**Advanced Engineering
Thermodynamics**

Advanced Engineering Thermodynamics

Adrian Bejan

*Department of Mechanical Engineering
and Materials Science
Duke University
Durham, North Carolina*

WILEY

A WILEY-INTERSCIENCE PUBLICATION

JOHN WILEY & SONS

New York · Chichester · Brisbane · Toronto · Singapore

ADRIAN BEJAN
Professor of Mechanical Engineering
Duke University
Durham, NC 27706
U.S.A.

Other Wiley books by Adrian Bejan:

Entropy Generation through Heat and Fluid Flow, Wiley, 1982, 248 pages.
  (*Solutions Manual*, Wiley, 1984, 50 pages,
  available from the author.)

Convection Heat Transfer, Wiley, 1984, 477 pages.
  (*Solutions Manual*, Wiley, 1984, 218 pages,
  available from the publisher and from the author.)

Library of Congress Cataloging-in-Publication Data:

Bejan, Adrian, 1948–
  Advanced engineering thermodynamics.

  "A Wiley-Interscience Publication."
  Bibliography: p.
  1. Thermodynamics. I. Title.
TJ265.B425   1988   621.402'1   88-5509
ISBN 0-471-83043-7

Printed in the United States of America

10 9 8 7 6 5 4 3 2 1

To my Cristina and Teresa

Preface

I have assembled in this book the notes prepared for my advanced class in engineering thermodynamics, which is open to students who have had previous contact with the subject. I decided to present this course in book form for the same reasons that I organized my own notes for use in the classroom. Among them is my impression that the teaching of engineering thermodynamics is dominated by an abundance of good introductory treatments differing only in writing style and quality of graphics. For generation after generation, engineering thermodynamics has flowed from one textbook into the next, essentially unchanged. Today the textbooks describe a seemingly "classical" engineering discipline, that is, a subject void of controversy and references, one in which the step-by-step innovations in substance and teaching method have been long forgotten.

Traveling back in time to rediscover the history of the discipline and looking into the future for new frontiers and challenges are activities abandoned by all but a curious few. This situation presents a tremendous pedagogical opportunity at the graduate level, where the student's determination to enter the research world comes in conflict with the undergraduate view that thermodynamics is boring and dead as a research arena. The few textbooks that qualify for use at the graduate level have done little to alleviate this conflict. On the theoretical side, the approach preferred by these textbooks has been to emphasize the abstract reformulation of classical thermodynamics into a sequence of axioms and corollaries. The pedagogical drawback of overemphasizing the axiomatic approach in engineering is that engineers do not live by axioms alone, and that the axiomatic reformulation seems to change from one revisionist author to the next. Of course, there is merit in the simplified phrasing and rephrasing of any theory: this is why a comparative presentation of various axiomatic formulations is a component of the present treatment. However, I see additional merit in proceeding to show how the theory can guide us through the everexpanding maze of contemporary problems. Instead of emphasizing the discussion of equilibrium states and relations among their properties, I see more value in highlighting irreversible processes, especially the kind found in practical engineering systems.

With regard to the presentation of engineering thermodynamics at the graduate level, I note a certain tendency to emphasize physics research developments and to deemphasize engineering applications. I am sure that the engineering student–his[†] sense of self esteem–has not been well served by the implication that the important and interesting applications are to be found only outside the domain chosen by him for graduate study. If he, like Lazare and Sadi Carnot two centuries earlier, sought to improve his understanding of what limits the "efficiency" of machines, then he finished the course shaking his head wondering about the mechanical engineering relevance of, say, negative absolute temperatures.

These observations served to define my objective in designing the present treatment. My main objective is to demonstrate that engineering thermodynamics is an active and often controversial field of research, and to encourage the student to invest his creativity in the future growth of the field. That there is opportunity for research in engineering thermodynamics is amply documented by Liu and Wepfer's recent survey of publications in just one portion of the field, namely, in the area of second-law analysis [1][‡]. The explosion of interest in this area is illustrated in Fig. P.1, which is reproduced from Ref. 2; it is further accentuated by the publication of no less than seven monographs and textbooks in the three years preceding Liu and Wepfer's survey [3–9].

The other considerations that have contributed to defining the objective of the present treatment are hinted at by the title *Advanced Engineering Thermodynamics*. The focus is being placed on "engineering" thermodynamics, that is, on that segment of classical thermodynamics that addresses the production of mechanical power and refrigeration in the field of engineering practice. I use the word "thermodynamics" in spite of the campaign fought on behalf of "thermostatics" as the better name for the theory whose subjects are either in equilibrium or, at least, in local equilibrium (more on this later, pp. 69–72). I must confess that I feel quite comfortable using the word "thermodynamics" in the broad sense intended by its creator, William Thomson (Lord Kelvin): this particular combination of the Greek words *therme* (heat) and *dynamis* (power) is a most appropriate name[§] for the field that united the "heat" and "work" lines of activity that preceded it (Table 1.2, pp. 31–33).

[†]Masculine pronouns are used throughout this treatment only for succinctness. They are intended to refer to both males and females.
[‡]Numbers in square brackets indicate references listed at the end of each chapter.
[§]The appetite for the "thermostatics" nomenclature is stimulated by comparisons with the dynamics/statics differentiation that is practiced in the field of mechanics: I believe that the contemporary mechanics meaning of "dynamics" is being mistakenly viewed as the *origin* of "-dynamics" in "thermodynamics."

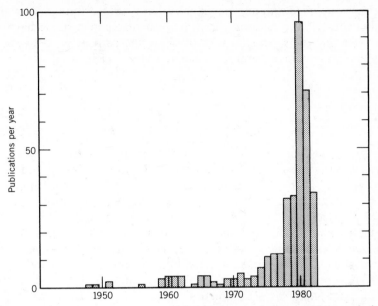

Figure P.1   Research publication activity in the area of second-law analysis [2].

Finally, I view this as an "advanced" course in engineering thermodynamics because it is the natural outcome of my own interaction with the research arena and with students who were previously acquainted with the subject of classical thermodynamics. There are at least two ways in which every subject can be advanced by a second course such as this. One is a "horizontal" expansion into the more remote fields intersected by the subject; the other is a "vertical" expansion, that is, a deepening of our understanding of the most basic concepts that define the subject. In the present treatment, I have followed the second approach because I see it as a more effective means of conveying a bird's-eye-view of engineering thermodynamics. An exhaustive coverage of the horizontal type already exists in the "handbooks"; and justice to each peripheral domain can be done only in specialized courses such as compressible fluid dynamics, combustion, turbomachinery, refrigeration and air conditioning, cryogenics, etc.

I have followed the vertical approach in order to make a statement of what I consider effective as a pedagogical tool. Although it has become fashionable to associate completeness and volume with "goodness," in this course I have made a conscious effort to focus on the structure of the field. I invite the research student to make his own contributions to this structure. For this last reason, the more applied segments of the present treatment are dominated by the topics that have attracted my own interest as a researcher.

To summarize, the combined research and pedagogical mission of this effort is to take a second look at the field, to make this view accessible in a one-semester course taken by individuals whose initial understanding of the subject is by no means homogeneous. Depth is provided through a comparative discussion of the various ways in which the fundamentals have been stated over the years, and by reestablishing the connection between fundamentals and contemporary research trends such as the "exergy" methodology.

*   *   *

The preceding words are the true preface because I wrote them in 1984, as I was starting the research for this book. I was then in the middle of a sabbatical leave at the University of Western Australia, which happened to be my first official assignment as a professor at Duke. Upon my arrival at Duke, I decided to use my enhanced freedom for the purpose of bettering my research and my life in general. Thinking in depth about engineering thermodynamics was one result of that decision. The fact that large numbers of thermal engineers continued to regard the field as mature is precisely why I picked engineering thermodynamics as a treatise topic: I not only saw merit in questioning the established point of view, but I also knew that a true research frontier is, quite often, the territory overlooked by the crowd.

As I look back at the past 3 to 4 years, I see a most gratifying project, a constant source of intellectual pleasure and new ideas. This project forced me to think on my own about those areas—the gaps—of which I knew the least. It challenged me to be creative and produce my own version of what fits best in any particular blank area. Overall, this book helped me diversify and enrich my research, which is why during this period I was able personally to take steps in new directions, such as the axiomatic formulation of classical thermodynamics (chapter 2), the graphic condensation of the relations between thermodynamic properties (chapters 4 and 6), the design of power plants for maximum power (chapter 8), the theory of the ideal conversion of solar radiation (chapter 9), and the design of refrigeration plants for maximum refrigeration effect per unit time (chapter 10). And, relative to engineering thermodynamics as a whole, this book gave me the opportunity to assemble in the same place many of the modern as well as the long-forgotten references. I also used every opportunity to do what I like best—produce original graphics.

Working on this book has been recreational. I did most of my thinking while walking through the Duke Forest between my West Campus office and our house in the Forest Hills section of Durham. I spent many hours consulting the truly exceptional collection of books of the libraries of Duke University. Ours is one university that from its early days in the 1800s invested in the important things. I made also many trips to the Library of Congress in Washington, DC, where, while reading the original writings, I had a chance to use the French, German, Latin, and Russian I learned in school.

The main contributor to the rewarding atmosphere of this project was Mary. I have benefited from her wisdom, sense of strategy and intellectual honesty during all my projects, big and small. This time, however, her participation transcended a number of much more important projects: the birth of child, the move from Colorado to North Carolina (via Western Australia!), and the triumphant completion of her PhD in business administration at the University of California, Berkeley. What I owe her is best condensed in the dedication that opens my *Convection Heat Transfer*.

I also benefited from my year-long association with Dr. Peter Jany of the Technical University of Munich, who generously contributed a most up-to-date section on critical-point phenomena in chapter 6. I will always remember the many conversations in which we compared notes on American engineering versus the German version, which had so much influence in Central Europe and Russia.

I recognize also the contribution made by Linda Hayes, who not only typed the manuscript, but also volunteered her rare talent of organization and sense of symmetry to the raw material that I have produced. Her work can be viewed directly in the *Solutions Manual*, which is available as a separate book. This manual can be obtained by writing to Wiley-Interscience (605 Third Avenue, New York, NY 10158) or directly to me.

At various stages, I was helped by old friends, colleagues in academia, and new students. Ren Anderson, Shigeo Kimura, Dimos Poulikakos, and Osvair V. Trevisan kept me in touch with their respective corners of the frontier and the literature. I am very grateful to my thermodynamics colleagues at Duke, Prof. C. M. Harman, Prof. E. Elsevier, and Prof. J. B. Chaddock, for commenting critically on early versions of the manuscript. While using those early drafts in the classroom, I collected many useful suggestions from the students, among whom I must mention: J. Gottwald, J. L. Lage, P. A. Litsek, A. Mahajan, D. P. Mendivil, M. Wang, Z. Xia, and Z. Zhang. Looking ahead, I will appreciate it very much if users of this book will write to call my attention to the imperfections that may have slipped into the final version.

ADRIAN BEJAN

Durham, North Carolina
October 1987

REFERENCES

1. Y. A. Liu and W. J. Wepfer, Theory and applications of second law analysis: A bibliography, in R. Gaggioli, ed., *Thermodynamics: Second Law Analysis*, Vol. II, American Chemical Society Symposium Series, ACS, Washington, DC, 1983, Chapter 18.

2. A. Bejan, Second law analysis: The method for maximizing thermodynamic

efficiency in thermal systems, invited position paper presented at the ASME–NSF Workshop on Research Goals and Priorities in Thermal Systems, Ft. Lauderdale, FL, April 25–27, 1984; published in W. O. Winer, A. E. Bergles, C. J. Cremers, R. H. Sabersky, W. A. Sirignano and J. W. Westwater, *Research Needs in Thermal Systems*, ASME, New York, 1986.

3. J. E. Ahern, *The Exergy Method of Energy System Analysis*, Wiley, New York, 1980.

4. M. V. Sussman, *Availability (Exergy) Analysis*, Mulliken House, Lexington, MA, 1981.

5. R. N. S. Rathore and W. F. Kenney, *Thermodynamic Analysis for Improved Energy Efficiency*, AIChE Today Series, AIChE, New York, 1980.

6. A. Bejan, Second law analysis in heat transfer and thermal design, in *Advances in Heat Transfer*, Vol. 15, pp. 1–58, 1982, Chapter 1.

7. M. J. Moran, *Availability Analysis: A Guide to Efficient Energy Use*, Prentice-Hall, Englewood Cliffs, NJ, 1982.

8. A. Bejan, *Entropy Generation through Heat and Fluid Flow*, Wiley, New York, 1982.

9. J. D. Seader, *Thermodynamic Efficiency of Chemical Processes*, Manual 1 in E. P. Gyftopoulos, ed., *Industrial Energy-Conservation*, MIT Press, Cambridge, MA, 1982.

Acknowledgments

I want to thank some very special individuals who, through their actions, have had a most positive effect on my work and my morale during the past three years. I add their names to the lists I treasure in *Entropy Generation through Heat and Fluid Flow* and in *Convection Heat Transfer*.

Prof. R. M. Hochmuth, Duke University
Prof. S. Kakac, University of Miami
Dr. J. H. Kim, Electric Power Research Institute
Prof. R. Kumar, Clemson University
Prof. L. F. Milanez, University of Campinas, Brazil
Prof. S. Sengupta, University of Miami
Prof. E. M. Sparrow, University of Minnesota
Prof. K. Vafai, Ohio State University

I am very grateful to the Lord Foundation of North Carolina for providing me with partial support during the summer of 1986.

A.B.

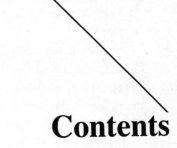

Contents

7   CHEMICALLY REACTIVE SYSTEMS   343

APPENDIX

AUTHOR INDEX

SUBJECT INDEX

ABOUT THE AUTHOR

1

The First Law of Thermodynamics

ELEMENTS OF THERMODYNAMICS TERMINOLOGY

The first step in the thermodynamic analysis of anything must be the definition of the entity that is being subjected to analysis. We refer to this entity (collection of matter, region in space) as a *system*. Although the need for executing this first move and for doing it once and unambiguously is obvious, in tackling certain engineering problems, the temptation is great to assume the system definition itself obvious and omit it. This tendency is in many cases a source of confusion if not outright error. Paradoxical differences between two results claimed by two experts who attack the same problem are often explained by the realization that the two problem solvers were mentally addressing different systems. As we will see very soon, the need for a precise system definition becomes most critical in the act of determining the location and magnitude of thermodynamic irreversibility.

The tendency to forget to define the system is due mainly to tradition, not to incompetence. In considerably older disciplines such as solid-body mechanics, the system is indeed obvious, as the mere sketching of a solid body focuses the attention of both problem solver and critic. In fluid mechanics and heat transfer, the system is again understood once the boundary conditions necessary for solving the Navier–Stokes equations are specified. However, even in fluid mechanics and heat transfer, the unambiguous specification of a unique system becomes a necessity if the analyst bases his method on order of magnitude or *scale analysis* [1–3], that is, if he replaces the Navier–Stokes equations with approximate algebraic statements that cannot be subjected to boundary conditions. For the same reasons, the origin of confusion and error in scale analysis can be traced to the tradition of "routine" fluid mechanics analysis, where it seems that everybody is expected to know and understand all the unspoken assumptions on which the mathematical model is based.

1

To define a system for the purpose of analysis means also to identify crisply the system's *environment*, or *surroundings*. The environment is the portion of matter or region in space that resides outside the system selected for analysis. What differentiates between the two entities—the system and its environment—is the surface called *boundary*. Now, one very important defining feature that sometimes falls prey to the same forces of tradition is that the boundary is a surface, not another system (note that the thickness of a surface is mathematically zero, therefore, the boundary can neither contain matter, nor can it fill a volume in space). Said another way, the value of a property that is measured at a point on the surface called "boundary" must be shared by both the system and the environment because, after all, the system and the environment are in contact at this point.

In order to see the importance of this observation, consider the heat transfer interaction Q between two fluid masses whose absolute temperatures are different, $T_H > T_L$. The thermal conductivities of the two fluids (or their states of agitation) are such that each fluid can be regarded as isothermal, in other words, the temperature drop $T_H - T_L$ occurs through the wall. If the wall thickness is small relative to the size of the fluid masses (e.g., the skin of a hot-air balloon in flight), it is of course tempting to regard the wall as the boundary between, say, the system (T_H) and the environment (T_L). This unfortunate choice is shown in Fig. 1.1. Its drawback is that unlike the heat transfer interaction Q, the entropy transfer Q/T is not conserved as it passes through the boundary. If made, this choice serves as permanent source of confusion: the unexperienced analyst has trouble deciding whether to use T_H or T_L in the denominator of Q/T, and the engineering component that is responsible for the generation of entropy (S_{gen}) is effectively hidden from view. The capricious augmentation of entropy transfer through "boundaries" of the kind shown in Fig. 1.1 helps perpetuate the mystery that surrounds the concepts of entropy, entropy transfer, and entropy generation.

Proper ways to select a system boundary between the two fluid masses T_H and T_L are exhibited in Figs. 1.2a–c. The common feature of these alternatives is that the temperature varies continuously across each boundary, hence, both the heat transfer interaction and the entropy transfer interaction are conserved. In Fig. 1.2a, the wall is situated outside the system, and for this reason, the entropy generation due to the system–environment heat transfer interaction can be termed "external." The opposite choice is made in Fig. 1.2b, where the entropy-generation effect is "internal" relative to the system. Figure 1.2c shows that the system can be divided further into subsystems, if the precise identification of the source of entropy generation (the wall) is one of the objectives of the thermodynamic analysis.

The observation that properties must vary continuously across the surface chosen as boundary is general. The discontinuity of absolute temperature

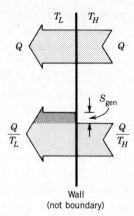

Figure 1.1 The discontinuity of entropy transfer through an improper "boundary."

was used in Fig. 1.2 only for the purpose of illustration: it is instructive to think of "boundaries" across which other properties vary discontinuously, and to imagine the analytical difficulties that are triggered by this decision.

The boundary and the types of interactions that are present at the boundary play an important role in the structure (organization) of the analysis devoted to solving a certain problem. One feature that must be recognized at an early stage is whether the boundary is crossed by the flow of mass. A system defined by a boundary impermeable to mass flow is a *closed system*. Most of the systems discussed in connection with the establishment of relationships between thermodynamic properties at equilibrium

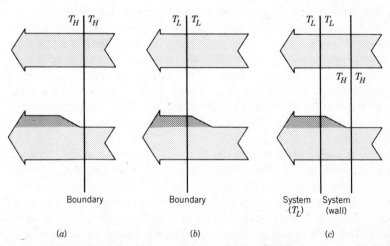

Figure 1.2 The continuity of temperature, heat transfer, and entropy transfer through a proper boundary.

are closed systems (chapters 4, 6, and 7). Conversely, systems whose
defining boundaries can be crossed by the flow of mass are recognized as
open systems, or *flow systems*. The engineering thermodynamics of open
systems tends to rely on a special terminology; for example, the thermodynamic system itself is usually referred to as the *control volume*, the system
boundary is the *control surface*, and the particular patches of the boundary
that are crossed by mass flow are the *inlet* or *outlet ports*.

The condition, or the being, of a thermodynamic system at a particular
point in time is described by an ensemble of quantities called *thermodynamic properties*. We refer to the condition described by properties as
state. One important issue that deserves emphasis is that not all the
quantities that the analyst calculates in connection with a certain system are
thermodynamic properties. Thermodynamic properties are only those quantities whose numerical values do not depend on the history of the system, as
the system evolves between two different states. Quantities such as pressure
and temperature are properties because their values depend strictly on the
instantaneous condition during which they are measured. Examples of
quantities that are not thermodynamic properties are work, heat and mass
transfer interactions, entropy transfer interaction, and entropy generation
(Fig. 1.2), lost available work, lost exergy, etc.

The thermodynamic properties that we encounter in engineering are
quite numerous: it seems that each generation has added to the list one or
more new properties that proved to be useful relative to the engineering
challenges faced by the generation. Some properties can be measured
directly (e.g., pressure, temperature, volume), whereas others can be derived based on such measurements (e.g., internal energy, entropy, enthalpy,
exergy). Thermodynamic properties whose values depend on the size of the
system are called *extensive properties* (e.g., volume, entropy, internal energy). *Intensive properties* are those whose values do not depend on the size
of the system, for example, pressure and temperature. The collection of all
the intensive properties of a system constitutes the *intensive state*.

A certain *phase* of a system is the collection of all the parts of the system
that have the same intensive state and the same per-unit-mass values of the
extensive properties. For example, liquid droplets dispersed in a liquid–
vapor mixture in equilibrium have the same pressure, temperature, specific
volume, specific enthalpy, etc.: taken together, the droplets represent the
liquid phase.

Finally, in engineering thermodynamics, we use the concept of *process* as
a one-word reference to the change of state from an initial state to a final
state. To know the process means to know not only the end states but also
the *interactions* experienced by the system while in communication with its
environment (e.g., work transfer, heat transfer, entropy transfer, mass
transfer). The *path* of the process is the history, or the succession of states,
followed by the system from the initial to the final state. Stressing again the
fundamental difference between thermodynamic properties and quantities

that are not, note that the changes in quantities that are not properties depend not only on the end states but also on the path.

The thermodynamic *cycle* is a special process in which the final state coincides with the initial state. Starting with Sadi Carnot's 1824 memoir [4], the concept of cycle evolved into a key concept in the field of power engineering, and into a vehicle for logical deduction in thermodynamics theory.

In this section, we reviewed only the most essential terms that are necessary for conducting a review of engineering-thermodynamics fundamentals. The complete nomenclature is considerably more extensive, however, I prefer to continue its discussion as we review the theoretical developments that one by one defined the need for the introduction of new terms.

THE FIRST LAW FOR CLOSED SYSTEMS

There are two principles of classical thermodynamics that must be stressed in an engineering review such as this. One principle has to do with the equivalence of work transfer and heat transfer as possible forms of energy interactions. This principle is encapsulated in the First Law of Thermodynamics that, in Max Planck's words, "is nothing more than the principle of the conservation of energy applied to phenomena involving the production or absorption of heat" [5]. The second principle is the inherent irreversibility of all processes that occur in nature. It is the irreversibility, or the generation of entropy, that prevents man from extracting the most possible work from various processes, and from doing the most with the work that is already at his disposal. This second principle is summarized by the Second Law of Thermodynamics.

It is an integral part of engineering tradition to discuss the first law first and the second law second. This ordering of the discipline is based apparently on views—both questionable—that the first law is older than the second law, and that the concept of internal energy defined by the first law is somehow easier to grasp than the concept of entropy introduced by the second law. The first view is fueled by the misinterpretation of statements of the kind quoted from Max Planck in the preceding paragraph: what is relatively older than the second law is the principle of conservation of energy known in mechanics, not the First Law of Thermodynamics. The first law and the second law emerged together out of the writings of William John MacQuorn Rankine, Rudolph Clausius, and William Thomson (Lord Kelvin) in the early 1850s [consult Ref. 6]: they *had to* emerge together in order to resolve the conflict between Sadi Carnot's theory that assumed the conservation of "caloric," and the growing evidence that work through friction can serve as an endless source of caloric. The second view—the feeling that internal energy is easier to understand than entropy—is again fueled by the engineer's relative familiarity with the aging concept of mechanical energy, not with internal energy.

The questioning of tradition aside, in this treatment, I start also with the first law because, above all, this is a review of the student's first encounter with engineering thermodynamics, not a review of the history of the subject. Note that a number of captivating historical accounts already exist in book form [6–9] or in certain prefaces and introductions that convey some of the historical flavor [10–12]. Further observations on the historical development of the First Law of Thermodynamics are placed at the end of this chapter.

Consider the closed system shown schematically in Fig. 1.3: if this system experiences a change of state from the initial state (1) to the final state (2), the First Law of Thermodynamics requires

$$Q_{1-2} \quad - \quad W_{1-2} \quad = \quad \underbrace{E_2 - E_1} \qquad (1.1)$$

$$\underbrace{\underset{\text{transfer}}{\text{Heat}} \qquad \underset{\text{transfer}}{\text{Work}}}_{\substack{\text{energy interactions} \\ \text{(Nonproperties)}}} \qquad \underset{\substack{\text{change} \\ \text{(Property)}}}{\text{Energy}}$$

in other words, the difference between the net heat input Q_{1-2} and the net work output W_{1-2} represents the change in the thermodynamic property called energy. The first law proclaims the existence of energy as a thermodynamic property.

Evident from eq. (1.1) and Fig. 1.3 is the use of the "heat-engine sign convention" [13], whereby the heat transfer into the system and the work transfer out of the system are considered positive. This is a "heat-engine"

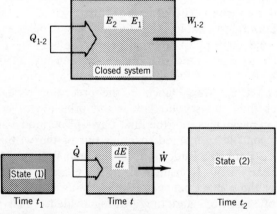

Figure 1.3 Graphic statements of the First Law of Thermodynamics for closed systems.

sign convention because the purpose of a heat engine as a closed thermo-dynamic system is to perform work on its environment. This sign convention is used consistently throughout the present treatment.

The energy change $E_2 - E_1$ depends only on the end states, whereas the energy interactions Q_{1-2} and W_{1-2} depend on the end states and on the path of the process that links the end states. This important distinction is stressed with reference to the concept of property under each term appearing in eq. (1.1). Another way to stress this difference is to use a different notation for the infinitesimal increments in work transfer and heat transfer relative to the exact differential notation that applies to the infinitesimal change in E. For this reason, the first law for a process between two states situated infinitely close to one another is often written as

$$\delta Q - \delta W = dE \tag{1.2}$$

In the same notation, the net energy interactions that appear on the left side of eq. (1.1) are

$$Q_{1-2} = \int_1^2 \delta Q \quad \text{and} \quad W_{1-2} = \int_1^2 \delta W \tag{1.3}$$

The peculiar notation "δ" may not be the ideal way to emphasize the difference between energy interactions and energy change. The alternative used by Truesdell [14] consists of introducing the concept of time in the description of the process (see the bottom of Fig. 1.3). In this new description, state (1) is the condition of the system at time t_1, state (2) is the condition at time t_2, and the net energy interactions Q_{1-2} and W_{1-2} are the time integrals

$$Q_{1-2} = \int_{t_1}^{t_2} \dot{Q} \, dt \quad W_{1-2} = \int_{t_1}^{t_2} \dot{W} \, dt \tag{1.4}$$

Quantities $\dot{Q}$ and $\dot{W}$ are the instantaneous heat transfer rate and the mechanical-power output, respectively (note that the $\dot{Q}$ and $\dot{W}$ notations are already used routinely in the analysis of open systems in steady flow). By using the notation of eqs. (1.4), the First Law of Thermodynamics for a closed system can be written on a per-unit-time basis as

$$\dot{Q} - \dot{W} = \frac{dE}{dt} \tag{1.5}$$

Another way of stressing the path dependence of Q_{1-2} and W_{1-2} (or $\dot{Q}$ and $\dot{W}$) is presented graphically in Fig. 1.4 [see also Ref. 15]. The system can proceed from state (1) to state (2) along an infinity of paths, for example, along paths (A) and (B) in Fig. 1.4. Assuming for the purpose of

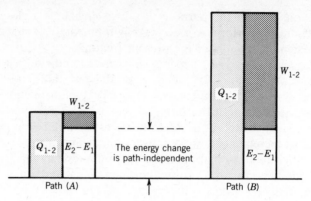

Figure 1.4   The path dependence of the energy transfer interactions Q_{1-2} and W_{1-2}.

illustration that paths (A) and (B) are such that Q_{1-2}, W_{1-2}, and $E_2 - E_1$ are all positive, the first-law statement (1.1) implies the stacking of the three building blocks shown in Fig. 1.4. While the difference $Q_{1-2} - W_{1-2}$ matches $E_2 - E_1$ along both paths, the sizes of Q_{1-2} and W_{1-2} vary from one path to the next. In particular, if the process executed by the closed system is a cycle, the first-law statement (1.1) reduces to

$$\oint \delta Q - \oint \delta W = 0 \tag{1.6}$$

in other words, the white blocks in Fig. 1.4 shrink to zero thickness. This statement stresses again the difference between energy-change and energy transfer interactions: the latter depend on the path followed by the cycle. We shall return to the diagram of Fig. 1.4 in the early part of chapter 4.

WORK TRANSFER INTERACTIONS

We continue with a review of the three concepts linked by the first law—work transfer, heat transfer, and energy change—and highlight some features that tend to be overlooked in the course of problem solving. The work transfer interactions encountered most often in classical engineering thermodynamics are those associated with the displacement of the system's boundary in the presence of forces that act on the boundary. If $\mathbf{F}$ is the force experienced by the system at a certain point on its boundary (i.e., the force exerted by the environment on the system), and if $d\mathbf{r}$ is the infinitesimal displacement of the point of application, then the infinitesimal work transfer interaction is

$$\delta W = -\mathbf{F} \cdot d\mathbf{r} \tag{1.7}$$

Discussed already in connection with Fig. 1.3 is the convention that the work transfer interaction is considered positive when the system does work on its environment, in other words, when the boundary displacement occurs against the force felt by the system, $\cos(\mathbf{F}, d\mathbf{r}) < 0$.

Two features must be present simultaneously if a system is to experience a work transfer interaction with its environment: (1) a force must be present on the boundary, and (2) the point of application of this force (hence, the boundary) must move. The mere presence of forces on the boundary, without the displacement or the deformation of the boundary, does not amount to work transfer. Likewise, the occurrence of boundary displacement without a force opposing or driving this motion does not mean work transfer. For example, in the "free expansion" of a gas into an evacuated space, the gas as a closed system does not experience work transfer because the pressure is zero on the moving boundary.

One special application of expression (1.7) is fundamental to the analytical description of the relations between the thermodynamic properties of a substance in equilibrium (chapter 4). The same special form is also used routinely (however, only as an approximate engineering model) to estimate the work transfer when a batch of "working fluid" expands or contracts in a cylinder and piston apparatus. With reference to the left side of Fig. 1.5, when the system is in equilibrium, pressure P is uniform throughout the system; therefore, $(-\mathbf{F} \cdot d\mathbf{r})$ can be replaced by $P\, dV$ in eq. (1.7):

$$\delta W_{\text{rev}} = P\, dV \tag{1.8}$$

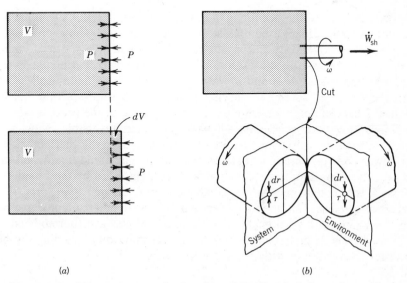

(a)   (b)

Figure 1.5 Observations on the location of $P\, dV$ and shaft work transfer.

The discussion of the subscript "rev," which stands for "reversible," is postponed until chapter 2, where we review the concepts associated primarily with the Second Law of Thermodynamics. One point to stress here is that if eq. (1.8) is to be used to evaluate the work transfer transmitted through the movement of a piston, then the pressure P at the boundary must be *known* at any instant during the volume change. In general, this means that the analyst has to first solve the complete equations that govern convection inside the expanding fluid (i.e., the mass, momentum, and energy equations) in order to calculate the value of P versus time right on the moving boundary. However, engineering thermodynamics occupies a domain much narrower than convection: the engineer is taught to rely on eq. (1.8) to estimate the work transfer as the area under the $P(V)$ curve on Watt's famous "indicator diagram."[†] Implicit in this learned fact is that the expansion is slow enough so that the state of the fluid batch can be represented at all times by a single point in the two-dimensional plane P–V. In engineering thermodynamics then, the requirement that P of eq. (1.8) must be known on the moving boundary is considerably more restrictive: the pressure P must also be instantaneously *uniform* throughout the expanding system.

The sufficiently slow process to which eq. (1.8) applies is often called *quasistatic*, and the states along the path of such a process are sometimes referred to as *quasistatic states*. It is unfortunate that over the years this terminology has served as a steady source of confusion—not that the word "quasistatic" is that difficult to grasp (literally, it means "seemingly static," which is an appropriate description for a process that is infinitely slow), but because different schools of thermodynamics have attached different meanings to the word.

If we go back to the beginning of this century, we find that one influential author (Carathéodory) used quasistatic to describe an adiabatic process that happens infinitely slowly so that it "can be regarded as a series of equilibrium states" (Carathéodory's formulation of classical thermodynamics is discussed on pages 39–41 and 77–89). Carathéodory was quite explicit in his argument that in systems in which rate-dependent processes (internal friction was his example) converge to zero as the pace of the process becomes infinitely slow, the quasistatic adiabatic process is reversible. Now, if we think of the simple systems that are of concern to us in the thermodynamics of power and refrigeration—batches of common gases and liquids, in other words, the same simple systems that were contemplated by the founders of classical thermodynamics—then, adiabatic or not, *any quasistatic process is a reversible process*. Carathéodory's special interest in adiabatic processes was essential only to his axiomatic reconstruction of classical thermodynamics, which consists of first ruling out heat transfer and, later, defining the heat transfer interaction as a derived concept (Table 1.3, p. 39).

[†]The mechanism for drawing this diagram was conceived by John Southern (see Fig. 2.1).

The competing interpretation of the words "quasistatic process" stops at the literal translation and stresses that an infinitely slow process is not necessarily a reversible process. There are two factors that fueled the emergence of this newer interpretation. First, it was the effort to generalize the science of thermodynamics in order to cover systems (bodies) whose internal constitution differs from that of the simple systems of the classical calorimetry–thermostatics line. Carathéodory noted that the quasistatic processes executed by substances in which the internal friction effect does not converge to zero are not reversible, and that such substances would require in fact a new kind of thermodynamics.

The second reason is that by employing a mechanics-sounding name (quasi*static*) in a discussion that deliberately avoids the concept of heat transfer, Carathéodory left the impression that the infinite slowness implied by the word quasistatic refers only to the time scale of mechanical effects, say, to the time needed for the pressure to become uniform inside a working cylinder and piston expander. Obviously, if heat transfer takes place through the cylinder wall lined by the system boundary, then temperature gradients and irreversibility will be present inside the system (e.g., the system on the left side of the boundary surface shown in Fig. 1.2a). Since the effect of thermal diffusion has a time scale that, generally speaking, is not the same as the scale of viscous slowdown or the scale of the imposed volume change, then the process is not a sequence of equilibrium states even though it be slow enough to be called quasistatic [for more on this, see Ref. 12, pp. 16 and 77]. Therefore, according to the newer interpretation, the concept of a reversible process is more restrictive than the concept of a quasistatic process: all the reversible processes are quasistatic, however, not all the quasistatic processes are reversible.

As a summary to the two competing interpretations, the best I can do is to warn the engineer that the potential for confusion exists and that, because of this potential, the best course is to avoid using the word "quasistatic." If the process is sufficiently slow so that it can be viewed as a sequence of equilibrium states, then the process is reversible. If, for any reason, the intermediate states visited during the process cannot be regarded as equilibrium states (i.e., if each state cannot be represented as one point in a two-dimensional plane such as the $P–v$ diagram of Fig. 2.2), then the process is not reversible.

Another mode of work transfer that is very common in engineering applications is the shaft work W_{sh} transmitted through a shaft crossed by the system boundary. The origin of this work transfer mode may seem mysterious to the problem solver in view of the exclusive use of the $P\,dV$ work transfer mode in thermodynamics, and also in view of definition (1.7) and the fact that the boundary surface does not move. In order to clarify this issue, Fig. 1.5b shows the cut made by the boundary through the shaft. In the shaft cross-section that is attached to the system, the point of application of each shear stress vector τ moves as the shaft turns. By integrating the work done by each shear stress over the cross-section, i.e., by applying definition (1.7), it is easy to show that the infinitesimal work transfer δW_{sh} is equal to the angular displacement times the torque with which the environment opposes the turning of the shaft (or that the shaft power output $\dot{W}_{sh}$ is equal to the angular speed ω times the same torque).

Based on this discussion of Fig. 1.5*b* and on the coincidence that in English the words "shaft" and "shear" admit the same abbreviation, the symbol W_{sh} can also be used as notation for "shear work transfer." An example of work transfer associated with shear forces occurs in the derivation of the First Law of Thermodynamics for an infinitesimally small control volume in a flow field [e.g., Ref. 1, pp. 8–16]. An example of shear forces that although present along the system's boundary do not account for any shear work transfer is the distribution of shear stresses caused by fluid friction against a rigid wall that confines a certain fluid in motion. In this case, the work transfer is zero because the point of application of each shear force is stationary.

In this section, we focused on some of the modes in which work transfer interactions enter the constitution and analysis of engineering thermodynamics systems. Since a large segment of engineering thermodynamics deals with the convertibility of heat transfer into mechanical work, the above observations were directed at mechanical work transfer interactions for which the classical definition (1.7) is adequate. A more general definition of work transfer that applies also to electrical and magnetic work interactions was formulated by Hatsopoulos and Keenan [Ref. 11, p. 22]: "Work is an interaction between two systems such that what happens in each system at the interaction boundary could be repeated while the sole effect external to each system was the change in level of a weight." Analogous definitions can be formulated in terms of the energy stored in a translational spring [16] or in another conservative mechanical system. Hatsopoulos and Keenan refer to a footnote in Gibbs' second paper [17] as the origin of the idea behind their general definition of work transfer interactions. Along the same line, it is worth noting that since the early eighteenth century, it has been standard engineering practice to evaluate the capacity of a machine or engine in terms of the height to which it could raise a given weight [18]. Indeed, the weight lifted to a height was the British engineer's common unit of "duty" in the description of the early steam engines [Ref. 9, p. 162].

Related to the issue of generalizing the concept of work transfer, it is worth mentioning another generalization that is used in contemporary thermodynamics. The concept of *reversible* work transfer can be envisioned not only in the context of systems that expand or contract quasistatically, eq. (1.8), but also for systems that can experience other modes of work transfer. A collection of such work transfer interactions is presented in Table 1.1 next to examples of very simple mechanical and electrical systems whose energy storage capability is the subject of the discussion that ends this section (and is the reason for constructing the table). It is sufficient to note at this point that for systems that are capable to experience more than one work transfer interaction, eq. (1.8) can be replaced by

$$\delta W_{\text{rev}} = -\sum_i Y_i \, dX_i \qquad (1.8')$$

The terms Y_i and X_i are *the generalized forces* and *the generalized displacements* (or *deformation* coordinates), respectively. (From the first column of Table 1.1 and eq. (1.8), it should be clear that the units of these quantities are not necessarily those of force and displacement.) In this somewhat abstract terminology, it is the negative of the pressure P that plays the role of generalized force in the reversible work done by closed systems that expand quasistatically.

HEAT TRANSFER INTERACTIONS

The First Law of Thermodynamics does not distinguish between heat transfer and work transfer as two possible forms of energy interaction between a system and its environment. Indeed, the role of the first law is to place the heat transfer interaction and the work transfer interaction on an equal footing. The equivalence of the two interactions is made especially evident by their traditional appearance on the same side of the equal sign in eq. (1.1), as written by Planck [Ref. 5, p. 45]. In the earliest analytical statements of the first law made by Clausius and adopted by contemporary engineers, the work transfer and the energy change terms appear on the same side of the equal sign [19, 20]. Clausius' arrangement of the terms was adopted also by Poincaré in his thermodynamics course taught in 1888–1889 at the Faculty of Sciences of the University of Paris [21]. These early works and their appearance are worth keeping in mind: like Planck's and Zeuner's courses among the physicists and engineers educated in the German language, Poincaré's course emerged as a dominant factor in the practice and rewriting of thermodynamics in the first half of the twentieth century.

The fundamental distinction between heat transfer and work transfer is brought to light by the Second Law of Thermodynamics: heat transfer is the energy interaction accompanied by entropy transfer, whereas work transfer is the energy interaction that takes place in the absence of entropy transfer. If this way of distinguishing δQ from δW sounds a bit abstract, it is simply because the second law and related concepts such as entropy transfer are usually not practiced in the problems proposed during a first course in engineering thermodynamics. Nevertheless, it is a rigorous definition that has the additional benefit that it draws attention to the existence of entropy transfer interactions of type $\delta Q/T$, where T is the thermodynamic (absolute) temperature of the boundary crossed by δQ (Fig. 1.2).

Another way to think of heat transfer is to say that it is the energy interaction whose effect on the system and the environment cannot be reproduced such that the sole effect external to each system is the change in level of a weight. This alternative description relies on the definition of work transfer, and stresses the fundamental difference between the two modes of energy interaction. In this sense, it is related to the definition given in the preceding paragraph in terms of the presence or absence of entropy transfer.

The intuitively more appealing description of heat transfer preferred by most engineering treatments of thermodynamics originated with Poincaré's course [21]: it amounts to the statement that heat transfer is the energy interaction driven by the temperature difference between the system and its environment. The same view has been held from the beginning in the not too distant field of heat transfer engineering. The phenomenological treatment of heat transfer initiated by Poincaré makes sense because it appeals to the familiarity of modern man with the concepts of heating and temperature. Yet, the phenomenological description has been criticized by some who favor a precise definition of each word that appears in the thermodynamics language. For example, since Poincaré's course appears to suggest that, on the one hand, the heat transfer is the interaction that is driven by a temperature difference, and, on the other hand, that the temperature difference is left without a definition, Hatsopoulos and Keenan and a few after them called Poincaré's description of heat transfer a "triviality" [Ref. 11, p. XXII]. This evaluation, I feel is far too simplistic, for the challenge faced so successfully by Poincaré was not to "define" terminology to an audience already familiar with the essence of classical thermodynamics, rather, it was to communicate and explain a new theory that had gelled only two decades earlier and was still unknown to waves of would-be inventors. To do this, Poincaré used (or misused) very effectively the terminology of his time.

No matter how rigorous the treatment and how strong the desire to pin the definition of each new concept on the definitions of older concepts, sooner or later the engineer must speak of thermal equilibrium and temperature. To review what is meant by thermal equilibrium, consider two closed systems whose boundaries are such that both systems cannot experience work transfer (e.g., two arbitrary amounts of air sealed in rigid containers, where "arbitrary" means that the mass, volume, and pressure of each system are not specified). If two systems of this kind are positioned close to one another, it is generally observed that changes are induced in both systems. In the case of the air-filled containers of the above example, these changes can be documented by recording the air pressure versus time. It is commonly observed that there exists a time interval beyond which the changes triggered by the proximity of the two systems cease. In general, the condition of the closed system is said to be one of *equilibrium* when, after a sufficiently long period, changes cease to occur inside the system. In particular, when the closed system is incapable of experiencing work transfer interactions, the long-time condition illustrated in the above example is one of *thermal equilibrium* [22].

Let (A) and (B) be the closed systems that interact and reach thermal equilibrium in the preceding example. The same experiment can be repeated using system (A) and a third system (C), which is also closed and unfit for work transfer. It is also a matter of common experience that if systems (B) and (C) are individually in thermal equilibrium with system (A), then, when placed in direct communication, systems (B) and (C) do not undergo any changes as time passes. This second observation can be summarized as follows: if systems (B) and (C) are separately in thermal equilibrium with a third system, then they are in thermal equilibrium with each other. It was stressed more than a century ago by Maxwell that this summarizing statement carries the weight of physical law. After Maxwell's death, in fact more than half a century after the formulation and labeling of the First and Second Laws of Thermodynamics, this view has come to be recognized as the *Zeroth Law of Thermodynamics*. The zeroth law was first formulated and labeled in 1931 by Fowler [23, 24].

Each law of thermodynamics can be thought of as a way to define a new system property, for example, the internal energy via the first law and the entropy via the second law. In this sense, the zeroth law defines the thermodynamic property called *temperature*. Returning to the vast experimental evidence on which the zeroth law and the science of thermometry are based, we recognize as temperature the property whose numerical value determines whether the system is in thermal equilibrium with another system. Two systems are in thermal equilibrium when their temperatures are identical.

The temperature of a system is measured by placing it in thermal communication with a special system (a test system) called a *thermometer*. The thermometer has to be sufficiently smaller than the actual system so that the heat transfer interaction en route to thermal equilibrium is negligible from the point of view of the system. The thermometer, on the other hand, is designed so that the same heat transfer interaction leads to measurable effects such as changes in volume or electrical resistance.

The development of the science concerning the measuring of temperature (the science of thermometry) has a long history that is tightly connected to that of calorimetry, caloric theory, and classical thermodynamics (see Table 1.2 later in this chapter). The calibration of thermometers and the adoption of certain temperature scales is very much part of this history. Traditionally, calibration consisted of agreeing on two easy-to-reproduce states of the thermometer: following a suggestion made in 1701 by Newton that the interval between the freezing point of water and the human body temperature be a scale of 12 degrees, the most often used states were the thermal equilibrium with a mixture of ice and water at atmospheric pressure and the thermal equilibrium with a batch of water boiling at atmospheric pressure [25]. These traditional scales, named in order after Fahrenheit,[†] Réaumur,[‡] and Celsius,[§] are said to be based on two fiducial points (literally, on two points based on firm faith). In view of the arbitrariness of the material that fills the thermometer, the temperature measurements recorded on the traditional scales are recognized nowadays as *empirical temperatures*.

The temperature scales in use today are all based on the concept of *thermodynamic temperature* defined in terms of the Second Law of Thermodynamics (p. 59). Following the Tenth General Conference on Weights and Measures (1954), we use the four thermodynamic temperature scales shown in Fig. 1.6. These scales are based on only one fiducial point, namely, the triple point of water: on the Kelvin scale, the numerical value assigned to this point is 273.16. More on the reasoning behind these scales and the distinction between, on the one hand, the *absolute temperatures* recorded on the Kelvin and Rankine scales and, on the other, the temperatures of the new, one-point Celsius and Fahrenheit scales is found in chapter 2. Of problem-solving interest at this point are the relations that affect the conversion of one thermodynamic temperature into another:

$$T(°C) = T(K) - 273.15$$

$$T(R) = \tfrac{9}{5} T(K)$$

$$T(°F) = T(R) - 459.67 \tag{1.9}$$

$$T(°C) = \tfrac{5}{9}[T(°F) - 32]$$

The relative size of the divisions of these scales is also highlighted in Fig. 1.6:

$$1 \text{ R or } 1°F = \tfrac{5}{9}(1 \text{ K or } 1°C) \tag{1.10}$$

[†]Gabriel Daniel Fahrenheit (1686–1736), German instrument maker native of Danzig (today, Gdansk) and long-time resident of Holland, invented the mercury-in-glass thermometer in 1714. He assigned the number 0 to the mercury level corresponding to the thermal equilibrium of a mixture of ice and common salt, and the number 96 to the level corresponding to the temperature of the human body. He found that on the same scale, the freezing and boiling points of water correspond to numbers 32 and 212, respectively.

[‡]René Antoine Ferchault de Réaumur (1683–1757) was a leading physicist, engineer, and naturalist. In thermodynamics, he is remembered for inventing in 1731 the alcohol thermometer and the Réaumur temperature scale, on which the freezing point of water is 0 degrees and the boiling point 80 degrees. His fascinating career included the study of gold-bearing rivers, turquoise mines, forests, insects, crayfish, Chinese porcelain, opaque glass, the composition and manufacture of iron and steel, and methods for tinning iron.

[§]Anders Celsius (1701–1744) professor of astronomy at the University of Uppsala, proposed in 1742 the centigrade scale on which the freezing and boiling of water at atmospheric pressure occur at 0°C and 100°C, respectively.

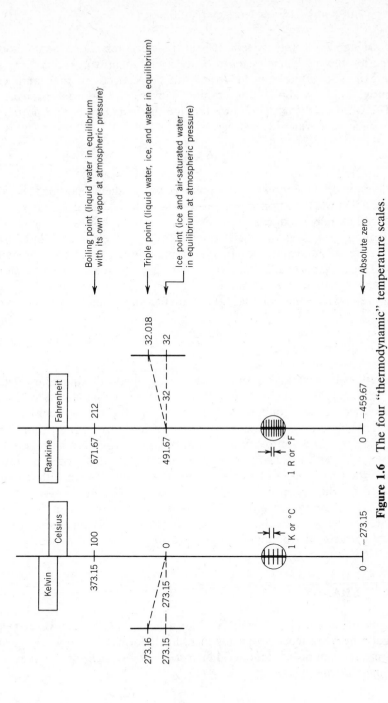

Figure 1.6 The four "thermodynamic" temperature scales.

Boiling point (liquid water in equilibrium with its own vapor at atmospheric pressure)

Triple point (liquid water, ice, and water in equilibrium)

Ice point (ice and air-saturated water in equilibrium at atmospheric pressure)

Absolute zero

Kelvin   Celsius   Rankine   Fahrenheit

373.15 ─┼─ 100   671.67 ─┼─ 212

273.16 ─┼──
273.15 ─┼─ 0   491.67 ─┼─ 32   32.018
                                          32

1 K or °C   1 R or °F

0 ─ ─273.15   0 ─ ─459.67

Returning now to the concept of heat transfer interaction, the method of engineering thermodynamics relies on two additional words, adiabatic and diathermal, that effectively do away with the concept of time and, consequently, build a wall between thermodynamics and heat transfer engineering. The word *adiabatic*[†] is used to describe the boundary or that portion of the boundary for which we can write

$$\dot{Q} = 0 \qquad (1.11)$$

regardless of the magnitude of the temperature gradient one might be able to measure in the direction normal to the boundary. This concept was first introduced by Laplace in caloric theory [Ref. 8, p. 6], however, in engineering thermodynamics, it was made popular by Rankine's book [26] and the creative use of adiabatic lines in the graphic description of steam-engine cycles. The second word, *diathermal*,[‡] refers to a boundary across which the temperature gradient is zero even in the presence of heat transfer. If n is the direction normal to the boundary, for a diathermal boundary, we can write

$$\frac{\partial T}{\partial n} = 0 \qquad (1.12)$$

The adiabatic boundary model does not invalidate the boundary definition rule discussed in connection with Fig. 1.2: the temperature varies continuously across the boundary surface, however, the thermal conductivity of the local material is so low, or the time of observation is so short, that $\dot{Q} = 0$ is a very good approximation of the energy transferred as heat across the boundary. To model a boundary as adiabatic or diathermal means to compare the time scale of the process executed by the system with the time that elapses if the system and its environment are allowed to reach thermal equilibrium. If the process time scale is considerably shorter than the time to thermal equilibrium, the boundary can be modeled as adiabatic. In the opposite extreme, the boundary approaches the diathermal model.

ENERGY CHANGE

The right side of eq. (1.1) is shorthand for a general expression whose terms distinguish between macroscopically identifiable forms of energy storage and the form that cannot be identified macroscopically (which for this reason is called *internal energy*):

[†]From the Greek word *adiabatos* (not to be passed, impossible to pass).
[‡]From the Greek words *dia* (through) and *therme* (hot), or *thermotis* (heat).

$$E_2 - E_1 = U_2 - U_1 + \tfrac{1}{2}mV_2^2 - \tfrac{1}{2}mV_1^2 + mgz_2 - mgz_1 + (E_2 - E_1)_i \qquad (1.13)$$

| Energy | Internal energy | Kinetic energy | Gravitational potential energy | Other macroscopic forms of energy storage (Table 1.1) |

Whether all these terms have to be included in the composition of $E_2 - E_1$ depends on the system selected for analysis. In the thermodynamics of power and refrigeration sytems, which is the main focus of the present treatment, the three components that usually enter the analysis are the internal energy, the kinetic energy, and the gravitational potential energy. In thermostatics, i.e., the study of finite-size batches of substances or mixtures of substances in equilibrium, only the internal energy change term is relevant.

The term that accounts for forms of macroscopic energy storage, $(E_2 - E_1)_i$, depends on the constitution (makeup, construction) of the system. A listing of the simplest possible expressions for such terms is compiled in Table 1.1, which is based on examples drawn from early mechanical and electrical engineering courses [for a unified treatment of these examples, see Refs. 16, 27]. The important notion to recognize and always keep in mind in connection with any of the expressions listed for $(E_2 - E_1)_i$ is that each is based on assuming the existence of a particular *constitutive relation*. The examples listed in Table 1.1 are by far the simplest because the cited constitutive relations are independent of one another, and each energy storage term $(E_2 - E_1)_i$ can be increased or decreased only through a characteristic energy transfer interaction listed in the δW column. Such energy interactions and forms of energy storage can be described as *uncoupled* [16]. Not listed in Table 1.1 are examples of *coupling*, that is, the existence of two or more energy transfer interaction that can affect the same mode of energy storage. Energy-conversion systems are primary examples of coupled behavior: for example, the electromechanical energy of an electric motor can be changed through shaft work transfer, electrical work transfer, and through a combination of shaft work transfer and electrical work transfer. In the systems encountered regularly in power and refrigeration engineering, the existence of the internal energy U as a thermodynamic property is a sign of coupled thermodynamic behavior, because the system's internal energy can be changed through work transfer, heat transfer, and through a combination of work and heat transfer.

The general decomposition of energy change revealed by eq. (1.13) is also a hint of the historical development of the concept of energy and its terminology. It was Leibnitz who first discussed the conservation of the sum of the kinetic and potential energies, using the name *vis viva* (live force) for mV^2 and *vis mortua* (dead force) for mgz. The same conservation idea was

TABLE 1.1 Examples of Simple (Uncoupled) Forms of Energy Storage and Corresponding Work Transfer Interactions

Macroscopic Forms of Energy Storage, $(E_2 - E_1)_i$, Eq. (1.13)	Relation Assumed in Writing Each $(E_2 - E_1)_i$ Expression	Infinitesimal Work Transfer Interaction, δW, Eq. (1.8)	Notation
Kinetic, translational: $\dfrac{1}{2}mV_2^2 - \dfrac{1}{2}mV_1^2$	$F = m\dfrac{dV}{dt}$	$-F\,dx$	(a)
Kinetic, rotational: $\dfrac{1}{2}J\omega_2^2 - \dfrac{1}{2}J\omega_1^2$	$T = J\dfrac{d\omega}{dt}$	$-T\,d\theta$	(b)
Spring, translational: $\dfrac{1}{2}kx_2^2 - \dfrac{1}{2}kx_1^2$	$F = kx$	$-F\,dx$	(c)

Spring, rotational:	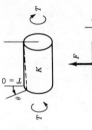 (d)	$-T\,d\theta$	$T = K\theta$	$\frac{1}{2}K\theta_2^2 - \frac{1}{2}K\theta_1^2$
Gravitational spring (or constant-force translational spring):	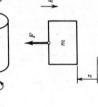 (e)	$-F\,dz$	$F = mg$	$mgz_2 - mgz_1$
Electrical capacitance:	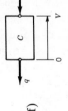 (f)	$-V\,dq$	$V = \dfrac{q}{C}$	$\frac{1}{2}\dfrac{q_2^2}{C} - \frac{1}{2}\dfrac{q_1^2}{C}$
Electrical inductance:	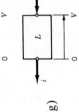 (g)	$-V\,dq$	$V = L\dfrac{di}{dt}$	$\frac{1}{2}Li_2^2 - \frac{1}{2}Li_1^2$

21

implicit in Galileo Galilei's earlier formula for the velocity of a free-falling body, $V = (2gs)^{1/2}$, where s is the travel measured downward from the position of rest. The *vis-viva* theory entered the realm of fluid mechanics in 1738 through Daniel Bernoulli's famous treatise on hydrodynamics [28], and, in an isolated earlier instance, through Torricelli's 1644 formula for the discharge velocity of a fluid driven by its own weight through an orifice. The "internal energy" term and the symbol U come from the works of Clausius [19, 29] and Rankine [26], although the words "inner work," "internal work," and "intrinsic energy" were also used by their engineering contemporaries (e.g., Zeuner [20]). The name "energy," which in thermodynamics was proposed by William Thomson in 1852, had been coined in 1807 by Thomas Young, the discoverer of the phenomenon of optical interference [30]. Additional highlights of the history of first-law concepts are given in Table 1.2 (p. 31) and in the closing sections of this chapter.

Example 1.1. Consider a rigid and evacuated container (bottle) of volume V that is surrounded by the atmosphere (T_0, P_0). At some point in time, the neck valve of the bottle opens, and atmospheric air gradually flows in. The wall of the bottle is thin and conductive enough so that the trapped air and the atmosphere eventually reach thermal equilibrium. In the end, the trapped air and the atmosphere are also in mechanical equilibrium, because the neck valve remains open.

We are asked to determine the net heat transfer interaction that takes place through the wall of the bottle during the entire filling process. The challenge consists of solving the problem using the first-law statement for *closed systems*, eq. (1.1).

As the closed system in this example, we identify the total air mass that eventually rests inside the bottle:

$$m = \frac{RT_0}{P_0 V} \tag{a}$$

The final state of the system is represented by the properties (T_0, P_0, V). Next, we try to visualize the position of the air mass m in the beginning of the process: that mass resides outside the bottle, and its temperature and pressure are atmospheric. Using eq. (a) and the $PV = mRT$ equation of state, we learn that the original volume occupied by m outside the bottle is also equal to V.

Using (1) and (2) for the beginning and the end of the process by which the closed system m moves inside the bottle, the first law provides the equation with which to calculate the unknown Q_{1-2}:

$$Q_{1-2} - W_{1-2} = U_2 - U_1 \tag{b}$$

Since we are treating the air mass as an ideal gas in which $T_1 = T_2 = T_0$, we note that $U_2 - U_1 = mc_v(T_2 - T_1) = 0$, i.e.,

$$Q_{1-2} = W_{1-2} \tag{c}$$

Finally, we calculate the work transfer interaction by noting that *two* portions of the boundary of system m move during the process: first, the interface between m and the rest of the atmosphere, and, second, the interface between m and the evacuated space. The pressure along these two surfaces are P_0 and 0, respectively, which means that a work transfer interaction is associated only with the movement of the first interface:

$$W_{1-2} = \int_1^2 P\,dV = P_0 \int_1^2 dV \tag{d}$$

The volume integral $\int_1^2 dV$ represents the volume swept by the interface as it is being pushed inward by the atmosphere. The size of the volume integral is $-V$, where the "V" comes from the volume originally occupied by m outside the bottle (note that V must be swept by the interface entirely if m is to end up in the bottle), and where the "$-$" comes from the fact that the interface moves in the direction of the forces applied by the ambient on the moving boundary [review the work transfer definition (1.7)]. In conclusion

$$Q_{1-2} = -P_0 V < 0 \tag{e}$$

In other words, the physical sense of the heat transfer through the bottle wall is such that the atmosphere acts as heat sink. In absolute terms, the heat transfer rejected to the atmosphere matches the work done by the atmosphere for the purpose of "extruding" m through the neck valve. We consider this problem again in Example 1.2.

THE FIRST LAW FOR OPEN SYSTEMS

In engineering thermodynamics, we rely most often on a generalization of the first-law statement that in the preceding section was reviewed in the context of closed systems. The generalization is almost as old as the statement of the first law for closed systems: it consists of allowing for mass flow across certain portions of the system boundary and writing the equivalent of eqs. (1.1) or (1.5) for an open system (control volume). The basic assumptions that make this classical restatement of the first law possible become clearer if we review the main steps involved in extending eq. (1.1) to open systems.

Figure 1.7 shows the main features of an open system, namely, heat transfer interactions per unit time, $\dot{Q}$; work transfer interactions per unit time, $\dot{W}$; and portions of the boundary that are crossed by the flow of mass. For simplicity, the figure shows only one of each type of boundary crossing, one inlet port labeled "in," and one outlet port labeled "out." The open system, or the control volume, is the rectangular region contained between the inlet and outlet ports, in other words, the stationary dashed lines labeled "in" and "out" belong to the boundary of the open system. Note also that the work transfer rate term $\dot{W}$ refers to any mode of combination of modes of work transfer, $P\,dV/dt$, $\dot{W}_{sh}$, $\dot{W}_{electrical}$, $\dot{W}_{magnetic}$, etc.

Since the first-law statement (1.1) applies strictly to closed systems, we must first identify a system with a fixed mass inventory that is unambiguously related to the open system of interest. If M_{open} is the mass inventory of the open system at a certain point in time t, then we can think of the fixed mass inventory M_{closed} that at time t "flows" through the control volume. According to Fig. 1.7, the relationship between M_{open} and M_{closed} is

$$M_{closed} \text{ (constant)} = M_{open,t} + \Delta M_{in} = M_{open,(t+\Delta t)} + \Delta M_{out} \quad (1.14)$$

For the process from state (1) (time t) to state (2) (time $t + \Delta t$) executed by the closed system, the First Law of Thermodynamics (1.1) reads

$$E_{closed,(t+\Delta t)} - E_{closed,t} = \dot{Q}\,\Delta t - \dot{W}\,\Delta t + (P\,\Delta V)_{in} - (P\,\Delta V)_{out} \quad (1.15)$$

The last two terms appearing on the right side account for the $P\,dV$-type work transfer associated with the deformation of the closed system from time t to time $t + \Delta t$. Note that in each term, P is the local pressure, that is, the pressure in the immediate vicinity of the port. Relations similar to eqs. (1.14) express the relative size of the energy inventories of closed and open systems:

$$E_{closed,\,t} = E_{open,t} + \Delta E_{in} \quad (1.16)$$

$$E_{closed,(t+\Delta t)} = E_{open,(t+\Delta t)} + \Delta E_{out} \quad (1.17)$$

Furthermore, the ΔEs and ΔVs can be rewritten in terms of their per-unit-mass counterparts e and v as

$$(\Delta E)_{in,out} = (e\,\Delta M)_{in,out} \quad \text{and} \quad (\Delta V)_{in,out} = (v\,\Delta M)_{in,out} \quad (1.18)$$

Like the port pressure P, the specific energy and volume (e and v,

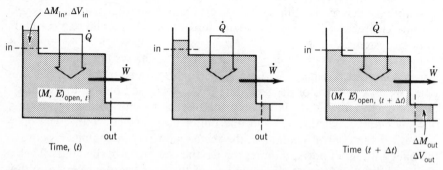

Figure 1.7 The flow of a closed system (the shaded area) through the space occupied by an open system, and the conversion of the first-law statement for closed systems into a statement valid for open systems.

respectively) are properties of the intensive state of the fluid that crosses the boundary at time t. Combining eqs. (1.15)–(1.17) for the purpose of eliminating the terms that refer to the energy inventory of the closed system (E_{closed}), we obtain

$$\frac{1}{\Delta t}\left(E_{open,(t+\Delta t)} - E_{open,t}\right) = \dot{Q} - \dot{W} + \left[(e + Pv)\frac{\Delta M}{\Delta t}\right]_{in} - \left[(e + Pv)\frac{\Delta M}{\Delta t}\right]_{out}$$

(1.19)

Invoking the limit $\Delta t \to 0$, writing $\dot{m}$ for the mass flowrate $\Delta M/\Delta t$, dropping the subscript "open" from the energy inventory of the control volume, and allowing for the existence of more than one inlet port and outlet port, we arrive at the most general statement of the First Law of Thermodynamics for an open system:

$$\frac{dE}{dt} = \dot{Q} - \dot{W} + \sum_{in} \dot{m}(e + Pv) - \sum_{out} \dot{m}(e + Pv)$$

(1.20)

What makes this statement more general than the per-unit-time version of the first law for closed systems, eq. (1.5), is the presence of the terms $\dot{m}(e + Pv)$: these terms represent the energy transfer associated with the flow of mass across the system boundary. Finally, in the absence of macroscopic forms of energy storage other than kinetic and gravitational, the specific energy e can be decomposed into $(u + \frac{1}{2}V^2 + gz)$, eq. (1.13). The result of this decomposition is that the specific *enthalpy*

$$h = u + Pv$$

(1.21)

shows up explicitly in the terms accounting for energy transfer via mass flow:

$$\frac{dE}{dt} = \dot{Q} - \dot{W} + \sum_{in} \dot{m}(h + \frac{1}{2}V^2 + gz) - \sum_{out} \dot{m}(h + \frac{1}{2}V^2 + gz)$$

(1.22)

In the fields of gas dynamics and compressible fluid mechanics, the group $(h + \frac{1}{2}V^2)$ is recognized as the local *stagnation enthalpy* of the flowing fluid. In 1966, Kestin proposed an engineering generalization of the enthalpy concept under the name *methalpy* [symbol $h°$; Ref. 24, p. 223]:

$$h° = e + Pv = h + \frac{1}{2}V^2 + gz$$

(1.23)

which is intended to mean "beyond enthalpy" or "transcending enthalpy" [note the Greek word *meta* (beyond)].

Following a procedure that is analogous to the closed system to open system transformation illustrated in Fig. 1.7, the first-law statement (1.22) can be generalized further by considering the class of open systems where the inflows and the outflows are not restricted to penetrating discrete patches (ports) on the control surface. If $\mathscr{A}$ is the closed control surface that contains the control volume $\mathscr{V}$, the First Law of Thermodynamics reads

$$\int_{\mathscr{V}} \frac{\partial(\rho e)}{\partial t}\, d\mathscr{V} = -\int_{\mathscr{A}} \mathbf{q} \cdot \mathbf{n}\, d\mathscr{A} - \dot{W} - \int_{\mathscr{A}} \rho h^{\circ} \mathbf{v} \cdot \mathbf{n}\, d\mathscr{A} \qquad (1.24)$$

In this expression, $\mathbf{q}$ and $\mathbf{v}$ represent the heat-flux vector and the velocity vector, respectively, at the points that make up the control surface. The unit vector $\mathbf{n}$ is normal to the control surface and points outward. The specific energy e and methalpy h° are the local properties of the material that resides inside the volume element $d\mathscr{V}$ and along the area element $d\mathscr{A}$ (this observation is amplified on p. 71).

The first-law statement for a control volume of point size situated inside $\mathscr{V}$ is obtained by transforming the surface integrals of eq. (1.24) into volume integrals via the divergence theorem:

$$\frac{\partial}{\partial t}(\rho e) = -\nabla \cdot \mathbf{q} - \rho h^{\circ} \nabla \cdot \mathbf{v} - w''' \qquad (1.25)$$

or, using the mass-continuity statement shown later, eq. (1.29),

$$\rho \frac{\partial e}{\partial t} = -\nabla \cdot \mathbf{q} - P\nabla \cdot \mathbf{v} - w''' \qquad (1.26)$$

In these expressions, w''' represents the contribution made by the point-size system to the overall work transfer rate $\dot{W}$ delivered by the finite-size control volume $\mathscr{V}$, in other words, $\dot{W} = \int_{\mathscr{V}} w'''\, d\mathscr{V}$ (note the definition of positive $\dot{W}$, Fig. 1.3). In the continua studied in the field of conduction heat transfer, w''' usually accounts for the negative of the volumetric rate of electrical power dissipation q''' [31]. In the fluid media encountered in convection heat transfer, w''' accounts for both $-q'''$ and the negative of the work done via viscous forces on the point-size control volume [Ref. 1, pp. 8–11].

The $\Delta t \to 0$ limit that led to the first-law statement (1.22) can be invoked also in connection with the second of equations (1.14) to yield the *mass-conservation* statement:

$$\frac{dM}{dt} = \sum_{\text{in}} \dot{m} - \sum_{\text{out}} \dot{m} \qquad (1.27)$$

This equation spells out the basic difference between open systems and closed systems (in the latter, the $\dot{m}$s are all zero and the mass inventory M is a constant). In the language of eq. (1.24), that is, for a control volume $\mathscr{V}$ enclosed by a permeable control surface $\mathscr{A}$, the mass-conservation equation is

$$\int_V \frac{\partial \rho}{\partial t} \, d\mathcal{V} = -\int_{\mathcal{A}} \rho \mathbf{v} \cdot \mathbf{n} \, d\mathcal{A} \tag{1.28}$$

The corresponding statement for a control volume of point size is

$$\frac{\partial \rho}{\partial t} = -\rho \nabla \cdot \mathbf{v} \tag{1.29}$$

Important in engineering applications is a special class of open systems whose inventories of mass (M), energy (E), and entropy $(S$, chapter 2$)$ are time-independent. Such systems are said to operate in the *steady-state* or the *stationary regime*. The equations that govern their operation are simpler because time derivatives such as dE/dt and dM/dt in eqs. (1.22) and (1.27) vanish. It is important to keep in mind that the constancy of the M, E, and S inventories in time does not mean that the mass, energy, and entropy are distributed uniformly through the space occupied by the open system. The steady state should not be confused with the spatial uniformity of the intensive state.

The First Law of Thermodynamics for open systems and its enthalpy-based presentation illustrate admirably the aging of engineering thermodynamics into a discipline that threatens to lose sight of its origins. The first law for open systems was first stated by Gustav Zeuner as part of the analysis of flow systems that operate in the steady state: he made this result known primarily through his technical thermodynamics treatise, whose first German edition was published in 1859 [Ref. 20, pp. 225–231]. Equally impressive is that Zeuner saw and stressed the important role played by the first law in fluid mechanics next to the other equations that in his time were recognized as the pillars of fluid and gas dynamics [32]. Zeuner's name never made it into fluid mechanics vocabulary; more surprising is that it disappeared from engineering thermodynamics beginning with the turn of the century.[†] The most recent reference I can find in connection with

---

[†]About the forget-first-the-engineer syndrome, Rankine wrote in 1859:

> ... the improvers of the mechanical arts were neglected by biographers and historians, from a mistaken prejudice against practice, as being inferior in dignity to contemplation; and even in the case of men such as Archytas [an ancient Greek philosopher] and Archimedes, who combined practical skill with scientific knowledge, the records of their labours that have reached our time give but vague and imperfect accounts of their mechanical inventions, which are treated as matters of trifling importance in comparison with their philosophical speculations. The same prejudice, prevailing with increased strength during the middle ages, and aided by the prevalence of the belief in sorcery, rendered the records of the progress of practical mechanics, until the end of the fifteenth century, almost a blank. Those remarks apply, with peculiar force, to the history of those machines called PRIME MOVERS ... [Ref. 26, p. xv].

Which is why Rankine—the engineer and cofounder of classical thermodynamics (next to Clausius and Kelvin)—is almost never mentioned by the philosophes.

"Zeuner's formula" is in Stodola's treatise on steam turbines, first published in German in 1903 [33]. Zeuner's statement of the first law for steady flow and the argument on which its derivation was based are present in virtually every engineering thermodynamics treatise of the twentieth century. On this background of namelessness, it is not surprising that in 1965 Hatsopoulos and Keenan give credit to Keenan's 1941 treatise [34] for providing a rigorous proof for a truth that to some segments of the peer group must have seemed known, natural, and of spontaneous origin [Ref. 11, p. XXIV].

Another noteworthy example of death and forgetting in the world of engineering thermodynamics has to do with the invention of the word "enthalpy." First, it is interesting that the widespread use of the term in engineering was triggered by the work of another professor from the old University of Dresden, Richard Mollier (the other influential Dresden figure had been Gustav Zeuner). Mollier recognized the importance of the group $(u + Pv)$ in the first-law analysis of steam turbines, next to entropy (s) in second-law analysis. He presented graphically and in tabular form the properties of steam as the now famous enthalpy–entropy chart (the Mollier chart, h–s) [35]. Mollier referred to the group $u + Pv$ as "heat contents" and "total heat" and labeled it "i." G. A. Goodenough, famous professor of thermodynamics at the University of Illinois, called i "thermal potential" and "thermal head" [36]. The symbol i was used until about twenty years ago in the engineering thermodynamics taught in German, Russian, and in the languages of the other Europe. Mollier's contribution is not the discovery of the group $u + Pv$—this group was known already as Gibbs' "heat function for constant pressure" [symbol χ; Ref. 17, p. 92]—rather, it is the invention of an important graphical tool whose impact on the efficiency of slide-rule calculations in thermal design is beyond question.

At least from the point of view of North American engineers, Mollier's "total heat" appears to have been replaced spontaneously by the term "enthalpy" somewhere in the 1930s. Some authors explain the correct pronunciation of enthalpy[†] [e.g., Ref. 37], however, the originator of this terminology is not mentioned. The term "enthalpy" was coined by Kamerlingh-Onnes [38], professor at the University of Leiden, otherwise famous for having been the first to liquefy helium and to discover the phenomena of superconductivity and superfluidity. Part of the mystery that persists in the wake of Kamerlingh-Onnes' innovations is due to the limited circulation enjoyed by his original writings, for which he used Dutch as language and the bulletin of his own low-temperature laboratory as journal [39].

Example 1.2.   Consider again the problem stated in Example 1.1, this time in the more natural context of *open systems* (after all, this phenomenon is the classical "filling" process). The object is to determine the heat transfer interaction that crosses the bottle wall during the filling process.

[†]Accent on the second syllable; from the Greek word *enthalpein* (to heat).

As "open system," we choose the space contained by the bottle. The system has one inlet port (the neck valve), and the operation of the system is definitely unsteady (the system accumulates mass during the process). The mass-conservation equation and the first law require at any instant that

$$\frac{dM}{dt} = \dot{m} \tag{a}$$

$$\frac{dU}{dt} = \dot{Q} + \dot{m}h_0 \tag{b}$$

where M and U are the instantaneous inventories of mass and internal energy of the system. Symbols $\dot{Q}$, $\dot{m}$, and h_0 stand for the instantaneous heat transfer rate into the system, the instantaneous inlet flowrate, and the enthalpy of atmospheric air, $h_0(T_0, P_0)$ = constant. The unknown in this problem is the integral

$$Q_{1-2} = \int_1^2 \dot{Q}\, dt \tag{c}$$

where 1 and 2 denote the start and finish of the filling operation, respectively. Combining eqs. (a)–(c), it is easy to show that

$$Q_{1-2} = U_2 - U_1 - h_0(M_2 - M_1) \tag{d}$$

or, since the open system is initially evacuated ($U_1 = 0$, $M_1 = 0$),

$$Q_{1-2} = U_2 - M_2 h_0 \tag{e}$$

Finally, we note that $U_2 = M_2 u_0$, where u_0 is the specific internal energy of air at T_0 and P_0 (recall that $T_2 = T_0$ and $P_2 = P_0$). Combining eq. (e) with the definition of enthalpy,

$$h_0 = u_0 + P_0 v_0 \tag{f}$$

and noting that $V = M_2 v_0$, we arrive at the same answer as in Example 1.1:

$$Q_{1-2} = -P_0 M_2 v_0 = -P_0 V \tag{g}$$

Comparing the two methods of deriving this answer, my own impression is that the open-system analysis of Example 1.2 is more direct. However, the closed-system approach used in Example 1.1 reveals not only the size and sign of Q_{1-2}, but also the physical meaning of the expression $P_0 V$. We discuss this physical meaning again in Example 2.2.

The combined message of Examples 1.1 and 1.2 is that there is more than one way in which to pursue the solution to a given problem.

HISTORICAL BACKGROUND

The review presented so far emphasized the main concepts associated with the first law and those items that are most likely to lead to confusion in the process of analyzing an engineering problem. I was unable to discuss these points without drawing attention to their historical background: I believe that the effort to understand the pioneers (their personality, research methodology, fights, victories, and disappointments en route to "making it") deserves an emphasis that has been generally avoided in engineering thermodynamics teaching and research. If we are to speak exotic words such as energy, enthalpy and entropy, then the best teachers of this language can only be its inventors.

We develop a better understanding of the meaning of the first law by looking at its position against the development of engineering science in general. A number of highlights are presented in Table 1.2 by recording first a new concept or discovery, the individual responsible for it, and my best estimate of the time frame (usually, the year of publication of the in-novator's main opus). The historic record is so vast that any condensation of the type exhibited here reflects first the writer's bias and incomplete knowledge of history. In the present display, an engineering bias was used intentionally to organize these events in two columns (or "currents") whose confluence is marked by the emergence of thermodynamics in the mid-1800s.

On the practical side, the "work" line refers to man's preoccupation with mechanisms that *transmit* the mechanical power derived from animal, hydraulic, aeolian, or combustible origins, and with machines that *produce* mechanical power while consuming fuel (atmospheric-pumping engines, steam engines). On the practical side of man's preoccupation with hotness (the "heat" line, Table 1.2), we recall the measurement of temperature, quantity of heat, and—in more recent times—the rate of heat propagation (heat transfer). One very important practical aspect of the "heat" line was the recording of the changes undergone by various substances under the influence of heating and cooling, for example, the dilation of thermometric fluids and "permanent gases." Through such experiments, it was discovered that the state of a certain batch is determined by *two independent properties* in addition to the mass of the batch that was being studied. It is this early work that provides the empirical foundation for what we now call *state principle* and for the analytical and tabular summaries known as *equations of state* (chapter 4).

The confluence of the "work" and "heat" lines was accompanied on a theoretical level by the coexistence of two views on the nature of hotness: (i) the mechanical theory holding that heat is the manifestation of motion (live force, p. 19) at the molecular level, and (ii) the material or caloric theory maintaining that the caloric fluid contained in a substance is uniquely defined if the state is specified. A key word in describing the evolution of

TABLE 1.2 Highlights in the Conceptual Development of Classical Thermodynamics

The *Work* Line	The *Heat* Line
Machines	*Thermometry*
The 12th century: gunpowder is brought from China to Europe, marking the beginning of the technology of firearms. From Manchester, which as an intellectual environment had played a leading role in the birth of thermodynamics, Osborne Reynolds remarked that "the combustion, in the form of the cannon is the oldest form of heat engine." A similar view had been advocated earlier by Amontons and Daniel Bernoulli	Galilei's barothermoscope (1592): a glass bulb filled with air and having a downward stem dipped into a pool of mercury
	Sealed-stem thermometers filled with alcohol; stem calibrated in thousandths parts of bulb volume (Grand Duke Ferdinand II of Tuscany, 1654)
The 13th, 14th, and 15th centuries: the proliferation of water-driven machines, air bellows, water pumps, irrigation, and the draining of mines	The air thermometer: a volume of air confined by a column of mercury as indicator (Amontons, late 1600s)
The technology and study of pumps (Stevinus, 1586; della Porta, 1601)	The mercury-in-glass thermometer (Fahrenheit, 1714; the empirical Fahrenheit, Réaumur, and Celsius scales are described on p. 16)
	Calorimetry
The use of mathematical analysis in mechanics, the motion under the influence of gravity, the first instrument for measuring temperature (Galilei, 1623; see also "thermometry" in the adjacent column)	The elasticity of a gas, PV = constant at constant T (Boyle, 1660, Mariotte, 1679, both preceded by Towneley, Boyle's student)
The barometer, the orifice velocity of a fluid driven by its own weight (Torricelli, 1644)	The phlogiston theory: phlogiston is a substance without weight, odor, color, or taste that is contained by all flammable bodies and is given off during burning (advanced by Becher; extended and made popular by Stahl in the late 1600s)
A basic understanding of the origins of atmospheric pressure (Pascal, 1648)	The constancy of temperature during phase change (Newton, 1701; observed also by Amontons)
The invention, demonstration, and popularization of the air (vacuum) pump (Otto von Guericke, 1654 and later. Noteworthy is his 1672 book in which he makes popular the idea that the weight of the atmosphere can be put to work: the famous woodcut known as the "Magdeburg	The foundations of quantitative calorimetry, the concepts of "quantity of heat," "latent heat," the discovery of CO_2, called "fixed air" (Black, late 1700s)

TABLE 1.2 (Continued)

The *Work* Line	The *Heat* Line
hemispheres" shows two eight-horse teams trying to pull apart two 36-cm-diameter hemispheres from which the air had been evacuated—an exaggerated image that invited the work on atmospheric-pumping engines, e.g., Huygens, 1657)	The discovery of oxygen by Priestley (1774), who called it "dephlogisticated air" (later it was named "oxygen" by Lavoisier, who by explaining combustion, discredited the phlogiston theory; Priestley also discovered sulfur dioxide and ammonia)
Captain Thomas Savery builds the first atmospheric engine (1698): The development of heat engines in the prethermodynamics era continues in Fig. 2.1	The latent heat of fusion of ice and the concept of "specific heat" (Wilcke, 1772, 1781)
From Mechanics to Machine Science	Lavoisier and Laplace publish *Memoire sur la Chaleur* (1783): a systematic foundation for the science of calorimetry, the heat conservation axiom ("all variations in heat, real or apparent, which a system of bodies undergoes in changing state are reproduced in inverse order when the system returns to its final state"), the calorimetric measurement of specific heat, heat of reaction, etc. In 1789, Lavoisier publishes "*Traité élémentaire de chimie*: a system of chemistry in which the caloric fluid (*calorique*) is chosen as one of the simple substances or elements. The material or caloric theory of heat becomes established
The *vis-viva* theory or the conservation of live force, the method of infinitesimal calculus along with the system of notation that was universally adopted (Leibnitz, 1684; Newton's calculus was published three years later)	The gas law $V \sim T$ at constant P (Gay-Lussac, 1802; he also discovered that $\Delta U = 0$ at constant T in gases)
The law of universal gravitation, the three laws of motion, calculus presented in geometric terms (Newton, 1687)	The law of partial pressures in gas mixtures (Dalton, 1805); Avogadro's law (1811)
The conservation of live force in hydraulics, the kinetic-molecular theory of gases (Daniel Bernoulli, 1738)	The discovery of "critical temperature" (Cagnard Latour, 1810s)
The mathematical foundations of inviscid fluid flow (Euler, 1755)	The approximate character of Boyle's law for real gases, the careful measurement of the specific heat and thermal expansion coefficient of gases, liquids, and solids (Regnault, mid-1800s)
The gravitational field theory, the mathematics of thermal diffusion (Laplace, 1785 and later)	
The law governing friction (Coulomb, in a 1781 prize-winning paper)	
The equations of analytical mechanics (Lagrange, 1788)	
The foundations of descriptive geometry (Monge, 1795)	
The beginnings of a science of machines (mechanisms), the "Carnot principle" of avoiding shocks, percussion and	

The proportionality between cooling rate and body-surroundings temperature difference (Newton, 1701)

Comparative measurement of thermal conductivity (Ingen Housz, 1785, 1789; Count Rumford, 1786 and later)

Convection as a principal heat transfer mechanism through clothing (Rumford, 1797; the word "convection" was coined by Prout in 1834)

The proportionality between heat transfer rate and temperature gradient, also the distinction between the thermal conductivity coefficient and the coefficient in Newton's law of cooling (Biot, 1804)

Fourier formulates the partial differential equation for time-dependent heat conduction (1807): in today's language, this ranks as the first analytical formulation of the first law (in the context of zero-work processes, Table 1.3). The field of heat transfer continues to develop along the pure "heat" line into the late 1900s, when it is reunited with the field of engineering thermodynamics [13]

---

turbulent flow in order to achieve maximum efficiency or continuity in the transmission of mechanical power (Lazare Carnot, 1783; he also defines the concept of "moment of activity," that in 1829 was named "work," independently by Coriolis and Poncelet)

The *École Polytechnique* is established in 1795: under its influence and through the teachings of some of its first graduates, the study of machines becomes central to engineering education everywhere (e.g., courses by Navier, 1826, Coriolis, 1829, Poncelet 1829)

The "dynamic unit" or "dynamode," as the work required to raise one kilogram to a height of one meter (Hachette, 1811); the "calorie" was defined as the quantity of heat required to raise the temperature of 1 kilogram of water by 1°C (Clément, 1826)

The assault on the conservation of caloric doctrine (Count Rumford, 1798 and later; Sir Humphry Davy, 1799)

The "heat" equivalent of "work," or traditionally, the "mechanical equivalent of heat" (the theoretical line: Mayer, 1842 and later, also Séguin, 1839, Holtzmann, 1845; the experimental line: Joule, 1843 and later, enriched by Violle, 1870, Rowland, 1879, Hirn, and others)

The "first law" as an integral part of the new science of "thermodynamics" (Clausius, 1850 and later; Rankine, 1850 and later; Kelvin, 1851 and later)

these two views is "coexistence," which means that the mechanical and caloric theories were not mutually exclusive and that they were accepted together as complementary. Note that the success and wide acceptance of the newer theory (the caloric theory) was the result of the great service that this theory rendered to the quantitative fields of thermochemistry and "heat engineering."

However, no theory is perfect and forever. One respect in which the caloric theory failed—the generation of heat through friction—was well known in the 1700s, and was certainly known by Lavoisier and Laplace. This particular limitation of the caloric theory was assaulted in a series of papers started in 1798 by Count Rumford [40, 41] based on exhaustive and otherwise approximate observations of the heat generated by a drill during the boring of a cannon.[†] This theme was advocated also by Sir Humphry Davy, who communicated that he was able to induce the melting of two blocks of ice by rubbing them against each other. Although the correctness of Davy's

---

[†]Count Rumford's 1798 paper begins with advice to all researchers to keep their eyes open:

> It frequently happens, that in the ordinary affairs and occupations of life, opportunities present themselves of contemplating some of the most curious operations of nature; and very interesting philosophical experiments might often be made, almost without trouble or expence, by means of machinery contrived for the mere mechanical purposes of the arts and manufacturer.

> I have frequently had occasion to make this observation; and am persuaded, that a habit of keeping the eyes open to every thing that is going on in the ordinary course of the business of life has oftener led, as it were by accident, or in the playful excursions of the imagination, put into action by contemplating the most common appearances, to useful doubts and sensible schemes for investigating and improvement, than all the more intense meditations of philosophers, in the hours expressly set apart for study [Ref. 40, p. 80].

About the experiment in which water boiled as a result of frictional heating, he wrote:

> At 2 hours 20 minutes it was at 200°(F); and at 2 hours 30 minutes it ACTUALLY BOILED!

> It would be difficult to describe the surprise and astonishment expressed by the countenances of the by-standers, on seeing so large a quantity of cold water heated, and actually made to boil, without any fire.

> Though there was, in fact, nothing that could justly be considered as surprising in this event, yet I acknowledge fairly that it afforded me a degree of childish pleasure, which, were I ambitious of the reputation of a *grave philosopher*, I ought most certainly rather to hide than to discover [Ref. 40, p. 92].

His final and famous conclusion on the origin of the observed heating effect was

> It is hardly necessary to add, that any thing which any *insulated* body, or system of bodies, can continue to furnish *without limitation*, cannot possibly be a *material substance*: and it appears to me to be extremely difficult, if not quite impossible, to form any distinct idea of any thing, capable of being excited, and communicated, in the manner the heat was excited and communicated in these experiments, except it be MOTION [Ref. 40, p. 99].

communication has been questioned [42], and despite the view expressed by some that Rumford had a gift for exaggeration, the vociferous attack on one shortcoming of the caloric theory played an important role in the developments that were to take place in the 1840s and 1850s. The Rumford–Davy theme did not win many converts at the turn of the century, however, it was used conveniently and successfully by the founders of thermodynamics (Joule even attributed to Rumford the honor of having first measured the heat equivalent of work—this, in order to draw the establishment's attention to his own measurements [43]).

Even more important is that the Rumford–Davy line of questioning contributed to preparing an audience for the theory that was to arrive. Note in this regard Sadi Carnot's *Manuscript Notes* papers, which were saved and revealed first in 1871 by Sadi's brother Hippolyte and which had been written most likely around the time of his 1824 memoir [44]. In these notes, we find that Sadi Carnot had questioned the conservation of caloric doctrine and decided, "Heat is nothing more than motive power, or, in other words, the motion that has changed form. Wherever motive force is produced, there is always production of heat in a quantity precisely proportional to the quantity of motive power destroyed. Conversely: wherever there is destruction of heat, there is production of motive power." He concludes that "one can state as a general thesis that the motive power is an invariable quantity in nature; that it is never, properly speaking, produced or destroyed. In fact, it changes form, i.e., it produces sometimes one kind of motion, sometimes another, but it is never exhausted" [Ref. 21, p. 51; my translation]. After this unambiguous statement of the future principle of the conservation of energy (the first law), he reports an estimate of the heat equivalent of one unit of motive power, which today amounts to 3.7 joules/calorie, i.e., only 12 percent below the modern value of 4.186 joules/calorie.

Sadi Carnot's unpublished notes do not in any way detract from the credit that Mayer and Joule deserve for publishing similar views first and then for *fighting the battle to have them accepted*. The notes show, however, that Sadi Carnot's thinking was well ahead of the views expressed in his published memoir and, quite possibly, that the conservation of caloric doctrine was losing ground among some of his peers. One of these, Emile Clapeyron, wrote in 1834 that "a quantity of mechanical action, and a quantity of heat which can pass from a hot body to a cold body, are quantities of the same nature, and that it is possible to replace the one by the other; in the same manner as in mechanics a body which is able to fall from a certain height and a mass moving with a certain velocity are quantities of the same order, which can be transformed one into the other by physical means" [Ref. 6, pp. 36–51].

The first clear-cut theory expressing that "heat" and "work" are equivalent and that their respective units are convertible was published independently by Mayer in May 1842 [45] and Joule in August 1843 [46]. This dual approach to such a great step is a perfect example of how differently two individuals can think, and a very strong case for free access to the marketplace of ideas as the best recipe for scientific progress. An important item in the history of the first law is the fact that both Mayer and Joule had difficulty in getting their papers published and in being taken seriously by their established contemporaries.

Mayer was the theoretician, the man obsessed by the idea: he conceived it in circumstances that even today appear removed from the thermodynamics scene (more on this sortly), and then relied on the contemporary state of knowledge in order to support its validity. Joule, on the other hand, was the ultimate experimentalist: he first discovered in his measurements that the heat generated by electrical resistances is proportional to the mechanical power required to generate the electrical power. He then recognized the importance of this proportionality and drew the revolutionary conclusion that a *universal* proportionality must exist between the two effects (work and heat). Only to polish this idea and to convince the skeptics (e.g., the Royal Society), he produced a series of nakedly simple experiments whose message proved impossible to refute. From the point of view of mechanical engineers, the most memorable among these experiments was the heating of a pool of water by an array of paddle wheels driven by falling weights.

Mayer's entrance on this controversial stage was considerably different. A medical doctor by training, he was serving as surgeon on a ship sailing through the East Indies when he observed that the blood of European sailors showed a brighter color of red, i.e., a smaller rate of oxidation. This was in July 1840. He attributed this observation to the high temperatures near the equator, i.e., to the lower metabolic rate that is needed to maintain the body temperature. Mayer, however, went beyond the connection between the chemical energy contained in food and the rejection of body heat to the ambient: he saw the energy of food as the common source of both body heat and muscular work, in other words, he saw intuitively that heat and work have similar origins and are interconvertible.

It was pointed out by Epstein [47] that the field of physiology had already seen a "connection" between heat and work in the form of an outmoded theory of respiration that claimed that body heat is generated by the friction of air in the air passages of the lungs. The theory was being displaced at the time by Lavoisier's conclusive theory of oxidation. Epstein's research of the physiology literature and of the healthy intellectual milieu of Manchester (of which Cardwell [9] writes so well) goes on to suggest that the physiologists' connection between heat and work was also known to Joule.[†]

[†]Revealing is Tisza's one-sentence dismissal of the suggestion that Joule's work and ideas may have had something to do with physiology and Count Rumford [Ref. 10, p. 25]: this is just one of the many subtle attempts to banish the thought that the pioneers might have been something other than physicists, i.e., engineers (Sadi Carnot, Clapeyron, Séguin, Rankine, and the many engine builders of Scotland and Cornwall, Fig. 2.1), medical doctors (Mayer, Helmholtz), public servants (Fourier, Lazare Carnot), and—we should not forget— military men (Lazare and Sadi Carnot, Count Rumford, Helmholtz). The symbol of this diverse group is Joule himself—a man who received no formal education, i.e., a veritable amateur.

Mayer was very clear about the meaning of his theory: "We must find out how high a particular weight must be raised above the surface of the earth in order that its falling power may be equivalent to the heating of an equal weight of water from 0° to 1°C" [Ref. 45, p. 240]. He reasoned that an amount of gas has to be heated more at constant pressure than at constant volume, because at constant pressure, it is free to dilate and do work against the atmosphere; in today's notation, we would write $mc_P \Delta T - mc_v \Delta T = P_{atm} \Delta V$, where ΔV is the volume increment associated with ΔT and $P = P_{atm}$. Using the c_P and c_P/c_v constants known in his time, he estimated the left side of the equation in calories, while the right side was known in mechanical units. He established the equivalence between these units numerically by listing "365m" as the answer to the question quoted earlier in this paragraph (this number corresponds to 3.58 joules/calorie, i.e., it is nearly the same as Sadi Carnot's estimate of p. 35). Worth noting is that if we use Clapeyron's equation $Pv = RT$ in Mayer's argument previously, we arrive at "$c_P - c_v = R$": this classical relation between the specific heats of an ideal gas is recognized by some as *Mayer's equation*.

Most if not all the credit for convincing the skeptics and putting the heat/work equivalence on the books belongs to Joule, whose experiments found a strong and very influential supporter in William Thomson.[†] Mayer received recognition for his theoretical contribution later, thanks to the efforts of Tyndall, Helmholtz, and, among engineers, Zeuner. Helmholtz's letter to Tait on behalf of Mayer [49, 50] is, in retrospect, an important statement on scientific progress and a discoverer's troubles with the establishment in general:

> I must say that to me the discoveries of Kirchhoff in this area (radiation and absorption) appear to be one of the most instructive cases in the history of science, precisely because so many others had previously been so close to making the same discoveries. Kirchhoff's predecessors in this field were related to him in roughly the same way in which, with respect to the conservation of force, Robert Mayer, Colding, and Seguin were related to Joule and William Thomson.

[†]In a note dated 1885, Joule wrote:

> It was in the year 1843 that I read a paper "On the Calorific Effects of Magneto-Electricity and the Mechanical Value of Heat" to the Chemical Section of the British Association assembled at Cork. With the exception of some eminent men . . . the subject did not excite much general attention; so that when I brought it forward again at the meeting in 1847, the chairman suggested that . . . I should not read my paper, but confine myself to a short verbal description of my experiments. This I endeavoured to do, and discussion not being invited, the communication would have passed without comment if a young man had not risen in the section, and by his intelligent observation created a lively interest in the new theory. The young man was William Thomson, who had two years previously passed the University of Cambridge with the highest honour, and is now probably the foremost scientific authority of the age . . . [48].

With respect to Robert Mayer, I can, of course, understand the position you have taken in opposition to him; I cannot, however, let this opportunity pass without stating that I am not completely of the same opinion. The progress of the natural sciences depends always upon new inductions being formed out of available facts, and upon the consequence of these inductions, insofar as they refer to new facts, being compared with reality through the use of experiments. There can be no doubt concerning the necessity of this second undertaking. This part of science often requires a large amount of work and great ingenuity, and we are obligated in the highest degree to those who do it well. The fame of discovery, however, remains with those who have found the new idea; the later experimental verification is quite a mechanical occupation. Further, we cannot demand unconditionally that the person who discovers a new idea also be obligated to carry out the second part of the work. If this were the case, we would have to reject the greatest part of the work of all mathematical physicists. William Thomson, for example, produced a number of theoretical papers concerning Carnot's law and its consequences before he performed a single experiment, and it would not occur to any one of us to treat these lightly.

Robert Mayer was not in a position to conduct experiments; he was repulsed by the physicists with whom he was acquainted (this also happened to me several years later); it was only with difficulty that he could find space for the publication of his first condensed formulation of this principle. You must know that as a result of this rejection he at last became mentally ill. It is now difficult to set oneself back into the modes of thought of that period, and to make clear to oneself how absolutely new the whole idea seemed at that time. I should imagine that Joule too must have fought for a long time in order to gain recognition for his discovery.

Thus, although no one can deny that Joule did much more than Mayer, and although one must admit that in the first publications of Mayer there were many things that were unclear, still I believe that one must accept that Mayer formulated this idea, which determined the most important recent progress in the natural sciences, independently and completely. His reward should not be lessened because at the same time another man in another country and under other conditions made the same discovery and, to be sure, carried it through afterwards better than he did.

THE STRUCTURED PRESENTATION OF THE FIRST LAW

The objective of this chapter has been to review the various concepts of the first law and to reestablish the connection between this law and the historical background on which it was formulated. In line with a general effort to outline the development of the engineering side of classical thermo-dynamics, I find it appropriate to end the chapter by reviewing some of the more recent attempts of streamlining and structuring the presentation of the first law. In this respect, an article written long ago by Keenan and Shapiro [51] provides useful reading. Summarized in Table 1.3 are three formula-

TABLE 1.3 Alternatives for the Structured Presentation of the First Law and Its Concepts[a]

Structure	Poincaré [21]	Carathéodory [52]	Keenan and Shapiro's Second Method [51]
Primary concepts	Work transfer Temperature Heat transfer	Work transfer Adiabatic boundary	Temperature Heat transfer Zero-work boundary
The first law	$\oint \delta W = \oint \delta Q$	$\int_1^2 \delta W_{\text{adiabatic}} = f(1,2)$	$\int_1^2 \delta Q_{\text{zero-work}} = f(1,2)$
The definition of the infinitesimal change in property E	$\delta Q - \delta W$	$-\delta W_{\text{adiabatic}}$	$\delta Q_{\text{zero-work}}$
Derived definition of heat transfer	—	$\delta W - \delta W_{\text{adiabatic}}$	—
Derived definition of work transfer	—	—	$\delta Q - \delta Q_{\text{zero-work}}$
Other derived concepts	Adiabatic boundary Zero-work boundary	Heat transfer Temperature Zero-work boundary	Work transfer Adiabatic boundary

[a]After Ref. 51.

tions that have been proposed in connection with the first law for closed systems. Each formulation reveals the same structure, which begins with the selection of the smallest number of understandable notions that serve as primary concepts for the remainder of the scheme. Additional concepts are later defined using the primary concepts and an appropriate statement of the first law.

Poincaré's Scheme

The first structured presentation of this kind was made by Poincaré [21]. As primary concepts, he chose the heat transfer, the temperature, and the experience condensed in the Zeroth Law of Thermodynamics. For the definition of the concept of heat transfer interaction (heat, in the thermodynamics language of the late 1800s), he relied heavily on the science of calorimetry, which—it is worth noting—is a science that serves us well to this day despite the impression that is fueled by reading too much in the celebrated failure of certain aspects of the caloric theory. The heat transfer interaction was defined as the system–environment interaction made possible by the inequality of temperatures between the two systems. The size of the heat transfer interaction was measured by counting the number of auxiliary standard systems that must be placed in thermal communication with the system of interest. The function of the standard system is to undergo a temperature excursion between two standard temperature levels under the influence of the heat transfer interaction that is to be measured.

As foundation for the first law, Poincaré chose the numerous measurements of the mechanical equivalent of heat pioneered by Joule and continued by Rowland, Violle, and Hirn, and then stated that the work transfer equals the heat transfer during a complete cycle. Finally, Poincaré defined the difference between heat transfer and work transfer as a new quantity called internal energy, and relied on the first law to prove that the internal energy is a thermodynamic property.

Carathéodory's Scheme

Listed in the third column of Table 1.3 is another influential approach, which relies on the concepts of work transfer interaction and adiabatic boundary as primary concepts [52]. The work transfer interaction is defined by reference to the mechanical concept of work (see the general definition in terms of changing the level of a weight, p. 12). While avoiding any reference to thermal concepts that would certainly sound more appealing and natural, the adiabatic wall is defined by the special feature that the "equilibrium of a body enclosed by (this wall) is not disturbed by any external process as long as no part of the wall is moved (distance forces being excluded in the whole consideration)" [53].

The next step in the Carathéodory structure is the first-law statement that the work transfer interaction experienced by a closed system surrounded by an adiabatic boundary depends only on the end states of the process. One consequence of this statement is that the adiabatic work transfer interaction is a thermodynamic property, hence, the definition of the property E:

$$dE = -\delta W_{\text{adiabatic}} \tag{1.30}$$

Under the Carathéodory scheme, the heat transfer interaction becomes a derived concept, as δQ is defined simply as the sum $dE + \delta W$. The temperature becomes also a derived concept: two bodies are said to have the same temperature when the heat transfer interaction δQ (or $dE + \delta W$) is zero in the absence of an adiabatic wall.

Keenan and Shapiro's Second Scheme

Although the Poincaré and Carathéodory presentations cover very well the range from the most successful to the most abstract, they do not represent all the possibilities of introducing the first law and its concepts. In their 1947 article, Keenan and Shapiro [51] proposed two alternate routes, the second of which is summarized in the fourth column of Table 1.3. The primary concepts according to this alternative are the temperature, the heat transfer interaction, and the zero-work boundary. The temperature and the heat transfer are defined in the same way as Poincaré's method. The concept of zero-work boundary, on the other hand, is an instrument required to get to the first law while avoiding any reference to work transfer (note here the relationship between this idea and Carathéodory's). In Keenan and Shapiro's definition, "a zero-work wall is a wall which is motionless (except in the absence of force at the wall) and through which pass no moving force fields and no electrical currents."

The first law amounts to the statement that the heat transfer interaction experienced by a system surrounded by a zero-work boundary depends only on the end states of the process. Energy, then, is the name given to this thermodynamic property:

$$dE = \delta Q_{\text{zero-work}} \tag{1.31}$$

As a derived concept, the work transfer interaction is defined as the difference $\delta Q - dE$, or as

$$\delta W = \delta Q - \delta Q_{\text{zero-work}} \tag{1.32}$$

Finally, the adiabatic boundary is the one for which we can write $\delta Q = 0$ (see the discussion that follows eq. (1.12) earlier in this chapter).

It is fitting that we close this chapter with a discussion of Keenan and Shapiro's second scheme, which draws attention to a very special set of circumstances in which "heat" is indeed "conserved," eq. (1.31). This observation begs us to look back at what preceded modern thermodynamics and to appreciate a little more the *legitimacy* of the caloric theory as a milestone in man's search for truth and in the development of his language. Spending some time to study this theory directly is very helpful [e.g., Refs. 54, 55]. Of course, no theory is perfect and forever, which is why there is something rotten in the contemporary trend of portraying the caloric thinking as the equivalent of "failure" and "bad thermodynamics." A theory that was created by the minds of Lavoisier and Laplace—the theory that was misused so creatively by Sadi Carnot—couldn't have been all bad!

SYMBOLS

$\mathcal{A}$	surface
C	capacitance
e	specific energy [J/kg]
E	energy [J]
f	function
$\mathbf{F}$	force
g	gravitational acceleration
h	specific enthalpy [J/kg], [eq. (1.21)]
H	enthalpy [J]
$h°$	methalpy [eq. (1.23)]
i	electric current
J	moment of inertia
k	constant of translational spring
K	constant of rotational spring
L	inductance
m, M	mass [kg]
$\dot{m}$	mass flowrate [kg/s]
P	pressure
q	electric charge
$\mathbf{q}$	heat flux vector
Q	heat transfer interaction [J]
$\dot{Q}$	heat transfer rate [W]
$\mathbf{r}$	position vector
t	time
T	temperature; also torque (Table 1.1)
u	specific internal energy [J/kg]
U	internal energy [J]
$\mathbf{v}$	velocity vector
V	volume; also velocity and voltage (Table 1.1)

$\mathcal{V}$	volume in space
w'''	work transfer rate per unit volume [W/m^3]
W	work transfer interaction [J]
$\dot{W}$	work transfer rate [W]
X_i	generalized deformation coordinates
Y_i	generalized forces
z	elevation
θ	angular position
ρ	density [kg/m^3]
ω	angular speed
$(\ )_H$	high
$(\ )_L$	low
$(\ )_{rev}$	reversible
$(\ )_{sh}$	shaft, shear

REFERENCES

1. A Bejan, *Convection Heat Transfer*, Wiley, New York, 1984.
2. A. Bejan, The method of scale analysis: Natural convection in fluids, in S. Kakac, W. Aung, and R. Viskanta, eds., *Natural Convection: Fundamentals and Applications*, Hemisphere, Washington, DC, 1985.
3. A. Bejan, The method of scale analysis: Natural convection in porous media, in S. Kakac, W. Aung, and R. Viskanta, eds., *Natural Convection: Fundamentals and Applications*, Hemisphere, Washington, DC, 1985.
4. S. Carnot, *Reflections on the Motive Power of Fire, and on Machines Fitted to Develop that Power*, Bachelier, Paris, 1824; also in E. Mendoza, ed., *Reflections on the Motive Power of Fire and Other Papers*, Dover, New York, 1960; abbreviated in Ref. [6].
5. M. Planck, *Treatise on Thermodynamics*, 3rd ed., translated by A. Ogg, Dover, New York, 1945, p. 40.
6. J. Kestin, ed., *The Second Law of Thermodynamics*, Part I, Dowden, Hutchinson & Ross, Stroudsburg, PA, 1976.
7. C. Truesdell, *The Tragicomedy of Classical Thermodynamics*, International Centre for Mechanical Sciences, Udine, Courses and Lectures, No. 70, Springer-Verlag, Berlin and New York, 1983.
8. C. Truesdell, *Rational Thermodynamics*, 2nd ed., Springer-Verlag, New York, 1984, pp. 1–48.
9. D. S. L. Cardwell, *From Watt to Clausius*, Cornell University Press, Ithaca, NY, 1971.
10. L. Tisza, *Generalized Thermodynamics*, MIT Press, Cambridge, MA, 1966, pp. 5–29.
11. G. N. Hatsopoulos and J.H. Keenan, *Principles of General Thermodynamics*, Wiley, New York, 1965, pp. XV–XLII.
12. M. Modell and R. C. Reid, *Thermodynamics and Its Applications*, Prentice-Hall, Englewood Cliffs, NJ, 1974, pp. 1–8.

13. A. Bejan, *Entropy Generation Through Heat and Fluid Flow*, Wiley, New York, 1982, p. 3.

14. C. Truesdell, Irreversible heat engines and the second law of thermodynamics, *Invited Paper*, *Lett. Heat Mass Transfer*, Vol. 3, 1976, pp. 267–290.

15. A. Bejan, Engineering thermodynamics, in M. Kutz, ed., *Mechanical Engineers' Handbook*, Wiley, New York, 1986, Chapter 54.

16. E. G. Cravalho and J. L. Smith, Jr., *Engineering Thermodynamics*, Pitman, Boston, MA, 1981, p. 92.

17. J. W. Gibbs, *The Collected Works of J. Willard Gibbs*, Vol. I, Longmans, Green, New York, 1928, p. 51.

18. C. C. Gillispie, *Lazare Carnot, Savant*, Princeton University Press, Princeton, NJ, 1971, p. 110.

19. R. Clausius, On the moving force of heat, and the laws regarding the nature of heat itself which are deducible therefrom, *Philos. Mag.*, Ser. 4, Vol. 2, 1851, pp. 1–20, 102–119; On the application of a theorem of the equivalence of transformations to the internal work of a mass of matter, *ibid.*, Vol. 24, 1862, pp. 81–97, 201–213; also in Ref. 6, pp. 98, 142).

20. G. Zeuner, *Technical Thermodynamics*, 1st Engl. ed., translated by J. F. Klein, Van Nostrand, New York, 1907.

21. H. Poincaré, *Thermodynamique*, Georges Carré, Paris, 1892, pp. 66–68.

22. S. Petrescu and V. Petrescu, *Principiile Termodinamicii*, Editura Tehnica, Bucharest, 1983, pp. 29–35.

23. R. H. Fowler and E. A. Guggenheim, *Statistical Thermodynamics*, 3rd impression, Cambridge University Press, London and New York, 1951, p. 56.

24. J. Kestin, *A Course in Thermodynamics*, revised printing, Vol. I, Hemisphere, Washington, DC, 1979, p. 40.

25. I. Newton, Scale graduum Caloris Descriptiones & signa, *Philos. Trans. Soc.*, Vol. 8, 1701, pp. 824–829.

26. W. J. M. Rankine, *A Manual of the Steam Engine and Other Prime Movers*, 12th ed., revised by W. J. Millar, Charles Griffin & Co., London, 1888.

27. J. L. Shearer, A. T. Murphy, and H. H. Richardson, *Introduction to System Dynamics*, Addison-Wesley, Reading, MA, 1967, pp. 46–49.

28. Daniel Bernoulli, *Hydrodynamics*, translated by T. Carmody and H. Kobus, Dover, New York, 1968; Johann Bernoulli, *Hydraulics*, translated by T. Carmody and H. Kobus, Dover, New York, 1968.

29. R. Clausius, *Mechanical Theory of Heat*, edited by T. A. Hirst, John Van Voorst, London, 1867.

30. M. von Laue, *History of Physics*, Translated by R. Oesper, Academic Press, New York, 1950, p. 83.

31. V. S. Arpaci, *Conduction Heat Transfer*, Addison-Wesley, Reading, MA, 1966, p. 33.

32. G. Zeuner, *Das Locomotiven-Blasrohr* (The Locomotive Blast-pipe. Experimental and theoretical investigations on the production of draft by steam jets and on the suction action of liquid jets in general), Verlag von Meyer & Zeller, Zurich, 1863.

33. A. Stodola, *Steam Turbines* (with an appendix on gas turbines and the future of heat engines), translated by L. C. Loewenstein, Van Nostrand, New York, 1905.

34. J. H. Keenan, *Thermodynamics*, Wiley, New York, 1941, Chapters IV, XVI, and XVIII.

35. R. Mollier, *The Mollier Steam Tables and Diagrams* (Extended to the critical pressure), translated by H. Moss, Pitman, London, 1927 (1st German ed. published in 1906).

36. G. A. Goodenough, *Principles of Thermodynamics*, 3rd ed., Henry Holt & Co., New York, 1932, p. 85 (first published in 1911).

37. P. J. Kiefer and M. J. Stuart, *Principles of Engineering Thermodynamics*, Wiley, New York, 1930, p. 52.

38. A. W. Porter, The generation and utilization of cold. A general discussion—general introduction, *Trans. Faraday Soc.*, Vol. 18, 1922–1923, pp. 139–143.

39. F. A. Freeth, "H. Kamerlingh Onnes, 1853–1926," *Nature (London)*, Vol. 117, 1926; also in *Smithson. Inst., Ann. Rep.*, 1926, pp. 533–535.

40. Count of Rumford, Benjamin, An inquiry concerning the source of heat which is excited by friction, *Philos. Trans. R. Soc., London*, 1798, pp. 80–102.

41. S.C. Brown, *Benjamin Thompson: Count Rumford*, MIT Press, Cambridge, MA, 1979.

42. E. N. da C. Andrade, Humphry Davy's experiments on the frictional development of heat, *Nature (London)*, March 9, 1935, pp. 359, 360.

43. J. P. Joule, *The Scientific Papers of James Prescott Joule*, Dawson's, London, 1963, p. 299.

44. E. Mendoza, Contributions to the study of Sadi Carnot and his work, *Arch. Int. Hist. Sci.*, Vol. 12, 1959, pp. 377–396.

45. J. R. Mayer, Remarks on the forces of inorganic nature, *Ann. Chem. Pharm.*, Vol. 42, 1842, pp. 233–240.

46. J. P. Joule, On the calorific effects of magneto-electricity, and on the mechanical value of heat, *Philos. Mag.*, Ser. 3, Vol. 23, 1843.

47. P. S. Epstein, *Textbook of Thermodynamics*, Wiley, New York, 1937, pp. 27–34.

48. J. P. Joule, *Joint Scientific Papers of James Prescott Joule*, Taylor & Francis, London, 1887, p. 215.

49. P. G. Tait, *Sketch of Thermodynamics*, Edmonston & Douglas, Edinburgh, 1868, pp. v–vii.

50. H. von Helmholtz, *Selected Writings of Hermann von Helmholtz*, edited by R. Kahl, Wesleyan University Press, Middleton, CT, 1971, pp. 52–53.

51. J. H. Keenan and A. H. Shapiro, History and exposition of the laws of thermodynamics, *Mech. Eng.*, Vol. 69, 1947, pp. 915–921.

52. C. Carathéodory, Untersuchungen über die Grundlagen der Thermodynamik, *Math. Ann (Berlin)*, Vol. 67, 1909, pp. 355–386; an English translation can be found in Ref. 6, pp. 229–256.

53. M. Born, *Natural Philosophy of Cause and Chance*, Oxford University Press, London and New York, 1949, pp. 34, 35; also in Ref. 6, pp. 294, 295.

54. J. B. Emmett, On the chemical phenomena of heat, *Ann. Philos.*, Vol. 9, 1817, pp. 421–430.

55. S. C. Brown, The caloric theory of heat, *Am. J. Phys.*, Vol. 18, 1950, pp. 367–373.

56. J. B. Chaddock, Private communication, Duke University, Durham, NC, September 1985.

PROBLEMS

1.1 One invention that revolutionized the design of the early steam engine was the *principle of expansive operation* (see Hornblower and Watt, Fig. 2.1). According to the "old design," high-pressure steam was admitted from the boiler throughout the work-producing stroke of the piston, (i)–(f). During this process, the cylinder pressure remained practically constant and equal to the steam supply pressure, P_1. Before the piston could be returned to its original position (i), the high-pressure steam that filled the V_2 volume had to be exhausted into the atmosphere.

The "new design" consisted of cutting off the admission of steam at some intermediate volume V_1 and allowing the gas m_1 trapped inside the cylinder to work "expansively" (i.e., to expand as a closed system) as the piston completed its work-producing stroke. The purpose of the steam cutoff feature was to lower the final cylinder pressure to the atmospheric level in order to avoid the costly discharge of high-pressure steam.

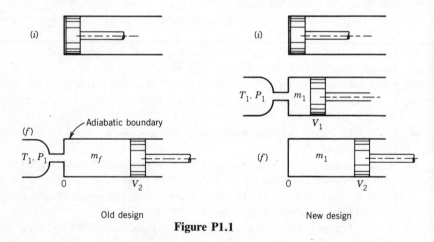

Old design   New design

Figure P1.1

As an engineer, you may be interested in evaluating the relative goodness of the new design, where "goodness" can be measured as the work produced during one full stroke W_{i-f} divided by the total amount of (T_1, P_1) gas drawn from the gas supply in order to execute that stroke. In order to develop an answer analytically, you can make the following simplifying assumptions:

- Instead of steam, the fluid that enters and expands in the cylinder is an ideal gas with known constants R and c_v.
- Initially, the piston touches the bottom of the cylinder, $V_i = 0$.
- The heat transfer through the wall of the cylinder is negligible.

(a) Consider first the old design and calculate in order:
- the work delivered during one stroke, W_{i-f}
- the final temperature inside the cylinder, T_f
- the final ideal gas mass m_f admitted into the cylinder at state (f)
- the goodness ratio W_{i-f}/m_f

(b) Consider next the new design, where the connection between the gas supply and the cylinder stays open only between $V=0$ and $V=V_1$. The expansion from $V=V_1$ to $V=V_2$ can be modeled as reversible and adiabatic. Calculate in order:
- the ideal gas mass m_1 trapped in the cylinder at $V=V_1$ and later
- the work delivered during one stroke, W_{i-f}
- the goodness ratio W_{i-f}/m_1

(c) Show that the goodness ratio of the new design (b) is greater than the goodness ratio of the old design (a).

1.2 One kilogram of H_2O (the "system") is being heated in a rigid container. In the initial state, the system is a mixture of liquid water and steam of temperature $T_1 = 100°C$ and vapor quality $x_1 = 0.5$. The total heat transfer interaction experienced by the system from state (1) to state (2) is $Q_{1-2} = 2199.26$ kJ.
(a) Draw the path of the process on a P–v diagram.
(b) Pinpoint the final state (2), i.e., determine *two* independent properties of the system at state (2).
(c) Determine T_2 and P_2.
(d) Comment on the dependence of the system's internal energy on temperature at states near state (2). For example, can H_2O be modeled as an ideal gas in the vicinity of state (2)?

1.3 An amount of ideal gas m (with known constants R and c_v) is confined by means of two diathermal diaphragms and one floating piston in three compartments, as shown in the drawing. Each compartment contains one-third of the amount of ideal gas, however, the pressure differs from one compartment to the next (pressures P_{1A}, P_{1B}, and P_{1C} are known). The three compartments are initially in thermal equilibrium at temperature T_1.

Consider next the adiabatic process $(1) \rightarrow (2)$ triggered when the two diaphragms are punctured. In the end [at state (2)], the ideal gas is characterized by a unique pressure P_2 and a unique temperature T_2. Determine the final temperature T_2, and show that your conclusion can be shaped as follows:

$$\frac{T_2}{T_1} = \text{function}\left(\frac{R}{c_v}, \frac{P_{1A}}{P_{1B}}, \frac{P_{1A}}{P_{1C}} \right)$$

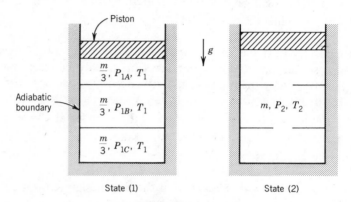

Figure P1.3

1.4 During its warm-up phase and before the pressure-release safety valve opens, a pressure cooker can be modeled as a constant-volume rigid container that is being heated at a rate $\dot{Q}$. Let m be the mass of the liquid–water vapor mixture at some point in time t, when the quality of the mixture is x and the pressure is P. Assuming that $\dot{Q}$, x, and P are known, determine analytically the rate of pressure increase in the pressure cooker (dP/dt) as a function of $\dot{Q}$, m, x and various properties of the saturated states f (liquid) and g (gas) that correspond to pressure P.

1.5 The air from a glass tube has been completely evacuated before the tip of the tube was fused and sealed. You are invited to break the tip of the tube, wait a second or two, and dip the open end of the tube into a beaker containing some water. Do you expect the water to rise into the glass tube, or do you expect air bubbles to come out of the tube into the water? Explain the basis for your expectations [56].

1.6   A rigid container with volume V is originally filled with water of temperature T_1. A hose is then connected to an inlet port and hotter water of temperature T_2 is pumped into the container at a mass flow rate $\dot{m}_{in}$, while another venting port allows the displaced water to flow out, the pressure being constant.

Assume that inside the container the incoming hot water mixes quickly with the water inside, so that the temperature is to a good approximation uniform. The water can be viewed as an incompressible liquid with constant specific volume v_w and with $du = c\, dT$, where the specific heat c is given.

   (a) Treating the container as an open system, derive an expression for the temperature T inside the container as a function of time. Neglect effects due to gravity and kinetic energy.

   (b) Suppose the container has a volume of $1\, m^3$ and $v_w = 10^{-3}\, m^3/kg$. If $T_1 = 10°C$ and $T_2 = 40°C$, what hot water mass must be pumped into the container in order to raise the inside temperature from 10 to 20°C?

1.7   A rigid and well-insulated container holds 1 kg of dry saturated steam at $P_1 = 5\,MPa$. At some point in time, a tiny crack develops in the uppermost region of the container. Through this crack, some of the H_2O escapes slowly into the atmosphere until the pressure inside the container falls to $P_2 = 1.5\,MPa$. This leakage process is slow enough so that at any instant the H_2O inventory is in a state of equilibrium. Any liquid that forms during this process accumulates at the bottom of the container, whereas the escaping H_2O is always gaseous.

   (a) Calculate the vapor quality at the final equilibrium state (2). Note that between 5 MPa and 1.5 MPa, the specific enthalpy of dry saturated steam is practically constant, $h_g \cong 2800\, kJ/kg$.

   (b) Calculate the ratio of the volumes occupied by vapor and liquid in the final state.

2

The Second Law of Thermodynamics

In this chapter, we turn our attention to the relatively more controversial part of engineering thermodynamics—the second law and the concepts of irreversibility, entropy, entropy generation, and the special features of system evolution that are associated with the name "Carnot." That we use the second law while speaking such words is not the result of a careful process of streamlining and purification: on the contrary, it is the fingerprint of chance and randomness in the course of human events, in this case, a record of the individuality of the pioneers. I think it is clear that what we inherited from Rankine, Clausius, and Kelvin is not in optimum shape, otherwise the writing of every new treatise on thermodynamics would not arouse so much passion, especially in relation to the formulation of the second law.

To review the historical development of second law concepts by merely retracing the meandering course followed by the pioneers is certainly one way to conduct a review. However, I feel that a more effective approach is to review the student's own encounter with the second law and to try to demystify certain aspects of this experience by reestablishing the connection between what the textbooks teach and what we actually inherited from the pioneers. It is only after this student-centered review that we discuss some of the more recent and more abstract reformulations of the second law and, for that matter, classical thermodynamics.

THE SECOND LAW FOR CLOSED SYSTEMS

The starting point is the physical law, i.e., the narrative or mathematical summary of physical observations that consistently reinforce older observations of the same phenomenon. Here, we have a choice, because whereas the impact of the second law is universal, the circumstances in which the law makes itself evident vary widely with regard to complexity. Historically, the

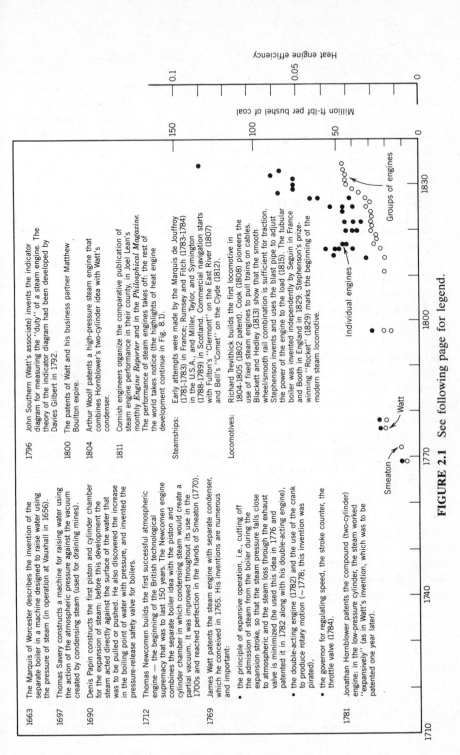

FIGURE 2.1 See following page for legend.

second law started with Sadi Carnot's contemplation of the development of fire engines (Fig. 2.1)—a class of systems that even today cannot be described as "simple." So instead of retracing Sadi Carnot's steps—this has created enough confusion among his contemporaries—we can look back at a class of simpler examples, namely, the experiments of Count Rumford and Joule. We note that in each experiment, the apparatus (a closed system) absorbed work and rejected heat to the ambient, i.e., to the only heat reservoir with which it could communicate.

Cycle in Contact with One Heat Reservoir

The essence of the second law is that in the above experiments, the apparatuses *received* work and *rejected* heat: it was never the other way around; in fact, all the attempts to construct a heat engine that would operate cyclically as a closed system while in possible contact with a single heat reservoir have failed.

Planck summarized these observations by writing that: "It is impossible to construct an engine which will work in a complete cycle, and produce no effect except the raising of a weight and the cooling of a heat-reservoir" [1]. A similar statement had been made in 1851 by William Thomson (Lord Kelvin): "It is impossible, by means of inanimate material agency, to derive mechanical effect from any portion of matter by cooling it below the temperature of the coldest of the surrounding objects," which is followed by the footnote, "If this axiom be denied for all temperatures, it would have to be admitted that a self-acting machine might be set to work and produce mechanical effect by cooling the sea or earth, with no limit but the total loss of heat from the earth and sea, or, in reality, from the whole material world" [2].

One difference between Planck's and Kelvin's statements is that Planck makes explicit reference to a cycle and to a *heat reservoir* as a system whose temperature does not change while experiencing a heat transfer interaction. Another difference is that Kelvin speaks words that betray the lingering influence of the caloric theory, namely, the idea that heat is stored in a body. Of course, Kelvin came to understand better than most the limitation of caloric theory with regard to the conservation of caloric fluid. However,

---

Figure 2.1 Highlights in the development of steam engines in the period before Sadi Carnot. The engine performance data are from a compilation by D. S. L. Cardwell, *From Watt to Clausius*, Cornell University Press, Ithaca, NY, 1971.

On the ordinate of Fig. 2.1, I estimated the modern equivalent, called "heat engine efficiency," $\eta = \dot{W}/\dot{Q}_H$, noting that one British imperial bushel equals 2219.36 in.^3, the calorific value of 1 lb of coal is $15{,}225 \text{ Btu/lbm}$ and the density of anthracite is roughly 1600 kg/m^3.

The relative efficiency of engines—individual versus groups—is not unlike that of people.

like Sadi Carnot's essay a quarter of a century earlier, Kelvin's terminology is an interesting example of how, in order to communicate a new theory, the proponent has no choice but to misuse the language of the old theory ("Without a constant misuse of language there can not be any discovery, any progress" [3]).

By recalling the heat-engine sign convention of Fig. 1.3, the impossibility described by Kelvin and Planck reduces to the analytical statement:

$$\oint \delta W \leq 0 \tag{2.1}$$

or, in view of the first law for a cycle executed by a closed system, eq. (1.6),

$$\oint \delta Q \leq 0 \tag{2.2}$$

As we proceed with the generalization of these statements to cover systems and processes of increasing complexity, it is important to keep in mind the origin of eqs. (2.1) and (2.2), and that these early statements apply only to a closed system that executes an integral number of cycles while in communication with one heat reservoir.

Cycle in Contact with Two Heat Reservoirs

Traditionally, the next step in the direction of more complex cycles is made using the concept of a *reversible cycle*, introduced by Sadi Carnot in 1824. This concept became known as the "Carnot cycle" perhaps under the influence of Emile Clapeyron [4], who in 1834 translated into analysis and graphics the message of Carnot's essay, in this way rereleasing a theory that had been totally ignored despite adequate exposure in 1824 [5].

The modern graphical interpretation of the Carnot cycle is shown in Fig. 2.2 either as an ideal gas working in the $P-v$ plane, which is one of the

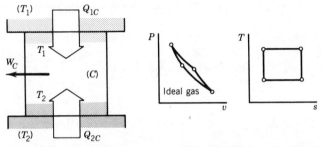

Figure 2.2   A closed system executing a reversible (Carnot) cycle while in communication with two heat reservoirs (T_1) and (T_2). Note that no assumption is being made regarding the relative size of (T_1) and (T_2) and the sense of the cycle on the $P-v$ and $T-s$ planes.

drawings made originally by Clapeyron (the other was done in the T–v plane), or in terms of an unspecified working fluid on the T–s diagram, as done by Gibbs in his first paper [6]. The cycle consists of a total of four processes, an alternating sequence of constant-temperature and adiabatic changes of volume. The analytical details of this cycle (the relationship between the properties of the working fluid, the volume, pressure, and temperature ratios, and the net magnitude of each energy interaction) are not important at this stage. Important is the notion that each state visited by the system during the cycle is one of *uniform* pressure, temperature, and specific volume (otherwise, it would be impossible to draw the cycle on the P–v plane). Physically, this means that the system expands and contracts at such a slow rate that at each point in time, the state of the system is one of equilibrium.

In particular, an isothermal volume change is slow enough so that the system is in thermal equilibrium with the heat reservoir. Under such limiting conditions, the system can execute the same cycle in the reverse sense, i.e., it can pass through the same sequence of equilibrium states in reverse order. Therefore, if W_C is the net work produced by the Carnot cycle, and if Q_{1C} and Q_{2C} are the respective heat transfer interactions with reservoirs T_1 and T_2 during the same cycle (interactions Q_{1C} and Q_{2C} are assumed positive when absorbed by the system, Fig. 1.3), then the energy interactions of the *reversed* Carnot cycle are

$$(W_C, Q_{1C}, Q_{2C})_{\text{reversed}} = -(W_C, Q_{1C}, Q_{2C}) \tag{2.3}$$

Consider now the object of this subsection—the step by which the second law represented by eq. (2.1) is extended to cover the cyclical operation of closed systems in contact with two heat reservoirs. Figure 2.3a shows a closed system (A) executing an unspecified cycle while in contact with the heat reservoirs (T_1) and (T_2), with $T_1 \neq T_2$, discussed already in connection with Fig. 2.2 and the concept of a reversible cycle. The net energy interactions experienced by this unspecified cycle, W, Q_1, and Q_2, are related through the first law:

$$Q_1 + Q_2 = W \tag{2.4}$$

In addition, the Kelvin–Planck version of the second law places a constraint on the signs of the two heat transfer interactions (Q_1, Q_2). For the sake of clarity, let us assume that W is positive:

$$W > 0 \tag{2.5}$$

Examining the signs of Q_1 and Q_2, we consider the three options:

$$\text{(i)} \quad Q_1 < 0 \quad \text{and} \quad Q_2 < 0 \tag{2.6}$$

$$\text{(ii)} \quad Q_1 > 0 \quad \text{and} \quad Q_2 > 0 \tag{2.7}$$

$$\text{(iii)} \quad Q_1 Q_2 < 0 \tag{2.8}$$

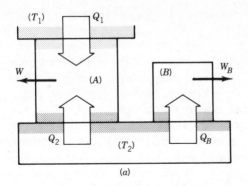

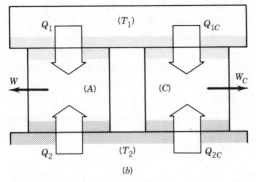

Figure 2.3 The translation of the Kelvin–Planck impossibility statement into the second law for any closed system that executes a cycle while in communication with two heat reservoirs.

and note immediately that option (i) is disallowed by the first law and the assumption that W is positive, eqs. (2.4) and (2.5).

Next, we reason that if option (ii) is possible, then after the unspecified cycle (W, Q_1, Q_2) runs its course, we can place one of the heat reservoirs [for example, reservoir (T_2), Fig. 2.3a] in communication with an auxiliary system (B) that executes one cycle characterized by interactions Q_B and W_B. Since the cycle of system (B) is executed in contact with only one heat reservoir, the Second Law of Thermodynamics (2.2) requires

$$Q_B < 0 \qquad (2.9)$$

One admissible value for Q_B is $-Q_2$, because according to option (ii), Q_2 is assumed positive. Therefore, if we set

$$Q_B = -Q_2 \qquad (2.10)$$

it follows also that the (T_2) reservoir completes a cycle at the end of the cycles executed by systems (A) and (B) [note that the net energy change for the (T_2) heat reservoir is $-Q_2 - Q_B = 0$, hence the completion of a cycle]. The punch line to this argument is that if systems (A), (B), and the heat reservoir (T_2) complete cycles, then the composite system $(A) + (B) + (T_2)$ executes a cycle. And, since this composite system makes contact with only one heat reservoir, (T_1), the second law (2.2) requires $Q_1 < 0$, which contradicts the first assumption of option (ii).

In conclusion, option (ii) constitutes a violation of the second law, and the only alternative left is option (iii). The heat transfer interactions Q_1 and Q_2 must be of *opposite sign*, $Q_1 Q_2 < 0$. It is true that in the discussion of Fig. 2.3a, we wrote $Q_2 > 0$ and, in the end, $Q_1 < 0$; however, this does not diminish the generality of the conclusion $Q_1 Q_2 < 0$. We have the choice of reconstructing the argument based on a different version of Fig. 2.3a, in which system (B) would make contact with the (T_1) heat reservoir: the conclusion reached along this second route is the same, $Q_1 Q_2 < 0$ (Problem 2.1).

Returning to the most general cycle that can be executed in contact with two heat reservoirs, Fig. 2.3a, we known now that regardless of the actual sign of $T_2 - T_1$,[†] the heat transfer interactions (Q_1, Q_2) cannot have the same sign. Let Q_1 be the positive heat transfer interaction, in other words, let us convene that indices 1, 2 serve to identify the sign of the algebraic value taken by the heat interaction Q:

$$Q_1 > 0 \quad \text{and} \quad Q_2 < 0 \tag{2.11}$$

Note further that Q_1 is *strictly* positive (and Q_2 strictly negative), because $Q_1 = 0$ is not allowed as long as $W \neq 0$. Consider now the arrangement in Fig. 2.3b: the unspecified cycle (A) and the Carnot cycle (C) of Fig. 2.2 share the two heat reservoirs (T_1) and (T_2). We have the freedom to size the Carnot cycle and to select its sense in such a way that

$$Q_1 + Q_{1C} = 0 \tag{2.12}$$

which means that the (T_1) heat reservoir also executes a cycle. Since the composite system $(A) + (C) + (T_1)$ completes a cycle while making contact with only one heat reservoir, (T_2), we invoke the second law (2.2) one more time to write

$$Q_2 + Q_{2C} \leq 0 \tag{2.13}$$

Dividing this inequality through Q_1 or $(-Q_{1C})$, which are both positive

---

[†]We made no assumption regarding the relative sizes of T_1 and T_2: we only required $T_1 \neq T_2$ in order to be able to talk about *two* heat reservoirs or, better yet, two *temperature* reservoirs.

according to eqs. (2.11) and (2.12), we obtain an inequality between two positive ratios:

$$\frac{(-Q_2)}{Q_1} \geq \frac{Q_{2C}}{(-Q_{1C})} \tag{2.14}$$

This inequality represents the Second Law of Thermodynamics for a cycle executed by a closed system in contact with two heat reservoirs (more useful versions of this result involve the concept of thermodynamic temperature, which is discussed shortly). At this stage, eq. (2.14) allows us to conclude that the positive ratio obtained by dividing the absolute value of the negative heat transfer interaction by the value of the positive heat transfer interaction cannot be smaller than a certain limiting value. The limiting case corresponds to the equal sign in eq. (2.14):

$$\frac{(-Q_2)}{Q_1} = \frac{Q_{2C}}{(-Q_{1C})} \tag{2.14'}$$

which, in combination with eq. (2.12) and the first-law statements $W = Q_1 + Q_2$ and $W_C = Q_{1C} + Q_{2C}$, translates into

$$(W_C, Q_{1C}, Q_{2C}) = -(W, Q_1, Q_2) \tag{2.15}$$

The meaning of the limiting case highlighted by the equal sign in the second law statement (2.14) becomes clear if we compare eq. (2.15) with eq. (2.3): in the limiting case, the cycle executed by the unspecified system (A) is the reverse of a Carnot cycle (C), and vice versa. The equal sign in the second law (2.14) is associated with any *reversible* cycle that is executed by the unspecified system (A). Therefore, the second law can be stated by making reference only to the system of interest:

$$\frac{-Q_2}{Q_1} \geq \left(\frac{-Q_2}{Q_1}\right)_{\text{rev}} \tag{2.16}$$

where the subscript "rev" stands for "reversible," in this case, for "reversible cycle executed by an arbitrary closed system in contact with two heat reservoirs."

The question that poses itself at this juncture is: "How small is the number that serves as lower bound for the ratio $(-Q_2)/Q_1$?" The analytical reply that follows appears to have been first given by Poincaré [7]: the same argument is found in some of the more influential treatises on modern engineering thermodynamics, for example, in Refs. 8–10.

According to eq. (2.16), the lower bound $(-Q_2/Q_1)_{\text{rev}}$ is independent of the cycle design (the sequence of processes) and the working fluid, because these items were intentionally left unspecified in the definition of system (A). Therefore, the critical value $(-Q_2/Q_1)_{\text{rev}}$ can only be a function of parameters T_1 and T_2, since the existence of two *different* heat reservoirs was assumed in the derivation of eq. (2.16). Writing

$$\left(\frac{-Q_2}{Q_1}\right)_{\text{rev}} = f(T_1, T_2) \tag{2.17}$$

we note that f is an unknown function and that T_1 and T_2 are two different numbers obtained by reading the scale of one thermometer (this scale can be even one of the traditional temperature scales reviewed as an introduction to the current one-point scales compared in Fig. 1.6). The unknown function $f(T_1, T_2)$ has a special property that surfaces while invoking the definition (2.17) in connection with two additional closed systems that execute cycles while in contact with pairs of heat reservoirs (Figs. 2.4*b* and *c*):

$$\left(\frac{-Q_3}{Q_1}\right)_{\text{rev}} = f(T_1, T_3) \tag{2.18}$$

$$\left(\frac{-Q_3}{-Q_2}\right)_{\text{rev}} = f(T_2, T_3) \tag{2.19}$$

In writing eqs. (2.17)–(2.19), we observe the convention that the numerator and denominator appearing on the left side represent, respectively, the absolute value of the negative heat transfer interaction and the positive heat transfer interaction. Dividing eq. (2.18) by (2.19), and using eq. (2.17) to eliminate $(-Q_2/Q_1)_{\text{rev}}$, yields

$$f(T_1, T_2) = \frac{f(T_1, T_3)}{f(T_2, T_3)} \tag{2.20}$$

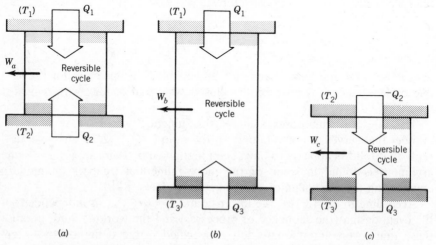

Figure 2.4 Graphic summary of the argument leading to eq. (2.21′).

Finally, since the left side of eq. (2.20) does not depend on the constant T_3, we conclude that the analytical form of $f(T_1, T_2)$ has to be

$$f(T_1, T_2) = \frac{\psi(T_1)}{\psi(T_2)} \qquad (2.21)$$

or, letting $\phi = 1/\psi$,

$$\left(\frac{-Q_2}{Q_1}\right)_{rev} = \frac{\phi(T_2)}{\phi(T_1)} \qquad (2.21')$$

Equation (2.21) can now be generalized to express the ratio between the heat transfer interactions of a reversible cycle that absorbs one unit of energy (Q_0) from a reference heat reservoir of empirical temperature θ_0, and rejects the amount $(-Q)$ to an arbitrary reservoir (θ):

$$\phi(\theta) = \phi(\theta_0)\left(\frac{-Q}{Q_0}\right)_{rev} \qquad (2.22)$$

Here the older temperature symbol θ (introduced by Poisson and adopted by Fourier [11]) is used instead of the modern symbol T in order to emphasize the observation that the heat reservoir temperatures can be compared and found different on any empirical temperature scale. The measurement of $(-Q/Q_0)_{rev}$ while running any reversible cycle between reservoirs (θ) and (θ_0) reveals the relationship between function ϕ and the particular scale θ, subject to the adoption of a numerical value for the constant $\phi(\theta_0)$. The numerical values obtained in this manner for function ϕ constitute the *thermodynamic temperature scale* (symbol T), in other words, $\phi \equiv T$ or

$$T = T_0\left(\frac{-Q}{Q_0}\right)_{rev} \qquad (2.23)$$

A geometric interpretation of the thermodynamic temperature scale definition (2.23) is presented in the *wedge of minimum Q* diagram of Fig. 2.5. The constant T_0 has been assigned the value 273.16 on the Kelvin temperature scale (or 491.69 on the Rankine scale); this constant represents the thermodynamic temperature of a heat reservoir in thermal equilibrium with a batch of ice, water vapor, and liquid water in equilibrium. Therefore, the current temperature scales displayed side by side in Fig. 1.6 are all based on one fiducial point (the triple point of water). The need for adopting one-point scales as opposed to traditional two-point scales is most critical in the field of low-temperature physics: an entertaining description of these difficulties was provided by Giauque [12], who drew an analogy between the traditional two-point temperature scales and the idea of redefining the measurement of weights using two reference weights.

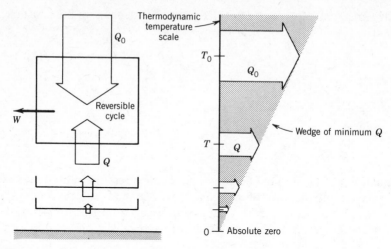

Figure 2.5 Reversible cycle Q measurements and the geometric construction of a thermodynamic temperature scale T (the "wedge of minimum Q" diagram).

When you again note that the ratio $(-Q/Q_0)_{\text{rev}}$ is not affected by the identity of the closed system and fluid that executes the reversible cycle, the chief advantage of the thermodynamic scale (T) is that it is universal [the distribution of numbers on an empirical (θ) scale depends on the choice of thermometric fluid]. The observation that a reversible cycle and an equation like eq. (2.22) can serve as basis for a universal temperature scale is Kelvin's contribution [13]. Of course, any monotonic function $\phi(T)$ can be used to define a thermodynamic temperature scale T in accordance with eq. (2.22). The special choice $\phi \equiv T$ made in eq. (2.23) is not inherited from Kelvin: this choice has prevailed in classical thermodynamics, partly because of the *coincidence* that ϕ equals θ if the reversible cycle uses an ideal gas as working fluid. It is not difficult to show that one of the properties of the ideal gas "Carnot cycle" sketched in the P–v plane of Fig. 2.2 is

$$\frac{-Q_{2C}}{Q_{1C}} = \frac{\theta_{2,\text{ideal gas}}}{\theta_{1,\text{ideal gas}}} \tag{2.24}$$

where $\theta_{\text{ideal gas}}$ is the ideal gas temperature defined by

$$\theta_{\text{ideal gas}} = \frac{Pv}{R} \tag{2.25}$$

Comparing eq. (2.24) with eq. (2.21'), we see $\phi = \theta_{\text{ideal gas}}$; hence,

$$\theta_{\text{ideal gas}} = T \tag{2.26}$$

This last equality is purely a coincidence because the thermodynamic scale T and the ideal gas scale $\theta_{\text{ideal gas}}$ are two entirely different concepts. The latter is an old concept from thermometry and calorimetry: note Guillaume Amontons' invention of the air thermometer, its use in the 1700s and 1800s, and the discovery of the empirical laws of Towneley, Boyle, and Mariotte ($Pv = \text{constant}$ at constant temperature), and Gay-Lussac ($v \sim \theta_{\text{ideal gas}}$ at constant pressure), Table 1.2. These two laws were combined and written for the first time as eq. (2.25) by Clapeyron [4] in his analytical interpretation of Sadi Carnot's essay (which is why "$Pv = RT$" is recognized by some as the *Clapeyron equation*—not to be confused with the *Clausius–Clapeyron relation* of page 262). The concepts of "absolute temperature" and "absolute zero" seem to go back to Amontons himself, who by measuring the air pressure–temperature relationship at constant volume (thus anticipating the work of both Charles and Gay-Lussac) reasoned that the air pressure must become zero at a finite temperature [14]. In Clapeyron's writings, the place of $\theta_{\text{ideal gas}}$ is occupied by $[\theta(°C) + 267]$, where $\theta(°C)$ is the Celsius temperature measured on the two-point empirical scale of the time, and where $-267°C$ is the absolute zero. On the current Celsius scale, which is defined in terms of a one-point thermodynamic scale, absolute zero corresponds to $-273.15°C$ (see eq. (1.9) and Fig. 1.6).

The reward for introducing the concept of a thermodynamic temperature scale is that we can now write $(-Q_2/Q_1)_{\text{rev}} = T_2/T_1$ and the second-law statement (2.16) becomes

$$\frac{Q_1}{T_1} + \frac{Q_2}{T_2} \leq 0 \tag{2.27}$$

This new statement is general despite the assumption $W > 0$ made early in the derivation of eq. (2.14). The second law expressed as eq. (2.27) is certainly insensitive to whether we use Q_1 or Q_2 as a label for the positive of the two heat transfer interactions under option (iii), eq. (2.11). Now, if the net work transfer interaction is negative, we find that admissible are two options: (i) $Q_1 < 0$ and $Q_2 < 0$, and (iii) $Q_1 Q_2 < 0$ (Problem 2.2). Option (i) is obviously covered by eq. (2.27). Option (iii) and the notation Q_2 for the negative of the two heat transfer interactions produce an analysis identical to the segment contained between eqs. (2.11) and (2.27). The second-law statement (2.27) is therefore independent of the sign of W.

Cycle in Contact with Any Number of Heat Reservoirs

Having just extended the second law from the realm of cycles in contact with only one heat reservoir, eq. (2.2), to cycles in contact with two heat reservoirs, we are entitled to ask the question: "Is the last statement of the second law, eq. (2.27), a generalization of the Kelvin–Planck statement?"

The answer is clearly, "yes," because we can use the label Q_1 for the heat transfer integral and rewrite the Kelvin–Planck statement (2.2) as

$$\frac{Q_1}{T_1} \leq 0 \tag{2.28}$$

where T_1 is the thermodynamic temperature of the lone heat reservoir. Comparing eq. (2.28) with eq. (2.27), we see the beginning of a pattern that takes us to the second law for any cycle executed while in contact with any number of heat reservoirs (n):

$$\frac{Q_1}{T_1} \leq 0 \tag{2.28}$$

$$\frac{Q_1}{T_1} + \frac{Q_2}{T_2} \leq 0 \tag{2.27}$$

$$\vdots$$

$$\frac{Q_1}{T_1} + \frac{Q_2}{T_2} + \cdots + \frac{Q_n}{T_n} \leq 0 \tag{2.29}$$

The validity of the general statement (2.29) can be proven in a number of ways: as an alternative to the proof seen perhaps in a first course in engineering thermodynamics, in the present treatment I rely on the method of mathematical induction. With the validity of eq. (2.28) supported by a wealth of physical observations, and with the validity of eq. (2.27) demonstrated in the preceding subsection, I assume that the statement corresponding to n heat reservoirs, eq. (2.29), is correct. If, based on this last assumption, I can show that the statement for $n+1$ heat reservoirs is correct,

$$\sum_{i=1}^{n+1} \frac{Q_i}{T_i} \leq 0 \tag{2.30}$$

then the assumed eq. (2.29) is valid. Consider for this purpose, a system (A) that executes a cycle while in thermal communication with $n+1$ heat reservoirs, $(T_1), (T_2), \ldots, (T_n), (T_{n+1})$, Fig. 2.6. No assumption is made in connection with the direction (sign) of each of the heat transfer interactions, $Q_1, Q_2, \ldots, Q_n, Q_{n+1}$. We did assume that the second law for cycles in contact with n heat reservoirs is valid; therefore, in order to use eq. (2.29), we have to return the reservoir (T_{n+1}) to its original state by placing it in contact with a reversible cycle (C) such that

$$Q_{n+1} + Q_{(n+1),C} = 0 \tag{2.31}$$

Invoking now the second law for the composite system $(A) + (T_{n+1}) + (C)$, which completes a cycle while in contact with n heat reservoirs,

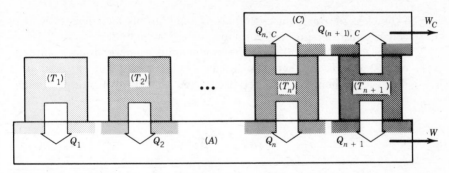

Figure 2.6 A device for extending the generality of the second law to cycles executed by a closed system while in communication with any number of heat reservoirs.

$$\sum_{i=1}^{n} \frac{Q_i}{T_i} + \frac{Q_{n,C}}{T_n} \leq 0 \qquad (2.32)$$

and using eq. (2.31) and $Q_{n,C}/T_n + Q_{(n+1),C}/T_{n+1} = 0$, we write in order:

$$\sum_{i=1}^{n} \frac{Q_i}{T_i} - \frac{Q_{(n+1),C}}{T_{n+1}} \leq 0 \qquad (2.33)$$

$$\sum_{i=1}^{n} \frac{Q_i}{T_i} + \frac{Q_{n+1}}{T_{n+1}} \leq 0 \qquad (2.34)$$

to arrive at eq. (2.30). Equation (2.34) completes the inductive proof that the same body of evidence that supports the Kelvin–Planck statement of the second law supports also eq. (2.29).

The final step in this line of generalization amounts to considering the possibility of a continuous variation of system boundary temperature as the cycle is executed in contact with an infinite sequence of heat reservoirs, each contributing a heat transfer interaction of size δQ or $\dot{Q}\,dt$. In such cases, eq. (2.29) is written as

$$\oint \frac{\delta Q}{T} \leq 0 \qquad (2.35)$$

with the understanding that T represents the Kelvin or Rankine thermodynamic temperature of that portion of the system boundary that is instantly crossed by the heat transfer interaction δQ. It is also understood that the equal sign of eq. (2.35) refers to reversible cycles:

$$\oint \frac{\delta Q_{rev}}{T} = 0 \qquad (2.36)$$

i.e., to the class of cycles that, if reversed, exhibit only a change in the sign of the energy transfer interactions, not in their magnitude.

The concept of entropy as a thermodynamic property follows directly from eq. (2.36): if the net change in $\delta Q_{rev}/T$ is zero at the end of a reversible cycle, then $\delta Q_{rev}/T$ represents the change in a thermodynamic property S:

$$dS = \frac{\delta Q_{rev}}{T} \tag{2.37}$$

This new property was described explicitly and named *entropy* by Clausius[†] in 1865; however, the same property was discovered and used earlier by Rankine [16], who called it "thermodynamic function," labeled it ϕ instead of S, and regarded eq. (2.37) as "the general equation of thermodynamics."

Process in Contact with Any Number of Heat Reservoirs

The cycles discussed until now are a restricted class of processes. Let the change between states (1) and (2) represent an arbitrary process, that is, a process whose path and energy interactions along the path are left unrestricted. Thinking of only that family of paths that are reversible (a path is reversible if it can be a part of a reversible cycle), we integrate eq. (2.37) and obtain what amounts to a definition for end-to-end entropy change:

$$S_2 - S_1 = \int_1^2 \frac{\delta Q_{rev}}{T} \tag{2.38}$$

The arbitrary process $(1) \rightarrow (2)$ can certainly be a part of a cycle $(1) \rightarrow (2) \rightarrow (1)$, where the return process $(2) \rightarrow (1)$ takes place along a reversible path: rewriting eq. (2.35) for this cycle and using the definition (2.38), we have, in order:

[†] . . . I have felt it more suitable to take the names of important scientific quantities from the ancient languages in order that they may appear unchanged in all contemporary languages. Hence I propose that we call S the *entropy* of the body after the Greek word ἡ τροπή, meaning "transformation." I have intentionally formed the word *entropy* to be as similar as possible to the word *energy*, since the two quantities that are given these names are so closely related in their physical significance that a certain likeness in their names has seemed appropriate [15].

Note: In modern Greek, the word τροπη means "turn," "change," not "transformation." In ancient Greek, the words ἐντροπή or ἐντροπία meant "turning inward," "twist" (private communication by Dr. S. Monti-Pouagare).

$$\int_1^2 \frac{\delta Q}{T} + \int_2^1 \frac{\delta Q_{\text{rev}}}{T} \leq 0 \tag{2.39}$$

$$\underbrace{\int_1^2 \frac{\delta Q}{T}}_{\substack{\text{Entropy} \\ \text{transfer} \\ \text{(Nonproperty)}}} \leq \underbrace{S_2 - S_1}_{\substack{\text{Entropy} \\ \text{change} \\ \text{(Property)}}} \tag{2.40}$$

The Second Law of Thermodynamics for a process, eq. (2.40), states that—algebraically—the entropy transfer never exceeds the entropy change. A measure of the strength of the inequality sign in eq. (2.40) (i.e., a new definition) is the *entropy generation*, or *entropy production* S_{gen}, a quantity that is never negative:

$$S_{\text{gen}} = S_2 - S_1 - \int_1^2 \frac{\delta Q}{T} \geq 0 \tag{2.41}$$

Like the entropy transfer interaction, the entropy generation is path-dependent.

The engineering significance of the concept of entropy generation is the focus of chapter 3 and much of the advanced engineering topics included in the present treatment. At this stage in the review of the second law, it is worth noting that the entropy generation term is not known by only one name and symbol in the thermodynamics literature. Alternative notations are P_S (entropy production [17]), ΔS_{net} (net entropy increase [18]), S_{irr} (produced entropy [19]), θ (entropy produced or generated; Ref. 9, p. 587), σ (created entropy [20]), S_{prod} (entropy production [21]), and ΔS_c (entropy creation [22]). In his 1865 paper [15], in an equation identical to eq. (2.41), Clausius wrote N instead of S_{gen}. In his terminology, N represented the "uncompensated transformation," which today might translate into "entropy imbalance" [note Clausius' translation of the word entropy (page 64 footnote) and his alternative view that S represents the "transformation" content of the body]. We are reminded by Denbigh [20] that the practice of writing a symbol to quantize the generation of entropy during an irreversible process was first used extensively by the Belgian physicist de Donder (his work is discussed in the beginning of Chapter 7).

There is an obvious need for consistency in how we label and use the entropy generation term. The symbol S_{gen}, which is adopted in the present treatment, has been preferred by a number of textbooks published in the 1980s [23–25] and by numerous research articles.

Example 2.1.   Two bodies of water of masses m_1 and m_2 and temperature T_1 and T_2 serve as "instantaneous heat reservoirs" for a reversible heat engine that operates in an integral number of cycles. These heat reservoirs are said to be instantaneous because their temperatures change as the engine bleeds heat transfer from one and rejects heat transfer to the other. The object of this example is to determine the final equilibrium temperature reached by the two bodies of water, T_∞ and the total work transfer delivered by the heat engine to an external user.

As thermodynamic "system," we choose the aggregate system containing the two bodies of water and the engine sandwiched between them. The aggregate system exchanges only work transfer with its environment. The "process" in this problem is the reversible engine cycle (or integral number of cycles) in which the two water masses come to equilibrium. Writing (i) and (f) for the initial and final states of the aggregate system, respectively, the first law and the second law for the aggregate system are

$$-W_{i-f} = U_f - U_i \tag{a}$$

$$S_{\text{gen},i-f} = S_f - S_i = 0 \tag{b}$$

The second law (b) is an "equation" because the $(i) \rightarrow (f)$ process is executed reversibly. Note also the absence of the heat transfer "Q" term from eq. (a) and the entropy transfer "Q/T" term from eq. (b): these interactions are absent because the aggregate system boundary is adiabatic.

Next, we focus on the net internal energy and entropy changes of the aggregate system, and express these in terms of the changes experienced by the three subsystems:

$$U_f - U_i = (U_f - U_i)_{m_1} + (U_f - U_i)_{\text{engine}} + (U_f - U_i)_{m_2}$$

$$= m_1 c(T_\infty - T_1) + 0 + m_2 c(T_\infty - T_2) \tag{c}$$

$$S_f - S_i = (S_f - S_i)_{m_1} + (S_f - S_i)_{\text{engine}} + (S_f - S_i)_{m_2}$$

$$= m_1 c \ln \frac{T_\infty}{T_1} + 0 + m_2 c \ln \frac{T_\infty}{T_2} \tag{d}$$

The second term in each of these decompositions is zero because the engine operates cyclically. The terms that account for property changes in the water bodies have been estimated invoking the incompressible liquid model, whereby

$$du = c\, dT \quad \text{and} \quad ds = c\, \frac{dT}{T} \tag{e}$$

and where c is the lone specific heat of water as an incompressible liquid.

Combining eq. (d) with the second law (b), we first find the final equilibrium temperature:

$$T_\infty = T_1^\alpha T_2^{1-\alpha} \tag{f}$$

where $\alpha = m_1/(m_1 + m_2)$. The first law (a) in combination with eqs. (c) and (f) delivers the expression for the total work output of the engine:

$$W_{i-f} = m_1 c T_1 \left[1 - \left(\frac{T_2}{T_1} \right)^{1-\alpha} \right] + m_2 c T_2 \left[1 - \left(\frac{T_1}{T_2} \right)^\alpha \right] \tag{g}$$

In the limit $\alpha \rightarrow 0$ when m_1 is negligibly small relative to m_2, these results become

$$T_\infty = T_2 \quad \text{and} \quad W_{i-f} = m_1 c(T_1 - T_2) \tag{h}$$

The work output in this limit is equal to the internal energy drop experienced by m_1, as its temperature decreases from T_1 to T_2.

The method that was employed in solving the above problem is not unique. A considerably more laborious alternative consists of analyzing "m_1," the "engine," and "m_2" as separate systems. For each such system, one can write two equations, namely, the first law and the second law in the reversible limit. The problem reduces to solving a system of six algebraic equations for a total of six unknowns: T_∞, W_{i-f}, the heat transfer and entropy transfer interactions between m_1 and the engine, and, finally, the heat transfer and entropy transfer interactions between the engine and m_2. In contrast to this more laborious route, the "aggregate system" formulation that was used in the above solution produced only two equations for the only two declared unknowns, T_∞ and W_{i-f}.

THE SECOND LAW FOR OPEN SYSTEMS

The next and final step in the generalization of the second-law statements discussed until now consists of applying one more time the closed system to open system transformation illustrated already via Fig. 1.7. With reference to the same drawing, we write:

$$S_{\text{closed},t} = S_{\text{open},t} + \Delta S_{\text{in}} \tag{2.42}$$

$$S_{\text{closed},(t+\Delta t)} = S_{\text{open},(t+\Delta t)} + \Delta S_{\text{out}} \tag{2.43}$$

and

$$(\Delta S)_{\text{in,out}} = (s\,\Delta M)_{\text{in,out}} = (s\dot{m})_{\text{in,out}}\,\Delta t \tag{2.44}$$

Labeling with ΔS_{gen} the entropy generated from time t to $t + \Delta t$, eq. (2.41) yields

$$\Delta S_{\text{gen}} = S_{\text{open},(t+\Delta t)} - S_{\text{open},t} - \frac{\dot{Q}_i}{T_i}\,\Delta t + (\dot{m}s)_{\text{out}}\,\Delta t - (\dot{m}s)_{\text{in}}\,\Delta t \geq 0 \tag{2.45}$$

Invoking the limit $\Delta t \to 0$, dropping the subscript "open," and considering the existence of any number of spots with heat transfer (i) and mass flow (in, out) on the control surface, we obtain:

$$\underbrace{\dot{S}_{\text{gen}}}_{\substack{\text{Entropy} \\ \text{generation} \\ \text{rate}}} = \underbrace{\frac{dS}{dt}}_{\substack{\text{Rate of} \\ \text{entropy} \\ \text{accumulation} \\ \text{inside the} \\ \text{control} \\ \text{volume}}} - \underbrace{\sum_i \frac{\dot{Q}_i}{T_i}}_{\substack{\text{Entropy} \\ \text{transfer} \\ \text{rate} \\ \text{(via heat} \\ \text{transfer)}}} + \underbrace{\sum_{\text{out}} \dot{m}s - \sum_{\text{in}} \dot{m}s}_{\substack{\text{Net entropy} \\ \text{flow rate out} \\ \text{of the} \\ \text{control volume} \\ \text{(via mass flow)}}} \geq 0 \tag{2.46}$$

Absent from the above statement of the second law and from the corresponding statement for closed systems, eq. (2.41), are terms that contain the
work transfer interaction W. This observation complements the distinction
made between work transfer and heat transfer on page 13: work transfer is
the energy interaction that is not accompanied by entropy transfer.

Equation (2.46) represents the Second Law of Thermodynamics for a
multiport system: an alternative statement can be made using the terminology of eqs. (1.24) and (1.28):

$$\dot{S}_{gen} = \int_{\mathcal{V}} \frac{\partial (\rho s)}{\partial t} \, d\mathcal{V} + \int_{\mathcal{A}} \frac{1}{T} \mathbf{q} \cdot \mathbf{n} \, d\mathcal{A} + \int_{\mathcal{A}} \rho s \mathbf{v} \cdot \mathbf{n} \, d\mathcal{A} \geq 0 \quad (2.47)$$

where $\mathcal{V}$ represents the volume occupied by the open system; $\mathcal{A}$, the control
surface; $\mathbf{q}$, the heat flux vector; $\mathbf{v}$, the velocity vector; and $\mathbf{n}$, the unit vector
that is locally normal to $\mathcal{A}$ and points away from the system. The meaning of
the thermodynamic properties that appear under various integral signs in
eqs. (1.24), (1.28), and (2.47) is discussed in the next section.

The second-law statement for a point-size control volume is obtained
directly from eq. (2.47) by transforming the surface integrals in accordance
with the divergence theorem, and by invoking the mass-continuity equation
(1.29). The resulting statement is

$$s'''_{gen} = \rho \frac{\partial s}{\partial t} + \nabla \cdot \left(\frac{q}{T} \right) \geq 0 \quad (2.48)$$

The volumetric rate of entropy generation s'''_{gen} is expressed in $W/K/m^3$, and
is defined as $\dot{S}_{gen} = \int_{\mathcal{V}} s'''_{gen} \, d\mathcal{V}$.

> **Example 2.2.**   We are now in a position to study the application of the second
> law to the unsteady filling process discussed already in Examples 1.1 and 1.2. We
> adopt the open-system approach, where the system is the space V confined by the
> bottle wall. In the beginning, the system is evacuated ($M_1 = 0$, $U_1 = 0$, $S_1 = 0$),
> while in the end it is filled with air at atmospheric conditions ($T_2 = T_0$, $P_2 = P_0$).
>
> We question whether the filling process is reversible. To answer this question,
> we calculate the entropy generated during the entire process:
>
> $$S_{gen} = \int_1^2 \dot{S}_{gen} \, dt \quad (a)$$
>
> where the instantaneous entropy generation rate is
>
> $$\dot{S}_{gen} = \frac{dS}{dt} - \dot{m} s_0 - \frac{\dot{Q}}{T_0} \geq 0 \quad (b)$$
>
> where S is the instantaneous entropy inventory of the system; $\dot{m}$, instantaneous
> inlet flow rate; and $\dot{Q}$, the instantaneous heat transfer rate (defined positive when
> directed into the system). The inlet specific entropy s_0 is evaluated at atmospheric
> conditions. The temperature T_0 that is used as a denominator in eq. (b) indicates

that the temperature of the control surface crossed by $\dot{Q}$ is equal to the ambient temperature. In other words, if temperature differences develop during the filling process, they are all situated *inside* the control volume delineated by this surface.

Performing the time integral (a), using eq. (b) of this example and eqs. (a) and (g) of Example 1.2, yields

$$S_{gen} = S_2 - S_1 - s_0(M_2 - M_1) + \frac{P_0 V}{T_0} \qquad (c)$$

We note finally that $S_1 = 0$, $M_1 = 0$, and $S_2 = M_2 s_0$, which means that the entropy generation is definitely positive and that the filling process is irreversible:

$$S_{gen} = \frac{P_0 V}{T_0} > 0 \qquad (d)$$

Exactly the same S_{gen} expression is reached if the second law is applied to the process undergone by the closed system outlined in Example 1.1.

A subtle aspect of the conclusion centered around eq. (d) above is that S_{gen} is directly proportional to the work $P_0 V$ done by the atmosphere on the air mass that ultimately resides inside the bottle. As far as a potential external observer is concerned, the $P_0 V$ work is "lost": consequently, eq. (d) can be rewritten as

$$W_{lost} = T_0 S_{gen} \qquad (e)$$

which offers us a first glimpse at much more general result in engineering thermodynamics—the Gouy–Stodola theorem, eq. (3.7). The quantity called "entropy generation" is proportional to work that was, in principle at least, available, but was not tapped and delivered to an external user.

THE LOCAL THERMODYNAMIC EQUILIBRIUM MODEL

Looking back at the structure of the preceding two sections, we see that we adopted as "law" the Kelvin–Planck statement that the net heat transfer interaction of a closed system during a cycle in communication with one heat reservoir cannot be positive. We then generalized this statement one step at a time, by first relaxing the assumption of a certain number of heat reservoirs, next by considering arbitrary processes instead of cyclical ones, and finally by allowing the flow of mass and entropy across the system boundary. Having just reached the level of generalization represented by eq. (2.47) for the second law, eq. (1.24) for the first law, and eq. (1.28) for the law of mass conservation, it is appropriate to pause and question the applicability of these analytical conclusions to the quantitative study of real engineering systems and processes.

While avoiding specific references to heat engines and refrigeration machines and to the relative hotness of reservoirs (T_1) and (T_2), we showed that the equal sign in the second law corresponds to cycles that can be "reversed" in such a way that the net energy interactions retain their magnitudes and only change their signs, eq. (2.15). This limiting operation was then related to Carnot's description of a reversible cycle as a succession of equilibrium states assumed by a batch of fluid expanding and contracting in a cylinder and piston apparatus. Further on the connection between this requirement of "equilibrium" and the system's ability to "reverse" its energy interactions, recall that we can write $\delta W = P\,dV$ only for a change that can be viewed as a succession of equilibrium states. Note that the sign of $\delta W = P\,dV$ is *reversed* if the system visits the succession of equilibrium states in the *reverse* direction. It pays to remember also the parental relationship between the mechanics concept of equilibrium and the concept of *thermodynamic equilibrium*: a closed thermodynamic system is in equilibrium if it undergoes no further changes in the absence of "interactions" (energy, entropy, mass) with the environment.

If the equal sign in the analytical statement of the second law is associated with reversible processes that can be regarded as successions of equilibrium states, then it is natural to associate the inequality sign in the second law with processes that are not reversible and cannot be regarded as successions of equilibrium states: the words "irreversible" and "nonequilibrium" are used to describe real processes and states visited by systems whose nature departs from the limiting behaviour envisioned by Sadi Carnot (more on this vision is said at the end of this chapter). At this juncture, it is essential to recognize that the irreversible/nonequilibrium terminology and the routine use of the second-law statement with the inequality sign in it are not consistent with the conceptual foundation on which the analytical statement of the second law is built.

The concepts of entropy (S) and temperature $(T$ or $\theta)$ have been defined for a system in equilibrium. The entropy definition (2.38) refers to a succession of equilibrium states; likewise, the property "temperature" is introduced in order to describe quantitatively the relationship of thermal equilibrium between a system and its environment (see the discussion of the Zeroth Law of Thermodynamics, p. 15). Then, on what basis do we use equilibrium concepts in order to express the departure from reversible operation as (note the inequality sign)

$$\int_{\mathscr{V}} \frac{\partial(\rho s)}{\partial t}\,d\mathscr{V} + \int_{\mathscr{A}} \frac{1}{T}\,\mathbf{q}\cdot\mathbf{n}\,d\mathscr{A} + \int_{\mathscr{A}} \rho s \mathbf{v}\cdot\mathbf{n}\,d\mathscr{A} > 0 \qquad (2.47')$$

The assumption that is implicit in writing eq. (2.47'), and is only rarely acknowledged in the field of engineering thermodynamics, is that even when the system operates irreversibly, each of its infinitesimally small subcompartments is in a state of thermodynamic equilibrium. The assumption of *local thermodynamic equilibrium* means that if we consider as "sample" any infinitesimally small subcompartment of mass Δm and volume ΔV, and if we instantly isolate this sample, the state of the sample is such that no changes (e.g., pressure) are observed as we follow the evolution of the sample in time. Again, when we talk about "isolating" the sample, we think of encasing it in a zero-work, zero-heat transfer, and zero-mass transfer boundary. At equilibrium, the properties that describe the intensive state of the sample (T, P, $\Delta V/\Delta m$, $\Delta E/\Delta m$, $\Delta S/\Delta m$, etc.) are related to each other in the same manner as the properties of a finite-size batch of the substance in equilibrium (T, P, v, e, s, etc.). These important relationships form the subject of chapters 4, 6, and 7.

It follows that the properties ρ, e, and s that appear under the volume integral sign in eqs. (1.24), (1.28), and (2.47) refer to the instantaneous local equilibrium state of the infinitesimally small sample whose volume is $d\mathcal{V}$. Likewise, the properties ρ, h°, s, and T that appear in the surface integral terms of the same equations refer to the instantaneous equilibrium state of each sample that resides on the system boundary or control surface: the cut made by the system boundary through each sample gives birth to the area element $d\mathcal{A}$. This serves to remind the analyst that the properties $(h^\circ, s)_{\text{in}}$ and $(h^\circ, s)_{\text{out}}$ that appear in the first law and the second law for multiport systems, eqs. (1.22) and (2.46), respectively, represent the equilibrium properties of the samples that instantly pass through the inlet and outlet ports. The assumption that at every point in time there is only one sample that passes through one port whose cross-sectional area is finite, is recognized more succinctly as the *bulk flow model*.

The local thermodynamic equilibrium assumption turns out to be a very good modeling decision in the majority of systems that form the subject of engineering thermodynamics. The reason for its success is that the time scales of the rate processes that make the aggregate system behave irreversibly are considerably greater than the local relaxation time (the time needed for the redistribution of internal energy among the various molecular states, i.e., the time of readjustment via molecular collisions [26]). The study of systems that depart from the local thermodynamic equilibrium model takes us into the relatively newer field of *irreversible*, or *nonequilibrium*, *thermodynamics*. An introduction to this field is presented in chapter 12.

One begins to appreciate the vastness of the territory covered by the heat- and fluid-flow phenomena that subscribe to the local thermodynamic equilibrium model by reviewing the mathematical foundations of established fields such as fluid mechanics and heat transfer. The equations that account for mass conservation, force balances, and the first law at every point in the flow and temperature field contain as unknowns the density (specific volume), pressure, and specific energy [27]. The problem is routinely closed by assuming the existence of *two equations of state*, one for specific volume as a function of intensive properties (P, T), and the other for specific energy (or enthalpy) in terms of the same intensive properties: the essence of this assumption is that relations of type $v(P, T)$ and $u(P, T)$ can be measured *separately* by studying the behavior of finite-size *equilibrium* batches of the same substance that instantaneously occupies the sample for which the Navier–Stokes equations have been written. It is important to keep in mind, however, that the adoption of the local thermodynamic equilibrium model does not mean that the finite-size system that contains the flow is itself in a state of equilibrium.

THE ENTROPY MAXIMUM AND ENERGY MINIMUM PRINCIPLES

Of special relevance to the language of engineering thermodynamics and the issue of "stability" discussed later in chapter 6 are the implications of the first and second laws when applied to *isolated systems*. An isolated system is a closed system that is incapable of both heat transfer and work transfer interactions. Consider then a closed system whose boundary is simultaneously a zero-work surface (e.g., the inner surface of a rigid wall that is not penetrated by a rotating shaft, energized power cables, etc.) and an adiabatic surface: if such a system undergoes an infinitesimally small change of state, the first law (1.2) and the second law (2.40) require

$$dU = 0 \qquad\qquad (2.49)$$

$$dS \geq 0 \qquad\qquad (2.50)$$

A simplifying assumption that is also a smooth introduction to the thermodynamics nomenclature of simple systems (chapter 4) is that the energy change (dE) is equal to the change in internal energy (dU) in eq. (2.49).

The obvious message of eqs. (2.49) and (2.50) is that during an arbitrary change of state, the energy of the isolated system is fixed, whereas the entropy increases or, at best, remains constant. This conclusion is recognized as the *principle of entropy increase* or, alternatively, *entropy maximum principle*. The second name makes more sense if we think of all the changes of state that could possibly be experienced by an isolated system: since in the course of each change the entropy cannot decrease, at the end of the series of possible changes, the entropy reaches its greatest algebraic value.

Although this conclusion is analytically straightforward, the practically inclined may wonder whether the processes discussed above are real. Indeed, it is possible for an isolated system—one that is being "left alone" by its neighbors—to undergo changes of state? Experience shows us that an isolated system can behave in one of two ways:

(i) Its state remains unchanged regardless of the observation time interval: this state is said to be one of *stable equilibrium*.
(ii) Its state changes by chance or is triggered by a "disturbance" that is sufficiently weak to qualify as a zero-energy interaction.

The classical example of isolated system behavior of type (ii) is provided by Joule's *free-expansion* experiment [28], in which the gas that originally was held pressurized in half of the system's total volume was suddenly

THE ENTROPY MAXIMUM AND ENERGY MINIMUM PRINCIPLES   73

allowed to expand "freely"and fill the evacuated half. Joule measured the equilibrium temperatures before and after the free expansion, found that they were practically the same, and, invoking the first law (2.49), told the world that the internal energy of room-temperature "permanent gases" is only a function of temperature. In the present terminology, Joule's system was an "isolated system" because its mass inventory was fixed (closed system), the total volume was defined by two rigid-wall bottles connected through a valve (zero-work boundary), and the free-expansion process was finished before any meaningful heat transfer would have crossed the wall (adiabatic boundary). This isolated system underwent a change that was triggered by the opening of the valve between the two bottles, i.e., by an event whose occurrence did not violate the zero-work or zero-heat transfer description of the system boundary (assuming, of course, that the work needed to open the valve was negligible). The change of state was triggered by a sudden modification in the internal geometry of the isolated system.

As a practical example of isolated-system behavior of type (ii), the description of Joule's experiment can be condensed in the following words: the system was initially in a state of *constrained equilibrium*, a change of state took place when the lone *internal constraint* was removed, and, eventually, the system settled in a state of equilibrium without any internal constraints. Experience teaches us that this final state is one of stable equilibrium.

The above observations can be generalized to make room for more complicated systems whose internal constitution depends initially on an arbitrary number of internal constraints. The geometry of the isolated system at equilibrium is described by a number of *deformation* parameters, X_k. For example, the geometry of the gas–vacuum system designed by Joule is described by only two deformation parameters, the volume occupied by gas, V_g, and the volume occupied by vacuum, V_v; if V is the total volume of the isolated system, the (V_g, V_v) set of values changes from $(V/2, V/2)$ to $(V, 0)$ as the internal constraint is removed.

The (X_k) set of deformation parameters of the general system is shown attached to the vertical axis of the three-dimensional construction in Fig. 2.7: the (X_k) values change as the internal constitution of the system changes with each removal of an internal constraint. The drawing is meant to suggest that the bottom plane corresponds to the *configuration* in which the system is free of internal constraints, in other words, the bottom plane is the locus of stable equilibrium states. It follows that the attainment of maximum entropy in the limit of zero internal constraints is represented by the curve drawn in the constant-U plane of Fig. 2.7. To say that after the removal of all possible internal constraints an isolated system reaches its entropy maximum is to make *two* analytical statements about the vicinity of a stable equilibrium state in an isolated system:

$$(dS)_U = 0 \tag{2.51}$$

$$(d^2S)_U < 0 \tag{2.52}$$

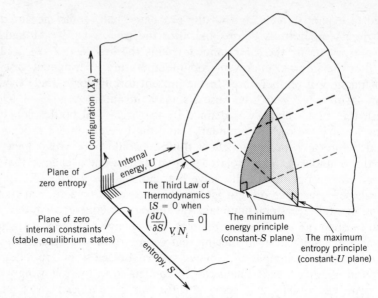

Figure 2.7 The entropy maximum and energy minimum principles for a closed system incapable of work transfer.

The subscript "U" serves as a reminder that the energy does not change as the isolated system approaches stable equilibrium.

Another way of looking at the second law and the behavior of closed systems is by focusing on systems that are only incapable of work transfer interactions (note that in the preceding discussion, the "isolated system" was incapable of both δW and δQ). When $\delta W = 0$, the first and second laws for an infinitesimal change of state require

$$\delta Q = dU \tag{2.53}$$

$$\delta Q \le T \, dS \tag{2.54}$$

or, if we eliminate δQ,

$$dU \le T \, dS \tag{2.55}$$

We conclude that in the constant-S plane of Fig. 2.7, in which eq. (2.55) states

$$dU \le 0 \tag{2.56}$$

the energy does not increase as the system frees itself of internal constraints. The stable equilibrium state that is approached at constant entropy is also a state of minimum energy; hence,

$$(dU)_S = 0 \tag{2.57}$$

$$(d^2U)_S > 0 \tag{2.58}$$

The *energy minimum principle* of eqs. (2.56)–(2.58) is illustrated by the curve drawn in the constant-S plane of Fig. 2.7. Together with the maximum-entropy principle discussed earlier, the energy minimum principle describes a surface in the S, U, (X_k) space: the main feature of this surface is that near the (U, S) plane of stable equilibrium states, it is perpendicular to that plane. For the sake of completeness, Fig. 2.7 shows that its line of intersection with the plane of stable equilibrium states is normal to the plane of zero entropy. This last feature is the graphic version of the *Third Law of Thermodynamics*, which is discussed on pp. 582–583 in the present treatment.

Although the entropy maximum and energy minimum principles are nothing more than reformulations of the two laws in the realm of two special classes of closed systems, historically, they have been very effective in conveying to the scientific world the essence of the two laws. The energy minimum principle showed that the emerging science of thermodynamics was consistent with the field of mechanics, in which a principle of potential-energy decrease was well-established. The entropy maximum principle drew and continues to draw attention from fields traditionally removed from engineering and physics, ranging from economics [29] to biology [30]. This principle received maximum publicity in the hands of its creator, Rudolf Clausius, who wrote [31]:

1. The energy of the universe is constant.
2. The entropy of the universe strives to attain a maximum value.

One hundred years later, it can be said with some justification that Clausius exaggerated in extending his principle from an isolated system to an evolving concept such as the "universe." However, considering the challenges faced by Clausius in his own time, it is evident that he was very enthusiastic about the strength and generality of his own theoretical conclusions, and that exaggeration (not apology) is the natural manifestation of enthusiasm. Had Clausius bent down to the skeptics, by using an apologetic tone that emphasized the limitations of his theory rather than its far-reaching consequences, it is unlikely that we would be using the entropy language in today's thermodynamics.

Clausius' contribution as a greater promoter of thermodynamics, continues today, both directly and indirectly. On the indirect plane, for example, his writings attracted to thermodynamics individuals like the young Gibbs,[†] who spent the years 1866–1869 in Europe [32]: we may disagree on

---

[†]Clausius' words on the energy and entropy of the universe were used by Gibbs as the opening to his paper on the equilibrium of heterogeneous substances [Ref. 32, p. 55]: "Die Energie der Welt ist constant. Die Entropie der Welt strebt einem Maximum zu."

the greatness of Gibbs' work, however, it is hard to imagine today's thermodynamics language without Gibbs' terminology in it.

As an example of Clausius' influence today, I note the emergence of a highly peculiar "anti-Clausius" movement among some thermodynamicists of the other Europe (e.g., Bazarov [33]). The members of this movement, the self-titled "progressives" who see Clausius' theory of the entropy death of the universe as a threat to the communist state, seem to take pleasure in calling Clausius—the true revolutionary—a "reactionary"!

Example 2.3. The presentation of the entropy maximum and energy minimum principles in Fig. 2.7 suffers from the abstract flavor of terms such as "internal constraints" and "constrained equilibrium states." The purpose of this example is to illustrate the physical meaning of Fig. 2.7 and its terminology. Consider the closed system of total mass $4m$ shown in Fig. 2.8. A certain incompressible liquid fills each of the four compartments that are separated by means of three rigid, impermeable, and adiabatic partitions. The initial state (A) of the closed system $(4m)$ is one of constrained equilibrium, as each compartment is at a different temperature. For the sake of concreteness, let us select the initial temperature distribution:

$$T_1 = T_0 , \qquad T_2 = 2T_0 , \qquad T_3 = 3T_0 , \quad \text{and} \quad T_4 = 4T_0 \qquad (a)$$

where T_0 is a reference absolute temperature. The properties U and S of incompressible liquids are functions of T only (see Table 4.7); if, in addition, c is constant, the initial internal energy and entropy of the $(4m)$ system can be expressed analytically as

$$U_A - U_0 = \sum_{i=1}^{4} mc(T_i - T_0) = 6mcT_0 \qquad (b)$$

$$S_A - S_0 = \sum_{i=1}^{4} mc \ln \frac{T_i}{T_0} = 3.18mc \qquad (c)$$

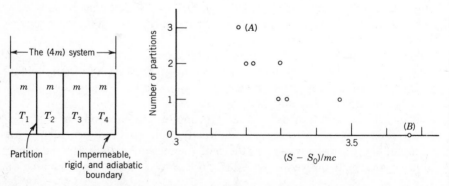

Figure 2.8 Example of how the total entropy of an isolated system tends toward a maximum (supremum) as the internal constraints disappear.

If one or more internal partitions are removed,[†] the $(4m)$ system evolves in a U-constant plane of the kind that is in Fig. 2.7. Given a sufficiently long time interval after the removal of a single partition, the $(4m)$ system settles into a new equilibrium state that, in general, is constrained. The energy and entropy of the $(4m)$ system and the compartment-by-compartment temperature distribution of a new equilibrium state can be determined quantitatively (Problem 2.6). The results are summarized by plotting the total entropy S on the abscissa and the number of remaining constraints on the ordinate. Each equilibrium state—constrained or unconstrained—is represented by a point in this plane (Fig. 2.8).

This exercise teaches us first that there is more than one sequence in which the partitions can be removed. The actual numerical calculations show that the removal of each partition leads to a net increase in the total entropy of the $(4m)$ system. Second, the unconstrained stable equilibrium state (B) is unique, and the total entropy of this state is greater than that of any other constrained equilibrium state that precedes it.

The simplicity of Fig. 2.8 is the result of having assumed that the initial compartments are of equal size and that the initial temperature distribution is given by eqs. (a). Even if we hold the number of original partitions fixed (i.e., equal to 3), we have the freedom to repeat this exercise by changing the $T_1, \ldots, T_4$ values, and by choosing any combination of original compartments of unequal size. In the end, when all these additional exercises are completed, the territory contained between points A and B will be covered by a much denser population of constrained equilibrium states. The conclusion that is not affected by the size of the population of intermediate constrained equilibrium states is that the entropy of the lone unconstrained equilibrium state (B) is *greater* than in any preceding state. For this reason, the appropriate name for the numerical value S_B is "supremum" rather than "maximum." At the end of a similar series of exercises using a closed system that evolves at contant S instead of constant U, we would see that a better name for the smallest U value reached in a state of stable equilibrium is "infimum" rather than "minimum."

In view of all these observations, the superstructural features of Fig. 2.7 (the ordinate axis and the three-dimensional surface whose tangent plane is vertical at a point that represents a stable equilibrium state) are acts of pure symbolism. These features are meant to summarize in three dimensions the main characteristics of an endless list of considerably more detailed pictures of the type shown in Fig. 2.8.

CARATHÉODORY'S TWO AXIOMS

Shortly after the turn of the century, Carathéodory proposed a two-axiom condensation of the essence of the two laws [34]. There were important reasons for such a step, for example, some mathematicians' lack of familiarity with the heat-engine cycle arguments of the Carnot–Clausius line, and the need for analytical thermodynamics that would make the subject applicable

[†]To remove a partition in this example means to assume only that the partition is no longer adiabatic.

to systems more general than the power/refrigeration examples contemplated by the pioneers. The effort to mathematize and "generalize" thermodynamics continues: for a critical review of successes and failures in this adjacent arena, the reader is directed to Truesdell's Rational Thermodynamics [11]. Yet, a study of Carathéodory's point of view can also play an instructive role in engineering thermodynamics, because it forces the reader to rediscover in a few abstract words the combined message of the entire first two chapters.

Reworded by Sears [35], the two axioms read as follows:

Axiom I. The work is the same in all adiabatic processes that take a system from a given initial state to a given final state.

Axiom II. In the immediate neighborhood of every state of a system, there are other states that cannot be reached from the first by an adiabatic process.

In the second scheme of Table 1.3, we saw already how axiom I is used to define "energy" as a thermodynamic property. The change in energy is taken as equal to the negative of the adiabatic work transfer, which is a unique quantity when the two end states are specified:

$$-W_{1-2,\text{adiabatic}} = U_2 - U_1 \tag{2.59}$$

In the same table, the heat transfer interaction is defined as the difference between the actual work transfer and the adiabatic work transfer associated with the given end states, $\delta Q = \delta W - \delta W_{\text{adiabatic}}$, which means $\delta Q - \delta W = dU$. Relevant to the analytical part of this section is the δQ expression for a reversible change of state, which is obtained using eq. (1.8'),

$$\delta Q_{\text{rev}} = dU - \sum_i Y_i \, dX_i \tag{2.60}$$

The second axiom is used to prove sequentially the existence of "reversible and adiabatic surfaces," the property "entropy," and, finally, the property "thermodynamic temperature." Carathéodory's original presentation of this sequence has become the subject of a sustained process of simplification that is designed to improve its accessibility [35–43]. An important step towards establishing the equivalence between axiom II and classical statements of the second law was made by Landsberg [41], who showed that axiom II follows from the Kelvin–Planck statement, eq. (2.1). His argument refers to the thermodynamic phase space in Fig. 2.9, in which an equilibrium state is represented by a point (the two volume coordinates V_A and V_B illustrate two of the deformation coordinates X_i).

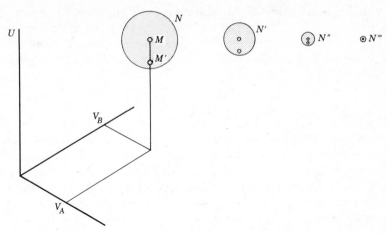

Figure 2.9 Graphic example of the connection between axiom II and the Kelvin–Planck statement of the second law.

Let (M) be an equilibrium state that possesses a neighborhood (N). We make the assumption that all the points that belong to (N) can be reached adiabatically from (M), in other words, we assume that axiom II is not true. Consider next the equilibrium state (M') that belongs to (N) and has the same configuration (V_A, V_B) as the first equilibrium state (M). We choose the state (M') such that $U_M > U_{M'}$.

We can think of the cycle $(M') \rightarrow (M) \rightarrow (M')$, in which the first leg $(M') \rightarrow (M)$ is executed in the absence of work transfer. The zero-work process $(M') \rightarrow (M)$ takes place as V_A and V_B are held constant and as the closed system is placed in thermal communication with a heat reservoir whose temperature is at least as high as θ_M. Note here the use of the concept of empirical temperature, as the θ values are the readings proved by a thermometer. The return process $(M) \rightarrow (M')$ is carried out adiabatically: that such a process is always possible, was assumed at the start of this discussion.

The cycle $(M') \rightarrow (M) \rightarrow (M')$ described above is characterized by positive net heat transfer $Q > 0$ and—if we invoke the first law—positive net work transfer. The net heat transfer is positive because during the only nonadiabatic portion of the cycle, the work transfer is zero while the energy increases (recall the assumption $U_M > U_{M'}$). Since the cycle $(M') \rightarrow (M) \rightarrow (M')$ is said to be executed by a closed system in communication with one heat reservoir, the $Q > 0$ conclusion violates the Kelvin–Planck statement of the second law (2.2). Therefore, relying on the Kelvin–Planck statement, we conclude that the assumption that all the points in (N) are adiabatically accessible from (M) is erroneous. In other words, in every neighborhood (N) of every point (M) in the thermodynamic phase space, there exist points that are adiabatically inaccessible from (M). This conclusion is analogous to axiom II. We can repeat this argument for a succession of neighborhoods (N'), (N''), (N'''), . . . that are progressively closer to state (M), Fig. 2.9, to show that the "immediate neighborhood" mentioned in axiom II contains adiabatically inaccessible states (M') that are *infinitesimally close* to (M).

Reversible and Adiabatic Surfaces

It is useful to illustrate the connection between axiom II and classical conclusions that are accessible based on the much simpler graphics of engineering thermodynamics. Consider the batch of fluid that expands adiabatically in the piston and cylinder apparatus in Fig. 2.10. The piston rod is connected to an external apparatus that can serve as either a work absorber or work producer. Our closed system is "the fluid." We follow its expansion from V_1 to V_2 in the energy–volume plane and, invoking the first law for the adiabatic process $(1) \rightarrow (2)$, we write

$$-W_{1-2} = U_2 - U_1 \tag{2.61}$$

This statement places no restriction on the sign of $U_2 - U_1$, yet we may question whether state (2) can land just anywhere on the vertical line $V = V_2$ of the (U, V) plane.

The energy change $U_2 - U_1$ can be negative, zero, or positive, because it is controlled by the work transfer apparatus with which our system com-

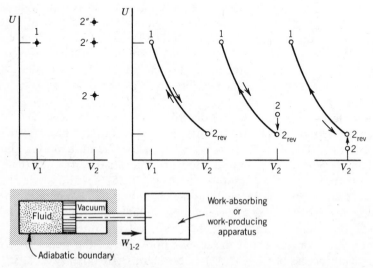

Figure 2.10 The uniqueness of state (2_{rev}), which can be reached adiabatically and reversibly from state (1).

municates. The energy decreases when the apparatus opposes the lateral progress of the rod, $U_2 - U_1 < 0$. In the special case where the rod is physically detached from the external apparatus, the expansion of the fluid is "free" and the net energy change is zero, $U_{2'} - U_1 = 0$ (note that the special process $(1) \rightarrow (2')$, Fig. 2.10, is attained in Joule's free-expansion experiment [28]). And, finally, the end-state energy $U_{2''}$ can rise above U_1 if, following the free expansion, the external apparatus does work on the fluid. One way of achieving this effect is by rotating the piston rod around its axis while the fluid shear integrated over the piston face opposes the rotation. In this manner, the external apparatus does work on the system, from state $(2')$ to $(2'')$.

In spite of the seemingly arbitrary location of state (2) on the vertical line $V = V_2$, we should expect some restrictions. Otherwise, given the fact that the energy drop must equal the adiabatic work output, eq. (2.61), it would mean that the adiabatic expansion from V_1 to V_2 could release any amount of work, no matter how great. A certain ordering of the final states (2), (2'), (2''), etc. becomes visible if we assume the existence of a special end-state (2_{rev}), whose noteworthy property is that it can also serve as the initial state for an adiabatic process that terminates at state (1) [note that state (2_{rev}), like states (2), (2'), and (2'') seen in the first (U, V) plane of Fig. 2.10, can be reached adiabatically from state (1)]. In other words, we assume the existence of a state (2_{rev}) that is adiabatically accessible from state (1) and from which state (1) is itself adiabatically accessible. The assumed two-way accessibility is indicated by the two arrows along the solid $(1)-(2_{rev})$ curve in the second (U, V) plane of Fig. 2.10.

We question whether an ordinary state (2) [adiabatically accessible from state (1)] is situated above or below state (2_{rev}) on the $V = V_2$ line. Assuming first that $U_2 > U_{2_{rev}}$, we can picture the cycle $(1) \rightarrow (2) \rightarrow (2_{rev}) \rightarrow (1)$ shown in the third (U, V) plane of Fig. 2.10. The cycle proceeds clockwise, because state (2) is adiabatically accessible from state (1) and state (1) is adiabatically accessible from state (2_{rev}). The net heat transfer interaction for this cycle is negative, $Q < 0$, as the only heat transfer occurs with a decrease in U during the zero-work process $(2) \rightarrow (2_{rev})$. We conclude that the positioning of state (2) above state (2_{rev}) on the $V = V_2$ line is consistent with the Kelvin–Planck statement of the second law.

Imagine now a sequence of states (2) that are situated progressively closer to state (2_{rev}), in other words, a sequence of cycles $(1) \rightarrow (2) \rightarrow (2_{rev}) \rightarrow (1)$, whose net heat transfer interaction approaches zero. In the limiting case, where state (2) coincides with state (2_{rev}), the net heat transfer is zero and, according to the terminology introduced on pp. 53–54, the cycle $(1) \rightarrow (2_{rev}) \rightarrow (1)$ is said to be *reversible*. In the present case, the two processes that make up the cycle are also adiabatic, hence, the name *reversible and adiabatic* for the solid line that symbolizes the processes $(1) \rightarrow (2_{rev})$ and $(2_{rev}) \rightarrow (1)$ in Fig. 2.10. The fact that the $(1)-(2_{rev})$ path is a "line" is discussed later in this subsection.

The possibility of a state (2) being located under state (2_{rev}) on the $V = V_2$ line is ruled out by the second law of Kelvin and Planck. We can assume the situation shown in the fourth (U, V) plane of Fig. 2.10, in which $U_2 < U_{2_{rev}}$. Since state (2) is adiabatically accessible from (1), the three-process cycle $(1) \rightarrow (2) \rightarrow (2_{rev}) \rightarrow (1)$ would have to be executed counterclockwise. The net heat transfer interaction would be positive in order to account for the energy increase from state (2) to state (2_{rev}) during the zero-work process $(2) \rightarrow (2_{rev})$. Such a conclusion, however, would violate the Kelvin–Planck statement.

The discussion centered around Fig. 2.10 can be summarized by saying that the states of volume V_2 that are accessible adiabatically from state (1) cannot be located under state (2_{rev}) in the (U, V) plane (or, simply, $U_2 \geq U_{2_{rev}}$). Conversely, all the states of volume V_2 that are situated under state (2_{rev}) in the (U, V) plane are not accessible adiabatically from state (1). It follows that the special state called (2_{rev}) is a *unique* point on the $V = V_2$ line, because it divides the vertical line into adiabatically accessible and adiabatically inaccessible points. Another way of concluding that the point $(U_{2_{rev}}, V_2)$ is unique is by starting with the assumption that two states of type (2_{rev}) can exist on the $V = V_2$ line, and arriving at a contradiction of the Kelvin–Planck statement (Problem 2.3).

The volume V_2 used in the preceding discussion is arbitrary. Indeed, for any value that we choose to assign to V_2, we can identify a unique point (2_{rev}) on the $V = V_2$ line in the (U, V) plane. It is not difficult to show that this conclusion—the uniqueness of point (2_{rev})—holds also for $V_2 < V_1$ (Problem 2.4), not just for the adiabatic expansions theorized in connection with Fig. 2.10. Taken together, all the points "(2_{rev})" that are reversibly and adiabatically accessible from point (1) form a curve that passes through point (1). The function $U(V)$ that represents this *reversible and adiabatic curve* is single-valued, otherwise we would be able to locate more than one (2_{rev}) point on a $V = $ constant line (Fig. 2.11).

The initial state [point (1)] considered in Fig. 2.10 is also arbitrary. This means that through any point $[(1), (1'), \text{etc.}]$, we can draw one reversible and adiabatic curve, and that the (U, V) plane is covered by a family of such curves. Two reversible and adiabatic curves cannot intersect: if they did, then, at constant V, there would exist two states that would be accessible reversibly and adiabatically from the point of intersection (see the right drawing of Fig. 2.11). The reversible and adiabatic curve drawn through point (1) divides the (U, V) field into a lower-half plane that is adiabatically inaccessible from (1), and an upper-half plane whose points (equilibrium states) are all accessible from (1) via adiabatic albeit *irreversible* processes.

These conclusions can be generalized by considering a closed system whose equilibrium configuration depends on more than one parameter [35]. Figure 2.12 shows a closed system that can be deformed by varying independently the two compartment volumes V_A and V_B. The "system" contains the fluids (A) and (B), whose respective pressures (P_A, P_B) are, in

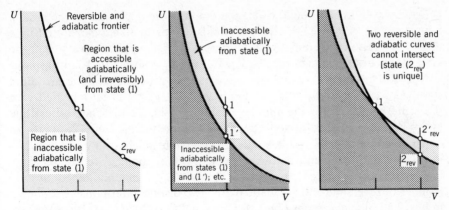

Figure 2.11 The uniqueness of the reversible and adiabatic curve that passes through state (1), or the fact that two reversible and adiabatic curves do not intersect.

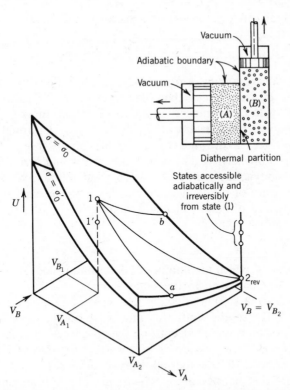

Figure 2.12 The uniqueness of the reversible and adiabatic surface that passes through a point in the (U, V_A, V_B) space, and the family of nonintersecting "$\sigma =$ constant" surfaces.

general, different. At equilibrium, the empirical temperature of the system is uniform ($\theta_A = \theta_B = \theta$), as the two fluids are permanently separated by a diathermal (also rigid, impermeable) partition. The equilibrium states of the system appear as points in the (U, V_A, V_B) space, or in the (θ, V_A, V_B) space.

The Kelvin–Planck sorting of the adiabatic processes linking an initial state $(U, V_A, V_B)_1$ and a final configuration $(V_A, V_B)_2$ can be repeated with the same results as for the simpler example of Fig. 2.10. The vertical line $(V_A = V_{A_2}, V_B = V_{B_2})$ is found to be divided by the unique state (2_{rev}) into an upper semiinfinite line of states that are accessible adiabatically and irreversibly from state (1), and a lower segment whose points are adiabatically inaccessible from state (1). States (1) and (2_{rev}) can be joined by any number of reversible and adiabatic paths: two distinct alternatives are shown in Fig. 2.12, namely, paths $(1)–(a)–(2_{rev})$ and $(1)–(b)–(2_{rev})$. It is shown in Problem 2.5 that by following the system's evolution along each path, we obain the same final temperature $\theta_{2_{rev}}$ (or final energy $U_{2_{rev}}$); in other words, we arrive at the same final point on the reversible and adiabatic surface. This argument leads in the end to the existence of a unique *reversible and adiabatic surface* that passes through a given point in the (U, V_A, V_B) space, hence, the family of nonintersecting surfaces in Fig. 2.12:

$$\sigma(U, V_A, V_B) = \sigma_0, \text{constant} \tag{2.62}$$

Each constant σ_0 denotes a reversible and adiabatic surface. We are free to choose the σ_0 values such that the sequence σ_0', σ_0'', σ_0''', . . . increases monotonically with the sequence of empirical temperatures θ', θ'', θ''', . . . of the states of fixed configuration $[(1'), (1''), (1'''), . . .]$ through which each constant-σ surface is drawn. If we would choose a σ_0 sequence that decreases monotonically as the θs increase, we would arrive eventually at negative thermodynamic temperatures. In what follows, we regard the σ_0 value as a measure of the "empirical entropy," in the same way that the θ values are a record of empirical temperatures.

Entropy (S)

The general case in which the number of deformation coordinates (X_i) is not specified can be dealt with in the same manner, that is, by repeating any number of times the system-selection leap executed by switching from Fig. 2.10 to Fig. 2.12. In the end, this line of reasoning would lead to the existence of constant-empirical entropy hypersurfaces $\sigma(U, X_i) = \sigma_0$ in the hyperspace (U, X_i). The existence of constant-σ surfaces can be used to prove the existence of the thermodynamic properties of "entropy" and "temperature." For the sake of mental visualization in three dimensions, we illustrate these steps in the context of the two-volume system defined in Fig. 2.12: this presentation is based largely on Sears' simplification of Carathéodory's approach [40].

Since σ varies monotonically with θ, the state of the $(A) + (B)$ system of Fig. 2.12 can be specified either in terms of (θ, V_A, V_B) or (σ, V_A, V_B). Any state-dependent quantity such as U can then be regarded as a function of σ, V_A, and V_B; hence, in an infinitesimal change of state:

$$dU = \left(\frac{\partial U}{\partial \sigma}\right)_{V_A, V_B} d\sigma + \left(\frac{\partial U}{\partial V_A}\right)_{\sigma, V_B} dV_A + \left(\frac{\partial U}{\partial V_B}\right)_{\sigma, V_A} dV_B \qquad (2.63)$$

In the event that this process occurs along a *reversible* path, eq. (2.63) assumes the form given already in eq. (2.60), namely,

$$dU = \delta Q_{rev} - P_A \, dV_A - P_B \, dV_B \qquad (2.64)$$

Note here the identification of $-P_A$ and $-P_B$ as the "generalized forces" Y_i discussed on page 13. Subtracting eq. (2.64) from eq. (2.63), we obtain

$$0 = \left(\frac{\partial U}{\partial \sigma}\right)_{V_A, V_B} d\sigma - \delta Q_{rev} + \left[\left(\frac{\partial U}{\partial V_A}\right)_{\sigma, V_B} + P_A\right] dV_A$$
$$+ \left[\left(\frac{\partial U}{\partial V_B}\right)_{\sigma, V_B} + P_B\right] dV_B \qquad (2.65)$$

Since σ, V_A and V_B can be varied independently, from eq. (2.65) we deduce the following identities:

$$\delta Q_{rev} = \left(\frac{\partial U}{\partial \sigma}\right)_{V_A, V_B} d\sigma \qquad (2.66)$$

$$-P_A = \left(\frac{\partial U}{\partial V_A}\right)_{\sigma, V_B} \quad \text{and} \quad -P_B = \left(\frac{\partial U}{\partial V_B}\right)_{\sigma, V_A} \qquad (2.67)$$

The first of these identities can be rewritten as

$$\delta Q_{rev} = \lambda \, d\sigma \qquad (2.68)$$

where $\lambda = (\partial U/\partial \sigma)_{V_A, V_B}$ is an unknown function of three properties, for example, $\lambda(\sigma, V_A, V_B)$, $\lambda(\theta, V_A, V_B)$, and $\lambda(\theta, \sigma, V_A)$. We show next that λ can only be a function of θ and σ, and that the form of this function is the product $\lambda = \phi(\theta)f(\sigma)$.

Applied to the closed system (A) alone (i.e., to the fluid on the left side of the diathermal partition, Fig. 2.12), the analysis contained between eqs. (2.63)–(2.68) concludes with

$$\delta Q_{rev,A} = \lambda_A \, d\sigma_A \qquad (2.69)$$

where λ_A is a function of (θ, σ_A), or of (θ, V_A), etc. For the adjacent system, we write similarly

$$\delta Q_{\text{rev},B} = \lambda_B \, d\sigma_B \tag{2.70}$$

with $\lambda_B = \lambda_B(\theta, \sigma_B)$. Influenced by our previous encounter with classical thermodynamics, where the Clausius entropy (S) of the aggregate system $(A) + (B)$ is the sum of the entropies of the two subsystems, $S_A + S_B$, we ask whether the *empirical* entropy of the aggregate system is only a function of the empirical entropies of the subsystems:

$$\sigma = \sigma(\sigma_A, \sigma_B) \tag{2.71}$$

Note that the question is not whether σ is equal to the sum $\sigma_A + \sigma_B$, but whether σ depends only on σ_A and σ_B. The answer falls out of eqs. (2.68)–(2.70), which, combined into the statement that the heat transferred to $(A) + (B)$ must be the sum of the heat transfer interactions experienced by (A) and (B) individually,

$$\delta Q_{\text{rev}} = \delta Q_{\text{rev},A} + \delta Q_{\text{rev},B} \tag{2.71'}$$

yields

$$\lambda \, d\sigma = \lambda_A \, d\sigma_A + \lambda_B \, d\sigma_B \tag{2.72}$$

or

$$d\sigma = \frac{\lambda_A}{\lambda} \, d\sigma_A + \frac{\lambda_B}{\lambda} \, d\sigma_B \tag{2.73}$$

In view of eq. (2.73), we conclude that σ is constant whenever σ_A and σ_B are held constant, or that σ is a function of σ_A and σ_B only. As partial derivatives of $\sigma(\sigma_A, \sigma_B)$, the two ratios, λ_A/λ and λ_B/λ, are also functions of only σ_A and σ_B:

$$\frac{\lambda_A}{\lambda} = f_1(\sigma_A, \sigma_B) \tag{2.74}$$

$$\frac{\lambda_B}{\lambda} = f_2(\sigma_A, \sigma_B) \tag{2.75}$$

Equation (2.74) suggests that since λ_A is not dependent on V_B, the denominator λ cannot be a function of V_B either. Similarly, eq. (2.75) states that λ is also independent of V_A. We are left with the possibility [43]

$$\lambda(\theta, \sigma_A, \sigma_B), \; \lambda_A(\theta, \sigma_A), \; \lambda_B(\theta, \sigma_B)$$

which, by virtue of eqs. (2.74) and (2.75), means

$$\frac{\partial}{\partial \theta}\left(\frac{\lambda_A}{\lambda}\right) = 0 \quad \text{and} \quad \frac{\partial}{\partial \theta}\left(\frac{\lambda_B}{\lambda}\right) = 0 \tag{2.76}$$

The two-equation system (2.76) is the same as

$$\frac{\partial}{\partial \theta}(\ln \lambda) = \frac{\partial}{\partial \theta}(\ln \lambda_A) = \frac{\partial}{\partial \theta}(\ln \lambda_B) \tag{2.76'}$$

where the three terms are, in order, functions of $(\theta, \sigma_A, \sigma_B)$, (θ, σ_A) and (θ, σ_B). However, since σ_A and σ_B can be varied independently of θ, eqs. (2.76') imply that each term must equal the same function of θ, for example,

$$\frac{\partial}{\partial \theta}(\ln \lambda) = \phi_1(\theta) \tag{2.77}$$

or

$$\lambda = f(\sigma) \exp\left[\int \phi_1(\theta)\, d\theta\right] \tag{2.77'}$$

or, finally,

$$\lambda = f(\sigma)\phi(\theta) \tag{2.78}$$

The reversible heat transfer expression (2.68) is therefore equal to $\delta Q_{rev} = \phi(\theta)f(\sigma)\, d\sigma$, and, recognizing $f(\sigma)\, d\sigma$ as the total differential of a function S named "entropy," we finally arrive at

$$\delta Q_{rev} = \phi(\theta)\, dS \tag{2.79}$$

This result is very similar to eq. (2.37): it remains to show that the temperature function $\phi(\theta)$ is the same as the thermodynamic temperature T defined via eq. (2.23).

Thermodynamic Temperature

Consider the family of cycles in the three-dimensional space of Fig. 2.13: each cycle $(a) \rightarrow (b) \rightarrow (c) \rightarrow (d) \rightarrow (a)$ consists of a reversible and isothermal process in the $\theta = \theta_a$ plane, a reversible and adiabatic process in the $S = S_2$ plane, a reversible and isothermal process in the reference temperature plane $\theta = \theta_0$, and, finally, a reversible and adiabatic process in the plane $S = S_1$. The constants θ_a, S_1, and S_2 are arbitrary (unspecified). Writing now $Q_{rev,a}$ for the net heat transfer input to the system during the $(a) \rightarrow (b)$ process, eq. (2.79) yields

$$Q_{rev,a} = \phi(\theta_a)(S_2 - S_1) \tag{2.80}$$

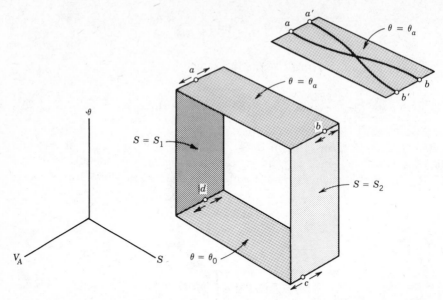

Figure 2.13 The family of reversible cycles for recovering the concept of thermo-dynamic temperature from the line of reasoning started by Carathéodory's axioms.

Similarly, the heat transferred to the system during the $(c) \rightarrow (d)$ process is

$$Q_{\text{rev},0} = \phi(\theta_0)(S_1 - S_2) \tag{2.81}$$

Together, eqs. (2.80) and (2.81) produce

$$\phi(\theta_a) = \phi(\theta_0)\left(\frac{Q_{\text{rev},a}}{-Q_{\text{rev},0}}\right) \tag{2.82}$$

which, analytically, is the same as eq. (2.22). We can then use eq. (2.82) in the same way as on p. 59, to define the *thermodynamic temperature* scale $T \equiv \phi$; hence,

$$\delta Q_{\text{rev}} = T \, dS \tag{2.37'}$$

One final observation concerns the special role of the thermodynamic temperature factor T vis-à-vis δQ_{rev}. According to eqs. (2.60) and (2.64), δQ_{rev} is a *linear differential form*, called also *Pfaffian form*, or *Pfaffian*. It is shown in the Appendix that this form can be integrated only when certain integrability conditions are met, which is to say that in general δQ_{rev} is not an exact differential (this observation is stressed also by the "δ" notation). The thermodynamic temperature then serves as an *integrating denominator* for the linear differential form, because the result of dividing δQ_{rev} by T is always an exact differential, dS.

The Two Parts of the Second Law

This concludes the review of the second law for closed systems using as a starting point Carathéodory's axioms. We saw that the existence of reversible and adiabatic surfaces enables us to recover on a purely mathematical basis the concepts of entropy, thermodynamic temperature, and, in particular, the second law for a reversible process, $\delta Q_{rev} = T\,dS$. The geometric beginnings of this line of reasoning highlight the two-part nature of the second law:

(i) the uniqueness of the reversible and adiabatic surface that passes through a given point (M) [e.g., state (1), Figs. 2.11 and 2.12], and,

(ii) the fact that the states that can be reached adiabatically and *irreversibly* from (M) are all situated on the same side of the reversible and adiabatic surface drawn through (M).

These two parts can be rediscovered in the equality and the inequality that are traditionally placed on top of each other into a single statement, such as in eqs. (2.1) or (2.40).

A HEAT TRANSFER MAN'S TWO AXIOMS

Carathéodory's axiomatic encapsulation of the essence of the first law and the second law represents a mechanistic point of view, because the first primary concept chosen for this formulation (work transfer, Table 1.3) is a concept borrowed from the field of mechanics. Taken together, the two axioms and the primary concepts deliver the remaining pieces of engineering thermodynamics as we learn it, namely, the concepts of heat transfer, entropy, and thermodynamic temperature. One develops a better feel for what this axiomatic formulation accomplishes by rethinking the idea of combining a few primary concepts with two axioms, this time using the very language that Carathéodory tried so hard to avoid.

Imagine a world in which man was considerably more at ease with thermal effects and their quantitative analysis than with mechanics, in other words, a world in which "temperature," "heat transfer," and "zero-work boundary" were the obvious choices as primary concepts. Note that the same choice was made also by Keenan and Shapiro in their second scheme (Table 1.3). Derived concepts that follow immediately from this choice are the concepts of adiabatic boundary and work transfer. The definitions of these derived concepts are discussed in the closing pages of chapter 1.

The members of this make-believe society had already established the convention that "temperature" always means the ideal gas thermometer temperature (the "absolute temperature") and labeled it T. Note here that the adoption of the ideal gas temperature as the true thermodynamic temperature scale is practiced sometimes as a pedagogically effective move in introductory engineering thermodynamics courses (e.g., Cravalho and Smith [23]), despite the subtle conceptual difference between $\theta_{\text{ideal gas}}$ and T (pp. 60–61).

The fact that these people were not comfortable with mechanical things does not mean in any way that they were backward. On the contrary, their mathematics was the same as ours despite trivial differences in language, notation, and ways of measuring. For example, it was considered proper to measure the size of a closed system in terms of its "entropy" (labeled S): this method had been perfected two thousand years earlier by their own "old Greeks," so that even the children could tell that the entropy of the closed system of Fig. 2.14 is forever equal to the sum of the entropies of its two subsystems:

$$S = S_A + S_B \tag{2.83}$$

In short, these people were about as conversant in entropy measurement and accounting as you and I are today when it comes to measuring, say, "volume."

Incidentally, the people I am describing learned the concept of volume also (in college), however, they also learned to fear this concept along with

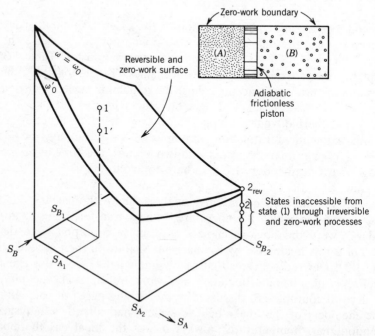

Figure 2.14 A family of nonintersecting reversible and zero-work surfaces in the (U, S_A, S_B) space.

everything else geometric or mechanical. Contributing to this fear was the fact that the name "volume" sounded alien: it had been invented by a classics-loving professor, who made it up by piecing together two interesting and otherwise very dead words. Feared as it was, the word "volume" was quite popular among the practitioners of name dropping, especially at conferences on pure heat and congressional hearings on all sorts of nongeometric topics. There were even passionate references to the "volume death" of the universe, when, in fact, no one remembered the professor's original theory with the same name.

Relatively late in its scientific development, this make-believe society learned to attach great importance to the quantity $T\,dS$, which was called "reversible heat transfer interaction" and labeled δQ_{rev}. The word "reversible" was used as a reminder of the special state of tranquility and temperature uniformity (T) that reigned during the entropy change dS. At ease with measuring both T and S, these people had no trouble calculating δQ_{rev} using eq. (2.37).

Consider now the implications of the following two axioms, which—it is easy to see—are my entropy-seeing society's alternative to Carathéodory's axioms [44]:

Axiom I′. The heat transfer is the same in all zero-work processes that take a system from a given initial state to a given final state.

Axiom II′. In the immediate neighborhood of every state of a system, there are other states that cannot be reached from the first via a zero-work process.

The implications of axiom I′ are listed already in the rightmost column of Table 1.3, namely, the existence of a thermodynamic property (function of state) called energy:

$$Q_{1-2,\text{zero-work}} = U_2 - U_1 \qquad (2.84)$$

and the definition of work transfer:

$$\delta W = \delta Q - \delta Q_{\text{zero-work}} = \delta Q - dU \qquad (2.85)$$

For an infinitesimally small reversible change of state, we can now write

$$\delta W_{\text{rev}} = -dU + \delta Q_{\text{rev}} \qquad (2.86)$$

and, with special reference to the two-fluid system in Fig. 2.14,

$$\delta W_{\text{rev}} = -dU + T_A\,dS_A + T_B\,dS_B \qquad (2.87)$$

In that example, the two compartments are separated by an adiabatic wall (piston) that can slide without encountering any resistance from the cylinder wall. The "empirical pressure" π is measured with a locally available pressure gage (not a standard instrument). Measurements show that the value of π is the same on both sides of the sliding partition.

It should be obvious that the analysis contained between eqs. (2.61) and (2.82) can be redone for the example of Fig. 2.14 in order to define the rigorously derived concepts of "volume" (V) and thermodynamic "pressure" (P). This analysis is outlined in Ref. [44]. The first conclusion to emerge from axiom II' is the existence of *reversible and zero-work surfaces*:

$$\omega(U, S_A, S_B) = \omega_0, \text{constant} \qquad (2.88)$$

This conclusion replaces eq. (2.62) of the original analysis. The value ω_0 of the particular surface that passes through a given point (1) in the (U, S_A, S_B) space can be called the *empirical volume* of the system at state (1). One finds also that eq. (2.88) represents a family of nonintersecting surfaces. Furthermore, all the states (2) that can be reached via zero-work processes from state (1) are situated on the same side of the reversible and zero-work surface that passes through state (1).

The next conclusion is that U can be considered a function of ω, S_A, and S_B, eq. (2.88); hence,

$$dU = \left(\frac{\partial U}{\partial \omega}\right)_{S_A, S_B} d\omega + \left(\frac{\partial U}{\partial S_A}\right)_{\omega, S_B} dS_A + \left(\frac{\partial U}{\partial S_B}\right)_{\omega, S_A} dS_B \qquad (2.89)$$

Comparing this differential with eq. (2.87), we identify, among other things,

$$\delta W_{\text{rev}} = \Lambda \, d\omega \qquad (2.90)$$

where the coefficient Λ is itself a function of (ω, S_A, S_B), or (π, S_A, S_B), or (π, ω, S_A), etc.:

$$\Lambda = -\left(\frac{\partial U}{\partial \omega}\right)_{S_A, S_B} \qquad (2.91)$$

Assuming that $\Lambda = \Lambda(\pi, \omega, S_A)$ and reconstructing the analysis presented earlier between eqs. (2.69) and (2.78), we find that Λ can only be a function of π and ω, and that this function is a product

$$\Lambda = \Phi(\pi) F(\omega) \qquad (2.92)$$

The details of this proof are omitted for the sake of conciseness: note, for example, that to "reconstruct" the earlier analysis means to begin with $\delta W_{\text{rev}, A} = \Lambda_A \, d\omega_A$ in place of eq. (2.69), and so on. In the end, eqs. (2.90) and (2.92) together imply the existence of an exact differential $dV = F(\omega) \, d\omega$, so that

$$\delta W_{rev} = \Phi(\pi) \, dV \tag{2.93}$$

The new function V can be named *volume* to distinguish it from the empirical volume function ω (note that the function ω represents the sequence of monotonically increasing or decreasing numbers assigned to the surfaces that pass through points (1), (1), (1″), ... in Fig. 2.14). The analogy between this last conclusion and the naming of entropy in eq. (2.79) is evident.

The final step consists of constructing a reversible cycle that is a sequence of four processes:

(i) Reversible and isobaric expansion in the $\pi = \pi_a$ plane (to be concrete, we assume $V_2 > V_1$)
(ii) Reversible and zero-work depressurization in the $V = V_2$ plane
(iii) Reversible and isobaric compression in the $\pi = \pi_0$ plane
(iv) Reversible and zero-work pressurization in the $V = V_1$ plane

The four planes that house this cycle are shown in Fig. 2.15. In most cases, i.e., when common working fluids are used, the reversible and zero-work depressurization process (ii) means a process of zero-work reversible *cooling*. Note further that the heat transfer interaction during this process [as well as during the heating process (iv)] is of type $\delta Q_{rev} = T \, dS$, i.e., that the system makes contact with a sequence of heat reservoirs whose tempera-

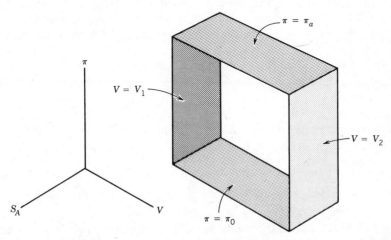

Figure 2.15 The four-plane locus of a family of reversible cycles for recovering the concept of thermodynamic pressure from axioms I′ and II′.

tures match the continuous variation of the system temperature T. Applying eq. (2.93) to the two isobaric processes, we have

$$\delta W_{rev,a} = \Phi(\pi_a)(V_2 - V_1) \tag{2.94}$$

$$\delta W_{rev,0} = \Phi(\pi_0)(V_1 - V_2) \tag{2.95}$$

and, dividing side by side,

$$\Phi(\pi_a) = \Phi(\pi_0)\left(\frac{\delta W_{rev,a}}{-\delta W_{rev,0}}\right) \tag{2.96}$$

This relationship can be used to define the *thermodynamic pressure* scale $P \equiv \Phi$ in the same way that eqs. (2.22) and (2.82) served as definitions for thermodynamic temperature. The thermodynamic pressure scale P is based on one fiducial point, so that, as a matter of convention, the factor $\Phi(\pi_0)$ is a constant [e.g., $\Phi(\pi_0) = 0.6113$ kPa if the fiducial point is the triple point of water]. Relative to the linear differential form (2.87), the thermodynamic pressure P serves as the integrating denominator.

Reviewing the progress made after axioms I′ and II′, we note the three-step definition of the concepts of reversible and zero-work surface, volume, and, finally, thermodynamic pressure. The expression for the infinitesimal *reversible* work transfer interaction is $P\,dV$, which means that the glossary of words possessed now by the heat transfer-biased individual is as complete as the glossary of the mechanic who just finished digesting Carathéodory's two axioms.

The perfect symmetry between the classical (historical) construction of thermodynamics and the present, heat transfer-based reconstruction is outlined in Table 2.1. The concepts assembled in this table are only the frontispiece of the voluminous subject of thermodynamics known here (the second column) or in the heat transfer-minded world described earlier (the third column). The table can be continued downward by identifying one by one the (derived) analytical highlights of our own thermodynamics, and then working out the equivalent analytical fact or statement that must be entered in the third column.

As an example of how Table 2.1 might be continued, consider the so-called "two parts" of the classical second law, which were highlighted at the end of the preceding section. The same two parts can be identified in the third column of the table as well. They are

(i′) the uniqueness of the reversible and zero-work surface drawn through state (1), Fig. 2.14, and

(ii′) the fact that all the states that are accessible via zero-work processes from state (1) are all situated on the same side of the constant-ω (or constant-V) surface drawn through (1).

TABLE 2.1   The Parallel Thermodynamics Structure of the Classical Scheme due to Carathéodory and the Heat Transfer–Based Reconstruction [44]

Structure	Classical Scheme, Rooted in Mechanics	Heat Transfer-Based Reconstruction
Primary concepts	Pressure (force) Volume (displacement) Work transfer Adiabatic boundary	Temperature[a] Entropy Heat transfer[a] Zero-work boundary[a]
The first law	$\int_1^2 \delta W_{\text{adiabatic}} = f(1,2)$	$\int_1^2 \delta Q_{\text{zero-work}} = f(1,2)^a$
The definition of energy change, dE	$-\delta W_{\text{adiabatic}}$	$\delta Q_{\text{zero-work}}{}^a$
Derived definition of heat transfer	$\delta W - \delta W_{\text{adiabatic}}$	
Derived definition of work transfer		$\delta Q - \delta Q_{\text{zero-work}}{}^a$
Derived concepts	Heat transfer Zero-work boundary Reversible and adiabatic surface Entropy Temperature	Work transfer[a] Adiabatic boundary[a] Reversible and zero-work surface Volume Pressure
The second law for a closed system executing an integral number of cycles while in communication with no more than one "reservoir"	"$\oint \delta Q > 0$ is impossible" or "$\oint \delta W > 0$ is impossible"	"$\oint \delta W < 0$ is impossible" "$\oint \delta Q < 0$ is impossible"
Type of reservoir	Temperature (T_0)	Pressure (P_0)

[a] Entries found also in Keenan and Shapiro's second method for stating the first law (Table 1.3).

The immediate question is whether the "accessible" states referred to in part (ii′) are situated above or below the reversible and zero-work surface. The equivalent question in the second column of the table is settled by invoking the Kelvin–Planck statement of the second law. This means that in order to be able to differentiate between the two sides of the constant-ω surface, we must first state the equivalent of the "Kelvin–Planck second law" that would fit in the third column of Table 2.1. Looking again at the vast evidence on which the second law is based, the equivalent of the Kelvin–Planck statement must be

> The net work transfer interaction of a closed system that executes a cycle while in communication with no more than one pressure reservoir cannot be negative, i.e.,

$$\oint \delta W < 0 \text{ is impossible}$$

Had this statement not been true, man would have opted a long time ago for letting the atmospheric pressure reservoir alone do work for him.

CONCLUDING REMARKS

Despite an already impressive record of articles, books, public debates, and lost friendships, the second law continues to be a preferred topic in modern research—this, as an activity distinct from the modern engineering interest in second-law implications in thermal design (e.g., chapters 3, 5, and 11). Among the modern works devoted to the second law, I note the comparative discussion and the ordering of the various alternative formulations that have been proposed. No less than nine such alternatives are discussed by Montgomery [45] and seven by Huang and Clothier [46]. The latter place special emphasis on a formulation proposed originally by Hatsopoulos and Keenan [47] based on the body of empirical evidence that concerns the behavior of isolated systems (see also Huang [48]). The evidence can be summarized in two statements (the two laws), first, that the energy of the isolated system is forever conserved, and, second, that the system's ability to do useful work (for us) can either deteriorate or, at best, stay the same. The connection between this second statement and the second law written here as a principle of entropy increase for isolated systems (p. 72) becomes clear after applying the concept of nonflow exergy to isolated systems (Problem 3.5).

Another modern development is the questioning of the view that all the second-law formulations are equivalent [11]. For example, we are brought up to believe (1) that with regard to constructing heat-engine cycles, the second law boils down to having access to at least two heat reservoirs, and (2) that when only two reservoirs are involved, the net effect of the engine cycle is the emission of heat transfer by the warm reservoir and the absorption of heat transfer by the cold one. Professor Truesdell draws attention to a class of heat-engine cycles [49]—the so-called "strange" or unusual Carnot cycles [50]—that make us think twice about the generality of such teachings. Shown in Fig. 2.16 is the trace left in the T–V plane by the reversible cycle of a closed system that consists of a quantity of cold water. The noteworthy property of cold water is that its density reaches a well-defined maximum as the temperature varies isobarically. At atmospheric

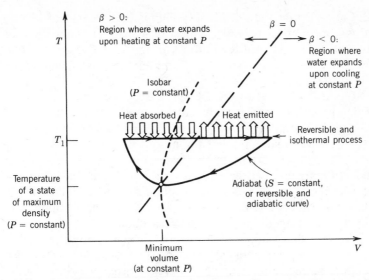

Figure 2.16 A reversible cycle executed by a closed system consisting of a batch of liquid water near the temperature of its density maximum (after Truesdell [49]).

pressure, the temperature of maximum density is very close to 4°C: this feature governs the circulation pattern of cold water in nature, both in pure liquid layers [51] and in liquid-saturated porous media [52].

One interesting consequence of the anomalous behavior of cold water is that the reversible and adiabatic lines are cup-shaped, as shown in Fig. 2.16. The Carnot cycle, then, consists of only two processes, a reversible and adiabatic volume change connected to a reversible and isothermal volume change. If the sense of the cycle is as indicated in Fig. 2.16, then during the first part of the reversible and isothermal expansion process, the system absorbs heat from the lone heat reservoir (T_1). During the second part of the process, a heat transfer interaction of exactly the same magnitude as the first occurs in the opposite direction: the net heat transfer interaction for the entire cycle is zero, in other words, the proposed heat-engine cycle does not deviate from the Kelvin–Planck statement of the second law. Nevertheless, contrary to the rules of thumb listed in the preceding paragraph, in Fig. 2.16, we have an example of a heat-engine cycle that proceeds while in contact with only one heat reservoir. Furthermore, at the end of each cycle, the net effect is not the emission of heat transfer by the reservoir that serves as heat source.

Finally, there is considerable material still being written about Sadi Carnot and the origins of what later became the second law. There is little disagreement on the greatness of Sadi Carnot's intuitive description of a "limiting" cycle that consists of a succession of equilibrium states, and of his claim that the efficiency of such a cycle depends only on the temperatures of the two heat reservoirs and not on the choice of working fluid. Unfortunately, Sadi Carnot's premature death of cholera in the epidemic of 1832 and the belated discovery of his theory left most of us with the impression that his revolutionary ideas materialized out of thin air. I use this as an opportunity to draw attention to a growing body of literature that paints an entirely different picture [5, 53–56].

Sadi Carnot[†] belonged to one of the most remarkable families that the worlds of engineering and government have ever known. The concept of reversibility in cyclical operation was formulated first by his father, Lazare Carnot, as an essential condition for maximizing the efficiency of purely mechanical energy converters. In his treatise on engineering mechanics [57], Lazare Carnot argued that the efficiency of a "machine" is maximum when violent effects such as percussion and turbulence (when fluid machinery is involved) are avoided. Lazare Carnot referred to this limiting and most efficient regime of operation as "geometric motion."

It is also true that with so little material evidence left after Sadi's death and the atmosphere of political disrepute that surrounded Lazare's final years, much of the apparently "filial" relationship between Sadi's heat-engine theory and Lazare's theory of mechanisms can only form the subject of educated speculation. Interesting reading in this direction is provided by Refs. 53 and 54, which focus on Sadi's adolescence and engineering studies, when Lazare's occupation was that of recording secretary for the *Institut de France*. In that capacity, Lazare Carnot had to examine firsthand and comment on a number of inventions that dealt with heat engines. At the end of such reading, Sadi Carnot's analogy between the fall of a water stream through a work-producing water wheel and the fall of caloric through a work-producing heat engine emerges as an understandable product of the intellectual environment in which he was raised. His vision that the temperature differences between the heat engine and the heat reservoirs must be avoided emerges as a very powerful generalization of Lazare Carnot's principle of avoiding the free fall of water upstream and downstream of the waterwheel.

These speculations address only one aspect of the father–son relationship. Lazare Carnot was a revolutionary in more than just engineering. A firm believer in the concept of Republic, he resigned from government in 1800 after only five months as Napoleon's first Minister of War. Retired, he devoted his time to engineering science and to raising his children. So, if today, we regard Sadi Carnot's thinking as "revolutionary," perhaps we are paying also a tribute to Lazare Carnot's presence as father and role model.

[†]Nicolas Léonard Sadi Carnot (1796–1832) was the son of Lazare Nicolas Marguerite Carnot (1753–1823), a military engineer and one of the leaders of the French Revolution. In French history, Lazare is remembered as the "Great Carnot" and the "Organizer of Victory." Lazare's younger son, Hippolyte Carnot (1801–1888) enjoyed a successful political career that culminated with the position of senator for life. It was Hippolyte who in 1871 published Sadi's *Manuscript Notes*. Hippolyte's own son, named also Sadi Carnot (1837–1894), was President of the Third Republic from 1887 until his assassination in 1894.

SYMBOLS

$\mathcal{A}$	surface
c	specific heat of incompressible substance
f, F	functions
$h°$	methalpy
m, M	mass [kg]
$\dot{m}$	mass flowrate [kg/s]
P	pressure
$\mathbf{q}$	heat-flux vector
Q	heat transfer interaction
R	ideal gas constant
s	specific entropy [J/kgK]
s'''_{gen}	volumetric rate of entropy generation [W/m^{3}K]
S	entropy [J/K]
S_{gen}	entropy generation [J/K]
$\dot{S}_{gen}$	entropy-generation rate [W/K]
T	temperature
U	internal energy
v	specific volume
$\mathbf{v}$	velocity vector
V	volume
$\mathcal{V}$	region in space
W	work transfer interaction
W_{lost}	lost or destroyed work
θ	empirical temperature
λ	coefficient [eq. (2.68)]
Λ	coefficient [eq. (2.90)]
π	empirical pressure
ρ	density
σ	empirical entropy
ϕ, Φ	functions
ψ	function
ω	empirical volume [eq. (2.88)]
$(\ )_C$	Carnot
$(\ )_{rev}$	reversible
$(\ )_0$	reference quantity

REFERENCES

1. M. Planck, *Treatise on Thermodynamics*, 3rd ed., translated by A. Ogg, Dover, New York, 1945, p. 89 (1st German ed. published in 1897).
2. Lord Kelvin, *Mathematical and Physical Papers of William Thomson*, Vol. 1, Cambridge University Press, 1882, p. 179; from the article "On the dynamical theory of heat, with numerical results deduced from Mr. Joule's equivalent of a

thermal unit, and M. Regnault's observations on steam," *Trans. R. Soc. Edinburgh*, March 1851, and *Philos. Mag.*, Ser. 4, 1852, reprinted in Ref. 6 of Chapter 1, pp. 106–132.

3. P. Feyerabend, *Against Method*, Verso, London, 1978, p. 27.

4. E. Clapeyron, *Memoir on the Motive Power of Heat*, originally published in the *Journal de l'École Polytechnique*, Vol. 14, 1834; translated in E. Mendoza, ed., *Reflections on the Motive Power of Fire and Other Papers*, Dover, New York, 1960, and in Ref. 6 of Chapter 1.

5. E. Mendoza, Contribution to the study of Sadi Carnot and his work, *Arch. Int. Hist. Sci.*, Vol. 12, 1959, pp. 377–396.

6. J. W. Gibbs, Graphical methods in the thermodynamics of fluids, *Trans. Conn. Acad. Arts Sci.*, Vol. 2 (April–May), 1873; reprinted in *The Collected Works of J. Willard Gibbs*, Vol. I, Longmans, Green, New York, 1928, Fig. 3, p. 10.

7. H. Poincaré, *Thermodynamique*, Georges Carré, Paris, 1892, pp. 136–138.

8. J. H. Keenan, *Thermodynamics*, Wiley, New York, 1941, pp. 74–76.

9. J. Kestin, *A Course in Thermodynamics*, revised printing, Vol. I, Hemisphere, Washington, DC, 1979, pp. 429–433.

10. V. A. Kirilin, V. V. Sychev, and A. E. Sheindlin, *Engineering Thermodynamics*, translated by S. Semyonov, Mir Publishers, Moscow, 1976, pp. 74–87.

11. C. Truesdell, *Rational Thermodynamics*, 2nd ed., Springer-Verlag, New York, 1984, p. 44.

12. W. F. Giauque, A proposal to redefine the thermodynamic temperature scale: With a parable of measures to improve weights, *Nature (London)*, Vol. 143, 1939, pp. 623–626.

13. William Thomson (Lord Kelvin), On an absolute thermometric scale founded on Carnot's theory of the motive power of heat, and calculated from Regnault's observations, *Proc. Cambridge Philos. Soc.*, June 5, 1848, and *Philos. Mag.*, Ser. 3, October 1848; also in *Mathematical and Physical Papers of William Thomson*, Vol. I, Cambridge University Press, 1882, pp. 100–106.

14. K. Mendelssohn, *The Quest for Absolute Zero*, McGraw-Hill, New York, 1966, pp. 10, 11.

15. R. Clausius, On different forms of the fundamental equations of the mechanical theory of heat and their convenience for application, translated by R. B. Lindsay, pp. 162–193 of Ref. 6, Chapter 1; first presented to the *Züricher naturforschende Gesellschaft* on April 24, 1865, and printed in the *Quarterly Journal of the Gesellschaft*, Vol. 10, p. 1, in *Phys. (Leipzig)*, Ser. 2, Vol. 125, 1865, p. 313, and in *Journal de Liouville*, Ser. 2, Vol. 10, p. 361.

16. W. J. M. Rankine, On the thermal energy of molecular vortices, *Trans. R. Soc. Edinburgh*, Vol. 25, 1869, pp. 557–566; also, On the hypothesis of molecular vortices, or centrifugal theory of elasticity, and its connexion with the theory of heat, *Philos. Mag.*, Ser. 4, No. 67, November 1855, pp. 354–363; No. 68, December 1855, pp. 411–420.

17. W. C. Reynolds and H. C. Perkins, *Engineering Thermodynamics*, 2nd ed., McGraw-Hill, New York, 1977, p. 156.

18. R. E. Sonntag and G. J. Van Wylen, *Introduction to Thermodynamics*, 2nd ed., Wiley, New York, 1982, p. 217.

19. H. D. Baehr, *Thermodynamik*, 3rd ed., Springer-Verlag, Berlin, 1973, p. 114.

20. K. G. Denbigh, The second law efficiency of chemical processes, *Chem. Eng. Sci.*, Vol. 6, No. 1, 1956, pp. 1–9.

21. W. Z. Black and J. G. Hartley, *Thermodynamics*, Harper & Row, New York, 1985, p. 295.

22. R. W. Haywood, *Equilibrium Thermodynamics for Engineers and Scientists*, Wiley, New York, 1980, p. 141.

23. E. G. Cravalho and J. L. Smith, Jr., *Engineering Thermodynamics*, Pitman, Boston, MA, 1981, p. 360.

24. A. Bejan, *Entropy Generation Through Heat and Fluid Flow*, Wiley, New York, 1982, p. 4.

25. J. R. Howell and R. O. Buckius, *Fundamentals of Engineering Thermodynamics*, McGraw-Hill, New York, 1986, Chapter 7.

26. W. G. Vincenti and C. H. Kruger, Jr., *Introduction to Gas Dynamics*, Wiley, New York, 1965, Chapter vii.

27. A. Bejan, *Convection Heat Transfer*, Wiley, New York, 1984, Chapter 1.

28. J. P. Joule, On the changes of temperature produced by the rarefaction and condensation of air, *Philos. Mag.*, Ser. 3, Vol. 26, May 1845; also in *The Scientific Papers of James Prrescott Joule*, Taylor & Francis, London, 1884, pp. 172–189; J. S. Ames, *The Free Expansion of Gases, Memoirs by Gay-Lussac, Joule and Joule and Thomson*, Harper's Scientific Memoirs, Harper & Brothers, New York, 1898, pp. 17–30.

29. N. Georgescu-Roegen, *The Entropy Law and the Economic Process*, Harvard University Press, Cambridge, MA, 1971.

30. H. J. Morowitz, *Entropy for Biologists*, Academic Press, New York, 1970.

31. R. Clausius, *Abhandlungen über die mechanische Wärmetheorie*, Vol. II, Vieweg, Braunschweig, 1867, p. 44.

32. H. A. Bumstead, Biographical sketch in *The Collected Works of J. Willard Gibbs*, Vol. I, Longmans, Green, New York, 1928, p. xiv.

33. I. P. Bazarov, *Thermodynamics*, translated by F. Immirizi and edited by A. E. J. Hayes, Pergamon, Oxford, 1964, pp. 74–78.

34. C. Carathéodory, Untersuchungen über die Grundlagen der Thermodynamik, *Math. Ann. (Berlin)*, Vol. 67, 1909, pp. 355–386.

35. F. W. Sears, Modified form of Carathéodory's second axiom, *Am. J. Phys.*, Vol. 34, 1966, pp. 665–666.

36. H. A. Buchdahl, The concepts of classical thermodynamics, *Am. J. Phys.*, Vol. 28, 1960, pp. 196–201.

37. P. T. Landsberg, On suggested simplifications of Carathéodory's thermodynamics, *Phys. Status Solidi*, Vol. 1, 1961, pp. 120–126.

38. L. A. Turner, Simplification of Carathéodory's treatment of thermodynamics, *Am. J. Phys.*, Vol. 28, 1960, pp. 781–786.

39. L. A. Turner, Simplification of Carathéodory's treatment of thermodynamics. II, *Am. J. Phys.*, Vol. 30, 1962, pp. 506–508.

40. F. W. Sears, A simplified simplification of Carathéodory's treatment of thermodynamics, *Am. J. Phys.*, Vol. 31, 1963, pp. 747–752.

41. P. T. Landsberg, A deduction of Carathéodory's principle from Kelvin's principle, *Nature (London)*, Vol. 201, 1964, pp. 485–486.

42. J. Kestin, A simple, unified approach to the first and second laws of thermodynamics, *Pure Appl. Chem.*, Vol. 22, 1970, pp. 511–518.

43. A. B. Pippard, *Elements of Classical Thermodynamics*, Cambridge University Press, London and New York, 1964, pp. 39–40.

44. A. Bejan, Heat transfer-based reconstruction of the concepts and laws of classical thermodynamics, *J. Heat Transfer*, Vol. 110, 1988, pp. 243–249.

45. S. R. Montgomery, *Second Law of Thermodynamics*, Pergamon, Oxford, 1966.

46. F. F. Huang and R. F. Clothier, *Let us De-mystify the Concept of Entropy*, ASEE Paper No. 3257, Presented at the 87th Annual Conference of the American Society for Engineering Education, Louisiana State University, Baton Rouge, June 25–28, 1979.

47. G. N. Hatsopoulos and J. H. Keenan, *Principles of General Thermodynamics*, Wiley, New York, 1965, p. 382.

48. F. F. Huang, *Engineering Thermodynamics: Fundamentals and Applications*, MacMillan, New York, 1976, p. 58.

49. C. Truesdell, Perpetual motion consistent with classical thermodynamics, *Atti Accad. Sci. Torino, Cl. Sci. Fiz., Mat. Nat.*, Vol. 114, 1980, pp. 433–436.

50. J. S. Thomsen and T. J. Hartka, Strange Carnot cycles; thermodynamics of a system with a density extremum, *Am. J. Phys.*, Vol. 30, 1962, pp. 26–33, 388–389.

51. K. R. Blake, D. Poulikakos, and A. Bejan, Natural convection near 4°C in a horizontal water layer heated from below, *Phys. Fluids*, Vol. 27, 1984, pp. 2608–2616.

52. K. R. Blake, A. Bejan, and D. Poulikakos, Natural convection in a water saturated porous layer heated from below, *Int. J. Heat Mass Transfer*, Vol. 27, 1984, pp. 2355–2364.

53. M. J. Klein, Carnot's contribution to thermodynamics, *Phys. Today*, August 1974, pp. 23–28.

54. T. S. Kuhn, Sadi Carnot and the Cagnard engine, *Isis*, Vol. 52, 1961, pp. 567–574.

55. M. V. Sussman, *Availability (Exergy) Analysis*, Mulliken House, Lexington, MA, 1981.

56. C. C. Gillispie, *Lazare Carnot Savant*, Princeton University Press, Princeton, NJ, 1971.

57. L. Carnot, *Essai sur les Machines en Général*, Dijon, 1783.

58. M. W. Zemansky, *Heat and Thermodynamics*, 5th ed., McGraw-Hill, New York, 1968, p. 126.

PROBLEMS

2.1   Consider the most general case of a closed-system cyclical operation while in communication with two heat reservoirs (Fig. 2.3a). Relying on the Kelvin–Planck statement of the second law, eq. (2.2), show that $Q_1 Q_2 < 0$ regardless of the sign of Q_2. (Note that in the text, the proof that Q_1 and Q_2 must be of opposite sign is based on the assumption that $Q_2 < 0$: this assumption must be relaxed in order to show that the $Q_1 Q_2 < 0$ conclusion is general.)

2.2   The Second Law of Thermodynamics for a cycle executed by a closed system in contact with two heat reservoirs, eq. (2.27), was derived in the text based on the assumption that the net work transfer interaction W is positive. Show that eq. (2.27) holds for any value of W.

2.3   Consider the class of adiabatic expansion processes discussed in conjunction with the closed system of Fig. 2.10. Assume that two distinct states (2_{rev}) and $(2'_{rev})$ can be reached reversibly and adiabatically from state (1); for example, assume

$$U_{2'_{rev}} > U_{2_{rev}} \qquad \text{and} \qquad V_{2'_{rev}} = V_{2_{rev}}$$

Rely on the Kelvin–Planck second-law statement to prove that such an assumption is physically impossible. Can more than two states of type (2_{rev}) be found at $V = V_2$? In other words, is the state (2_{rev}) unique?

2.4   Consider the adiabatic work transfer process executed by the closed system (the fluid) of Fig. 2.10 when its volume *decreases* from V_1 to V_2. Using a line of reasoning that parallels the discussion based on Fig. 2.10 in the text, show that the end state (2_{rev}) that is accessible reversibly and adiabatically from state (1) is represented by a unique point on the $V = V_2$ line of the (U, V) plane.

2.5   Consider the two-fluid closed system of Fig. 2.12 and assume that each fluid charge can be modeled as an ideal gas. Let $(m, R, c_v)_A$ and $(m, R, c_v)_B$ be the respective masses, ideal gas constants, and specific heat constants of the two fluids. The initial state $(\theta, V_A, V_B)_1$ and the final configuration $(V_A, V_B)_2$ are given. The state of the system changes reversibly and adiabatically along two distinct paths, $(1)–(a)–(2_{rev})$ and $(1)–(b)–(2_{rev})$ (Fig. 2.12). The $(1)–(a)–(2_{rev})$ path, for example, is a reversible and adiabatic process in which the first part is characterized by $V_B = V_{B_1}$ (constant) and the second part by $V_A = V_{A_2}$ (constant).

Determine the final temperature reached by the system along both paths, $\theta_{2_{rev}}^{(a)}$ and $\theta_{2_{rev}}^{(b)}$, and show that $\theta_{2_{rev}}^{(a)} \equiv \theta_{2_{rev}}^{(b)}$. With this result in hand, comment on how many states of known configuration can be reached reversibly and adiabatically from a given initial state.

2.6   Determine one by one the exact location of the points that appear in the two-dimensional plot of Fig. 2.8 (see also Example 2.3). How many different paths can the system follow from state (A) to state (B)?

2.7   As an example of the evolution of an isolated system towards a state of maximum entropy as one or more internal constraints are removed, consider the quantity of ideal gas (m, R, c_v) trapped in the piston and cylinder chamber shown in part (a). The initial volume is V_1; the initial pressure P_1 is greater than the pressure that would be necessary to support the piston (mass M, area A); therefore, in the initial state, the piston is held in place with a pin. The space $(V_2 - V_1)$ located above the piston is evacuated.

   The pin is removed, the piston rises, and enough time passes until a new equilibrium state is reached. Assuming that the piston comes to rest against the top wall of the cavity, determine the critical piston mass that differentiates between the final states shown in (b) and (c). First, note that the "system" that is "isolated" has two parts, the gas m and the piston M. (What are the energy transfer interactions experienced by this system during the expansion?) In the case shown in (b), determine analytically the final temperature T_2 (note that V_2 is given). In the case shown in (c), determine analytically the final volume V_3. Show *analytically* that in both cases, the entropy of the "system" increases.

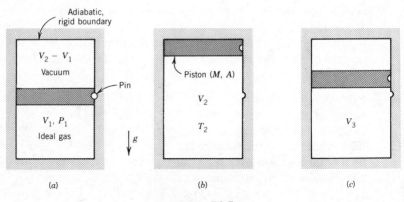

Figure P2.7

2.8   As a second example of changes caused by the removal of an internal constraint, consider an insulated piston and cylinder chamber in which the weight of the piston (mass M, area A) controls the pressure of the quantity of ideal gas $(m/2, R, c_v)$ situated in the compartment immediately under the piston. The lower compartment contains an equal quantity of the same ideal gas $(m/2, R, c_v)$: the temperature and volume of this second quantity are known (T_1 and V_1). The two quanities of ideal gas are initially in thermal equilibrium across a rigid, motionless, leakproof, and thermally conducting partition. The space above the piston is and remains evacuated.

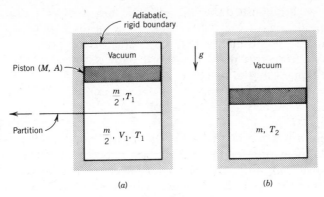

Figure P2.8

The partition is removed to the side (the work required to make this change can be assumed negligible), the two quantities of ideal gas become one (m, R, c_v), and, after enough time, the piston settles in a new position. Assume that the cylinder chamber is sufficiently tall so that the piston never touches the ceiling [drawing (b)]. Analytically determine the final temperature T_2 of the ideal gas. For the aggregate system $(m + M)$ as an isolated thermodynamic system, show that during the change caused by the removal of the partition, the entropy increases while the energy remains constant.

2.9   A long time ago, the specific heat at constant volume of gases (c_v) was determined indirectly from c_P measurements and from a special measurement of the c_P/c_v ratio, which is described below. A small amount of gas is held in a rigid bottle fitted with a valve. Initially [state (1)], the valve is closed, the gas temperature is atmospheric (T_a), and the gas pressure is slightly above atmospheric (P_1). The valve is then opened and closed immediately [state (2)]: during this incident, enough of the bottled gas escapes and the pressure inside the bottle drops to the atmospheric level (P_a). The experimenter also notices that during the same incident, temperature T_2 of the bottled gas drops, $T_2 < T_a$. Waiting so that the bottled gas comes to thermal equilibrium with the atmosphere [state (3)], the experimenter finds that the gas pressure climbs to a final level P_3 above atmospheric.

Show that if the expansion of the gas that remains trapped in the bottle is modeled as "reversible" and "adiabatic," and if the gas is modeled as "ideal," then the ratio c_P/c_v is a function of only P_1/P_a and P_3/P_a. In other words, show that the ratio of specific heats can be calculated from the measurement of three pressures, P_1, P_3, and P_a. This classic technique is Gay-Lussac's modification of an experiment conceived by Clement and Desormes [58].

2.10 It is well established that the adiabatic and quasistatic volume change of an ideal gas (R, c_v) in a cylinder and piston apparatus follows the path $PV^k = \text{constant}$, where $k = c_p/c_v$. Real processes, however, depart from this path; the actual path is named *polytropic* and is represented by the function

$$PV^n = \text{constant}$$

where n is a constant $(n \neq k)$. One possible reason for this departure is the heat transfer that takes place between the ideal gas and the massive wall of the cylinder. Your job as a problem solver is to show the manner in which the constant exponent n is affected by the heat capacity of the cylinder wall.

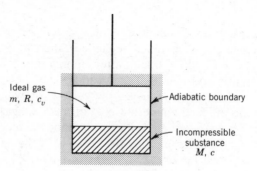

Figure P2.10

The simplest model that retains the effect of gas–wall heat transfer is presented in the drawing. The mass of the ideal gas is m, the mass of the cylinder wall is M, and the specific heat of the wall material is c. Consider now the expansion from an initial volume V_1 and pressure P_1 to a final volume V_2, and assume that at any instant during this process, the ideal gas and the wall material are in mutual thermal equilibrium. The expansion process is sufficiently slow so that $\delta W = P\, dV$. Furthermore, the combined system (ideal gas and cylinder material) does not exchange heat with its surroundings.

(a) Show that the path of the process is $PV^n = \text{constant}$, where

$$n = 1 + \frac{R/c_v}{1 + Mc/mc_v}$$

(*Hints*: Write the first law for an infinitesimally short change of state; express the instantaneous gas–cylinder heat transfer interaction in terms of the energy change experienced by the cylinder material.)

(b) How large (or how small) should the wall heat capacity be if the path is to approach $PV^k = $ constant?

(c) Evaluate the entropy change dS for the combined system (ideal gas and cylinder material) during the infinitesimal change from V to $V + dV$.

(d) Invoke the Second Law of Thermodynamics in order to decide whether the process executed by the combined system is reversible or irreversible.

2.11 A basic question in the design of boilers and pressurized-water reactor vessels refers to estimating the time interval in which the vessel would lose all its liquid, following an accident in which the lower part of the vessel develops a leak. If the vessel contains originally a mixture of saturated water liquid and saturated water vapor at a pressure higher than atmospheric, then the loss of liquid causes the depressurization of the vessel. This blowdown process is accompanied by the vaporization of some of the liquid before the liquid is evicted entirely through the bottom leak. The phenomenon is quite complicated: in general, the liquid–vapor interface is crossed by both heat transfer and mass transfer. A water–carbon dioxide analog of the discharge of the boiling water mixture is available to anyone who chooses to experiment with an old-fashioned seltzer bottle.

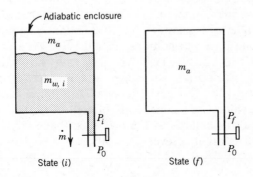

Figure P2.11

We develop an understanding of the phenomenon by focusing on the behavior of the air–water model shown in the drawing. The rigid vessel (volume V) contains in the initial state (i) a quantity of air, m_a, and a quantity of water, $m_{w,i}$. The initial pressure of this composite system, P_i, is higher than the atmospheric pressure P_0. The initial volume occupied by air, $V_{a,i}$, is known. The air is modeled as an ideal gas with known constants R and c_v, whereas the water behaves as an incompressible liquid of density ρ_w and constant specific heat c. The

water–air interface is modeled as impermeable to mass transfer, i.e.,
it is assumed that the discharge process is fast enough so that the
diffusion of water into air (and vice versa) is negligible.

The rupture of the lower part of the vessel is symbolized by a valve
that opens at time $t = 0$ and stays open until time $t = t_f$, when all the
water is ejected ("f" stands for the final state when the vessel
contains only air, at a pressure that is still higher than atmospheric).
Fluids engineering studies of the turbulent liquid flow through the
valve suggest that the instantaneous liquid ejection flowrate can be
assumed to vary as $\dot{m}(t) = C[P(t) - P_0]^{1/2}$, where C is a known
constant.

(a) Determine the discharge time t_f by modeling the water–air interface adiabatic and the expansion of the air batch m_a reversible.
Determine t_f analytically by assuming that throughout the
$(i) \rightarrow (f)$ process P is much greater than P_0.

(b) Derive the equations for calculating t_f numerically in the case
when the water–air interface is modeled as diathermal (i.e., when
the instantaneous temperature T is the same for air and water
throughout V). Assume that the expansion is again sufficiently
slow so that the instantaneous pressure P is uniform throughout
the vessel.

2.12 Newcomen's atmospheric pumping engine operated on the principle
summarized as states $(1) \rightarrow (2) \rightarrow (3)$. Initially in state (1), the piston
and cylinder apparatus contains m kilograms of saturated steam at
atmospheric pressure, $P_1 = 0.10135\,\text{MPa}$. An amount of tap water Δm
(at atmospheric conditions, $P_0 = 0.10135\,\text{MPa}$ and $T_0 = 298.15\,\text{K}$) is
introduced into the cylinder, and, with the piston still locked in its
uppermost position, the charge attains a new equilibrium represented
by state (2). The new pressure P_2 is lower than atmospheric, the
actual value of P_2 depending, of course, on the amount of tap water
Δm. Let us assume that Δm is such that the pressure falls to about
half of its original level, $P_2 = 50\,\text{kPa}$.

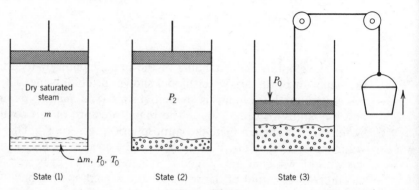

State (1)   State (2)   State (3)

Figure P2.12

The work-producing stroke of the engine takes place as the atmosphere drives the piston (assumed massless) into the cylinder until the pressure inside the cylinder rises back to atmospheric level, $P_3 = 0.10135$ MPa. An appropriate mechanism, which originally was devised for raising water from coal mines, delivers the difference between the work done by the atmosphere and the work needed to compress the contents of the cylinder. The process $(2) \rightarrow (3)$ can be modeled as adiabatic and quasistatic.

(a) Determine the quality at state (2) and the ratio $\Delta m / m$.

(b) Determine the quality at state (3) and the work required to compress the liquid–vapor mixture from state (2) to state (3).

(c) The piston and cylinder apparatus is returned to state (1) by discharging the saturated liquid collected at state (3), $m_{f,3}$, and by replenishing the dry saturated steam amount $m_{g,3}$ in order to reestablish the original charge m. Calculate the amount of steam added, $m - m_{g,3}$, i.e., the amount that is consumed during one complete cycle.

(d) Determine the work delivered during one cycle.

2.13 The two-chamber apparatus shown in the drawing contains atmospheric air at temperature T_0 and pressure P_0. The two chambers contain equal amounts of air; in other words, their initial volumes are equal $(V_{A1} = V_{B1} = V_1)$. The air may be modeled as an ideal gas whose constants R and c_v are known. The cylinder wall and the two frictionless leakproof pistons are adiabatic. The bottom of the cylinder (the right wall of chamber B) is a diathermal boundary through which B communicates with the atmospheric temperature reservoir (T_0).

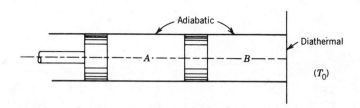

Figure P2.13

Starting from the state (1) described above, the air in both chambers is compressed by the slow movement of the left piston. The process ends at state (2), where the pressure level P_2 is measured and recorded (note that $P_{A2} = P_{B2} = P_2$).

(a) Determine analytically the final volumes V_{A2} and V_{B2} and the final temperature of chamber A, T_{A2}.

(b) Determine the work transfer interaction experienced by the closed system represented by the air trapped in chamber A.

(c) Determine the work transfer interaction experienced by the aggregate closed system represented by all the trapped air (chamber $A + B$).

(d) Apply the second law to the process $(1) \rightarrow (2)$ executed by the aggregate system $(A + B)$, and determine whether the process is reversible.

2.14 The function of the cylinder and piston apparatus shown in the drawing is to lift a weight by using the compressed air stored in the reservoir of pressure P_R and atmospheric temperature T_0. In the initial state the piston touches the bottom of the cylinder. Throughout the lifting process compressed air is being admitted into the cylinder through a valve. The function of the valve is to lower the pressure of the inflowing air from P_R to the pressure that is maintained in the cylinder by the load itself, P_L. The air that resides inside the cylinder is in thermal communication with the ambient and, as a result, its temperature is constant and equal to T_0.

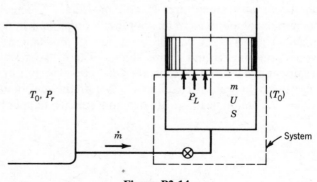

Figure P2.14

Consider the open system indicated by the dashed box in the figure. Calculate the entropy generated inside it during a finite-excursion lifting process. Calculate also the total heat transfer interaction (Q_0) between this system and the ambient reservoir (T_0).

3

The Two Laws Combined:
The Destruction of Exergy

In this chapter, we focus on the most essential engineering implication of the thermodynamics laws discussed until now, namely, the close relationship that exists between irreversibility (entropy generation) and the one-way destruction of available work. This relationship is essential because the field of engineering thermodynamics is the result of our interest in "work" as a commodity, for example, in extracting work from various sources and in accomplishing the most with the work that is already in our possession. Beginning with this chapter, we will see that the many efficiency-maximization rules learned in the study of various devices and processes are nothing more than special manifestations of one general theorem. The lasting message of this theorem is that in the field of energy engineering, the losses can be measured in units of entropy generation as currency.

There are additional reasons for highlighting the lost-work theorem in an advanced course. At the practical level, the relationship between entropy generation and lost available work is an ideal instrument with which to correlate the seemingly unrelated engineering applications of thermodynamics. Furthermore, this relationship bridges the gap between the problems encountered by the student in the classroom and the real problems faced by the engineer in practice. At a theoretical level, the concept of destroyed available work reminds all of us that the two laws function simultaneously, in spite of the problem-solving tradition that invites us to ignore the second law. The concepts that form the subject of this chapter have their origin in the *simultaneous* invocation of the first law and the second law. This important feature tends to be obscured by the label "second-law analysis" that is often placed on the evaluation of lost available work, and on the minimization of entropy generation. Understood in the intended sense, however, the term "second-law analysis" is quite effective in reminding the analyst that his work is incomplete unless the second law is also a part of the analysis.

LOST AVAILABLE WORK (LOST EXERGY)

It is for the sake of conciseness that we begin by deriving the lost-work theorem in a general setting, rather than traveling again on the road from simple to complex as we did in the treatment of the second law (chapter 2). The physical meaning of the general results is illustrated in the latter parts of this chapter by focusing on simpler and more familiar classes of systems and processes.

Consider first the multiport system in Fig. 3.1. At a certain point in time, the system can be in thermal contact with any number of heat reservoirs of temperatures T_i $(i = 0, 1, 2, \ldots, n)$. We will soon see that a special role in the functioning of an engineering installation is played by the *atmosphere*, which in Fig. 3.1 is represented by the temperature and pressure reservoir (T_0, P_0). The work transfer rate $\dot{W}$ represents any combination of possible modes of work transfer $(P\,dV/dt, \dot{W}_{\text{shear}}, \dot{W}_{\text{electrical}}, \dot{W}_{\text{magnetic}})$. One possible work transfer interaction of the $P\,dV/dt$ type is the work done against the atmosphere, while the atmosphere acts as a pressure reservoir, $P_0\,dV/dt$: this possible mode is illustrated in Fig. 3.1 because it is essential to the discussion that concludes this section.

With reference to the open system defined in Fig. 3.1, the first law and the second law are written as

$$\frac{dE}{dt} = \sum_{i=0}^{n} \dot{Q}_i - \dot{W} + \sum_{\text{in}} \dot{m}h^{\circ} - \sum_{\text{out}} \dot{m}h^{\circ} \tag{3.1}$$

$$\dot{S}_{\text{gen}} = \frac{dS}{dt} - \sum_{i=0}^{n} \frac{\dot{Q}_i}{T_i} - \sum_{\text{in}} \dot{m}s + \sum_{\text{out}} \dot{m}s \geq 0 \tag{3.2}$$

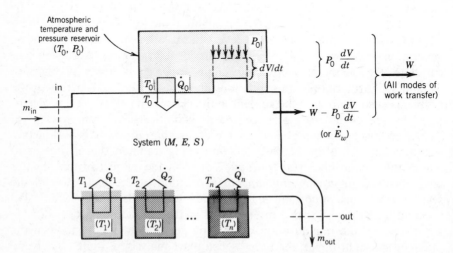

Figure 3.1 Open system in communication with the atmosphere and n additional heat reservoirs.

where the methalpy symbol $h°$ is shorthand notation for the generalized enthalpy group $(h + V^2/2 + gz)$, eq. (1.23).

Next, we consider the possibility of changing the design (the internal functioning) of the system for the purpose of maximizing the work transfer rate $\dot{W}$. Of long-term engineering interest is the common characteristic of all the changes that consistently lead to increases in $\dot{W}$. We must recognize, however, that since the First Law of Thermodynamics is an equation, the wish to see changes in $\dot{W}$ means to allow the variation of at least one other term in eq. (3.1). Let us assume that the heat transfer interaction with the atmosphere, $\dot{Q}_0$, varies as $\dot{W}$ is maximized. In other words, let us assume that all the other interactions that are specified around the system (heat transfer rates $\dot{Q}_1, \ldots, \dot{Q}_n$, inflows and outflows of methalpy and entropy) are fixed by design, and that only $\dot{Q}_0$ "floats" in order to balance the changes in $\dot{W}$. The choice of $\dot{Q}_0$ as the interaction that floats in the wake of design changes is consistent with the role that is assigned traditionally to the rate of heat rejection to the atmosphere in design of power and refrigeration systems.

If we eliminate $\dot{Q}_0$ between the first law (3.1) and the entropy generation-rate definition (3.2), we find that the work transfer rate $\dot{W}$ depends explicitly on the degree of thermodynamic irreversibility of the system, $\dot{S}_{gen}$,

$$\dot{W} = -\frac{d}{dt}(E - T_0 S) + \sum_{i=1}^{n}\left(1 - \frac{T_0}{T_i}\right)\dot{Q}_i$$
$$+ \sum_{in} \dot{m}(h° - T_0 s) - \sum_{out} \dot{m}(h° - T_0 s) - T_0 \dot{S}_{gen} \qquad (3.3)$$

Furthermore, since according to the second law (3.2), the entropy generation rate $\dot{S}_{gen}$ cannot be negative, the first four terms on the right side of eq. (3.3) represent algebraically an upper bound for $\dot{W}$. This upper bound is reached when the system operates reversibly ($\dot{S}_{gen} = 0$). In this manner, we identify the first four terms on the right side of eq. (3.3) as the work transfer rate in the limit of reversible operation:

$$\dot{W}_{rev} = -\frac{d}{dt}(E - T_0 S) + \sum_{i=1}^{n}\left(1 - \frac{T_0}{T_i}\right)\dot{Q}_i$$
$$+ \sum_{in} \dot{m}(h° - T_0 s) - \sum_{out} \dot{m}(h° - T_0 s) \qquad (3.4)$$

Equation (3.3) can be rewritten briefly as

$$\dot{W} = \dot{W}_{rev} - T_0 \dot{S}_{gen} \qquad (3.5)$$

or, invoking the second law (3.2) one more time, as

$$\dot{W}_{rev} - \dot{W} = T_0 \dot{S}_{gen} \geq 0 \qquad (3.6)$$

The engineering conclusion that follows from combining the first law with the second law is this: whenever a system operates irreversibly, it destroys work at a rate that is proportional to the system's rate of entropy generation. The work destroyed through thermodynamic irreversibility, $\dot{W}_{rev} - \dot{W}$, is appropriately called *lost available work*: this terminology was first used prominently by Kestin [1–3]. The proportionality between lost available work and entropy generation, or between their respective rates,

$$\dot{W}_{lost} = T_0 \dot{S}_{gen} \tag{3.7}$$

constitutes the *lost-work theorem*. The same result is remembered by many as the *Gouy–Stodola theorem*, in memory of the first two thermodynamicists who were at least partially successful in convincing their contemporaries that eq. (3.7) deserves attention [4, 5]. The history of the concept of lost work is about as old as the history of engineering thermodynamics: remarks on this topic can be found in Refs. 1–3 and 6.

A few observations regarding the sign of $\dot{W}$, $\dot{W}_{rev}$, and $\dot{W}_{lost}$ are needed, because there is a conceptual difference between $\dot{W}$ (or $\dot{W}_{rev}$) and $\dot{W}_{lost}$. The work transfer rate $\dot{W}$ and its limiting value $\dot{W}_{rev}$ can take both positive and negative values, depending on whether the system is designed to produce or absorb work. For this reason, the second law inequality (3.6), or simply

$$\dot{W}_{rev} \geq \dot{W} \tag{3.8}$$

must be recognized as correct in an algebraic sense. In other words, regardless of the sign of the values assigned to $\dot{W}$ and $\dot{W}_{rev}$, the position of $\dot{W}_{rev}$ always falls to the right of $\dot{W}$ on the work transfer rate axis in Fig. 3.2. The lost available work, on the other hand, can never be negative. This fundamental distinction between lost available work and work transfer interaction is to be kept in mind, especially in view of the observation made earlier (p. 4) that work transfer and lost available work are "similar" in the sense that both are not thermodynamic properties of the system. The work transfer and lost available work depend on the path (design, constitution, functioning) of the system, however, whereas the system can be designed for both positive and negative $\dot{W}$s, the second law prevents the designer from even dreaming of negative values for $\dot{W}_{lost}$.

The analytical content of this section can be summarized in terms of two conclusions, first, that the entropy generated by the system is a measure of the available work that has been destroyed, and, second, that by invoking the reversible limit and using eq. (3.4), it is possible to evaluate the upper limit to the work transfer rate of which the system might be capable. An engineering question that arises in connection with the second conclusion is whether $\dot{W}_{rev}$ (or, for that matter, $\dot{W}$) is entirely available for consumption. The answer depends on whether the atmospheric pressure reservoir P_0 is part of the environment and whether the system experiences a change in volume while being resisted or aided by this pressure reservoir. In cases

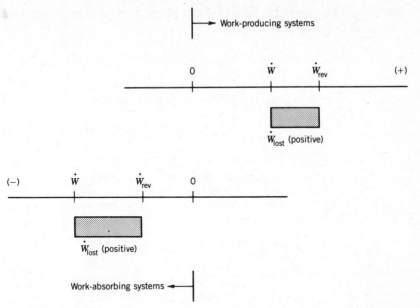

Figure 3.2 $\dot{W}$ and $\dot{W}_{rev}$ can be either positive or negative, whereas $\dot{W}_{lost}$ can only have one sign.

where the atmospheric pressure reservoir P_0 exchanges work with the system (Fig. 3.1), the fraction of $\dot{W}$ that is transferred to the atmosphere is $P_0\,dV/dt$, while the remainder constitutes the rate of *available work*, $\dot{E}_W$,

$$\dot{E}_W = \dot{W} - P_0\frac{dV}{dt}$$

$$= -\frac{d}{dt}(E + P_0V - T_0S) + \sum_{i=1}^{n}\left(1 - \frac{T_0}{T_i}\right)\dot{Q}_i$$

$$+ \sum_{in}\dot{m}(h^\circ - T_0s) - \sum_{out}\dot{m}(h^\circ - T_0s) - T_0\dot{S}_{gen} \qquad (3.9)$$

In most of the flow systems that are of engineering interest, the atmospheric work $P_0\,dV/dt$ is absent, and $\dot{E}_W$ is simply equal to $\dot{W}$. [The $\dot{E}_W$ notation was chosen in order to be consistent with the "exergy" nomenclature that has been developed for the right-hand terms of eq. (3.11); see p. 223.]

The work transfer rate to the atmosphere $P_0\,dV/dt$ can be positive or negative, depending on whether the system expands or contracts while in contact with the P_0 reservoir. Consequently, the absolute value of $\dot{W}$ is not necessarily greater than the absolute value of $\dot{E}_W$. A classical engineering system in which $|\dot{W}| < |\dot{E}_W|$ is Newcomen's atmospheric pumping engine, where the work-producing stroke occurs when the system shrinks under the "weight" of the atmosphere (Problem 2.12).

Finally, in the reversible limit, we can identify an algebraic ceiling value for the available work transfer rate,

$$(\dot{E}_W)_{\text{rev}} = \dot{W}_{\text{rev}} - P_0 \frac{dV}{dt} \tag{3.10}$$

which can be evaluated by combining the definition (3.10) with eq. (3.4):

$$(\dot{E}_W)_{\text{rev}} = \underbrace{-\frac{d}{dt}(E + P_0 V - T_0 S)}_{} + \underbrace{\sum_{i=1}^{n}\left(1 - \frac{T_0}{T_i}\right)\dot{Q}_i}_{}$$

Maximum delivery of useful (available) mechanical power	Accumulation of nonflow exergy	Exergy transfer via heat transfer

$$+ \underbrace{\sum_{\text{in}} \dot{m}(h° - T_0 s)}_{} - \underbrace{\sum_{\text{out}} \dot{m}(h° - T_0 s)}_{} \tag{3.11}$$

Intake of flow exergy via mass flow	Release of flow exergy via mass flow

Under each of the five types of terms in eq. (3.11) is the terminology usually attached to them in modern engineering thermodynamics. This terminology is introduced on a case-by-case basis in the next three sections. The algebraically maximum rate of available work delivery $(\dot{E}_W)_{\text{rev}}$ emerges as

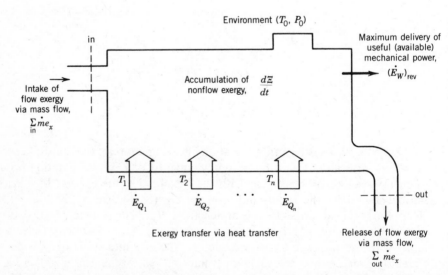

Figure 3.3 The exergy accounting "balance" that rules the open system of Fig. 3.1 in the reversible operation limit.

the difference between the net flow of exergy into the control volume and the net flow of exergy out of the control volume. This state of exergy "balance," which exists only hypothetically in the reversible operation limit, is illustrated in the exergy accounting diagram of Fig. 3.3. Note that Fig. 3.3 is the exergy flow network that rides on top of Fig. 3.1 in the reversible operation limit.

The concept of lost available work—defined already as the difference between the ceiling value $\dot{W}_{rev}$ and the actual work transfer rate $\dot{W}$—can be defined alternatively as the difference between the corresponding available work quantities, Fig. 3.4,

$$\dot{W}_{lost} = (\dot{E}_W)_{rev} - \dot{E}_W = (\dot{E}_W)_{lost} \tag{3.12}$$

The relationships between work, available work, and lost available work (or lost exergy) are summarized geometrically in Fig. 3.5, where it has been assumed that all the work quantities (not just $\dot{W}_{lost}$) are positive.

Equations (3.11) and (3.12) point out two important directions in engineering thermodynamics:

(i) the estimation of the theoretically ideal operating conditions of a proposed installation, in particular, the maximum mechanical power output for engines or the minimum mechanical power requirement for refrigerators, eq. (3.11), and

(ii) the estimation and minimization of lost available work or entropy generation through improved thermal design, eqs. (3.12) and (3.7).

The first direction is the common thread of all "exergy analyses," as illustrated in this chapter and chapters 5, 7 to 11. The second direction is a relatively new point of view in applied engineering thermodynamics, a modern brand of thermal design that would be better named "entropy generation minimization" or "thermodynamic design." The growth of this activity forms the subject of a 1982 monograph [6] and chapter 11 in the present treatment.

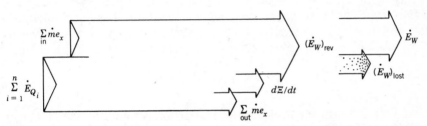

Figure 3.4 Alternative to the exergy accounting of Fig. 3.3, showing how the lost exergy ruins the balance between exergy inflow and exergy outflow.

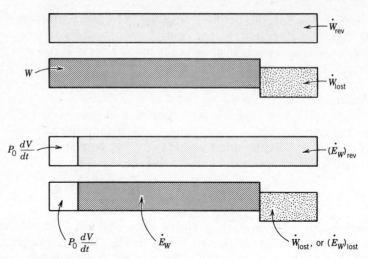

Figure 3.5 The relationship between work transfer $\dot{W}$, available work $\dot{E}_W$, and lost available work $\dot{W}_{\text{lost}}$ or $(\dot{E}_W)_{\text{lost}}$.

CYCLES

The meaning of the lost-work theorem becomes clearer if we take another look at the classical topic of heat engines and refrigerators, this time from the engineering position of trying to avoid the destruction of available work. This step serves also as an introduction to the more detailed and advanced applications treated in chapter 8. We begin with the observation that we are dealing with a class of relatively uncomplicated systems, namely, closed systems that operate in an integral number of cycles. For this class, the ceiling value of the available power (3.11) reduces to

$$(\dot{E}_W)_{\text{rev}} = \sum_{i=1}^{n} \left(1 - \frac{T_0}{T_i}\right) \dot{Q}_i \tag{3.13}$$

We learn that a certain heat transfer interaction (Q_i) can affect the system's ability to produce work if the temperature of the boundary crossed by $\dot{Q}_i$ differs from the atmospheric reservoir temperature, $T_i \neq T_0$. Generalizing the "E" (exergy) notation used in eq. (3.9), we can speak of the *available work* *(exergy) content of the heat transfer interaction* $(\dot{Q}, T, T_0)$ and label it

$$\dot{E}_Q = \dot{Q}\left(1 - \frac{T_0}{T}\right) \tag{3.14}$$

Note that although the heat transfer interaction is fully described by $\dot{Q}$, the available work (exergy) content requires the specification of three quantities, $\dot{Q}$, T, and T_0. For many years, the quantity $\dot{Q}(1 - T_0/T)$ has been recognized as the *availability* of $\dot{Q}$ [7, 8]. By using the $\dot{E}_Q$ notation, the lost-work theorem for closed systems that operate cyclically can be written as

$$\dot{W}_{\text{lost}} = \sum_{i=1}^{n} (\dot{E}_Q)_i - \dot{E}_W \qquad (3.15)$$

The per-unit-time notation ($\dot{}$) drops out if the analysis refers to a complete cycle or an integral number of cycles.

Heat-Engine Cycles

Dating back to Sadi Carnot's memoir, the simplest representation of heat-engine operation is in terms of cycles executed in contact with two temperature reservoirs. Figure 3.6 shows the *physical* sense of the three energy interactions: the function of the device (the closed system) is to produce work by absorbing heat from a high-temperature reservoir and rejecting heat to a low-temperature reservoir. The simultaneous occurrence of heat absorption and rejection is a consequence of the Second Law of Thermodynamics, as demonstrated by the proof to option (iii), eq. (2.8). The graphic definition of the physical sense of the energy interactions, Fig. 3.6, means that in the following analysis, Q_H, Q_L, and W represent positive numbers of joules per cycle.

With reference to the heat engine as a closed system, the first law and the second law state

$$Q_H - Q_L - W = 0 \qquad (3.16)$$

$$S_{\text{gen}} = \frac{Q_L}{T_L} - \frac{Q_H}{T_H} \geq 0 \qquad (3.17)$$

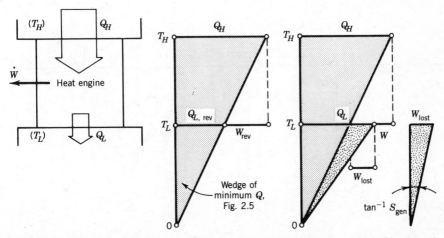

Figure 3.6 Triangular diagram showing the occurrence of lost available work and entropy generation during a heat-engine cycle (after Bejan [9]).

The same statements are made graphically by the triangular diagram [9] presented twice in Fig. 3.6. Next, we apply the lost-work theorem (3.15) by identifying T_L as the temperature that plays the role of T_0, or Q_L as the energy interaction that "floats" as W changes [review the assumption that preceded eq. (3.3)]. Equation (3.15) reduces to

$$W_{\text{lost}} = E_{Q_H} - E_W = Q_H\left(1 - \frac{T_L}{T_H}\right) - W \qquad (3.18)$$

Equation (3.18) is the subject of the second triangular diagram of Fig. 3.6: the lost-work theorem, $W_{\text{lost}} = T_L S_{\text{gen}}$, is an integral feature of this drawing, as the destroyed work increases in proportion to the departure from the limit of reversible operation. The "efficiency" of the work-producing device is related, of course, to the destruction of available work inside the device. We define the *relative efficiency* (or *utilization factor*) as [2]

$$\eta_{\text{II}} = \frac{E_W}{(E_W)_{\text{rev}}} = 1 - \frac{T_L S_{\text{gen}}}{(E_W)_{\text{rev}}} \qquad (3.19)$$

The value of η_{II} increases from 0 to 1 as the engine approaches its reversible limit. The indicator defined by eq. (3.19) is recognized also as the *second-law efficiency* of the heat engine. The subscript "II" is used to draw attention to the difference between this newer figure of merit and the traditional ("first-law") efficiency:

$$\eta_{\text{I}} = \frac{W}{Q_H} = \eta_{\text{II}}\left(1 - \frac{T_L}{T_H}\right) \qquad (3.20)$$

whose range is $0 \leq \eta_{\text{I}} \leq (1 - T_L/T_H)$. A comparative view of the first-law and second-law efficiencies is presented in Fig. 3.7. In particular, the shaded

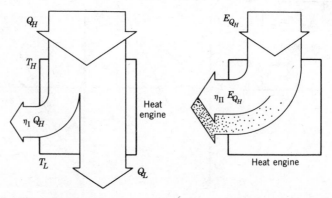

Figure 3.7 The relationship between the first-law efficiency (also called "heat-engine" efficiency) and the second-law efficiency during a heat-engine cycle.

area in the right side of Fig. 3.7 shows the destruction of exergy during the heat-engine cycle. The same drawing also shows that the exergy transfer associated with Q_L is always zero.

Refrigeration Cycles

According to the simplest model, a refrigeration cycle is analyzed as a closed system in communication with two heat reservoirs, namely, the cold space (T_L) from which the cycle extracts the refrigeration load (Q_L), and the room-temperature ambient (T_H) to which the refrigerator rejects heat (Q_H). In the drawing on the left side of Fig. 3.8, the arrows indicate the physical sense of the energy interactions, therefore, W, Q_H, and Q_L represent positive numerical values. The laws of thermodynamics require

$$Q_L - Q_H + W = 0 \qquad (3.21)$$

$$S_{\text{gen}} = \frac{Q_H}{T_H} - \frac{Q_L}{T_L} \geq 0 \qquad (3.22)$$

The triangular diagrams of Fig. 3.8 show graphically the content of the two laws [9], in particular, the fact that in refrigerators, the work requirement and the rejected heat increase as the entropy generation increases. The identification of Q_H as the heat transfer interaction that "floats" as W and S_{gen} are minimized is the preliminary step in using the lost-work formula (3.15); we obtain

$$W_{\text{lost}} = E_{Q_L} - E_W = Q_L\left(1 - \frac{T_H}{T_L}\right) - (-W) \qquad (3.23)$$

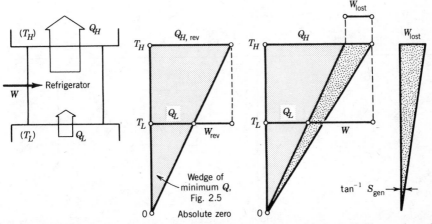

Figure 3.8 Triangular diagram showing the occurrence of lost available work and entropy generation during a refrigeration cycle (after Bejan [9]).

and note that the E_{Q_L} term is negative. A more instructive version of eq. (3.23) is

$$W = Q_L\left(\frac{T_H}{T_L} - 1\right) + W_{\text{lost}} \tag{3.23'}$$

where the term $Q_L(T_H/T_L - 1)$ represents the fixed amount of exergy $(-E_{Q_L})$ that must be deposited in the T_L-cold space. The work delivered to the refrigerating machine (W) must account for both $(-E_{Q_L})$ and the available work lost because of the irreversibility of the cycle. This last observation is illustrated better in the drawing on the right side of Fig. 3.9.

The relative efficiency, or utilization factor, for the refrigeration cycle is defined as the ratio of the minimum exergy (work) requirement divided by the actual exergy input:

$$\eta_{\text{II}} = \frac{(-E_W)_{\text{rev}}}{(-E_W)} = \frac{(-E_{Q_L})}{(-E_{Q_L}) + T_H S_{\text{gen}}} \tag{3.24}$$

This second-law efficiency can vary from 0 to 1. The traditional ("first-law") figure of merit for the same cycle is the coefficient of performance:

$$\text{COP} = \frac{Q_L}{W} = \frac{\eta_{\text{II}}}{T_H/T_L - 1} \tag{3.25}$$

whose values can vary from 0 to $(T_H/T_L - 1)^{-1}$. Figure 3.9 displays side by side the energy flow and the exergy flow through the same refrigeration machine. Again, we note the zero exergy associated with the heat rejected to the ambient.

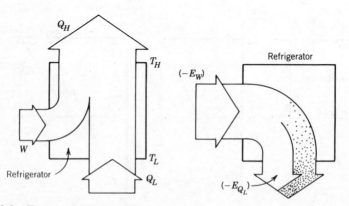

Figure 3.9 Energy conservation versus exergy destruction during a refrigeration cycle.

To deposit the exergy amount $(-E_{Q_L})$ in the cold space is the job of the refrigerator: this exergy output is steadily destroyed in the leaky thermal insulation that separates the (T_L) space from the (T_H) ambient (Problem 3.1). It is easy to show that the exergy transferred to the cold space $(-E_{Q_L})$ can be recovered fully as useful work by operating a reversible heat engine between (T_H) and (T_L), so that the heat rejected by the engine to (T_L) matches the refrigeration load Q_L pulled by the refrigerator out of the cold space (T_L).

Heat-Pump Cycles

As a final example of the application of the exergy accounting formula (3.13), consider a heat-pump cycle whose function is to deliver the heat transfer Q_H to the interior of a building (T_H) warmer than the ambient (T_L). With reference to the left side of Fig. 3.10, in which Q_H, Q_L, and W are all positive numbers, the first law is written exactly as in eq. (3.21). This is why the left sides of Figs. 3.9 and 3.10 are identical. The exergy-flow diagram, however, is different because in a heat-pump cycle, Q_L is the heat transfer interaction with the ambient, that is, the interaction that "floats" as the degree of irreversibility of the cycle (S_{gen}) varies. Equation (3.15) yields in this case:

$$W_{\text{lost}} = \left(1 - \frac{T_L}{T_H}\right)(-Q_H) - (-W) \qquad (3.26)$$

which means that the work requirement W must always exceed a theoretical threshold value:

$$W = \left(1 - \frac{T_L}{T_H}\right)Q_H + W_{\text{lost}} \qquad (3.26')$$

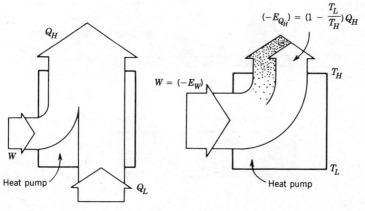

Figure 3.10 Energy conservation versus exergy destruction during a heat-pumping cycle.

The first term on the right side of eq. (3.26′) represents the exergy deposited by the heat-pump cycle into the (T_H) space. This exergy term can be labeled $(-E_{Q_H})$ as done on the right side of Fig. 3.10: this notation is consistent with the $\dot{E}_Q$ definition (3.14), in which $\dot{Q}$ is considered positive when entering the system. Note that $(-E_{Q_H})$ is positive and so is $(-E_W)$.

The second-law efficiency of the heat-pump cycle is calculated by dividing the minimum work requirement by the actual work:

$$\eta_{\rm II} = \frac{(-E_W)_{\rm rev}}{(-E_W)} = \frac{(-E_{Q_H})}{(-E_{Q_H}) + T_L S_{\rm gen}} \tag{3.27}$$

Since $(-E_{Q_H})$ and $T_L S_{\rm gen}$ are positive quantities, the $\eta_{\rm II}$ ratio takes values in the interval $[0, 1]$. The seond-law efficiency is to be distinguished from the classical coefficient of performance of the heat pump:

$$\mathrm{COP} = \frac{Q_H}{W} = \frac{\eta_{\rm II}}{1 - T_L/T_H} \tag{3.28}$$

which can vary from 0 to $(1 - T_L/T_H)^{-1}$.

A graphic summary of the first-law and second-law figures of merit of heat engines, refrigerators, and heat pumps is presented in Fig. 3.11. The

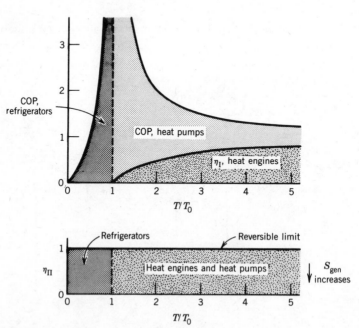

Figure 3.11 The range of values taken by the first-law and second-law efficiencies of heat engines, refrigerators, and heat pumps.

top figure was drawn after Radcenco et al. [10]. In each case, the upper boundary of the domain represents the limit of reversible operation. The T_0 denominator in the abscissa is the ambient temperature, with which all the devices referred to in the figure are assumed to be in contact (i.e., $T_0 = T_L$ in heat engines and heat pumps, and $T_0 = T_H$ in refrigerators).

The bottom of Fig. 3.11 shows the range of values taken by the second-law efficiencies of refrigerators, heat engines, and heat pumps. The upper boundary in this second drawing ($\eta_{II} = 1$) is the common limit of reversibility for all these devices. It is clear from this figure and from eqs. (3.20), (3.25), and (3.28) that the second-law efficiency is a way of normalizing the traditional figures of merit so that they all equal 1 in the limit of reversible operation.

NONFLOW PROCESSES

Consider next a process $(1) \to (2)$ executed by a *closed* system while in contact with $n + 1$ temperature reservoirs $(T_i, i = 0, 1, \ldots, n)$. Special among these reservoirs is (T_0), because its heat transfer interaction Q_0 is assumed to vary in response to changes in the degree of irreversibility of the process, S_{gen}. The available work E_W delivered by the closed system during the process $(1) \to (2)$ is obtained by integrating eq. (3.9) from $t = t_1$ to $t = t_2$:

$$E_W = A_1 - A_2 + \sum_{i=1}^{n} (E_Q)_i - T_0 S_{gen} \qquad (3.29)$$

where the *nonflow availability* (A, a) is shorthand notation for [11]

$$\begin{aligned} A &= E - T_0 S + P_0 V \\ a &= e - T_0 s + P_0 v \end{aligned} \qquad (3.30)$$

The available work terms $(E_Q)_i$ associated with the heat transfer interactions other than Q_0 have been defined in eq. (3.14). The nonflow availability A is a thermodynamic property of the system as long as T_0 and P_0 are fixed.

The most frequent application of eq. (3.29) is in the evaluation of the most work that would become available as the closed system comes to thermal and mechanical equilibrium with the atmosphere, during a process in which the atmosphere is the only temperature reservoir with which the system can interact. By replacing the subscript "2" with "0" to indicate the final state, and noting that the last two terms drop out from eq. (3.29), the maximum available work reduces to

$$(E_W)_{rev, T_0 \text{ only}} = A - A_0 \qquad (3.31)$$

The difference $A - A_0$ has been named the *nonflow exergy* [12] of the system and given the symbol (Ξ, ξ),

$$\Xi = A - A_0 = E - E_0 - T_0(S - S_0) + P_0(V - V_0)$$
$$\xi = a - a_0 = e - e_0 - T_0(s - s_0) + P_0(v - v_0)$$

$$(3.32)$$

As indicated by the notation on the left side of eq. (3.31), the nonflow energy is the reversible work delivered by a fixed-mass system during a process in which the atmosphere (T_0) is the only temperature reservoir available. The final state of equilibrium with the atmosphere, $(\ )_0$, is called the *restricted dead state* of the system. In the present treatment of closed systems, the "restricted" dead state is a state of only mechanical and thermal equilibrium with the atmosphere $(P = P_0, T = T_0)$. The concepts of equilibrium and dead state will be generalized in chapter 5, after the formal introduction of chemical equilibrium.

The numerical evaluation of "exergy" quantities $(E_Q, \Xi$, and, as shown in the next section, $E_x)$ depends on the numerical values assigned to the constants T_0 and P_0. There is an engineering need to standardize the calculation of exergies, i.e., to ensure that the calculated numerical values do not change from year to year and from country to country. The (T_0, P_0) constants that have been used most frequently are those that traditionally have been associated with standard atmospheric conditions [13], namely, 25°C and 1 atm, or

$$T_0 = 298.15 \text{ K} \qquad \text{and} \qquad P_0 = 0.101325 \text{ MPa} \qquad (3.33)$$

According to the recommendations of a recent thermodynamics workshop [14], the above constants can be regarded as the *standard environmental state* for all future exergy calculations.

A simple way to illustrate the meaning of nonflow exergy calculations is by considering a fixed mass of *incompressible substance* at temperature T and pressure P, such that $T \neq T_0$ and $P \neq P_0$. The extent to which this closed system can serve as "source" of available work is determined from eqs. (3.31) and (3.32), that is, by considering the reversible process by which the system reaches its restricted dead state. The specific nonflow exergy ξ is obtained by recalling that for an incompressible substance

$$du = c \, dT$$
$$dv = 0$$
$$ds = \frac{c}{T} \, dT$$

$$(3.34)$$

where the lone specific heat c is, at best, a function of temperature. Assuming that c is constant and that $de = du$, eq. (3.32) yields

$$\xi = cT_0\left(\frac{T}{T_0} - 1 - \ln\frac{T}{T_0}\right) \tag{3.35}$$

Figure 3.12 shows in dimensionless form the relationship between non-flow exergy on temperature. The incompressible substance can serve as a source of exergy as long as $T \neq T_0$. A hot fixed-mass system contains available work because it can serve as a high-temperature reservoir to a heat-engine cycle that rejects heat to the (T_0) reservoir. Likewise, a cold mass contains available work because, theoretically, it can serve as a low-temperature reservoir to a heat-engine cycle that absorbs heat from the "high"-temperature reservoir (T_0). We see again the image conveyed by the right diagram of Fig. 3.9: to refrigerate a system to a temperature that is lower than that of the ambient is to deposit exergy in the cold system.

We can examine the same issue in the context of a fixed-mass system that contains an *ideal gas* whose temperature and pressure differ from the environmental conditions. We evaluate the specific nonflow exergy by combining the definition (3.32) with the ideal gas relations:

$$du = c_v\, dT$$

$$dv = R\, d\left(\frac{T}{P}\right) \tag{3.36}$$

$$ds = \frac{c_P}{T}\, dT - \frac{R}{P}\, dP$$

Treating the specific heats $c_v(T)$ and $c_P(T)$ as constants, and assuming again $de = du$, we obtain

$$\xi = c_v T_0\left(\frac{T}{T_0} - 1 - \frac{c_P}{c_v}\ln\frac{T}{T_0}\right) + RT_0\left[\left(\frac{T}{T_0}\right)\left(\frac{P_0}{P}\right) - 1 - \ln\frac{P_0}{P}\right] \tag{3.37}$$

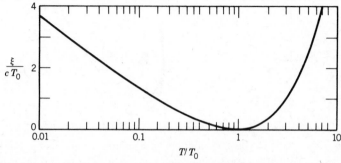

Figure 3.12 The specific nonflow exergy of an incompressible liquid.

The nonflow exergy of an ideal gas depends on both temperature and pressure in a way that, in three dimensions, resembles a bucket-shaped ξ surface whose minimum ($\xi = 0$) is located at $T = T_0$ and $P = P_0$. The ideal gas system possesses positive nonflow exergy anywhere else in the (T, P) domain. For example, if we set $T = T_0$ in eq. (3.37), we arrive at an exergy–pressure relationship that is analytically similar to the exergy–temperature relationship for an incompressible substance, eq. (3.35). Consequently, in Fig. 3.12, we can relabel the ordinate as ξ/RT_0 and the abscissa as P_0/P and use the same graph as an illustration of the nonflow exergy of an ideal gas whose temperature is the same as the environmental temperature, $\xi(T_0, P)$. A potential for doing useful work is therefore present at both above-atmospheric and below-atmospheric pressure levels. We draw similar conclusions if instead of holding T constant in eq. (3.37), we set P equal to P_0.

The convexity of the three-dimensional surface $\xi(T, P)$ can be shown analytically by considering the limit $(T \to T_0, P \to P_0)$ in which eq. (3.37) becomes

$$2\, \frac{\xi}{c_v T_0} = \tau^2 + \frac{R}{c_v}\, (\pi - \tau)^2 \tag{3.38}$$

where $\tau = (T - T_0)/T_0$, and $\pi = (P - P_0)/P_0$. Near point (T_0, P_0) in the

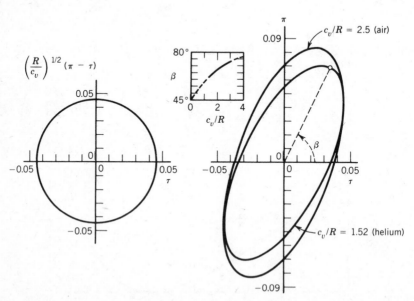

Figure 3.13 The shape of constant nonflow exergy lines for an ideal gas in the limit $T \to T_0$, $P \to P_0$. All curves correspond to $\xi/c_v T_0 = 10^{-3}$. Note also the dimensionless notations $\tau = (T - T_0)/T_0$ and $\pi = (P - P_0)/P_0$.

(T, P) plane, the lines of constant nonflow exergy constitute a family of ellipses whose axes of symmetry are inclined relative to the $T-P$ system of coordinates (Fig. 3.13). The complete nonflow exergy diagram for a particular ideal gas, $\xi(T, P)$, can be drawn using eq. (3.37) directly, as shown in the case of air $(c_P/c_v = 1.4)$ by Brodianskii [15] and Moran [16].

STEADY-FLOW PROCESSES

By far the most frequent engineering application of the two laws is in the realm of "steady-flow processes," that is, in the analysis of installations that can be modeled as open systems, operating steadily or periodically, if the period-averaged behavior does not change from one period to the next. For such systems, the "combined" law (3.9) reduces to

$$\dot{E}_W = \sum_{i=1}^{n} (\dot{E}_Q)_i + \sum_{\text{in}} \dot{m}b - \sum_{\text{out}} \dot{m}b - T_0 \dot{S}_{\text{gen}} \qquad (3.39)$$

where the new property (B, b) is the *flow availability* measured at each port [11]:

$$B = H° - T_0 S$$
$$b = h° - T_0 s \qquad (3.40)$$

Since in many cases the bulk methalpy $h°$ of each stream is equal to the bulk enthalpy h, it is important to bear in mind the difference between the flow availability property $h - T_0 s$ and the specific Gibbs free energy $h - Ts$ (chapter 4). Equation (3.39) states that the available work or exergy (E_W) delivered by the flow system is equal to the *net* flow of exergy into the system (via heat transfer and fluid flow) minus the exergy destroyed through thermodynamic irreversibility.

As a first application of eq. (3.39), consider the class of systems in which the streams $\dot{m}$ do not mix as they flow through the apparatus. The simplest examples of this kind are the single-stream shaft work components encountered in power and refrigeration cycles, namely, expanders, turbines, pumps, and compressors. Two-stream or multistream heat exchangers in which the streams do not mix also belong to this special class of systems. By attaching the subscript $(\ )_k$ to the inlet/outlet properties of each of the r unmixed streams, eq. (3.39) can be rewritten as

$$\dot{E}_W = \sum_{i=1}^{n} (\dot{E}_Q)_i + \sum_{k=1}^{r} [(\dot{m}b)_{\text{in}} - (\dot{m}b)_{\text{out}}]_k - T_0 \dot{S}_{\text{gen}} \qquad (3.41)$$

hence,

$$\dot{E}_W = \sum_{i=1}^{n} (\dot{E}_Q)_i + \sum_{k=1}^{r} [(\dot{m}e_x)_{in} - (\dot{m}e_x)_{out}]_k - T_0 \dot{S}_{gen} \qquad (3.42)$$

where the new property (E_x, e_x) is the *flow exergy* of each fluid [12]:

$$E_x = B - B_0 = H° - H_0° - T_0(S - S_0)$$
$$e_x = b - b_0 = h° - h_0° - T_0(s - s_0) \qquad (3.43)$$

The flow exergy e_x is defined as the difference between the flow availability of the stream (b) and that of the same stream at its restricted dead state, that is, the flow availability evaluated at standard environmental conditions, $b_0 = h_0° - T_0 s_0$. Note also that the transition from eq. (3.41) to eq. (3.42) is made by invoking the steady-state conservation of mass for each stream, eq. (1.27); hence, the reminder that eq. (3.42) applies to systems in which each stream k flows unmixed through the apparatus (in general, it is assumed that each stream can carry a different substance or mixture of substances).

In the very special case where all the flows into and out of the control volume carry the same substance, subscript k loses its usefulness and eq. (3.42) becomes

$$\dot{E}_W = \sum_{i=1}^{n} (\dot{E}_Q)_i + \sum_{in} \dot{m}e_x - \sum_{out} \dot{m}e_x - T_0 \dot{S}_{gen} \qquad (3.44)$$

In this case, the number of inlet ports need not be equal to the number of outlet ports (e.g., the mixing of two water streams in a steam ejector and the mixing of two helium gas streams on the low-pressure side of the main heat exchanger of a helium liquefier, chapter 10). Note further that eq. (3.44) can be obtained directly from eq. (3.39) by invoking the steady-state conservation of the lone substance processed by the apparatus.

Either in flow availability or flow exergy formulation, the above statements consistently draw attention to the destruction of exergy in steady-flow systems ($T_0 \dot{S}_{gen}$). Figure 3.14 shows the linkage of four steady-flow components in the classical Rankine cycle sandwiched between a high (flame) temperature T_H and the atmospheric reservoir temperature T_0. The energy interactions experienced by the pump, heater, turbine, and condenser are indicated by arrows that point in the correct physical direction; therefore, as in the usual treatment of this cycle, in the present discussion, $\dot{W}_p$, $\dot{Q}_H$, $\dot{W}_t$, and $\dot{Q}_0$ represent positive numerical values. Each component processes a single stream; in the case of the heater, for example, eq. (3.42) or (3.43) yields

$$0 = \dot{E}_{Q_H} + \dot{m}e_{x,2} - \dot{m}e_{x,3} - T_0 \dot{S}_{gen,heater} \qquad (3.45)$$

We rewrite this result as

$$\underbrace{\dot{E}_{Q_H} + \dot{m}e_{x,2}}_{\text{Exergy inflow}} = \underbrace{\dot{m}e_{x,3}}_{\substack{\text{Exergy} \\ \text{outflow}}} + \underbrace{T_0 \dot{S}_{\text{gen,heater}}}_{\substack{\text{Destroyed} \\ \text{exergy}}} \qquad (3.45')$$

to show the meaning of the confluence and branching out of the exergy streams in the "heater" part of the exergy (bottom) diagram of Fig. 3.14. The exergy flowrate into the heater, $\dot{E}_{Q_H} + \dot{m}e_{x,2}$, is partly transmitted to the

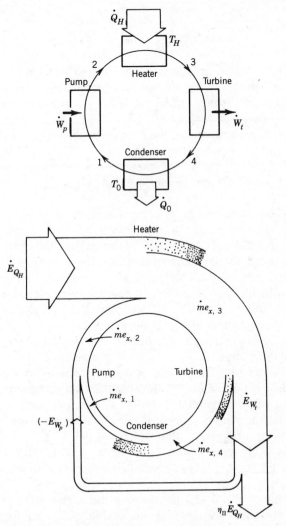

Figure 3.14 The conversion and partial destruction of exergy around a simple Rankine cycle. Top: the traditional notation and energy-interaction diagram. Bottom: the exergy flow diagram and the definition of second-law efficiency.

next flow component, $\dot{m}e_{x,3}$, and partly destroyed due to the inherent irreversibility of boilers (we concentrate on this feature and on the minimization of power cycle irreversibilities in chapters 8 and 9).

In the corresponding analysis of the turbine, we find that the heat transfer exergy terms $(\dot{E}_Q)_i$ are all zero, because turbines are adequately represented by the adiabatic steady-flow system model [however, even in cases where the turbine leaks heat at a significant rate $\dot{Q}_t$ to the ambient, the exergy flow associated with $\dot{Q}_t$ would be by definition zero, eq. (3.14)]. The right side of the exergy flow diagram of Fig. 3.14 shows that the turbine converts part of its exergy input $\dot{m}e_{x,3}$ into shaft work:

$$\dot{E}_{W_t} = \dot{W}_t = \dot{m}e_{x,3} - \dot{m}e_{x,4} - T_0 \dot{S}_{\text{gen,turbine}} \tag{3.46}$$

The remaining fraction is either passed on to the condenser or is lost through irreversibility. Clearly, the fraction $\dot{W}_t / \dot{m}e_{x,3}$ is related to the classical concept of turbine efficiency, $\dot{W}_t / \dot{W}_{t,\text{rev}}$: we examine this relationship in greater detail in chapter 8.

As a steady-flow component, the condenser distinguishes itself as having the simplest exergy flow equation:

$$0 = \dot{m}e_{x,4} - \dot{m}e_{x,1} - T_0 \dot{S}_{\text{gen,condenser}} \tag{3.47}$$

because the heat transfer exergy term associated with $\dot{Q}_0$ is by definition zero. The condenser destroys a significant portion of the stream's exergy, primarily as a result of the heat transfer across the finite temperature difference between the condenser tubes and the atmosphere. Even though the theoretical job of the condenser is to bring the stream to thermal equilibrium with the ambient, the exit temperature of the condensate (T_1) is, in general, greater than T_0, which is why the exit exergy $e_{x,1}$ is shown finite in the exergy wheel of Fig. 3.14.

Finally, by going through the pump, the $\dot{m}e_{x,1}$ stream is augmented somewhat by the exergy put into the circuit as pump work, $\dot{E}_{W_p} = -\dot{W}_p$, and returned to the heater. The pump, of course, adds its own contribution to the overall destruction of exergy through the circuit; however, in order to keep the exergy-flow diagram of Fig. 3.14 simple, the pump $T_0 \dot{S}_{\text{gen}}$ contribution is not shown. Another decision aimed at making the diagram more readable was to exaggerate the width of the $(-\dot{E}_{W_p})$ arrow: in most steam-turbine cycle calculations, the pump power requirement and its irreversibility ($T_0 \dot{S}_{\text{gen,pump}}$) are minuscule compared with the corresponding quantities for the turbine.

As a summary, it is useful to compare the exergy diagram of Fig. 3.14 with the power cycle exergy diagram of Fig. 3.7, and to recognize the agreement between the two diagrams with regard to the *external* exergy interactions $\dot{E}_{Q_H}$ and $\dot{E}_W$ ($= \eta_{\text{II}} \dot{E}_{Q_H}$). The exergy analysis of each steady-flow component of the power cycle allows us to determine the *distribution* of irreversibility among the cycle components.

The component-by-component contribution to the destruction of exergy shown in the refrigeration cycle of Fig. 3.9 can be determined similarly. Figure 3.15 shows the internal circulation and successive destruction of exergy in a vapor-compression cycle: seen from the outside, this exergy-flow diagram is analogous to the right side of Fig. 3.9. An internal examination of the refrigeration cycle shows that the destruction of exergy is the combined effect of four irreversible steady-flow components, the compressor, the condenser, the expansion valve, and the evaporator. The detailed study and minimization[†] of the irreversibility associated with each of these components forms the subject of chapter 10.

The flow exergy analyses that led to Figs. 3.14 and 3.15 could have been carried out just as well in terms of flow availability, that is, based on eq. (3.39). The use of flow exergy offers a special advantage in connection with a generic question in the thermodynamics of power engineering, namely, "What is the available work (exergy) content of a stream that is not in thermal and mechanical equilibrium with the environment?" Note that this question is the steady-flow counterpart of the nonflow exergy questions answered in Figs. 3.12 and 3.13. The answer is simply

$$(\dot{E}_W)_{\mathrm{rev}, T_0 \,\mathrm{only}} = \dot{m} e_x \tag{3.48}$$

because the would-be outlet flow exergy $\dot{m} e_{x,0}$ is by definition zero, eq. (3.43). It is assumed that the environmental temperature reservoir is the

---

[†] I use this opportunity to draw attention to an important document on research needs in thermal systems engineering [17], to which I was honored to contribute early versions of Figs. 3.14 and 3.15. This document is a summary of an extensive study initiated by the American Society of Mechanical Engineers and the National Science Foundation [18]. Of interest here is that the topic of *Thermodynamic Optimization and Design Methodology* was identified as an important research priority:

> Research should be done on methods of applying second-law analysis to achieve thermodynamic optimization of thermal systems. An analysis of all major systems and components to identify ideal targets for thermodynamic optimization should be undertaken. Development of a design methodology for thermal systems, similar to that for mechanical systems, is necessary. These methods should be primarily concerned with thermal processes, as well as with alternatives that take into account economic and first- and second-law constraints on thermal performance. Such methodology is important for both the evaluation of existing systems and the development of new thermal processes [Ref. 17, p. 46].

Another priority item identified in this document is the research on *Modeling, Simulation, and Optimization*:

> Innovative methodologies are needed in the design, operation, and adaptive control of thermal systems, all of which could lead to improved forecasting of loads and resources. *The methodologies should explicitly include the second law in optimizing systems and subsystems* [emphasis added; Ref. 17, p. 45].

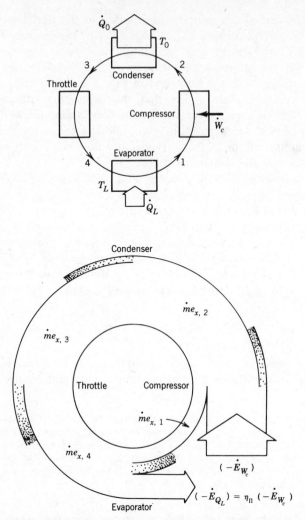

Figure 3.15 The conversion and partial destruction of exergy around a simple vapor-compression refrigeration cycle. Top: the traditional notation and energy-interaction diagram. Bottom: the exergy flow diagram and the definition of second-law efficiency.

only one with which the stream can make thermal contact as it produces available work and reaches its restricted dead state (T_0, P_0).

In conclusion, the answer to the generic question formulated in the preceding paragraph is none other than the stream's original "specific flow exergy." This answer explains the modern interest in calculating e_x for various working fluids and constructing in this way an entirely new generation of thermodynamic property diagrams. Well-known flow exergy and flow availability diagrams have been published by Keenan [19], Reistad [20], and Thirumaleshwar [21], among others.

The specific flow exergy assumes a particularly simple form in the case of an *incompressible liquid* (Problem 3.4):

$$e_x = cT_0\left(\frac{T}{T_0} - 1 - \ln\frac{T}{T_0}\right) + v(P - P_0) \tag{3.49}$$

where the lone specific heat c and the specific volume v are both constant. Looking back at the expression obtained for the nonflow exergy of the same incompressible liquid, eq. (3.35), we note the relationship

$$e_x = \xi + v(P - P_0) \tag{3.50}$$

This means that the flow energy chart of an incompressible liquid is obtained by replacing ξ with e_x on the ordinate in Fig. 3.12 and by shifting the cup-shaped curve vertically by an amount (i.e., a "distance" measured in J/kg) that corresponds to the value of the $v(P - P_0)$ term. The e_x chart is then a family of nonintersecting isobaric curves that have the same shape as the curve shown already in Fig. 3.12. Unlike the nonflow exergy ξ, which cannot be negative, the flow exergy e_x is negative in a certain, finite (T, P) domain at sufficiently low pressures.

The flow exergy of an *ideal gas* has an equally simple expression,

$$e_x = c_P T_0\left(\frac{T}{T_0} - 1 - \ln\frac{T}{T_0}\right) + RT_0 \ln\frac{P}{P_0} \tag{3.51}$$

where the ideal gas model is represented by the known constants c_P and R. The e_x chart that can be built based on eq. (3.51) will have the same features as the chart described in the preceding paragraph, except that this time, the vertical position of each cup-shaped isobaric curve is controlled by the size of the pressure term $RT_0 \ln(P/P_0)$. The flow exergy of an ideal gas can be negative, unlike the nonflow exergy listed in eq. (3.37).

MECHANISMS OF ENTROPY GENERATION AND EXERGY DESTRUCTION

As a preview of the irreversibility minimization methods described in the more applied segments of the present treatment, in this section, we identify three common design features that always contribute to the irreversibility of the installation in which they are present. The mission of this section is to provide a feel for what causes entropy generation, and to teach how to identify those system components that contribute the most to the overall irreversibility of the installation.

Heat Transfer Across a Finite Temperature Difference

Consider the first drawing shown in Fig. 3.16, in which two systems of different temperatures (T_H, T_L) experience the heat transfer interaction $\dot{Q}$. An important observation is that the two systems do not communicate directly, because if they did, they would have a common boundary; hence, $T_H \equiv T_L$ (review the message of Fig. 1.2). Sandwiched between the system (T_H) and the system (T_L) is a third system called the "temperature gap." The heat transfer $\dot{Q}$ enters and leaves the temperature gap system undiminished, which is why the temperature gap is not recognized as a separate thermodynamic system in the field of heat transfer engineering.

With reference to the temperature-gap system, the second law shows that the entropy generated by the system is finite as long as the temperature difference $T_H - T_L$ is finite:

$$\dot{S}_{gen} = \frac{\dot{Q}}{T_L} - \frac{\dot{Q}}{T_H} = \frac{\dot{Q}}{T_L T_H} (T_H - T_L) \geq 0 \qquad (3.52)$$

If T_L happens to be greater than T_H, then the physical direction of $\dot{Q}$ must be the opposite of the direction in Fig. 3.16, so that the second-law $\dot{S}_{gen} \geq 0$ is always satisfied.

We can also view the temperature-gap system as a closed system that operates steadily while in communication with two heat reservoirs. Designating (T_L) as reference (ambient) reservoir, eq. (3.15) in this case reduces to

$$\dot{W}_{lost} = \left(1 - \frac{T_L}{T_H}\right)\dot{Q} \qquad (3.53)$$

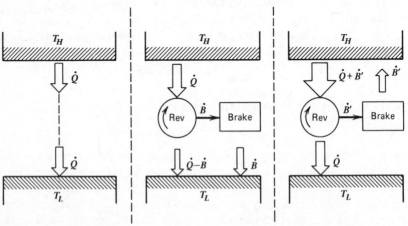

Figure 3.16 The destruction of useful work in the temperature gap crossed by a heat transfer interaction [22].

or, in view of eq. (3.52), $\dot{W}_{lost} = T_L \dot{S}_{gen}$. The physical meaning of eq. (3.53) is illustrated in the second and third drawings of Fig. 3.16 [22]. We see that the transfer of heat from T_H to T_L is thermodynamically equivalent to a reversible engine that operates between (T_H) and (T_L) and dissipates its entire work output into a brake, which in turn rejects heat either to (T_H) or (T_L). The work output of the reversible engine occupies the right side of eq. (3.53). Putting all these observations together, we conclude that the heat transfer across a finite temperature gap always contributes to the destruction of useful work in the greater system that contains the temperature gap.

This is an appropriate place for repeating an observation made back in 1982 [Ref. 6, p. 99], namely, that the simple entropy generation expression (3.52) has the power to *unify* the two seemingly antagonistic sectors of thermal design practice. In very broad terms, the contemporary thermal design problems fall into two large categories:

(1) heat transfer augmentation problems, and

(2) thermal insulation problems

The first category contains problems in which the heat transfer rate is usually prescribed by design, whereas the objective is to minimize the temperature difference between the heat-exchanging entities, $\Delta T = T_H - T_L$. According to eq. (3.52), however, this effort of enhancing the thermal contact is essentially an entropy-generation minimization effort.

In a thermal insulation problem, on the other hand, the objective is to minimize the heat leak $\dot{Q}$ when the temperature difference ΔT is usually fixed. Unlike heat transfer augmentation problems, in thermal insulation

TABLE 3.1   The Two Main Problems in Thermal Design, as Two Distinct Approaches to Entropy Generation Minimization [6]

Problem		Temperature Difference, ΔT	Heat Transfer Rate, $\dot{Q}$	Entropy Generation Rate, $\dot{S}_{gen} = \dot{Q}\,\Delta T/T^2$
1. Heat transfer augmentation (thermal contact enhancement)		Reduced	Fixed	Reduced
2. Thermal insulation (heat leak minimization)		Fixed	Reduced	Reduced

problems, an intimate thermal contact between (T_H) and (T_L) is considered repulsive. Yet, eq. (3.52) shows that to solve a thermal insulation problem is one more effort of minimizing entropy generation. Table 3.1 summarizes the conclusion that two seemingly opposite camps of thermal design engineers, one devoted to enhancing thermal contact and the other to preventing it, solve their individual problems in ways that *universally* lead to reductions in entropy generation.

Flow with Friction

An equally common feature that serves as source of entropy generation is the effect of fluid friction in various ducts and flow networks. In order to see the direct connection between frictional pressure drop and thermodynamic irreversibility, consider the steady and adiabatic flow of a pure substance through a short segment of pipe. The mass-conservation equation and the first and second laws require:

$$\dot{m}_{in} = \dot{m}_{out} = \dot{m} \tag{3.54}$$

$$h_{in} = h_{out} \tag{3.55}$$

$$\dot{S}_{gen} = \dot{m}(s_{out} - s_{in}) \geq 0 \tag{3.56}$$

We are interested in relating the entropy-generation rate to the pressure drop along the pipe segment; therefore, we note that for a fluid of fixed chemical composition, $dh = T\,ds + v\,dP$ (Table 4.3, the enthalpy representation), in other words,

$$ds = \frac{1}{T}\,dh - \frac{v}{T}\,dP \tag{3.57}$$

The first law (3.55) shows that the bulk enthalpy h does not change along the stream: combining this idea with eq. (3.57), we can calculate the entropy-generation rate (3.56) with the pressure integral:

$$\dot{S}_{gen} = \dot{m} \int_{out}^{in} \left(\frac{v}{T}\right)_{h=\text{constant}} dP \tag{3.58}$$

Since the integrand v/T is positive and since the direction of integration is "upstream" (i.e., toward higher pressures), we conclude that the entropy-generation rate is finite as soon as the pressure drop $\Delta P = P_{in} - P_{out}$ is finite. It is a known fact in fluids engineering and convection heat transfer that, regardless of the flow regime, the pressure drop is finite when both $\dot{m}$ and the length of the duct are finite [23]. The pressure drop is the result of (it is balanced by) the frictional shear stress integrated over the internal surface of the duct, hence, the name "flow with friction" for the entropy-generation mechanism discussed here.

The one-to-one relationship between entropy generation rate and pressure drop is even more visible in the two limits of thermodynamic behavior in fluids. In the case of an *ideal gas*, eq. (3.58) becomes [6]

$$\dot{S}_{gen} = \dot{m}R \ln \frac{P_{in}}{P_{out}} \qquad (3.59)$$

and for pressure drops that are much smaller than the local absolute pressure,

$$\dot{S}_{gen} \cong \dot{m}R \frac{\Delta P}{P_{in}} \qquad (\Delta P \ll P_{in}) \qquad (3.60)$$

The corresponding expressions for an *incompressible liquid* are [6]

$$\dot{S}_{gen} = \dot{m}c \ln \frac{T_{out}}{T_{in}} \qquad (3.61)$$

$$\dot{S}_{gen} = \dot{m} \frac{v}{T_{in}} \Delta P \qquad \left(\Delta P \ll \frac{c}{v} T_{in}\right) \qquad (3.62)$$

All these expressions show that it takes both flow ($\dot{m}$) and pressure drop (ΔP) to have entropy generation in duct flow, much in the way that $\dot{Q}$ and ΔT *together* contribute to generating entropy in a heat transfer device (Table 3.1).

The duct segment example treated until now can be viewed also as a system that steadily destroys flow exergy. Since the flow is steady and adiabatic, eq. (3.44) reduces to

$$
\begin{aligned}
T_0 \dot{S}_{gen} &= \dot{m}(e_{x,in} - e_{x,out}) \\
&= \dot{m}[(h - T_0 s)_{in} - (h - T_0 s)_{out}]
\end{aligned} \qquad (3.63)
$$

which says exactly the same thing as eqs. (3.55) and (3.56) combined. The flow exergy decreases in the downstream direction, and the decrease is proportional to both ΔP and $\dot{m}$.

The "flow with friction" acts as an entropy-generation mechanism in all flow configurations, not just in the *internal flow* configuration discussed until now. As a most general example of *external flow*, consider the entropy generated by an object (strut, fin, vehicle body) that is swept by a much larger reservoir of fluid whose velocity relative to the object is U_∞. The object experiences a net drag force F_D. The analysis is greatly simplified by choosing the fluid reservoir (the "rest of the world," more accurately) as a closed system [24]. Next, we note that the only energy interaction experienced by this closed system is the work done by the entity that drags the object through the reservoir. The system is otherwise encased in a constant-volume adiabatic enclosure. The first law and the second law for this very large closed system are

$$F_D U_\infty = \frac{dU}{dt} \tag{3.64}$$

$$\dot{S}_{\text{gen}} = \frac{dS}{dt} \geq 0 \tag{3.65}$$

The incremental changes in the fluid reservoir's internal energy and entropy are related by (Table 4.3):

$$dU = T_\infty \, dS - P_\infty \, dV \tag{3.66}$$

where (T_∞, P_∞) are the temperature and pressure of the reservoir, respectively, and where $dV = 0$. Eliminating dU and dS between eqs. (3.64)–(3.66) yields

$$\dot{S}_{\text{gen}} = \frac{F_D U_\infty}{T_\infty} \tag{3.67}$$

In conclusion, an external flow configuration that is characterized by a net drag force F_D and a relative (so-called "free-stream") velocity U_∞ is also the locus of thermodynamic irreversibility. The rate of entropy generation is proportional to the mechanical power invested in dragging the object through the fluid reservoir, $F_D U_\infty$, that is, to the rate of lost work.

Mixing

A third class of system design features that consistently serve as sources of entropy generation can be loosely grouped together under the general title of "mixing processes." The examples of mixing are as diverse as the number and type of dissimilar entities that are being mixed. Even when the mixing process involves only two substances, there can be two types of mixing apparatuses that can be modeled as adiabatic and zero-work, namely, a "flow" device in which two streams are mixed and a "nonflow" device in which mixing occurs between two batches. Furthermore, the two substances that engage in this relatively simple subset of mixing processes can be "dissimilar" because of their respective temperatures, or pressures, or chemical constitution (chemical potentials), or any combination of these properties. The examples alluded to in Table 3.2 illustrate the diversity of mixing processes, hence, the impossibility of covering the entire domain in the limited space that is provided by the present treatment.

Consider as an example the mixing of two streams that carry the same substance into a third, mixed stream. At the joining of the three streams is a control volume with two inlets labeled (1) and (2) and one outlet labeled (3). The control surface is obviously a zero-work transfer surface; assuming further that the surface is also adiabatic, the steady-state statements for mass conservation and the first law and the second law are

TABLE 3.2 The Diversity of Adiabatic and Zero-Work Mixing Processes That Can Be Constructed with Two Nonreacting Dissimilar Substances

Type of Mixing Device	Type of Dissimilarity		
	Thermal	Mechanical	Chemical
Nonflow	● Ref. 6, p. 17		
Flow	● Problem 3.6	● Treated in the present subsection	● Ref. 6, p. 40
	● Treated in the present subsection		

$$\dot{m}_1 + \dot{m}_2 = \dot{m}_3 \tag{3.68}$$

$$\dot{m}_1 h_1 + \dot{m}_2 h_2 = \dot{m}_3 h_3 \tag{3.69}$$

$$\dot{S}_{gen} = -\dot{m}_1 s_1 - \dot{m}_2 s_2 + \dot{m}_3 s_3 \tag{3.70}$$

How the entropy-generation rate depends on the degree of dissimilarity between the two inflowing streams becomes visible if we place the analysis in the limit of small changes or "marginal mixing" ($\Delta T \ll T$, $\Delta P \ll P$, $\Delta v \ll v$, etc.). In other words, we are assuming that a differential form like $dh = T\,ds + v\,dP$ can be integrated to obtain an approximate linear relationship between the enthalpy changes that appear in the first law (3.69) and the entropy changes that add up to the entropy-generation rate (3.70). For example, between states (1) and (3), we can write approximately

$$h_3 - h_1 \cong T(s_3 - s_1) + v(P_3 - P_1) \tag{3.71}$$

so that, taken together, eqs. (3.68)–(3.70) yield

$$\frac{1}{\dot{m}_3}\dot{S}_{gen} \cong x\left[\frac{1}{T}(h_3 - h_1) - \frac{v}{T}(P_3 - P_1)\right]$$
$$+ (1-x)\left[\frac{1}{T}(h_3 - h_2) - \frac{v}{T}(P_3 - P_2)\right] \tag{3.72}$$

Is this last expression, x represents the fraction contributed by the $\dot{m}_1$ stream to the final mixed stream $\dot{m}_3$:

$$x = \frac{\dot{m}_1}{\dot{m}_3} \quad \text{hence} \quad 1 - x = \frac{\dot{m}_2}{\dot{m}_3} \tag{3.73}$$

Two special applications of eq. (3.72) are particularly instructive, the case where the mixing streams carry the same *ideal gas* (or ideal gas mixture) with known constants R and c_P:

$$\frac{\dot{S}_{\text{gen}}}{\dot{m}c_P} \cong x(1-x)\left(\frac{T_1-T_2}{T_1}\right)^2 + x\,\frac{R}{c_P}\left(\frac{P_1-P_3}{P_3}\right) + (1-x)\,\frac{R}{c_P}\left(\frac{P_2-P_3}{P_3}\right) \geq 0$$

$$(3.74)$$

and the case of an *incompressible liquid* with known constants v and c:

$$\frac{\dot{S}_{\text{gen}}}{\dot{m}c} \cong \frac{1}{2}\,x(1-x)\left(\frac{T_1-T_2}{T_1}\right)^2 + x\,\frac{v}{cT_1}\left(\frac{P_1-P_3}{P_3}\right) + (1-x)\,\frac{v}{cT_1}\left(\frac{P_2-P_3}{P_3}\right)$$

$$\geq 0 \qquad\qquad\qquad\qquad\qquad\qquad\qquad\qquad\qquad\qquad (3.75)$$

The symmetry between these last two results is obvious. The first term on the right side expresses the irreversibility associated with the initial temperature mismatch between streams (1) and (2). And it does not matter which of these two streams is the warmer one: the thermal mixing irreversibility increases as the inflowing temperature difference squared. An additional feature of the first term is its dependence on the mass fraction x. This first term reaches its maximum value at $x = \frac{1}{2}$; in other words, the thermal mixing is maximized when neither of the streams overwhelms the other.

The remaining two terms on the right side of eqs. (3.74) and (3.75) account for the pressure drops experienced by both streams as they pass through the three-part joint. In many cases, the exit pressure P_3 is dictated by one of the two original streams (e.g., $P_3 = P_2$), so that only one of the pressure mixing terms contributes to entropy generation. Examples of tee-shaped joints that mix two ideal gas streams abound in the flow circuitry of advanced liquid-helium temperature refrigeration cycles (chapter 10).

Finally, we note that each pressure mixing term depends also on the mass fraction x, which means that in general the maximum entropy-generation rate of the mixing joint depends on all three effects (one ΔT, and two ΔPs). Highlighting only the x dependence, we note that eqs. (3.74) and (3.75) are both of the form

$$\sigma \cong x(1-x)t^2 + xp_1 + (1-x)p_2 \qquad\qquad (3.76)$$

hence, the mass fraction for maximum entropy-generation rate:

$$x \cong \frac{1}{2} + \frac{p_1-p_2}{2t^2} \qquad\qquad (3.77)$$

The meaning of the shorthand notation (t, p_1, p_2) changes from ideal gases (3.74) to incompressible liquids (3.75). In all cases, however, the absolute values of t, p_1, and p_2 are considerably smaller than one—this, as a result of the linearization assumption adopted in eq. (3.71).

SYMBOLS

a	specific nonflow availability [J/kg], [eq. (3.30)]
A	nonflow availability [J], [eq. (3.30)]
b	specific flow availability [J/kg], [eq. (3.40)]
B	flow availability [J], [eq. (3.40)]
c	lone specific heat of incompressible liquid
c_P	specific heat at constant pressure
c_v	specific heat at constant volume
COP	coefficient of performance of refrigeration cycles [eq. (3.25)] and heat-pump cycles [eq. (3.28)]
e	specific energy [J/kg]
E	energy [J]
E_Q	exergy associated with the heat transfer interaction Q [eq. (3.14)]
E_W	exergy associated with the work transfer interaction W (note: $E_W = W$)
e_x	specific flow exergy [J/kg], [eq. (3.43)]
E_x	flow exergy [J], [eq. (3.43)]
F_D	drag force
$h°$	specific methalpy [J/kg], [eq. (1.23)]
$H°$	methalpy [J]
$\dot{m}$	mass flowrate
p	shorthand notation [eqs. (3.76)–(3.77)]
P	pressure
P_0	pressure of the environment as a pressure reservoir
Q	heat transfer interaction
$\dot{q}$	heat transfer rate (Table 3.1)
$\dot{Q}$	heat transfer rate
R	ideal gas constant
s	specific entropy
S	entropy
S_{gen}	entropy generation
$\dot{S}_{gen}$	entropy-generation rate
t	time
t	shorthand notation [eqs. (3.76)–(3.77)]
T	absolute temperature
T_0	temperature of the environment as a temperature reservoir
U	internal energy
U_∞	speed of fluid reservoir
v	specific volume
V	volume
W	work transfer interaction
$\dot{W}$	work transfer rate (mechanical power)
$\dot{W}_{lost}$	rate of work destruction; the same as $(\dot{E}_W)_{lost}$
x	mass fraction [eq. (3.73)]

$\delta T, \Delta T$	temperature difference (Table 3.1)
ΔP	pressure drop
η_I	first-law efficiency [eq. (3.20)]
η_{II}	second-law efficiency [eqs. (3.19), (3.24), and (3.27)]
ξ	specific nonflow exergy [J/kg], [eq. (3.32)]
Ξ	nonflow exergy [J], [eq. (3.32)]
π	dimensionless pressure [defined under eq. (3.38)]
σ	function [eq. (3.76)]
τ	dimensionless temperature [defined under eq. (3.38)]
$(\ )_H$	high
$(\ )_{in}$	inlet
$(\ )_L$	low
$(\ )_{out}$	outlet
$(\ )_{rev}$	reversible
$(\dot{\ })$	per unit time
$(\ )_\infty$	properties of the fluid reservoir

REFERENCES

1. J. Kestin, *Available Work in Geothermal Energy*, Report No. CATMEC/20, Division of Engineering, Brown University, Providence, RI, July 1978.

2. J. Kestin, Availability, the concept and associated terminology, *Energy*, Vol. 5, 1980, pp. 679–692.

3. J. Kestin, Available work in geothermal energy, in J. Kestin, R. DiPippo, H. E. Khalifa, and D. J. Ryley, eds., *Sourcebook on the Production of Electricity from Geothermal Energy*, DOE/RA/28320-2, U.S. Department of Energy, Washington, DC, 1980, Chapter 3.

4. G. Gouy, Sur l'énergie utilisable, *Journal de Physique*, Vol. 8, 1889, pp. 501–518.

5. A. Stodola, *Steam Turbines* (with an appendix on gas turbines and the future of heat engines), translated by L. C. Loewenstein, Van Nostrand, New York, 1905, p. 402.

6. A. Bejan, *Entropy Generation Through Heat and Fluid Flow*, Wiley, New York, 1982, p. 32.

7. G. A. Goodenough, *Principles of Thermodynamics*, 3rd ed., Henry Holt & Co., New York, 1932, pp. 50–51.

8. E. F. Miller, *Notes on Heat Engineering*, Technology Press (now MIT Press), Cambridge, MA, 1931, p. 30.

9. A. Bejan, Graphic techniques for teaching engineering thermodynamics, *Mech. Eng. News*, 1977, pp. 26–28.

10. V. Radcenco, S. Porneala, and A. Dobrovicescu, *Procese in Instalatii Frigorifice*, Editura Didactica si Pedagogica, Bucharest, 1983, p. 21.

11. J. H. Keenan, *Thermodynamics*, Wiley, New York, 1941, pp. 284–293.

12. Z. Rant, Exergie ein neues Wort für "technische Arbeitsfähigkeit," *Forsch. Ingenieurwes.*, Vol. 22, 1956, p. 36.

13. R. E. Sonntag and G. J. Van Wylen, *Introduction to Thermodynamics*, Wiley, New York, 1982, p. 460.

14. T. J. Kotas, Private communication (Proposed Nomenclature for the Exergy Method of Thermodynamic Analysis, Thermodynamics Workshop, Cambridge, England, September 20–21, 1984), September 19, 1985.

15. V. M. Brodianskii, *Eksergeticheskii Metod Termodinamicheskogo Analiza*, Energia, Moscow, 1973, p. 32.

16. M. J. Moran, *Availability Analysis: A Guide to Efficient Energy Use*, Prentice-Hall, Englewood Cliffs, NJ, 1982, p. 115.

17. W. O. Winer, A. E. Bergles, C. J. Cremers, R. H. Sabersky, W. A. Sirignano, and J. W. Westwater, Needs in thermal systems, *Mech. Eng.*, Vol. 108, August 1986, pp. 39–46.

18. W. O. Winer, A. E. Bergles, C. J. Cremers, R. H. Sabersky, W. A. Sirignano, and J. W. Westwater, *Research Needs in Thermal Systems*, Am. Soc. Mech. Eng., New York, 1986.

19. J. H. Keenan, A steam chart for second-law analysis, *Mech. Eng.*, Vol. 54, March 1932, pp. 195–204.

20. G. M. Reistad, A property diagram to illustrate irreversibilities in the R12 refrigeration cycle, *ASHRAE Trans.*, Vol. 78, Part II, 1972, pp. 97–101.

21. M. Thirumaleshwar, Exergy method of analysis and its application to a helium cryorefrigerator, *Cryogenics*, Vol. 19, 1979, pp. 355–361.

22. A. Bejan, A general variational principle for thermal insulation system design, *Int. J. Heat Mass Transfer*, Vol. 22, 1979, pp. 219–228.

23. A. Bejan, *Convection Heat Transfer*, Wiley, New York, 1984, pp. 76, 257.

24. D. Poulikakos and A. Bejan, Fin geometry for minimum entropy generation in forced convection, *J. Heat Transfer*, Vol. 104, 1982, pp. 616–623.

PROBLEMS

3.1 Consider a thermal insulation device sandwiched between a cold space (T_L) and the ambient (T_H), i.e., between the same entities that border the refrigerator of Figs. 3.8 and 3.9. Heat leaks from (T_H) to (T_L) through the device, at exactly the same rate that it is removed from the cold space by the refrigerator. Show that the exergy delivered by the refrigerator to the cold space is destroyed fully in the thermal insulation.

3.2 Show that exergy deposited in the cold space (T_L) by the refrigerator of Fig. 3.9 can be retrieved as useful work. Show also that the exergy deposited into the building (T_H) by the heat pump of Fig. 3.10 can be retrieved as useful work.

3.3 Show that sufficiently close to the standard environmental state (T_0, P_0), the lines of constant nonflow exergy for an ideal gas have an elliptical shape. Show also that the axes of each ellipse form an acute angle β with the $T = T_0$ and $P = P_0$ axes, and that β is a function of only c_v/R.

3.4 Derive the formulas for the specific flow exergy of an incompressible liquid and of an ideal gas, eqs. (3.49) and (3.51). Since these formulas are quite similar, focus on only one of them and describe the (T, P) domain in which the flow exergy is negative. What is the physical meaning of a negative flow exergy value?

3.5 Show that the entropy maximum principle encountered in the study of isolated systems, eq. (2.50), is equivalent to the statement that the nonflow exergy of an isolated system is destined to decrease or, at best, remain unchanged.

3.6 Consider the adiabatic and zero-work mixing of two streams $(\dot{m}_1, \dot{m}_2)$ that carry the same ideal gas (R, c_P). The two inflowing streams and the resulting mixed stream are all at the same pressure. The original temperatures of the inflowing streams (T_1, T_2) are such that the absolute temperature ratio $\tau = T_2/T_1$ is not necessarily approximately equal to one. Show that the entropy generated during this flow mixing process is maximum when the mass fraction $x = \dot{m}_1/(\dot{m}_1 + \dot{m}_2)$ is equal to [6]

$$x = \frac{\tau - 1 - \tau \ln \tau}{(1 - \tau) \ln \tau}$$

3.7 A new vehicle propulsion scheme calls for the use of liquid nitrogen as "fuel." The details of the power cycle are not the issue here (basically, the liquid nitrogen is heated in contact with the atmosphere, pressurized, and expanded through the turbine that drives the vehicle). You are asked to calculate the maximum work that could theoretically be derived from the liquid nitrogen fuel. During each refueling stop, the driver purchases a Dewar vessel (a bottle) containing $0.05 \, m^3$ of liquid nitrogen at atmospheric pressure. He leaves in exchange a used bottle, that is, a bottle containing gaseous N_2 at atmospheric pressure and temperature. The properties of nitrogen as saturated liquid at 1 atm are

$$v = (1.24)10^{-3} \, m^3/kg$$

$$h = -121.5 \, kJ/kg$$

$$s = 2.85 \, kJ/kg/K$$

The corresponding properties of nitrogen at atmospheric temperature and pressure (300 K, 1 atm) are

$$v = 0.49 \, m^3/kg$$

$$h = 172.1 \, kJ/kg$$

$$s = 6.25 \, kJ/kg/K$$

4

Single-Phase Systems

The purpose of this chapter is to outline an important analytical segment that serves as skeleton for much of the classical thermodynamics that is used in engineering. The most recognizable features are the numerous analytical, tabular, and graphic relationships that have been established among thermodynamic properties. One engineering objective is to sort these relationships and summarize them for future reference in the course of specialized applications. Another objective is to formally introduce the concept of chemical equilibrium and to widen the applicability of the combined-law exergy concepts of chapter 3.

From the outset, we restrict the discussion to a special class of systems that are, on one hand, general enough to serve adequately as models in many engineering circumstances and, on another, are sufficiently uncomplicated to lend themselves to concise analytical treatment. Consider a system that is not subjected to the influence of gravitational, electrical, and magnetic fields, and inertial forces. The system is sufficiently large so that surface (capillarity) effects can be neglected. Furthermore, the system is macroscopically homogeneous and isotropic; in other words, it does not contain any internal constraint (e.g., adiabatic, impermeable, or rigid and immobile partitions; if the system is solid, then it is in a state of uniform, hydrostatic compression). Following Gibbs [1], the "simplest kind of system" defined above will be called *simple system*.

Each of the defining features of the simple system contributes toward simplifying the analysis that follows in this chapter (we will comment on each simplifying feature at its introductory point in the analysis). At this point, however, it is instructive to review the contents of Table 4.1 and develop a feeling for the kind of physical system that fits the simple-system description. Shown in the table are common examples of "batches" of substances that are encountered in engineering applications. Each batch is characterized by a state of internal equilibrium, because of the absence of internal constraints. If we examine the chemical composition of each sample, we can divide Table 4.1 into single-component sustances (the upper half) and multicomponent substances (the lower half). The feature that

TABLE 4.1 Samples of Substances in Internal Equilibrium: the Distinction Between "Homogeneous" and "Heterogeneous," and Between "Single-Component" and "Multicomponent," as a Way of Visualizing the Meaning of "Simple System" and "Pure Substance"

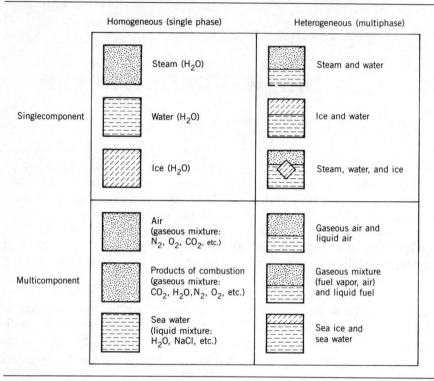

separates the simple systems from the examples of Table 4.1 is the requirement that the system be "macroscopically homogeneous and isotropic." Therefore, only the single-phase examples shown on the left of Table 4.1 qualify as simple systems in the sense intended in the definition. The right part of the table shows examples of macroscopically heterogeneous (multiphase) substances, which, as a class, form the subject of chapter 6.

The horizontal and vertical alignment of the examples of Table 4.1 serves an additional purpose by showing the relationship between the Gibbsian thermodynamics terminology and the concept of *pure substance*, which is a central concept in engineering thermodynamics. In Table 4.1, pure substances are all the samples in which the chemical composition does not vary (spatially) through the sample. Therefore, we recognize as pure substances the single-component examples that occupy the upper half of the table. In addition, the examples placed in the lower left quarter of the table (multicomponent, homogeneous) qualify also as pure substances. The only group that does not accept the pure-substance label is the one that fits in the lower right quarter of the table (multicomponent, heterogeneous). For example, in a batch of gaseous air in equilibrium with liquid air, the proportions of N_2, O_2, CO_2, etc. in the liquid phase are not the same as in the gaseous phase.

EQUILIBRIUM CONDITIONS

A simple system can be open or closed. We chose an *open* simple system because the closed-system conclusions can be derived in one small step from the more general conclusions reached in the study of open systems. We examine once again the implications of the combined first and second laws that opened chapter 3, this time in the context of the simple system defined by the control volume of Fig. 4.1. The system is in thermal contact with the temperature reservoir (T_0), while its volume change is being resisted or assisted by the pressure reservoir (P_0). Noting the filial position of Fig. 4.1 relative to Fig. 3.1, we write the first law and the second law as in eqs. (3.1) and (3.2). This time, however, we integrate the resulting equations over the infinitesimal time interval dt:

$$dU = \delta Q - \delta W + \sum_{i=1}^{n} h_{0,i}\, dm_i \qquad (4.1)$$

$$\delta S_{\text{gen}} = dS - \frac{\delta Q}{T_0} - \sum_{i=1}^{n} s_{0,i}\, dm_i \geq 0 \qquad (4.2)$$

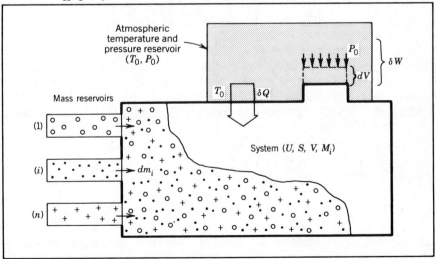

Figure 4.1 Open system in communication with n mass reservoirs and the atmospheric temperature and pressure reservoirs.

Equation (4.1) already reflects the use of the simple-system model, namely, the ruling out of gravity and inertia ($E = U$, $h° = h$). In accordance with the boundary-selection rule stressed in Fig. 1.2, the local temperature of the boundary crossed by δQ is equal to the known temperature of the reservoir, T_0. At this stage in the analysis, we say nothing about the system temperature T except that it is equal to T_0 right on the boundary crossed by δQ (later, we invoke the notion of equilibrium in the absence of internal constraints to conclude that the temperature of the simple system, T, must be uniform and equal to T_0).

We are ultimately interested in the relationship between the chemical composition of the simple system and other properties such as its energy U, volume V, pressure P, and temperature T. With this objective in mind, in Fig. 4.1, we picture a system that exchanges mass with an unspecified number of mass reservoirs (i). Each of the mass reservoirs contains a single chemically pure substance that is also present in the chemical composition of the system. The thermodynamic properties of each mass reservoir are known (e.g., $h_{0,i}$ and $s_{0,i}$). The mass flows between the system and each mass reservoir occur through boundary patches that can be modeled as adiabatic and rigid (zero-work); in addition, each boundary patch of this kind is permeable only to the chemical species of the respective mass reservoir. We follow the tradition of calling this type of mass transfer boundary a *semipermeable membrane*, even though the name "selectively permeable membrane" would be more accurate [after all, the membrane is assumed to be fully permeable to the flow of species (i), not half-permeable].

The chemical composition of the system is described by the masses of the individual chemical species found at time t inside the control volume, M_i ($i = 1, 2, \ldots, n$). Each M_i represents the number of kilograms of species (i). The same chemical composition can be expressed in terms of the number of moles of each species, N_i ($i = 1, 2, \ldots, n$): this alternative description will be adopted later in this section. As a unit of quantity of matter (i.e., not as unit of mass), 1 mole is the amount of substance that contains as many elementary entities (molecules, in this case) as there are in 0.012 kg of carbon-12. That special number of entities is *Avogadro's number*, 6.023×10^{23}.

Ruling out the occurrence of any chemical reaction inside the system, we invoke the mass-conservation principle (1.27) for each species and, integrating the resulting equations over the time interval dt, write

$$dM_i = dm_i \qquad (i = 1, 2, \ldots, n) \tag{4.3}$$

We are now in a position to combine the three governing principles represented by eqs. (4.1)–(4.3), and cast the result in the following form:

$$dU = T_0\, dS - \delta W + \sum_{i=1}^{n} (h_{0,i} - T_0 s_{0,i})\, dM_i - T_0\, \delta S_{\text{gen}} \tag{4.4}$$

This particular combination of the laws of thermodynamics is analogous to the statement that led earlier to the lost-work theorem, eq. (3.3).

In the present analysis, we pursue the emerging relationship between system property increment such as dU, dS, and dM_i. With reference to Fig. 4.2, we consider two infinitesimally close equilibrium states (a) and $(a + da)$ represented by the properties $(U, S, M_i,$ etc.) and $(U + dU, S + dS, M_i + dM_i,$ etc.), respectively. The system can evolve from (a) to $(a + da)$ in an infinite number of ways: among the possible paths $(a) \rightarrow (a + da)$, only the reversible ones can be traced as lines (sequences of equilibrium states) in the horizontal plane of Fig. 4.2. Regardless of the irreversibility level of each path, the end states of all paths are represented by the points (a) and $(a + da)$ in the horizontal plane.

In the *reversible* limit, the entropy generation δS_{gen} is zero, whereas the work transfer interaction assumes the form $\delta W_{rev} = P_0 \, dV$ (note the use of P_0 as the pressure measured along the moving boundary, Fig. 4.1). The other modes of reversible work transfer, which are summarized in eq. (1.8′) and Table 1.1, are ruled out by the definition of "simple system." Therefore, eq. (4.4) can be rewritten as

$$dU = T_0 \, dS - P_0 \, dV + \sum_{i=1}^{n} (h_{0,i} - T_0 s_{0,i}) \, dM_i \qquad (4.5)$$

According to the preceding paragraph and Fig. 4.2, eq. (4.5) holds for both reversible and irreversible paths $(a) \rightarrow (a + da)$. Equation (4.5) is simply a relation between the property increments $(dU, dS,$ etc.) that separate two neighboring equilibrium states. Opting for the molar description of the chemical composition of our simple system, we have

$$dU = T_0 \, dS - P_0 \, dV + \sum_{i=1}^{n} \mu_{0,i} \, dN_i \qquad (4.5')$$

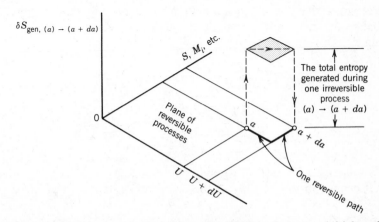

Figure 4.2 Reversible and irreversible paths linking the same initial and final equilibrium states, (a) and $(a + da)$.

where each coefficient $\mu_{0,i}$ is the *chemical potential* of species (i) supplied by the mass reservoir (i), Fig. 4.1. The chemical potential notation is shorthand for the partial molal Gibbs free energy ($\bar{h}_{0,i} - T_0 \bar{s}_{0,i}$), that is, for the per-mole equivalent of the ($\bar{h}_{0,i} - T_0 \bar{s}_{0,i}$) coefficients that appear in eq. (4.5).

The important conclusion at this stage is that the energy of a simple system can be affected independently by changes in its entropy, volume, and chemical composition. The coefficients that appear on the right side of eq. (4.5′) are fixed by the selection (specification) of the reservoirs with which the system interacts during the process $(a) \rightarrow (a + da)$. Each mass reservoir acts as a *chemical potential reservoir* ($\mu_{0,i}$) in the same way that the atmosphere acts as a pressure reservoir (P_0) and temperature reservoir (T_0).

Consider applying the analysis of eqs. (4.1)–(4.5′) to the "external" system that contains the temperature, pressure, and chemical potential reservoirs of Fig. 4.1. We continue to identify the properties of the external system with the subscript "0." This time, however, we assume that the original system (U, S, V, N_i) is characterized by one temperature T, one pressure P, and one chemical potential μ_i for each of the species that leak out of it through the semipermeable membranes. In other words, we assign the values (T, P, μ_i) to the boundary segments crossed by heat, work, and mass interactions. Repeating the analysis, we obtain in place of eq. (4.5′)

$$dU_0 = T \, dS_0 - P \, dV_0 + \sum_{i=1}^{n} \mu_i \, dN_{0,i} \qquad (4.6)$$

Finally, we examine under what conditions the reservoir parameters (T_0, P_0, $\mu_{0,i}$) match the corresponding parameters of the original system (T, P, μ_i). Consider for this purpose the isolated aggregate system indicated in Fig. 4.1: this new system includes the original system and the external system. We identify the properties of the aggregate system with the subscript "Σ," and recognize that the aggregate system's boundary is impermeable, rigid, adiabatic, and zero-work. Analytically, these boundary features amount to writing

$$dN_{\Sigma,i} = dN_i + dN_{0,i} = 0 \qquad (i = 1, 2, \ldots, n)$$

$$dV_\Sigma = dV + dV_0 = 0 \qquad\qquad\qquad (4.7)$$

$$dU_\Sigma = dU + dU_0 = 0$$

Combining eqs. (4.5′)–(4.7) to eliminate $dN_{0,i}$, dV_0, and dU_0, we evaluate the net change in the entropy of the aggregate system[†]:

[†]The net change dS_Σ is also the net entropy generated inside the aggregate system.

$$dS_\Sigma = dS + dS_0$$

$$= \left(\frac{1}{T_0} - \frac{1}{T}\right) dU + \left(\frac{P_0}{T_0} - \frac{P}{T}\right) dV + \sum_{i=1}^{n} \left(\frac{\mu_i}{T} - \frac{\mu_{0,i}}{T_0}\right) dN_i \quad (4.8)$$

In accordance with the entropy maximum principle, eq. (2.51), and Fig. 2.7, dS_Σ is zero if the aggregate system is internally in a state of unconstrained equilibrium: in this case, eq. (4.8) requires, in order,

$$T = T_0$$
$$P = P_0 \quad\quad\quad\quad\quad (4.9)$$
$$\mu_i = \mu_{0,i} \quad (i = 1, 2, \ldots, n)$$

Since the two constituent parts of the aggregate system (the external system plus the original system) are not separated by a rigid, adiabatic, or impermeable partition, eqs. (4.9) constitute the equilibrium conditions for our original simple system and its environment. At equilibrium, the temperature, pressure, and n chemical potentials of an open simple system are exactly the same as their corresponding environmental values.

The aggregate system argument that culminated with the entropy maximum principle and eqs. (4.9) can be used any number of times to show that a simple system is internally in equilibrium if its T, P, and μ_i values are the same throughout the system. This can be shown by first isolating the simple system and dividing it mentally into two subsystems, (A) and (B), that are not separated by a rigid, impermeable, and adiabatic partition. The equilibrium conditions resulting from this first step are $T_A = T_B$, $P_A = P_B$, and $\mu_{A,i} = \mu_{B,i}$. Next, subsystem (A) can be divided into $(A1)$ and $(A2)$ to show that $T_{A1} = T_{A2}$, $P_{A1} = P_{A2}$, $\mu_{A1,i} = \mu_{A2,i}$, etc.

THE FUNDAMENTAL RELATION

In the remainder of this chapter, we consider simple systems in equilibrium with their respective environments. In view of the equilibrium conditions (4.9), the equilibrium version of eq. (4.5′) for an open system is

$$dU = T\,dS - P\,dV + \sum_{i=1}^{n} \mu_i\,dN_i \quad\quad\quad (4.10)$$

The physical meaning of each of the $(n+2)$ terms that appear on the right side of eq. (4.10) is illustrated in Fig. 4.3. The reversible heat transfer interaction δQ_{rev} is associated with the passing of an amount of entropy dS from the temperature reservoir (T) through a boundary whose temperature is equal to T. The reversible work transfer term $\delta W_{\text{rev}} = P\,dV$ takes place as the pressure reservoir and the system match pressures at the moving

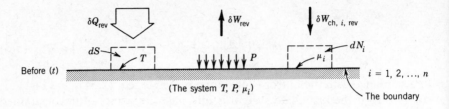

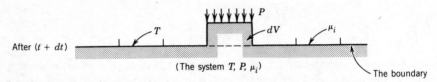

Figure 4.3   The meaning of the terms on the right side of the fundamental relation (4.10): reversible heat transfer, reversible work transfer, and reversible chemical work transfer.

boundary (Fig. 1.5). A new feature of Fig. 4.3 is the *reversible chemical work transfer* $\delta W_{ch,i,rev}$, which is shorthand notation for each of the $\mu_i\, dN_i$ terms of eq. (4.10). The passing of dN_i moles of (i) from the chemical potential reservoir (μ_i) through the system boundary has the effect of increasing the internal energy of the system in the same way as δQ_{rev} and $-\delta W_{rev}$. An important assumption is that the $(n+2)$ reservoirs, (T), (P), and $(\mu_1), \ldots, (\mu_n)$, are large enough so that their respective intensities $(T, P, \mu_1, \ldots, \mu_n)$ remain *constant* throughout the infinitesimal change of state invoked in writing eq. (4.10).

The Energy Representation

The analytical form of the combined law (4.10) proclaims the existence of a function of $(n+2)$ variables:

$$U = U(S, V, N_1, \ldots, N_n) \tag{4.11}$$

where

$$T = \left(\frac{\partial U}{\partial S}\right)_{V, N_1, \ldots, N_n} \tag{4.12}$$

$$-P = \left(\frac{\partial U}{\partial V}\right)_{S, N_1, \ldots, N_n} \tag{4.13}$$

$$\mu_i = \left(\frac{\partial U}{\partial N_i}\right)_{S, V, N_1, \ldots, N_{i-1}, N_{i+1}, \ldots, N_n} \quad (i = 1, \ldots, n) \tag{4.14}$$

Equation (4.11) is recognized as the *fundamental equation* in energy representation, or the energetic fundamental equation. It describes a surface on which every stable equilibrium state is represented by a point, as shown symbolically in the horizontal plane of Fig. 4.2. As first partial derivatives of the $U(S, V, N_1, \ldots, N_n)$ surface, the thermodynamic temperature, pressure, and n chemical potentials are themselves functions of S, V, and N_i:

$$T = T(S, V, N_1, \ldots, N_n) \tag{4.15}$$

$$P = P(S, V, N_1, \ldots, N_n) \tag{4.16}$$

$$\mu_i = \mu_i(S, V, N_1, \ldots, N_n) \quad (i = 1, 2, \ldots, n) \tag{4.17}$$

Equations (4.15)–(4.17) are the $(n + 2)$ *equations of state* of the simple system. Since all the equations of state can be obtained by differentiating the fundamental relation, we note at this point that the information conveyed by eqs. (4.15)–(4.17) is already available in the fundamental relation. Later, we show that the reverse is also true, i.e., that the fundamental relation and the $(n + 2)$ equations of state (taken together) are equivalent.

The Entropy Representation

An alternative formulation of the fundamental equation and the equations of state begins with rewriting eq. (4.10) as

$$dS = \frac{1}{T} dU + \frac{P}{T} dV - \sum_{i=1}^{n} \frac{\mu_i}{T} dN_i \tag{4.18}$$

hence,

$$S = S(U, V, N_1, \ldots, N_n) \tag{4.19}$$

$$\frac{1}{T} = \left(\frac{\partial S}{\partial U}\right)_{V, N_1, \ldots, N_n} \tag{4.20}$$

$$\frac{P}{T} = \left(\frac{\partial S}{\partial V}\right)_{U, N_1, \ldots, N_n} \tag{4.21}$$

$$-\frac{\mu_i}{T} = \left(\frac{\partial S}{\partial N_i}\right)_{U, V, N_1, \ldots, N_{i-1}, N_{i+1}, \ldots, N_n} \quad (i = 1, \ldots, n) \tag{4.22}$$

Equation (4.19) is the fundamental equation in entropy representation, or the so-called entropic fundamental relation. Each of the $(n + 2)$ coefficients $(T^{-1}, P/T, -\mu_i/T)$ is a function of U, V, and N_i: in the manner of eqs. (4.15)–(4.17), these relations constitute the $(n + 2)$ equations of state in entropy representation.

Either in energy or entropy representation, the fundamental equation serves as origin for all the relations that exist between properties at equilibrium. Since only some of these properties and relations can be measured with relative ease (e.g., pressure, temperature, specific heat at constant pressure, and coefficient of thermal expansion), the fundamental equation makes the connection between direct measurements and properties that prove absolutely essential in first-law and second-law calculations (e.g., internal energy, entropy, enthalpy, and exergy). The pivotal role played by the fundamental equation is the reason why it has become fashionable to begin the treatment of simple systems in equilibrium with the *postulate* that a relation of type (4.11) or (4.19) exists. Witness in this regard Callen's Postulate I, which states: "There exist particular states (called equilibrium states) of simple systems that, macroscopically, are characterized completely by the internal energy U, the volume V, and the mole numbers N_1, $N_2, \ldots, N_n$ of chemical components" [2]. Although it certainly would have been expedient to start out by postulating eq. (4.11), in an engineering treatment, I find it much more appropriate to trace the roots of the fundamental relation to the first law and the second law, that is, to eqs. (4.1) and (4.2).

Extensive Versus Intensive Properties

A remarkable feature of the fundamental relation (4.11) is that absolutely all variables that appear in it are *extensive* properties of the system. If, for example, a simple system (Λ) is made up of λ identical "unit'" subsystems in mutual equilibrium, the property X of the unit subsystem is "extensive" if the corresponding property of the system (Λ) obeys the relation $X_\Lambda = \lambda X$. With reference to the properties that appear in eq. (4.11), we write

$$U_\Lambda = \lambda U, \quad S_\Lambda = \lambda S, \quad V_\Lambda = \lambda V, \quad N_{\Lambda,i} = \lambda N_i \qquad (4.23)$$

and, invoking eq. (4.11) itself,

$$U_\Lambda(S_\Lambda, V_\Lambda, N_{\Lambda,i}) = \lambda U(S, V, N_i) \qquad (4.24)$$

In these statements, N_i is used simply as an abbreviation for the sequence $N_1, N_2, \ldots, N_n$.

System (Λ) has its own combined law (4.10):

$$dU_\Lambda = T_\Lambda \, dS_\Lambda - P_\Lambda \, dV_\Lambda + \sum_{i=1}^{n} \mu_{\Lambda,i} \, dN_{\Lambda,i} \qquad (4.25)$$

Therefore, we can ask what relationship exists between the coefficients (T_Λ, P_Λ, $\mu_{\Lambda,i}$) and the corresponding properties of the unit system (T, P, μ_i). By definition, we have

$$T_\Lambda = \left(\frac{\partial U_\Lambda}{\partial S_\Lambda}\right)_{V_\Lambda, N_{\Lambda,i}} = \left[\frac{\partial(\lambda U)}{\partial(\lambda S)}\right]_{\lambda V, \lambda N_i} = \left(\frac{\partial U}{\partial S}\right)_{V, N_i} \qquad (4.26)$$

which, according to eq. (4.12), means

$$T_\Lambda(\lambda S, \lambda V, \lambda N_i) = T(S, V, N_i) \qquad (4.27)$$

The temperature of the aggregate system (Λ) is therefore the same as the temperature of any of its units; in other words, T_Λ is independent of the size of the system (independent of λ). This conclusion is the same as the one described in the last paragraph of the preceding section. The operations listed in eq. (4.26) can be repeated for P_Λ and $\mu_{\Lambda,i}$ to conclude that

$$P_\Lambda(\lambda S, \lambda V, \lambda N_i) = P(S, V, N_i) \qquad (4.28)$$

$$\mu_{\Lambda,i}(\lambda S, \lambda V, \lambda N_i) = \mu_i(S, V, N_i) \qquad (i = 1, 2, \ldots, n) \qquad (4.29)$$

Properties such as T, P, and μ_i are recognized as *intensive properties* or *intensities*. Looking back at the structure of the combined law (4.10), we note that the places of partial differential coefficients are all occupied by intensive properties. The fundamental equation (4.11), on the other hand, is *exclusively* a relation among extensive properties. Finally, each of the $(n + 2)$ equations of state (4.15)–(4.17) contains at least one intensive property.

The Euler Equation

An important relation between all the extensive and intensive properties discussed until now follows directly from eq. (4.24). Differentiating both sides with respect to parameter λ, we obtain

$$\left(\frac{\partial U_\Lambda}{\partial S_\Lambda}\right)_{V_\Lambda, N_{\Lambda,i}} \frac{\partial S_\Lambda}{\partial \lambda} + \left(\frac{\partial U_\Lambda}{\partial V_\Lambda}\right)_{S_\Lambda, N_{\Lambda,i}} \frac{\partial V_\Lambda}{\partial \lambda} + \sum_{i=1}^{n} \left(\frac{\partial U_\Lambda}{\partial N_{\Lambda,i}}\right)_{S_\Lambda, V_\Lambda, N_{\Lambda,j,(j \neq i)}} \frac{\partial N_{\Lambda,i}}{\partial \lambda} = U$$
$$(4.30)$$

or

$$T_\Lambda S - P_\Lambda V + \sum_{i=1}^{n} \mu_{\Lambda,i} N_i = U \qquad (4.30')$$

and, since T, P, and the μ_is are intensive properties [see eqs. (4.27)–(4.29)]:

$$U = TS - PV + \sum_{i=1}^{n} \mu_i N_i \qquad (4.30'')$$

or, in entropy representation:

TABLE 4.2 Alternatives to Reconstructing the Information Content of the Fundamental Relation

I	The fundamental equation, $U = U(S, V, N_i)$ or $S = S(U, V, N_i)$			
II	The Euler equation;		The $(n + 2)$ equations of state	
III	The Euler equation;	Only $(n + 1)$ equations of state;	The Gibbs–Duhem relation [to deduce the $(n + 2)$th equation of state through integration]	One undetermined constant of integration

$$S = \frac{1}{T} U + \frac{P}{T} V - \sum_{i=1}^{n} \frac{\mu_i}{T} N_i \qquad (4.30''')$$

In thermodynamics, eq. $(4.30'')$ is recognized as the *Euler equation* because it is an application of Euler's Theorem on Homogeneous Functions[†] [3] to $U(S, V, N_1, \ldots, N_n)$ as a homogeneous function of degree one. The Euler equation justifies an earlier claim that, taken together, the $(n + 2)$ equations of state are equivalent to the fundamental relation (p. 155). When T, P, and μ_i $(i = 1, \ldots, n)$ are known as functions of only $(S, V, N_1, \ldots, N_n)$, these functions can be substituted into eq. $(4.30'')$ in order to recover the fundamental relation (4.11). This relationship of equivalence is illustrated graphically as alternatives I and II in Table 4.2.

The Gibbs–Duhem Relation

Reviewing the structure developed so far, we note that the combined law (4.10) is a relation among the infinitesimal changes of $(n + 3)$ extensive properties, where n is the number of constituent species in the chemical composition of the system. The Euler equation $(4.30'')$, on the other hand, is a relation among the $(n + 3)$ extensive properties themselves (i.e., not among the changes dU, dS, etc.) and the system's $(n + 2)$ intensive proper-

---

[†]A function $z = f(x, y)$ is said to be homogeneous of degree k in a region $\mathcal{R}$ if the relation

$$f(\lambda x, \lambda y) = \lambda^k f(x, y)$$

holds identically for every point (x, y) that belongs to $\mathcal{R}$ and for every λ from a certain neighborhood of $\lambda = 1$. Note that according to eqs. (4.23) and (4.24), the function $U(S, V, N_i)$ is homogeneous of degree $k = 1$. According to the general theorem, the following relation also holds in $\mathcal{R}$:

$$x \frac{\partial f}{\partial x} + y \frac{\partial f}{\partial y} = k f(x, y)$$

ties. A third relation that follows from the preceding two is obtained by first differentiating the Euler equation:

$$dU = T\,dS + S\,dT - P\,dV - V\,dP + \sum_{i=1}^{n} \mu_i\,dN_i + \sum_{i=1}^{n} N_i\,d\mu_i \quad (4.31)$$

and by subtracting eq. (4.31) from eq. (4.10). The result is the *Gibbs–Duhem relation* [Ref. 1, p. 88] that corresponds to the energy representation of the fundamental relation:

$$S\,dT - V\,dP + \sum_{i=1}^{n} N_i\,d\mu_i = 0 \qquad (4.32)$$

What distinguishes this relation from the fundamental relation (4.10) is that all the infinitesimal changes represent changes in intensive properties. Furthermore, only $(n+1)$ of the total of $(n+2)$ intensities can be changed independently of one another: if the changes in $(n+1)$ intensive properties are specified, the change in the $(n+2)$th intensive property results from the Gibbs–Duhem eq. (4.32). The intensive properties that can be varied independently are the *degrees of freedom* of the system; consequently, a simple system containing a total of n species possesses $(n+1)$ degrees of freedom. For example, in the case of a chemically pure substance (i.e., in a single-species system, $n = 1$), only two of the three intensive properties T, P, and μ can be varied independently. The change in chemical potential is fixed by the changes in temperature and pressure:

$$d\mu = -\frac{S}{N}\,dT + \frac{V}{N}\,dP \qquad (4.33)$$

which also means that if the system evolves at constant T *and* constant P, then its chemical potential does not change. Continuing with the same example, we note that the Euler equation for $n = 1$ reduces to

$$U = TS - PV + \mu N \qquad (4.34)$$

hence,

$$\mu = \frac{G}{N} = \bar{g} \qquad (4.34')$$

where $G = U - TS + PV$ is the Gibbs free energy displayed in Table 4.3 later in this chapter, and where $\bar{g}$ is the molal Gibbs free energy expressed in joules per mole. In conclusion, at constant T and P, the molal Gibbs free energy of the single-species system is constant also.

An alternative version of the Gibbs–Duhem relation is obtained by using instead of eq. (4.10), the fundamental relation in entropy representation, eq. (4.18). The result is

$$U\,d\!\left(\frac{1}{T}\right) + V\,d\!\left(\frac{P}{T}\right) - \sum_{i=1}^{n} N_i\,d\!\left(\frac{\mu_i}{T}\right) = 0 \qquad (4.35)$$

Again, it is worth noting that all the infinitesimal changes refer to intensive properties in the entropy representation, namely, T^{-1}, P/T, and $\mu_1/T, \ldots, \mu_n/T$. It can be shown that eqs. (4.35) and (4.32) are analogous, that is, eq. (4.35) can be derived from eq. (4.32), and vice versa (Problem 4.1).

The Gibbs–Duhem relation concludes the most basic analytical statements that can be derived based on the first law and the second law in the study of simple systems. A graphic summary of the value of these analytical results is presented in Table 4.2. The equivalence between alternatives I and II as conveyers of maximum thermodynamic information was discussed in connection with the Euler equation. Alternative III is made possible by the Gibbs–Duhem relation: when only $(n + 1)$ equations of state are known, the missing equation of state can be obtained by integrating the Gibbs–Duhem relation once. Because of this operation, alternative III differs from either I and II through one undetermined constant of integration.

Example 4.1.   An effective way of illustrating the technique of constructing the fundamental relation on the basis of experimental measurements is to consider the case of a *single-component ideal gas*. We seek the fundamental relation in the entropy representation, $S = S(U, V, N)$, where N is the number of moles of the single chemical species (e.g., O_2, N_2, CO_2, or H_2O). Guided by Table 4.2, we identify the appropriate Euler equation, eq. (4.30'''):

$$S = \left(\frac{1}{T}\right)U + \left(\frac{P}{T}\right)V - \left(\frac{\mu}{T}\right)N \qquad (a)$$

and the equations of state that have been produced by experiments. In the case of ideal gases (either single-component or multi-component), the equations of state are two:

$$PV = N\bar{R}T \qquad (b)$$

$$U = U(T) \qquad (c)$$

The first is the Clapeyron relation in which $\bar{R}$ is the universal ideal gas constant:

$$\bar{R} = 8.314\ \text{J/mol K} \qquad (d)$$

while the second is the result of calorimetric measurements (e.g., Joule's free-expansion experiment). In sufficiently narrow temperature intervals, the second equation of state is well approximated by the line

$$U = N\bar{c}_v T \qquad (e)$$

where $\bar{c}_v$ [J/mol K] is a constant. Figure 4.4 shows that the constant-$\bar{c}_v$ approximation is particularly appropriate for monatomic gases. Certain segments of the

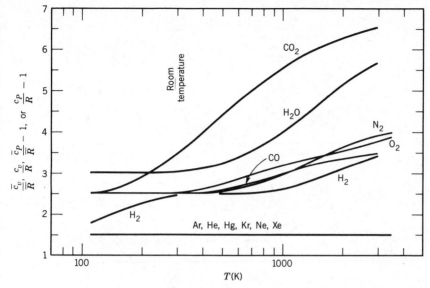

Figure 4.4   The c_v/R ratio of several ideal gases as a function of temperature.

engineering-thermodynamics literature use the name "perfect gas" for the ideal gas that is represented by eqs. (b) and (e), with $\bar{c}_v$ = constant. The ideal gas whose specific heats (c_v and c_p) are functions of temperature, i.e., the gas represented by eqs. (b) and (c), is called "semiperfect."

The two equations of state, (e) and (b), deliver the first two coefficients that appear on the right side of eq. (a):

$$S = \left(\frac{N\bar{c}_v}{U}\right)U + \left(\frac{N\bar{R}}{V}\right)V - \left(\frac{\mu}{T}\right)N \tag{a'}$$

The third equation of state, for μ/T, follows from the Gibbs–Duhem relation in entropy representation, eq. (4.35):

$$N\,d\left(\frac{\mu}{T}\right) = U\,d\left(\frac{1}{T}\right) + V\,d\left(\frac{P}{T}\right) = U\,d\left(\frac{N\bar{c}_v}{U}\right) + V\,d\left(\frac{N\bar{R}}{V}\right) \tag{f}$$

Integrating from a reference condition where $U = U_0$ and $V = V_0$, yields the third equation of state:

$$\frac{\mu}{T} = -\bar{c}_v\,\ln\frac{U}{U_0} - \bar{R}\,\ln\frac{V}{V_0} + \bar{C} \tag{g}$$

where $\bar{C}$ is an undetermined constant of integration (option III, Table 4.2). Combining this last result with eq. (a'), we obtain the entropic fundamental relation:

$$S(U, V, N) = S_0 + N\bar{c}_v\,\ln\frac{U}{U_0} + N\bar{R}\,\ln\frac{V}{V_0} \tag{h}$$

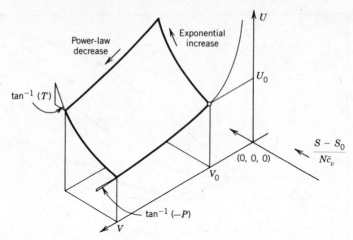

Figure 4.5 The surface representing the fundamental relation for a single-component ideal gas.

where $S_0 = N(\bar{c}_v + \bar{R} - \bar{C})$ is another name for the undetermined constant. The corresponding relation in the energy representation is

$$U(S, V, N) = U_0 \left(\frac{V}{V_0}\right)^{-\bar{R}/\bar{c}_v} \exp\left(\frac{S - S_0}{N\bar{c}_v}\right) \tag{i}$$

Either in entropy representation or energy representation, the fundamental relation constitutes a surface in the four-dimensional space (U, S, V, N). As an extensive property, however, the entropy scales with the size of the system (N); therefore, the fundamental relation can be visualized as a surface in the three-dimensional space $(U, S/N, V)$. Figure 4.5 shows the shape of the surface represented by eqs. (h) or (i) in the reduced three-dimensional space; the same figure shows also the geometric significance of T and $(-P)$.

LEGENDRE TRANSFORMS

Either in energetic or entropic representation, the variables that appear in the fundamental relation are all extensive properties of the simple system, eqs. (4.11) and (4.19). The intensive properties $(T, P, \mu_1, \ldots, \mu_n)$ can be derived from the fundamental relation in the same way that the various slopes of the internal energy hypersurface $U(S, V, N_1, \ldots, N_n)$ can be measured geometrically at any given point in the $(n + 3)$-dimensional frame $(U, S, V, N_1, \ldots, N_n)$. The preponderance of chemical- and mechanical-engineering applications in which systems undergo changes while in thermal (T), mechanical (P), or thermomechanical $(T$ and $P)$ equilibrium with the ambient reservoir of constant temperature and pressure has created a demand for alternatives to the all-extensive presentation of thermodynamic relations, i.e., for alternative fundamental relations in which T, or P, or T and P together appear as independent variables. The demand for such alternatives is stimulated also by the fact that temperature and pressure are relatively easy to measure in the laboratory.

We focus on three particular alternatives to the internal energy fundamental relation (4.11), namely, the *enthalpy* fundamental relation:

$$H = H(S, P, N_1, N_2, \ldots, N_n) \tag{4.36}$$

the *Helmholtz free-energy* fundamental relation:

$$F = F(T, V, N_1, N_2, \ldots, N_n) \tag{4.37}$$

and, finally, the *Gibbs free-energy* fundamental relation:

$$G = G(T, P, N_1, N_2, \ldots, N_n) \tag{4.38}$$

Note the appearance of T and P, alone or together, as coordinates of the thermodynamic space in which these new fundamental relations are defined. The three alternatives listed above are certainly not the only ones that can be constructed. Of special interest in the present treatment is the equivalence between any two such alternatives, in other words, the *transformation* that guarantees that *no information is lost from the fundamental relation* as we travel from the $U(S, V, N_1, \ldots, N_n)$ representation to, for example, $H(S, P, N_1, \ldots, N_n)$ and vice versa.

The transformation is best understood in terms of its geometrical implications [4, 5]. Consider the curve

$$y = y(x) \tag{4.39}$$

which in the x–y frame of Fig. 4.6 is represented by a certain alignment of points (x, y). In what follows, we regard $y = y(x)$ as the two-dimensional version of the fundamental relation $U = U(S, V, N_1, \ldots, N_n)$; therefore, the wish to replace one of the extensive properties $(S, V, N_1, \ldots, N_n)$ with an intensive property (a slope of the U surface), reduces in this case to the problem of replacing x with dy/dx as the independent variable in a new analytical expression that represents *the same* curve. We seek the new function

$$\eta = \eta(\zeta) \tag{4.40}$$

where

$$\zeta = \frac{dy}{dx} \tag{4.41}$$

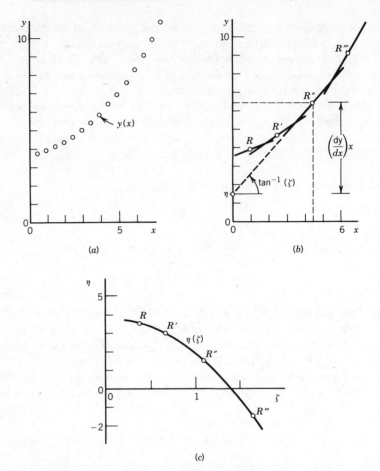

Figure 4.6 Representing a curve as a sequence of points (a), or as the envelope to a family of lines (b), leading to the transformed curve (c).

such that to each (x, y) point corresponds only one point (ζ, η), and vice versa. The actual analytical form of $\eta(\zeta)$ follows from the observation that the curve that originally was drawn as a sequence of special points $[x, y(x)]$ can also be drawn as the envelope of the straight lines that have the slope (dy/dx) while passing through each special point $[x, y(x)]$. In other words, the curve can be drawn by means of all its tangent lines, as shown in the second x–y frame of Fig. 4.6b.

To graphically "describe" the curve means to vary the cut (intercept) on the ordinate (η) in a certain way relative to the slope of each line (ζ), so that the original curve is the one that emerges as the envelope of all the tangent lines. This special relationship between η and ζ follows from the condition that each straight line that passes through the point $[x, y(x)]$ must be tangent to the original $y = y(x)$ curve, Fig. 4.6b:

$$y = \left(\frac{dy}{dx}\right)x + \eta \qquad (4.42)$$

or, using the definition (4.41),

$$\eta(\zeta) = y(x) - \zeta x \qquad (4.43)$$

Eliminating x and y between eqs. (4.39), (4.41), and (4.43) yields the new function $\eta = \eta(\zeta)$, which is shown in Fig. 4.6c. The elimination is possible as long as the new variable ζ is not independent of x, i.e., when

$$\frac{d\zeta}{dx} \neq 0 \qquad \text{or} \qquad \frac{d^2 y}{dx^2} \neq 0 \qquad (4.44)$$

If the original function is a straight line, $d^2 y/dx^2 = 0$, the tangents of Fig. 4.6b would fall on top of each other and the entire $[x, y(x)]$ curve would collapse into a single point in the $\zeta-\eta$ plane: in this extreme, to each (ζ, η) would correspond more than one point in the $x-y$ plane, and the $\eta(\zeta)$ function defined in eq. (4.40) does not exist. It turns out that in simple systems, the condition (4.44) is satisfied as long as the equilibrium is stable [e.g. eq. (6.3), p. 239].

The function $\eta(\zeta)$ is the *Legendre transform* of $y(x)$. The scale drawings of Fig. 4.6 stress that although graphically (and, also, analytically) $y(x)$ and $\eta(\zeta)$ can be quite different, they do represent the same curve. Said another way, the original relation and the Legendre transform contain the same information. Analytically, the transformation consists of using eqs. (4.41) and (4.43) for the purpose of eliminating x and y as variables. One distinguishing feature of this transformation is that by applying it one more time to the Legendre transform $\eta(\zeta)$, we recover the original function $y(x)$. For example, if $\hat{\eta}(\hat{\zeta})$ is the Legendre transform of $\eta(\zeta)$, then eqs. (4.41) and (4.43) require

$$\hat{\zeta} = \frac{d\eta}{d\zeta} \qquad (4.45)$$

$$\hat{\eta} = \eta - \hat{\zeta}\zeta \qquad (4.46)$$

However, according to eq. (4.43), the derivative $d\eta/d\zeta$ is equal to $-x$; therefore, eqs. (4.45) and (4.46) become

$$\hat{\zeta} = -x \qquad (4.45')$$

$$\hat{\eta} = y \qquad (4.46')$$

Except for the minus sign in eq. (4.45'), the successive application of two Legendre transformations leads back to the original function $y(x)$. We can substitute eqs. (4.45′) and (4.46′) into (4.45) and (4.46) and call the resulting set *the inverse Legendre transformation*:

$$-x = \frac{d\eta}{d\zeta} \tag{4.47}$$

$$y = \eta - (-x)\zeta \tag{4.48}$$

The symmetry between the inverse transformation listed above and the direct transformation of eqs. (4.41) and (4.43) is evident. If $d^2\eta/d\zeta^2 \neq 0$, we can eliminate η and ζ between eqs. (4.40), (4.47), and (4.48) to recover the original relation $y = y(x)$.

The preceding discussion can easily be extended to a function of $(n+2)$ variables:

$$y = y(x_1, x_2, \ldots, x_{n+1}, x_{n+2}) \tag{4.49}$$

such as the energetic fundamental relation $U = U(S, V, N_1, \ldots, N_n)$. The y function has $(n+2)$ first partial derivatives:

$$\zeta_k = \frac{\partial y}{\partial x_k} = \zeta_k(x_1, x_2, \ldots, x_{n+1}, x_{n+2}) \qquad (k = 1, 2, \ldots, n+2) \tag{4.50}$$

while each of these derivatives is a function of the same $(n+2)$ variables x_1, $x_2, \ldots, x_{n+2}$. Based on the argument presented in Fig. 4.6, the effect of x_k on y can be presented as a curve [i.e., an alignment of points according to eq. (4.49)] in the $x_k - y$ plane, or as the envelope of all the lines drawn tangent to that curve. Let $\eta^{(1)}$ be the y intercept that corresponds to the tangent of slope ζ_k; in place of eq. (4.43), we write

$$\eta^{(1)}(x_1, \ldots, x_{k-1}, \zeta_k, x_{k+1}, \ldots, x_{n+2}) = y - \zeta_k x_k \tag{4.51}$$

The function $\eta^{(1)}$ is the *first Legendre transform* of the original function y with respect to x_k [6]. The superscript "(1)" indicates that only *one* of the original $(n+2)$ variables was replaced by the corresponding partial derivative. Obviously, there are $(n+2)$ first Legendre transforms, as ζ_k can replace any single one of the original variables $x_1, x_2, \ldots, x_{n+2}$. In the examples of "first Legendre transforms" that conclude this section, Table 4.3, we see that the role of ζ_k is played once by $(-P)$ in the construction of the enthalpy fundamental relation, and another time by T in the Helmholtz fundamental relation. Finally, we note that the first Legendre transform $\eta^{(1)}$ emerges as a function of $(x_1, \ldots, x_{k-1}, \zeta_k, x_{k+1}, \ldots, x_{n+2})$ by eliminating y and x_k between eqs. (4.49)–(4.51): this elimination is possible if $\partial^2 y/\partial x_k^2 \neq 0$.

If we are interested in replacing not one x_k by its corresponding slope ζ_k, but two, then we can simply apply the first Legendre transformation once more to the $\eta^{(1)}$ function of eq. (4.51). We obtain

$$\eta^{(2)}(x_1, \ldots, \zeta_k, \ldots, \zeta_l, \ldots, x_{n+2}) = y - \zeta_k x_k - \zeta_l x_l \qquad (4.52)$$

where ζ_l is the slope $\partial y/\partial x_l$ that now occupies the place of the second variable (x_l) that had to be eliminated. Considering the relationship between eqs. (4.52) and (4.51), the superscript "(2)" is shorthand notation for $[(\ )^{(1)}]^{(1)}$, and stresses that *two* of the original variables have been replaced by corresponding partial derivatives. Following Beegle et al. [6], we refer to the function $\eta^{(2)}$ as the *second Legendre transform* of y with respect to x_k and x_l. Depending on the selection of the (x_k, x_l) pair, the original function y can have a maximum of $(n + 1)(n + 2)/2$ second Legendre transforms. Table 4.3 shows only one example of second Legendre transform, namely, the construction of the Gibbs free-energy fundamental relation, in which the chosen (ζ_k, ζ_l) pair is $(T, -P)$. Legendre transforms of higher order can be developed by continuing the procedure represented by the step from eq. (4.51) to eq. (4.52) (Problem 4.4).

In conclusion, the fundamental relation can be transformed without losing any information by replacing one or more of its original arguments with the corresponding first partial derivatives. In the case of first Legendre transforms, for example, the procedure consists of writing three equations: eq. (4.49), eq. (4.50) once for the lone variable x_k that is to be replaced, and, finally, eq. (4.51). Eliminating y and x_k between these three equations yields, in principle, the first Legendre transform $\eta^{(1)}(x_1, \ldots, \zeta_k, \ldots, x_{n+2})$ listed on the left side of eq. (4.51).

In Table 4.3, the enthalpy and Helmholtz free-energy fundamental relations are developed as first Legendre transforms of $U = U(S, V, N_1, \ldots, N_n)$. The Gibbs free-energy fundamental relation can be viewed as a second Legendre transform of $U = U(S, V, N_1, \ldots, N_n)$, as done in Table 4.3, or as a first Legendre transform of either $H = H(S, P, N_1, \ldots, N_n)$ or $F = F(T, V, N_1, \ldots, N_n)$. The Gibbs free-energy fundamental relation $G = G(T, P, N_1, \ldots, N_n)$ is particularly useful in view of the presence of easily measurable properties (T, P) as arguments.

Example 4.2. As an application of the steps outlined in Table 4.3, consider the task of transforming the fundamental relation of a single-component ideal gas from the energy representation $U(S, V, N)$, Fig. 4.5, to the enthalpy representation $H(S, P, N)$. The analysis consists of writing *three* equations, namely, the original fundamental relation deduced in Example 4.1:

$$U(S, V, N) = U_0 \left(\frac{V}{V_0}\right)^{-\bar{R}/\bar{c}_v} \exp\left(\frac{S - S_0}{N\bar{c}_v}\right) \qquad (a)$$

the partial derivative $(-P)$ that is to replace the extensive variable (V) in the new representation:

$$-P = \left(\frac{\partial U}{\partial V}\right)_{S,N} = -\frac{\bar{R}}{\bar{c}_v} V_0^{\bar{R}/\bar{c}_v} U_0 V^{-\bar{c}_P/\bar{c}_v} \exp\left(\frac{S - S_0}{N\bar{c}_v}\right) \qquad (b)$$

and the Legendre transform definition:

TABLE 4.3 Legendre Transforms of the Fundamental Relation: the Enthalpy, the Helmholtz Free-Energy, and the Gibbs Free-Energy Fundamental Relations ($N_i = N_1, N_2, \ldots, N_n$)

I	Original relation, $y(x_1, \ldots, x_{n+2})$	(*) $U = U(S, V, N_i)$
II	Extensive properties (x_k, x_l), to be eliminated	V
III	Intensive properties (ξ_k, ξ_l), used as replacements; corresponding equation(s) of state	$-P$ (*) $P = P(S, V, N_i)$
IV	First Legendre transform, $\eta^{(1)}$, fundamental relation, name, and way of deriving it	(*) $H = U - (-P)V = U + PV = H(S, V, N_i)$ $H = H(S, P, N_i)$ Enthalpy Eliminate U and V between the equations marked with (*) above
V	Second Legendre transform, $\eta^{(2)}$, fundamental relation, name, and way of deriving it	
VI	Differential form of the transform; in place of the combined law (4.10)	$dH = T\,dS + V\,dP + \sum_{i=1}^{n} \mu_i\, dN_i$
VII	The $(n+2)$ equations of state	$5T = \left(\dfrac{\partial H}{\partial S}\right)_{P, N_i} = T(S, P, N_i)$ $V = \left(\dfrac{\partial H}{\partial P}\right)_{S, N_i} = V(S, P, N_i)$ $\mu_i = \left(\dfrac{\partial H}{\partial N_i}\right)_{S, P, N_{j(j\neq i)}} = \mu_i(S, P, N_i)$ $(i = 1, 2, \ldots, n)$

I (*) $U = U(S, V, N_i)$ (*) $U = U(S, V, N_i)$

II S S, V

III T $T, -P$

(*) $T = T(S, V, N_i)$ (*) $T = T(S, V, N_i)$

(*) $P = P(S, V, N_i)$

IV (*) $F = U - TS = F(S, V, N_i)$

$F = F(T, V, N_i)$

Helmholtz free energy

Eliminate U and S between the equations marked with (*) above

V (*) $G = U - TS - (-P)V = U - TS + PV = G(S, V, N_i)$

$G = G(T, P, N_i)$

Gibbs free energy

Eliminate U, S, and V between the equations marked with (*) above

VI

$$dF = -S\,dT - P\,dV + \sum_{i=1}^{n} \mu_i\,dN_i$$

$$dG = -S\,dT + V\,dP + \sum_{i=1}^{n} \mu_i\,dN_i$$

VII

$$S = -\left(\frac{\partial F}{\partial T}\right)_{V,N_i} = S(T, V, N_i)$$

$$S = -\left(\frac{\partial G}{\partial T}\right)_{P,N_i} = S(T, P, N_i)$$

$$P = -\left(\frac{\partial F}{\partial V}\right)_{T,N_i} = P(T, V, N_i)$$

$$V = \left(\frac{\partial G}{\partial P}\right)_{T,N_i} = V(T, P, N_i)$$

$$\mu_i = \left(\frac{\partial F}{\partial N_i}\right)_{T,V,N_j(j\neq i)} = \mu_i(T, V, N_i)$$

$$\mu_i = \left(\frac{\partial G}{\partial N_i}\right)_{T,P,N_j(j\neq i)} = \mu_i(T, P, N_i)$$

$(i = 1, 2, \ldots, n)$ $(i = 1, 2, \ldots, n)$

$$H = U + PV = \text{etc.} \ldots = H(S, V, N) \tag{c}$$

Note that the three equations (a)–(c) correspond to eqs. (4.49)–(4.51), or to steps I, III, and IV in Table 4.3, where the corresponding equations are marked with (∗). The elimination of U and V between the three equations listed above yields

$$H(S, P, N) = H_0 \left(\frac{P}{P_0}\right)^{\bar{R}/\bar{c}_P} \exp\left(\frac{S - S_0}{N\bar{c}_P}\right) \tag{d}$$

where $\bar{c}_P = \bar{c}_v + \bar{R}$. The new constants P_0 and H_0 are defined as

$$P_0 = \frac{U_0}{V_0} \quad \text{and} \quad H_0 = U_0\left[\left(\frac{\bar{c}_v}{\bar{R}}\right)^{\bar{R}/\bar{c}_P} + \left(\frac{\bar{R}}{\bar{c}_v}\right)^{\bar{c}_v/\bar{c}_P}\right] \tag{e}$$

It is shown in Fig. 4.7 that the enthalpy fundamental relation $H(S, P, N)$ represents a surface whose main features can be drawn in the reduced three-dimensional space $(H, S/N, P)$. Similar surfaces can be constructed for the other representations of the fundamental equation of a single-component ideal gas (Problem 4.3). Examined side by side, Figs. 4.5 and 4.7 show the geometric meaning of the first Legendre transformation: one of the slopes of the original fundamental relation ($-P$ in Fig. 4.5) replaces its corresponding directional argument (V in Fig. 4.5), and becomes a new argument in the transformed relation (Fig. 4.7). The equations of state for T, V, and μ as functions of (S, P, N) follow immediately from eq. (d):

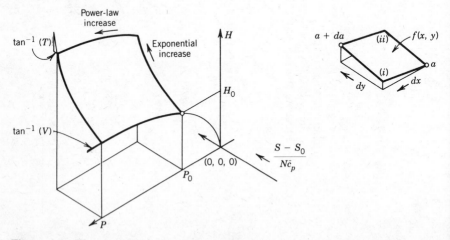

Figure 4.7 The enthalpy fundamental surface for a single-component ideal gas, or the first Legendre transform of the surface of Fig. 4.5.

$$T(S, P, N) = \frac{H_0}{N\bar{c}_P} \left(\frac{P}{P_0}\right)^{\bar{R}/\bar{c}_P} \exp\left(\frac{S - S_0}{N\bar{c}_P}\right) \qquad \text{(f)}$$

$$V(S, P, N) = \frac{H_0\bar{R}}{P_0\bar{c}_P} \left(\frac{P}{P_0}\right)^{-\bar{c}_v/\bar{c}_P} \exp\left(\frac{S - S_0}{N\bar{c}_P}\right) \qquad \text{(g)}$$

$$\mu(S, P, N) = -\frac{H_0}{N} \left(\frac{S - S_0}{N\bar{c}_P}\right)\left(\frac{P}{P_0}\right)^{\bar{R}/\bar{c}_P} \exp\left(\frac{S - S_0}{N\bar{c}_P}\right) \qquad \text{(h)}$$

Regarding N as constant, we can identify T and V geometrically as the S-slope and the P-slope of the enthalpy fundamental relation (Fig. 4.7). It is not difficult to verify that the equations of state (f)–(h) are consistent with the original equations of state that were used to construct the energy fundamental relation in Example 4.1. In other words, eqs. (f)–(h) can be used to recreate eqs. (b), (c), and (g) of Example 4.1.

RELATIONS BETWEEN PROPERTIES AND THEIR DERIVATIVES

In Examples 4.1 and 4.2, we invoked the limiting case of ideal gas behavior to show how the fundamental relation and one of its Legendre transforms can serve as "summary" for the empirical information stored in the equations of state. The practical function of such a summary is that it allows us to deduce from it the changes in properties that, generally, are not measured in the laboratory (e.g., entropy). We were able to derive the fundamental relation for an ideal gas in closed form only because the ideal gas equations of state are very simple. For most other substances, the equations of state suggested by measurements are sufficiently complicated to rule out the possibility of deriving the fundamental relation analytically. In these cases, the challenge remains the same, namely, to calculate the changes in certain properties while measuring the least number of some of the most accessible properties. The connection between the two groups of properties was established a long time ago through a long list of "relations" that, ultimately, have their origin in the geometric features of the surface represented by the fundamental relation. To provide a bird's-eye view of these relations is the objective of this section.

Maxwell's Relations

The calculus theorem on the change of order of differentiation in functions of two or more variables states that, given a function $f(x, y)$, the mixed derivatives

$$\frac{\partial^2 f}{\partial x \, \partial y} \quad \text{and} \quad \frac{\partial^2 f}{\partial y \, \partial x} \qquad (4.53)$$

are equal at a point (x_0, y_0) if they are continuous. This is also known as the theorem on the interchangeability of mixed derivatives [Ref. 3, p. 446]. The geometric meaning of the equality of $\partial^2 f / \partial x \, \partial y$ and $\partial^2 f / \partial y \, \partial x$ is illustrated in the upper right figure of Fig. 4.7. When the equality holds, we can move from one point (a) on the surface and arrive at the *same* neighboring point $(a + da)$ on the surface, either by stepping first in x and second in y [option (i)], or by stepping first in y and second in x [option (ii)]. If we recall that in Fig. 4.5, the role of first derivatives of the fundamental surface is played by T and $-P$, it becomes clear that the above mixed derivatives theorem implies the existence of an equality between the first derivatives of T and $-P$.

The energy fundamental relation $U = U(S, V, N_1, \ldots, N_n)$ presents U as a function of $(n + 2)$ variables. We can form a certain number of relations of the type

$$\frac{\partial^2 f}{\partial x \, \partial y} = \frac{\partial^2 f}{\partial y \, \partial x} \tag{4.54}$$

by replacing f with U, and the pair (x, y) with any pair of the $(n + 2)$ arguments of the fundamental relation. The U function has a total of $(n + 1)(n + 2)$ mixed derivatives, hence, only $(n + 1)(n + 2)/2$ equations of type (4.54). In the case of a single-component system $(n = 1)$, for example, we obtain the three relations listed in the first block of Table 4.4. These

TABLE 4.4   Maxwell's Relations for a Single-Component System

$dU = T \, dS - P \, dV + \mu \, dN$	$dH = T \, dS + V \, dP + \mu \, dN$
$\left(\dfrac{\partial T}{\partial V}\right)_{S,N} = -\left(\dfrac{\partial P}{\partial S}\right)_{V,N}$	$\left(\dfrac{\partial T}{\partial P}\right)_{S,N} = \left(\dfrac{\partial V}{\partial S}\right)_{P,N}$
$\left(\dfrac{\partial T}{\partial N}\right)_{S,V} = \left(\dfrac{\partial \mu}{\partial S}\right)_{V,N}$	$\left(\dfrac{\partial T}{\partial N}\right)_{S,P} = \left(\dfrac{\partial \mu}{\partial S}\right)_{P,N}$
$-\left(\dfrac{\partial P}{\partial N}\right)_{S,V} = \left(\dfrac{\partial \mu}{\partial V}\right)_{S,N}$	$\left(\dfrac{\partial V}{\partial N}\right)_{S,P} = \left(\dfrac{\partial \mu}{\partial P}\right)_{S,N}$
$dF = -S \, dT - P \, dV + \mu \, dN$	$dG = -S \, dT + V \, dP + \mu \, dN$
$\left(\dfrac{\partial S}{\partial V}\right)_{T,N} = \left(\dfrac{\partial P}{\partial T}\right)_{V,N}$	$-\left(\dfrac{\partial S}{\partial P}\right)_{T,N} = \left(\dfrac{\partial V}{\partial T}\right)_{P,N}$
$-\left(\dfrac{\partial S}{\partial N}\right)_{T,V} = \left(\dfrac{\partial \mu}{\partial T}\right)_{V,N}$	$-\left(\dfrac{\partial S}{\partial N}\right)_{T,P} = \left(\dfrac{\partial \mu}{\partial T}\right)_{P,N}$
$-\left(\dfrac{\partial P}{\partial N}\right)_{T,V} = \left(\dfrac{\partial \mu}{\partial V}\right)_{T,N}$	$\left(\dfrac{\partial V}{\partial N}\right)_{T,P} = \left(\dfrac{\partial \mu}{\partial P}\right)_{T,N}$

relations are headed by the differential form of the fundamental relation, which shows that the intensities T, $-P$, and μ play the role of first derivatives. Similar three-equation sets are obtained for the Legendre transforms of the fundamental relation, as shown under dH, dF, and dG in Table 4.4.

As a second example, we list the mixed derivative equalities for a *closed system*, that is, for a simple system whose chemical composition (N_1, N_2, ..., N_n) is fixed. Referring once again to the analysis centered on Fig. 4.1, the closed system is in a state of *restricted equilibrium* with its environment, because out of the total of ($n + 2$) intensities of the system, only T and P are the same throughout the system and the environment. Writing $dN_1 = 0, \ldots, dN_n = 0$, the differential forms of the U, H, F, and G fundamental equations reduce to (Table 4.3):

$$dU = T\,dS - P\,dV \tag{4.55u}$$

$$dH = T\,dS + V\,dP \tag{4.55h}$$

$$dF = -S\,dT - P\,dV \tag{4.55f}$$

$$dG = -S\,dT + V\,dP \tag{4.55g}$$

Note that in a state of restricted equilibrium (zero mass transfer), the system possesses only two modes of reversible energy transfer interactions, $T\,dS$ and $-P\,dV$, eq. (4.55u). Therefore, we can write only one equality of mixed derivatives for each of the four differential forms listed above:

$$\left(\frac{\partial T}{\partial V}\right)_S = -\left(\frac{\partial P}{\partial S}\right)_V \tag{4.56u}$$

$$\left(\frac{\partial T}{\partial P}\right)_S = \left(\frac{\partial V}{\partial S}\right)_P \tag{4.56h}$$

$$\left(\frac{\partial S}{\partial V}\right)_T = \left(\frac{\partial P}{\partial T}\right)_V \tag{4.56f}$$

$$-\left(\frac{\partial S}{\partial P}\right)_T = \left(\frac{\partial V}{\partial T}\right)_P \tag{4.56g}$$

where, for the sake of brevity, the constant mole numbers (N_1, N_2, ..., N_n) have been omitted from the usual subscript notation.

The four relations (4.56) were identified by Maxwell [7] based on a very interesting geometric construction involving the intersection of two $T =$ constant lines with two $S =$ constant lines in the (P, V) plane, that is, without using the fundamental relation differentials and, certainly, without invoking the calculus theorem on the interchangeability of mixed derivatives. In a section-ending footnote, Maxwell also pointed out that his four geometric equalities "may be concisely expressed in the language of differential calculus" as eqs. (4.56u)–(4.56g). Among these, the last two relations are especially useful because, as shown next, their right-hand terms can be calculated directly from volume expansivity and isothermal compressibility measurements. Equations (4.56) and their more numerous counterparts for systems with more than two modes of reversible energy transfer (e.g., Table 4.4) are appropriately recognized as *Maxwell's relations*. The reader should note however that Maxwell did not discover eq. (4.54), i.e., that the impression conveyed by some loosely worded contemporary treatises is historically inaccurate.

Relations Measured During Special Processes

The empirical information on how the system properties vary as the system shifts to a new state of equilibrium is stored in a series of relations whose analytical outlook is descriptive of the experiments that generated them. We continue to focus on the behavior of a system whose equilibrium with the environment is restricted to the equality of temperature and pressure only. Through direct (P, V, T) measurements that go back to the prethermodynamics era of thermometry and calorimetry, the most accessible relation has been the one between pressure, temperature, and volume, or, on a per-unit-mass basis, the $P = P(v, T)$ surface. A separate class of relations has been developed by combining the changes measured during special processes with the requirements posed by the first law and the second law. Figure 4.8 shows a graphic summary of the special processes that support the relations presented below.

(a) Reversible Heating at Constant Volume (Fig. 4.8a). By measuring the heat transfer δQ_{rev} and the temperature change dT, it is possible to calculate and record *the specific heat at constant volume*:

$$c_v = \left(\frac{\delta Q_{rev}}{m\,dT}\right)_v \quad \text{or} \quad \bar{c}_v = \left(\frac{\delta Q_{rev}}{N\,dT}\right)_v \tag{4.57}$$

which, in general, depends on both T and v (or T and P, etc.). Invoking the first law and the second law for this infinitesimal change, $dU = \delta Q_{rev}$ and $\delta Q_{rev} = T\,dS$, the c_v definition (4.57) translates into two important relations:

$$c_v = \left(\frac{\partial u}{\partial T}\right)_v \quad \text{and} \quad c_v = T\left(\frac{\partial s}{\partial T}\right)_v \tag{4.58}$$

With c_v known as a function of T and v, in eqs. (4.58), we recognize two relations between, on the one hand, properties and changes in properties that are measured directly, and, on the other, changes in internal energy and entropy. The specific-heat relation is particularly simple in ideal gases, where c_v is only a function of temperature or, in sufficiently narrow temperature intervals, a constant (Fig. 4.4).

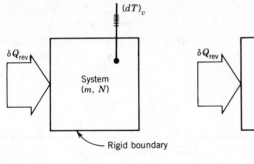

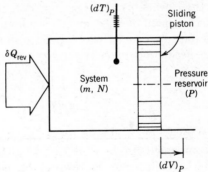

(a) Reversible heating at constant volume

(b) Reversible heating at constant pressure

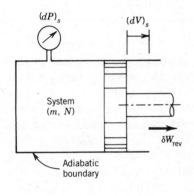

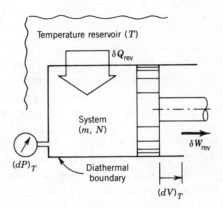

(c) Reversible and adiabatic change of volume

(d) Reversible and isothermal change of volume

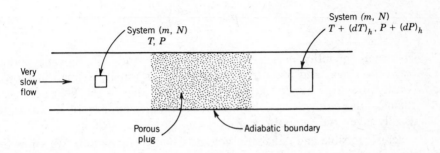

(e) Constant—enthalpy expansion

Figure 4.8 Five special processes for the direct measurement of relations between properties.

(b) Reversible Heating at Constant Pressure (Fig. 4.8b). Similarly, the measurement of δQ_{rev} and dT while the pressure of the system is maintained constant by contact with a pressure reservoir (P) permits the calculation of *the specific heat at constant pressure*:

$$c_P = \left(\frac{\delta Q_{rev}}{m \, dT}\right)_P \quad \text{or} \quad \bar{c}_P = \left(\frac{\delta Q_{rev}}{N \, dT}\right) \qquad (4.59)$$

The reversible heating process executed by the closed system of Fig. 4.8b is accompanied also by work transfer, $\delta W_{rev} = P \, dV$. At constant pressure, the first law allows us to rewrite the heat transfer as $\delta Q_{rev} = dU + P \, dV = (dH)_P$. In addition, the second law states that δQ_{rev} is equal to $T \, dS$; therefore, the per-unit-mass c_P definition (4.59) can be restated as

$$c_P = \left(\frac{\partial h}{\partial T}\right)_P \quad \text{or} \quad c_P = T\left(\frac{\partial s}{\partial T}\right)_P \qquad (4.60)$$

Assuming that the function $c_P(T, P)$ is known from experiment, eqs. (4.60) provide the means to calculate the changes in enthalpy and entropy associated with a change in temperature when the pressure is held constant. In the case of ideal gas behavior, Fig. 4.4 and Mayer's equation (p. 37) show that c_P is at most a function of temperature. The specific-heat terminology and methodology, like the state principle (p. 30), are reminders of how much of the present-day thermodynamics is the contribution of the calorimetrists and caloric theorists of the late 1700s and the 1800s—this, in spite of the exuberance displayed even today by some who proclaim the death of the caloric theory. The concept of "specific heat," as distinct from that of "quantity of heat," was introduced in 1772 by Wilcke [8].

An important measurement that is made possible by the constant-pressure process of Fig. 4.8b is *the volumetric coefficient of thermal expansion*, or *the volume expansivity*:

$$\beta = \frac{1}{v} \left(\frac{\partial v}{\partial T}\right)_P \qquad (4.61)$$

The same quantity occurs frequently in the study of buoyancy-induced convection heat and mass transfer, where it is usually defined in terms of density, $\beta = -\rho^{-1}(\partial \rho / \partial T)_P$ [9]. Note further that in the field of physicochemical thermodynamics, the volumetric coefficient of thermal expansion is usually given the symbol α.

(c) Reversible and Adiabatic Volume Change (Fig. 4.8c). We know from the second law (especially from the discussion presented on pp. 80–87) that during a reversible and adiabatic expansion, the entropy remains constant while measurable properties such as P and v change. Viewing the entropy as a function of P and v,

$$ds = \left(\frac{\partial s}{\partial P}\right)_v dP + \left(\frac{\partial s}{\partial v}\right)_P dv \tag{4.62}$$

and invoking the second law $(ds = 0)$, we obtain

$$0 = \left(\frac{\partial s}{\partial P}\right)_v dP + \left(\frac{\partial s}{\partial v}\right)_P dv \tag{4.63}$$

The partial derivatives can be replaced using the first two of Maxwell's relations (4.56):

$$0 = -\left(\frac{\partial v}{\partial T}\right)_s dP + \left(\frac{\partial P}{\partial T}\right)_s dv \tag{4.64}$$

In other words, during the isentropic volume change, P and v are related by

$$0 = \frac{dP}{P} + k \frac{dv}{v} \tag{4.65}$$

where the so-called *isentropic expansion exponent* k is shorthand notation for

$$k = -\frac{v}{P}\left(\frac{\partial P}{\partial v}\right)_s \tag{4.66}$$

In general, k is a function of both P and v (or T and P); eq. (4.65) suggests that the pressure and volume vary locally as

$$Pv^k = \text{constant} \tag{4.67}$$

during an infinitesimal expansion [note eq. (4.67) as the origin of the name "exponent" that is used for k].

In conclusion, the measurement of k provides a relation between P and V during a constant-entropy reversible expansion. This relation is particularly important in engineering thermodynamics, in view of the role of reference model played by the reversible and adiabatic expansion in the analysis of work transfer components of power and refrigeration systems. The work delivered by the system during a reversible and adiabatic expansion from a state (P_1, V_1) to a sufficiently close[†] end state (P_2, V_2) is found by combining the local path (4.67) with $\delta W_{\text{rev}} = P\,dV$:

$$W_{\text{rev},1-2} \cong \frac{P_1 V_1}{k-1}\left[1 - \left(\frac{V_1}{V_2}\right)^{k-1}\right] \tag{4.68}$$

Figure 4.9 shows that in general k is a function of two variables. The $k = \text{constant}$ lines were drawn based on the data presented in Ref. 10. Only

---

[†]Sufficiently close so that k can be treated as constant.

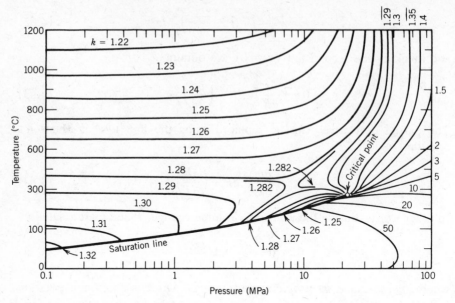

Figure 4.9 The isentropic expansion coefficient (k) of steam.

in the limit of extremely low pressures, that is, in the limit of ideal gas behavior does k become a function of temperature only. This observation is emphasized further by Fig. 4.10, which shows the isentropic exponent of the common ideal gases reviewed earlier in Fig. 4.4. Both Fig. 4.10 and Fig. 4.4 are based on the numerical data presented for gases at low pressures in Ref. 11. It is important to distinguish between the isentropic expansion exponent k and specific-heat ratio c_P/c_v: as shown in Problem 4.5, in general k is not the same as c_P/c_v:

$$k = -\frac{v}{P}\left(\frac{\partial P}{\partial v}\right)_T \frac{c_P}{c_v} \qquad (4.69)$$

The isentropic exponent k equals c_P/c_v only in an ideal gas limit, as noted on the ordinate of Fig. 4.10 and in Table 4.5.

   (d) Reversible and Isothermal Volume Change (Fig. 4.8d). During this process, the system is maintained at constant temperature by contact with the temperature reservoir (T). Measuring the change in pressure associated with a given change in volume, we calculate the *isothermal compressibility*:

$$\kappa = -\frac{1}{v}\left(\frac{\partial v}{\partial P}\right)_T \qquad (4.70)$$

where the new symbol κ should not be confused with the isentropic exponent k. The isothermal compressibility is generally a function of T and

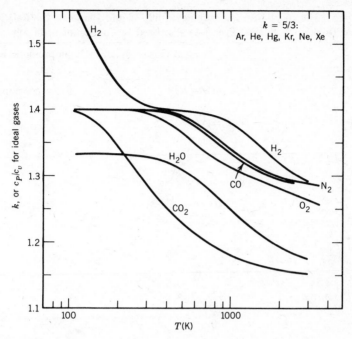

Figure 4.10 The effect of temperature on the isentropic expansion coefficient of several ideal gases (note that $k = c_p/c_v$ for ideal gases).

P that bridges the gap between the ideal gas and incompressible-liquid limits listed in Table 4.5. The κ relation (4.70) and the β relation (4.61) record the measurement of two slopes of the $v(T, P)$ surface. In terms of these measurements, the last two of Maxwell's relations (4.56) state

$$\left(\frac{\partial s}{\partial v}\right)_T = \frac{\beta}{\kappa} \tag{4.71f}$$

$$-\left(\frac{\partial s}{\partial P}\right)_T = \beta v \tag{4.71g}$$

The variation of pressure with volume during the constant-temperature expansion [i.e., the path $P(v)$ of the process] is found by considering the total differential of the function $T(P, v)$:

$$dT = \left(\frac{\partial T}{\partial P}\right)_v dP + \left(\frac{\partial T}{\partial v}\right)_P dv \tag{4.72}$$

which, since $dT = 0$, yields

TABLE 4.5 The Ideal Gas and Incompressible-Liquid Limits of the Relations Measured During the Special Processes Discussed in Connection with Fig. 4.8

Relation for	Ideal Gas Limit	Incompressible-Liquid Limit
$P = P(v, T)$	$Pv = RT$	$v = \text{constant}$
$c_v = \left(\dfrac{\partial u}{\partial T}\right)_v = T\left(\dfrac{\partial s}{\partial T}\right)_v$	$c_v(T)$	$c(T)$
$c_P = \left(\dfrac{\partial h}{\partial T}\right)_P = T\left(\dfrac{\partial s}{\partial T}\right)_P$	$c_P(T) = c_v - R$	$c(T)$
$\beta = \dfrac{1}{v}\left(\dfrac{\partial v}{\partial T}\right)_P$	T^{-1}	0
$k = -\dfrac{v}{P}\left(\dfrac{\partial P}{\partial v}\right)_s$	c_P/c_v	∞
$\kappa = -\dfrac{1}{v}\left(\dfrac{\partial v}{\partial P}\right)_T$	P^{-1}	0
$k_T = -\dfrac{v}{P}\left(\dfrac{\partial P}{\partial v}\right)_T$	1	∞
$\mu_J = \left(\dfrac{\partial T}{\partial P}\right)_h$	0	$-\dfrac{v}{c}$

$$0 = \frac{dP}{P} + k_T \frac{dv}{v} \tag{4.73}$$

The *isothermal-expansion exponent* k_T is clearly

$$k_T = \frac{v}{P}\left(\frac{\partial T}{\partial v}\right)_P \left(\frac{\partial P}{\partial T}\right)_v \tag{4.74}$$

or, invoking the cyclical relation (Appendix),

$$k_T = -\frac{v}{P}\left(\frac{\partial P}{\partial v}\right)_T = (\kappa P)^{-1} \tag{4.74'}$$

Like κ, the isothermal exponent k_T depends on both T and P. For sufficiently small departures from a given state, k_T can be regarded as constant; therefore, the *local* path suggested by eq. (4.73) is

$$Pv^{k_T} = \text{constant} \tag{4.75}$$

and, by analogy with eq. (4.68), the reversible and isothermal work output is

$$W_{rev,1-2} \cong \frac{P_1 V_1}{k_T - 1}\left[1 - \left(\frac{V_1}{V_2}\right)^{k_T - 1}\right] \tag{4.76}$$

The ideal gas limit of this last result is

$$\lim_{k_T \to 1} (W_{rev,1-2}) = P_1 V_1 \ln \frac{V_2}{V_1} \tag{4.77}$$

Finally, comparing eqs. (4.69) and (4.74'), we reach the interesting conclusion that the two expansion exponents and the two specific heats are not independent [12]:

$$\frac{k}{k_T} = \frac{c_P}{c_v} \tag{4.78}$$

(e) Constant-Enthalpy Expansion (Fig. 4.8e). The variation of tempera-ture with pressure at constant enthalpy can be measured in the classical experiment conceived by Thomson and Joule [13]. A fluid flows sufficiently slowly through a fine porous plug (flow restriction), such that both upstream and downstream of the plug any finite-size sample taken out of the fluid is in a state of equilibrium. If the duct that houses the flow and the plug is bounded by an adiabatic surface, the first law for steady flow guarantees that the enthalpy downstream of the plug is the same as upstream. The measure-ment of temperatures and pressures upstream and downstream of the flow restriction is recorded in the form of the *Joule–Thomson coefficient*:

$$\mu_J = \left(\frac{\partial T}{\partial P}\right)_h \tag{4.79}$$

where, in general, μ_J is a function of both T and P (Fig. 4.11). Note also that the μ_J notation has nothing to do with Gibbs' chemical potentials μ_i.

The Joule–Thomson coefficient and, especially, its sign are particularly relevant in the field of low-temperature refrigeration. When the μ_J value of a fluid is positive, it means that the flow and depressurization of a stream across a throttle is accompanied by a drop in temperature, i.e., that the resulting stream can be used as a refrigerant. An incompressible liquid, on the other hand, warms up while being throttled adiabatically, as indicated by the negative μ_J listed in Table 4.5.

These features are evident also in Fig. 4.11, which shows the "inversion curves" $\mu_J(T, P) = 0$ for the two most common low-temperature refriger-ants, gaseous nitrogen and helium-4. These curves were drawn from the tabulated data of Jacobsen et al. [14] and McCarthy [15]. The Joule–Thomson expansion proceeds from right to left along the dashed line shown in the figure. On the right side of the inversion curve ($\mu_J < 0$), the expansion brings about a temperature increase, meaning that only the $\mu_J > 0$ domain bordered by the inversion curve contains (T, P) states from which the temperature can drop via a throttling process.

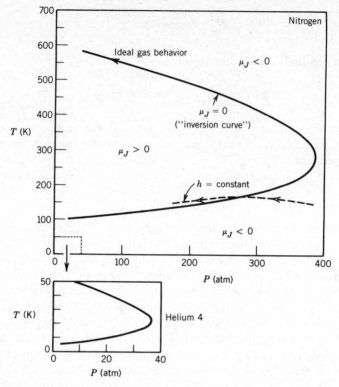

Figure 4.11 The $T(P)$ inversion curves ($\mu_J = 0$) for nitrogen and helium-4.

Bridgman's Table

The relationships discussed until now in this section were put to use in 1914 by P. W. Bridgman—famous professor of mathematics and natural philosophy at Harvard University during the first half of this century [16, 17]. His objective was to express all the first partial derivatives of the most frequently used properties:

$$P, T, v, s, u, h, f, \text{ and } g$$

in terms of the measured $P = P(v, T)$ surface and three directional slopes (first partial derivatives) that are the easiest to measure in the laboratory, namely,

$$c_P, \beta, \text{ and } \kappa$$

Table 4.6 is a reproduction of Bridgman's table for first partial derivatives. In order to construct the most concise table possible, Bridgman made use of the peculiar notation "$(\partial x_j)_{x_i}$," whose meaning will become clear

TABLE 4.6 Bridgman's Table of Relations for First Partial Derivatives (after Ref. 16); the Property Listed in the Square Brackets, [x], Indicates that the Respective Group of Relations is either for $[\partial(\ )]_x$ or $-(\partial x)_{(\ )}$

[P] $(\partial T)_P = -(\partial P)_T = 1$

$(\partial v)_P = -(\partial P)_v = \beta v$

$(\partial s)_P = -(\partial P)_s = \dfrac{c_P}{T}$

$(\partial u)_P = -(\partial P)_u = c_P - \beta P v$

$(\partial h)_P = -(\partial P)_h = c_P$

$(\partial f)_P = -(\partial P)_f = -s - \beta P v$

$(\partial g)_P = -(\partial P)_g = -s$

[T] $(\partial v)_T = -(\partial T)_v = \kappa v$

$(\partial s)_T = -(\partial T)_s = \beta v$

$(\partial u)_T = -(\partial T)_u = \beta T v - \kappa P v$

$(\partial h)_T = -(\partial T)_h = -v + \beta T v$

$(\partial f)_T = -(\partial T)_f = -\kappa P v$

$(\partial g)_T = -(\partial T)_g = -v$

---

[v] $(\partial s)_v = -(\partial v)_s = \beta^2 v^2 - \dfrac{\kappa v c_P}{T}$

$(\partial u)_v = -(\partial v)_u = T\beta^2 v^2 - \kappa v c_P$

$(\partial h)_v = -(\partial v)_h = T\beta^2 v^2 - \beta v^2 - \kappa v c_P$

$(\partial f)_v = -(\partial v)_f = \kappa v s$

$(\partial g)_v = -(\partial v)_g = \kappa v s - \beta v^2$

---

[s] $(\partial u)_s = -(\partial s)_u = \beta^2 v^2 P - \dfrac{\kappa v c_P P}{T}$

$(\partial h)_s = -(\partial s)_h = -\dfrac{v c_P}{T}$

$(\partial f)_s = -(\partial s)_f = \beta v s + \beta^2 v^2 P - \dfrac{\kappa v c_P P}{T}$

$(\partial g)_s = -(\partial s)_g = \beta v s - \dfrac{v c_P}{T}$

---

[u] $(\partial h)_u = -(\partial u)_h = P\beta v^2 + \kappa v c_P P - v c_P - PT\beta^2 v^2$

$(\partial f)_u = -(\partial u)_f = s T \beta v - \kappa v c_P P - \kappa v s P + PT\beta^2 v^2$

$(\partial g)_u = -(\partial u)_g = \beta v^2 P + \beta v s T - v c_P - \kappa v s P$

---

[h] $(\partial f)_h = -(\partial h)_f = (s - v\beta P)(v - v\beta T) - \kappa v c_P P$

$(\partial g)_h = -(\partial h)_g = \beta v s T - v(s + c_P)$

---

[g] $(\partial f)_g = -(\partial g)_f = \kappa v s P - v s - \beta v^2 P$

shortly. The most general way of writing one of the first partial derivatives that can be constructed with the eight properties listed above is

$$\left(\frac{\partial x_j}{\partial x_k}\right)_{x_i}$$

where x_j, x_k, and x_i can represent eight, seven, and six properties, respectively. This means that the total number of first partial derivatives that can be evaluated via Table 4.6 is $(8)(7)(6) = 336$: compare 336 with 28, which is the total number of lines in Table 4.6, and you begin to appreciate the merit of Bridgman's notation "$(\partial x_j)_{x_i}$." In principle, the wanted partial derivative $(\partial x_j/\partial x_k)_{x_i}$ can be found by dividing two other derivatives, say

$$\left(\frac{\partial x_j}{\partial x_k}\right)_{x_i} = \frac{(\partial x_j/\partial \alpha_i)_{x_i}}{(\partial x_k/\partial \alpha_i)_{x_i}} \tag{4.80}$$

where α_i is any other property, not necessarily one of the original eight, P, $T, \ldots, g$. Bridgman's notation $(\partial x_j)_{x_i}$ is shorthand for $(\partial x_j/\partial \alpha_i)_{x_i}$, *only in the sense that the wanted derivative can be obtained by dividing the quantities listed as $(\partial x_j)_{x_i}$ and $(\partial x_k)_{x_i}$ in Table 4.6:*

$$\left(\frac{\partial x_j}{\partial x_k}\right)_{x_i} = \frac{(\partial x_j)_{x_i}}{(\partial x_k)_{x_i}} \tag{4.81}$$

Note that the purpose of the table is not the determination of the auxiliary variables α_i that were used in the construction of the table. Bridgman's table is to be used always in association with eq. (4.81), that is, by dividing the appropriate two expressions (lines) from the table in order to generate the wanted partial derivative.

Example 4.3.   Consider the problem of calculating the Joule–Thomson coefficient in terms of the experimental information listed in Table 4.6. We write, in order,

$$\mu_J = \left(\frac{\partial T}{\partial P}\right)_h = \frac{(\partial T/\partial \alpha_h)_h}{(\partial P/\partial \alpha_h)_h} = \frac{(\partial T)_h}{(\partial P)_h} \tag{a}$$

and, by dividing the eleventh and fifth lines of Table 4.6, we obtain

$$\mu_J = \frac{v - \beta v T}{-c_P} = \frac{v}{c_P}(\beta T - 1) \tag{b}$$

As a second example, consider deducing the value of c_v from measurements of the $P = P(v, T)$ surface and c_P, β, and κ. Again, we start with the wanted derivative and express it in terms of Bridgman's special notation:

$$c_v = \left(\frac{\partial u}{\partial T}\right)_v = \frac{(\partial u)_v}{(\partial T)_v} \qquad \text{(c)}$$

Dividing the fifteenth and eighth lines of Table 4.6, we have

$$c_v = \frac{T\beta^2 v^2 - \kappa v c_P}{-\kappa v} = c_P - \frac{\beta^2}{\kappa} vT \qquad \text{(d)}$$

The validity of the two identities discovered above, eqs. (b) and (d), can be tested easily against the ideal gas and incompressible-liquid limits summarized in Table 4.5. As demonstrated in Problem 4.6, the work of proving eqs. (b) and (d) is considerably more challenging in the absence of Bridgman's table.

Shaw's Use of Jacobians

We close this section on relations between properties and their derivatives, by outlining a more general and pedagogically important method that was proposed by Shaw in 1935 [18]. The method relies on the concept of a functional determinant or Jacobian, which, for the functions of two variables $x(\alpha, \beta)$ and $y(\alpha, \beta)$, is labeled $\partial(x, y)/\partial(\alpha, \beta)$ and is defined by the expression [Ref. 3, p. 456]:

$$\frac{\partial(x, y)}{\partial(\alpha, \beta)} = \begin{vmatrix} \left(\dfrac{\partial x}{\partial \alpha}\right)_\beta & \left(\dfrac{\partial x}{\partial \beta}\right)_\alpha \\ \left(\dfrac{\partial y}{\partial \alpha}\right)_\beta & \left(\dfrac{\partial y}{\partial \beta}\right)_\alpha \end{vmatrix} = \left(\frac{\partial x}{\partial \alpha}\right)_\beta \left(\frac{\partial y}{\partial \beta}\right)_\alpha - \left(\frac{\partial x}{\partial \beta}\right)_\alpha \left(\frac{\partial y}{\partial \alpha}\right)_\beta \qquad (4.82)$$

Of special interest are the following properties of the Jacobian:

$$\frac{\partial(x, y)}{\partial(\alpha, \beta)} = -\frac{\partial(y, x)}{\partial(\alpha, \beta)} \qquad (4.83)$$

$$\frac{\partial(x, y)}{\partial(\alpha, \beta)} = \left[\frac{\partial(\alpha, \beta)}{\partial(x, y)}\right]^{-1} \qquad (4.84)$$

$$\frac{\partial(x, y)}{\partial(\alpha, \beta)} = \frac{\partial(x, y)}{\partial(z, w)} \frac{\partial(z, w)}{\partial(\alpha, \beta)} \qquad (4.85)$$

where the last two equations are similar to the reciprocal and chain rules encountered in the manipulation of partial derivatives (see Appendix).

We focus on the task of evaluating a first partial derivative, $(\partial x/\partial z)_y$, and note that this derivative can be written in Jacobian notation as

$$\left(\frac{\partial x}{\partial z}\right)_y = \frac{\partial(x, y)}{\partial(z, y)} \qquad (4.86)$$

It is easy to verify the validity of eq. (4.86) by invoking the original definition (4.82). Next, we apply the chain rule (4.85) and conclude that

$$\left(\frac{\partial x}{\partial z}\right)_y = \frac{\partial(x, y)}{\partial(\alpha, \beta)} \bigg/ \frac{\partial(z, y)}{\partial(\alpha, \beta)} \tag{4.87}$$

The promising end of this analytical path is that using eq. (4.87), we can express any derivative $(\partial x/\partial z)_y$ only in terms of the partial derivatives of x, y, and z with respect to the same two properties, α and β. The derivatives with respect to α and β appear on the right side of eq. (4.87), after invoking twice the Jacobian definition (4.82).

Partial derivatives of the type $(\partial x/\partial z)_y$ occur naturally in the description of simple systems of fixed composition (e.g., Tables 4.5 and 4.6). The generalization of the rule (4.87) for the case where x is a function of any number of properties $(z, y, u, v, \ldots)$ reads

$$\left(\frac{\partial x}{\partial z}\right)_{y,u,v,\ldots} = \frac{\partial(x, y, u, v, \ldots)}{\partial(\alpha, \beta, \gamma, \delta, \ldots)} \bigg/ \frac{\partial(z, y, u, v, \ldots)}{\partial(\alpha, \beta, \gamma, \delta, \ldots)} \tag{4.88}$$

Note that the number of auxiliary properties $(\alpha, \beta, \gamma, \delta, \ldots)$ must match the number of properties on which x depends, namely $(z, y, u, v, \ldots)$.

The use of Jacobians in the transformation (reduction) of certain thermodynamic derivatives forms the subject of a very instructive paper that was published recently by Somerton and Arnas [19]. Their study is recommended.

Example 4.4. Consider again the Joule–Thomson coefficient identity of the preceding example:

$$\mu_J = \frac{v}{c_P}(\beta T - 1) \tag{a}$$

and let us prove it, using this time the Jacobian expression (4.87). The Joule–Thomson coefficient is a partial derivative of type $(\partial x/\partial z)_y$; therefore, we immediately write

$$\mu_J = \left(\frac{\partial T}{\partial P}\right)_h = \frac{\partial(T, h)}{\partial(\alpha, \beta)} \bigg/ \frac{\partial(P, h)}{\partial(\alpha, \beta)} \tag{b}$$

Next, we select the auxiliary properties α and β using the right side of the target identity (a) as a guide. Since c_P is present on the right side of eq. (a), and since $c_P = (\partial h/\partial T)_P$, we conclude that all the partial derivatives that form on the right side of eq. (b) must be taken with respect to T or P. Therefore, setting $\alpha = T$ and $\beta = P$ in eq. (b), we have

$$\mu_J = \frac{\partial(T, h)}{\partial(T, P)} \bigg/ \frac{\partial(P, h)}{\partial(T, P)}$$

$$= -\left(\frac{\partial h}{\partial P}\right)_T \bigg/ \left(\frac{\partial h}{\partial T}\right)_P$$

$$= -\frac{1}{c_P}\left(\frac{\partial h}{\partial P}\right)_T \tag{c}$$

At this stage, we reconstructed only the c_P denominator of the right side of eq. (a). To show that $(\partial h/\partial P)_T$ is in fact the same as $v(1 - \beta T)$ is to express the $(\partial h/\partial P)_T$ solely in terms of (P, v, T) information. From the differential expression for enthalpy, Table 4.3, we write that for fixed-composition simple systems:

$$dh = T \, ds + v \, dP \tag{d}$$

hence,

$$\left(\frac{\partial h}{\partial P}\right)_T = T\left(\frac{\partial s}{\partial P}\right)_T + v$$

$$= -T\left(\frac{\partial v}{\partial T}\right)_P + v$$

$$= v(-\beta T + 1) \tag{e}$$

The last two steps in eq. (e) consisted of applying Maxwell's relation (4.56g) and the volume expansivity definition (4.61). Putting eqs. (c) and (e) together completes the reconstruction of the target identity (a).

Example 4.5. The analysis sandwiched between eqs. (d) and (e) in the preceding example is an important sequence of steps that often occur together, as a building block, in the construction of more complicated relations between properties. Table 4.7 shows one group of such relations, namely, the relations that permit the calculation of du, dh, and ds strictly from (P, v, T) and specific-heat information. These relations form the analytical foundation of the many numerical tables of thermodynamic properties.

The challenge that was faced by the problem solver in the second half of

TABLE 4.7 The Calculation of Changes in Internal Energy, Enthalpy, and Entropy Based on the Information Provided by the $P = P(v, T)$ Surface and Specific-Heat Measurements (after Ref. 20)

Closed Simple System[a] (Fixed Composition)	Ideal Gas Limit	Incompressible-Liquid Limit
$du = c_v \, dT + \left(\dfrac{\beta}{\kappa} T - P\right) dv$	$du = c_v \, dT$	$du = c \, dT$
$dh = c_P \, dT + (1 - \beta T)v \, dP$	$dh = c_P \, dT$	$dh = c \, dT + v \, dP$
$ds = \dfrac{c_P}{T} dT - \beta v \, dP$	$ds = \dfrac{c_P}{T} dT - \dfrac{R}{P} dP$	$ds = \dfrac{c}{T} dT$
$\quad = \dfrac{c_v}{T} dT + \dfrac{\beta}{\kappa} dv$	$\quad = \dfrac{c_v}{T} dT + \dfrac{R}{v} dv$	
$\quad = \dfrac{c_P}{v\beta T} dv + \dfrac{\kappa c_v}{\beta T} dP$	$\quad = \dfrac{c_P}{v} dv + \dfrac{c_v}{P} dP$	

[a]The coefficients c_v, c_P, β, and κ are defined in Table 4.5.

Example 4.4 was to express an (enthalpy) partial derivative in terms of (P, v, T) and specific-heat information. The sequence of steps that followed was this:

(i) the (enthalpy) derivative was first replaced by an entropy derivative, using the appropriate differential form listed in Table 4.3, and

(ii) the entropy derivative was replaced by a partial derivative involving only P, v, and T, using one of the last two of Maxwell's relations (4.56).

We rediscover these steps as we attempt to rederive the formulas listed in the left column of Table 4.7. For example, the internal energy formula is one of the type

$$du = \left(\frac{\partial u}{\partial T}\right)_v dT + \left(\frac{\partial u}{\partial v}\right)_T dv \tag{a}$$

where the first partial derivative on the right side is clearly c_v. In order to evaluate the $(\partial u/\partial v)_T$ derivative in terms of (P, v, T) information, we execute steps (i) and (ii) by replacing the word (enthalpy) with (internal energy):

(i)
$$du = T\,ds - P\,dv \tag{b}$$

$$\left(\frac{\partial u}{\partial v}\right)_T = T\left(\frac{\partial s}{\partial v}\right)_T - P \tag{c}$$

(ii)
$$\left(\frac{\partial u}{\partial v}\right)_T = T\left(\frac{\partial P}{\partial T}\right)_v - P \tag{d}$$

Equations (a) and (d) contain the wanted result:

$$du = c_v\,dT + \left[T\left(\frac{\partial P}{\partial T}\right)_v - P\right] dv \tag{e}$$

The du expression in the first line of Table 4.7 is the result of an additional step in which $(\partial P/\partial T)_v$ is expressed in terms of expansivity (β) and compressibility (κ) by invoking the cyclical relation:

$$\left(\frac{\partial P}{\partial T}\right)_v\left(\frac{\partial T}{\partial v}\right)_P\left(\frac{\partial v}{\partial P}\right)_T = -1 \tag{f}$$

The derivation of the dh and ds formulas of Table 4.7 proceeds along a similar course. Note that step (i) disappears in the derivation of ds.

GEOMETRIC REPRESENTATIONS OF THERMODYNAMIC RELATIONS

There have been a number of atempts aimed at using the power of graphics to "condense" some of the analytical facts of equilibrium thermodynamics, the most famous of these being Koenig's [21] thermodynamic mnemonic square, which was made popular under Max Born's name by Tisza [5] and

Callen [2]†. Koenig and Born's "thermodynamic square" is shown in Fig. 4.12. Listed on the four sides of the square are alternatives to the fundamental relation in energy representation (U, H, F, G). The two corners that terminate one side indicate the variables of which the particular side quantity is a function. For example, the left side of the square suggests an internal energy fundamental relation of the type $U = U(S, V)$. It is evident that the thermodynamic square refers to closed simple systems, that is, to systems whose constituent mole numbers remain fixed.

Another feature of the thermodynamic square is the upward sense of the arrows of the two diagonals. These arrows help memorize the sign of the two terms that appear on the right side of total differentials such as $dU = T\,dS - P\,dV$. If the diagonal arrow points away from a corner, then the differential of the property listed in that corner appears with the plus sign (note in the preceding example, the plus sign of dS on the right side of $dU = T\,dS - P\,dV$). On the other hand, when the diagonal arrow points toward the corner, the differential of the corner variable enters with the negative sign on the right side of the differential expression for the respective side quantity (note the minus sign of dV on the right side of $dU = T\,dS - P\,dV$ or, looking at the top edge of the square, the minus signs of both dT and dV on the right side of $dF = -S\,dT - P\,dV$).

A third group of relations that are condensed in the thermodynamic square of Fig. 4.12 is the group of Maxwell's relations seen already in eqs. (4.56u)–(4.56g). These relations can be read off the square by looking only at the corner properties. For example, in order to read eq. (4.56u), we recognize the two right triangles that share the side labeled "U," namely, triangles TVS and PSV. These two triangles are isolated in the first part of Fig. 4.13, which was rotated by 90 degrees counterclockwise relative to its original position in Fig. 4.12. The vertical sides of the triangles list the partial derivatives involved in the Maxwell relation, in such a way that the

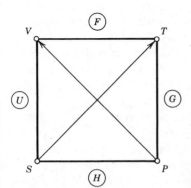

Figure 4.12 Koenig and Born's thermodynamic square.

---

†The omission of Koenig's name is highly peculiar, since Koenig published his discovery and Born did not.

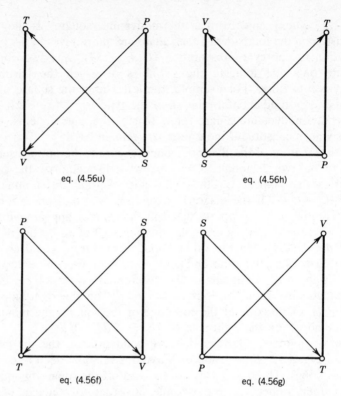

eq. (4.56u)   eq. (4.56h)

eq. (4.56f)   eq. (4.56g)

Figure 4.13 Deducing Maxwell's relations (4.56u)–(4.56g) from the thermodynamic square of Fig. 4.12.

properties seen at the upper corners of the triangles (i.e., T and P) assume the role of numerators in the Maxwell relations (i.e., $\partial T/\partial V$ and $\partial P/\partial S$). Furthermore, the third corner of each triangle—the corner that has not been used yet—indicates the property that is held constant in the respective partial derivative [i.e., $(\partial T/\partial V)_S$ and $(\partial P/\partial S)_V$]. Finally, when the two diagonals point in different directions—one up and the other down, as in the first drawing of Fig. 4.13—it means that the two partial derivatives identified so far enter with opposite signs in the desired Maxwell's relation. This rule is confirmed by the minus sign seen on the right side of eq. (4.56u). The remaining relations, eqs. (4.56h)–(4.56g), can be reconstructed visually by reading the remaining three drawings in Fig. 4.13.

The fact that the preceding relations can be read off from relatively simple geometric constructions drawn on a plane is due to the simplicity of the closed system that was considered until now. A first step in the direction of more complicated systems is to consider the case of a single-component system whose number of moles N can vary. The twelve Maxwell-type relations that apply to such a system have been listed in Table 4.4. The geometric construction that does for this system what the thermodynamic square did for a closed system is the thermodynamic octahedron discovered independently (and with slight differences) by Branch and Hayashi [22], Koenig [23], and Reddy [24], Fig. 4.14. The use of this device in the classroom was discussed in detail by Reddy [24], who showed that in addition to the four thermodynamic potentials presented in Table 4.4,

$$U(S, V, N)$$
$$H(S, P, N)$$
$$F(T, V, N)$$
$$G(T, P, N)$$

one could construct four additional Legendre transforms, which here are labeled as

$$U_\mu(S, V, \mu)$$
$$H_\mu(S, P, \mu)$$
$$F_\mu(T, V, \mu)$$
$$G_\mu(T, P, \mu)$$

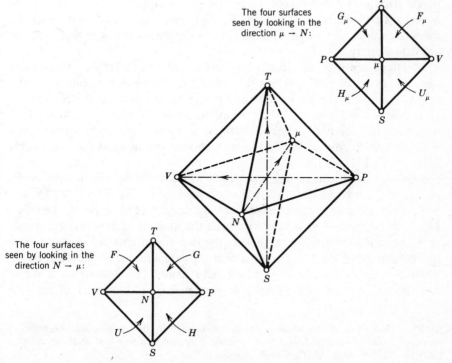

The four surfaces seen by looking in the direction $\mu \to N$:

The four surfaces seen by looking in the direction $N \to \mu$:

Figure 4.14 The thermodynamic octahedron.

In particular, $F_\mu(T, V, \mu)$ is the grand canonical potential discussed in Problem 4.2. The G_μ function, on the other hand, is identically equal to zero based on the Euler equation, $G_\mu = U - TS + PV - \mu N \equiv 0$. Figure 4.14 shows that each face of the octahedron represents one of the eight potentials listed above. The six vertices represent the possible arguments (T, S, P, V, μ, N).

There are three squares that appear embedded in the octahedron: each is a thermodynamic square of the type illustrated in Fig. 4.12. For example, the square whose plane is normal to the direction $N \rightarrow \mu$ is precisely the Koenig–Born thermodynamic square. The diagonal arrows drawn in these square cross-sections have the same meaning as in Koenig and Born's square. Consider, for example, the triangular face labeled "U" in Fig. 4.14: the three vertices of this triangle indicate that U must be a function of S, V, and N. Furthermore, since a diagonal arrow arrives at the "V" corner and since diagonal arrows depart from "S" and "N," the correct form of the U differential is $dU = T\,dS - P\,dV + \mu\,dN$. The Maxwell relations of the system represented by Fig. 4.14 can be identified by looking into each of the square cuts at a time and applying the rules illustrated in Fig. 4.13. This procedure is described further in Refs. 22 and 24.

The construction in space of a geometric summary for the thermodynamic relations of Table 4.4 was pursued also by Fox [25]. In place of the octahedron of Fig. 4.14, Fox proposes a "cuboctahedron," that is, a solid object with a total of 14 faces, 6 squares, plus 8 equilateral triangles. Worth noting is that both the octahedron (Fig. 4.14) and the cuboctahedron can be manufactured easily and used as models in the classroom or in the course of individual study.

Finally, I use this opportunity to draw attention to a highly compact and imaginative bit of new graphics that continues the modern trend[†] toward a geometrization of the methodology of research. This new geometric point of view belongs to Professor Frank Weinhold of the University of Wisconsin, Madison. His method is best illustrated in terms of the two-dimensional diagram constructed in Fig. 4.15, even though the method is considerably more general [28–34].

Consider again the closed system for which the differential form dU of eq. (4.55u) reveals the two pairs of *conjugate properties*, (T, S) and $(-P, V)$. Each pair corresponds to one mode of reversible energy transfer, Fig. 4.3; the number of such pairs equals the number of degrees of freedom of the system (two in this case, hence, the two dimensions of Fig. 4.15). One special feature of the (T, S) pair is that the temperature T cannot decrease when S increases while V is being held fixed. This feature is a direct consequence of the second law, as shown in the discussion of internal

---

[†]Another illustration of this trend is the progress made by the "large-scale instantaneous structure" point of view in the fields of turbulent flow and turbulent heat and mass transfer [see, for example, Refs. 26, 27; Ref. 9, chapters 6–8].

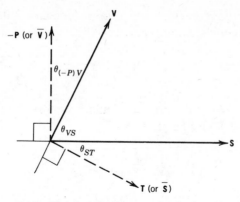

Figure 4.15   Weinhold's diagram for a closed system.

thermal stability of simple systems (pp. 244–246). The $(-P, V)$ pair behaves similarly, in that P cannot decrease when V decreases at constant S.

Let (Y_i, X_i) represent either of the pairs of conjugate properties discussed above. It is said that the "consistent" response of each Y_i to changes in its conjugate X_i (and vice versa) is *positive*. With each pair (Y_i, X_i), we can associate a vector $\mathbf{X}_i$ whose length measures the degree of such responsiveness:

$$|\mathbf{X}_i| = \left[\left(\frac{\partial X_i}{\partial Y_i}\right)_{Y_j}\right]^{1/2} \tag{4.89}$$

Figure 4.15 shows the two vectors that can be drawn based on this definition. Their respective lengths are

$$|\mathbf{S}| = \left[\left(\frac{\partial S}{\partial T}\right)_P\right]^{1/2} = \left(\frac{c_P}{T}\right)^{1/2} \tag{4.90}$$

$$|V| = \left[-\left(\frac{\partial V}{\partial P}\right)_T\right]^{1/2} = (\kappa V)^{1/2} \tag{4.91}$$

These lengths are not equal, in fact, they are measured in terms of different units. The size of the angle θ_{VS} formed by these two vectors is an issue to which we return at the end of this section.

Next, we can draw a new set of vectors $\bar{\mathbf{X}}_i$, such that each $\bar{\mathbf{X}}_i$ is perpendicular to $\mathbf{X}_j$. In Fig. 4.15, this would mean that $\bar{\mathbf{S}}$ is perpendicular to $\mathbf{V}$, and that $\bar{\mathbf{V}}$ is perpendicular to $\mathbf{S}$. The length of each of these perpendicular vectors can be chosen such that the scalar product $\mathbf{X}_i\bar{\mathbf{X}}_i$ is equal to 1, i.e.,

$$|\mathbf{X}_i||\bar{\mathbf{X}}_i| \cos \theta_{X_i \bar{X}_i} = 1 \tag{4.92}$$

It turns out that the $\bar{\mathbf{X}}_i$ vector defined in this manner is the conjugate of $\mathbf{X}_i$ [33]; in other words, $\bar{\mathbf{X}}_i$ is the same as the vector $\mathbf{Y}_i$ of length

$$|\mathbf{Y}_i| = \left[\left(\frac{\partial Y_i}{\partial X_i}\right)_{X_j}\right]^{1/2} \tag{4.93}$$

In Fig. 4.15, the role of the "perpendicular" vectors $\mathbf{Y}_i$ (or $\bar{\mathbf{X}}_i$) is played by the vectors labled $\mathbf{T}$ and $(-\mathbf{P})$, where

$$|\mathbf{T}| = \left[\left(\frac{\partial T}{\partial S}\right)_V\right]^{1/2} = \frac{T}{c_v} \tag{4.94}$$

$$|-\mathbf{P}| = \left[-\left(\frac{\partial P}{\partial V}\right)_S\right]^{1/2} = \left(\frac{kP}{V}\right)^{1/2} \tag{4.95}$$

The angles formed between $\mathbf{S}$ and $\mathbf{T}$ and between $\mathbf{V}$ and $(-\mathbf{P})$ are determined by invoking eq. (4.92):

$$\cos\theta_{ST} = \left(\frac{c_v}{c_P}\right)^{1/2} \tag{4.96}$$

$$\cos\theta_{(-P)V} = (\kappa kP)^{-1/2} \tag{4.97}$$

Finally, the in-between angle θ_{VS} is found by recognizing the relationship between it and either of its adjacents:

$$\cos\theta_{VS} = (1 - \sin^2\theta_{VS})^{1/2} = (1 - \cos^2\theta_{ST})^{1/2} = (1 - c_v/c_P)^{1/2} \tag{4.98}$$

Starting with the consideration of internal stability ("positive" response) that via eq. (4.89) was built into the geometric construction of Fig. 4.15, Weinhold's diagram summarizes a good portion of the analytical coverage of closed-system equilibrium thermodynamics. A number of relations between properties and their derivatives can be read directly off the diagram:

Geometric relations	**Thermodynamic relations**	
$\theta_{ST} = \theta_{(-P)V}$	$\dfrac{c_P}{c_v} = \kappa kP$	(4.99a)
$\cos^2\theta_{ST} \le 1$	$c_P \ge c_v$	(4.99b)
$\cos^2\theta_{(-P)V} \le 1$	$k \ge (\kappa P)^{-1}$	(4.99c)

This diagram is only a small part of a much more general geometric approach to the thermodynamics of simple systems in equilibrium. For example, the Weinhold diagram of a binary mixture (a three-degrees-of-freedom system) assumes the shape of a star with six rays (vectors) in a three-dimensional space [33].

PARTIAL MOLAL PROPERTIES

In the presentation of property relations of the preceding sections, we assumed that the system composition is fixed, and focused on the effects associated with changes in temperature and pressure. In this section, we adopt the reverse point of view and focus on a simple system whose temperature and pressure are held fixed, for example, by contact with the atmospheric reservoir (Fig. 4.1). We are particularly interested in the effect of composition variations on the extensive properties of the system.

Let X represent an extensive property of the system. Assuming that X depends on T, P, and composition $(N_1, N_2, \ldots, N_n)$, we have

$$dX = \left(\frac{\partial X}{\partial T}\right)_{P,N_1,\ldots,N_n} dT + \left(\frac{\partial X}{\partial P}\right)_{T,N_1,\ldots,N_n} dP$$

$$+ \sum_{i=1}^{n} \left(\frac{\partial X}{\partial N_i}\right)_{T,P,N_j} dN_i \qquad (j \neq i) \qquad (4.100)$$

The *partial molal* value of X with respect to the ith constituent is by definition [35, 36]

$$\bar{x}_i = \left(\frac{\partial X}{\partial N_i}\right)_{T,P,N_j} \qquad (j \neq i) \qquad (4.101)$$

so that at constant temperature and pressure

$$dX = \sum_{i=1}^{n} \bar{x}_i \, dN_i \qquad \text{(constant } T, P) \qquad (4.102)$$

The partial molal property $\bar{x}_i$ describes the change caused in the extensive property X by the addition of one unit of constituent "i," while holding T, P, and all the other mole numbers fixed. It is easy to see that at constant T and P, the function X is first-order homogeneous in $(N_1, \ldots, N_n)$: invoking Euler's Theorem on Homogeneous Functions (p. 158), we conclude that

$$X = \sum_{i=1}^{n} \bar{x}_i N_i \qquad \text{(constant } T, P) \qquad (4.103)$$

A related result is obtained by differentiating eq. (4.103) and subtracting it side by side from eq. (4.102):

$$\sum_{i=1}^{n} N_i \, d\bar{x}_i = 0 \qquad \text{(constant } T, P) \qquad (4.104)$$

The partial molal properties $\bar{x}_i$ should not be confused with the more popular notation used for *mole fraction* (x_i), whose definition is

$$x_i = \frac{N_i} {N} \quad \text{where } N = \sum_{i=1}^{n} N_i; \quad \text{hence } \sum_{i=1}^{n} x_i = 1 \qquad (4.105)$$

Using mole fractions (x_i) instead of numbers of moles (N_i) in eq. (4.104), we can also write

$$\sum_{i=1}^{n} x_i \, d\bar{x}_i = 0 \qquad \text{(constant } T, P) \qquad (4.106)$$

An entirely different molal quantity can be defined by dividing X by the total number of moles present in the system:

$$\bar{x} = \frac{X}{N} \qquad (4.107)$$

To distinguish it from the partial molal properties defined earlier, we regard $\bar{x}$ as the *molal* or *proper molal* [35] quantity associated with the extensive property X. Again, the $\bar{x}$ notation should not be confused with the classical mole fraction notation x_i. Finally, from eqs. (4.103) and (4.107), we learn that in single-component systems $(n = 1)$, the partial molal and proper molal quantities associated with property X are equal:

$$\bar{x} = \bar{x}_i \qquad (n = 1) \qquad (4.108)$$

In order to see the difference between partial molal and proper molal quantities, consider a binary mixture whose composition is described by the mole fractions $x_1 = x$ and $x_2 = 1 - x$. According to eq. (4.102), we have

$$dX = \bar{x}_1 \, dN_1 + \bar{x}_2 \, dN_2 \qquad \text{(constant } T, P) \qquad (4.109)$$

and, after dividing by N:

$$d\bar{x} = \bar{x}_1 \, dx_1 + \bar{x}_2 \, dx_2 = (\bar{x}_1 - \bar{x}_2) \, dx \qquad \text{(constant } T, P) \qquad (4.110)$$

or

$$\left(\frac{\partial \bar{x}}{\partial x} \right)_{T,P} = \bar{x}_1 - \bar{x}_2 \qquad (4.111)$$

On the other hand, dividing eq. (4.103) by N, we obtain directly

$$\bar{x} = (x)\bar{x}_1 + (1 - x)\bar{x}_2 \qquad (4.112)$$

Solving eqs. (4.111) and (4.112) for $\bar{x}_1$ and $\bar{x}_2$, we see that in general the partial molal quantities are not the same as the proper molal quantity $\bar{x}$:

$$\bar{x}_1 = \bar{x} + (1-x)\left(\frac{\partial \bar{x}}{\partial x}\right)_{T,P} \tag{4.113}$$

$$\bar{x}_2 = \bar{x} - x\left(\frac{\partial \bar{x}}{\partial x}\right)_{T,P} \tag{4.114}$$

These results have a simple graphic interpretation. The curved solid line in Fig. 4.16 shows the manner in which the proper molal $\bar{x}$ might vary with the composition. The left extremity of the diagram corresponds to a pure system that contains only the $i = 1$ component. The state of the system is represented by the point M of unspecified concentration x. Equations (4.113) and (4.114) show that the partial molal quantities $\bar{x}_1$ and $\bar{x}_2$ are equal to the cuts made by the dash-line tangent on the $x = 1$ and $x = 0$ side lines, respectively. When point M moves to the left, the discrepancy between $\bar{x}_1$ and $\bar{x}$ vanishes as the system approaches the single-component description, eq. (4.108). Of course, the same effect is observed as M approaches the right extremity of the diagram.

A premier example of an extensive property of type X (T, P, $N_1, \ldots, N_n$) is the Gibbs free energy G. Comparing the μ_i equations of state (the bottom of the fourth column in Table 4.3) with the definition (4.101), we see the identity between chemical potentials and partial molal Gibbs free energies:

$$\bar{g}_i = \mu_i \tag{4.115}$$

This identity was commented on earlier in the context of single-component systems, eq. (4.34'). Since $\bar{g}_i$ is also a function of (T, P, $N_1, \ldots, N_n$), we can write [36]

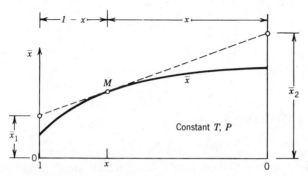

Figure 4.16 The relationship between a proper molal quantity ($\bar{x}$) and the partial molal quantities ($\bar{x}_1, \bar{x}_2$) of a binary mixture.

$$d\bar{g}_i = d\mu_i = \left(\frac{\partial\mu_i}{\partial T}\right)_{P,N_1,\,\ldots,\,N_n} dT + \left(\frac{\partial\mu_i}{\partial P}\right)_{T,N_1,\,\ldots,\,N_n} dP$$

$$+ \sum_{j=1}^{n}\left(\frac{\partial\mu_i}{\partial N_j}\right)_{T,P,N_k} dN_j \quad (k \neq j) \tag{4.116}$$

and note that

$$\left(\frac{\partial\mu_i}{\partial T}\right)_{P,N_1,\,\ldots,\,N_n} = -\left(\frac{\partial S}{\partial N_i}\right)_{T,P,N_j} \quad (j \neq i) \tag{4.117}$$

$$\left(\frac{\partial\mu_i}{\partial P}\right)_{T,N_1,\,\ldots,\,N_n} = \left(\frac{\partial V}{\partial N_i}\right)_{T,P,N_j} \quad (j \neq i) \tag{4.118}$$

Equations (4.117) and (4.118) are two equalities of type (4.54) written for the Gibbs free energy: according to the partial molal property definition (4.101), the right sides of these two equations are $(-\bar{s}_i)$ and $\bar{v}_i$, respectively. We reach the important conclusion that

$$d\bar{g}_i = d\mu_i = -\bar{s}_i\,dT + \bar{v}_i\,dP + \sum_{j=1}^{n}\left(\frac{\partial\mu_i}{\partial N_j}\right)_{T,P,N_k} dN_j \quad (k \neq j) \tag{4.119}$$

which for a single-component system reduces to

$$d\bar{g}_i = d\mu_i = -\bar{s}_i\,dT + \bar{v}_i\,dP \tag{4.119'}$$

Other relations between partial molal quantities and between proper molal quantities can be obtained through the consistent application of definitions (4.101) and (4.107):

$H = G + TS$	$\bar{h}_i = \bar{g}_i + T\bar{s}_i$	$\bar{h} = \bar{g} + T\bar{s}$	(4.120h)
$U = H - PV$	$\bar{u}_i = \bar{h}_i - P\bar{v}_i$	$\bar{u} = \bar{h} - P\bar{v}$	(4.120u)
$F = U - TS$	$\bar{f}_i = \bar{u}_i - T\bar{s}_i$	$\bar{f} = \bar{u} - T\bar{s}$	(4.120f)

Worth keeping in mind is that in these expressions, H, U, F, and their molal counterparts are considered to be functions of $(T, P, N_1, \ldots, N_n)$.

IDEAL GAS MIXTURES

The analytical treatment of a mixture of ideal gases that itself behaves like an ideal gas rests on Dalton's 1802 law, which states that "any gas is a vacuum to any other gas mixed with it" [37]. Let P, V, N, and T represent the pressure, volume, total number of moles, and temperature, respectively, of an ideal gas mixture. One way to interpret Dalton's law is to imagine that the ith component fills the entire volume V while at the mixture temperature T. The pressure attained by the ith component in these circumstances is the *partial pressure* P_i, which is *defined* as

$$P_i = N_i \frac{\bar{R}T}{V} \qquad (4.121)$$

Comparing this quantity with the pressure of the mixture $(P = N\bar{R}T/V)$, we conclude that the sum of all the partial pressures defined by eq. (4.121) equals the pressure of the "mixture" system:

$$P = \sum_{i=1}^{n} P_i \qquad (4.122)$$

We note also that the relative partial pressure of the ith component is the same as its mole fraction:

$$\frac{P_i}{P} = \frac{N_i}{N} = x_i \qquad (4.123)$$

Equation (4.122) constitutes *the law of additivity of pressures* [35], as the first part of Gibbs' analytical reformulation of Dalton's law—now recognized generally as the *Gibbs–Dalton Law* [37]. The second part of the Gibbs–Dalton law is that an extensive mixture property such as U, H, and S is equal to the sum of the corresponding properties of each ideal gas component (U_i, H_i, S_i) that would fill the mixture volume V at the mixture temperature T:

$$U = \sum_{i=1}^{n} U_i = \sum_{i=1}^{n} N_i \bar{u}_i \qquad \bar{u} = \sum_{i=1}^{n} x_i \bar{u}_i \qquad (4.124u)$$

$$H = \sum_{i=1}^{n} H_i = \sum_{i=1}^{n} N_i \bar{h}_i \qquad \bar{h} = \sum_{i=1}^{n} x_i \bar{h}_i \qquad (4.124h)$$

$$S = \sum_{i=1}^{n} S_i = \sum_{i=1}^{n} N_i \bar{s}_i \qquad \bar{s} = \sum_{i=1}^{n} x_i \bar{s}_i \qquad (4.124s)$$

A special feature of the ideal gas mixtures considered here is that the molal quantities $\bar{u}$, $\bar{u}_i$, $\bar{h}$, and $\bar{h}_i$ are functions of T only, whereas the molal entropies depend also on pressure, $\bar{s}_i = \bar{s}_i(T, P_i)$.

It is the additivity property exhibited by the "energies" of eqs. (4.124u) and (4.124h) that makes the gaseous mixture an ideal one. This additivity property allows us to deduce also the specific heats of the ideal gas mixture:

$$\bar{c}_v = \left(\frac{\partial \bar{u}}{\partial T} \right)_{V, N_1, \dots, N_n} = \sum_{i=1}^{n} x_i \bar{c}_{v,i} \qquad (4.125)$$

$$\bar{c}_P = \left(\frac{\partial \bar{h}}{\partial T} \right)_{P, N_1, \dots, N_n} = \sum_{i=1}^{n} x_i \bar{c}_{P,i} \tag{4.126}$$

where the molal heats of each component are defined as

$$\bar{c}_{v,i} = \left(\frac{\partial \bar{u}_i}{\partial T} \right)_V \quad \text{and} \quad \bar{c}_{P,i} = \left(\frac{\partial \bar{h}_i}{\partial T} \right)_P \tag{4.127}$$

Of particular importance in chemical energy calculations is the Gibbs free energy of the ideal gas mixture, which follows from eqs. (4.124):

$$G = H - TS = \sum_{i=1}^{n} N_i (\bar{h}_i - T\bar{s}_i)$$

$$= \sum_{i=1}^{n} N_i \mu_i \tag{4.128}$$

Note here that the chemical potential of the ith component is equal to the partial molal Gibbs free energy evaluated at mixture temperature (T) and the component partial pressure (P_i):

$$\mu_i = \bar{g}_i(T, P_i) \tag{4.129}$$

Focusing on the ith component alone, eq. (4.119′) implies

$$d\mu_i = d\bar{g}_i = -\bar{s}_i \, dT + \bar{v}_i \, dP \tag{4.130}$$

In order to calculate the partial molal volume $\bar{v}_i$, we use the definition (4.101) and $H = U + PV$:

$$\bar{v}_i = \left(\frac{\partial V}{\partial N_i} \right)_{T,P,N_j} = \frac{1}{P} \left[\frac{\partial}{\partial N_i} (H - U) \right]_{T,P,N_j} \quad (j \neq i) \tag{4.131}$$

According to eqs. (4.124u) and (4.124h), this result translates into

$$\bar{v}_i = \frac{1}{P} (\bar{h}_i - \bar{u}_i) \tag{4.132}$$

where, if we view the ith constituent as an ideal gas of pressure P_i, temperature T, and volume V:

$$N_i \bar{h}_i - N_i \bar{u}_i = P_i V = N_i \bar{R} T \tag{4.133}$$

It follows that the group $(\bar{h}_i - \bar{u}_i)$ is equal to $\bar{R}T$ and that the partial molal volume of eq. (4.132) is simply

$$\bar{v}_i = \frac{\bar{R}T}{P} \tag{4.134}$$

Before proceeding with the chemical potential formula (4.130), in which $\bar{v}_i$ is known, it is worth recognizing a different point of view in the description of an ideal gas mixture of ideal gas components. We define the *partial volume* V_i of constituent "i" as

$$V_i = N_i \frac{\bar{R}T}{P} \tag{4.135}$$

i.e., as the volume that would be occupied by the entire quantity of "i" if its temperature and pressure would match the T and P of the mixture. We see immediately that

$$V = \sum_{i=1}^{n} V_i \tag{4.136}$$

and that

$$\frac{V_i}{V} = \frac{N_i}{N} = x_i \tag{4.137}$$

Equation (4.136) represents *Amagat's law of the additivity of volumes.* This law and eq. (4.135) can be substituted into the $\bar{v}_i$ definition (4.131) in order to obtain directly eq. (4.134).

We are now in a position to integrate eq. (4.130) at constant temperature, from the reference pressure P to the partial pressure P_i:

$$\mu_i = \mu_i^{(x_i=1)}(T, P) + \bar{R}T \ln \frac{P_i}{P} \tag{4.138}$$

where $\mu_i^{(x_i=1)}(T, P)$ is the chemical potential of the ith constituent in the limit where it is *alone* in the system at the temperature T and pressure P, that is, when the system is a "single-component" system. From eq. (4.138), we learn that the chemical potential of an ideal gas constituent of an ideal gas mixture, μ_i, is a function of only three properties, the temperature and pressure of the mixture and the mole fraction of the particular constituent (P_i/P or x_i):

$$\mu_i(T, P, x_i) = \mu_i^{(x_i=1)}(T, P) + \bar{R}T \ln x_i \tag{4.139}$$

In addition, the effect of the mole fraction x_i is felt through the natural logarithm $\ln x_i$, which is negative. It means that the chemical potential $\mu_i(T, P, x_i)$ decreases monotonically as the constituent "i" gradually disappears from the mixture, that is, as the mole of fraction x_i decreases. Conversely, $\mu_i^{(x_i=1)}$ represents the ceiling value of μ_i at constant T and P.

Systems where constituents obey a μ_i expression of type (4.139) belong to a more general class of systems that in chemical thermodynamics are called *ideal systems* [38]. The ideal gas mixture considered in this section is just one example of ideal system behavior. Another example of mixtures that obey eq. (4.139) is the class of extremely dilute solutions: these systems fall outside the scope of the present treatment of engineering thermodynamics principles and applications.

REAL GAS MIXTURES

Mixtures of real gases depart from the behavior listed in eqs. (4.138) and (4.139), especially as the pressure P increases. At sufficiently high pressures, the ith chemical potential is the considerably more complicated function suggested by the $\mu_i = \mu_i(T, P, N_1, N_2, \ldots, N_n)$ equation of state presented in the fourth column of Table 4.3.

The special μ_i form that in the case of real gas mixtures replaces the eqs. (4.138) and (4.139) of ideal gas mixtures is easier to grasp if we focus first on a single-component system. If, in addition, this single-component system behaves as an ideal gas, we can combine eqs. (4.130) and (4.134) to write that during an isothermal change:

$$d\mu = d\bar{g} = \frac{\bar{R}T}{P}\, dP \qquad (4.140)$$

in other words,

$$d\mu = d\bar{g} = \bar{R}T\, d(\ln P) \qquad \text{(constant } T\text{, single ideal gas)} \qquad (4.141)$$

If the single-component gaseous system does not behave as an ideal gas, its isothermal change in μ (or $\bar{g}$) does not obey the special function of T and P shown on the right side of eq. (4.141). The actual function of T and P that represents this change can be structured in a way that mimics the structure of eq. (4.141), namely,

$$d\mu = d\bar{g} = \bar{R}T\, d(\ln f) \qquad \text{(constant } T\text{, single nonideal gas)} \qquad (4.142)$$

where $f(T, P)$ is now the new property called *fugacity*.[†] This nomenclature was introduced in 1901 by G. N. Lewis[‡] [39] in order to describe the

---

[†]The name "fugacity" is patterned after the Latin feminine noun *fuga* (flight, the act of running away) in order to suggest the escaping tendency of the nonideal gas. In fact, before naming it fugacity, G. N. Lewis labeled the same quantity *escaping tendency* and gave it the symbol ψ [40].

[‡]Gilbert Newton Lewis (1875–1946) taught chemistry at the Massachusetts Institute of Technology (1907–1912) and the University of California, Berkeley (1912–1946). At Berkeley, he was also dean of the college of chemistry. Worth noting among his many achievements is his and H. C. Urey's 1923 discovery of heavy water (D_2O, deuterium oxide).

Observation: Students of heat and mass transfer may be familiar with the *Lewis number* notation $Le = \alpha/D$, where α and D are the thermal and mass diffusivities of the medium. This terminology was established in honor of another MIT professor, Warren K. Lewis, who taught chemical engineering.

property that in the case of a nonideal gas plays a role analogous to that of pressure in the limit of ideal gas behavior. Recognizing $P \to 0$ as the ideal gas limit, eqs. (4.141) and (4.142) require

$$\lim_{P \to 0} \left(\frac{f}{P} \right) \to 1 \qquad (4.143)$$

This is another way of saying that the fugacity of an ideal gas is equal to the pressure of the ideal gas system.

The departure of the fugacity function of a real gas is directly related to the departure of the shape of the $P(\bar{v}, T)$ surface from the shape approached by this surface in the ideal gas limit, $P\bar{v} = \bar{R}T$. This can be shown analytically by integrating eq. (4.142) along an isothermal path from the *reference* state (T, P^*) to the arbitrary state (T, P) [36]:

$$\bar{g}(T, P) - \bar{g}^*(T, P^*) = \bar{R}T \ln \frac{f}{f^*} \qquad \text{(constant } T) \qquad (4.144)$$

where $f^* = f^*(T, P^*)$. However, since in general $d\bar{g} = -\bar{s}\, dT + \bar{v}\, dP$, the isothermal-change difference $(\bar{g} - \bar{g}^*)$ shown above is also equal to

$$\bar{g}(T, P) - \bar{g}^*(T, P^*) = \left(\int_{P^*}^{P} \bar{v}\, dP \right)_T \qquad (4.145)$$

Putting eqs. (4.144) and (4.145) together, we see that

$$\ln f = \ln f^* + \frac{1}{\bar{R}T} \left(\int_{P^*}^{P} \bar{v}\, dP \right)_T \qquad (4.146)$$

and, after subtracting $\ln P$ from both sides:

$$\ln \frac{f}{P} = \ln \frac{f^*}{P^*} - \ln \frac{P}{P^*} + \left(\int_{P^*}^{P} \frac{\bar{v}}{\bar{R}T}\, dP \right)_T$$

$$= \ln \frac{f^*}{P^*} + \left[\int_{P^*}^{P} \left(\frac{\bar{v}}{\bar{R}T} - \frac{1}{P} \right) dP \right]_T \qquad (4.147)$$

Taking this last conclusion to the limit $P^* \to 0$ and recognizing that in this limit $\ln(f^*/P^*) \to 0$, we learn finally that

$$\ln \frac{f}{P} = \left[\int_{0}^{P} \left(\frac{\bar{v}}{\bar{R}T} - \frac{1}{P} \right) dP \right]_T \qquad (4.148)$$

or, in term of the specific volume $v(\text{m}^3/\text{kg})$ and the particular ideal gas constant of the gaseous system, $R\ (\text{J/kg K})$:

$$\ln \frac{f}{P} = \left[\int_{0}^{P} \left(\frac{v}{RT} - \frac{1}{P} \right) dP \right]_T \qquad (4.149)$$

It is easy to see that in the case of an ideal gas, the integrand is zero; therefore, $f = P$ regardless of P. Just like the fugacity function, the *fugacity coefficient* f/P is a function of both T and P. The fugacity coefficient can be either smaller or greater than 1, as illustrated by the three Honari–Brown generalized fugacity coefficient charts [41] reproduced in Fig. 4.17. The reduced temperature and pressure coordinates employed in these figures are defined as

$$T_r = \frac{T}{T_c} \quad \text{and} \quad P_r = \frac{P}{P_c} \tag{4.150}$$

where T_c and P_c are the conditions at the critical point. The theoretical basis for the use of these dimensionless coordinates is presented in detail on pages 275–299 of chapter 6. At this point, it helps to study the shape and location of the f/P curves in order to see one more time the locus of ideal gas behavior, namely, the limit of low pressures and relatively high temperatures.

Consider now the chemical potential of the gaseous constituent "i" in a mixture. Reexamining eq. (4.138), we see that in the ideal gas limit, the change in μ_i (or $\bar{g}_i$) at constant temperature has the special form

$$d\mu_i = d\bar{g}_i = \bar{R}T \, d(\ln P_i) \quad \text{(constant } T\text{, ideal gas)} \tag{4.151}$$

Following the same argument that gave us eq. (4.142), the isothermal change in the chemical potential of a constituent in a nonideal gas mixture is

$$d\mu_i = d\bar{g}_i = \bar{R}T \, d(\ln f_i) \quad \text{(constant } T\text{, nonideal gas)} \tag{4.152}$$

where f_i is the fugacity of the ith constituent in that particular mixture. In the limit of ideal gas behavior, $P \to 0$, the fugacity of the constituent becomes the same as its partial pressure in the sense of Dalton (that is, a partial pressure in a mixture that behaves as an ideal gas, $P_i = x_i P$):

$$\lim_{P \to 0} \frac{f_i}{x_i P} = 1 \tag{4.153}$$

Integrating at constant temperature as the overall system pressure increases from the reference P^* to P, eq. (4.152) yields

$$\mu_i = \mu_i^* + \bar{R}T \ln \frac{f_i}{f_i^*} \tag{4.154}$$

In this expression μ_i^* and f_i^* represent the chemical potential and fugacity, respectively, of "i" if the mixture conditions are T and P^*. This form can be compared now with eq. (4.138) to see the similarity in their construction. We shall use these forms in the treatment of applications involving nonreacting and reacting gaseous mixtures (chapters 5 and 7, respectively).

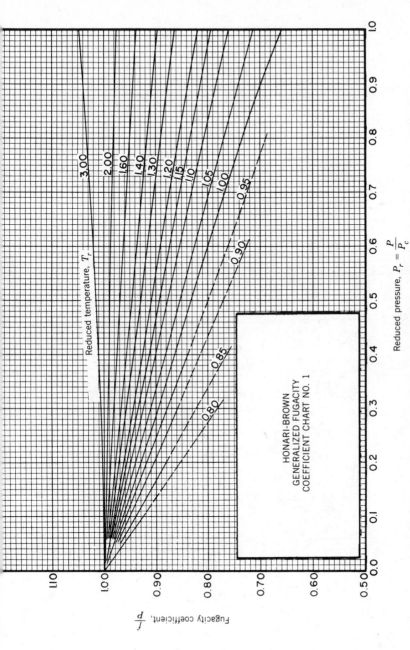

Figure 4.17 The Honari–Brown generalized fugacity-coefficient charts [41] (courtesy of Professor Edward F. Obert, University of Wisconsin, Madison). Charts 2 and 3 on following pages.

205

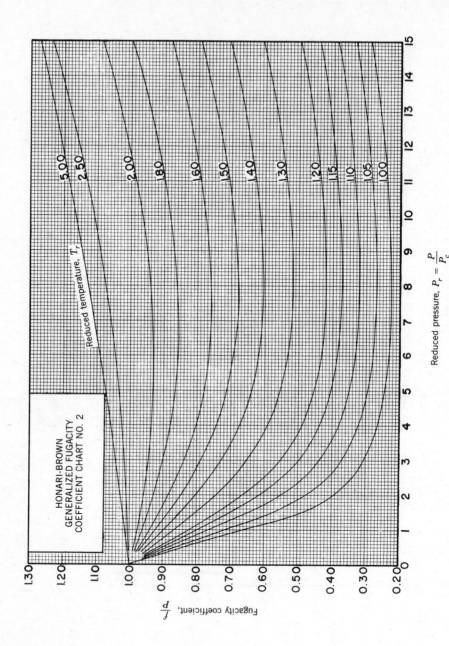

Figure 4.17 Continued.

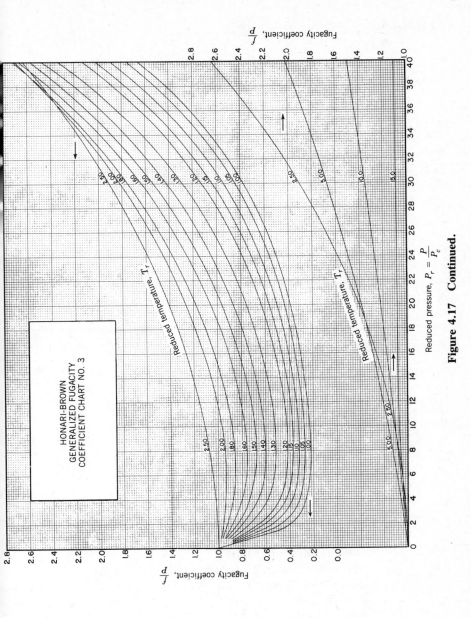

Figure 4.17 Continued.

207

SYMBOLS

c	lone specific heat of an incompressible substance
$c_v, \bar{c}_v$	specific heat at constant volume [J/kg K, J/mol K]
$c_P, \bar{c}_P$	specific heat at constant pressure [J/kg K, J/mol K]
f	fugacity [eqs. (4.142) and (4.143)]
$f(x, y)$	function
$f, \bar{f}, F$	Helmholtz free energy [J/kg, J/mol, J]
F_μ	grand canonical potential $[F_\mu(T, V, \mu)]$
$g, \bar{g}, G$	Gibbs free energy [J/kg, J/mol, J]
G_μ	total Legendre transform $[G_\mu(T, P, \mu) \equiv 0]$
$h, \bar{h}, H$	enthalpy [J/kg, J/mol, J]
H_μ	thermodynamic potential $[H_\mu(S, P, \mu)]$
k	isentropic expansion coefficient [eq. (4.66)]; equal to c_P/c_v only in the ideal gas limit
k_T	isothermal expansion coefficient [eq. (4.74)]; equal to 1 only in the ideal gas limit
m, M	mass [kg]
N	number of moles
P	pressure
P_i	partial pressure [eq. (4.121)]
P^*	reference pressure
Q	heat transfer interaction
δQ_{rev}	reversible heat transfer interaction
R	ideal gas constant
$\bar{R}$	universal ideal gas constant [8.314 J/mol K]
$s, \bar{s}, S$	entropy [J/kg K, J/mol K, J/K]
S_{gen}	entropy generation [J/K]
T	temperature
$u, \bar{u}, U$	internal energy [J/kg, J/mol, J]
U_μ	thermodynamic potential $[U_\mu(S, V, \mu)]$
$v, \bar{v}, V$	volume [m³/kg, m³/mol, m³]
V_i	partial volume [eq. (4.135)]
W	work transfer interaction
δW_{rev}	reversible work transfer interaction
$\bar{x}$	proper molal property [eq. (4.107)]
x_i	mole fraction [eq. (4.105)]
$\bar{x}_i$	partial molal property [eq. (4.101)]
x_i, x_j, x_k	sample properties employed in Bridgman's scheme
x, y, z, u, v	sample properties employed in Shaw's method
$y(x)$	original function (Fig. 4.6)
α	sample property employed in Bridgman's scheme
$\alpha, \beta, \gamma, \delta$	sample properties employed in Shaw's method
β	volumetric coefficient of thermal expansion [eq. (4.61)]
$\eta(\zeta)$	Legendre transform of $y(x)$ (Fig. 4.6)

$\eta^{(1)}$	first Legendre transform
$\eta^{(2)}$	second Legendre transform
κ	isothermal compressibility [eq. (4.70)]
λ	number
μ	chemical potential [J/mol]
μ_J	Joule–Thomson expansion coefficient [eq. (4.79)]
$(\ )_c$	critical-point properties
$(\ )_i$	property of the ith constituent
$(\ )_r$	reduced properties [eqs. (4.150)]
$(\ )_0$	property of an environmental reservoir
$(\ )^*$	reference state
$(\ )_\Lambda$	property of system (Λ)
$(\ )_\Sigma$	property of system (Σ)

REFERENCES

1. J. W. Gibbs, *The Collected Works of J. Willard Gibbs*, Vol. I, Longmans, Green, New York, 1928, p. 62.

2. H. B. Callen, *Thermodynamics*, Wiley, New York, 1960, p. 12.

3. K. Rektorys, ed., *Survey of Applicable Mathematics*, MIT Press, Cambridge, MA, 1969, pp. 454–455.

4. R. Courant and D. Hilbert, *Methods of Mathematical Physics*, Vol. 2, Wiley (Interscience), New York, 1962, p. 32.

5. L. Tisza, *Generalized Thermodynamics*, MIT Press, Cambridge, MA, 1966, pp. 235–241.

6. B. L. Beegle, M. Modell, and R. C. Reid, Legendre transforms and their application in thermodynamics, *AIChE J.*, Vol. 20, 1974, pp. 1194–1200.

7. J. C. Maxwell, *Theory of Heat*, 10th ed., with corrections and additions (1891) by Lord Rayleigh, Longmans, Green, London, 1904, pp. 165–169.

8. J. C. Wilcke, *Proc. R. Soc. Stockholm*, 1772.

9. A. Bejan, *Convection Heat Transfer*, Wiley, New York, 1984, p. 113.

10. J. H. Keenan, F. G. Keyes, P. G. Hill, and J. G. Moore, *Steam Tables*, Wiley, New York, 1969, p. 124.

11. J. H. Keenan and J. Kaye, *Gas Tables*, Wiley, New York, 1966.

12. R. W. Haywood, *Equilibrium Thermodynamics for Engineers and Scientists*, Wiley, New York, 1980, p. 284.

13. W. Thomson and J. P. Joule, On the thermal effects of fluids in motion, *Philos. Trans. R. Soc. London*, 1853, p. 357; see also J. P. Joule, *Joint Scientific Papers of James Prescott Joule*, Taylor & Francis, London, 1887, pp. 231–245.

14. R. T. Jacobsen, R. B. Stewart, R. D. McCarthy, and H. J. M. Hanley, Thermophysical properties of nitrogen from the fusion line to 3500 R (1944 K) for pressures to 150,000 psia (10342×10^5 N/m^2), *NBS Tech. Note (U.S.)*, No. 648, December 1973.

15. R. D. McCarthy, Thermophysical properties of Helium-4 from 2 to 1500 K with pressures to 1000 atmospheres, *NBS Tech. Note (U.S.)*, No. 631, November 1972.

16. P. W. Bridgman, A complete collection of thermodynamic formulas, *Phys. Rev.*, Vol. 3, 1914, pp. 273–281.

17. P. W. Bridgman, *A Condensed Collection of Thermodynamic Formulas*, Dover, New York, 1961.

18. A. N. Shaw, The derivation of thermodynamical relations for a simple system, *Philos. Trans. R. Soc. London, Ser. A*, Vol. 234, 1935, pp. 299–328.

19. C. W. Somerton and Ö. A. Arnas, On the use of Jacobians to reduce thermodynamic derivatives, *Int. J. Mech. Eng. Educ.*, Vol. 13, No. 1, 1985, pp. 9–18.

20. A. Bejan and H. M. Paynter, *Solved Problems in Thermodynamics*, Mechanical Engineering Department, Massachusetts Institute of Technology, Cambridge, MA, 1976, p. 10-5.

21. F. O. Koenig, Families of thermodynamic equations. I, The method of transformations by the characteristic group, *J. Chem. Phys.*, Vol. 3, 1935, pp. 29–35.

22. M. C. Branch and A. K. Hayashi, Thermodynamic derivatives, *Int. J. Mech. Eng. Educ.*, Vol. 12, No. 1, 1984, pp. 25–34.

23. F. O. Koenig, Families of thermodynamic equations. II. The case of eight characteristic functions, *J. Chem. Phys.*, Vol. 56, 1972, pp. 4556–4562.

24. R. P. Reddy, Octahedral mnemonic model for thermodynamic relations, *Mech. Eng. News*, Vol. 23, No. 1, February 1986, pp. 3–6.

25. R. F. Fox, The thermodynamic cuboctahedron, *J. Chem. Educ.*, Vol. 53, 1976, pp. 441–442.

26. A. Bejan, Buckling flows: A new frontier in fluid mechanics, in T. C. Chawla, ed., *Annual Review of Numerical Fluid Mechanics and Heat Transfer*, Vol. I, Hemisphere, Washington, DC, 1987, pp. 262–304.

27. A. Bejan, Buckling flows: A new frontier in convection heat transfer. Opening lecture at the 2nd Latin American Congress on Heat and Mass Transfer, São Paulo, Brazil, May 12–15, 1986, *Latin Am. J. Heat Mass Transfer*, Vol. 10, 1987, pp. 83–103.

28. F. Weinhold, Metric geometry of equilibrium thermodynamics, *J. Chem. Phys.*, Vol. 63, No. 6, 1975, pp. 2479–2483.

29. F. Weinhold, Metric geometry of equilibrium thermodynamics. II. Scaling, homogeneity, and generalized Gibbs–Duhem relations, *J. Chem. Phys.*, Vol. 63, No. 6, 1975, pp. 2484–2487.

30. F. Weinhold, Metric geometry of equilibrium thermodynamics. III. Elementary formal structure of a vector-algebraic representation of equilibrium thermodynamics, *J. Chem. Phys.*, Vol. 63, No. 6, pp. 2488–2495.

31. F. Weinhold, Metric geometry of equilibrium thermodynamics. IV. Vector-algebraic evaluation of thermodynamic derivatives, *J. Chem. Phys.*, Vol. 63, No. 6, pp. 2496–2501.

32. F. Weinhold, Geometric representation of equilibrium thermodynamics, *Acc. Chem. Res.*, Vol. 9, 1976, pp. 236–240.

33. F. Weinhold, Thermodynamics and geometry, *Phys. Today*, Vol. 29, 1976, pp. 23–30.

34. F. Weinhold, Metric geometry of equilibrium thermodynamics. V. Aspects of heterogeneous equilibrium, *J. Chem. Phys.*, Vol. 65, 1976, pp. 559–564.

35. E. A. Guggenheim, *Thermodynamics*, 7th ed., North-Holland Publ., Amsterdam, 1985, p. 20.

36. J. S. Hsieh, *Principles of Thermodynamics*, McGraw-Hill, New York, 1975, p. 137.

37. D. B. Spalding and E. H. Cole, *Engineering Thermodynamics*, 3rd ed., Edward Arnold, London, 1973, p. 324.

38. I. Prigogine and R. Defay, *Chemical Thermodynamics*, translated by D. H. Everett, Longmans, Green, London, 1954.

39. G. N. Lewis, The law of physico-chemical change, *Proc. Am. Acad. Arts Sci.*, Vol. 37, 1901, pp. 49–69.

40. G. N. Lewis, *Proc. Am. Acad. Arts Sci.*, Vol. 36, 1900, p. 145; also *Z. Phys. Chem.*, Vol. 35, 1900, p. 343.

41. E. F. Obert, Personal communication, March 4, 1987; the charts are from Honari's M.S. thesis done under the supervision of Prof. G. Brown, Chemical Engineering Department, Northwestern Technological Institute, 1955, based on Nelson's data (L. C. Nelson and E. F. Obert, Generalized pvT properties of gases, *Trans. ASME*, Vol. 76, 1954, pp. 1057–1066).

42. H. W. Liepmann and A. Roshko, *Elements of Gasdynamics*, Wiley, New York, 1957, p. 50.

PROBLEMS

4.1 Derive the Gibbs–Duhem relation in entropy representation, eq. (4.35), by relying on the Euler equation and the entropic fundamental relation (4.18). Show that the same relation can be derived directly from the Gibbs–Duhem relation in energy representation, eq. (4.32).

4.2 Consider a single-component system whose energetic fundamental relation $U(S, V, N)$ is known. Following the example of the second Legendre transform given in Table 4.3, outline the construction of the so-called *grand canonical potential* function $F_\mu(T, V, \mu)$, in which T and μ replace S and N as independent variables in the fundamental relation [Ref. 2, p. 100]. Determine $F_\mu(T, V, \mu)$ analytically for the single-component ideal gas system treated in Example 4.1.

4.3 Following the steps outlined in Table 4.3, derive analytical expressions for the Helmholtz free-energy fundamental relation of a single-component ideal gas, $F(T, V, N)$, and for the Gibbs free-energy fundamental relation of the same substance, $G(T, P, N)$.

4.4   Continuing the procedure represented by the step from eq. (4.51) to eq. (4.52) in the text, we can construct the (k)th Legendre transform:

$$\eta^{(k)} = y - \sum_{i=1}^{k} \zeta_i x_i$$

in which the x_is are k of the original $(n + 2)$ arguments of function y, while the ζ_is represent the corresponding first derivatives that replace the x_is. The ultimate example of this kind is the total Legendre transform $\eta^{(n+2)}$, in which all the original arguments are replaced by the corresponding first derivatives [6]. Show that the *total* Legendre transform of the energy fundamental relation $U = U(S, V, N_1, \ldots, N_2)$ is identically equal to zero.

4.5   Prove the validity of eq. (4.69), in other words, show that

$$\left(\frac{\partial P}{\partial v}\right)_T \frac{c_P}{c_v} = \left(\frac{\partial P}{\partial v}\right)_s$$

Hint: Use the entropic relations for c_P and c_v, eqs. (4.58) and (4.60), and the cyclical relation (see the Appendix).

4.6   Without using the Bridgman Table 4.6, prove the validity of the following relations:

(a) $\left(\dfrac{\partial s}{\partial v}\right)_T = \dfrac{\beta}{\kappa}$

(b) $\left(\dfrac{\partial s}{\partial P}\right)_T = -\beta v$

(c) $c_v = c_P - \dfrac{v}{\kappa}\beta^2 T$

(d) $\left(\dfrac{\partial h}{\partial P}\right)_T = v(1 - \beta T)$

(e) $\mu_J = \dfrac{v}{c_P}(\beta T - 1)$

(f) $\left(\dfrac{\partial T}{\partial P}\right)_s = \dfrac{\beta}{c_P} vT$

(g) $\left(\dfrac{\partial c_P}{\partial P}\right)_T = -T\left(\dfrac{\partial^2 v}{\partial T^2}\right)_P$

4.7   Use the formulas assembled in Bridgman's Table 4.6 in order to prove the identity

$$\frac{\partial^2 u}{\partial v^2} - \left(\frac{\partial^2 u}{\partial s\,\partial v}\right)^2 \Big/ \left(\frac{\partial^2 u}{\partial s^2}\right) = \frac{\partial^2 f}{\partial v^2}$$

where, according to Table 4.3 for closed systems, $u = u(s, v)$ and $f = f(T, v)$.

4.8 Reconstruct the relations listed in the left column of Table 4.7 by following the two-step procedure highlighted in Example 4.5. (Use Bridgman's Table 4.6 only to verify the accuracy of your results.)

4.9 The two equations of state for compressed liquid water near its density maximum can be written approximately as

$$v(T, P) = v_0[1 + \lambda(T - T_0 + aP)^2 - \kappa_0 P]$$

$$(c_P)_{P=0 \text{ atm}} = c_0 - b(T - T_0)$$

where v_0, λ, T_0, a, κ_0, c_0, and b are known constants (see the Appendix).

(a) Use these two equations to show that the specific heat at pressures other than atmospheric is

$$c_P(T, P) = c_0 - b(T - T_0) - 2\lambda v_0 TP$$

(b) Choosing as reference entropy $s(T_0, 0) = 0$, show that

$$s(T, P) = (c_0 + bT_0) \ln \frac{T}{T_0} - b(T - T_0)$$

$$- 2\lambda v_0 P(T - T_0) - \lambda v_0 aP^2$$

and that the reversible and adiabatic curves must appear cup-shaped in the T–P plane.

(c) Derive the fundamental relation in the Gibbs free-energy representation, $g = g(T, P)$. As reference, choose $g(T_0, 0) = 0$.

(d) Show that for an incompressible liquid with constant specific heat, the fundamental relation derived above reduces to

$$g(T, P) = c_0(T - T_0) - c_0 T \ln \frac{T}{T_0} + v_0 P$$

4.10 A batch of a certain substance (a simple system) is held in a constant-volume, rigid container. The system experiences an infinitesimal change of state during which its pressure rises. The question is whether during this constant-volume pressurization process, the system is being cooled or heated. Show that the constant-volume pressurization is the result of heating if the system's properties vary such that

$$\frac{\kappa}{\beta} c_v > 0$$

where c_v, κ, and β are the specific heat at constant volume, the isothermal compressibility, and the volumetric coefficient of thermal expansion, respectively.

4.11 The speed with which small pressure waves travel through a compressible fluid is the *speed of sound*, a, which is defined by [42]

$$a^2 = \left(\frac{\partial P}{\partial \rho}\right)_s$$

where ρ is the density of the fluid, $\rho = 1/v$. Demonstrate the validity of the following relations:

(a) $a^2 = \dfrac{v c_P}{\kappa c_v}$

(b) $a = (kRT)^{1/2}$, for an ideal gas.

4.12 The $P = P(v, T)$ equation of state of a van der Waals gas is

$$P = \frac{RT}{v - b} - \frac{a}{v^2}$$

in which R, a, and b are three constants.

(a) Prove that if c_v is also constant, the "caloric" equation of state $u = u(T, v)$ of the same gas is

$$u = u_0 + c_v(T - T_0) + a\left(\frac{1}{v_0} - \frac{1}{v}\right)$$

where $u_0 = u(T_0, v_0)$.

(b) Derive the expression for the entropy function $s = s(T, v)$, which is valid under the same conditions.

(c) Combining this last result with $dh = T\,ds + v\,dP$, derive the corresponding expression for enthalpy, $h = h(T, v)$.

5

Exergy Generalized

The objective of this chapter is to provide a few breathing moments for digesting all the analytical facts that have been condensed in chapter 4. We accomplish this by focusing on a special class of energy-engineering problems.

The formal introduction of the concept of chemical equilibrium in the preceding chapter allows us now to reexamine the basic exergy problem of chapter 3, this time in the more general and diverse world of systems that can reach not only thermal and mechanical equilibrium, but also chemical equilibrium with their respective environments. The energy-engineering mission of this chapter then is to establish in quantitative terms the theoretical limit to the production of useful work in processes in which systems can experience not only heat transfer and work transfer, but also mass transfer interactions with their environments.

NONFLOW SYSTEMS

Consider a batch of a certain mixture containing $N_1, N_2, \ldots, N_n$ moles of n different constituents. We shall refer to this batch as the "mixture" system. The initial equilibrium state of the system is characterized by a temperature (T), pressure (P), and n chemical potentials ($\mu_1, \mu_2, \ldots, \mu_n$) that differ from the corresponding intensities of the environment (T_0, P_0, $\mu_{0,1}$, $\mu_{0,2}, \ldots, \mu_{0,n}$). In other words, the batch system and the environment are not in mutual thermal, mechanical, and chemical equilibrium.

The question that can be formulated at this stage is: "What is the maximum useful work that could be produced as the system and the environment reach equilibrium?" The same question was studied in the context of closed systems in chapter 3 (see the section on "Nonflow Processes"), where equilibrium with the environment meant only $T = T_0$ and $P = P_0$. The new element that justifies the resurrection of the nonflow exergy question at this time is that the equilibrium now is not only thermal and mechanical, but also chemical.

Figure 5.1 shows the initial and final states of a process $(1) \rightarrow (2)$ that is designed to bring our original system in a state of equilibrium with the environment. Note the equality between the system intensities and the environmental intensities in the final state. Note also that all the extensive properties change from state (1) to state (2), including the mole numbers N_i $(i = 1, 2, \ldots, n)$. These numbers change as each of the n constituents diffuses through its own semipermeable membrane that makes up the system–environment boundary. It is due to these mass transfer interactions from state (1) to state (2) that the mixture functions as an *open* system, albeit in a nonflow apparatus.

The equations that account for the conservation of each constituent and for the first and second laws are

$$\frac{dN_i}{dt} = \dot{N}_i \qquad (i = 1, 2, \ldots, n) \tag{5.1}$$

$$\frac{dU}{dt} = \dot{Q}_0 - \dot{W} + \sum_{i=1}^{n} \dot{N}_i \bar{h}_{0,i} \tag{5.2}$$

$$\dot{S}_{\text{gen}} = \frac{dS}{dt} - \frac{\dot{Q}_0}{T_0} - \sum_{i=1}^{n} \dot{N}_i \bar{s}_{0,i} \geq 0 \tag{5.3}$$

In these equations, $\bar{h}_{0,i}$ and $\bar{s}_{0,i}$ represent the molal enthalpy and entropy of the ith constituent that crosses the mixture–environment boundary. Integrating in time from state (1) to state (2), and splitting $\dot{W}$ again into work done on the environment $P_0 \, dV/dt$ plus a useful component $\dot{E}_W$ (also called "exergy"), eqs. (5.1)–(5.3) yield

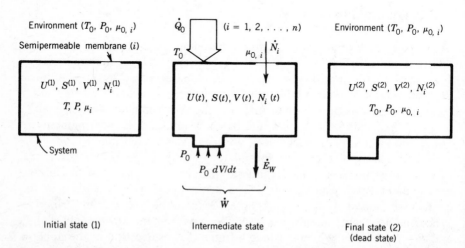

Figure 5.1 Nonflow system, reaching thermal, mechanical, and chemical equilibrium with the atmosphere.

$$N_i^{(2)} - N_i^{(1)} = \int_1^2 \dot{N}_i \, dt \qquad (i = 1, 2, \ldots, n) \tag{5.4}$$

$$U^{(2)} - U^{(1)} = Q_0 - E_w - P_0(V^{(2)} - V^{(1)}) + \sum_{i=1}^n \bar{h}_{0,i} \int_1^2 \dot{N}_i \, dt \tag{5.5}$$

$$S_{gen} = S^{(2)} - S^{(1)} - \frac{Q_0}{T_0} - \sum_{i=1}^n \bar{s}_{0,i} \int_1^2 \dot{N}_i \, dt \tag{5.6}$$

Eliminating Q_0 and the n integrals of type $\int_1^2 \dot{N}_i \, dt$, we obtain

$$E_W = U^{(1)} - T_0 S^{(1)} + P_0 V^{(1)} - \sum_{i=1}^n \mu_{0,i} N_i^{(1)}$$

$$- \left[U^{(2)} - T_0 S^{(2)} + P_0 V^{(2)} - \sum_{i=1}^n \mu_{0,i} N_i^{(2)} \right] - T_0 S_{gen} \tag{5.7}$$

where $\mu_{0,i} = \bar{g}_{0,i} = \bar{h}_{0,i} - T_0 \bar{s}_{0,i}$. Finally, we note that since T_0, P_0, and the $\mu_{0,i}$s are the intensive properties of the mixture in the final state (Fig. 5.1), the Euler equation (4.30″) assures us that the terms collected between the brackets in eq. (5.7) add up to zero:

$$U^{(2)} - T_0 S^{(2)} + P_0 V^{(2)} - \sum_{i=1}^n \mu_{0,i} N_i^{(2)} = 0 \tag{5.8}$$

The first four terms on the right side of eq. (5.7) represent the maximum useful work that could be extracted[†] as the mixture and the environment come to equilibrium at the end of a *reversible* process. Dropping the "(1)" superscripts, we conclude that

$$(E_W)_{rev} = U - T_0 S + P_0 V - \sum_{i=1}^n \mu_{0,i} N_i \tag{5.9}$$

where $(U, S, V, N_1, N_2, \ldots, N_n)$ are the original extensive properties of the mixture system. We reexamine this conclusion in the last paragraph of this section [see eq. (5.9′)].

It is useful to compare this last result with the chapter 3 conclusion that the maximum work to be extracted from a closed system that reaches only thermal and mechanical equilibrium with the ambient is the nonflow exergy Ξ, or, according to eq. (3.32),

$$\Xi = U - U^* - T_0(S - S^*) + P_0(V - V^*) \tag{5.10}$$

The new notation $( )^*$ indicates properties associated with the *restricted dead state* of the mixture system, which is the state where only the

[†]Or the minimum work that would have to be invested.

temperature and pressure match the corresponding environmental values. Invoking Euler's equation at the restricted dead state:

$$U^* = T_0 S^* - P_0 V^* + \sum_{i=1}^{n} \mu_i^* N_i \qquad (5.11)$$

we note further that the restricted dead-state chemical potentials $\mu_i^*(T_0, P_0)$ are not necessarily equal to the chemical potentials $(\mu_{0,i})$ enjoyed by the mixture in its ultimate (proper) dead state, when its equilibrium with the environment is thermal, mechanical, *and* chemical.

Combining eqs. (5.9)–(5.11), we learn that

$$(E_W)_{\text{rev}} = U - U^* - T_0(S - S^*) + P_0(V - V^*) + \sum_{i=1}^{n} (\mu_i^* - \mu_{0,i})N_i \quad (5.12)$$

or, in view of eq. (5.10),

$$(E_W)_{\text{rev}} = \Xi + \Xi_{\text{ch}} \qquad (5.13)$$

in which Ξ_{ch} represents the *nonflow chemical exergy* defined by

$$\Xi_{\text{ch}} = \sum_{i=1}^{n} (\mu_i^* - \mu_{0,i})N_i \qquad (5.14)$$

The maximum available work from the nonflow system, $(E_W)_{\text{rev}}$, emerges as the sum of two contributions, first, the nonflow exergy Ξ released en route to the restricted dead state, plus the chemical exergy Ξ_{ch} released as the mixture reaches chemical equilibrium with the environment, while the mixture temperature and pressure before and after this last process are fixed at T_0 and P_0, respectively.

In order to distinguish Ξ from Ξ_{ch}, the former is also recognized as the *nonflow thermomechanical exergy* (or the *nonflow physical exergy*) of the original fixed-mass and fixed-composition system $(U, S, V, N_1, N_2, \ldots, N_n)$ [1–9]. This name enforces the observation that useful work of algebraically maximum size Ξ is released as the system and the environment reach only thermal and mechanical equilibrium. With the same idea in mind, the sum $(\Xi + \Xi_{\text{ch}})$ listed in eq. (5.13) can be viewed as the *nonflow thermomechanical and chemical exergy* [8], or, more succinctly, the *total exergy* Ξ_t of the original mixture batch:

$$\Xi_t = \Xi + \Xi_{\text{ch}} \qquad (5.13')$$

This quantity is also known as the nonflow *essergy* of the mixture, after Professor Evans' seminal doctoral thesis on the generalization of the concepts of availability and exergy [10]. Note finally that the total nonflow exergy Ξ_t is nothing but a new name for $(E_W)_{\text{rev}}$ of eq. (5.9), in other words, that

$$\Xi_t = U - T_0 S + P_0 V - \sum_{i=1}^{n} \mu_{0,i} N_i \qquad (5.9')$$

FLOW SYSTEMS

Consider next the companion problem of calculating the maximum useful power that could be extracted from a stream as it steadily reaches thermal, mechanical, and chemical equilibrium with the environment. A batch of the fluid that flows into the control volume has the initial composition (N_1, $N_2, \ldots, N_n$). Figure 5.2 shows that the inflowing stream amounts to the superposition of n streams with molal flowrates $\dot{N}_{i,\text{in}}$ ($i = 1, 2, \ldots, n$). Inside the control volume, the mixture and the environment exchange mass, so that, in general, the outflowing mixture has a different composition, $\dot{N}_{i,\text{out}}$ ($i = 1, 2, \ldots, n$). The steady-state conservation of each constituent is guaranteed by writing directly ($\dot{N}_{i,\text{out}} - \dot{N}_{i,\text{in}}$) for the flowrate of constituent (i) through the ith semipermeable membrane at the boundary between the control volume and the environment.

The steady-state first-law and the second-law statements for the control volume are

$$\dot{E}_W = \dot{Q}_0 + \sum_{i=1}^{n} (\bar{h}_i \dot{N}_i)_{\text{in}} - \sum_{i=1}^{n} (\bar{h}_i \dot{N}_i)_{\text{out}} + \sum_{i=1}^{n} \bar{h}_{0,i}(\dot{N}_{i,\text{out}} - \dot{N}_{i,\text{in}}) \quad (5.15)$$

$$\dot{S}_{\text{gen}} = -\frac{\dot{Q}_0}{T_0} - \sum_{i=1}^{n} (\bar{s}_i \dot{N}_i)_{\text{in}} + \sum_{i=1}^{n} (\bar{s}_i \dot{N}_i)_{\text{out}} - \sum_{i=1}^{n} \bar{s}_{0,i}(\dot{N}_{i,\text{out}} - \dot{N}_{i,\text{in}}) \quad (5.16)$$

where, keeping up with the notation used in a related problem in chapter 3 (see the section on "Steady-Flow Processes"), $\dot{E}_W$ is the exergy delivery rate or the mechanical power $\dot{W}$ put out by the control volume as an open system. Eliminating $\dot{Q}_0$ between eqs. (5.15) and (5.16), and writing $\mu_{0,i}$ instead of ($\bar{h}_{0,i} - T_0 \bar{s}_{0,i}$), we obtain

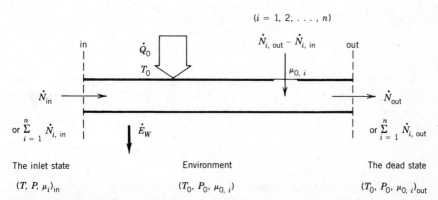

Figure 5.2 Steady-flow apparatus, in which a mixture stream is brought into thermal, mechanical, and chemical equilibrium with the environment.

$$\dot{E}_W = \sum_{i=1}^{n} [(\bar{h}_i - T_0\bar{s}_i)_{\text{in}} - \mu_{0,i}]\dot{N}_{i,\text{in}}$$

$$- \sum_{i=1}^{n} [(\bar{h}_i - T_0\bar{s}_i)_{\text{out}} - \mu_{0,i}]\dot{N}_{i,\text{out}} - T_0\dot{S}_{\text{gen}} \qquad (5.17)$$

The outflowing mixture is in thermal, mechanical, and chemical equilibrium with the ambient:

$$(\bar{h}_i - T_0\bar{s}_i)_{\text{out}} = \bar{g}_{0,i} = \mu_{0,i} \qquad (5.18)$$

which means that all the terms in the second summation of eq. (5.17) vanish. We are left with the conclusion that the maximum-exergy delivery rate occurs when the flow apparatus functions reversibly:

$$(\dot{E}_W)_{\text{rev}} = \bar{h}\dot{N} - T_0\bar{s}\dot{N} - \sum_{i=1}^{n} \mu_{0,i}\dot{N}_{i,\text{in}} \qquad (5.19)$$

where $\dot{N}$ is the total flowrate measured at the inlet:

$$\dot{N} = \sum_{i=1}^{n} \dot{N}_{i,\text{in}} \qquad (5.20)$$

and where $\bar{h}$ and $\bar{s}$ are the molal enthalpy and entropy, respectively, of the inflowing mixture:

$$\bar{h} = \frac{1}{\dot{N}} \sum_{i=1}^{n} \bar{h}_{i,\text{in}}\dot{N}_{i,\text{in}} \qquad \bar{s} = \frac{1}{\dot{N}} \sum_{i=1}^{n} \bar{s}_{i,\text{in}}\dot{N}_{i,\text{in}} \qquad (5.21)$$

On a per-unit-$\dot{N}$ basis, the maximum-exergy delivery rate (5.19) can be written as

$$\frac{(\dot{E}_W)_{\text{rev}}}{\dot{N}} = \sum_{i=1}^{n} (\bar{h}_{i,\text{in}} - T_0\bar{s}_{i,\text{in}} - \mu_{0,i})x_i$$

$$= \bar{h} - T_0\bar{s} - \sum_{i=1}^{n} \mu_{0,i}x_i \qquad (5.22)$$

in which x_i represents the mole fraction of the ith constituent in the inflowing mixture, $\dot{N}_{i,\text{in}}/\dot{N}$. The right side of eq. (5.22) lists the *total molal flow exergy* $\bar{e}_t$ of the mixture stream $\dot{N}$:

$$\bar{e}_t = \bar{h} - T_0\bar{s} - \sum_{i=1}^{n} \mu_{0,i}x_i \qquad (5.22')$$

where the word "total" reminds us that the stream is brought to thermal, mechanical, and chemical equilibrium with the environment (the dead state). We can also refer to this quantity as the molal *thermomechanical and chemical flow exergy*.

This last expression can be compared directly with the flow exergy delivered as the stream reaches the restricted dead state (T_0, P_0), eq. (3.43): on a per-unit-$\dot{N}$ basis, that result is written as

$$\bar{e}_x = \bar{h} - \bar{h}^* - T_0(\bar{s} - \bar{s}^*) \tag{5.23}$$

where $(\ )^*$ indicates properties evaluated at the restricted dead state. Note further that $\bar{h}^*$ and $\bar{s}^*$ are defined according to eqs. (5.21):

$$\bar{h}^* = \sum_{i=1}^{n} \bar{h}_i^* x_i \qquad \bar{s}^* = \sum_{i=1}^{n} \bar{s}_i^* x_i \tag{5.24}$$

Combining eqs. (5.22) and (5.23), we find that the total or thermomechanical and chemical flow exergy is the sum of two contributions:

$$\bar{e}_t = \bar{e}_x + \bar{e}_{ch} \tag{5.25}$$

where $\bar{e}_{ch}$ is the molal *chemical flow exergy* released as the bulk state of the stream changes from the restricted dead state to the dead state:

$$\bar{e}_{ch} = \sum_{i=1}^{n} (\mu_i^* - \mu_{0,i}) x_i \tag{5.26}$$

Note that the step that takes us from eqs. (5.22) and (5.23) to eq. (5.25) involves also the use of the definition

$$\bar{h}^* = \bar{u}^* + P_0 \bar{v}^* \tag{5.27}$$

and the Euler equation corresponding to the restricted dead state:

$$\bar{u}^* = T_0 \bar{s}^* - P_0 \bar{v}^* + \sum_{i=1}^{n} \mu_i^* x_i \tag{5.28}$$

In conclusion, the total or thermomechanical and chemical flow exergy $\bar{e}_t$ can be divided into the *thermomechanical* (or *physical*) *flow exergy* $\bar{e}_x$ of chapter 3 plus the chemical flow exergy $\bar{e}_{ch}$ defined by eq. (5.26). The physical flow exergy accounts for the maximum useful work delivered to an external user as the stream reaches the restricted dead state (T_0, P_0), whereas the chemical flow exergy represents the maximum useful work associated with the stream's transition from the restricted dead state to the ultimate, proper, dead state $(T_0, P_0, \mu_{0,1}, \mu_{0,2}, \ldots, \mu_{0,n})$.

Note finally that the restricted dead-state chemical potentials μ_i^*, which are shorthand for $(\bar{h}^* - T_0 \bar{s}_i^*)$, have the same meaning as in eq. (5.11). The chemical flow exergy $\bar{e}_{ch}$ is identical to the molal counterpart of the nonflow chemical exergy of eq. (5.14):

$$\bar{\xi}_{ch} = \frac{\Xi_{ch}}{N} = \bar{e}_{ch} \tag{5.29}$$

where N is the total number of moles in the initial state of the nonflow system of Fig. 5.1, $N = \Sigma_{i=1}^{n} N_i^{(1)}$.

GENERALIZED EXERGY ANALYSIS

We are now in a position to generalize the exergy analysis that opened chapter 3. The difference is that this time the open system can also experience mass transfer interactions with the environment (Fig. 5.3). The present system embodies the features introduced piecemeal via Figs. 5.1 and 5.2 with the general configuration of Fig. 3.1, in which the actual numbers of inlet ports, outlet ports, and heat reservoirs are not specified. It is not difficult to prove that the first-law and the second-law analyses of Fig. 5.3 combine into (Problem 5.1)

$$\dot{E}_W = -\frac{d\Xi_t}{dt} + \sum_{l=1}^{p} (\dot{E}_Q)_l + \sum_{j=1}^{q} (N\bar{e}_t)_j - \sum_{k=1}^{r} (N\bar{e}_t)_k - T_0 \dot{S}_{gen} \tag{5.30}$$

where the js and ks refer to inlet ports and outlet ports, respectively. The

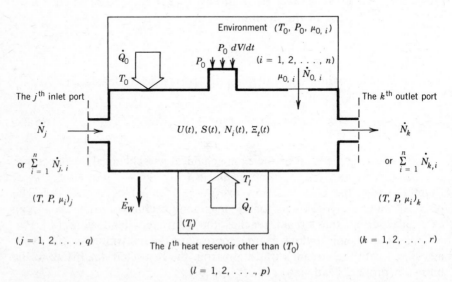

Figure 5.3 General open system exchanging heat, work, and mass transfer with the environment.

TABLE 5.1   Summary of "Exergy" Names and Symbols

Name	Nonflow Symbol	Nonflow Definition	Flow Symbol	Flow Definition
Exergy (total, or thermomechanical and chemical)	$\Xi_t, \bar{\xi}_t$	Eq. (5.13')	E_t, e_t	Eq. (5.22')
Thermomechanical exergy[a] (or physical exergy)	Ξ, ξ	Eq. (3.32)	E_x, e_x	Eq. (3.43)
Chemical exergy	$\Xi_{ch}, \bar{\xi}_{ch}$	Eq. (5.14)	$E_{ch}, \bar{e}_{ch}$	Eq. (5.26)

[a]The subscript "tm" could be attached to each thermomechanical exergy quantity (e.g., Moran [8]); in the present treatment, the thermomechanical exergies have been assigned the simplest and most popular symbols because they are the exergies that appear most frequently in the analysis of power and refrigeration systems.

exergy associated with each heat transfer interaction, $(\dot{E}_Q)_i$, has been defined in eq. (3.14). The "nonflow" and "flow" conclusions, (5.13) and (5.25), respectively, emerge as two special cases of the general result listed above.

One widely acknowledged source of difficulty for the student of "second-law" or "exergy" analysis is the proliferation of exergy-type names and symbols given to what are, in fact, different (i.e., self-standing) concepts and quantities. Table 5.1 is offered here both as a summary of the six-name exergy language of chapters 3 and 5, and as a quick reference guide for the problem solver. The relationship between the total, physical, and chemical exergies is illustrated further in Fig. 5.4 for nonflow systems and Fig. 5.5 for flow systems. Note that the labeling in both drawings is correct only if there are two or more components in the mixture, $n \geq 2$ (recall that the number of degrees of freedom is $n + 1$). In the case of a single-component batch, the lone chemical potential cannot vary independently of T and P (see p. 159). An example of how a single-component substance reaches the dead state prescribed by the ambient is given in the subsection on "The total flow exergy of liquid water."

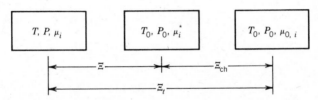

Figure 5.4 The relationship between *nonflow* total (Ξ_t), physical (Ξ), and chemical (Ξ_{ch}) exergies.

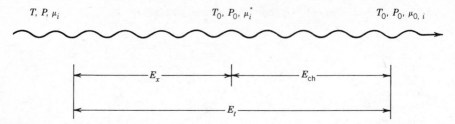

Figure 5.5 The relationship between *flow* total (E_t), physical (E_x), and chemical (E_{ch}) exergies.

AIR-CONDITIONING APPLICATIONS

We have an opportunity to practice using the new concepts of this chapter by focusing on the thermodynamics of air-conditioning processes and systems. The working fluid in this class of applications is atmospheric air, which, as a first approximation, can be viewed as a two-component mixture of dry air (a) and water vapor (v). The objective of most air-conditioning installations is to bring the humid air ($a + v$) mixture to a state in which the temperature and composition differ from the conditions found in the actual (environmental) atmospheric air. It is for this reason that the concepts of total exergy and chemical exergy play a crucial role in assessing the true thermodynamic merit of this class of applications.

The Dry Air and Water Vapor Mixture

The classical way of describing the thermodynamic properties of humid air is to treat it as an ideal gas mixture of two components (a, v) that individually exhibit ideal gas behavior (review the two parts of the Gibbs–Dalton Law, pp. 198–202). For the nearly environmental temperature and pressure ranges covered by most air-conditioning calculations, the appropriate ideal gas constants of these two components are the values corresponding to $T \cong 300$ K and the low-pressure limit [11, 12]

Dry air	Water vapor	
$R_a = 0.287$ kJ/kg K	$R_v = 0.461$ kJ/kg K	
$c_{P,a} = 1.003$ kJ/kg K	$c_{P,v} = 1.872$ kJ/kg K	(5.31)
$M_a = 28.97$ kg/kmol	$M_v = 18.015$ kg/kmol	

The state of any mixture of dry air and water vapor is pinpointed by three properties, for example, by the temperature T, the pressure P, and the composition. The latter consists of specifying one of the two mole fractions, x_a or x_v, because their sum is always equal to 1:

$$x_a + x_v = 1 \tag{5.32}$$

As a self-standing activity in thermal engineering, the field of air conditioning and psychrometry[†] developed its own terminology for describing the composition of a humid air mixture—this, as an alternative to speaking in terms of x_a and x_v. One way is to specify the mass ratio called *specific humidity* or *humidity ratio*:

$$\omega = \frac{m_v}{m_a} \tag{5.33}$$

which represents the number of kilograms of water that correspond to 1 kilogram of dry air in the given mixture. Another way is to specify the mole fraction ratio:

$$\tilde{\omega} = \frac{x_v}{x_a} \tag{5.34}$$

which represents the number of moles of water corresponding to 1 mole of dry air in the mixture. The proportionality between ω and $\tilde{\omega}$ is derived easily by writing

$$\tilde{\omega} = \frac{N_v}{N_a} = \frac{m_v/M_v}{m_a/M_a} = \frac{28.97}{18.015}\,\omega$$
$$= 1.608\omega \tag{5.35}$$

The relations between a ratio such as $\tilde{\omega}$ (or ω) and the individual mole fractions are [8]

$$x_a = \frac{1}{1 + \tilde{\omega}} \qquad x_v = \frac{\tilde{\omega}}{1 + \tilde{\omega}} \tag{5.36}$$

The other way of specifying the composition of the dry air and water vapor mixture is to report the *relative humidity*:

$$\phi = \frac{x_v \ [\text{in the actual mixture } (T, P)]}{x_v \ (\text{in the saturated mixture at the same } T \text{ and } P)} \tag{5.37}$$

This definition makes most sense while looking at the T–s diagram of the water vapor that exists in the mixture, Fig. 5.6. The isobars in this diagram

---

[†]Psychrometry is the science concerning the measurement of the moisture content of atmospheric air. Its name comes from the Greek word *psychros* (cold).

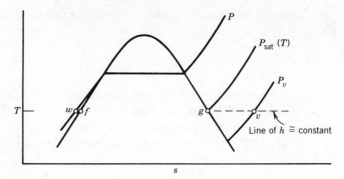

Figure 5.6 The T–s diagram for water, showing the position of the water vapor (v) state (T, P_v) and the liquid water (w) state (T, P).

are lines of constant partial pressure of water vapor, P_v. The diagram shows that when T is held fixed, the partial pressure cannot exceed the ceiling value represented by $P_{sat}(T)$. In the limit $P_v = P_{sat}(T)$, the humid-air mixture is said to be "saturated." Recalling that

$$x_v = \frac{P_v}{P}$$
(5.38)

the relative humidity can be defined also as

$$\phi = \frac{P_v}{P_{sat}(T)}$$
(5.39)

where T is the temperature of the dry air and water vapor mixture of interest. The fact that ϕ is an alternative to the job of specifying the moisture content of humid air is stressed further by the following "conversion" relations:

$$\phi = \frac{\omega P_a}{0.622 P_{sat}(T)} \qquad (P_a = P - P_v)$$
(5.40)

$$\phi = \frac{\omega}{\omega + 0.622} \frac{P}{P_{sat}(T)}$$
(5.41)

$$\omega = \frac{0.622}{\dfrac{P}{\phi P_{sat}(T)} - 1}$$
(5.42)

Figure 5.7 summarizes these relationships for the case where the pressure of the mixture is atmospheric, $P = 1$ atm. In this case, the state of the mixture is specified by only two parameters, namely, the temperature T and one "composition" parameter (ϕ, ω, or $\tilde{\omega}$). The figure shows that if T is constant, the relative humidity increases monotonically with ω (or $\tilde{\omega}$), and that ω is always proportional to $\tilde{\omega}$.

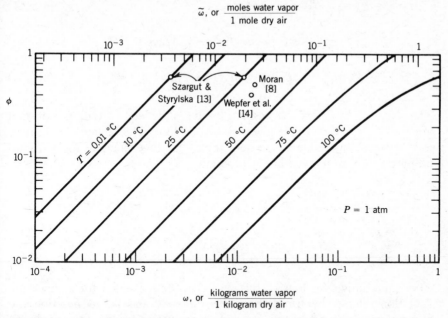

Figure 5.7 The relationship between relative humidity and the humidity ratio for humid air at atmospheric pressure.

The Total Flow Exergy of Humid Air

If we are to apply eq. (5.30) to a system that processes one or more streams of humid air in the steady state, what we need is a way to calculate the total exergy ($\bar{e}_t$) of each stream. This problem was dealt with by Szargut and Styrylska [13] and Wepfer et al. [14], and it has been summarized by Moran [8]. The concept of "exergy" requires the specification of a dead state for the mixture under consideration. In the present case, the dead state is set by the atmospheric conditions T_0, P_0, and ϕ_0 (or ω_0, or $\tilde{\omega}_0$), which in the following analysis are being assumed fixed and different from the corresponding properties of the given mixture (T, P, ϕ).

The total flow exergy per mole of a humid air mixture is deduced from eqs. (5.23)–(5.26):

$$\bar{e}_t = x_a[\bar{h}_a - \bar{h}_a^* - T_0(\bar{s}_a - \bar{s}_a^*) + \mu_a^* - \mu_{0,a}]$$
$$+ x_v[\bar{h}_v - \bar{h}_v^* - T_0(\bar{s}_v - \bar{s}_v^*) + \mu_v^* - \mu_{0,v}] \qquad (5.43)$$

Looking inside the first pair of square brackets, we make use of the ideal gas model and the notion that the ( )* superscript indicates only mechanical and thermal equilibrium with the ambient, that is, properties evaluated at T_0 and P_0:

$$\bar{h}_a - \bar{h}_a^* = \bar{h}_a(T) - \bar{h}_a^*(T_0) = \bar{c}_{P,a}(T - T_0) \qquad (5.44)$$

$$\bar{s}_a - \bar{s}_a^* = \bar{s}_a(T, P) - \bar{s}_a^*(T_0, P_0) = \bar{c}_{P,a} \ln \frac{T}{T_0} - \bar{R} \ln \frac{P}{P_0} \qquad (5.45)$$

$$\bar{\mu}_a^* - \bar{\mu}_{0,a} = \bar{\mu}_a^*(T_0, P_0, x_a) - \bar{\mu}_{0,a}(T_0, P_0, x_{0,a}) = \bar{R}T_0 \ln \frac{x_a}{x_{0,a}} \qquad (5.46)$$

Note that eq. (5.46) is the result of integrating eq. (4.130) at constant temperature, for which $\bar{v}_i$ is given by eq. (4.134). The contents within the second pair of square brackets of eq. (5.43) can be evaluated similarly and, putting all the results together, we obtain

$$\bar{e}_t = (x_a\bar{c}_{P,a} + x_v\bar{c}_{P,v})T_0\left(\frac{T}{T_0} - 1 - \ln \frac{T}{T_0}\right) + \bar{R}T_0 \ln \frac{P}{P_0}$$

$$+ \bar{R}T_0\left(x_a \ln \frac{x_a}{x_{0,a}} + x_v \ln \frac{x_v}{x_{0,v}}\right) \qquad (5.47)$$

Two alternative versions of this result are better suited for engineering calculations. First, by using the mole ratios $\tilde{\omega}$ and $\tilde{\omega}_0$ to describe the composition of the actual and dead-state mixtures, eqs. (5.47) and (5.36) yield [8]

$$\bar{e}_t = \frac{\bar{c}_{P,a} + \tilde{\omega}\bar{c}_{P,v}}{1 + \tilde{\omega}} T_0\left(\frac{T}{T_0} - 1 - \ln \frac{T}{T_0}\right) + \bar{R}T_0 \ln \frac{P}{P_0}$$

$$+ \bar{R}T_0\left(\ln \frac{1 + \tilde{\omega}_0}{1 + \tilde{\omega}} + \frac{\tilde{\omega}}{1 + \tilde{\omega}} \ln \frac{\tilde{\omega}}{\tilde{\omega}_0}\right) \qquad (5.48)$$

The second alternative is to report the total flow exergy *per kilogram of dry air*:

$$e_t = (c_{P,a} + \omega c_{P,v})T_0\left(\frac{T}{T_0} - 1 - \ln \frac{T}{T_0}\right) + (1 + \tilde{\omega})R_aT_0 \ln \frac{P}{P_0}$$

$$+ R_aT_0\left[(1 + \tilde{\omega}) \ln \frac{1 + \tilde{\omega}_0}{1 + \tilde{\omega}} + \tilde{\omega} \ln \frac{\tilde{\omega}}{\tilde{\omega}_0}\right] \qquad (5.49)$$

The step from eq. (5.47) to eq. (5.49) is based on the observation that to 1 mole of mixture correspond $M_a x_a$ kilograms of dry air, and that the quantity $\bar{e}_t$ of eq. (5.47) represents total flow exergy per mole of mixture.

The total flow exergy of a stream of *dry air* can be written immediately by setting $\tilde{\omega} = 0$ and $\omega = 0$ in eq. (5.49):

$$e_{t,a} = c_{P,a}T_0\left(\frac{T}{T_0} - 1 - \ln \frac{T}{T_0}\right) + R_aT_0 \ln \frac{P}{P_0} + R_aT_0 \ln (1 + \tilde{\omega}_0) \qquad (5.50)$$

Worth recognizing in this expression are the first two terms, which represent the thermomechanical flow exergy encountered in chapter 3. The last term is the chemical exergy or the maximum work that could theoretically be harvested as the dry-air stream (already at T_0 and P_0) becomes as humid as the ambient.

The Total Flow Exergy of Liquid Water

In many processes involving air and water vapor mixtures, one or more streams carry condensed water (e.g., the water supply to an adiabatic saturator, the condensed water dripping out of an air dehumidifier). The total flow exergy of liquid water cannot be deduced as a particular case of eq. (5.49), as was done for a stream of pure dry air in eq. (5.50). The reason for this is the ideal gas model invoked in eqs. (5.44)–(5.46), to which liquid water does not subscribe. Instead, we must begin with the beginning, namely, with the equivalent of eq. (5.22) for liquid water (w) as a single-component substance [14]:

$$\bar{e}_{t,w} = \bar{h}_w(T, P) - T_0 \bar{s}_w(T, P) - \mu_{0,w} \tag{5.51}$$

in which

$$\mu_{0,w} = \bar{h}_0(T_0, P_{0,w}) - T_0 \bar{s}_0(T_0, P_{0,w}) \tag{5.52}$$

The pressure $P_{0,w}$ is the partial pressure of water in the ambient, which in the preceding subsection would have been labeled $P_{0,v}$. We can write, therefore,

$$P_{0,w} = x_{0,v} P_0 = \frac{\tilde{\omega}_0}{1 + \tilde{\omega}_0} P_0 \tag{5.53}$$

Combining eqs. (5.51) and (5.52) and putting the result on a per-unit-mass basis (i.e., per kilogram of water) yields

$$e_{t,w} = h_w(T, P) - h_0(T_0, P_{w,0}) - T_0 s_w(T, P) + T_0 s_0(T_0, P_{0,w}) \tag{5.54}$$

The terms that appear on the right side can be related to the properties of respective neighboring states on the two-phase dome by applying the approximate relations presented in the Appendix, namely,

$$h_w(T, P) \cong h_f(T) + [P - P_{sat}(T)]v_f(T) \tag{5.55}$$

$$h_0(T_0, P_{w,0}) \cong h_g(T_0) \tag{5.56}$$

$$s_w(T, P) \cong s_f(T) \tag{5.57}$$

$$s_0(T_0, P_{0,w}) \cong s_g(T_0) - R_v \ln \frac{P_{w,0}}{P_{sat}(T_0)} \tag{5.58}$$

Noting further that $P_{w,0}/P_{sat}(T_0) = \phi_0$, the combined message of eqs. (5.54)–(5.58) becomes

$$e_{t,w} \cong h_f(T) - h_g(T_0) - T_0 s_f(T) + T_0 s_g(T_0)$$
$$+ [P - P_{sat}(T)] v_f(T) - R_v T_0 \ln \phi_0 \qquad (5.59)$$

The terms assembled in the first row add up to zero in the special case when the liquid water stream is already in thermal equilibrium with the ambient, $T = T_0$ [recall the relation $h_{fg}(T) = T s_{fg}(T)$]. In the second row, the first term is usually negligible when compared with the last term ($-R_v T_0 \ln \phi_0$).

An Evaporative Cooling Example

Consider as an example the steady-flow system in Fig. 5.8, whose job is to lower the temperature of a stream of dry air ($\dot{m}_a$) by mixing it with a trickle of water ($\dot{m}_w$). The latter evaporates and becomes part of the humid air mixture ($\dot{m}_a + \dot{m}_v$) that leaves the adiabatic mixing chamber. Clearly, the conservation of water in the steady state requires $\dot{m}_v = \dot{m}_w$.

There are two basic questions to focus on in this example. The first asks how much water ($\dot{m}_w$) is needed in order to lower the temperature of the outgoing mixture to a prescribed level T_2. The answer falls straight out of the first law for the control volume, which can be written sequentially as

$$\dot{m}_a h_a(T_1) + \dot{m}_w h_w(T_1, P_1) = \dot{m}_a h_a(T_2) + \dot{m}_v h_v(T_2, P_{v2}) \qquad (5.60)$$
$$h_a(T_1) + \omega h_w(T_1, P_1) = h_a(T_2) + \omega h_v(T_2, P_{v2}) \qquad (5.60')$$
$$h_a(T_1) + \omega h_f(T_1) \cong h_a(T_2) + \omega h_g(T_2) \qquad (5.60'')$$

The last step in this sequence consisted of using the approximations (5.55)–(5.56); in addition, the second term on the right side of eq. (5.55) was assumed negligible. In conclusion, the first-law analysis produces an explicit relationship for calculating the needed humidity ratio when the exit temperature is specified:

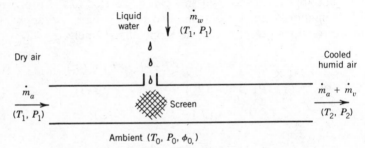

Figure 5.8 Adiabatic evaporative cooling process.

$$\omega \cong \frac{c_{P,a}(T_1 - T_2)}{h_g(T_2) - h_f(T_1)} \qquad (5.61)$$

The second question refers to how much exergy is being destroyed during the evaporative cooling process. For this, we use eq. (5.30). The control volume in this case does not deliver any exergy for direct use ($\dot{E}_w = 0$), the operation is steady ($d\Xi_t/dt = 0$), and there are no heat transfer interactions, i.e., all the $(\dot{E}_Q)_t$ terms are absent. Term for term, eq. (5.30) reduces to

$$0 = 0 + 0 + \dot{m}_a e_{t,a} + \dot{m}_w e_{t,w} - \dot{m}_a e_t - T_0 \dot{S}_{gen} \qquad (5.62)$$

where the last e_t is the per-kilogram-of-dry-air total flow exergy of the outflowing humid air, eq. (5.49). The rate of exergy destruction per kilogram of dry air is, therefore,

$$\frac{T_0 \dot{S}_{gen}}{\dot{m}_a} = \underbrace{e_{t,a} + \omega e_{t,w}}_{\substack{\text{Total flow exergy} \\ \text{arriving}}} - \underbrace{e_t}_{\substack{\text{Total flow exergy} \\ \text{leaving}}} \qquad (5.63)$$

Finally, the second-law efficiency of the evaporative cooler can be defined as the ratio:

$$\eta_{II} = \frac{\text{total flow exergy leaving}}{\text{total flow exergy arriving}} = \frac{e_t}{e_{t,a} + \omega e_{t,w}} \qquad (5.64)$$

Beyond this point, it makes more sense to continue this example numerically. Let the dead state be represented by the atmospheric conditions:

$T_0 = 25°C$ (298.15 K)

$P_0 = 1$ atm (0.101325 MPa) $\qquad (5.65)$

$\phi_0 = 0.6$, which means also $\omega_0 = 0.0119$, or $\hat{\omega}_0 = 0.0191$

Assume further that the pressure is atmospheric throughout the process:

$$P_1 = P_2 = P_0 \qquad (5.66)$$

and that the unmixed streams $\dot{m}_a$ and $\dot{m}_w$ are both at ambient temperature:

$$T_1 = T_0 \qquad (5.67)$$

Finally, let us assume that the function of the apparatus is to lower the temperature to

$$T_2 = 15°C \tag{5.68}$$

which, after solving eq. (5.61), means that

$$\omega = 0.00414, \text{ or } \tilde{\omega} = 0.00666 \tag{5.69}$$

The exergy analysis requires the evaluation of the following quantities, all for 1 kilogram of dry air:

$$e_{t,a} = R_a T_0 \ln (1 + \tilde{\omega}_0) = 1.619 \text{ kJ/kg} \tag{5.70}$$

$$\omega e_{t,w} \cong - \omega R_v T_0 \ln \phi_0 = 0.291 \text{ kJ/kg} \tag{5.71}$$

$$e_t = (c_{P,a} + \omega c_{P,v}) T_0 \left(\frac{T_2}{T_0} - 1 - \ln \frac{T_2}{T_0} \right)$$

$$+ R_a T_0 \left[(1 + \tilde{\omega}) \ln \frac{1 + \tilde{\omega}_0}{1 + \tilde{\omega}} + \tilde{\omega} \ln \frac{\tilde{\omega}}{\tilde{\omega}_0} \right]$$

$$= 0.631 \text{ kJ/kg} \tag{5.72}$$

The second-law efficiency (5.64) is then

$$\eta_{II} = \frac{0.631}{1.619 + 0.291} = 0.33 \tag{5.73}$$

meaning that two thirds of the original exergy inventory brought into the control volume are destroyed by the evaporative cooling process.

Observations of the Selection of Dead-State Conditions

An important characteristic of the exergy analysis just completed is that its numerical results are quite sensitive to the values assigned to the dead-state conditions. Whereas the choice of $P_0 = 1$ atm is obvious, a convention for the selection of T_0 and ϕ_0 is still lacking. Figure 5.7 shows the location of four different dead states that have been used in illustrative humid-air exergy calculations. The dead state chosen in the present example, eqs. (5.65), matches one that was used earlier by Szargut and Styrylska [13].

The selection of dead-state conditions is even more challenging in problems where the working fluids and the environment contain more than the two components discussed in this section. One class of problems of this kind are the chemically reactive flows treated in chapter 7. Considerable research is being devoted to standardizing the dead-state conditions of the natural, multicomponent environment, as demonstrated by the work of Ahrendts [15], Wepfer and Gaggioli [16], and Morris and Szargut [17].

SYMBOLS

$c_P, \bar{c}_P$	specific heat at constant pressure [J/kg K, J/mol K]
$\bar{e}_{ch}$	molal chemical exergy [J/mol], [eq. (5.26)]
e_t	total flow exergy per kilogram of dry air [J/kg], [eq. (5.49)]
$\bar{e}_t$	molal total flow exergy [J/mol], [eq. (5.22′)]
$e_{t,a}$	total flow exergy of dry air [J/kg], [eq. (5.50)]
$e_{t,w}$	total flow exergy of pure liquid water [J/kg], [eq. (5.59)]
$\bar{e}_x$	molal thermomechanical flow exergy [J/mol], [eq. (5.23)]
$\dot{E}_Q$	exergy transfer rate associated with the heat transfer interaction $\dot{Q}$ [W]
E_w	exergy delivered as useful work [J]
$\dot{E}_w$	rate of exergy delivery as useful power [W]
$\bar{g}$	molal Gibbs free energy [J/mol]
$\bar{h}$	molal enthalpy [J/mol]
m	mass [kg]
M	molecular weight [kg/kmol]
N	number of moles [mol]
$\dot{N}$	molal flowrate [mol/s]
P	pressure [Pa]
P_a	partial pressure of dry air [Pa]
$P_{sat}(T)$	saturation pressure of water vapor, corresponding to the humid-air temperature T [Pa]
P_v	partial pressure of water vapor [Pa]
Q	heat transfer interaction [J]
R	ideal gas constant [J/kg K]
$\bar{R}$	universal ideal gas constant [J/mol K]
$\bar{s}, S$	entropy [J/mol K, J/K]
$\dot{S}_{gen}$	entropy generation rate [W/K]
t	time
T	thermodynamic temperature
$\bar{u}, U$	internal energy [J/mol, J]
$\bar{v}, V$	volume [m³/mol, m³]
$\dot{W}$	mechanical power transfer [W]
x_i	mole fraction
η_{II}	second-law efficiency [eq. (5.64)]
μ	chemical potential [J/mol]
$\bar{\xi}_{ch}$	molal chemical exergy [J/mol], [eq. (5.29)]
Ξ	nonflow thermomechanical exergy [J], [eq. (5.10)]
Ξ_{ch}	chemical exergy [J], [eq. (5.14)]
Ξ_t	total nonflow exergy [J], [eq. (5.13′)]
ω	humidity ratio, specific humidity [eq. (5.33)]
$\tilde{\omega}$	mole-fraction ratio [eq. (5.34)]
$(\ )_a$	properties of dry air
$(\ )_f$	properties of saturated liquid

( )$_g$	properties of saturated vapor
( )$_{in}$	inlet
( )$_{out}$	outlet
( )$_{rev}$	reversible
( )$_v$	properties of water vapor
( )$_w$	properties of liquid water
( )$_0$	ambient (dead-state) properties
( )*	properties at thermomechanical equilibrium with the ambient
(˙)	flowrate

REFERENCES

1. J. Szargut and R. Petela, *Egzergia*, Wydawnictwa Naukowo-Techniczne, Warsaw, 1965.
2. I. Nerescu and V. Radcenco, *Exergy Analysis of Thermal Processes*, Editura Tehnica, Bucharest, 1970.
3. V. M. Brodianskii, *Eksergeticheskii Metod Termodinamicheskogo Analiza*, Energia, Moscow, 1973.
4. V. Radcenco, *Termodinamica Tehnica si Masini Termice—Procese Ireversibile*, Editura Didactica si Pedagogica, Bucharest, 1976.
5. V. Radcenco, *Optimization Criteria of (Irreversible) Thermal Processes*, Editura Tehnica, Bucharest, 1977.
6. J. Szargut, International progress in second law analysis, *Energy*, Vol. 5, 1980, pp. 709–718.
7. J. E. Ahern, *The Exergy Method of Energy System Analysis*, Wiley, New York, 1980.
8. M. J. Moran, *Availability Analysis: A Guide to Efficient Energy Use*, Prentice-Hall, Englewood Cliffs, NJ, 1982.
9. T. J. Kotas, *The Exergy Method of Thermal Plant Analysis*, Butterworth, London, 1985.
10. R. B. Evans, A proof that essergy is the only consistent measure of potential work (for chemical systems), PhD thesis, Thayer School of Engineering, Dartmouth College, Hanover, NH, June 1969.
11. A. Bejan, Engineering thermodynamics, in M. Kutz, ed., *Mechanical Engineers' Handbook*, Wiley, New York, 1986, Chapter 54, pp. 1530–1548.
12. R. E. Sonntag and G. J. Van Wylen, *Introduction to Thermodynamics*, 2nd ed., Wiley, New York, 1982, p. 729.
13. J. Szargut and T. Styrylska, Die exergetische Analyse von Prozessen der feuchten Luft, *Heiz.-Lueft-Haustech.*, Vol. 20, 1969, pp. 173–178.
14. W. J. Wepfer, R. A. Gaggioli, and E. F. Obert, Proper evaluation of available energy for HVAC, *ASHRAE Trans.*, Vol. 85, Part 1, 1979, pp. 214–230.
15. J. Ahrendts, Reference states, *Energy*, Vol. 5, 1980, pp. 667–677.
16. W. J. Wepfer and R. A. Gaggioli, Reference datums for available energy, *ACS Symp. Ser.*, Vol. 122, 1980, pp. 77–92.
17. D. R. Morris and J. Szargut, Standard chemical exergy of some elements and compounds on the planet Earth, *Energy*, Vol. 11, 1986, pp. 733–755.

PROBLEMS

5.1 Derive the most general exergy accounting statement (5.30), starting with the first-law and second-law statements for the control volume in Fig. 5.3. Keep in mind also the requirement that each constituent must be conserved. As a guide, use the analyses developed in the text with reference to Figs. 5.1 and 5.2.

5.2 The atmospheric air that surrounds us can be viewed as an ideal gas mixture at temperature T_0 and pressure P_0. At a particular location and time, the mole fractions of the ideal gas components of this mixture are

$$x_{0,N_2} = 0.7567 \qquad x_{0,H_2O} = 0.0303$$

$$x_{0,O_2} = 0.2035 \qquad x_{0,CO_2} = 0.0003, \text{ etc.}$$

(a) Consider a process by which 1 mole of O_2 is extracted (separated) from the atmospheric air mixture. The end result of this operation is a 1-mole batch of pure O_2 at T_0 and P_0. Determine the minimum work required to perform this task.

(b) Show that the work requirement calculated above is greater than the minimum work needed for extracting a corresponding batch of pure N_2.

5.3 An evacuated vessel of volume V (the "system") is surrounded by a fluid environment at temperature T_0, pressure P_0, and chemical potentials $\mu_{0,i}$ ($i = 1, \ldots, n$). What is the maximum work that one could theoretically extract from this two-system arrangement? Compare your answer with the conclusions reached in Examples 1.1 and 1.2.

5.4 (a) The results of the exergy analysis of air-conditioning systems are sensitive to the choice of the dead state (T_0, P_0, ϕ_0). In order to study this effect, repeat the analysis of Fig. 5.8 while assuming different values for the dead-state relative humidity. For example, assume $\phi_0 = 0.5$ and, later, $\phi_0 = 0.7$. Based on your results and eq. (5.73), sketch the way in which η_{II} depends on ϕ_0.

(b) The results of the same analysis depend also on the "effective" temperature of the process relative to the dead-state temperature T_0. In order to study this effect, reconstruct the curve $\eta_{II}(\phi_0)$ demanded in part (a), while assuming this time that the exit temperature of the humid air is $T_2 = 10°C$ [recall that in part (a) as well as in the text this temperature was $15°C$].

5.5 Consider the classical *adiabatic saturation process* in which a stream of humid air [state (1) in the drawing] is mixed adiabatically and isobarically with a stream of liquid water [state (2)] in order to become a saturated air + water vapor mixture at state (3) (note: $\phi_3 = 1$).

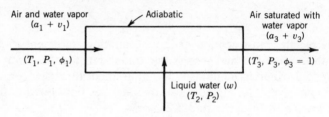

Figure P5.5

(a) Show that the adiabatic saturation temperature T_3 is a function of the inlet humidity ratio and that the $T_3(\omega_1)$ relation is obtained by combining

$$\omega_1 \cong \frac{c_{P,a}(T_3 - T_1) + \omega_3[h_g(T_3) - h_f(T_2)]}{h_g(T_1) - h_f(T_2)}$$

with the $\omega_3(T_3)$ relation expressed by eq. (5.42).

(b) Assume next that the inlet mixture is at environmental conditions $(T_1 = T_0, P_1 = P_0, \phi_1 = \phi_0)$ and these conditions are the same as the numerical values listed in eq. (5.65). Determine the exergy destroyed by the adiabatic saturator per kilogram of dry air. Calculate also the second law efficiency η_{II} of the apparatus, where η_{II} is defined by eq. (5.64).

5.6 An air and water vapor stream ($T_1 = 25°C$, $\phi_1 = 0.7$) is dehumidified in the two-part apparatus shown in the sketch. In the first part [state $(1) \rightarrow$ state (2)], the stream is cooled at constant (atmospheric) pressure to a low enough temperature T_2 such that the water stream $\dot{m}_w$ condenses and is collected as liquid water at atmospheric pressure and temperature T_2. In the second part of the apparatus, the remaining air and water vapor stream is heated isobarically back up to 25°C. The final relative humidity of the stream is $\phi_3 = 0.4$. Calculate in order:

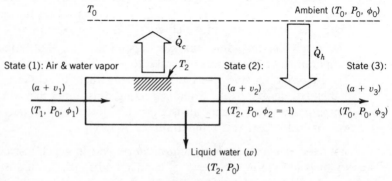

Figure P5.6

(a) the number of kilograms of condensate collected per kilogram of dry air of the original stream, $\dot{m}_w/\dot{m}_a$,

(b) the intermediate temperature T_2,

(c) the refrigeration rate required by the first part, $\dot{Q}_c/\dot{m}_a$,

(d) the heat transfer rate required by the second part, $\dot{Q}_h/\dot{m}_a$.

The refrigeration effect $\dot{Q}_c$ is provided by a reversible refrigerator operating beftween T_2 and the ambient temperature T_0. The heat transfer rate $\dot{Q}_h$ originates from T_0. Assuming further that the conditions of the ambient air and water vapor mixture are $T_0 = 25°C$, $P_0 = 1\,\text{atm}$, and $\phi_0 = 0.6$, calculate in order:

(e) the rate of exergy destruction in the first part of the apparatus,

(f) the rate of exergy destruction in the second part of the apparatus,

(g) the second law efficiency of the entire apparatus, and how the blame for its less-than-one value can be attributed to the individual irreversibilities of the two parts.

5.7 Show that the total *nonflow* exergy of a batch of humid air $(T, P, \tilde{\omega})$ relative to the ambient humid air mixture $(T_0, P_0, \tilde{\omega}_0)$ is given by

$$\xi_t = (c_{v,a} + \omega c_{v,v}) T_0 \left(\frac{T}{T_0} - 1 - \ln \frac{T}{T_0} \right)$$
$$+ (1 + \tilde{\omega}) R_a T_0 \left(\ln \frac{P}{P_0} - \ln \frac{T}{T_0} + \frac{T/T_0}{P/P_0} - 1 \right)$$
$$+ R_a T_0 \left[(1 + \tilde{\omega}) \ln \frac{1 + \tilde{\omega}_0}{1 + \tilde{\omega}} + \tilde{\omega} \ln \frac{\tilde{\omega}}{\tilde{\omega}_0} \right]$$

where ξ_t represents exergy *per kilogram of dry air* present in the mixture. From this result, deduce the total nonflow exergy of dry air, $\xi_{t,a}$.

5.8 Prove that the total nonflow exergy of pure liquid water (T, P) relative to the ambient humid air (T_0, P_0, ϕ_0) is given by

$$\xi_{t,w} = h_f(T) - h_f(T_0) - T_0[s_f(T) - s_f(T_0)]$$
$$+ (P_0 - P)v_f(T) - R_v T_0 \ln \phi_0$$

where $\xi_{t,w}$ represents exergy per kilogram of water.

6

Multiphase Systems

In this chapter, we focus on the properties of working fluids that are macroscopically heterogeneous or multiphase, as illustrated in the right half of Table 4.1. One approach to this subject is purely descriptive, as seen in the traditional coverage of two-phase and three-phase mixtures in a first course in engineering thermodynamics. An alternative—the approach taken in the present treatment—is to view a heterogeneous (multiphase) batch of substance as the aftermath of a process of thermodynamic instability by which an original single-phase (homogeneous, or simple) system splits into two or more homogeneous subsystems (phases) that coexist. This second approach, which is due to Gibbs [1], is a standard part of the backbone of physicochemical thermodynamics today. In the field of engineering thermodynamics, in which separation processes and chemically reacting systems receive less attention than pure substances, the merit of the Gibbsian stability approach is that it organizes into a geometric structure many of the empirical observation concerning the properties of multiphase systems in internal equilibrium.

THE ENERGY MINIMUM PRINCIPLE IN U, H, F, AND G REPRESENTATIONS

The issue of thermodynamic stability has its origins in the first law and the second law, or, more precisely, in the "entropy maximum" and "energy minimum" principles that were discussed in chapter 2. These two principles are combined in the graphic construction presented in Fig. 2.7. Each principle is in turn the combined result of the first law and the second law for an arbitrary (not necessarily reversible) process executed by a closed system. The words "minimum" and "maximum" refer to the manner in which the closed system settles into a state of stable equilibrium that is represented by a point in the horizontal plane of Fig. 2.7. The surface that extends above this plane is symbolism for the locus of constrained equilibrium states of the system. The graphic meaning of the entropy maximum and energy minimum principles is that at any point of intersection between this surface and the horizontal plane, the plane that is tangent to the surface is also perpendicular to the horizontal plane.

The objective of this introductory section is to restate the entropy maximum and energy minimum principles using the simple-system terminology developed in chapter 4. In the end, we will see that these energy-based principles have perfect equivalents in terms of enthalpy, Helmholtz free energy, and Gibbs free energy, in the same way that the information content of the fundamental relation in energy representation is conserved by the Legendre transforms H, F, and G of Table 4.3.

The Energy Minimum Principle

We begin with a few observations concerning the closed system of Fig. 2.7. Assuming that in the absence of internal constraints, the closed system is a simple system (i.e., macroscopically homogeneous or single-phase, etc., p. 148), its fundamental relation in energy representation is

$$U = U(S, V) \tag{6.1}$$

The mole numbers that describe the composition of the closed system, N_1, $N_2, \ldots, N_n$, are all fixed. The stable equilibrium states represented by the $U(S)$ curve in the horizontal plane of Fig. 2.7 are states that also belong to the fundamental surface given by eq. (6.1). It follows that the $U(S)$ curve corresponds to a fixed volume V. We can then draw not one but an entire family of symbolic surfaces in the three-dimensional space of Fig. 2.7, and the traces left by these surfaces in the horizontal plane will constitute a family of constant-V surfaces. This new aspect is shown in Fig. 6.1, which otherwise reproduces the main features of Fig. 2.7.

The positive slope and curvature of the $U(S)$ curves in the horizontal plane are tied to the internal stability conditions that form the subject of the next section:

$$\left(\frac{\partial U}{\partial S}\right)_V = T > 0 \tag{6.2}$$

$$\left(\frac{\partial^2 U}{\partial S^2}\right)_V = \frac{T}{mc_v} > 0 \tag{6.3}$$

where m is the mass inventory of the system.

The Enthalpy Minimum Principle

The closed-system fundamental relation in enthalpy representation is

$$H = H(S, P) \tag{6.4}$$

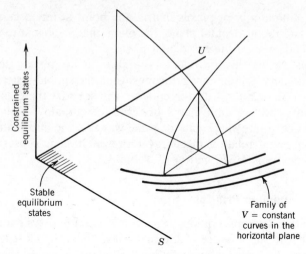

Figure 6.1 The energy minimum principle (fixed volume).

Guided by the relationship between eq. (6.1) and Fig. 6.1, we seek an enthalpy minimum by considering the class of constant-pressure processes. The graphic conclusion of this exercise is presented in Fig. 6.2.

The first and second laws for an infinitesimal change of state that occurs in the immediate vicinity of an unconstrained equilibrium state require

$$\delta Q - \delta W = dU \tag{6.5}$$

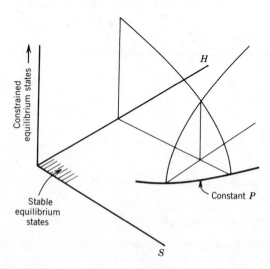

Figure 6.2 The enthalpy minimum principle (fixed pressure).

$$\delta Q \leq T\, dS \tag{6.6}$$

or, taken together,

$$\delta W \leq T\, dS - dU \tag{6.7}$$

Using $H = U + PV$, $\delta W = P\, dV$, and $dP = 0$, we find that the combined law (6.7) reduces to

$$dH \leq T\, dS \qquad \text{(at constant } P) \tag{6.8}$$

and, in particular,

$$dH \leq 0 \qquad \text{(at constant } S \text{ and } P) \tag{6.9}$$

and

$$dS \geq 0 \qquad \text{(at constant } H \text{ and } P) \tag{6.10}$$

In conclusion, of all the states that have the same pressure and entropy, the unconstrained equilibrium state is the one with the lowest (smallest) enthalpy. Equation (6.10) shows further that the entropy of the unconstrained equilibrium state reached under conditions of constant pressure and enthalpy is maximum. Both features—the enthalpy minimum and the entropy maximum—are included in the graphic construction of Fig. 6.2.

The trace left by the three-dimensional surface on the horizontal plane of Fig. 6.2 is a curve $H(S)$ that corresponds to a particular value of P. The slope and curvature of the $H(S)$ curve are both positive:

$$\left(\frac{\partial H}{\partial S}\right)_P = T > 0 \tag{6.11}$$

$$\left(\frac{\partial^2 H}{\partial S^2}\right)_P = \frac{T}{mc_P} > 0 \tag{6.12}$$

where, again, the curvature condition (6.12) is related to the stability of the simple system.

The Helmholtz Free-Energy Minimum Principle

In view of the structured presentation of the U and H minimum principles of Figs. 6.1 and 6.2, the derivation of the corresponding F minimum statement becomes a routine procedure whose details can be abbreviated. The starting point is the closed-system fundamental relation in F representation, Table 4.3,

$$F = F(T, V) \tag{6.13}$$

and the focus is on processes executed while the system is at uniform temperature T and in thermal equilibrium with the ambient temperature reservoir (T). Combining $F = U - TS$ with eq. (6.7) leads to

$$dF + P\,dV \leq 0 \qquad \text{(at constant } T\text{)} \tag{6.14}$$

with two special versions:

$$dF \leq 0 \qquad \text{(at constant } T \text{ and } V\text{)} \tag{6.15}$$

and

$$dV \leq 0 \qquad \text{(at constant } T \text{ and } F\text{)} \tag{6.16}$$

The first conclusion is that the unconstrained equilibrium state that is approached at constant temperature and constant volume is the state with minimum Helmholtz free energy. The companion conclusion is that the unconstrained equilibrium state that is reached under conditions of constant T and F is characterized by minimum volume. The F minimum and V minimum conclusions have been incorporated in Fig. 6.3. Note further the negative slope and positive curvature of the $F(V)$ trace left in the horizontal plane of equilibrium states:

$$\left(\frac{\partial F}{\partial V}\right)_T = -P < 0 \tag{6.17}$$

$$\left(\frac{\partial^2 F}{\partial V^2}\right)_T = \frac{1}{\kappa V} > 0 \tag{6.18}$$

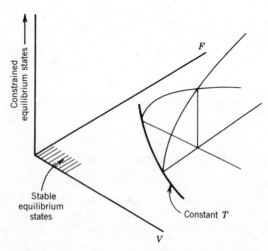

Figure 6.3 The Helmholtz free-energy minimum principle (fixed temperature).

The Gibbs Free-Energy Minimum Principle

The last of the Legendre transforms of Table 4.3 for closed systems reads

$$G = G(T, P) \tag{6.19}$$

where $G = U - TS + PV$. We concentrate on processes executed at constant and uniform temperature and pressure, for which the combined law (6.7) reads

$$dG \leq 0 \quad \text{(at constant } T \text{ and } P) \tag{6.20}$$

The Gibbs free energy of the unconstrained equilibrium state approached while in equilibrium with the temperature reservoir (T) and pressure reservoir (P) is, therefore, a minimum. This feature is illustrated in Fig. 6.4.

The aggregate contribution of this section is the identification of the minima and maxima that are reached by certain thermodynamic properties when a closed system settles into a stable (unconstrained) equilibrium state. Originally, I had intended to summarize these results in tabular form, however, a better overview is conveyed by the star-shaped diagram of Fig. 6.5. In this diagram, the minima are represented by open circles and the S maxima by the shaded circle. A property of the closed system reaches the indicated minimum or maximum when two other properties are held constant. The original property and the two that are being held fixed form a triangle. There are only four such triangles in the diagram, namely, USV, HPS, GTP, and FVT, one for each of the four figures 6.1–6.4. The solid lines that extend inward from the four energy corners (U, H, F, G) indicate the properties that were considered fixed in the construction of Figs. 6.1–6.4.

The star diagram of Fig. 6.5 illustrates also the structure of the Legendre transforms of Table 4.3, only in the special case of closed systems, of course.

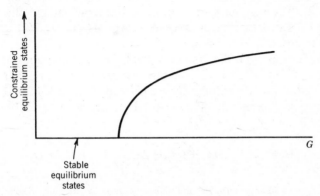

Figure 6.4 The Gibbs free-energy minimum principle (fixed temperature and pressure).

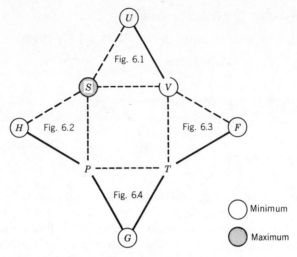

Figure 6.5 Star-shaped diagram summarizing the U, H, F, and G minimum and S maximum principles established for closed systems.

For example, the first Legendre transform from $U(S, V)$ to $H(S, P)$ corresponds to interchanging the position of diagonally opposed corners of the square portion of the diagram.

THE INTERNAL STABILITY OF A SIMPLE SYSTEM

Thermal Stability

The usefulness of the energy minimum principles outlined in the preceding section becomes apparent as we express analytically the thought that the equilibrium state of a simple system must be *stable*[†]. Consider as a first example, a simple system confined by a rigid, impermeable, and adiabatic boundary [2]. The system is therefore homogeneous (single-phase), closed and, during the process that is discussed next, its energy (U) and volume (V) remain constant, Fig. 6.6.

In the initial state (1), the system is divided by an adiabatic partition into two halves so that the left half is slightly warmer than the right. In Fig. 6.6, this is accounted for by writing $(U_L + \Delta U)$ and $(U_R - \Delta U)$ for the left and right energy inventories, so that the total energy at state (1) matches the final energy inventory $(U_L + U_R)$:

$$U_1 = (U_L + \Delta U) + (U_R - \Delta U) = U_L + U_R = U_2 \qquad (6.21)$$

[†]Otherwise, if the system is unstable and separates into two or more distinct subsystems (phases), the system ceases to be a "simple" system.

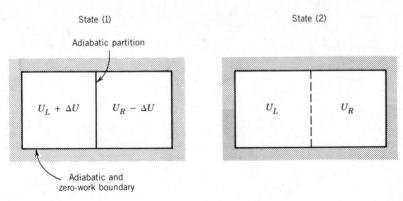

Figure 6.6 Isolated system for the study of internal thermal stability in a simple system.

The constancy of U during the constraint-removal process $(1) \rightarrow (2)$ is guaranteed by the adiabatic and zero-work boundary that defines the system. Note that this process belongs in the constant-U section taken through the three-dimensional surface in Fig. 6.1. Indeed, the object of the following analysis is to exploit the entropy maximum made visible by that section.

Let S_L and S_R represent the entropy inventories of the left and right halves, respectively, of the system in the final state (2); in other words, let

$$S_L = S_R = \tfrac{1}{2} S \tag{6.22}$$

The original entropy of the left half $(S_{L,1})$ can be estimated by noting the constant-volume process undergone by the left half as a system:

$$S_{L,1} = S_L + \left(\frac{\partial S_L}{\partial U} \right)_V (\Delta U) + \frac{1}{2} \left(\frac{\partial^2 S_L}{\partial U^2} \right)_V (\Delta U)^2 + \cdots + \tag{6.23}$$

A similar expression is obtained for the right-side entropy inventory by noting the internal energy defect registered on the right side in switching from state (2) to state (1):

$$S_{R,1} = S_R - \left(\frac{\partial S_R}{\partial U} \right)_V (\Delta U) + \frac{1}{2} \left(\frac{\partial^2 S_R}{\partial U^2} \right)_V (\Delta U)^2 - \cdots + \tag{6.24}$$

The entropy maximum of Fig. 6.1 suggests that the original total entropy $(S_{L,1} + S_{R,1})$ cannot exceed the final total entropy. The same idea is expressed by the statement that the entropy generated from state (1) to state (2) cannot be negative:

$$S_{\text{gen},1-2} = (S_L + S_R) - (S_{L,1} + S_{R,1}) \geq 0 \qquad (6.25)$$

which, using the "$\frac{1}{2}S$" notation indicated in eq. (6.22) becomes

$$S_{\text{gen},1-2} = -\frac{1}{2}\left(\frac{\partial^2 S}{\partial U^2}\right)_V (\Delta U)^2 \geq 0 \qquad (6.26)$$

Finally, if the entropy generation is to decrease steadily as the original left/right nonuniformity disappears, $\Delta U \rightarrow 0$, the second-order partial derivative must be negative and *finite*:

$$\left(\frac{\partial^2 S}{\partial U^2}\right)_V < 0 \qquad (6.27)$$

This last inequality can be rewritten in a number of ways:

$$\frac{\partial}{\partial U}\left(\frac{1}{T}\right)_V < 0 \qquad (6.27')$$

$$\frac{1}{T^2}\left(\frac{\partial T}{\partial U}\right)_V = \frac{1}{T^2 m c_v} > 0 \qquad (6.27'')$$

en route to the most essential conclusion that

$$c_v > 0 \qquad (6.28)$$

Therefore, in order for the process of Fig. 6.6 to proceed from state (1) to state (2) and not in the opposite direction, the specific heat at constant volume must be positive. In other words, if $c_v > 0$, the simple system of state (2) cannot split spontaneously into two thermally dissimilar regions. The inequality (6.28) emerges as a necessary condition for internal thermal stability.

Mechanical Stability

In this second example, we study the evolution of a system away from a state characterized by an internal pressure discontinuity, Fig. 6.7. In state (1), the system is divided into two slightly unequal parts by an off-center partition held in place with a locking mechanism. State (2) is one of unconstrained equilibrium: in this state, the partition floats exactly in the middle of the system. The partition is a diathermal piston that slides freely (without any resistance) during the process (1)→(2), that is, after the locking mechanism is disengaged. During this process, the temperature of the entire system is maintained uniform by contact with the temperature reservoir (T). Furthermore, the total volume of the system does not change:

$$V_1 = (V_L - \Delta V) + (V_R + \Delta V) = V_L + V_R = V_2 \qquad (6.29)$$

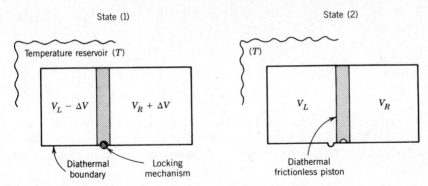

Figure 6.7 Isothermal process for the study of internal mechanical stability in a simple system.

According to the star-shaped summary of Fig. 6.5, the energy minimum principle that applies at constant T and V is that of minimum Helmholtz free energy. Let F_L and F_R be the final Helmholtz free-energy inventories of the identical left and right halves of the system at state (2):

$$F_L = F_R = \tfrac{1}{2}F \tag{6.30}$$

where F is the final inventory of the entire system. The original Helmholtz free energies of the two parts of the system ($F_{L,1}$ and $F_{R,1}$) can be related to the final values (F_L and F_R) by noting the constant-temperature volume changes experienced by each of the two parts:

$$F_{L,1} = F_L - \left(\frac{\partial F_L}{\partial V}\right)_T (\Delta V) + \frac{1}{2}\left(\frac{\partial^2 F_L}{\partial V^2}\right)_T (\Delta V)^2 - \cdots + \tag{6.31}$$

$$F_{R,1} = F_R + \left(\frac{\partial F_R}{\partial V}\right)_T (\Delta V) + \frac{1}{2}\left(\frac{\partial^2 F_R}{\partial V^2}\right)_T (\Delta V)^2 + \cdots + \tag{6.32}$$

The Helmholtz free energy at state (2) is a minimum only if

$$(F_L + F_R) - (F_{L,1} + F_{R,1}) \le 0 \tag{6.33}$$

or, using eqs. (6.30)–(6.32), if

$$\frac{1}{2}\left(\frac{\partial^2 F}{\partial V^2}\right)_T (\Delta V)^2 \ge 0 \tag{6.34}$$

Based on the same argument that earlier gave us eq. (6.27), we deduce that the second-order partial derivative must be finite:

$$\left(\frac{\partial^2 F}{\partial V^2}\right)_T > 0 \tag{6.35}$$

hence,

$$\frac{\partial}{\partial V}(-P)_T = \frac{1}{\kappa V} > 0 \tag{6.35'}$$

and, finally,

$$\kappa > 0 \tag{6.36}$$

We learn in this way that the internal mechanical stability of the system at state (2) is assured by a positive isothermal compressibility value κ, that is, by the system's property to contract upon pressurization at constant temperature.

Chemical Stability

The remaining question is what condition must be met for the simple system not to segregate spontaneously into two or more subsystems with different chemical composition. We consider for this purpose the process $(1) \rightarrow (2)$ triggered by the removal of the internal semipermeable membrane shown on the left side of Fig. 6.8. The membrane is impermeable only to the "i" species, and it is because of it that at state (1) the left half of the system holds more moles of "i" than the right half. The rectangular boundary of the system is impermeable to all the species, which means the total number of moles of "i" is conserved during the $(1) \rightarrow (2)$ process:

$$N_{i,1} = (N_{i,L} + \Delta N_i) + (N_{i,R} - \Delta N_i) = N_{i,L} + N_{i,R} = N_{i,2} \tag{6.37}$$

Held constant and uniform during the same process are the temperature and pressure of the system. This is the result of the intimate contact maintained with the (T) and (P) reservoirs, Fig. 6.8. We know from Figs. 6.4 and 6.5 that the energy minimum principle that rules this constant-T and

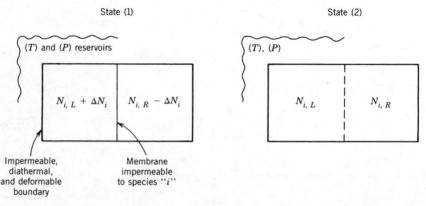

Figure 6.8 Constant-T, $-P$, and $-N_j$ process for the study of internal chemical stability in a simple system.

constant-P process is the G minimum principle. Labeling with G_L and G_R the Gibbs free energies of the two halves of the simple system in unconstrained equilibrium at state (2):

$$G_L = G_R = \tfrac{1}{2}G \tag{6.38}$$

the G minimum principle requires

$$(G_L + G_R) - (G_{L,1} + G_{R,1}) \leq 0 \tag{6.39}$$

where, at constant T, P, and N_j ($j \neq i$),

$$G_{L,1} = G_L + \left(\frac{\partial G_L}{\partial N_i}\right)_{T,P,N_j}(\Delta N_i) + \frac{1}{2}\left(\frac{\partial^2 G_L}{\partial N_i^2}\right)_{T,P,N_j}(\Delta N_i)^2 + \cdots + \tag{6.40}$$

$$G_{R,1} = G_R - \left(\frac{\partial G_R}{\partial N_i}\right)_{T,P,N_j}(\Delta N_i) + \frac{1}{2}\left(\frac{\partial^2 G_R}{\partial N_i^2}\right)_{T,P,N_j}(\Delta N_i)^2 - \cdots + \tag{6.41}$$

The G minimum trend (6.39) yields ultimately

$$\frac{1}{2}\left(\frac{\partial^2 G}{\partial N_i^2}\right)_{T,P,N_j}(\Delta N_i)^2 \geq 0 \qquad (j \neq i) \tag{6.42}$$

in other words,

$$\left(\frac{\partial^2 G}{\partial N_i^2}\right)_{T,P,N_j} > 0 \qquad (j \neq i) \tag{6.43}$$

Recalling the differential form for dG given in Table 4.3, eq. (6.43) means

$$\left(\frac{\partial \mu_i}{\partial N_i}\right)_{T,P,N_j} > 0 \qquad (j \neq i) \tag{6.43'}$$

The sign that at state (2) the system is chemically stable and that the process in Fig. 6.8 can only proceed in the (1)→(2) direction is the inequality (6.43'). The simple system is chemically stable at state (2) because of the property that the chemical potential of species "i" increases whenever a new quantity of that species is added to the mixture system under conditions of constant temperature, pressure, and mole numbers of species "j" ($j \neq i$).

THE CONTINUITY OF THE GASEOUS AND LIQUID STATES

What do the preceding stability conditions have to do with the thermodynamic description of multiphase systems in equilibrium? Is it not known already that the *P–v* diagram of a pure substance must have the features shown with solid lines in Fig. 6.9? One benefit of looking more closely at the gas–liquid transitions referred to in Fig. 6.9 is the learning of an interesting bit of history, for the isotherm map of Fig. 6.9 is far from having been the spontaneous product of a commonly shared point of view.

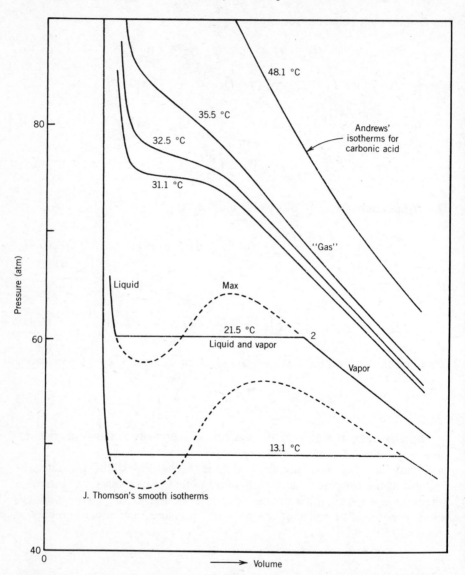

Figure 6.9   Schematic reproduction of Andrews' diagram and isotherms—the solid lines—and J. Thomson's continuous transition from gas to liquid through a region of unstable equilibrium—the dashed lines (this figure is a mirror image drawn after Ref. 5).

Andrews' Diagram and J. Thomson's Theory

The proper name for what we see in Fig. 6.9 is the *Andrews diagram* [3], for Dr. Thomas Andrews of Queen's College, Belfast, who put it together and did much of the work of clarifying the relationship between known concepts such as gas, vapor, liquid, and critical temperature. For a long time before

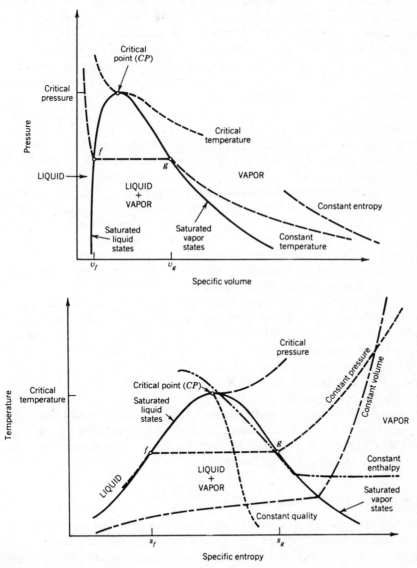

Figure 6.10 The modern *P–v* and *T–s* diagrams and terminology for a pure substance in gaseous and/or liquid form [4].

him, a "gas" and a "vapor" were thought to be different in the sense that only the latter could be transformed into liquid, say, by pressurization at constant temperature. Andrews showed that the difference between these common terms has to do with whether the temperature of the fluid in question is greater or lower than its critical temperature. When the temperature is greater than critical, the process of isothermal pressurization does not reveal the abrupt transition (separation) into liquid droplets and vapor.

While putting the known gas, vapor, liquid, and (liquid + vapor) states on the same map, Andrews drew also the essential conclusion that the gaseous and liquid forms of a pure substance are manifestations of "the same condition of matter" [3]. Furthermore, by locating the position of the (liquid + vapor) dome and by avoiding it, Andrews was able to demonstrate that "the gaseous and liquid forms of matter may be transformed into one another by a series of unbroken changes" [3].

The next important idea that appears to have been forgotten when "Andrews' diagram" became the modern "$P-v$ diagram" of Fig. 6.10 [4] is indicated by the humped dash lines in Fig. 6.9. If the gas can be transformed into liquid continuously by avoiding the two-phase dome (i.e., avoiding the abrupt changes marked by the beginning and the end of condensation), then, perhaps, the continuous-transition concept has a place even at lower, subcritical temperatures. After all, it is known that by respecting special rules of tranquility during the isothermal compression processes of Fig. 6.9, it is possible to raise the vapor pressure well above the known saturation pressure, before the formation of the first droplets of liquid. This effect is represented by the arc labeled (2)–(max) in Fig. 6.9, and forms the subject of the subsection on metastable states later in this chapter.

The idea that a continuous gas–liquid transition across the dome is theoretically conceivable and that only an instability effect prevents it from being part of the reality we observe was advanced by James Thomson [5], professor of mechanical engineering at the same Queen's College (Belfast) and brother of William Thomson (Lord Kelvin). He wrote that "it appears probable that, although there be a practical breach of continuity in crossing the line of boiling-points . . . , there may exist, in the nature of things, a theoretical continuity across this breach having some real and true significance. This theoretical continuity . . . must be supposed to be such as to have its various courses passing through conditions of pressure, temperature and volume in unstable equilibrium . . . ."

The van der Waals Equation of State

James Thomson's idea of fitting continuously the gas and liquid portion of Andrews' isotherms is shown with dash lines in Fig. 6.9. The next step is to express this continuous-fitting idea analytically and then to use the stability conditions of the preceding section in order to predict the abrupt transition from gas to liquid and vice versa. An analytical model that serves this purpose in an amazingly concise manner was constructed in 1873 by Johannes Diderik van der Waals [6]. Instead of the usual equation of state for an ideal gas ($Pv = RT$), van der Waals wrote

$$\left(P + \frac{a}{v^2}\right)(v - b) = RT \tag{6.44}$$

or

$$P = \frac{RT}{v - b} - \frac{a}{v^2} \tag{6.44'}$$

where a and b are two empirical constants. The b constant represents the minimum volume occupied by the substance in the limit $P \to \infty$. The additional pressure term a/v^2 accounts for the mutual attraction between molecules, under the assumption that the attraction forces are proportional to the density squared. That the van der Waals model bridges the gap between ideal gas behavior and incompressible-liquid behavior becomes evident as we take eq. (6.44) to the following two limits while T is finite:

$$\begin{aligned} \text{eq. (6.44)} &\to Pv = RT &&\text{as } v \to \infty \\ \text{eq. (6.44)} &\to v = b &&\text{as } P \to \infty \end{aligned} \tag{6.44''}$$

Figure 6.11 shows in dimensionless form the family of isotherms provided by the van der Waals model and how the shape of these curves matches what we saw in Andrews' diagram. First, the nondimensionalization of Fig. 6.11 consists of using the concepts of *reduced* pressure, temperature, and specific volume:

$$P_r = \frac{P}{P_c} \qquad T_r = \frac{T}{T_c} \qquad v_r = \frac{v}{v_c} \tag{6.45}$$

where P_c, T_c, and v_c represent the (P, T, v) values measured at the critical point CP. In terms of these new variables, eq. (6.44') reads

$$P_r = \frac{8T_r}{3v_r - 1} - \frac{3}{v_r^2} \tag{6.46}$$

A preliminary step in the transition from eq. (6.44') to eq. (6.46) is the "fitting" of the two-constant expression (6.44') to the critical-point data of an actual substance, Table 6.1. This is done by stating that the van der Waals isotherm has an inflexion at the critical point:

$$\left(\frac{\partial P}{\partial v}\right)_T = 0 \quad \text{and} \quad \left(\frac{\partial^2 P}{\partial v^2}\right)_T = 0 \quad \text{at } P = P_c, T = T_c, \text{ and } v = v_c \tag{6.47}$$

Solving the two-equation system (6.47) yields

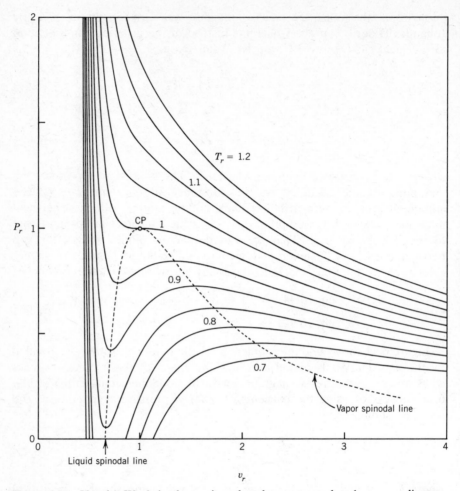

Figure 6.11 Van der Waals isotherms in reduced-pressure and -volume coordinates.

$$P_c = \frac{a}{27b^2} \qquad T_c = \frac{8a}{27bR} \qquad v_c = 3b \qquad (6.48)$$

In the pursuit of James Thomson's idea, it remains to show that the region covered by wavy isotherms in Fig. 6.11 is a domain of thermodynamic instability that accounts for the slope discontinuities (sharp corners) made evident by Andrews' solid lines (Fig. 6.9). The first observation to be made is that the fluid cannot exist in the area covered by van der Waals isotherms with positive slope:

$$\left(\frac{\partial P}{\partial v}\right)_T > 0 \qquad (6.49)$$

TABLE 6.1   Critical State Properties [4]

Fluid	Critical Temperature		Critical Pressure		Critical Specific Volume
	K	°C	MPa	atm	cm³/g
Air	133.2	−140	3.77	37.2	2.9
Alcohol (methyl)	513.2	240	7.98	78.7	3.7
Alcohol (ethyl)	516.5	243.3	6.39	63.1	3.6
Ammonia	405.4	132.2	11.3	111.6	4.25
Argon	150.9	−122.2	4.86	48	1.88
Butane	425.9	152.8	3.65	36	4.4
Carbon dioxide	304.3	31.1	7.4	73	2.2
Carbon monoxide	134.3	−138.9	3.54	35	3.2
Carbon tetrachloride	555.9	282.8	4.56	45	1.81
Chlorine	417	143.9	7.72	76.14	1.75
Ethane	305.4	32.2	4.94	48.8	4.75
Ethylene	282.6	9.4	5.85	57.7	4.6
Helium	5.2	−268	0.228	2.25	14.4
Hexane	508.2	235	2.99	29.5	4.25
Hydrogen	33.2	−240	1.30	12.79	32.3
Methane	190.9	−82.2	4.64	45.8	6.2
Methyl chloride	416.5	143.3	6.67	65.8	2.7
Neon	44.2	−288.9	2.7	26.6	2.1
Nitric oxide	179.3	−93.9	6.58	65	1.94
Nitrogen	125.9	−147.2	3.39	33.5	3.25
Octane	569.3	296.1	2.5	24.63	4.25
Oxygen	154.3	−118.9	5.03	49.7	2.3
Propane	368.7	95.6	4.36	43	4.4
Sulfur dioxide	430.4	157.2	7.87	77.7	1.94
Water	647	373.9	22.1	218.2	3.1

Equilibrium states are excluded from this region by the criterion of internal mechanical stability, eqs. (6.35′) and (6.36). In Fig. 6.11, this region is bordered by two dashed lines representing the solution to the first of eqs. (6.47). A three-dimensional version of Fig. 6.11 is shown in Fig. 6.12a in order to make more vivid the origin[†] of the *spinodal* name given to the two curves that border the instability domain.

The second observation is that the slope discontinuities associated with horizontal lines of type (f)–(g), Fig. 6.10, are part of a phenomenon that occurs at constant temperature and pressure. We are reminded by the star diagram of Fig. 6.5 that at constant T and P, the equilibrium of a simple

---

[†]The origin lies in the Latin word *spina*, which in this context means "spine." It also means "thorn," the thorny bone of a fish, etc.

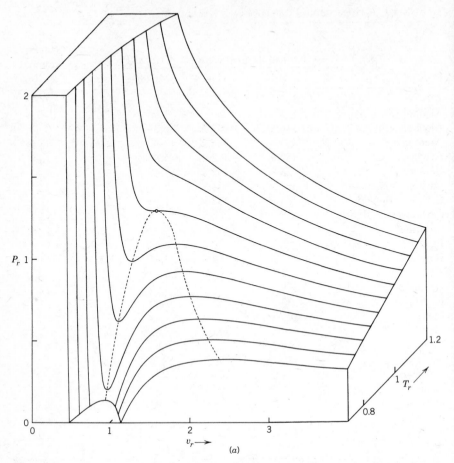

Figure 6.12   (a) The three-dimensional surface corresponding to the van der Waals equation of state.

system is characterized by minimum Gibbs free energy; therefore, we set out to investigate the variation of g [or μ, eq. (4.34′)] in the domain of wavy van der Waals isotherms.

One wavy isotherm from Fig. 6.11 has been isolated in Fig. 6.13. According to the closed-system version of the dG expression listed in Table 4.3, we can write

$$dg = -s\,dT + v\,dP \qquad (6.50)$$

This means that along the $T = T_0$ isotherm, the specific Gibbs free energy varies as

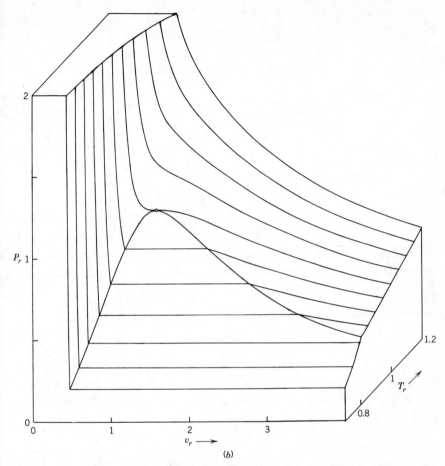

Figure 6.12 (b) The theoretically modified surface showing the two-phase region resulting from applying the equal-area rule described in Fig. 6.13.

$$g(T_0, P) = g_0 + \int_{P_0}^{P} v \, dP \qquad (6.51)$$

where $g_0 = g(T_0, P_0)$. The integration of eq. (6.50) starts from the point (0), where $T = T_0$ and $P = P_0$, and proceeds along the wavy portion of the isotherm passing most notably through the points labeled (max) and (min). The way in which the calculated $g(T_0, P)$ values vary with P is shown qualitatively on the right side of Fig. 6.13. From (0) to (max), g increases relatively fast because the integrand (v) of the pressure integral (6.51) is large. Between (max) and (min), g decreases because the pressure decreases (note that dP is negative). This portion of the isotherm is drawn with dash

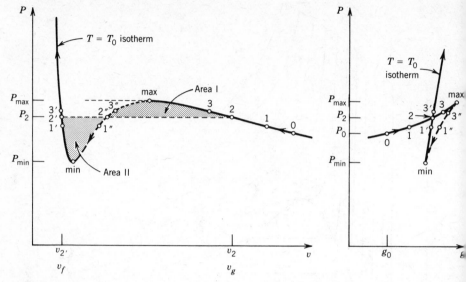

Figure 6.13 The equal-area rule for determining the pressure of the gas–liquid transition during isothermal compression.

line to remind us that the points along it are states of unstable equilibrium. Finally, the g values increase again as the pressure increases after turning the corner marked (min). This time, however, the rise of g is slow and almost linear in P, because the v integrand of the pressure integral (6.51) is small and almost constant.

What emerges in the P–g plane on the right side of Fig. 6.13 is a curve that crosses itself over at point (2) [see also Refs. 7, 8]. The two cusps of this curve correspond to the (max) and (min) points noted on the T_0 isotherm in the P–v plane. As the integral (6.51) sweeps the T_0 isotherm from right to left in the P–v plane, the T_0 isotherm is traveled in the clockwise direction in the P–g plane.

For any pressure P_1 in the range $P_{min} - P_2$, the g function calculated with eq. (6.51) has three values such that

$$g_1 < g_{1'} < g_{1''} \qquad \text{(at constant T and P)} \qquad (6.52)$$

Which of these values represents the true Gibbs free energy of the system is the same as asking which of the states, (1), (1'), and (1''), represents the equilibrium of the system in the P–v plane. State (1'') and $g_{1''}$ are ruled out by the instability argument built already around eq. (6.49). States (1) and (1') are both possible equilibrium states; however, in view of the G minimum principle and the fact that g_1 is less than $g_{1'}$, the system will opt for state (1) as stable equilibrium state. The choice is reversed in the case of pressure levels P_3 closer to P_{max}:

$$g_{3'} < g_3 < g_{3''} \qquad \text{(at constant } T \text{ and } P) \qquad\qquad (6.53)$$

as the system settles for state (3'). Looking at the crossed-over curve in the P–g plane, it is clear that the equilibrium state (the one with the lowest g) jumps from the first arc of the curve to the third at the intersection point labeled (2). This point corresponds to states (2), (2'), and (2") in the P–v plane.

The abrupt transition from gas to liquid during a process of isothermal compression can therefore be rationalized by invoking the principle of minimum Gibbs free energy at constant temperature and pressure. The discontinuity of volume measured between v_f and v_g along one of the horizontal isotherm segments in Andrews' diagram, Fig. 6.9 and 6.10 (top), is the same as the discrepancy between $v_{2'}$ and v_2 in the P–v plane of Fig. 6.13. The system can exist as a homogeneous system only at volumes larger than v_2 (as gas) and at volumes smaller than $v_{2'}$ (as liquid). If the system volume v is dictated by the design of the container wall in such a way that it falls between $v_{2'}$ and v_2:

$$v_{2'} < v < v_2 \qquad \text{(or } v_f < v < v_g) \qquad\qquad (6.54)$$

the system (mass $= m$) has the option to break up into two subsystems (m_f, m_g), where the specific volume of m_f is v_f and that of m_g is v_g. Put together, the two subsystems—the two phases—occupy the imposed overall volume (mv);

$$m_f v_f + m_g v_g = mv \qquad\qquad (6.55)$$

The mass fraction of the vapor phase in this assembly is the *quality* of the two-phase mixture (better said, the vapor quality):

$$x = \frac{m_g}{m_f + m_g} \qquad\qquad (6.56)$$

so that state (2) or (g) is represented by $x = 1$ and $T = T_0$ and state (2') or (f) by $x = 0$ and $T = T_0$. The liquid fraction or *moisture content* of the "$L + V$" mixture is equal to $(1 - x)$. The specific volume of the entire system is related to the specific volumes of the individual phases via the equation

$$v = v_f + x\underbrace{(v_g - v_f)}_{v_{fg}} \qquad\qquad (6.57\text{v})$$

which is obtained by combining eqs. (6.55) and (6.56). Worth noting is that equations of type (6.55) and (6.56) can be written for extensive properties other than specific volume, and that the u, h, and s equivalents of eq. (6.57v) are

$$u = u_f + x(u_g - u_f) = u_f + xu_{fg} \tag{6.57u}$$

$$h = h_f + x(h_g - h_f) = h_f + xh_{fg} \tag{6.57h}$$

$$s = s_f + x(s_g - s_f) = s_f + xs_{fg} \tag{6.57s}$$

Maxwell's Equal-Area Rule

The horizontal line $(2')$–(2) slices the hump of the wavy van der Waals isotherm at a particular pressure level P_2. We saw that this special (unique) pressure corresponds to the crossover point exhibited by the T_0 isotherm in the P–g plane. At this stage, we are in the position to draw an important conclusion and to ask a new question. The conclusion is that as long as the temperature T_0 is subcritical, there is a one-to-one relationship between T_0 and the pressure level of volume discontinuity, P_2. We shall return to this conclusion in subsequent pages of this chapter.

The new question is how to determine the condensation pressure P_2 based on the P–v diagram alone, that is, without having to construct the P–g diagram and to locate the crossover point (2). The answer consists of expressing in analytical form that states (2) and $(2')$ fall on top of each other on the P–g plane, namely,

$$g_{2'} = g_2 \tag{6.58}$$

In terms of pressure integrals, eq. (6.51), the above statement implies

$$g_0 + \left(\int_0^{2'} v \, dP \right)_{T=T_0} = g_0 + \left(\int_0^{2} v \, dP \right)_{T=T_0} \tag{6.59}$$

in other words,

$$\left(\int_2^{2'} v \, dP \right)_{T=T_0} = 0 \tag{6.60}$$

Now, looking at the wavy portion of the T_0 isotherm in the P–v plane of Fig. 6.13, we see that the arc (2)–$(2')$ is made up of four smaller arcs, (2)–$(\max)$, $(\max)$–$(2'')$, $(2'')$–$(\min)$, and $(\min)$–$(2')$. This observation allows us to rewrite eq. (6.60) as

$$\left(\int_2^{\max} v \, dP \right)_{T=T_0} + \left(\int_{\max}^{2''} v \, dP \right)_{T=T_0} + \left(\int_{2''}^{\min} v \, dP \right)_{T=T_0} + \left(\int_{\min}^{2'} v \, dP \right)_{T=T_0} = 0 \tag{6.61}$$

which is the same as saying

$$\underbrace{\left(\int_2^{\max} v\,dP\right)_{T=T_0} - \left(\int_{2''}^{\max} v\,dP\right)_{T=T_0}}_{\text{Area I}} = \underbrace{\left(\int_{\min}^{2''} v\,dP\right)_{T=T_0} - \left(\int_{\min}^{2'} v\,dP\right)_{T=T_0}}_{\text{Area II}} \qquad (6.61')$$

The graphic technique for locating the position of P_2 directly on the $P-v$ diagram consists of comparing Areas I and II shaded in Fig. 6.13. The wanted transition-pressure level P_2 is the one for which Areas I and II are equal. This equal-area rule is Maxwell's contribution[†] [9] to the line of thought reviewed here in the names of Andrews [3], Thomson [5], and van der Waals [6].

The equal-area rule can be used to slice off all the wavy isotherms of the van der Waals $P(v, T)$ surface presented in Fig. 6.12a. The result of this geometric construction is shown drawn to scale in Fig. 6.12b. The similarity between the projection of this theoretically modified surface on the P_r-v_r plane and the Andrews diagrams (Figs. 6.9 and 6.10, top) is the reward for having questioned the origin of the volume discontinuity and two-phase separation observed during isothermal compression at subcritical temperatures.

The flat-looking portion of the modified surface (the liquid + vapor dome of Fig. 6.12b) is what in space geometry is called a *ruled surface*. Projected on the P_r-T_r plane, the liquid + vapor dome appears as a line with positive curvature. This line is terminated by the critical point CP. The projection of a $P(v, T)$ surface on the $P-T$ plane was called the *Regnault diagram* in

---

[†]In the same paper, Maxwell offered these comments on the subject of research challenge and strategy:

In attempting the extension of dynamical methods to the explanation of chemical phenomena, we have to form an idea of the configuration and motion of a number of material systems, each of which is so small that it cannot be directly observed. We have, in fact, to determine, from the observed external actions of an unseen piece of machinery, its internal construction.

The method which has been for the most part employed in conducting such inquiries is that of forming a hypothesis, and calculating what would happen if the hypothesis were true. If these results agree with the actual phenomena, the hypothesis is said to be verified, so long, at least, as someone else does not invent another hypothesis which agrees still better with the phenomena.

The reason why so many of our physical theories have been built up by the method of hypothesis is that the speculators have not been provided with methods and terms sufficiently general to express the results of their induction in its early stages. They were thus compelled either to leave their ideas vague and therefore useless, or to present them in a form the details of which could be supplied only by the illegitimate use of the imagination.

In the meantime the mathematicians, guided by that instinct which teaches them to store up for others the irrepressible secretions of their own minds, had developed with the utmost generality the dynamical theory of a material system.

memory of one of the greatest experimentalists of the nineteenth century[t] [10].

The Clapeyron Relation

Translated into the modern terminology reviewed in Fig. 6.10, the crossover condition (6.58) reads

$$g_f = g_g \tag{6.62}$$

The position of states (f) and (g) is fixed as soon as their temperature *or* pressure is specified (recall the one-to-one relationship between T_0 and P_2 discovered in Fig. 6.13). For small variations in T (or P), eq. (6.62) yields, in order,

$$dg_f = dg_g \tag{6.63}$$

$$-s_f \, dT + v_f \, dP = -s_g \, dT + v_g \, dP \tag{6.64}$$

$$(s_g - s_f) \, dT = (v_g - v_f) \, dP \tag{6.65}$$

$$\frac{dP}{dT} = \frac{s_{fg}}{v_{fg}} \qquad \text{[for states of type } (f), (g), \text{ and } (f+g)] \tag{6.66}$$

The step from eq. (6.63) to eq. (6.64) required the use of eq. (6.50). The result of this brief analysis is that the slope of the unique curve $P(T)$, eq. (6.66), is related to the discontinuities registered by the entropy and the volume during the constant-T switch from (f) to (g). Since both s_{fg} and v_{fg} are positive, Fig. 6.10, the pressure of a two-phase mixture $(f+g)$ increases with the temperature.

An alternative to the dP/dT expression derived above is

$$\frac{dP}{dT} = \frac{h_{fg}}{Tv_{fg}} \qquad \text{[for states of type } (f), (g), \text{ and } (f+g)] \tag{6.67}$$

This alternative is based on the additional identity:

$$h_{fg} = Ts_{fg} \tag{6.68}$$

which comes from the analysis of a process of reversible isothermal expansion from (f) to (g):

---

[t]Henri Victor Regnault (1810–1878) was professor of chemistry at the École Polytechnique, Paris, and later at the Collège de France.

$$\underbrace{Q_{f\text{-}g} - W_{f\text{-}g}}_{mPv_{fg}} = \underbrace{U_g - U_f}_{mu_{fg}} \qquad \text{(first law)} \qquad (6.69)$$

$$Q_{f\text{-}g} = \underbrace{T(S_g - S_f)}_{ms_{fg}} \qquad \text{(second law)} \qquad (6.70)$$

where m is the total mass of the pure substance sample. The dP/dT relation (6.67) is recognized as either the *Clapeyron relation* or the *Clausius–Clapeyron relation*.

PHASE DIAGRAMS

The Gibbs Phase Rule

As a generalization of the observation that in the two-phase region of a pure substance P and T cannot be specified independently, consider the case of a simple system whose composition is described by the mole numbers of n components. We learned in the early part of chapter 4 that the simple system has a total of $(n + 2)$ intensive properties (T, P, μ_i) out of which only $(n + 1)$ can be specified independently. The simple system (pure phase) is said to have only $(n + 1)$ degrees of freedom because the $(n + 2)$ intensive properties are related through the Gibbs–Duhem relation (4.32). In the case of one phase, then, we reason that

Number of degrees of freedom	=	Number of intensive properties	−	Number of Gibbs–Duhem relations	(6.71)
$(n + 1)$		$(n + 2)$		(1)	

Consider now a post-instability situation where the n-component system is separated into p distinct phases all in equilibrium with one another. Each phase is a simple system of the kind discussed in the preceding paragraph. In chapter 4, we learned also that at equilibrium the $(T, P \mu_i)$ values of a simple system match those of its surroundings. It follows that if the p phases are in equilibrium with one another, they all share the same (T, P, μ_i) values, i.e., the same intensive properties. Therefore, the total number of intensive properties for the aggregate system (the assembly of p phases) is the same as for one individual phase, namely $(n + 2)$. The number of degrees of freedom of the aggregate system is equal to $(n + 2)$ minus the number of relations that exist between the $(n + 2)$ intensities. In the present case, that number is p because each phase has its own Gibbs–Duhem relation (4.32). The summary of this argument is the *Gibbs phase rule* [1] for calculating the number of degrees of freedom of an equilibrium mixture of n components coexisting as p phases:

$$\text{degrees of freedom} = n + 2 - p \qquad (6.72)$$

Single-Component Substances

As a first application of the Gibbs phase rule, consider the three-dimensional representation of the $P(v, T)$ surface seen already as a family of isotherms in Andrews' diagram (Fig. 6.9). In a single-component substance, $n = 1$, which means that when the system is in only one phase ($p = 1$), it has two degrees of freedom, T and P. The temperature and pressure can be varied independently along the single-phase regions marked "vapor," "liquid," and "solid" on the $P(v, T)$ surface of the "pure substance" diagram of Fig. 6.14. It was pointed out in the introduction to chapter 4 that the concept of a pure substance is slightly more general than that of a single-component substance. The defining feature of a pure substance is that its chemical composition is the same regardless of its form of aggregation

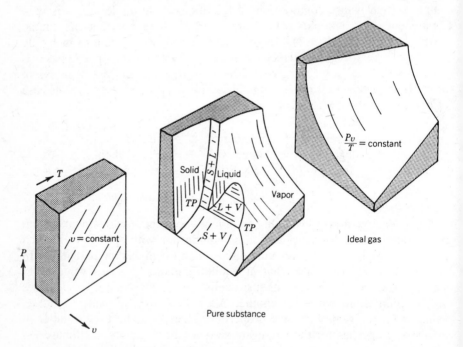

Figure 6.14 The $P(v, T)$ surface of a single-component substance that contracts upon freezing (the middle diagram), and the two extreme models of this surface, the incompressible-substance model (left) and the ideal gas model (right) [4].

(solid, liquid, vapor). Single-component substances meet this condition; therefore, they are all pure substances.

Two-phase mixture states are represented by the scooped-out regions "$L + V$" (liquid + vapor), "$S + V$" (solid + vapor) and "$S + L$" (solid + liquid) on the same $P(v, T)$ surface of the pure substance diagram in Fig. 6.14. A surface of this kind was discovered already while constructing the modified version of the van der Waals $P(v, T)$ surface, Fig. 6.12b. The distinctive feature of these regions of two-phase states is that their projections on the $P-T$ plane (i.e., on the Regnault diagram) are curves, not surfaces. The state of a two-phase mixture is pinpointed, for example, by specifying one intensive property (T or P) and the specific volume or any other property that depends on the two-phase composition of the mixture. The composition is described in terms of the quality x, eq. (6.56), which for a single-component two-phase mixture is also the mole fraction of the vapor phase:

$$x = \frac{N_g}{N_f + N_g} \qquad \text{(for a "}L + V\text{" mixture)} \qquad (6.73)$$

The mixing rule used in eqs. (6.57v)–(6.57s) for calculating the properties of a "$L + V$" mixture continues to hold for "$S + L$" and "$S + V$" mixtures. For example, the specific volume of a mixture of solid and liquid in equilibrium is

$$v = v_s + x_f(v_f - v_s) = v_s + x_f v_{sf} \qquad (6.74v)$$

where x_f represents the fraction occupied by liquid in the mixture. Likewise, the specific volume of a solid and vapor mixture is

$$v = v_s + x_g(v_g - v_s) = v_s + x_g v_{sg} \qquad (6.75v)$$

with x_g representing the vapor fraction and v_s the specific volume of saturated solid. Other properties of "$S + L$" and "$S + V$" mixtures can be calculated by extending the above rules in the manner indicated earlier in eqs. (6.57u)–(6.57s). Similarly, we can write a Clapeyron equation of type (6.66)–(6.67) to express the dP/dT slope of each region of two-phase states:

$$\frac{dP}{dT} = \frac{s_{sf}}{v_{sf}} = \frac{h_{sf}}{Tv_{sf}} \qquad \text{[for states of type } (s), (f), \text{ and } (s+f)] \quad (6.76)$$

$$\frac{dP}{dT} = \frac{s_{sg}}{v_{sg}} = \frac{h_{sg}}{Tv_{sg}} \qquad \text{[for states of type } (s), (g), \text{ and } (s+g)] \quad (6.77)$$

The collection of states in which the system contains all three phases in equilibrium is represented by the *triple-point* line "TP–TP" drawn parallel to the v direction in the middle diagram of Fig. 6.14. The triple-point line is the intersection of the three ruled surfaces "$S + L$," "$L + V$," and "$S + V$," and its projection on the P–T plane is a single point (the triple point "TP," Fig. 6.16). For a system whose state is situated on the triple-point line, we have $n = 1$ and $p = 3$; therefore, according to eq. (6.72), the number of degrees of freedom of the system is zero. Indeed, the temperature and pressure of "$S + L + V$" mixtures of the same chemical component are known constants or "fiducial points" on the T and P scales. Some of these constants are listed in Table 10.1 and, for water, in Fig. 6.15. The state of a

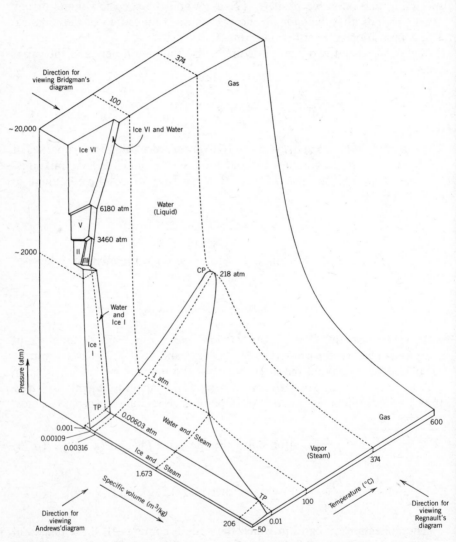

Figure 6.15 The $P(v, T)$ surface of water, showing the volume increase experienced upon freezing and the various ice transformations at very high pressures.

"$S + L + V$" mixture is pinpointed by specifying the composition (x_s, x_f, x_g), where

$$x_s + x_f + x_g = 1 \qquad (6.78)$$

The specific volume of the three-phase system is then

$$v = x_s v_s + x_f v_f + x_g v_g \qquad (6.79v)$$

where v_s, v_f, and v_g are known constants also.[†] Other specific extensive property values can be calculated similarly, by replacing v with u, h, s, etc. in eq. (6.79v).

The preceding observations acquire more meaning if we examine the actual $P(v, T)$ surface of the substance with which we are most familiar: Figure 6.15 shows the main features and dimensions of the $P(v, T)$ surface of water. One peculiarity that is emphasized by this figure is that the specific volume of ice I is greater than that of liquid water; in other words, water expands upon freezing. This feature is particularly visible along the triple-point line, where $v_s > v_f$. Another feature is the multitude of ice forms and associated transitions at pressures higher than 2000 atmospheres. Note the

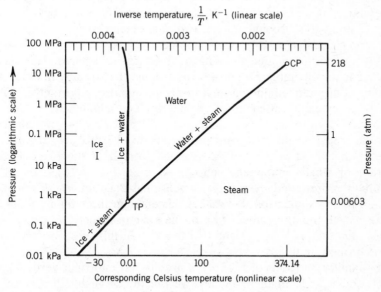

Figure 6.16 The P–T diagram of water, steam, and ice I at moderate pressures.

[†]In general, the specific volumes (or u, h, s, etc.) of the saturated single-phase states are functions of only T (or P); therefore, they are known constants at the fixed triple-point temperature (or pressure).

ruled surfaces representing two-phase mixtures and the triple-point lines along which some of these surfaces intersect. The transition from liquid water to ice VI, for example, shows that at sufficiently high temperatures and pressures, the process of solidification is accompanied by a reduction in volume. The high-pressure region of the $P(v, T)$ surface of water was mapped during the first decades of this century, most notably by Bridgman at Harvard [11, 12].

Viewed from the right (i.e., parallel to the v direction), the $P(v, T)$ surface of Fig. 6.15 reveals the Regnault diagram for water. The moderate pressure range of this diagram is presented in Fig. 6.16. One feature to note in this figure is the use of $\log P$ and $1/T$ as coordinates: this choice is responsible for making the "ice + steam" and "water + steam" curves appear almost linear (Problem 6.2).

Two-Component Mixtures

Placing $n = 2$ in the Gibbs phase rule (6.72) teaches us that the number of degrees of freedom of a two-component mixture is 3 for a single-phase system, 2 for a two-phase system, and 1 for a three-phase system. It means that in place of the areas that in the case of single-component substances represented single-phase systems, we must now use volumes to chart the domain of single-phase two-component mixtures. In place of $P(T)$ curves representing two-phase states of single-component substances, we now discover areas that account for the two-phase states of binary mixtures. In short, the phase diagrams of two-component systems have one more dimension relative to the diagrams of single-component substances. This accounts for their relative complexity and for the volume they occupy in fields where the subject of multicomponent mixtures is much more important than in this course, for example, in chemical engineering and metallurgy.

Consider a mixture of two components, A and B. The components are such that they can exist in any proportion in the mixture, regardless of whether the mixture is liquid or gaseous. Instead of the plane Regnault diagram of single-component substances (e.g., Fig. 6.16), the $(A + B)$ mixture is represented by a three-dimensional diagram, as shown in Fig. 6.17a. The new dimension that makes this diagram three-dimensional is the one that measures the composition of the system, namely, x_A or x_B. (Note at this point that $x_A + x_B = 1$ and that throughout this subsection, x_A and x_B represent mole fractions.)

The familiar portions of Fig. 6.17a are the left and right vertical planes, $x_A = 0$ and $x_A = 1$. In these planes, we see the Regnault diagrams of the pure components B and A, respectively. Each $P(T)$ curve in these planes is terminated by a critical point, $(CP)_B$ and $(CP)_A$. The states that reside above the two-phase curves labeled $(L + V)_B$ and $(L + V)_A$ in the lateral planes represent liquid states. The states situated below these curves are gaseous states. If we now look at the bulk of the diagram that covers the

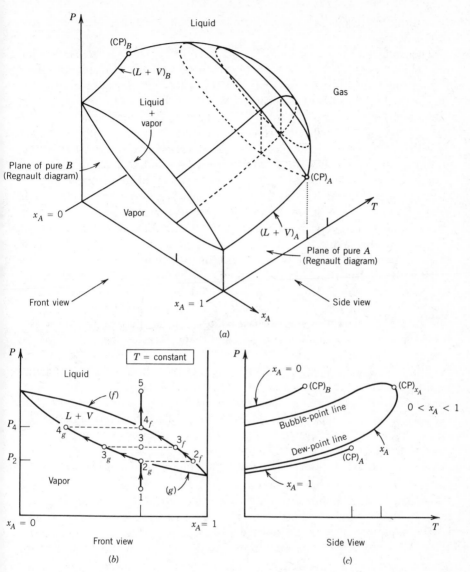

Figure 6.17 The single-phase and two-phase domains of a binary mixture, or the effect of composition (x_A) on the Regnault diagram.

range of intermediate compositions $0 < x_A < 1$, we see that the (T, P, x_A) space is divided into three distinct regions along the vertical, an upper high-pressure domain in which the binary mixture is a liquid, a lower domain where the mixture is a gas, and, finally, an intermediate space shaped like a mitten that houses all the "liquid + vapor" states of the mixture. Behind this intermediate space (i.e., at sufficiently high temperatures), the mixture is in gaseous form throughout the pressure range shown in the figure.

The shape of the intermediate space is visualized by three constant-temperature cuts. The second of these constant-temperature planes passes right through the critical point $(CP)_A$, but not through $(CP)_B$. This is due to the fact that the critical temperature of pure A happens to be higher than the critical temperature of pure B. The shape of the intermediate "liquid + vapor" domain is visualized further by a constant-x_A plane, the $P(T)$ trace of which is shown in Fig. 6.17c. The $P(T)$ curve seen in the constant-x_A cut is made up of two arcs—the bubble-point line and the dew-point line—joined at a critical point $(CP)_{x_A}$. The meaning of this terminology is discussed next. In conclusion, the mitten-shaped intermediate space sketched in the (T, P, x_A) space is bordered from above by the bubble-point surface and from below by the dew-point surface. These two surfaces are joined smoothly along a curve of critical points that runs from $(CP)_B$ to $(CP)_A$ around the high-temperature end of the "liquid + vapor" space.

How the intermediate "liquid + vapor" space accounts for the two-phase states of the $(A + B)$ mixture is illustrated in the pressure–composition diagram shown in Fig. 6.17b. Consider a process of isothermal pressurization beginning with the single-phase state (1). As the pressure increases from state (1) to state (2), the composition and form of aggregation of the mixture do not change. State (2) marks the appearance of the first droplets of liquid. The peculiar aspect of this incipient two-phase system is that the liquid phase (the droplets) and the vapor phase (the bulk of the system) have different compositions. The first droplets have the x_A value of state (2_f), while the surrounding vapor has the original composition, $x_{A,1} = x_{A,2_g}$. As the pressure continues to increase, the system evolves as a two-phase mixture in which the saturated vapor phase is a binary mixture of composition $x_{A,3_g}$ and where the saturated liquid phase is a binary mixture of composition $x_{A,3_f}$. Both $x_{A,3_g}$ and $x_{A,3_f}$ shift as P increases. The life of the two-phase system expires when the concentration of the liquid phase matches the original concentration of the gaseous system, $x_{A,4_f} = x_{A,1}$. The last bubbles that disappear from the system have a different composition, $x_{A,4_g}$. A further increase in pressure moves the state of the now single-phase mixture from (4_f) to (5), that is, along a line of constant concentration.

As summary to the above paragraph, we record that the system breaks up into "liquid + vapor" only in the pressure interval $P_2 - P_4$. The point (3) drawn along the constant-x_A line inside the two-phase region symbolizes the two-phase state in which the system contains liquid of type (3_f) and vapor of type (3_g). How much of (3_f) liquid and (3_g) vapor one finds at state (3) depends on the specific volumes of (3_f) liquid and (3_g) vapor, and on the overall (imposed) specific volume of the entire two-phase system.

A similar story can be told while looking at a constant-P cut through the mitten-shaped space of "liquid + vapor" states of Fig. 6.17a. One section of

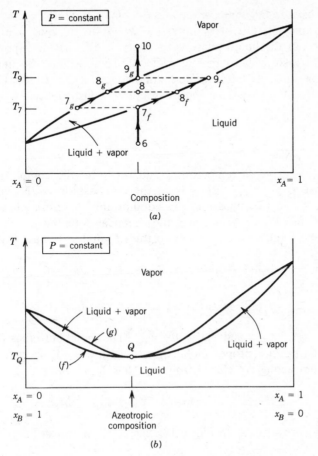

Figure 6.18 The two-phase domain of a binary mixture in the constant-P plane and the collapse of this domain at the point of azeotropic composition (Q).

this type is shown in Fig. 6.18a. The airfoil shape of the two-phase region has two sharp points because the constant-P plane happens to intersect the curves labeled $(L + V)_B$ and $(L + V)_A$ in Fig. 6.17a. A process in which the temperature rises isobarically proceeds along the vertical line 6–8–10, giving birth to two-phase mixtures only in the temperature interval T_7–T_9.

One conclusion that is worth stressing following this discussion is that, in general, at fixed T and P, the composition of the liquid phase differs from that of the vapor phase in a two-phase batch of a binary mixture. After all, this feature distinguishes the binary mixture from a pure substance. One exception from this conclusion is shown in 6.18b, where we see that the intermediate domain of two-phase states collapses to zero thickness at the point marked (Q). The composition where this collapse occurs is called the *azeotropic*[†] composition and the binary mixture $(x_A, x_B)_Q$ is called the azeotropic mixture or, simply the *azeotrope*. In such a mixture, the composition of the liquid phase is the same as that of the vapor phase.

Figure 6.18 shows also that at azetropic composition, the lines of saturated vapor state (g) and saturated liquid states (f) have zero slope as they become tangent at (Q). This feature is a direct consequence of the condition of azeotropy:

$$x_{A,f} = x_{A,g} \tag{6.80}$$

In general, at a given temperature and pressure, the composition of the liquid phase $(x_{A,f}, x_{B,f})$ differs from the composition of the vapor phase $(x_{A,g}, x_{B,g})$. The two phases are in equilibrium; therefore, they share not only the T and P values, but also the chemical potentials μ_A and μ_B. The Gibbs–Duhem relations that apply to the (f) and (g) phases are

$$\bar{s}_f \, dT - \bar{v}_f \, dP + x_{A,f} \, d\mu_A + x_{B,f} \, d\mu_B = 0 \tag{6.81}$$

$$\bar{s}_g \, dT - \bar{v}_g \, dP + x_{A,g} \, d\mu_A + x_{B,g} \, d\mu_B = 0 \tag{6.82}$$

where $x_{B,f} = 1 - x_{A,f}$, and $x_{B,g} = 1 - x_{A,g}$. The molal entropies $(\bar{s}_f, \bar{s}_g)$ and volumes $(\bar{v}_f, \bar{v}_g)$ are proper molal quantities in accordance with definition (4.107). Subtracting eq. (6.81) from eq. (6.82), we obtain

$$\bar{s}_{fg} \, dT - \bar{v}_{fg} \, dP + (x_{A,g} - x_{A,f})(d\mu_A - d\mu_B) = 0 \tag{6.83}$$

At constant pressure, as in Fig. 6.18, this conclusion translates into

$$\frac{dT}{dx_A} = \frac{x_{A,g} - x_{A,f}}{\bar{s}_{fg}} \frac{d}{dx_A}(\mu_B - \mu_A) \qquad \text{(constant } P\text{)} \tag{6.84}$$

In the case of an azeotropic composition, the right side of this equation vanishes because of eq. (6.80) and $\bar{s}_{fg} \neq 0$. It means that the slopes of the (f) and (g) curves must be zero at the point (Q) in the T–x_A plane of constant P. The (f) and (g) curves must also *touch* at (Q) because, taken together, eqs. (6.80) and (6.83) state that at (Q), the temperature is only a function of pressure. In conclusion, the azeotropy condition (6.80) is responsible for the zero slope and tangency of the (f) and (g) curves at point of azeotropic composition (Q).

It is easy to see that if the two-phase domain collapses to zero thickness at (Q) in the constant-P plane of Fig. 6.18b, it must also collapse in the constant-T plane, for example, in the P–x_A plane shown in Fig. 6.17b. In a constant-temperature plane, eq. (6.83) dictates

---

[†]From the Greek words *zeo* (to boil) and *atropos* (unchanging).

$$\frac{dP}{dx_A} = \frac{x_{A,g} - x_{A,f}}{\bar{v}_{fg}} \frac{d}{dx_A}(\mu_A - \mu_B) \qquad \text{(constant } T\text{)} \qquad (6.85)$$

which means zero slope dP/dx_A at the point where the azeotropic composition condition (6.80) is satisfied. The (f) and (g) curves that would be drawn in the constant-T plane would be tangent with zero slope at the point of azeotropic composition.

The phase diagrams presented in Figs. 6.17 and 6.18 are actually the simplest diagrams that can be drawn for mixtures of two chemical species. The relative simplicity of these diagrams stems from the assumption that the components A and B are miscible in all proportions. Complications arise when A and B are only partially miscible, for example, when liquid A and liquid B can mix at x_A values that do not cover the entire range $(0, 1)$.

Figure 6.19 shows a constant-P section through the phase surface of a mixture of two partially miscible liquids. In the limit $x_A \to 0$, liquid B mixes with small amounts of liquid A to form the mixed (homogeneous) liquid phase L_β. At the opposite edge of the diagram, liquid A mixes with small amounts of liquid B and forms a totally different (A-dominated) homogeneous liquid, L_α. Continuing to look at the low-temperature portion of Fig. 6.19, we see that at intermediate values of x_A, the system can only exist as an assembly of two distinct liquids (two phases), L_β and L_α. The properties of these two phases are those associated with the saturated liquid states that border the "$L_\beta + L_\alpha$" domain. For example, state (1) represents a two-phase mixture of saturated liquid L_β at state (1_β) and saturated liquid L_α at state (1_α).

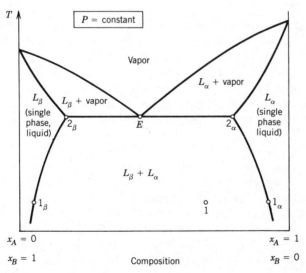

Figure 6.19 The constant-pressure phase diagram of a mixture of two partially miscible liquids.

The wing-shaped upper portion of Fig. 6.19 is caused also by the partial immiscibility of A and B as liquids. Note that at sufficiently high temperatures, gaseous A and gaseous B are miscible in all proportions. In the limit $x_A \to 0$, liquid A and liquid B are miscible; therefore, in that limit, Fig. 6.19 should have the same features as the constant-P diagrams seen already in Fig. 6.18. Indeed, directly above the single-phase domain L_β, we see the left extremity of the expected airfoil shape of the domain of liquid plus vapor states, in this case "L_β + vapor." Similarly, in the limit $x_A \to 1$, we rediscover the right end of the two-phase domain "L_α + vapor." The two domains "L_β + vapor" and "L_α + vapor" are bordered from above by the locus of saturated vapor states (dew-point line) and from below by the locus of saturated liquid states (bubble-point line). These two two-phase domains expire as the temperature decreases below the lowest temperature at which A and B exist as a gaseous mixture, T_E. If $T = T_E$ and if the composition x_A falls between that of states (2_β) and (2_α), then the system exists as a three-phase mixture, namely, liquid L_β of composition $x_{A,2_\beta}$, gaseous mixture of composition $x_{A,E}$, and liquid L_α of composition $x_{A,2_\alpha}$.

It is instructive to see the temperature–composition diagrams of miscible A and B (Fig. 6.18) as special cases of the diagram presented in Fig. 6.19. For example, if liquid A can mix[†] with increasingly larger amounts of liquid B, then the horizontal extent of the "L_α" domain in Fig. 6.19 increases toward the left, and so does the length of the airfoil-shaped domain "L_α + vapor." As this trend continues, Fig. 6.19 begins to look more and more like Fig. 6.18a. On the other hand, if the single-phase domains "L_β" and "L_α" become wider simultaneously, the two-phase liquid domain "$L_\beta + L_\alpha$" is eventually snuffed out and the resulting diagram looks more like Fig. 6.18b.

We can think also of a pair of liquids like water and mercury, which are practically immiscible in any proportion. For such a combination, the "L_β" and "L_α" domains of Fig. 6.19 shrink to zero thickness; in other words, the point labeled (2_β) resides now in the $x_A = 0$ edge of the diagram, while the point (2_α) lands on the $x_A = 1$ edge. The phase diagram of two immiscible liquids is therefore simpler than Fig. 6.19, in the sense that now the T–x_A plane is divided into only four subdomains: "$L_\beta + L_\alpha$," "L_β + vapor," "L_α + vapor," and "vapor."

The phase diagrams of the binary systems discussed in this subsection dealt only with liquid and gaseous mixtures. A completely analogous discussion can be presented for binary systems in liquid, solid, and two-phase (liquid + solid) states. If the components A and B are fully miscible as liquids but only partially miscible as solids, then the phase diagram looks the same as in Fig. 6.19, provided the "vapor" lable is replaced by "liquid," "L_β" by "S_β," and "L_α" by "S_α." Note that S_β and S_α refer to two distinct solid phases of A and B. The state (E) in this case is called the *eutectic point*

[†] In other words, if they can form a homogeneous liquid mixture—a phase.

and a mixture of A and B of composition $x_{A,E}$ is called the *eutectic mixture*. The melting temperature of the eutectic mixture, T_E, is lower than that of any other combination of the same components.[†]

CORRESPONDING STATES

Compressibility Factor

Perhaps the most useful by-product of van der Waals' analytical fitting of the $P(v, T)$ surface is the idea that in terms of the reduced variables (6.45), the pressure might be a unique function of volume and temperature for all single-component substances. This idea is suggested by eq. (6.46), which means

$$P_r = P_r(v_r, T_r) \tag{6.86}$$

It is best remembered as a principle or theory of *corresponding states*, by which all substances (pure, or mixtures) are expected to have the same pressure–volume–temperature equation of state when the variables P, v, and T are normalized with respect to the critical-point values P_c, v_c, and T_c. To any point (M_A) found on the $P(v, T)$ surface of substance A "corresponds" a particular point (M_B) from the $P(v, T)$ surface of substance B: both points, (M_A) and (M_B), are represented by a single point (M_r) on the surface expressed in reduced coordinates, $P_r(v_r, T_r)$.

We shall see that the theory works in the original sense suggested by eq. (6.86) only for one group of substances at a time, where each group contains substances whose molecular constitution is relatively similar. In practice, then, the simple two-argument expression (6.86) gives way to a three-argument expression in which the third argument accounts for the molecular identity (fingerprint) of each substance or group of similar substances.

Instead of using a reduced-pressure surface as in eq. (6.86), it has become customary to work with the dimensionless *compressibility factor Z* defined by

$$Z = \frac{Pv}{RT} \tag{6.87}$$

where R is the ideal gas constant of the particular substance whose $P(v, T)$ surface is being nondimensionalized. The compressibility factor measures the departure from ideal gas behavior, which is represented by $Z = 1$. Figure 6.20 shows that the compressibility factor of real gases is smaller than 1,

---

[†]This property is in fact summarized in the name "eutectic," which comes from the Greek word *eutektos* (easily fused, well-melted).

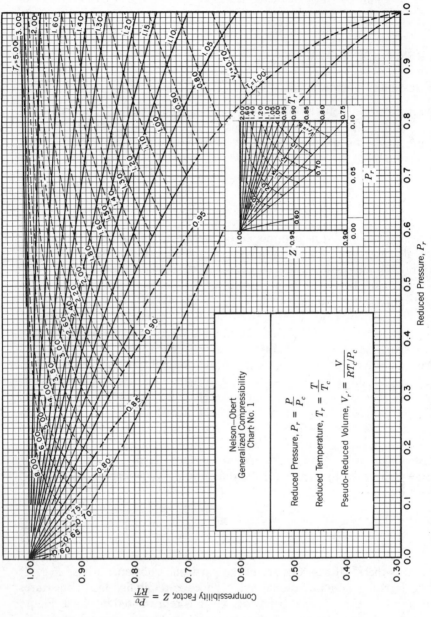

Figure 6.20 The Nelson–Obert compressibility charts. Chart No. 1: low-pressure region. (After Nelson and Obert [13], with permission from Professor Edward F. Obert, University of Wisconsin, Madison.)

Nelson—Obert
Generalized Compressibility
Chart No. 1

Reduced Pressure, $P_r = \dfrac{P}{P_c}$

Reduced Temperature, $T_r = \dfrac{T}{T_c}$

Pseudo-Reduced Volume, $V_r = \dfrac{V}{RT_c/P_c}$

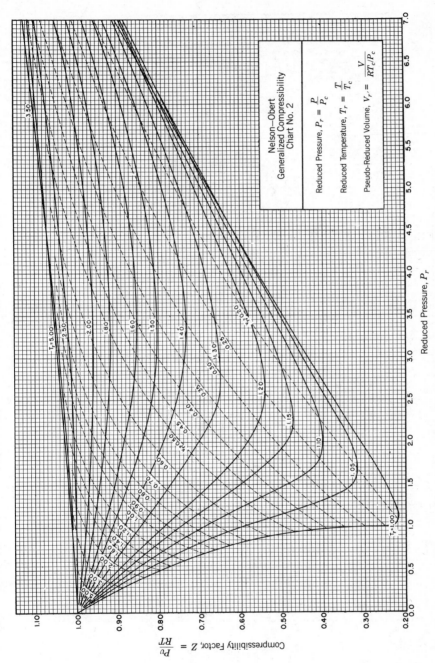

Figure 6.20 Continued. Chart No. 2: intermediate-pressure region.

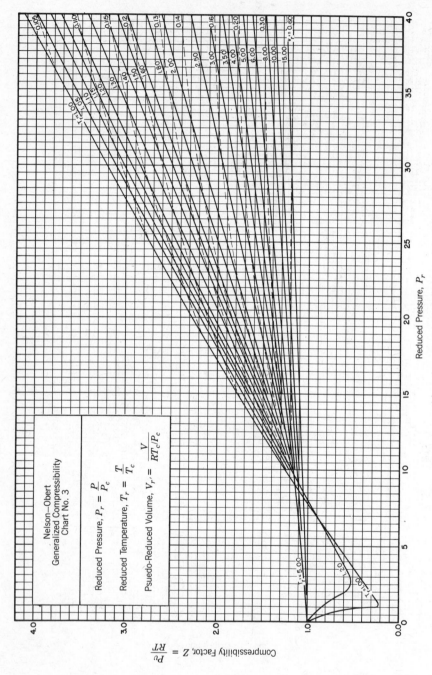

Figure 6.20 Continued. Chart No. 3: high-pressure region.

except at pressures and temperatures that are considerably higher than critical. In terms of reduced properties, the compressibility factor is equal to

$$Z = Z_c \frac{P_r v_r}{T_r} \tag{6.88}$$

where Z_c is the compressibility factor at the critical point:

$$Z_c = \frac{P_c v_c}{RT_c} \tag{6.87'}$$

Table 6.2 shows that substances can be categorized to some extent according to critical compressibility factor. Furthermore, since any single-component substance has a two-argument expression of type (6.86), by eliminating v_r between eqs. (6.88) and (6.86), we learn that Z_c may indeed play the role of the third (molecular) parameter in the corresponding states expression:

$$Z = Z(T_r, P_r, Z_c) \tag{6.89}$$

The description of the $P(v, T)$ surface is completed by eqs. (6.87) and (6.89). If we know two properties of a batch of a certain fluid (say, T and P), first we evaluate Z using eq. (6.89) so that we can calculate the remaining properties, for example, v from eq. (6.87) and h and s from eqs. (6.124) and (6.127), respectively, derived later in this section. Important then is to have access to the Z surface indicated in eq. (6.89). One approach

TABLE 6.2   Critical Compressibility Factors and Pitzer Acentric Factors[a]

Fluid	Z_c	ω	Fluid	Z_c	ω
Ammonia, NH_3	0.242	0.250	Hydrogen, H_2	0.305	−0.22
Argon, Ar	0.291	−0.004	Methane, CH_4	0.288	0.008
Butane, C_4H_{10}	0.274	0.193	Methyl chloride,		
Carbon dioxide,			CH_3Cl	0.268	0.156
CO_2	0.274	0.225	Neon, Ne	0.311	0.000
Carbon monoxide,			Nitric oxide, NO	0.25	0.607
CO	0.295	0.049	Nitrogen, N_2	0.29	0.040
Carbon			Octane, C_8H_{18}	0.259	0.394
tetrachloride,			Oxygen, O_2	0.288	0.021
CCl_4	0.272	0.194	Propane, C_3H_8	0.281	0.152
Chlorine, Cl_2	0.275	0.073	Sulfur dioxide,		
Ethane, C_2H_6	0.285	0.098	SO_2	0.268	0.251
Ethylene, C_2H_4	0.276	0.085	*van der Waals fluid*	*0.375*	*−0.302*
Helium-4, He	0.301	−0.387	Water, H_2O	0.229	0.344
Hexane, C_6H_{14}	0.260	0.296	Xenon, Xe	0.286	0.002

[a]Data extracted from a compilation by Reid et al. [14].

to the construction of this surface has been to ignore the effect of the third parameter [Z_c in eq. (6.89)] and to average the $Z(T_r, P_r)$ values calculated for a number of pure fluids. Figure 6.20 shows the generalized *Nelson–Obert Compressibility Charts* [13] that were obtained by averaging the Z information deduced from the $P(v, T)$ surfaces of 30 fluids.

The information projected on these charts is admittedly approximate, which is why it is better left in graphical form as opposed to tabular form or analytical expressions. The errors associated with using these charts are of the order of less than 5 percent for states that are not near the critical point. The charts are less accurate in the critical-point region. In any case, they are not to be used for strongly polar fluids (water, ammonia) as well as helium, nitrogen, and neon [14].

The constant-$v_{r'}$ lines drawn on the Z charts of Fig. 6.20 are for the *pseudo-reduced volume*:

$$v_{r'} = \frac{v}{RT_c/P_c} \tag{6.90}$$

This new reduced volume is nondimensionalized with respect to (RT_c/P_c) as opposed to the critical specific volume v_c used in eq. (6.45). The use of pseudo-reduced volumes is motivated by the fact that the measurement of T_c and P_c is considerably more accurate than that of v_c. Therefore, in order to recover the $P(v, T)$ information of a certain fluid from Fig. 6.20, the user must know the critical temperature, critical pressure, and ideal gas constant of the fluid of interest.

The generalized compressibility factor chart $Z(T_r, P_r)$ has been improved in a number of ways. One approach consisted of increasing the number of fluids in the process of averaging the reduced $P(v, T)$ information, which yielded Fig. 6.20. Another approach was motivated by the point of view that justice to the corresponding-states idea can only be done by taking into account the third argument (the molecular parameter) that is suggested by eq. (6.89). If Z_c is chosen as the molecular parameter, then a $Z(T_r, P_r)$ surface like that of Fig. 6.20 must exist for each group of fluids that have the same critical compressibility factor Z_c.

An alternative to using Z_c as third argument in a universal Z surface was proposed by Pitzer in 1955 [15]. This proposal amounts to differentiating between various fluids by means of the position occupied by each saturation-pressure curve (or locus of "$L + V$" states) on the dimensionless Regnault diagram P_r–T_r. Figure 6.21 shows three such curves. Note first that all these curves arrive at the same critical point (CP) because of our choice of nondimensionalizing the abscissa and the ordinate. The interesting feature brought out by this plot is the different slope of each curve. Pitzer attributed this difference to the degree of spherical symmetry of the molecular force field of each particular fluid (more on this is learned in a course on *molecular* thermodynamics [16]). It is found that fluids whose molecular

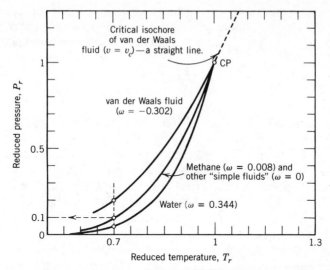

Figure 6.21 Regnault diagram in reduced coordinates for water, simple fluids, (e.g., methane) and the van der Waals fluid, showing the definition of the Pitzer acentric factor.

force fields exhibit spherical symmetry are represented by practically the same reduced saturation-pressure curve in the P_r–T_r plane. Examples of such fluids are the permanent gases with heavy molecules (argon, neon, xenon) and methane. Pitzer named this fluid class *simple fluids* and used the position of their $P_r(T_r)$ curve as reference. The departure from this reference position was then quantized in terms of the cut made by the $P_r(T_r)$ curve of another fluid on the vertical line $T_r = 0.7$ in Fig. 6.21. Worth noting is that the cut made by the $P_r(T_r)$ curve of a simple fluid on the same line occurs at $P_r = 0.1$.

Pitzer et al. [17] assigned a number ω—now called the *Pitzer acentric factor*—to each cut made by the $P_r(T_r)$ curves on the $T_r = 0.7$ vertical:

$$\omega = -\log_{10}(P_r) - 1.000 \qquad (\text{at } T_r = 0.7) \qquad (6.91)$$

such that $\omega = 0$ represents the class of simple fluids. Table 6.2 shows that each pure fluid has its own ω and that the value of ω departs from zero as the spherical symmetry of the molecular force field deteriorates. Strongly polar fluids (H_2O, NH_3) have high Pitzer acentric factors, which means that their two-parameter Z surfaces $Z(T_r, P_r)$ should differ from the reference Z surface of simple fluids.

The above proposal would amount to constructing a three-parameter family of charts:

$$Z = Z(T_r, P_r, \omega) \qquad (6.92)$$

Pitzer et al. [17] went one step further and assumed that the effect of ω is simply to correct the two-parameter compressibility factor chart of simple fluids, $Z^{(0)}(T_r, P_r)$. They assumed a linear relationship for this correction procedure:

$$Z(T_r, P_r, \omega) = Z^{(0)}(T_r, P_r) + \omega Z^{(1)}(T_r, P_r) \tag{6.93}$$

where the correction function $Z^{(1)}$ is also a function of only T_r and P_r. Fitting the linear expression (6.93) to the $P(v, T)$ surfaces of a large number of nonsimple fluids led to the calculation of the correction function $Z^{(1)}$. The companion operation of averaging the compressibility factor values of simple fluids produced the numerical information for the surface $Z^{(0)}$. Tables for $Z^{(0)}(T_r, P_r)$ and $Z^{(1)}(T_r, P_r)$ appeared first in Ref. 17. The most accurate set in use today appears to be that published in 1975 by Lee and Kesler [18]: this set is reproduced in the Appendix in Tables A.1 and A.2. Note in each table the zig-zag line that separates subcooled liquid states from superheated vapor states in the subcritical temperature range, $T_r < 1$. Highlighted in a box in the center of each table is the value that corresponds to the reduced critical point $T_r = P_r = 1$.

Coupled with Lee and Kesler's Tables A.1 and A.2, Pitzer et al.'s [17] linear decomposition of compressibility factor (6.93) is sufficiently accurate for engineering $P(v, T)$ calculations. Improvements of the corresponding states method of correlating $P(v, T)$ information can be sought in the direction suggested by Fig. 6.22. This figure shows that it takes at least two parameters to describe the molecular identity of a certain fluid; in other words, a one-to-one relation between the Z_c and ω values given in Table 6.2

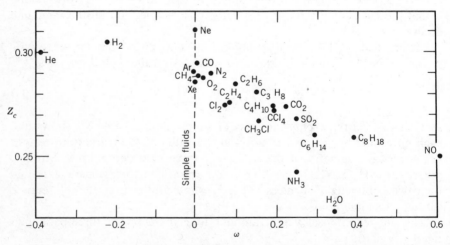

Figure 6.22 The position of the fluids of Table 6.2 in the plane $Z_c - \omega$.

does not exist. Using not one but two molecular parameters means constructing a four-parameter class of Z charts. This idea was put into practice by Hsieh [19], who viewed the compressibility factor as the function:

$$Z = Z(T_r, P_r, Z_c, \alpha_c) \qquad (6.94)$$

where α_c is the slope of the reduced saturation-pressure curve right at the critical point:

$$\alpha_c = \frac{dP_r}{dT_r} \qquad \text{(at } T_r = 1\text{)} \qquad (6.95)$$

According to Fig. 6.21, this slope should vary monotonically with ω; therefore, ω could be used instead of α_c in the four-parameter function (6.94). The dimensionless slope α_c has been introduced as the molecular parameter in corresponding states correlations by Riedel [20]. Hsieh displayed 30 nonpolar and slightly polar fluids in the Z_c–α_c plane and identified six groups of fluids according to their (Z_c, α_c) values. He then represented each group with a pair of group-averaged values (Z_c, α_c) and produced one two-dimensional chart $Z(T_r, P_r)$ for each of the six groups. These charts can be examined in Refs. 19 and 21.

Analytical $P(v, T)$ Equations of State

Van der Waals' impact on contemporary thermodynamics is illustrated also by the voluminous effort that has been devoted to the development of closed-form expressions for the $P(v, T)$ surface. The most compact expressions are those based on only two empirical constants: van der Waals' eq. (6.44) is one such example. A considerably more accurate two-constant expression[†] patterned after van der Waals' was proposed in 1949 by Redlich and Kwong [22]:

$$P = \frac{RT}{v - b} - \frac{a/T^{0.5}}{v(v + b)} \qquad (6.96)$$

Applying the critical-point conditions (6.47), we find that the new constants (a, b) are related to the critical-point coordinates via

$$a = 0.4275 \frac{R^2 T_c^{2.5}}{P_c} \qquad b = 0.08664 \frac{RT_c}{P_c} \qquad (6.97)$$

Equation (6.96) can be cast in dimensionless form as a cubic in Z:

[†]The most recent proposal of this kind is reviewed in Problem 6.20.

$$Z^3 - Z^2 + (A - B - B^2)Z - AB = 0 \qquad (6.98)$$

where A and B are now two dimensionless constants:

$$A = \frac{aP}{R^2 T^{2.5}} \qquad B = \frac{bP}{RT} \qquad (6.99)$$

Redlich and Kwong's equation of state was subjected to considerable testing, some of which is discussed by Reid et al. [14] and Miller [23]. This equation is preferred among the other two-constant equation of state proposals. The Redlich and Kwong equation of state and others like it can be used not only for pure fluids, but also for *mixtures* of fluids whose critical points are situated not too far from one another in the dimensional Regnault plane P–T. For example, in order to apply eq. (6.96) to a mixture, one must calculate first the "equivalent" critical temperature and pressure of the mixture. These values will be needed in eqs. (6.97). In Redlich and Kwong's case, the recommended rules for calculating the $(T_{c,m}, P_{c,m})$ coordinates of the fictitious critical point of the mixture are [14]

$$T_{c,m} = \left\{ \frac{[\sum_{i=1}^n x_i (T_{c,i}^{5/2}/P_{c,i})^{1/2}]^2}{\sum_{i=1}^n x_i (T_{c,i}/P_{c,i})} \right\}^{2/3} \qquad (6.100)$$

$$P_{c,m} = \frac{T_{c,m}}{\sum_{i=1}^n x_i (T_{c,i}/P_{c,i})} \qquad (6.101)$$

where x_i and $(T_{c,i}, P_{c,i})$ are the mole fraction and critical-point coordinates, respectively, of the ith component in the mixture.

Redlich and Kwong's equation of state was improved further by Soave [24], who replaced the $(a/T^{0.5})$ numerator with a temperature function $a(T)$ on the right side of eq. (6.96):

$$P = \frac{RT}{v - b} - \frac{a(T)}{v(v + b)} \qquad (6.102)$$

The temperature function $a(T)$ matches Redlich and Kwong's expression $(a/T^{0.5})$ only at the critical point:

$$a(T_c) = aT_c^{-0.5} \qquad (6.103)$$

where Redlich and Kwong's constant "a" is listed in eq. (6.97). Soave's complete temperature function is

$$a(T) = aT_c^{-0.5}\{1 + m[1 - (T/T_c)^{0.5}]\}^2 \qquad (6.104)$$

where m is a characteristic constant for each pure fluid. Soave was able to correlate m against the Pitzer acentric factor and obtained the following expression:

$$m = 0.48 + 1.574\omega - 0.176\omega^2 \tag{6.105}$$

Camporese et al. [25] showed that an even more accurate version of eq. (6.102) is one in which the temperature function $a(T)$ depends on two parameters, m and n,

$$a(T) = aT_c^{-0.5}\left[1 + \left(1 - \frac{T}{T_c}\right)\left(m + n\,\frac{T_c}{T}\right)\right] \tag{6.106}$$

Best values for the pair (m, n) are reported for 11 common refrigerants in Ref. 25.

It is instructive to review at this point the analytical form of other $P(v, T)$ expressions that have been proposed over the years, so that the reader can identify those special features that must enter in the construction of an equation of state. The following listing indicates also the proponent of each particular $P(v, T)$ expression. The constants that appear in these expressions should not be confused with the constants of the equations of state discussed until now.

Clausius [26]:

$$P = \frac{RT}{v - b} - \frac{a}{T(v + c)^2} \tag{6.107}$$

Lorentz [27]:

$$P = \frac{RT}{v^2}\,(v + b) - \frac{a}{v^2} \tag{6.108}$$

Dieterici [28]:

$$P = \frac{RT}{v - b}\,\exp\left(-\frac{a}{vRT}\right) \tag{6.109}$$

Generalized[†] Dieterici equation of state [29, 30]:

$$P_r = \frac{T_r}{2v_r - 1}\,\exp\left(2 - \frac{2}{v_r T_r^n}\right) \tag{6.110}$$

Generalized[‡] Redlich–Kwong equation of state [23]:

[†]Note that the case $n = 1$ represents the original Dieterici equation (6.109) in reduced coordinates.

[‡]The original Redlich–Kwong equation (6.96) corresponds to the case $n = 1/2$.

$$P_r = \frac{T_r}{Z_c(v_r - b/Z_c)} - \frac{a/T_r^n}{Z_c^2 v_r(v_r + b/Z_c)} \qquad (6.111)$$

Martin [31]:

$$P_r = \frac{\alpha T_r}{v_r - a} - \frac{b(c - T_r)}{(v_r + d)^2} \qquad (6.112)$$

Keyes and associates at MIT [e.g., Ref. 32]:

$$P = \frac{RT}{v - \beta \exp(-\alpha/v)} - \frac{a}{(v + l)^2} \qquad (6.113)$$

Kamerlingh Onnes and associates at the University of Leiden [e.g., Ref. 33]:

$$Pv = A(T) + \frac{B(T)}{v} + \frac{C(T)}{v^2} + \frac{D(T)}{v^3} \qquad (6.114)$$

A more complex expression patterned after van der Waals' and eqs. (6.107)–(6.109) was developed at MIT by Beattie and Bridgeman [34]

$$P = \frac{RT}{v^2}(1 - \epsilon)(v + B) - \frac{A}{v^2} \qquad (6.115)$$

where

$$A = A_0(1 - a/v) \qquad (6.115a)$$

$$B = B_0(1 - b/v) \qquad (6.115b)$$

$$\epsilon = c/vT^3 \qquad (6.115c)$$

The values of the five empirical constants (A_0, a, B_0, b, c) and the ideal gas constants of 18 fluids are presented in Table A.3, after Cravalho and Smith [35].

The Beattie–Bridgeman equation was modified by Benedict et al. [36–38] in order to improve its accuracy in the limit of high densities (small vs). This proposal is recognized now as the Benedict–Webb–Rubin equation of state:

$$P = \frac{RT}{v} + \left(B_0 RT - A_0 - \frac{C_0}{T^2}\right)\frac{1}{v^2} + (bRT - a)\frac{1}{v^3} + \frac{a\alpha}{v^6}$$

$$+ \frac{c}{v^3 T^2}\left(1 + \frac{\gamma}{v^2}\right)\exp\left(-\frac{\gamma}{v^2}\right) \qquad (6.116)$$

in which there are eight empirical constants $(A_0, B_0, C_0, a, b, c, \alpha, \gamma)$.

Table A.4 shows the values of these constants for the 12 pure hydrocarbons covered by Benedict et al.'s third paper [38]. The presentation of this table in SI units is due to Cravalho and Smith [35]. A more extensive compilation of constants for eq. (6.116) was produced by Cooper and Goldfrank [39].

For a mixture of the same hydrocarbons, Benedict et al. [38] suggest the following rules for calculating the eight constants of the mixture:

$$
A_0 = \left(\sum_{i=1}^{n} x_i A_{0,i}^{1/2} \right)^2 \qquad b = \left(\sum_{i=1}^{n} x_i b_i^{1/3} \right)^3
$$

$$
B_0 = \sum_{i=1}^{n} x_i B_{0,i} \qquad c = \left(\sum_{i=1}^{n} x_i c_i^{1/3} \right)^3
$$

$$
C_0 = \left(\sum_{i=1}^{n} x_i C_{0,i}^{1/2} \right)^2 \qquad \alpha = \left(\sum_{i=1}^{n} x_i \alpha_i^{1/3} \right)^3
$$

$$
a = \left(\sum_{i=1}^{n} x_i a_i^{1/3} \right)^3 \qquad \gamma = \left(\sum_{i=1}^{n} x_i \gamma_i^{1/2} \right)^2
$$

(6.117)

where x_i and the subscript "i" indicate the mole fraction and constants, respectively, of each component. The accuracy and range of the Benedict–Webb–Rubin equation of state were subsequently improved by Orye [40], who replaced the third constant C_0 with a smoothly varying function of temperature.

A look back at the equation-of-state coverage started with van der Waals' eq. (6.44) and ended with Benedict et al.'s eq. (6.116) reveals the following common characteristics of $P(v, T)$ expressions. One feature is the recovery of the ideal gas equation of state $Pv = RT$ in the "ideal gas" limit found in the third diagram of Fig. 6.14. In other words, the $P = P(v, T)$ function approaches RT/v either as P approaches zero at constant T or as T approaches infinity at constant P. Another feature is the inflexion with zero slope shown by the critical isotherm $(T = T_c)$ in the P–v plane, as was illustrated in Figs. 6.10 (top), 6.11, and 6.15. It means that the $P = P(v, T)$ function must satisfy the critical-point conditions stated in eqs. (6.47). And, speaking of the critical point, the slope of the constant-volume line that passes through the critical point in the P–T plane must equal the slope of the saturation-pressure curve right at the critical point (Problem 6.10). This particular feature is illustrated in Fig. 6.21 for van der Waals' equation of state.

A bird's-eye view of how two or more equations of state might compete for the user's preference is conveyed by Fig. 6.23. Projected on the T_r–P_r plane is the position of the "inversion" curves ($\mu_J = 0$) predicted by four different equations of state:

$$
\mu_J = \frac{RT^2}{Pc_P} \left(\frac{\partial Z}{\partial T} \right)_P = 0
$$

(6.118)

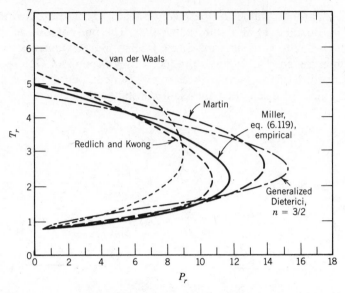

Figure 6.23   The average position of the Joule–Thomson inversion curve versus four predictions based on analytical equations of state (after Miller [23]).

plus the averaged [23] position of the actual $\mu_J = 0$ curves of several gases (CO_2, N_2, CO, CH_4, NH_3, C_3H_8, Ar, C_2H_4):

$$P_r \cong 24.21 - \frac{18.54}{T_r} - 0.825 T_r^2 \tag{6.119}$$

Note the similarity between Fig. 6.23 and Fig. 4.11. Among the four $P = P(v, T)$ models compared in Fig. 6.23, Redlich–Kwong's equation of state is the most accurate. Miller [23] noted also that since none of the four models predicts an inversion curve that falls exactly on top of the curve represented by eq. (6.119), the correct prediction of the position of the inversion curve emerges as a critical test in the evaluation of the relative goodness of future $P = P(v, T)$ models.

Calculation of Other Properties Based on $P(v, T)$ and Specific-Heat Information

One benefit of the corresponding-states correlation of the $P(v, T)$ surfaces of various fluids is the emergence of a common approach to the calculation of derived properties such as u, h, and s. At the basis of this methodology rest the differential expressions (du, dh, ds) listed in the left column of Table 4.7. These expressions show that in order to calculate the changes in these properties, we must know the $P(v, T)$ surface and *one* specific heat. In general, the specific heat is a function of both temperature and pressure, for example, $c_P(T, P)$. The most readily available specific heat information refers to the low-pressure "ideal gas" limit [41], where, as shown in Fig. 4.4, c_P depends only on temperature.

Coupled with the other ideal gas equation of state, $Pv = RT$, the ideal gas specific heat can be used to calculate the changes in the internal energy, enthalpy, and entropy of the fluid that behaves as an ideal gas. It is useful to identify with an asterisk the properties associated with the low-pressure limit:

$$u^*(T) = \lim_{P \to 0} u(T, P) \tag{6.120u}$$

$$h^*(T) = \lim_{P \to 0} h(T, P) \qquad c_P^*(T) = \lim_{P \to 0} c_P(T, P) \tag{6.120h}$$

$$s^*(T, P) = \lim_{P \to 0} s(T, P) \tag{6.120s}$$

where we see again that the ideal gas entropy is a function of both temperature and pressure (recall Table 4.7). The changes in u^*, h^*, and s^* are readily calculated using the relations listed in the second column of Table 4.7. These quantities are assumed known in the analysis that follows.

The final step in the pursuit of the actual (u, h, s) values at any state consists of calculating these values as *departures* from the corresponding properties known in the ideal gas limit. It is during this last step that the corresponding states formulation of the $P(v, T)$ information is extremely useful.

In order to illustrate the method, consider the calculation of enthalpy at a state (T, P), where the fluid does not necessarily behave as an ideal gas, $h(T, P)$. The *enthalpy-departure* function $[h(T, P) - h^*(T)]$ is obtained by integrating:

$$h(T, P) - h^*(T) = \left(\int_0^P dh \right)_T$$

$$= \int_0^P \left[v - T \left(\frac{\partial v}{\partial T} \right)_P \right] dP \tag{6.121}$$

where dh is the expression shown in the left column of Table 4.7. Replacing v with ZRT/P in the integrand and dividing by RT_c yields on the left side the dimensionless enthalpy-departure function:

$$\frac{h(T_r, P_r) - h^*(T_r)}{RT_c} = \int_0^{P_r} \left(-\frac{T_r^2}{P_r} \right) \left(\frac{\partial Z}{\partial T_r} \right)_{P_r} dP_r \tag{6.122}$$

Through the compressibility factor that appears now in the integrand, the dimensionless enthalpy-departure function depends on T_r, P_r, and the molecular parameter(s) used in the construction of function Z. If we rely on eq. (6.93) and Tables A.1 and A.2 in order to calculate $Z(T_r, P_r, \omega)$, then the right side of eq. (6.122) splits into two integrals such that the first represents the dimensionless enthalpy-departure function of simple fluids:

$$\frac{h - h^*}{RT_c} = \int_0^{P_r} \left(-\frac{T_r^2}{P_r} \right) \left(\frac{\partial Z^{(0)}}{\partial T_r} \right)_{P_r} dP_r + \omega \int_0^{P_r} \left(-\frac{T_r^2}{P_r} \right) \left(\frac{\partial Z^{(1)}}{\partial T_r} \right)_{P_r} dP_r$$

$$= \left(\frac{h - h^*}{RT_c} \right)^{(0)} + \omega \left(\frac{h - h^*}{RT_c} \right)^{(1)} \tag{6.123}$$

in other words,

$$\frac{h^* - h}{RT_c} = \left(\frac{h^* - h}{RT_c} \right)^{(0)} + \omega \left(\frac{h^* - h}{RT_c} \right)^{(1)} \tag{6.124}$$

Tables A.5 and A.6 contain Lee and Kesler's [18] values for the terms appearing on the right side of eq. (6.124).

As a second illustration of this method, consider the task of calculating the entropy at a certain T and P. We can relate the wanted $s(T, P)$ to the known value based on the ideal gas model, $s^*(T, P)$, by calculating the *entropy-departure* function:

$$s(T, P) - s^*(T, P) = \left(\int_0^P ds \right)_T - \left(\int_0^P ds^* \right)_T$$

$$= \left[-\int_0^P \left(\frac{\partial v}{\partial T} \right)_P dP \right]_T - \left[-\int_0^P \frac{R}{P} dP \right]_T \tag{6.125}$$

where ds and ds^* occupy the third line of the first and second columns of Table 4.7. By replacing again v with ZRT/P, the entropy-departure function assumes the following dimensionless form:

$$\frac{s^* - s}{R} = \int_0^{P_r} \left[Z - 1 + T_r \left(\frac{\partial Z}{\partial T_r} \right)_{P_r} \right] \frac{dP_r}{P_r} \tag{6.126}$$

or, seen through eq. (6.93),

$$\frac{s^* - s}{R} = \left(\frac{s^* - s}{R} \right)^{(0)} + \omega \left(\frac{s^* - s}{R} \right)^{(1)} \tag{6.127}$$

In conclusion, the dimensionless entropy-departure function $(s^* - s)/R$ can be calculated by combining the simple-fluid entropy-departure value read off Table A.7 with the first-order correction suggested by Table A.8. The final value of $(s^* - s)/R$ depends on T_r, P_r, and ω.

Another property that can be deduced from the generalized compressibility factor $Z(T_r, P_r, \omega)$ is the fugacity of a single-component fluid, or, more precisely, the *fugacity coefficient* f/P. The relationship between f/P and Z follows from the definition of fugacity (p. 202):

$$dg = RT \frac{df}{f} \qquad \text{(constant } T) \tag{6.128}$$

where $f/P \to 1$ as $P \to 0$. From eq. (6.50), the above definition means also that

$$RT \frac{df}{f} = v \, dP \qquad \text{(constant } T) \tag{6.129}$$

Subtracting $RT \, dP/P$ from both sides, we obtain

$$RT \, d\left(\ln \frac{f}{P}\right) = \left(v - \frac{RT}{P}\right) dP \qquad \text{(constant } T) \tag{6.130}$$

which can be integrated from $P = 0$, where $f/P = 1$,

$$RT \ln \frac{f}{P} = \int_0^P \left(v - \frac{RT}{P}\right) dP \qquad \text{(constant } T) \tag{6.131}$$

In terms of Z and reduced pressure and temperature, this last result reads

$$\ln \frac{f}{P} = \int_0^{P_r} (Z - 1) \, d(\ln P_r) \qquad \text{(constant } T_r) \tag{6.132}$$

It means that the fugacity factor is a function of T_r, P_r, and the molecular parameter ω. Applying one more time the decomposition of Z into $(Z^{(0)} + \omega Z^{(1)})$, eq. (6.93), it is possible to calculate f/P in two steps:

$$\log_{10} \frac{f}{P} = \left(\log_{10} \frac{f}{P}\right)^{(0)} + \omega \left(\log_{10} \frac{f}{P}\right)^{(1)} \tag{6.133}$$

Tables A.9 and A.10 list Lee and Kesler's [18] values for the (T_r, P_r)-dependent functions $(\ )^{(0)}$ and $(\ )^{(1)}$ listed on the right side of eq. (6.133). A more direct and less accurate approach is to read the fugacity coefficient directly off Fig. 4.17.

Saturated-Liquid and Saturated-Vapor States

The locus of saturated liquid and vapor states is an especially important curve contained in the $P(v, T)$ surface. Since the $P(v, T)$ surfaces of various fluids have been correlated through the generalized compressibility factor $Z(T_r, P_r, \omega)$ of eq. (6.93) and Tables A.1 and A.2, it should be possible to correlate the saturation-pressure[†] curves $P(T)$ of the same fluids in terms of reduced coordinates, $P_r(T_r, \omega)$. An effective correlation for saturation pressure was developed by Lee and Kesler [18]:

$$\ln P_r = 5.92714 - \frac{6.09648}{T_r} - 1.28862 \ln T_r + 0.169347 T_r^6$$

$$+ \omega\left(15.2518 - \frac{15.6875}{T_r} - 13.4721 \ln T_r + 0.43577 T_r^6\right) \quad (6.134)$$

which has the familiar form $(\ )^{(0)} + \omega(\ )^{(1)}$. Reasons for the particular analytical form of the two T_r functions appearing on the right side of eq. (6.134) are given in Problem 6.12.

The reduced saturation-pressure curve listed above satisfies three important conditions. First, it passes through the critical point, $T_r = P_r = 1$. Second, the curvature of the saturation-pressure curve right at the critical point is zero; in other words, the Riedel factor α_c is insensitive to temperature changes in the vicinity of the critical point ($d\alpha_c/dT_r = 0$ at $T_r = 1$). A third condition satisfied by Lee and Kesler's eq. (6.134) is the definition of the Pitzer acentric factor, eq. (6.91).

A simpler corresponding-states formula for saturation pressure was reported by Dong and Lienhard [42]:

$$\ln P_r = 5.37270\left(1 - \frac{1}{T_r}\right)$$

$$+ \omega(7.49408 - 11.18177 T_r^3 + 3.68769 T_r^6 + 17.92998 \ln T_r) \quad (6.135)$$

A more important advantage of this correlation is that it works very well in a considerably wider ω range, especially for fluids with large and *negative* Pitzer acentric factors (e.g., helium, cesium, hydrogen, mercury, potassium, sodium, van der Waals fluid). The agreement between the Dong–Lienhard correlation and measured saturation-pressure data is within 0.42 percent for fluids covering the range $-0.3 < \omega < 0.9$. Equation (6.135) satisfies only two of the three conditions satisfied by Lee and Kesler's eq. (6.134), namely, the ω definition (6.91) and the critical-point condition $P_r(T_r = 1) = 1$.

Another quantity of interest in the realm of the (f), (g), and $(f + g)$ states addressed in this subsection is the enthalpy of vaporization or condensation h_{fg}. This quantity can be deduced from the compressibility-factor surface Z and the saturation-pressure curve by invoking the Clapeyron relation (6.67):

[†]A more specialized notation for saturation pressure in the literature is P_{sat} or P_s. The same pressure level—the pressure of liquid and vapor states in equilibrium—was labeled P_2 in Fig. 6.13. In this chapter, the "sat" subscript is used only where it is absolutely necessary, for example, when the saturation pressure has to be distinguished from another pressure appearing in the same equation, as in eq. (6.143).

$$\frac{dP}{dT} = \frac{h_{fg}}{Tv_{fg}} \cong \frac{h_{fg}}{(RT^2/P)Z_{fg}} \tag{6.136}$$

where $Z_{fg} = Z_g - Z_f$. Note the use of the ideal gas model for the specific volume of the saturated vapor state (g), $v_g = RT/P$, in which P is the saturation pressure corresponding to T. The difference $(Z_g - Z_f)$ is available from generalized compressibility-factor charts (Fig. 6.20) or from the "saturation line" of Fig. 6.24. In many cases, Z_f is considerably smaller than 1; therefore, $Z_{fg} \cong Z_g$. The left side of eq. (6.136) can be calculated directly from eq. (6.134) or (6.135).

Corresponding-states correlations for $h_{fg}(T)$ have been sought also directly, that is, without relying on the Clapeyron relation and the Z charts. A classical example is Watson's empirical formula [43]:

$$\frac{h_{fg}(T)}{h_{fg}(T_1)} = \left(\frac{1 - T_r}{1 - T_{r,1}}\right)^{0.38} \tag{6.137}$$

which works particularly well at near-critical temperatures. A very effective correlation was proposed recently by Torquato and Stell [44] and Torquato and Smith [45]. The correlation is for the entire temperature range between

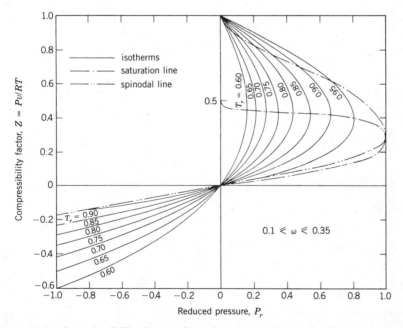

Figure 6.24 Compressibility-factor chart for saturated, metastable, and unstable states of nonpolar fluids in the range $0.1 \leq \omega \leq 0.35$ (from Dong and Lienhard [42]; courtesy of Professor John H. Lienhard, University of Houston).

the critical point (T_c) and the triple point (T_t), hence, the use of a new dimensionless temperature:

$$\tau = \frac{T_c - T}{T_c - T_t} \qquad (6.138)$$

instead of the reduced temperature T_r used by others. Figure 6.25 shows that the $h_{fg}(T)$ curves of six different fluids are correlated by a curve of type $\lambda(\tau)$, where λ is the result of nondimensionalizing h_{fg} by dividing it through the latent heat of vaporization at the triple point:

$$\lambda = \frac{h_{fg}(T)}{h_{fg}(T_t)} \qquad (6.139)$$

The solid-line curve drawn in Fig. 6.25 represents the latent heat data for water (and, approximately, the data for five other fluids). A good fit for this curve is [45]

$$\lambda = b_1 \tau^{1/3} + b_2 \tau^{0.79} + b_3 \tau^{1.208} + b_4 \tau + b_5 \tau^2 + b_6 \tau^3 \qquad (6.140)$$

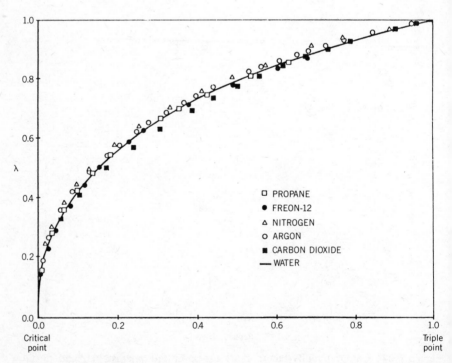

Figure 6.25 Correlation of the latent heat of vaporization of six fluids (after Torquato and Smith [45]).

where $b_1 = 0.60176$, $b_2 = 3.45913$, $b_3 = 4.62671$, $b_4 = -6.89614$, $b_5 =$ -1.10643, and $b_6 = 0.31522$.

In order to calculate h_{fg} from Fig. 6.25 or eq. (6.140), we must know the T_c, T_t, and $h_{fg}(T_t)$ values of the fluid of interest. These values are listed in Table A.11 along with the six constants of an accurate λ curve-fit for each fluid taken separately [44]:

$$\lambda = a_1 t^{1/3} + a_2 t^{0.79} + a_3 t^{1.208} + a_4 t + a_5 t^2 + a_6 t^3 \qquad (6.141)$$

In this expression, t is the reduced-temperature departure from the critical point:

$$t = \frac{T_c - T}{T_c} = 1 - T_r \qquad (6.142)$$

Listed in Table A.11 are also the maximum percentage deviation (δ_m) and the standard error (σ) associated with fitting eq. (6.141) to the $h_{fg}(T)$ data of each of the 20 fluids documented in the table.

Metastable States

The corresponding-states idea has preoccupied a relatively large number of thermodynamicists for more than 100 years. The many equations of state and Z surfaces discussed until now all refer to stable equilibrium states, especially to the large portion of the $P(v, T)$ surface that houses the superheated vapor states, i.e., the bridge between the saturated vapor states and the limiting region of ideal gas behavior. The intensive work on correlating the $P(v, T)$ properties of superheated vapor and $T_r > 1$ gaseous states is understandable in view of the predominance of such states in the analysis of actual power and refrigeration cycles. However, to concentrate exclusively on correlating the $P(v, T)$ relations that prevail outside the respective "$L + V$" domes of various fluids is to overlook the origin of much of this methodology, because, figuratively speaking, the "continuity-of-state" idea that preceded van der Waals originated from "inside" the dome (pp. 249–252).

All campaigns, no matter how glorious, return eventually to their place of origin. The corresponding-states theory is now rounded and steered back to its origins by those dedicated to correlating properties at the metastable states that are located inside the dome. Metastable states encountered already in this section are states (1') and (3) shown in Fig. 6.13. Relative to the long history of corresponding-states theory, the work on corresponding metastable states is very recent: this is best illustrated by the bibliography assembled in Lienhard et al.'s review [46]. For example, the first Z chart for the inner region of the dome was published in 1986 [42], Fig. 6.24.

Among the important practical issues that motivate the study of metastable states and, in particular, the pinpointing of the spinodal lines, is the question of how much a liquid can be heated above its saturation temperature before it has no choice but to boil. Figure 6.26 shows what can happen when a saturated liquid (f) is heated isobarically, at $P = P_f$. If the liquid is free of impurities that usually act as sites for the formation of the first vapor bubbles ("nucleation" sites), the temperature of the liquid can increase above the corresponding saturation temperature, $T_{sat}(P_f)$, while the liquid remains single-phase. We know from the stability considerations centered on Fig. 6.13 that this process cannot continue to temperatures that might exceed the liquid-spinodal temperature corresponding to P_f, namely $T_{sp}(P_f)$ in Fig. 6.26. The P–v region contained between the two spinodal lines is inaccessible. Of course, if the liquid batch is not clean enough and if the isobaric heating is not conducted carefully (without disturbances), the system starts to boil at a temperature lower than T_{sp}, that is, between T_{sat} and T_{sp}. The states visited by the system along the P_f isobar between T_{sat} and, possibly, T_{sp} are metastable (superheated) liquid states.

A similar phenomenon is encountered during the isobaric cooling of a clean batch of saturated vapor (g), Fig. 6.26. The system may survive as a single phase until it reaches the vapor spinodal line from the right. The

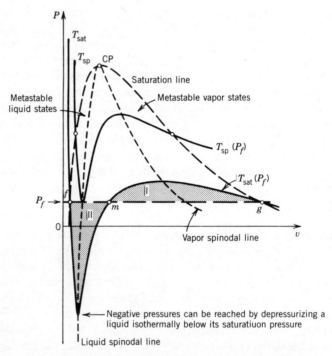

Figure 6.26 The maximum temperature (T_{sp}) that can be reached during the isobaric (P_f) heating of a batch of saturated liquid (after Lienhard et al. [46]).

low-temperature isotherm corresponding to the intersection of the P_f isobar with the vapor spinodal line is not shown in Fig. 6.26. The $P = P_f$ states found between (g) and this point of intersection are metastable (subcooled) vapor states. A "liquid" analog of this process of isobaric cooling into the domain of metastability is observed if one cools very carefully a batch of clean distilled water contained in a horizontal cylindrical enclosure (Fig. 6.27). The temperature of liquid water can drop below the freezing point (0°C), especially in the coldest layers[†] that gather near the top of the container. In one such experiment, the water temperature dropped as low as −2.7°C: after this point, knife-shaped ice crystals (dendritic ice) formed abruptly in the uppermost region of the container (Fig. 6.27) as the water temperature stabilized at 0°C [47].

The basic approach to locating and then correlating the positions of the spinodal lines and isotherms inside the saturation-line dome has been, first, to create a smooth fit between the $P(v, T)$ regions known on the left side and right side of the dome and, second to investigate the features of the wavy $P(v, T)$ surface that overlaps the dome. Very promising in this regard appears to be an original analytical form for a $P(v, T)$ equation of state, which was proposed by Shamsundar and Murali [46, 48]:

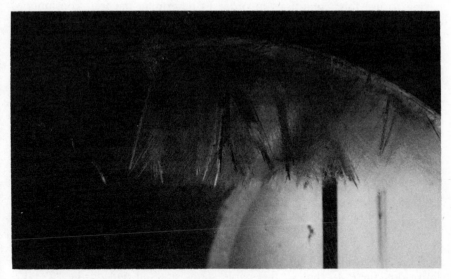

Figure 6.27 The formation of dendritic ice in the region once occupied by sub-cooled (metastable) liquid water, during a process of isobaric cooling by natural convection [47].

[†]These layers are also the lightest: recall the density maximum exhibited by liquid water (Problem 4.9).

$$\frac{P}{P_{sat}(T)} = 1 - \frac{[v - v_f(T)][v - v_m(T)][v - v_g(T)]}{[v + a(T)][v + b(T)][v + c(T)]} \qquad (6.143)$$

Evident in this expression are the three intersections of the $T_{sat}(P)$ isotherm with the P isobar: these points have been labeled (f), (m), and (g) in Fig. 6.26. The $P(v, T)$ surface that covers the dome is generated one isotherm at a time by first recognizing the well-documented values of $P_{sat}(T)$, $v_f(T)$, and $v_g(T)$. In the original application of eq. (6.143), Murali [48] set $c(T) = b(T)$ and, for each isotherm, determined the unknown constants (a, b, v_m) from three conditions: (1) the ideal gas limit, which must be reached as $P \rightarrow 0$, (2) the equal-area rule (6.60), shown again in Fig. 6.26, and (3) the condition that eq. (6.143) must predict correctly the (known) isothermal compressibility of saturated liquid. Fitted in this manner, Shamsundar and Murali's isotherms reproduced extremely well the known $P(v, T)$ surfaces of subcooled liquids and superheated vapors, that is, in addition to providing a smooth transition between the two.

A modified version of Shamsundar and Murali's method was used by Biney et al. [49] in order to determine, among other things, the location of the liquid spinodal line for water. The same version was used in the process of generating the compressibility-factor chart of Fig. 6.24. The modification consists of dropping Murali's assumption that $c(T) = b(T)$, which means having to use four conditions in order to determine the four constants of each Shamsundar–Murali isotherm (a, b, c, v_m). In addition to the three conditions listed in the preceding paragraph, Biney et al. used a high-pressure compressed-liquid condition, which is described in Ref. 49. Finally, instead of the $(v + a)(v + b)(v + c)$ expression used originally as the denominator in eq. (6.143), Biney et al. used $(v + a)(v^2 + bv + c)$.

The compressibility-factor chart of Fig. 6.24 is recommended for use in connection with all nonpolar substances in the Pitzer factor range $0.1 \le \omega \le 0.35$ and if $P_r < 0.8$ in the case of metastable liquid states and if $P_r < 0.9$ in the case of metastable vapor states. Within this domain, the Z value listed in Fig. 6.24 is ± 2 percent accurate [42]. Dong and Lienhard constructed also the complete Z chart for metastable water states: the reduced-pressure domain covered by this second chart is $-8 \le P_r \le 2$.

It is possible to combine the two equations of state provided by the $P(v, T)$ surface (6.143) and the specific-heat ("caloric") information known in the ideal gas limit, in order to construct the fundamental relation of a fluid batch as a simple system. Such a fundamental relation would cover the entire domain $P-v$, that is, the stable states situated outside the "$L + V$" dome and the metastable and unstable states positioned inside the dome. A fundamental relation of this kind was developed for water by Karimi and Lienhard [50, 51]. The $T(s, P)$ equation of state deduced from this fundamental relation is reproduced in Fig. 6.28. It is worth comparing this figure with the $T-s$ diagram of Fig. 6.10, in order to recognize the common features of the regions located outside the "$L + V$" dome.

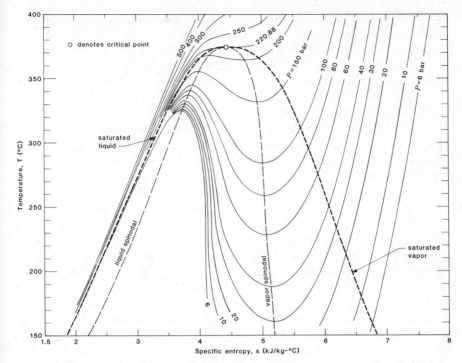

Figure 6.28 Karimi and Lienhard's $T(s, P)$ diagram for water, covering the stable, metastable, and unstable state regions (after Refs. 46 and 51; courtesy of Professor John H. Lienhard, University of Houston).

A more comprehensive picture of the modern work on metastability and corresponding states theory is painted by Skripov's monograph [52] and in Lienhard et al.'s review [46]. Additional research directions worthy of mention are the prediction of the relationship between surface tension and temperature by relying solely on $P(v, T)$ information, the prediction of the limits of homogeneous nucleation and isentropic-pressure undershoot, as well as the corresponding-states correlation of peak boiling heat flux [53, 54].

CRITICAL-POINT PHENOMENA
by Peter Jany[†]

In the early part of this chapter, the subject of phase transitions in gas–liquid systems has been discussed in detail. From Fig. 6.10, we learned

[†]Lehrstuhl A für Thermodynamik, Technische Universität München, Arcisstr. 21, D-8000 München 2, Federal Republic of Germany.

that the separation into two different phases (liquid and vapor) is impossible above a certain (critical) temperature T_c or pressure P_c. The phase transitions that occur in the vicinity of such critical points form the subject of the present section. Frequently, these phenomena are called "second-order" or "continuous" phase transitions, because not only the Gibbs free energy $g(P, T)$ becomes uniform throughout the coexisting phases [as in eq. (6.62)], but also their first partial derivatives. Phase transitions of this type are numerous, for example, the appearance of superfluidity in helium-4 below its λ-point, the separation into two phases of different concentrations in binary liquid mixtures below their consolute[†] point or in binary alloys below their order–disorder transition[‡] point. Another example is illustrated by the spontaneous magnetization of ferromagnets below their Curie point. In accordance with the main focus of this engineering treatment of thermodynamics, the emphasis of this section is placed on gas–liquid systems.

Critical-point phenomena have attracted attention from different fields and from individuals with diverging interests and points of view. For instance, in thermodynamics, the critical point (CP) distinguishes itself as the terminus of the vapor-pressure curve, Fig. 6.16, as a transition point in the system's number of degrees of freedom, eq. (6.72), or as a means of standardizing (nondimensionalizing) van der Waals' equation of state, eq. (6.45). In an attempt to describe the thermal behavior of different substances by equations of state, the engineer must face the fact that the derivatives of some variables disappear or diverge. This extreme behavior that is exhibited by certain thermophysical properties is the focus of considerable experimental work. Theoretical physicists are interested in the universality of phase transitions in different substances and physical systems. The coincidence of the coexistence curve and the spinodal line at the critical point (Fig. 6.28), which results in the simultaneous existence of stable, metastable, and unstable states, is a subject that preoccupies nonequilibrium physicists. In synergetics, the comparison of such seemingly divorced phenomena as phase transitions and turbulence has led to a better understanding of both phenomena. Experiments are already being conducted in space in order to eliminate the otherwise dominant effect of gravity on critical-point phenomena. And, back on earth, one of the most impressive experiments that the student can witness in a laboratory course is the passing of a fluid through its critical point, where boiling and raining become one and the same image.

[†]Such a point exists in some liquid mixtures, for example, oil and water, where the components become completely miscible in each other above a certain temperature (the temperature of this point) and separate into two phases of different concentrations below this temperature.

[‡]Such a point exists in some alloys, for example, β-brass, where the different atoms are randomly arranged above a certain temperature. Below this temperature, the atoms are distributed regularly in an alternating arrangement.

Basic Concepts

The preceding examples demonstrate the diversity of critical-point phenomena and the people and schools of thought associated with them. Numerous books and reviews have been published on this topic during the last decade [55–61]; therefore, a complete review of the field is not suited for this chapter-ending section. The objective of what follows is to introduce the research-minded engineering thermodynamicist to the main ideas that define the current research frontier in the field of critical-point phenomena. We begin with a review of some of the most basic terms and concepts.

Order Parameter. In the description of critical-point phenomena, it is common to define an order parameter Φ that corresponds to the actual physical system that is being considered. In principle, the only conditions to be satisfied by Φ are that it must be nonzero below the critical temperature T_c and zero above T_c. These features are found in Fig. 6.29a. A behavior of this kind is exhibited, for example, by the difference between the densities of the coexisting phases in gas–liquid systems, by the amplitude of the superfluid phase in helium-4, by the difference between the concentrations of the coexisting phases in binary liquid mixtures and alloys and, finally, by the spontaneous magnetization in ferromagnets.

Another common feature of these examples is shown in Fig. 6.29b. The logarithm of Φ varies linearly with the logarithm of the temperature difference $T_c - T$. The slope of the log Φ versus log $(T_c - T)$ line must be equal to β ($\cong 1/3$) for all the order parameters mentioned above. We return to this observation later in this section (p. 303).

Fluctuations. Even in states of macroscopic equilibrium at finite temperatures, the physical properties of a substance show deviations from their mean values. Classical thermodynamics neglects these "fluctuations" because of their smallness relative to the commonly recognized macroscopic

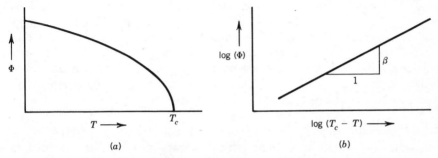

(a)   (b)

Figure 6.29   The temperature dependence of the order parameter.

scales. This assumption fails, however, when microscopic effects become important or when the fluctuations begin to grow rapidly.

Let us consider a volume V of a certain fluid in thermal equilibrium. The boundary that surrounds V is permeable. When V is microscopically small, the number of molecules within V will change continuously. The actual density of the V sample will fluctuate in time with an amplitude $\Delta\rho$ about a mean value ρ. The probability distribution of $\Delta\rho$ can be determined by means of statistical physics: an important result of this analysis is the mean (ensemble average) square density fluctuation $\langle\Delta\rho^2\rangle$ [62]:

$$\langle\Delta\rho^2\rangle = \frac{1}{V}\, k_B T\rho^2\kappa \qquad (6.144)$$

where k_B denotes the Boltzmann constant and κ the isothermal compressibility, $-v^{-1}(\partial v/\partial P)_T$. The increase of $\langle\Delta\rho^2\rangle$ with decreasing sample volume V or increasing disorder (temperature T) is obvious. The essential factor in relation to critical-point phenomena however is κ. It can be seen easily in Fig. 6.11 that the slopes of the isotherms $(\partial P/\partial v)_T$ become very small near (CP); therefore, the compressibility κ becomes large. In conclusion, in a microscopic volume of fixed size, the mean-square-density fluctuations become exceedingly large in the critical-point region.

Correlation Length. The correlation length ξ represents the length scale of the mean effective range of density fluctuations. This measure can be evaluated by means of eq. (6.144). Consider the case when the volume V of eq. (6.144) contains two coexisting phases, i.e., saturated liquid of density ρ_f and saturated vapor of density ρ_g. The natural limit for fluctuations in such a system will be attained if $\langle\Delta\rho^2\rangle$ becomes of the same order as $(\rho_f - \rho_g)^2$, which corresponds to the possibility of a spontaneous phase transition. The volume V associated with fluctuations of this order of magnitude is

$$V \sim \frac{k_B T\rho^2\kappa}{(\rho_f - \rho_g)^2} \qquad (6.144')$$

It will be shown later (p. 304) that $\rho^2\kappa$ and $(\rho_f - \rho_g)$ depend on simple powers of $(T_c - T)$ raised to specific exponents. In order to satisfy eq. (6.144'), V has to be a function of $(T_c - T)$. It can easily be deduced (Problem 6.15) that V increases with decreasing $(T_c - T)$ in exactly the same way as the third power of ξ. For that reason, ξ can be identified as a scale of length in which local fluctuations are correlated.

The Universality Hypothesis. All the fluids are expected to exhibit a number of common features near their critical point. Consider, for example, the following comparison of two typical length scales of molecular interactions. In classical fluids, the Lennard–Jones 6–12 potential indicates that the forces between molecules are relevant within a distance of about twice the diameter of hard cores of the molecules. For simple atoms and molecules, this diameter has a value of approximately $(3)10^{-10}$ to $(4)10^{-10}$ m [63]. Another typical length scale of molecular interactions in fluids near the critical point is the correlation length. This length depends strongly on $(T - T_c)$; for instance, at the critical isochore $(\rho = \rho_c)$ and $(T - T_c) = 0.01$ K, its value is somewhere between 10^{-7} and $(2)10^{-7}$ m regardless of the substance that is being examined (Problem 6.16). This second scale is more than two orders of magnitude larger than the first; in fact, the ratio between the two increases as $(T - T_c)$ decreases.

This comparison makes clear the fact that in the vicinity of the critical point, the range of interactions due to fluctuations exceeds the range of common fluid forces. The size and structure of an atom or molecule influence mainly short-range interactions. Therefore, the fluid behavior near the critical point becomes increasingly unrelated to the nature of the substance.

Critical-Point Exponents

From Fig. 6.29, we learn that the logarithm of the order parameter depends linearly on the logarithm of $(T_c - T)$. The slope β is called "critical-point exponent" and is expected to be practically independent of the actual substance and physical system. What follows is a listing of a few more critical-point exponents.

Power Laws. The singular behavior of some properties that diverge or vanish at the critical point can be described by a functional expression of the type

$$f(\tau) = A_{\lambda_1}|\tau|^{\lambda_1}(1 + A_{\lambda_2}|\tau|^{\lambda_2} + \cdots) \qquad (6.145)$$

where $\lambda_2 > 0$ and τ is the reduced temperature difference:

$$\tau = \frac{T - T_c}{T_c} \qquad (6.146)$$

The critical-point exponent λ_1 emerges by taking the limit:

$$\lambda_1 = \lim_{|\tau| \to 0} \frac{\ln f(\tau)}{\ln |\tau|} \qquad (6.147)$$

It is customary to retain only the first term in the parentheses of eq. (6.145). This simplifying procedure, of course, is valid only near the critical point, that is, in the limit $|\tau| \to 0$. It is nevertheless adequate for the purpose of determining λ_1 from experimental data. Regardless of whether this simplification is adopted, the power law (6.145) exhibits two additional features:

$$\lim_{\tau \to 0} f(\tau) = \begin{cases} 0 \text{ if } \lambda_1 > 0 \\ \infty \text{ if } \lambda_1 < 0 \end{cases} \tag{6.148}$$

The case $\lambda_1 = 0$ leads to special solutions that are not discussed here [see, for example, Ref. 61]. We continue with an outline of the most common simple power laws, for which we will employ the nomenclature of a gas–liquid system. These power laws, however, apply also to the descriptions of other systems as well.

The order parameter of the gas–liquid system is represented by the difference between the saturated-liquid density ρ_f and the saturated-vapor density ρ_g. The associated power law is

$$\frac{\rho_f - \rho_g}{2\rho_c} = A_\beta (-\tau)^\beta \tag{6.149}$$

which is defined for $\tau < 0$.

The dependence of the pressure P on ρ along the critical isotherm $T = T_c$ can be described for all values of ϕ as

$$\frac{P - P_c}{\rho_c R T_c} = A_\delta |\phi|^\delta \operatorname{sign}(\phi) \tag{6.150}$$

where ϕ is the reduced density difference:

$$\phi = \frac{\rho - \rho_c}{\rho_c} \tag{6.151}$$

The isothermal compressibilities of the two coexisting phases vary as

$$\left. \begin{array}{c} \left(\dfrac{\rho_f}{\rho_c}\right)^2 \kappa_f \\[2ex] \left(\dfrac{\rho_g}{\rho_c}\right)^2 \kappa_g \end{array} \right\} = A_{\gamma'} (-\tau)^{-\gamma'} \tag{6.152}$$

At supercritical temperatures $(T > T_c)$ along the critical isochore $(\rho = \rho_c)$, the compressibility varies as

$$\kappa = A_\gamma \tau^{-\gamma} \tag{6.153}$$

Along the same critical isochore, the specific heat at constant volume c_v behaves as

$$\frac{c_v}{R} = A_{\alpha'} (-\tau)^{-\alpha'} + B_{\alpha'} \qquad (\text{if } T < T_c) \tag{6.154}$$

$$\frac{c_v}{R} = A_\alpha \tau^{-\alpha} + B_\alpha \qquad (\text{if } T > T_c) \tag{6.155}$$

Both functions require background terms $(B_{\alpha'}, B_\alpha)$ as will be shown later.

The correlation length should be uniform for both coexisting phases when $T < T_c$:

$$\left.\begin{array}{c} \xi_f \\ \xi_g \end{array}\right\} = A_{\nu'}(-\tau)^{-\nu'} \tag{6.156}$$

however, at supercritical temperatures along the critical isochore, it behaves differently:

$$\xi = A_\nu \tau^{-\nu} \tag{6.157}$$

The power law for the correlation function $G(r)$ at the critical point completes these relations. The function G denotes the correlation of local densities with the distance r:

TABLE 6.3 Simple Power Laws for Several Thermodynamic Properties Near the Critical Point

Property	Quantity	Power Law	Domain τ	Domain ϕ
Density difference between saturated liquid and vapor	$\dfrac{\rho_f - \rho_g}{2\rho_c}$	$= A_\beta(-\tau)^\beta$	<0	ϕ_f, ϕ_g
Pressure	$\dfrac{P - P_c}{\rho_c RT_c}$	$= A_\delta \lvert\phi\rvert^\delta \operatorname{sign}(\phi)$	$=0$	arb.
Isothermal compressibility	$\left(\dfrac{\rho_f}{\rho_c}\right)^2 \kappa_f$	$= A_{\gamma'}(-\tau)^{-\gamma'}$	<0	ϕ_f
	$\left(\dfrac{\rho_g}{\rho_c}\right)^2 \kappa_g$		<0	ϕ_g
	κ	$= A_\gamma \tau^{-\gamma}$	>0	$=0$
Isochoric specific heat	$\dfrac{C_v}{R}$	$= A_{\alpha'}(-\tau)^{-\alpha'} + B_\alpha$	<0	$=0$
	$\dfrac{C_v}{R}$	$= A_\alpha \tau^{-\alpha} + B_\alpha$	>0	$=0$
Correlation length	$\dfrac{\xi_f}{\xi_g}$	$= A_{\nu'}(-\tau)^{-\nu'}$	<0 <0	ϕ_f ϕ_g
	ξ	$= A_\nu \tau^{-\nu}$	>0	$=0$
Correlation function	G	$\sim \dfrac{1}{r^{d-2+\eta}}$	$=0$	$=0$

$$G(r) \sim \frac{1}{r^{d-2+\eta}} \tag{6.158}$$

where d represents the dimensionality of the system; in this case, $d = 3$.

The independent variables in all these power laws, τ and ϕ, are defined according to eqs. (6.146) and (6.151). The factors of type "A" appearing in these relations are commonly called "critical-point amplitudes." The quantities labeled α, α', β, γ, γ', δ, ν, ν', and η are critical-point exponents. All the exponents are assumed positive, so that the convergence and divergence of the properties can be described by the signs of the powers.

A bird's-eye view of all power laws introduced above is presented in Table 6.3. Most of the experimental and theoretical research devoted to critical-point phenomena has focused on the determination of critical-point exponents. Before presenting some results, a few very useful relations between those exponents will be established.

Inequalities. Several inequalities between critical-point exponents were developed in the 1960s. Their derivation covers a broad spectrum, from relatively simple scale analysis to sophisticated statistical arguments, all of which are omitted here for the sake of brevity. Nine relations between the critical-point exponents introduced until now are presented in Table 6.4 [64–69]. They should be valid strictly with the sign "$\geq$." However, the scaling theories of Widom and Kadanoff derive the same relations without

TABLE 6.4   Relations Between Critical-Point Exponents

Inequality	Reference
(a) $\alpha' + 2\beta + \gamma' \overset{(>)}{=} 2$	Rushbrooke [64]
(b) $(1 + \delta)\beta \overset{(>)}{=} 2 - \alpha'$	Griffiths [65]
(c) $\gamma'(\delta + 1) \overset{(>)}{=} (2 - \alpha')(\delta - 1)$	Griffiths [66]
(d) $\gamma' \overset{(>)}{=} \beta(\delta - 1)$	
(e) $\nu'd \overset{(>)}{=} 2 - \alpha'$	Josephson [67]
(f) $\nu d \overset{(>)}{=} 2 - \alpha$	
(g) $(\delta - 1)d \overset{(>)}{=} (2 - \eta)(\delta + 1)$	Buckingham and Gunton [68]
(h) $\gamma'd \overset{(>)}{=} (2 - \eta)(2\beta + \gamma')$	
(i) $(2 - \eta)\nu \overset{(>)}{=} \gamma$	Fisher [69]

the greater-than option, as will be shown later. For that reason, in Table 6.4, the sign ">" is given in brackets.

Experimental findings suggest that the critical-point exponents α, γ, and ν are path-independent. If this assumption is true for one exponent, then the equations in Table 6.4 require these additional relations (Problem 6.17):

$$\alpha = \alpha' \tag{6.159a}$$

$$\gamma = \gamma' \tag{6.159b}$$

$$\nu = \nu' \tag{6.159c}$$

Theoretical Models

Beginning with the experimental discovery of the gas–liquid critical point (CP), commonly attributed to Andrews' 1869 paper [3], a considerable effort has been devoted to the theoretical description of critical-point phenomena. This effort continues today because of persisting discrepancies between experimental results and theoretical predictions. What follows is an introduction to some of the most important theoretical models and a discussion of their consequences regarding critical-point exponents. Emphasis is placed on the two models whose creators were awarded the Nobel prize in physics in 1910 and 1982, J. D. van der Waals and K. G. Wilson, respectively.

Van der Waals' Equation of State. Four years after the publication of Andrews' paper, J. D. van der Waals proposed in his dissertation [6] an equation of state, which was capable of describing for the first time analytically the gas–liquid phase transition. This equation was discussed thoroughly in an earlier section of this chapter. Its dimensionless form, eq. (6.46) can be rewritten into

$$\pi(2 - \phi) - 8\tau(\phi + 1) - 3\phi^3 = 0 \tag{6.160}$$

where τ and ϕ are defined by eqs. (6.146) and (6.151) and where

$$\pi = \frac{P - P_c}{P_c} \tag{6.161}$$

The expression $f(\pi, \tau, \phi) = 0$ given in eq. (6.160) is more appropriate for the study of critical-point exponents than the form $P_r = P_r(T_r, v_r)$ of eq. (6.46) or the frequently used form $f(\pi, \tau, v_r - 1)$, in which $v_r = (\phi + 1)^{-1}$.

Exponent δ: Along the critical isotherm $\tau = 0$, the pressure depends on the density as

$$\pi = \frac{3/2\phi^3}{1 - \phi/2} \tag{6.162}$$

which, for small ϕs, yields

$$\pi = \frac{3}{2}\,\phi^3\left(1 + \frac{\phi}{2}\right) \tag{6.163}$$

Comparing eq. (6.163) with the power law (6.150) yields immediately the critical-point exponent $\delta = 3$.

Exponent β: Assuming the existence of a symmetric coexistence curve, $\phi_f + \phi_g = 0$, the difference between the reduced-density differences of saturated liquid ϕ_f and saturated vapor ϕ_g can be derived directly from eq. (6.160) (Problem 6.18):

$$\phi_f - \phi_g = 4(-\tau)^{1/2} \tag{6.164}$$

In conclusion, the critical-point exponent for eq. (6.149) is $\beta = 1/2$.

Exponent γ: The definition of the isothermal compressibility in terms of π and ϕ reads:

$$\kappa = \frac{1}{P_c(\phi + 1)}\left(\frac{\partial \phi}{\partial \pi}\right)_\tau \tag{6.165}$$

The partial derivative appearing on the right side of eq. (6.165) can be obtained from eq. (6.160):

$$\left(\frac{\partial \phi}{\partial \pi}\right)_\tau = \frac{(2 - \phi)^2}{6(4\tau + 3\phi^2 - \phi^3)} \tag{6.166}$$

From eqs. (6.165) and (6.166), we find the κ value along the critical isochore ($\phi = 0, \tau > 0$):

$$\kappa = \frac{1}{6P_c}\,\tau^{-1} \tag{6.167}$$

According to eq. (6.153), then, the γ exponent is $\gamma = 1$. The more complicated derivation for the case when $\tau < 0$ yields the same exponent, $\gamma' = 1$.

Exponent α: Van der Waals' equation predicts a jump in c_v along the critical isochore ($\phi = 0, \tau = 0$), but no singularity, so that $\alpha = \alpha' = 0$. The results of the van der Waals model, $\alpha = \alpha' = 0$; $\beta = \frac{1}{2}$; $\delta = 3$; $\gamma = \gamma' = 1$ do not agree sufficiently well with the experimental findings, as will be seen later. Nevertheless, they are surprisingly good considering the relatively simple correction to the ideal gas behavior that is contributed by this model.

Landau's Mean-Field Theory. The following theories refer to the frequently mentioned ferromagnetic system rather than the gas–liquid system. If the temperature T is fixed, the magnetization $\mathcal{M}$ of a ferromagnet can be changed by applying a magnetic field $\mathcal{H}$. This effect is illustrated in Fig. 6.30*a*. The main feature is that even when $\mathcal{H} = 0$, "spontaneous" magnetization can occur if the system is cooled *below* a certain (critical) temperature T_c. This temperature is commonly referred to as the Curie temperature. Figure 6.30*b* shows the magnetization $\mathcal{M}$ versus T at zero field intensity, $\mathcal{H} = 0$. It is clear now that $\mathcal{M}$ fits the requirement of a critical-point order parameter. Other properties of the magnetic system find direct analogs in the gas–liquid system. Most of the results discussed next can easily be translated from one system to the other by replacing $\mathcal{M}$ and $\mathcal{H}$ with ρ and P, or vice versa. This analogy is illustrated further on pp. 575–577 of chapter 10.

The traditional approach to describing the magnetic behavior is to establish either the partition function $\mathcal{Z}$ as the probability distribution of energy levels of the elementary magnets, or the Helmholtz free energy F $[= -k_B T \ln \mathcal{Z}]$. Both provide information about all values of state. A surprisingly simple model [70] consists of expanding the Helmholtz free energy $F(T, \mathcal{M})^\dagger$ in a Taylor series about the critical point ($\mathcal{M} = 0$):

$$F(T, \mathcal{M}) = \sum_{i=0}^{\infty} c_i(T)\mathcal{M}^i \tag{6.168}$$

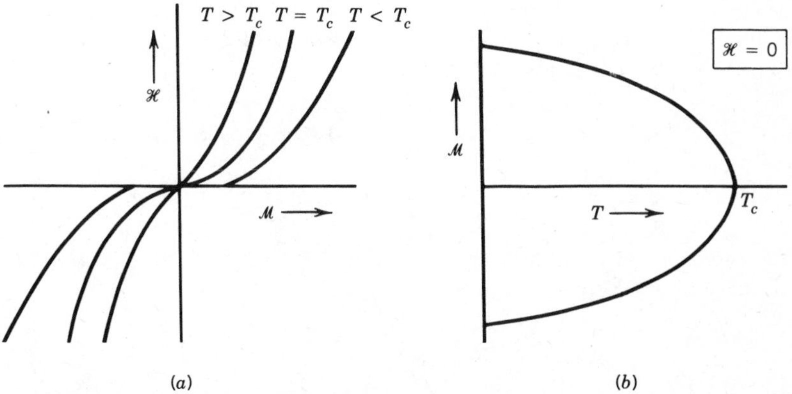

Figure 6.30 Characteristics of a ferromagnetic system: $\mathcal{H}(T, \mathcal{M})$ and $\mathcal{M}(T, \mathcal{H} = 0)$.

†Note that this function is analogous to the $F(T, V)$ function listed in eq. (6.13).

A number of simplifications in this series are straightforward. To begin with, the higher-order terms become negligible for small $\mathcal{M}$s, i.e., $c_6, c_7, \ldots = 0$. In the absence of a magnetic field, F cannot depend on the sign of $\mathcal{M}$; therefore, the odd powers must be dropped, $c_1 = c_3 = c_5 = 0$. Finally, in a way that is analogous to eqs. (6.17) and (6.18), the magnetic field $\mathcal{H}$ and the isothermal susceptibility K can be defined as

$$\mathcal{H}(T, \mathcal{M}) = \left(\frac{\partial F}{\partial \mathcal{M}}\right)_T \tag{6.169a}$$

$$K^{-1}(T, \mathcal{M}) = \left(\frac{\partial^2 F}{\partial \mathcal{M}^2}\right)_T \tag{6.169b}$$

From Fig. 6.30, it is obvious that $\mathcal{H} = 0$ solutions are possible only for $\mathcal{M} = 0$ if $T > T_c$ and for $\mathcal{M} \neq 0$ if $T < T_c$. This means that $c_2(T)$ and $c_4(T)$ must have different signs if $T < T_c$, and the same sign if $T > T_c$. On the other hand, the stability criterion $K > 0$ must be fulfilled. The simplest way to satisfy the two requirements identified above is to choose

$$c_2(T) = c_{2A}(T - T_c) \tag{6.170a}$$

$$c_4(T) = c_{4A} \tag{6.170b}$$

with $c_{2A}, c_{4A} > 0$. The resulting expession

$$F(T, \mathcal{M}) = c_0(T) + c_{2A}(T - T_c)\mathcal{M}^2 + c_{4A}\mathcal{M}^4 \tag{6.171}$$

leads to the critical-point exponents that are discussed next.

Exponent δ: By analogy with eq. (6.150), this exponent is associated with the relation $H \sim \mathcal{M}^\delta$ along the critical isotherm $T = T_c$. Combining eqs. (6.169a) and (6.171) leads to

$$\mathcal{H}(T, \mathcal{M}) = 2c_{2A}(T - T_c)\mathcal{M} + 4c_{4A}\mathcal{M}^3 \tag{6.172}$$

Considering the case $T = T_c$, we conclude that $\delta = 3$.

Exponent β: The relation that is analogous to eq. (6.149) is $\mathcal{M} \sim (-\tau)^\beta$. In the case $\mathcal{H} = 0$, eq. (6.172) yields

$$\mathcal{M}(\mathcal{H} = 0, T) = \pm\left[-\frac{c_{2A}}{2c_{4A}}(T - T_c)\right]^{1/2} \tag{6.173}$$

which means that at subcritical temperatures $(T < T_c)$, the exponent must be $\beta = 1/2$.

Exponent γ: According to eqs. (6.169b) and (6.171), the isothermal susceptibility can be expressed as

$$K^{-1}(T, \mathcal{M}) = 2c_{2A}(T - T_c) + 12c_{4A}\mathcal{M}^2 \qquad (6.174)$$

Along the "critical isochore" defined by $\mathcal{M} = 0$ and $T > T_c$, as well as along the "coexistence curve" $\mathcal{M} = \mathcal{M}(T)$ defined by eq. (6.173) and $T < T_c$, the susceptibility K is given by

$$K(T > T_c) = \frac{1}{2c_{2A}} (T - T_c)^{-1} \qquad (6.175a)$$

$$K(T < T_c) = -\frac{1}{4c_{2A}} (T - T_c)^{-1} \qquad (6.175b)$$

Comparing eqs. (6.175) and the power laws (6.152) and (6.153), we obtain the exponents $\gamma = \gamma' = 1$.

Exponent α: In a ferromagnetic system, this particular exponent is associated with the specific heat at constant (zero) magnetic field $\mathcal{H} = 0$. Landau's model predicts a jump in this value along $\mathcal{M} = 0$ at $T = T_c$ but no divergence, that is, $\alpha = \alpha' = 0$.

In summary, the critical-point exponents obtained based on this approach are $\alpha = \alpha' = 0$; $\beta = 1/2$; $\delta = 3$; and $\gamma = \gamma' = 1$. The perfect agreement with van der Waals' results is not an accident, because both models are based essentially on the same assumption. That assumption is that every elementary entity—magnet or molecule—interacts with all the others in the same way. All the theories based on this "mean-field" approximation necessarily predict the critical-point exponents given in Table 6.5.

Ising's Lattice Model. Less restrictive than the mean-field theories is Ising's model of interactions between site elements in a d-dimensional lattice. Every site contains one of two possible alternatives, for example, a positive versus a negative elementary magnet in a ferromagnetic system, a molecule of one component versus a molecule of the other component in a binary mixture, and, finally, a molecule versus a vacancy (lattice–gas model) in a gas–liquid system. All these examples exhibit a common degree of freedom regarding site occupation, $n = 1$. The number n is called the dimensionality of the order parameter.

Ising's model was derived originally for ferromagnetic systems, by assuming every site to be characterized by an elementary magnetic spin s equal to $+1$ or -1. If ϵ_{ij} is the energy interaction between the spins s_i and s_j, the total potential energy of the lattice is the Hamiltonian:

$$\mathbf{H} = -\sum_{n.n.} \epsilon_{ij} s_i s_j \qquad (6.176)$$

TABLE 6.5 Theoretical Critical-Point Exponents

Model	α'	α	β	γ'	γ	δ	η	ν'	ν	Reference
Mean field	0		$\frac{1}{2}$		1	3	0		$\frac{1}{2}$	Stanley [61]
Ising $d=2$		0	$\frac{1}{8}$		$\frac{7}{4}$	15	$\frac{1}{4}$		1	Stanley [61]
Ising $d=3$	$\frac{1}{16}$ $^{+0.16}_{-0.035}$	$\frac{1}{8}$ ±0.015	$\frac{5}{16}$ $^{+0.003}_{-0.006}$	$\frac{21}{6}$ $^{+0.03}_{-0.05}$	$\frac{5}{4}$ ±0.003	$\frac{26}{5}$ ±0.15	$\frac{1}{18}$ ±0.008	0.675 $^{+0.02}_{-0.03}$	$\frac{9}{14}$ ±0.025	Fisher [76]
		$\frac{1}{8}$	$\frac{5}{16}$		$\frac{5}{4}$	5	0.041 $^{+0.006}_{-0.003}$	0.625^a $^{+0.002}_{-0.001}$	0.638 $^{+0.002}_{-0.001}$	Domb [77]
		0.125 ± 0.020	0.312 ±0.002		1.250 ± 0.003	5.01^b	0.041 ±0.006	0.625^a ±0.002	0.638 ±0.002	Sengers [78]
Wilson		0.110 ± 0.0045	0.325 ±0.0015		1.241 ± 0.0020	4.82^b	0.031 ±0.004		0.630 ± 0.0015	Le Guillou and Zinn–Justin [84]

[a] From Table 6.4, line (e).
[b] From Table 6.4, line (b).

As a first approximation, the interactions can be limited to neighboring sites only, so that the sum in eq. (6.176) is evaluated only for the nearest neighboring (n.n.) pairs. The total number of summation terms depends on the dimensionality of the space, d. External fields are excluded from this consideration. The partition function

$$\mathscr{Z} = \sum_{\text{a.c.}} \exp\left(-\frac{\mathbf{H}}{k_B T}\right) \tag{6.177}$$

is then obtained by summing over all possible configurations (a.c.) of the lattice system (for example, N sites yield 2^N configurations). If the partition function $\mathscr{Z}$ can be obtained in closed form, then the derivation of all properties of state becomes straightforward.

It is interesting to take a brief look back at the history of this model (Brush [71]). The model was first introduced in 1920 by Wilhelm Lenz of Rostock University [72]. After Lenz became Professor at the University of Hamburg, one of his first students, Ernst Ising, was able to solve the problem for $d = 1$ in the course of his doctoral dissertation. Ising published his results in 1925 [73]. The model became famous as Ising's model, and no claim of Lenz as originator of the idea is known. Ising did not continue the work along this line. He was greatly disappointed by the fact that the model failed to predict ferromagnetic behavior for $d = 1$. Furthermore, the political developments in Germany prevented him from continuing his research; in fact, it took him several years to learn about advances that had been made based on his ideas. One advance took place in 1944 when Ernst Onsager [74] solved the model in $d = 2$ dimensions and showed that the model describes the phase transformation. In 1928, Werner Heisenberg proposed an extension of the model to three dimensional spins ($n = 3$) in order to improve the description of isotropic ferromagnets [75]. The discussion of the effects of d and n will be continued later in this section.

As summary, we conclude that Ising's model does not describe phase transitions for $d = 1$. Onsager's exact solution for $d = 2$ yields the critical-point exponents exhibited in Table 6.5. When compared with experimental findings, they turn out to be less accurate than the results of mean-field theories. An exact solution for $d = 3$ has not been obtained yet. Some approximate methods like series expansions for low or high temperatures and Padé approximations, however, permit the calculation of critical-point exponents within acceptable margins of uncertainty. A number of results that have been compiled in review papers [76–78] are also listed in Table 6.5. Until recently, these results were regarded as the best estimates.

Widom's Scaling Laws. The following description of critical-point phenomena is due to Widom [79], whose work leads to the so-called "scaling laws." The Gibbs free energy G is a function of temperature T and magnetic field $\mathscr{H}$, analogous to eq. (6.19). Widom's first assumption was to split $G(T, \mathscr{H})$ into two parts:

$$G(T, \mathcal{H}) = G_b(T, \mathcal{H}) + G_s(\tau, \mathcal{H}) \tag{6.178}$$

where $G_b(T, \mathcal{H})$ means a background part that is not influenced in any way by the critical-point singularity. The singular behavior near the critical point is accounted for entirely by $G_s(\tau, \mathcal{H})$, in which the T argument is replaced from now on by τ, eq. (6.146). Widom's second crucial assumption is that $G_s(\tau, \mathcal{H})$ changes its scale but not its basic functional form when approaching the critical point. In mathematical terminology, it is said that $G_s(\tau, \mathcal{H})$ fulfills the definition of a generalized homogeneous function (see footnote on page 158):

$$G_s(\lambda^{a_\tau}\tau, \lambda^{a_{\mathcal{H}}}\mathcal{H}) = \lambda G_s(\tau, \mathcal{H}) \tag{6.179}$$

for all the values of the scaling factor λ. The numbers $a_{\mathcal{H}}$ and a_τ are arbitrary.

Equation (6.179) provides direct information concerning the critical-point exponents. For instance, its partial derivative with respect to $\mathcal{H}$ yields

$$\lambda^{a_{\mathcal{H}}}\mathcal{M}(\lambda^{a_\tau}\tau, \lambda^{a_{\mathcal{H}}}\mathcal{H}) = \lambda\mathcal{M}(\tau, \mathcal{H}) \tag{6.180}$$

where $\mathcal{M} = -(\partial G_s/\partial \mathcal{H})_\tau$. For $\mathcal{H} = 0$, eq. (6.180) suggests

$$\mathcal{M}(\tau, 0) = \lambda^{a_{\mathcal{H}}-1}\mathcal{M}(\lambda^{a_\tau}\tau, 0) \tag{6.181}$$

Since eq. (6.181) must hold for any λ, we set $\lambda = (-1/\tau)^{1/a_\tau}$ and obtain

$$\mathcal{M}(\tau, 0) = (-\tau)^{(1-a_{\mathcal{H}})/a_\tau}\mathcal{M}(-1, 0) \tag{6.182}$$

Comparing this with the power law $\mathcal{M} \sim (-\tau)^\beta$ that corresponds to eq. (6.149), we obtain the critical-point exponent:

$$\beta = \frac{1 - a_{\mathcal{H}}}{a_\tau} \tag{6.183}$$

Similar expressions can be derived for other critical-point exponents, for example, δ (Problem 6.19), α', α, γ' and γ. Eliminating $a_{\mathcal{H}}$ and a_τ between these exponents leads to definite equations that relate these exponents, as shown in entries a–d in Table 6.4. The same elimination procedure yields, additionally, $\alpha' = \alpha$ and $\gamma' = \gamma$, validating once more eqs. (6.159a, b). Finally, since the six exponents $(\alpha', \alpha, \beta, \delta, \gamma', \gamma)$ can be expressed as functions of only $a_{\mathcal{H}}$ and a_τ, only two of these exponents can be specified independently.

Kadanoff's Scaling Laws. A similar scaling method was applied to Ising's lattice model in 1966 by Kadanoff [80]. Consider a d-dimensional lattice with "a" as the lattice constant (site distance) and ξ as the correlation length, such that $\xi \gg a$. The number of possible configurations [eq. (6.177)] in such a lattice can be reduced drastically when blocks of spins are considered rather than all spins. If every block contains l^d sites, it is assumed also that $1 \ll l \ll \xi/a$. Such a scaling change from a spin system to a block-spin system definitely requires a transformation of all the descriptive parameters. Kadanoff's approach was to state that the singular part of Gibbs free energy should change its scale but not its analytical form:

$$g_s(\tau', \mathcal{H}') = l^d g_s(\tau, \mathcal{H}) \tag{6.184}$$

Here $g_s(\tau, \mathcal{H})$ is the Gibbs free energy per site and $g_s(\tau', \mathcal{H}')$ is the Gibbs free energy per block in the transformed lattice. Kadanoff chose the transformed temperature and magnetic field to scale as

$$\tau' = l^{b_\tau} \tau \tag{6.185a}$$

$$\mathcal{H}' = l^{b_\mathcal{H}} \mathcal{H} \tag{6.185b}$$

therefore,

$$g_s(l^{b_\tau} \tau, l^{b_\mathcal{H}} \mathcal{H}) = l^d g_s(\tau, \mathcal{H}) \tag{6.186}$$

The above expression obeys the definition of a generalized homogeneous function, eq. (6.179). Setting $b_\tau = a_\tau d$ and $b_\mathcal{H} = a_\mathcal{H} d$ leads to the same scaling laws as in Widom's model. In addition, Kadanoff's scaling of Ising's lattice model permits the calculation of the critical-point exponents for the correlation length and correlation function. Note that ν, ν', and η can also be expressed as functions of only $a_\mathcal{H}$ and a_τ; therefore, only two of the nine critical-point exponents appearing in Tables 6.3 to 6.5 can be specified independently. Finally, Kadanoff's approach proves the validity of the equal signs in the relations $e-i$ of Table 6.4 and the relation $\nu = \nu'$ of eq. (6.159c).

Wilson's Renormalization Group Theory. Based on the preceding theoretical framework and, particularly, on Ising's model and Kadanoff's scaling, Wilson [81] was able to develop a new treatment of critical-point phenomena. This work sets the standard for current theoretical work on critical-point phenomena, which is why it was recognized through the Physics Nobel Prize in 1982 [82]. Wilson's main ideas—the application of renormalization group techniques and ϵ-expansions on critical-point phenomena—are highly sophisticated theoretical models. In this segment, we review a brief outline of his method.

The method of renormalization is essentially a scaling transformation of Ising's lattice model, one that is analogous to Kadanoff's approach. Since the analytical form of the Gibbs free energy is not known a priori, the scaling is applied to the partition function, eq. (6.177). In contrast to Ising's classical model, which takes into account energy interactions between nearest neighbors exclusively [eq. (6.176)], this model is able to consider also long-range interactions. The scaling transformation from a spin system to a new one in which every block-spin represents l^d spins of the old one requires a recalculation of the properties of state (e.g., temperature, magnetic field, energy of interaction) from nonlinear relations. Scaling is carried out as long as the lattice constant of the block-spin lattice remains smaller than the transformed correlation length ξ'. If the system is in a noncritical state, ξ' is finite and the limit of scaling is reached after a finite number of transformations. This process can also be viewed as an apparent departure of the system from the critical state (i.e., decreasing ξ'). If the system is in the critical state (i.e., infinite ξ), the transformations do not converge to a limit. Unchanged properties of state can be observed after every transformation of the block-spin system. In this scaling procedure, the critical point is distinguished as a fixed point.

The sequence of transformations represents the "renormalization group," a method that seeks to determine the fixed point. In this process, an influence of the dimensionality of space d can be detected in terms of a (small) deviation ϵ from the physical dimensionality $d = 4$, namely, $\epsilon = 4 - d$. All the results of this method depend on ϵ, particularly the critical-point exponents. For example, one finds perfect agreement between this approach and mean-field theories in the case $\epsilon = 0$, which corresponds to the imaginary dimensionality $d = 4$. The best critical-point exponent values are found in the limit of very small ϵs. Expansions in terms of ϵ yield critical-point parameters that are functions of ϵ (e.g., Wilson [83]). Table 6.5 shows also the exponents evaluated in this manner by Le Guillou and Zinn–Justin [84].

Another advantage of Wilson's approach is the explicit verification of the universality hypothesis, because all results are independent of lattice geometry. Only the dimensionality of space (d) and of the order parameter (n) are free parameters. Therefore, systems with fixed (d, n) belong to the same "universality class" and are characterized by equivalent critical-point exponents. All the systems with critical points mentioned in this book belong to the class $(d = 3, n = 1)$.

Finally, the shortcomings of Landau's mean-field theory can be explained by using Wilson's terminology [82]. In the statement $F(T, \mathcal{M})$, eq. (6.171), Landau implies the assumption of a mean magnetization $\mathcal{M}$ that is averaged over all fluctuations, which is analogous to the hydrodynamic theory. In statistical mechanics, however, $\mathcal{M}$ must be a weighted average over all the possible values if it is to contain information concerning the actual fluctuations. This averaging procedure shows, as a first approximation, that the coefficients c_2 and c_4 in eq. (6.170) depend not only on T but also on the scaling length $l \times a$. Wilson showed that this relatively simple modification of Landau's theory also yields critical-point exponents that depend on ϵ.

Quantitative Results

In this subsection, we focus on the problem of estimating the thermophysical properties that are influenced by critical-point singularities. The discussion revolves around two alternative descriptions of the relations between these properties, namely, equations of state and correlations of experimental results.

Equations of State. The classical equations of state presented earlier in this chapter (p. 283 ff.) fail to describe the property relations in an extended region around the critical point. Their analytical forms are incapable of accounting for the scaled nonanalytical behavior. For this reason, an effort has been made to develop special equations of state for the critical-point region. Kestin et al. [85], for example, describe the behavior of water by means of two distinct equations of state. One of the form $F(T, \rho)$ covers the noncritical region and the other of the form $(P/T)(1/T, \mu/T)$ describes the critical region. In these functions, F denotes the Helmholtz free energy and μ the chemical potential.

The development of an equation of state of this type for the critical region is described in detail by Levelt Sengers and Sengers [86]. An important feature of the potential $(P/T)(1/T, \mu/T)$ is that both independent variables are equal in the coexisting phases. In a way that is analogous to Widom's scaling model, the function P/T is split into a regular part $(P/T)_b$ and a nonanalytical part $(P/T)_s$, while it is assumed that the latter is exclusively influenced by the critical-point anomaly. Introducing the new variables r and θ in place of $1/T$ and μ/T [87]:

$$\frac{\mu}{T} \sim r^{\beta\delta}(1 - \theta^2) \tag{6.187a}$$

$$\frac{1}{T} + c\,\frac{\mu}{T} \sim r(1 - b^2\theta^2) \tag{6.187b}$$

the scaling of $(P/T)_s$ is

$$\left(\frac{P}{T}\right)_s \sim r^{2-\alpha}f_1(\theta) + r^{2-\alpha+\Delta}f_2(\theta) \tag{6.188}$$

Here, b and c are free parameters, whereas β, δ, and α are the well-known critical-point exponents. The new exponent Δ describes the scaling behavior of the potential. The main advantage of the parametric expressions (6.187a, b) is the exclusive influence of scaling behavior on the coordinate r, whereas $f_1(\theta)$ and $f_2(\theta)$ in eq. (6.188) represent purely analytic functions.

The equation of state is defined completely if the following parameters are known: the universal exponent Δ and two other critical-point exponents; the critical values ρ_c, P_c, and T_c; the five parameters required by the scaling;

and, finally, eight parameters for the regular part $(P/T)_b$. With this information, properties like c_P, c_v, κ, ρ, h_{fg}, vapor pressure, and velocity of sound can be derived. The entire equation of state that includes all the parameters can be found in Levelt Sengers et al. [88] for water and in Hastings et al. [89] for ethylene. The authors describe the range of validity of their equations as $-0.38 < \phi < 0.30$, $-(5)10^{-3} < \tau < (7)10^{-2}$ for water and $\doteq 10^{-2} < \tau < (7)10^{-2}$ for ethylene.

Although these equations are adequate for describing the thermodynamic behavior of fluids in the critical region in accordance with modern theories, two shortcomings continue to exist. First, there is considerable lack of information concerning several common fluids. As early as 1976, Levelt Sengers et al. [90] used the scaling of F to develop two equations of state for helium-3, helium-4, xenon, carbon dioxide, oxygen, and water. However, they were considerably less accurate than above-mentioned equations, especially with regard to caloric properties, and covered a much smaller range around the critical point. The second drawback is that the use of these equations is not that simple, particularly when using the parametric formulation, eqs. (6.187a, b). Quite often the information needed by the engineer consists of just one thermophysical property. More useful in these instances are the direct experimental results or their correlations.

Experimental Results. What follows is a presentation of a collection of experimental results concerning a number of thermophysical properties. The discussion of the experimental methods employed, the sources of error, and the accuracy of each result is omitted for the sake of brevity.

The main difficulty when carrying out experiments near the critical point stems from the earth's gravitational field, which produces density gradients over the height of the fluid sample. These gradients increase as the height and the isothermal compressibility increase. The latter was shown to increase rapidly towards the critical point; therefore, it can be responsible for the establishment of significant density gradients. The existence of these density variations was demonstrated experimentally as early as 1965 by Straub [91]. Looking at this effect from another point of view, the mean density prevails only in a fluid layer whose height decreases when approaching the critical point. The experimental limitations that are due to this effect are discussed thoroughly by Moldover et al. [92]. Regarding the measurement of ρ or κ, these authors report relations between the minimum distance away from the critical point, the maximum fluid sample height, and the maximum accuracy of the ultimate results. Caloric measurements (c_v) have to deal with the fact that the mean density is an average over all the local values, and that any reorganization of the density profile by changing the temperature distribution requires the transfer of energy. Finally, optical measurements depend also on the length of the path that light has to travel through the sample. This is due to the strong refraction effect caused by density gradients and by the increasing turbidity of the fluid near the critical point.

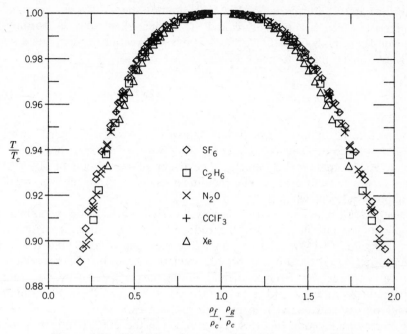

Figure 6.31 Densities of saturated liquid and vapor [93].

(a) *Densities of Saturated Liquid and Vapor.* The temperature dependence of the saturated-liquid density ρ_f and the saturated-vapor density ρ_g is represented by the characteristic parabola-shaped curve emerging in Fig. 6.31 [93]. In this figure, the densities and the temperatures are normalized with respect to the critical values ρ_c and T_c. This is why the data of five distinct fluids, sulphur hexafluoride (SF_6), ethane (C_2H_6), nitrous oxide (N_2O), monochlorotrifluoromethane ($CClF_3$ or Freon 13), and xenon (Xe), fall on practically the same curve. The critical-point properties of these substances are listed in Table 6.6.

TABLE 6.6 Numerical Parameters for Eqs. (6.149) and (6.189) and Figs. 6.37 and 6.38 [93]

Fluid	ρ_c (kg/m^3)	T_c (K)	A_1 Eq. (6.189)	A_β Eq. (6.149)	β Eq. (6.149)	D_T° (10^{-6} m^2/s) Figs. 6.37, 6.38
SF_6	741	318.67	0.722	1.868	0.338	0.201
C_2H_6	206	305.30	0.701	1.865	0.350	0.288
N_2O	457	309.60	0.730	1.882	0.348	0.194
$CClF_3$	581	301.93	0.797	1.997	0.358	0.178
Xe	1105	289.73	0.607	1.870	0.363	0.112

The shape of the curve of Fig. 6.31 invites the eye to see an axis of symmetry, which is commonly called "rectilinear diameter." According to the well-known law of Cailletet and Mathias, the rectilinear diameter should be only a weak linear function of T/T_c:

$$\frac{\rho_f + \rho_g}{2\rho_c} = 1 + A_1(-\tau) \tag{6.189}$$

This feature is supported by all data shown in Fig. 6.31. Coupled with the A_1 values listed in Table 6.6, eq. (6.189) represents the experimental results within approximately 0.2 percent. The order of magnitude of A_1 (~ 0.7) suggests a mean deviation of the rectilinear diameter from the $\rho/\rho_c = 1$ line of about 7 percent when $T/T_c = 0.9$.

The phase-transition theories discussed early in this section recommend a simple power law in $(-\tau)$ for the reduced difference between coexisting densities, eq. (6.149). The A_β and β values shown in Table 6.6 were determined based on the least-square deviation criterion. These values do not vary appreciably from one fluid to the next. Specifically, A_β ranges between 1.85 and 2.0 and β between 0.34 and 0.36. These experimental results depart noticeably from the critical-point exponent $\beta = 0.325$ predicted by renormalization theories, Table 6.5. On the other hand, they agree well with a number of other independent experimental reports [94–98]. The reason for the disagreement goes back to eqs. (6.145) and (6.147). The "real" critical-point exponents predicted by the theories refer to the limit $|\tau| \to 0$. Most experiments, however, do not reach this limit and, consequently, yield "apparent" critical-point exponents. An experiment whose objective is to test the theories has either to be restricted to a very small $|\tau|$ range or to be fitted by "extended scaling" expressions rather than by eqs. (6.149) and (6.189). This analytical procedure is not discussed further because even the simple power law (6.149) is capable of correlating the exhibited data in the relatively wide range $10^{-4} < |\tau| < 10^{-1}$ with an accuracy of approximately 1 percent. Additional experimental results concerning densities are cited in the review papers published by Levelt Sengers and Sengers [86], Greer and Moldover [97], Moldover [98], and Beysens [99].

(b) *Isochoric Specific Heat.* The temperature dependence of c_v along the critical isochore ($\rho = \rho_c$) is illustrated in Fig. 6.32. This experimental result—obtained by Straub et al. [100] using sulphur hexaflouride—shows not only a divergence at $T = T_c$, but also a jump as predicted by the theories. Both effects become even more visible when c_v is plotted against the logarithm of $|\tau|$, Fig. 6.33. The relationship between c_v/R and $\log|\tau|$ appears to be quite similar for both the subcritical and the supercritical temperature ranges. The power laws (6.154) and (6.155) have been added on Fig. 6.33. From their measurements, Straub et al. determined the parameters $\alpha = \alpha' = 0.0983$, $A_\alpha = 10.00$, $A_{\alpha'} = 18.29$, $B_\alpha = B_{\alpha'} = 2.60$, and $T_c = 318.733$ K for which the τ range is $(3.5)10^{-5} < |\tau| < (2)10^{-3}$.

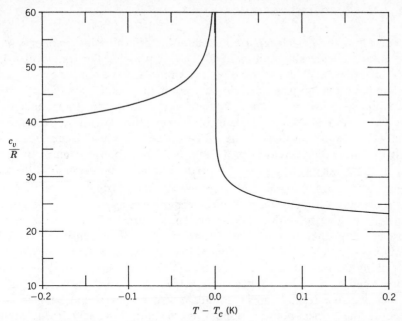

Figure 6.32 Isochoric specific heat of SF_6 along the critical isochore (drawn after Straub et al. [100]).

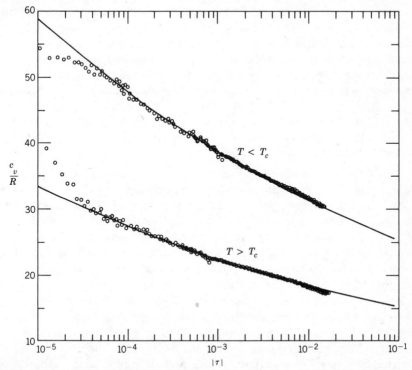

Figure 6.33 Isochoric specific heat measurements versus eqs. (6.154) and (6.155) (drawn after Straub et al. [100]).

For a long time, each reported value of the critical-point exponents α and α' caused considerable controversy. The reasons for the controversy may have been the smallness of the α and α' values and the serious experimental problems that creep up in caloric measurements. Additional measurements are available today for helium-3, helium-4, carbon dioxide, argon, and ethane (see Greer and Moldover [97] and Moldover [98]). All these experimental data yield α and α' values in the range 0.10–0.15. These values agree well with the theoretical predictions (Table 6.5), but cannot confirm or refute any of them except the $\alpha = \alpha' = 0$ prediction based on mean-field theories.

(c) *Correlation Length.* Unlike ρ and c_v, the correlation length ξ does not represent a directly measurable thermodynamic property. The correlation length must be deduced from system features that depend on it. The most common methods are the measurement of the intensity of scattered light or of the fluid turbidity. As far as I know, experiments were carried out only along the critical isochore ($\rho = \rho_c$, $T > T_c$). These experiments are well represented by the simple power law (6.157). A comprehensive compilation of the available results can be found in Van Leeuwen and Sengers [101]. Keeping the critical-point exponent ν constant, $\nu = 0.630$, these authors report the A_ν parameters of twelve fluids: helium-3, sulphur hexafluoride, xenon, neon, krypton, argon, nitrogen, carbon dioxide, isobutane, ethylene, heavy steam, and light steam. The A_ν values occupy the narrow range from $(1.3)10^{-10}$–$(2.6)10^{-10}$ m. Beysens [99] reports analogous results for eight binary-liquid mixtures: aniline + cyclohexane, isobutyric acid + water, triethylamine + water, nitroethane + isooctane, nitroethane + 3-methylpentane, nitrobenzene + n-hexane, 2–6 lutydine + water, and carbon tetrachloride + perfluoromethylcyclohexane. Their parameters A_ν are restricted to the range $(1.25)10^{-10}$–$(3.64)10^{-10}$ m. The critical-point exponent ν varies from 0.61 to 0.63. These experimental critical-point exponents cover not only a wide variety of fluid systems, but also agree very well with theoretical values given in Table 6.5.

(d) *Thermal Conductivity.* Until now, the emphasis has been placed on static (equilibrium) problems in critical-point phenomena. What follows is a discussion of three very important transport properties. Their behavior near the critical point is not anticipated fully by the theories described in the preceding subsection. In this case, it is necessary to combine the preceding theories with dynamic nonequilibrium considerations such as the mode-mode coupling approximation. These advances are well covered by the specialized literature [55, 56, 61, 102, 103].

The experimental study of the thermal conductivity k in the critical-point region requires sophisticated equipment and technique. This is one reason why for a long time people doubted whether k exhibits a critical-point anomaly. This question was settled in 1962 when Michels et al. [104] reported accurate measurements on carbon dioxide. These authors were able to observe an unambiguous enhancement of k near the critical point. Figure 6.34 shows the more recent results of Pittman et al. [105] for

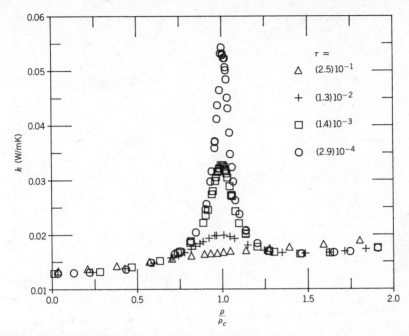

Figure 6.34 The thermal conductivity of helium-3 near the critical point (drawn after Pittman et al. [105]).

helium-3. The thermal conductivity k along several supercritical isotherms $(T > T_c)$ is plotted against the reduced density ρ/ρ_c. The sharp increase of k with decreasing $|\rho/\rho_c - 1|$ and decreasing τ is quite clear.

Figure 6.35 shows the thermal conductivity measured along the critical isochore ($\rho = \rho_c$, $T > T_c$). The two logarithmic scales allow us to see the relatively weak dependence on τ, which does not approach a simple power law. For this reason, k is frequently divided into two parts:

$$k(\rho, T) = k_b(\rho, T) + k_s(\rho, \tau) \qquad (6.190)$$

This suggestion is due to Sengers and Keyes [106], as an analog of the scaling idea expressed in eq. (6.178). The background term $k_b(\rho, T)$ does not undergo any anomaly at the critical point; it is determined by extrapolating into the critical-point region the measurements obtained far away from the critical point. The singular part $k_s(\rho, \tau)$, which exclusively describes the critical-point enhancement, is assumed to follow the simple power law:

$$k_s(\rho_c, \tau) = A_\psi \tau^{-\psi} \qquad (6.191)$$

along the critical isochore ($\rho = \rho_c$, $T > T_c$). Pittman et al. [105] estimated the background term for helium-3 to be $k_b(\rho_c, T) = (9.4 + 1.63 \times T) \times$

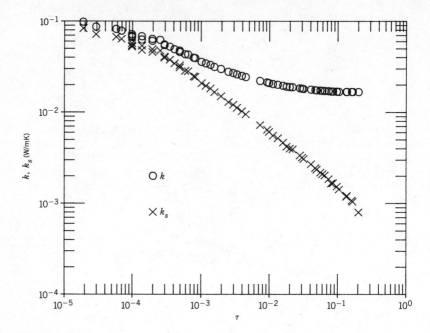

Figure 6.35 The thermal conductivity of helium-3 along the critical isochore (drawn after Pittman et al. [105]).

10^{-3} W/(mK). They used this estimate in order to determine from the experimental measurements the singular part $k_s(\rho_c, \tau)$, which is also shown in Fig. 6.35. Comparing this term with eq. (6.191) yields the parameters $A_\psi = (4.2)10^{-4}$ W/(mK) and $\psi = 0.57$, for which the τ range is $(5)10^{-4} < \tau < 0.1$. For carbon dioxide, on the other hand, Sengers and Keyes [106] had found $A_\psi = (3)10^{-3}$ W/(mK) and $\psi = 0.60$.

These experimental findings agree well with the theoretical predictions based on mode-mode coupling approximations (e.g., Sengers [107]). The main difficulty in comparing theoretical and experimental results stems from the need to know the correlation length, the shear viscosity, and, of course, the background contribution $k_b(\rho, T)$. Additional references on thermal-conductivity measurements near the critical point are Acton and Kellner [108] for helium-4, Prasad and Venart [109] for ethane, Weber [110] for oxygen, and Tufeu et al. [111] for ammonia and water.

(e) *Thermal Diffusivity.* Despite the enhancement of k, the thermal diffusivity[†] $D_T = k/(\rho c_P)$ decreases when approaching the critical point. This effect is due to the strongly diverging (increasing) isobaric specific heat

---

[†]The usual symbol for the thermal diffusivity α is not used here in order to avoid confusing it with the critical-point exponent α.

c_P. If we attempt to estimate $D_T(\tau)$, we find that it is convenient to define the new property (the "modified" thermal diffusivity):

$$D_T^* = \frac{k - k_b}{\rho(c_P - c_v)} \tag{6.192}$$

which employs $k - k_b = k_s$, eq. (6.191), rather than k, and $c_P - c_v \sim \kappa$, eq. (6.153), rather than c_P. In fact, D_T^* follows a simple power law along the critical isochore:

$$D_T^* = A_{\gamma - \psi} \tau^{\gamma - \psi} \tag{6.193}$$

The critical-point exponent $(\gamma - \psi)$ should be positive and approximately equal to ν. The main drawback of using eq. (6.193) to correlate the experimental results stems from the fact that k, k_b, c_P, and c_v would all have to be known. In most cases, however, they are not known; therefore, the following discussion is restricted to the "true" thermal diffusivity, that is, to the measurable value D_T.

Figure 6.36 [112] shows the thermal diffusivity of sulphur hexafluoride versus $|\tau|$. As $|\tau|$ decreases, the D_T values of saturated liquid and saturated vapor decrease in practically the same fashion. The thermal diffusivity

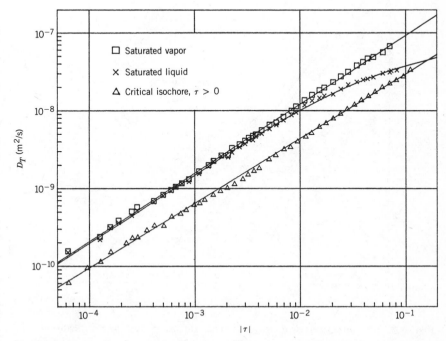

Figure 6.36 The thermal diffusivity of SF_6 near the critical point [112].

behaves differently along the critical isochore ($\rho = \rho_c, \tau > 0$). The three trends in Fig. 6.36 can reasonably be described by simple power laws $D_T(\tau)$. The resulting critical-point exponent values of 0.83 and 0.89 should not be confused with results from the theoretical predictable power law for the modified thermal diffusivity D_T^*, eq. (6.193).

Figure 6.37 shows a dimensionless thermal diffusivity versus ρ/ρ_c for several supercritical and subcritical isotherms and for saturated liquid and vapor. The ordinate shows the ratio D_T/D_T°, whose denominator has the following meaning. During the measurement of D_T in five different fluids (sulphur hexafluoride, ethane, nitrous oxide, monochlorotrifluoromethane, and xenon), Jany and Straub [113] found a universal functional dependence of D_T on τ and ρ/ρ_c. The molecular characteristics of each substance influence the magnitude of D_T but not the manner in which it is affected by τ and ρ/ρ_c. All the $D_T(\tau, \rho/\rho_c)$ surfaces collapse into a single dimensionless surface (Fig. 6.37) if D_T is divided by an empirical "specific-amplitude" constant D_T°, which is a characteristic of the fluid. Table 6.6 lists the D_T° values of the five fluids studied in Ref. 113. Jany and Straub developed in this way a universal dimensionless correlation for D_T/D_T° as a function of τ and ρ/ρ_c for a broad region around the critical point. The isotherms in Fig.

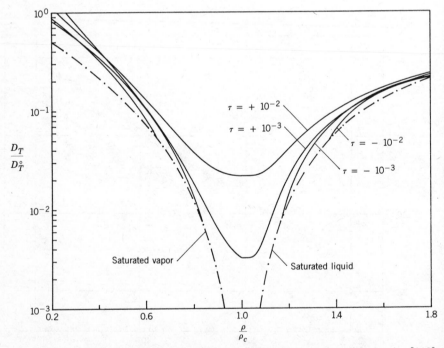

Figure 6.37 Dimensionless chart for thermal diffusivity near the critical point [113]: D_T° is listed in Table 6.6 and eq. (6.194).

6.37 are part of this correlation. The maximum departure between it and measurements is 10 percent. Worth noting is that the $D_T(\tau)$ chart of Fig. 6.36 can be nondimensionalized by dividing D_T by the D_T° value of SF_6 listed in Table 6.6.

Regarding the specific amplitude D_T°, Jany and Straub found that it is correlated adequately by

$$D_T^\circ = (8.3)10^{-6}\, M^{1/3}\rho_c^{-5/6} \qquad (6.194)$$

where M is the molecular mass in kg/kmol, ρ_c is the critical density in kg/m^3, and D_T° is expressed in m^2/s. This correlation and Fig. 6.37 allow the first-cut correlation of D_T for fluids other than the ones considered in Table 6.6.

An expanded ρ/ρ_c scale is used in Fig. 6.38, which shows how D_T/D_T° varies along the coexistence curve. This variation is contrasted with the simpler correlations that are known for ideal gases and perfect (incompressible) liquids. It is clear that these simpler models cannot be extrapolated into the critical region, although they agree well with the critical-region results at the left and right extremities of this region, $\rho/\rho_c \sim 0.2$ and $\rho/\rho_c \sim 2.0$. In

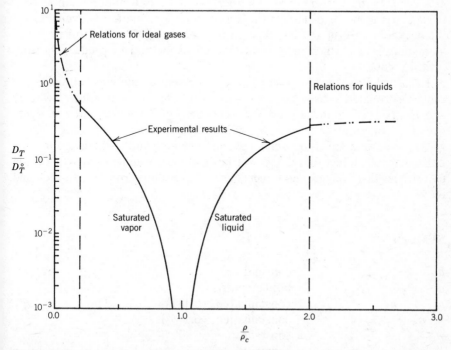

Figure 6.38 Comparison between the thermal diffusivity of ideal gases and incompressible liquids and the critical-region correlation of Fig. 6.37 and Ref. 113.

conclusion, the critical point influences D_T strongly in an unexpectedly broad region. Additional experimental results for thermal diffusivity have been reported by Ackerson and Straty [114] for methane, Grabner et al. [115] for methanol and acetonitrile, Tufeu et al. [116] for ammonia, and Reile et al. [117] for carbon dioxide and trifluoromethane.

(f) *Mass Diffusivity.* The preceding experimental results referred to the gas–liquid system. Corresponding results can also be developed for binary-liquid mixtures near their consolute critical point. A comprehensive review of recent experiments on this class of systems is provided by Beysens [99].

An additional effect that can occur in a two-component system is the phenomenon of mass diffusion. This effect is controlled by the diffusion coefficient or mass diffusivity D, which is expected to decrease when approaching the critical point. Sengers [107] presented recently an extensive study of this phenomenon. The mass diffusivity D can be separated into a background term D_b and a singular term D_s, in accordance with the scaling model shown in eq. (6.178). The D_s part is represented well along the "critical isochore" (constant critical concentration) by the power law:

$$D_s = A_\phi \tau^\phi \tag{6.195}$$

Assuming that D_s obeys the diffusion law of Stokes–Einstein ($D_s \sim k_B T / 6\pi\eta\xi$, with η as the shear viscosity), then the critical-point exponent ϕ is expected to be of order ν. No experimental information is available for the background term D_b. For this reason, Sengers [107] refers to Burstyn et al.'s study [118], which includes a theoretical method to predict D_b from viscosity data.

Even if the background contribution is neglected, the mass diffusivity D can be correlated satisfactorily by means of a simple power law. Swinney and Henry [119] found a power-law exponent of roughly 0.63 by studying five two-component mixtures: aniline + cyclohexane, isobutyric acid + water, n-hexane + nitrobenzene, phenol + water, and 3-methylpentane + nitroethane. The fact that this power-law exponent is quite close to ν (Table 6.6) proves that background effects have a lesser impact on D then they had on k.

SYMBOLS

a	constant
a	lattice constant
$a_1, \ldots, a_6$	constants [eq. (6.141)]
$a_\tau, a_{\mathcal{H}}$	constants [eq. (6.179)]
A	constant
A_0, A_1	constants

$A_\alpha, A_{\alpha'}, A_\beta, A_\gamma, A_{\gamma'},$ $A_{\gamma-\psi}, A_\nu, A_{\nu'}, A_\phi, A_\psi$	coefficients of powers in $|\tau|$ associated with critical-point exponents α, α', β, γ, γ', $\gamma - \psi$, ν, ν', ϕ, ψ
$A_{\lambda_1}, A_{\lambda_2}$	constants [eq. (6.145)]
b	constant
$b_1, \ldots, b_6$	constants [eq. (6.140)]
$b_\tau, b_{\mathscr{H}}$	constants [eq. (6.186)]
B	constant
B_0	constant
$B_\alpha, B_{\alpha'}$	constants [eqs. (6.155) and (6.154)]
c	constant
$c_0, c_1, \ldots$	constants [eq. (6.168)]
c_{2A}, c_{4A}	constants [eqs. (6.170a, b)]
c_P	specific heat at constant pressure
c_v	specific heat at constant volume
C	constant [eq. (6.114)]
C_0	constant [eq. (6.116)]
d	constant [eq. (6.112)]
d	dimensionality of a system
D	constant [eq. (6.114)]
D	mass diffusivity
D_T	thermal diffusivity
D_T°	normalization value of the thermal diffusivity [eq. (6.194)]
D_T^*	modified thermal diffusivity [eq. (6.192)]
f	fugacity [eq. (6.128)]
$f(\tau)$	function near the critical point [eq. (6.145)]
F	Helmholtz free energy
g	specific Gibbs free energy
G	Gibbs free energy
$G(r)$	correlation function [eq. (6.158)]
h	specific enthalpy
H	enthalpy
$\mathscr{H}$	magnetic field
$\mathbf{H}$	Hamiltonian [eq. (6.176)]
k	thermal conductivity
k_B	Boltzmann's constant
K	isothermal susceptibility
l	constant [eq. (6.113)]
l^d	number of sites in a d-dimensional block system
m	mass
m	parameter, [eqs. (6.104) and (6.106)]

M	molecular mass
$\mathcal{M}$	magnetization
n	dimensionality of the order parameter
n	number of components in a system [eq. (6.72)]
n	parameter [eqs. (6.106), (6.110), and (6.111)]
N	number of moles
p	number of phases in a system [eq. (6.72)]
P	pressure
Q	heat transfer interaction
r	distance
r	scaling parameter [eqs. (6.187a, b)]
R	ideal gas constant
s	specific entropy
s_i, s_j	spins at lattice sites i and j
S	entropy
S_{gen}	entropy generation
t	reduced-temperature departure [eq. (6.142)]
T	temperature
T_{sp}	liquid-spinodal temperature [Fig. 6.26]
u	specific energy
U	energy
v	specific volume
$v_{r'}$	pseudo-reduced volume [eq. (6.90)]
V	volume
W	work
x	mass fraction of the vapor phase [eq. (6.56)]
Z	compressibility factor [eq. (6.87)]
$\mathscr{Z}$	partition function [eq. (6.177)]
$Z^{(0)}, Z^{(1)}$	components of the compressibility factor [eq. (6.93)]
α	constant
α, α'	critical-point exponents [eqs. (6.155) and (6.154)]
α_c	slope of the reduced saturation-pressure curve at the critical point
β	constant [eq. (6.113)]
β	critical-point exponent [eq. (6.149)]
γ	constant [eq. (6.116)]
γ, γ'	critical-point exponents [eqs. (6.153) and (6.152)]
δ	critical-point exponent [eq. (6.150)]
Δ	critical-point exponent [eq. (6.188)]
$\Delta\rho$	density fluctuation
ϵ	constant [eq. (6.115)]
ϵ	deviation from physical dimensionality $d = 4$

ϵ_{ij}	energy interaction between spins s_i and s_j
η	critical-point exponent [eq. (6.158)]
η	shear viscosity
θ	scaling parameter [eqs. (6.187a, b)]
κ	isothermal compressibility
λ	dimensionless latent heat of vaporization [eq. (6.139)]
λ	scaling factor [eq. (6.179)]
λ_1, λ_2	constants [eq. (6.145)]
μ	chemical potential
μ_J	Joule–Thomson coefficient [eq. (6.118)]
ν, ν'	critical-point exponents [eqs. (6.157) and (6.156)]
ξ	correlation length
π	reduced-pressure difference [eq. (6.161)]
ρ	mass density
τ	dimensionless temperature [eq. (6.138)]
τ	reduced-temperature difference [eq. (6.146)]
ϕ	critical-point exponent [eq. (6.195)]
ϕ	reduced-density difference [eq. (6.151)]
Φ	order parameter
ψ	critical-point exponent [eq. (6.191)]
ω	Pitzer acentric factor [eq. (6.91)]
$(\ )_{A,B}$	property of component A, B
$(\ )_b$	quantity of noncritical background [eq. (6.178)]
$(\ )_c$	property at the critical point
$(\ )_f$	property along the saturated-liquid line
$(\ )_{fg}$	indicates a difference between saturated-liquid and saturated-vapor properties [eq. (6.57)]
$(\ )_g$	property along the saturated-vapor line
$(\ )_{i,j}$	property of components i and j
$(\ )_L$	quantity of the left half of the system
$(\ )_m$	property of a mixture
$(\ )_{max}$	maximum
$(\ )_{min}$	minimum
$(\ )_r$	reduced property
$(\ )_R$	quantity of the right half of the system
$(\ )_s$	property along the saturated-solid line
$(\ )_s$	quantity of the critical part of a property [eq. (6.178)]
$(\ )_{sat}$	property along the saturation line
$(\ )_t$	property at the triple point
$(\ )^t$	transformed property [eq. (6.184)]
$(\bar{\ })$	molal quantity
$(\ )^*$	property associated with the low-pressure limit

REFERENCES

1. J. W. Gibbs, *The Collected Works of J. Willard Gibbs*, Vol. I, Longmans, Green, New York, 1928, pp. 56–62.

2. E. A. Guggenheim, *Thermodynamics*, 7th ed., North-Holland Publ., Amsterdam, 1985, pp. 30–33.

3. T. Andrews, On the continuity of the gaseous and liquid states of matter, *Philos. Trans. R. Soc. London*, Vol. 159, 1869, pp. 575–590.

4. A. Bejan, Engineering thermodynamics, in M. Kutz, ed., *Mechanical Engineers' Handbook*, Wiley, New York, 1986, Chapter 54.

5. J. Thomson, Considerations on the abrupt change at boiling or condensation in reference to the continuity of the fluid state of matter, *Proc. R. Soc. London*, Vol. 20, 1871, pp. 1–8.

6. J. D. van der Waals, *Over de continuiteit van den gas en vloeistof toestand*, A. W. Sijthoff, Leiden, 1873; published in German as *Die Continuität des Gasförmigen und Flüssigen Zustandes*, Verlag von Johann Ambrosius Barth, Leipzig, 1881.

7. H. B. Callen, *Thermodynamics*, Wiley, New York, 1960, Chapter 9.

8. A. B. Pippard, *Elements of Classical Thermodynamics*, Cambridge University Press, London and New York, 1964, pp. 118–121.

9. J. C. Maxwell, On the dynamical evidence of the molecular constitution of bodies, *Nature (London)*, Vol. 11, 1875, pp. 357–359, 374–377.

10. M. Mott-Smith, *The Concept of Heat and Its Workings Simply Explained*, Dover, New York, 1962.

11. P. W. Bridgman, *The Physics of High Pressure*, Macmillan, New York, 1931.

12. P. W. Bridgman, *Collected Experimental Papers*, Vol. I, Harvard University Press, Cambridge MA, 1964, Paper 7; Vol. II, Paper 15.

13. L. C. Nelson and E. F. Obert, Generalized pvT properties of gases, *Trans. ASME*, Vol. 76, 1954, pp. 1057–1066.

14. R. C. Reid, J. M. Prausnitz and T. K. Sherwood, *The Properties of Gases and Liquids*, 3rd ed., McGraw-Hill, New York, 1977.

15. K. S. Pitzer, The volumetric and thermodynamic properties of fluids. I. Theoretical basis and virial coefficients, *J. Am. Chem. Soc.*, Vol. 77, 1955, pp. 3427–3433.

16. J. M. Prausnitz, *Molecular Thermodynamics of Fluid-Phase Equilibria*, Prentice-Hall, Englewood Cliffs, NJ, 1969.

17. K. S. Pitzer, D. Z. Lippmann, R. F. Curl, Jr., C. M. Huggins, and D. E. Petersen, The volumetric and thermodynamic properties of fluids. II. Compressibility factor, vapor pressure and entropy of vaporization, *J. Am. Chem. Soc.*, Vol. 77, 1955, pp. 3433–3440.

18. B. I. Lee and M. G. Kesler, A generalized thermodynamic correlation based on three-parameter corresponding states, *AIChE J.*, Vol. 21, 1975, pp. 510–527.

19. J. S. Hsieh, Four-parameter generalized compressibility charts for nonpolar fluids, *J. Eng. Ind.*, Vol. 88, 1966, pp. 263–273.

20. L. Riedel, Eine neue universelle Dampfdruckformel-Untersuchungen über eine Erweiterung des Theorems der übereinstimmenden Zustände. Teil 1, *Chem.-Ing.-Tech.*, Vol. 26, 1954, p. 83.

21. J. S. Hsieh, *Principles of Thermodynamics*, McGraw-Hill, New York, 1975, pp. 490–491.

22. O. Redlich and J. N. S. Kwong, On the thermodynamics of solutions. V. An equation of state. Fugacities of gaseous solutions, *Chem. Rev.*, Vol. 44, 1949, pp. 233–244.

23. D. G. Miller, Joule–Thomson inversion curve, corresponding states, and simpler equations of state, *Ind. Eng. Chem. Fundam.*, Vol. 9, 1970, pp. 585–589.

24. G. Soave, Equilibrium constants from a modified Redlich–Kwong equation of state, *Chem. Eng. Sci.*, Vol. 27, 1972, pp. 1197–1203.

25. R. Camporese, G. Bigolaro, and L. Rebellato, Calculation of thermodynamic properties of refrigerants by the Redlich–Kwong–Soave equation of state, *Int. J. Refrig.*, Vol. 8, No. 3, pp. 147–151.

26. R. Clausius, Ueber das Verhalten der Kohlensäure in Bezug auf Druck, Volumen und Temperatur, *Wiedemanns Ann.*, Vol. 9, 1880, pp. 337–357.

27. H. A. Lorentz, Ueber die Anwendung des Satzes vom Virial in der Kinetischen Theorie der Gase, *Wiedemanns Ann.*, Vol. 12, 1881, pp. 127–146; also *ibid.*, Suppl., pp. 660–661.

28. C. Dieterici, Ueber den Kritischen, *Wiedemanns Ann.*, Vol. 69, 1899, pp. 685–705; also Die Berechnung der Isothermen, *Ann. Phys.* (*Leipzig*), Ser. 4, Vol. 5, 1901, pp. 51–88.

29. A. W. Porter, On the inversion-points for a fluid passing through a porous plug and their use in testing proposed equations of state, *Philos. Mag.*, Ser. 6, Vol. 11, April 1906, pp. 554–568.

30. A. W. Porter, On the inversion-points for a fluid passing through a porous plug and their use in testing proposed equations of state. Part II. An examination of experimental data, *Philos. Mag.*, Ser. 6, Vol. 19, No. 114, June 1910, pp. 888–897.

31. J. J. Martin, Equations of state, *Ind. Eng. Chem.*, Vol. 59, No. 12, 1967, pp. 34–52.

32. F. G. Keyes and R. S. Taylor, The adequacy of the assumption of molecular aggregation in accounting for certain of the physical properties of gaseous nitrogen, *J. Am. Chem. Soc.*, Vol. 49, 1927, pp. 896–911.

33. H. Kamerlingh-Onnes and W. H. Keesom, Die Zustandgleichung, *Commun. Phys. Lab. Univ. Leiden*, Vol. 11, Suppl. 23, 1912.

34. J. A. Beattie and O. C. Bridgeman, A new equation of state for fluids, *Proc. Am. Acad. Arts Sci.*, Vol. 63, 1928, pp. 229–308.

35. E. G. Cravalho and J. L. Smith, Jr., *Engineering Thermodynamics*, Pitman, Boston, MA, 1981, p. 316.

36. M. Benedict, G. B. Webb, and L. C. Rubin, An empirical equation for thermodynamic properties of light hydrocarbons. I. Methane, ethane, propane and n-butane, *J. Chem. Phys.*, Vol. 8, 1940, pp. 334–345.

37. M. Benedict, G. B. Webb, and L. C. Rubin, An empirical equation for

thermodynamic properties of light hydrocarbons and their mixtures. II. Mixtures of methane, ethane, propane and n-butane, *J. Chem. Phys.*, Vol. 10, 1942, pp. 747–758.

38. M. Benedict, G. B. Webb, and L. C. Rubin, An empirical equation for thermodynamic properties of light hydrocarbons and their mixtures. Constants for twelve hydrocarbons, *Chem. Eng. Prog.*, Vol. 47, No. 8, 1951, pp. 419–422.

39. H. W. Cooper and J. C. Goldfrank, B–W–R constants and new correlations, *Hydrocarbon Process.*, Vol. 46, No. 12, 1967, pp. 141–146.

40. R. V. Orye, Prediction and correlation of phase equilibria and thermal properties with the BWR equation of state, *Ind. Eng. Chem. Process Des. Dev.*, Vol. 8, 1969, pp. 579–588.

41. J. H. Keenan and J. Kaye, *Gas Tables*, Wiley, New York, 1966.

42. W-G. Dong and J. H. Lienhard, Corresponding states correlation of saturated and metastable properties, *Can. J. Chem. Eng.*, Vol. 64, 1986, pp. 158–161.

43. K. M. Watson, Thermodynamics of the liquid state, *Ind. Eng. Chem.*, Vol. 35, No. 4, 1943, pp. 398–406.

44. S. Torquato and G. R. Stell, An equation for the latent heat of vaporization, *Ind. Eng. Chem. Fundam.*, Vol. 21, 1982, pp. 202–205.

45. S. Torquato and P. Smith, The latent heat of vaporization of a widely diverse class of fluids, *J. Heat Transfer*, Vol. 106, 1984, pp. 252–254.

46. J. H. Lienhard, N. Shamsundar, and P. O. Biney, Spinodal lines and equations of state: A review, *Nucl. Eng. Des.*, Vol. 95, 1986, pp. 297–314.

47. P. S. Beloff, Transient natural convection in a large-diameter horizontal cylinder, MS thesis, Department of Mechanical Engineering and Materials Science, Duke University, Durham, NC, 1986.

48. C. S. Murali, Improved cubic equations of state for polar and non-polar fluids, MEMS thesis, Department of Mechanical Engineering, University of Houston, TX, 1983.

49. P. O. Biney, W-G. Dong, and J. H. Lienhard, Use of a cubic equation to predict surface tension and spinodal limits, *J. Heat Transfer*, Vol. 108, 1986, pp. 405–410.

50. A. Karimi and J. H. Lienhard, Toward a fundamental equation for water in the metastable states, *High Temp.—High Pressures*, Vol. 11, 1979, pp. 511–517.

51. A. Karimi and J. H. Lienhard, *A Fundamental Equation Representing Water in the Stable, Metastable and Unstable States*, EPRI Report NP-3328, Electric Power Res. Inst., Palo Alto, CA, December 1983.

52. V. P. Skripov, *Metastable Liquids*, Halsted Press, Wiley, New York, 1974.

53. J. H. Lienhard, Corresponding states correlation of the spinodal and homogeneous nucleation limits, *J. Heat Transfer*, Vol. 104, 1982, pp. 379–381.

54. A. Sharan, J. H. Lienhard, and R. Kaul, Corresponding states correlations for pool and flow boiling burnout, *J. Heat Transfer*, Vol. 107, 1985, pp. 392–397.

55. C. Domb and M. S. Green, *Phase Transitions and Critical Phenomena*, Vol. 1, Academic Press, London, 1972; Vol. 2, 1973; Vol. 3, 1974; Vols. 5a, 5b, 6, 1976.

56. C. Domb and J. L. Lebowitz, *Phase Transitions and Critical Phenomena*, Vol. 7, Academic Press, London, 1983; Vol. 8, 1983; Vol. 9, 1984; Vol. 10, 1986.

57. M. Lévy, J.-C. Le Guillou, and J. Zinn-Justin, eds., *Phase Transition—Cargèse 1980*, Plenum, New York, 1982.

58. S.-K. Ma, *Modern Theory of Critical Phenomena*, Benjamin, Reading, MA, 1976.

59. P. Pfeuty and G. Toulouse, *Introduction to the Renormalization Group and to Critical Phenomena*, Wiley, London, 1977.

60. H. J. Raveché, ed., *Perspectives in Statistical Physics. Part II. Phase Transitions*, North-Holland Publ., Amsterdam, 1981, pp. 135–291.

61. H. E. Stanley, *Introduction to Phase Transitions and Critical Phenomena*, Oxford University Press, London and New York, 1971.

62. L. D. Landau and E. M. Lifshitz, *Statistical Physics*, Part I, 3rd ed., Pergamon, Oxford, 1980.

63. L. E. Reichl, *A Modern Course in Statistical Physics*, University of Texas Press, Austin, 1980.

64. G. S. Rushbrooke, On the thermodynamics of the critical region for the Ising problem, *J. Chem. Phys.*, Vol. 39, 1963, pp. 842–843.

65. R. B. Griffiths, Thermodynamic inequality near the critical point for ferromagnets and fluids, *Phys. Rev. Lett.*, Vol. 14, 1965, pp. 623–624.

66. R. B. Griffiths, Ferromagnets and simple fluids near the critical point: Some thermodynamic inequalities, *J. Chem. Phys.*, Vol. 43, 1965, pp. 1958–1968.

67. B. D. Josephson, Inequality for the specific heat. I. Derivation. II. Application to critical phenomena, *Proc. Phys. Soc., London*, Vol. 92, 1967, pp. 269–275, 276–284.

68. M. J. Buckingham and J. D. Gunton, Correlations at the critical point of the Ising model, *Phys. Rev.*, Vol. 178, 1969, pp. 848–853.

69. M. E. Fisher, Rigorous inequalities for critical-point correlation exponents, *Phys. Rev.*, Vol. 180, 1969, pp. 594–600.

70. L. Landau, Zur Theorie der Phasenumwandlungen. I and II, *Phys. Z. Sowjetunion*, Vol. 11, 1937, pp. 26–47, 545–555.

71. S. G. Brush, History of the Lenz–Ising model, *Rev. Mod. Phys.*, Vol. 39, 1967, pp. 883–893.

72. W. Lenz, Beitrag zum Verständnis der magnetischen Erscheinungen in festen Körpern, *Phys. Z.*, Vol. 21, 1920, pp. 613–615.

73. E. Ising, Beitrag zur Theorie des Ferromagnetismus, *Z. Phys.*, Vol. 31, 1925, pp. 253–258.

74. E. Onsager, Crystal statistics. I. A two-dimensional model with an order–disorder transition, *Phys. Rev.*, Vol. 65, 1944, pp. 117–149.

75. W. Heisenberg, Zur Theorie des Ferromagnetismus, *Z. Phys.*, Vol. 49, 1928, pp. 619–636.

76. M. E. Fisher, The theory of equilibrium critical phenomena, *Rep. Prog. Phys.*, Vol. 30, Part II, 1967, pp. 615–730.

77. C. Domb, Ising model, in Ref. 55, Vol. 3, pp. 357–484.

78. J. V. Sengers, Universality of critical phenomena in classical fluids, in Ref. 57, pp. 95–136.

79. B. Widom, Equation of state in the neighborhood of the critical point, *J. Chem. Phys.*, Vol. 43, 1965, pp. 3898–3905.

80. L. P. Kadanoff, Scaling laws for Ising models near T_c, *Physics*, Vol. 2, 1966, pp. 263–272.

81. K. G. Wilson, Renormalization group and critical phenomena. I. Renormalization group and the Kadanoff scaling picture. II. Phase-space cell analysis of critical behavior, *Phys. Rev. B*, Vol. 4, 1971, pp. 3174–3183, 3184–3205.

82. K. G. Wilson, The renormalization group and critical phenomena, *Rev. Mod. Phys.*, Vol. 55, 1983, pp. 583–600.

83. K. G. Wilson, Critical phenomena in 3.99 dimensions, *Physica (Amsterdam)*, Vol. 73, 1974, pp. 119–128.

84. J. C. Le Guillou and J. Zinn-Justin, Critical exponents from field theory, *Phys. Rev. B*, Vol. 21, 1980, pp. 3976–3998.

85. J. Kestin, J. V. Sengers, B. Kamgar-Parsi and J. M. H. Levelt Sengers, Thermophysical properties of fluid H_2O, *J. Phys. Chem. Ref. Data*, Vol. 13, 1984, pp. 175–205.

86. J. M. H. Levelt Sengers and J. V. Sengers, How close is "close to the critical point"?, in Ref. 60, Chapter 14, pp. 239–271.

87. P. Schofield, Parametric representation of the equation of state near a critical point, *Phys. Rev. Lett.*, Vol. 22, 1969, pp. 606–608.

88. J. M. H. Levelt Sengers, B. Kamgar-Parsi, F. W. Balfour, and J. V. Sengers, Thermodynamic properties of steam in the critical region, *J. Phys. Chem. Ref. Data*, Vol. 12, 1983, pp. 1–28.

89. J. R. Hastings, J. M. H. Levelt Sengers, and F. W. Balfour, The critical-region equation of state of ethene and the effect of small impurities, *J. Chem. Thermodyn.*, Vol. 12, 1980, pp. 1009–1045.

90. J. M. H. Levelt Sengers, W. L. Greer, and J. V. Sengers, Scaled equation of state parameters for gases in the critical region, *J. Phys. Chem. Ref. Data*, Vol. 5, 1976, pp. 1–51.

91. J. Straub, Dichtemessung am kritischen Punkt mit einer optischen Methode bei reinen Stoffen und Gemischen, Doctoral Thesis, Technical University Munich, 1965; Optische Bestimmung von Dichteschichtungen im kritischen Zustand, *Chem.-Ing.-Tech.*, Vol. 39, 1967, pp. 291–296.

92. M. R. Moldover, J. V. Sengers, R. W. Gammon, and R. J. Hocken, Gravity effects in fluids near the gas–liquid critical point, *Rev. Mod. Phys.*, Vol. 51, 1979, pp. 79–99.

93. P. Jany, Die Temperaturleitfähigkeit reiner Fluide im weiten Zustandsbereich um den kritischen Punkt, Doctoral Thesis, Technical University Munich, 1986.

94. D. Balzarini and K. Ohrn, Coexistence curve of sulfur hexafluoride, *Phys. Rev. Lett.*, Vol. 29, 1972, pp. 840–842.

95. W. Rathjen and J. Straub, Surface tension and refractive index of six refrigerants from triple point up to the critical point, in A. Cezairliyan, ed., *Thermophysical Properties*, 7th Symp., Am. Soc. Mech. Eng., New York, 1977, pp. 839–850.

96. K. Morofuji, F. Fujii, M. Uematsu, and K. Watanabe, Precise measurements of critical parameters of sulfur hexafluoride by laser interferometry, *Int. J. Thermophys.*, Vol. 7, 1986, pp. 17–28.

97. S. C. Greer and M. R. Moldover, Thermodynamic anomalies at critical points of fluids, *Annu. Rev. Phys. Chem.*, Vol. 32, 1981, pp. 233–265.

98. M. R. Moldover, Thermodynamic anomalies near the liquid–vapor critical point: A review of experiments, in Ref. 57, pp. 63–94.

99. D. Beysens, Status of the experimental situation in critical binary fluids, in Ref. 57, pp. 25–62.

100. J. Straub, R. Lange, K. Nitsche, and K. Kemmerle, Isochoric specific heat of sulfur hexafluoride at the critical point: Laboratory results and outline of a spacelab experiment for the D1-mission in 1985, *Int. J. Thermophys.*, Vol. 7, 1986, pp. 343–356.

101. J. M. J. Van Leeuwen and J. V. Sengers, Gravity effects on the correlation length in gases near the critical point, *Physica (Amsterdam)*, Vol. 128A, 1984, pp. 99–131.

102. L. P. Kadanoff and J. Swift, Transport coefficients near the liquid–gas critical point, *Phys. Rev.*, Vol. 166, 1968, pp. 89–101.

103. K. Kawasaki, Kinetic equations and time correlation functions of critical fluctuations, *Ann. Phys.*, Vol. 61, 1970, pp. 1–56.

104. A. Michels, J. V. Sengers, and P. S. Van der Gulik, The thermal conductivity of carbon dioxide in the critical region. II. Measurements and conclusions, *Physica (Amsterdam)*, Vol. 28, 1962, pp. 1216–1237.

105. C. E. Pittman, L. H. Cohen, and H. Meyer, Transport properties of helium near the liquid–vapor critical point. I. Thermal conductivity of ^{3}He, *J. Low Temp. Phys.*, Vol. 46, 1982, pp. 115–135; C. E. Pittman, The thermal conductivity of ^{3}He near the liquid–vapor critical point, Doctoral Thesis, Duke University, Durham, NC, 1981.

106. J. V. Sengers and P. H. Keyes, Scaling of the thermal conductivity near the gas–liquid critical point, *Phys. Rev. Lett.*, Vol. 26, 1971, pp. 70–73.

107. J. V. Sengers, Transport properties of fluids near critical points, *Int. J. Thermophys.*, Vol. 6, 1985, pp. 203–232.

108. A. Acton and K. Kellner, The low temperature thermal conductivity of ^{4}He. II. Measurements in the critical region, *Physica (Amsterdam)*, Vol. 103B, 1981, pp. 212–225.

109. R. C. Prasad and J. E. S. Venart, Thermal conductivity of ethane from 290 to 600 K at pressures up to 700 bar, including the critical region, *Int. J. Thermophys.*, Vol. 5, 1984, pp. 367–385.

110. L. A. Weber, Thermal conductivity of oxygen in the critical region, *Int. J. Thermophys.*, Vol. 3, 1982, pp. 117–135.

111. R. Tufeu, D. Y. Ivanov, Y. Garrabos, and B. Le Neindre, Thermal conductivity of ammonia in a large temperature and pressure range including the critical region, *Ber. Bunsenges. Phys. Chem.*, Vol. 88, 1984, pp. 422–427; R. Tufeu and B. Le Neindre, Thermal conductivity of steam from 250 to 510°C at pressures up to 95 MPa including the critical region, *Int. J. Thermophys.*, Vol. 8, 1987, pp. 283–292.

112. P. Jany and J. Straub, Thermal diffusivity of fluids in a broad region around the critical point, *Int. J. Thermophys.*, Vol. 8, 1987, pp. 165–180.

113. P. Jany and J. Straub, Thermal diffusivity of five pure fluids, *Chem. Eng. Commun.*, 1988 (to be published).

114. B. J. Ackerson and G. C. Straty, Rayleigh scattering from methane, *J. Chem. Phys.*, Vol. 69, 1978, pp. 1207–1212.

115. W. Grabner, F. Vesely, and G. Benesch, Rayleigh linewidth measurements on polar liquids in the critical region, *Phys. Rev. A*, Vol. 18, 1978, pp. 2307–2314.

116. R. Tufeu, A. Letaief, and B. Le Neindre, Turbidity, thermal diffusivity and thermal conductivity of ammonia along the critical isochore, in J. V. Sengers, ed., *Thermophysical Properties*, 8th Symp., Am. Soc. Mech. Eng., New York, 1982, pp. 451–457.

117. E. Reile, P. Jany, and J. Straub, Messung der Temperaturleitfähigkeit reiner Fluide und binärer Gemische mit Hilfe der dynamischen Lichtstreuung, *Waerme- Stoffuebertrag.*, Vol. 18, 1984, pp. 99–108.

118. H. C. Burstyn, J. V. Sengers, J. K. Bhattacharjee, and R. A. Ferrell, Dynamic scaling function for critical fluctuations in classical fluids, *Phys. Rev. A*, Vol. 28, 1983, pp. 1567–1578.

119. H. L. Swinney and D. L. Henry, Dynamics of fluids near the critical point: Decay rate of order-parameter fluctuations, *Phys. Rev. A*, Vol. 8, 1973, pp. 2586–2617.

120. H. M. Paynter, Simple veridical state equations for thermofluid simulation: generalization and improvements upon van der Waals, *J. Dyn. Syst., Meas., Control*, Vol. 107, 1985, pp. 233–234.

PROBLEMS

6.1 A closed system contains a single-phase substance that undergoes an infinitesimally small change at constant entropy (S) and pressure (P). The change is such that the entropy inventories of the two halves of the system become, respectively, $\frac{1}{2}(S + \delta S)$ and $\frac{1}{2}(S - \delta S)$, where S is the original (constant) entropy inventory of the entire system. Invoking the enthalpy minimum principle for constant S and P, show that the system is in a state of stable equilibrium if the system's temperature increases during heating at constant pressure; in other words, if

$$c_P > 0$$

Show also that the "positive c_P" requirement is consistent with the criteria for internal thermal and mechanical equilibria, $c_v > 0$ and $\kappa > 0$.

6.2 Show that if the vapor phase of a liquid–vapor mixture can be treated as an ideal gas whose constant R_g is known, the saturation pressure and temperature are related through

$$d(\ln P) = -\frac{h_{fg}}{R_g} d\left(\frac{1}{T}\right)$$

In other words, in the special temperature range in which h_{fg} is also a constant, the relationship between ($\ln P$) and ($1/T$) is graphically represented by a straight line (e.g., Fig. 6.16).

6.3   The specific heat "at saturation" of a liquid phase of mass m is defined as the ratio

$$c_f = \frac{\delta Q_{rev}/m}{dT}$$

where δQ_{rev} represents the heat transfer input responsible for raising the temperature of the liquid by the amount dT. Since $\delta Q_{rev}/m = T\,ds_f$, the definition for c_f can also be written as

$$c_f = T\,\frac{ds_f}{dT}$$

Treating s_f as a T function of the form $s_f[T, P(T)]$, show that in general

$$c_f = c_{P,f} - \beta_f v_f \frac{h_{fg}}{v_{fg}}$$

in other words, the specific heat at saturation is generally not the same as the specific heat at constant pressure of the liquid. Using the $(c_{P,f}, \beta_f, v_f, h_{fg}, v_{fg})$ data for saturated liquid water at 95°C, show that in that particular case, $c_f \cong c_{P,f}$.

6.4   In a way that parallels the first paragraph of the preceding problem, the specific heat at saturation of a gaseous phase is defined as

$$c_g = T\,\frac{ds_g}{dT}$$

Demonstrate that in general, c_g is not the same as the c_P of saturated vapor; specifically, show that

$$c_g = c_{P,g} - \beta_g v_g \frac{h_{fg}}{v_{fg}}$$

Assuming that the saturated vapor can be treated as an ideal gas, show further that

$$c_g \cong c_{P,g} - s_{fg}$$

6.5   Show that in the limit $v_r \to \infty$, the pressure along the vapor spinodal line of Fig. 6.11 decays as $1/v_r^2$.

6.6   The two-phase line separating the ice I and water domains in Fig. 6.16 appears to be perfectly vertical over the range of moderate pressures 1 kPa–1 MPa. This effect is due solely to the semilogarithmic plot that was chosen for drawing the figure (such a plot is recommended by Problem 6.2). In reality, the slope of the "ice + water" line is finite and negative throughout the pressure range: this slope is seen better by looking at the right edge of the "ice I" surface in Fig. 6.15. Estimate the slope dP/dT of the "ice + water" line by using the triple-point information listed on Fig. 6.15. Note, in addition, that the specific entropy of triple-point ice is -1221 J/kgK, whereas the specific entropy of triple-point liquid water is 0.00 J/kgK. What is the Celsius temperature of an equilibrium ice + water mixture whose pressure is 10 MPa?

6.7 The van der Waals equation of state can be nondimensionalized in the coordinate system centered at the critical point. The new dimensionless pressure, volume, and temperature are

$$\pi = \frac{P - P_c}{P_c} \qquad \phi = \frac{v - v_c}{v_c} \qquad \tau = \frac{T - T_c}{T_c}$$

Determine the new equation of state, $\pi = \pi(\phi, \tau)$, and the slope of the critical isochore ($\phi = 0$) in the pressure–temperature plane π–τ.

6.8 Invoke Maxwell's equal-area rule and determine numerically the reduced saturation-pressure curve $P_r(T_r)$ of a fluid whose $P(v, T)$ surface matches van der Waals' equation of state (6.46). From the $P_r(T_r)$ curve, deduce the values of ω and α_c of this "van der Waals fluid."

6.9 Determine the critical compressibility factor Z_c compatible with the Redlich and Kwong equation of state.

6.10 Demonstrate that the slope of the constant-volume line passing through the critical point (the "critical isochore," Fig. 6.21) must be the same as the slope of the saturation-pressure curve right at the critical point.

6.11 Prove that the position of the Joule–Thomson coefficient inversion curve ($\mu_J = 0$) in the T_r–P_r plane is given by the equation

$$\frac{\partial}{\partial T_r} [Z(T_r, P_r)] = 0$$

6.12 Sufficiently far from the critical point (i.e., to the right in Fig. 6.25), the enthalpy of vaporization is approximated well by a linear function of temperature. In the same range, the compressibility factor of saturated vapor, Z_g, can be regarded as constant. Show that the saturation-pressure curve that is compatible with these assumptions has the form

$$\ln P_r = \frac{c_1}{T_r} + c_2 \ln T_r + c_3$$

where c_1, c_2, and c_3 are three empirical constants.

6.13 Prove the validity of the identity (6.68),

$$h_{fg} = Ts_{fg}$$

by recognizing first that the Gibbs free energy at state (f) is the same as at state (g), and that in both states the system is single phase. In other words, exploit the relations between the properties of a simple system (Table 4.3), as an alternative to the derivation given in the text by eqs. (6.69)–(6.70).

6.14 Consider the process executed by a single-component system as its state (m) moves along the edge of the dome from state (f) to state (g). As indicated in the drawing, the initial and final temperature in this process is T_0. Prove that if dP_m/dT_m is practically constant, then

$$\int_f^g (T_m - T_0) \, ds_m = \frac{s_{fg}(T_0)}{v_{fg}(T_0)} \int_{T_0}^{T_c} v_{fg}(T) \, dT$$

in which the line integral appearing on the left side follows the edge of the dome, (f)–(m)–(g). Note further that this integral represents the melon slice-shaped area trapped between the dome and the $T = T_0$ base.

Show also that, in general (i.e., when dP_m/dT_m is not necessarily constant), we must have

$$\int_{T_0}^{T_c} s_{fg}(T) \, dT = \int_{P_0}^{P_c} v_{fg}(P) \, dP$$

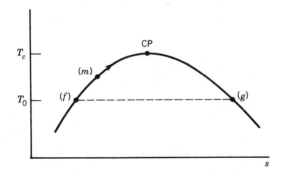

Figure P6.14

6.15 Assuming the mean-square-density fluctuations $\langle \Delta \rho^2 \rangle$ in eq. (6.144) of order $(\rho_f - \rho_g)^2$, show that $V \sim \xi_{f,g}^d$.

6.16 Estimate the order of magnitude of the correlation length along the critical isochore $(\rho = \rho_c)$ if, in order, $T - T_c = 0.1$ K, 0.01 K, and 0.001 K. Assume that T_c is of order 300 K.

6.17 Show that eqs. (6.159b) and (6.159c) are valid as soon as $\alpha = \alpha'$.

6.18 Express $(\phi_f - \phi_g)$ as a function of $(-\tau)$ using eq. (6.160) and the symmetry condition $\phi_f + \phi_g = 0$.

6.19 Invoking eq. (6.180), express the critical-point exponent δ as a function of $a_{\mathscr{H}}$. Note that δ is the exponent in the power law $M(\tau = 0, \mathscr{H}) \sim \mathscr{H}^\delta$.

6.20 Show that the van der Waals equation of state (6.46) can be written alternatively as

$$P_r = [1 + (\rho_r - 1)^3 M]T_r - A\rho_r^2(1 - T_r)$$

in which $\rho_r = v_r^{-1}$ and

$$M = (1 - \tfrac{1}{3}\rho_r)^{-1} \qquad \text{and} \qquad A = 3$$

In a recent paper Paynter [120] showed that if the parameters M and A are replaced by

$$M' = |\rho_r - 1|(1 - \tfrac{1}{4}\rho_r)^{-1} \qquad \text{and} \qquad A' = 7$$

the resulting $P_r = P_r(T_r, \rho_r)$ relation does a considerably better job of fitting the measured relations of actual fluids. Verify this conclusion by plotting the compressibility factor Z corresponding to Paynter's relation (plot Z versus P_r, as lines of constant temperature T_r). Compare your chart with the classical empirical chart of Fig. 6.20. Calculate also the critical point compressibility factor Z_c and the Riedel parameter α_c that correspond to Paynter's equation of state.

7

Chemically Reactive Systems

The present chapter places us at the crossroads between the elements of equilibrium thermodynamics reviewed in chapters 4 to 6 and the engineering applications that follow starting with chapter 8. Relative to the former, the present chapter recounts some of the extensions that have been built on Gibbs' equilibrium thermodynamics during the three or four decades around the turn of the century. These developments are what gave birth to the modern discipline of chemical thermodynamics. They are well illustrated by comparing, say, Gibbs' paper on the equilibrium of heterogeneous substances [1] with Lewis and Randall's [2] chemical thermodynamic treatise of 1923.

Relative to applied engineering topics such as the generation of power, the present chapter shows the thermodynamic origin of the many heat transfer inputs assumed and labeled "Q_H" in routine analyses of power cycles and plants. As such, this is a chapter of transition between the theoretical and the design-oriented parts of the present treatment. It is also a great opportunity for practicing one more time the general laws and theorems reviewed in chapters 1 to 3.

EQUILIBRIUM

Chemical Reactions

As a way of introducing some of the nomenclature needed in the analysis of combustion processes for power generation, let us examine the equilibrium of a mixture whose constituents can react chemically. The issue of chemical equilibrium formed the subject of most of chapters 4 and 6. The new feature of the mixture system considered now is the possibility that the interatomic bonds of the molecules of some or all the original chemical components of the system may break, leading in this way to the formation of a new generation of chemical components that coexist in the mixture. The "system" is in the upper-left corner of Fig. 7.1: it consists of a single-phase mixture surrounded by an impermeable boundary and maintained at constant temperature and pressure by contact with the reservoirs (T) and (P).

We refer to the original components whose molecules disintegrate as the *reactants* in the mixture (mole numbers, $N_{r1}, N_{r2}, \ldots$). The new components whose molecules are formed as the atoms of the original reactants change partners are recognized as the *products* in the mixture (mole numbers $N_{p1}, N_{p2}, \ldots$). These ideas are illustrated symbolically in Fig. 7.1: a *chemical reaction* is the process by which the mole numbers of the mixture change, making room at the same time for the mole numbers of the products.

The relationship between the depletion of the moles of reactants and the formation of moles of products is established on the basis of an atom accounting argument. For example, if the two reactants pictured in Fig. 7.1 are H_2 and CO_2, and if the two products are CO and H_2O, then it takes only one molecule of H_2 and one of CO_2 in order to form one molecule of CO and one of H_2O. We write the counting of the atoms in the form of a *chemical equation*:

$$H_2 + CO_2 \rightleftarrows CO + H_2O \tag{7.1}$$

where the two arrows indicate that at any *equilibrium* state of the type (ζ_e) shown in Fig. 7.1, the reaction proceeds steadily in both directions, so that at equilibrium, the net change in the mole numbers $(N_{H_2}, N_{CO_2}, N_{CO}, N_{H_2O})$ is zero. Equation (7.1) indicates that at equilibrium, the system contains certain amounts of H_2, CO_2, CO, and H_2O. It would be wrong to view eq. (7.1) as a statement that H_2 and CO_2 are converted completely into CO and H_2O.

Another chemical reaction example is

$$H_2 + \tfrac{1}{2}O_2 \rightleftarrows H_2O \tag{7.2}$$

Here we note first the $\frac{1}{2}$ coefficient of O_2 (i.e., it takes only one atom of oxygen to form one molecule of water), and, second, the fact that the number of products does not necessarily equal the number of reactants in the mixture.

A chemical equation can be expressed in general as

$$\nu_{r1}A_{r1} + \nu_{r2}A_{r2} + \cdots + \nu_{rm}A_{rm} \rightleftarrows \nu_{p1}A_{p1} + \nu_{p2}A_{p2} + \cdots + \nu_{pn}A_{pn} \tag{7.3}$$

where the A_rs and A_ps represent the chemical symbols of the m reactants and n products, and where the ν_rs and ν_ps are the respective *stoichiometric coefficients*.[†] These coefficients are the numbers that describe the propor-

---

[†]Stoichiometry means, literally, the counting of elements. The name is derived from the Greek words *stoicheion*, for "element" or "first principle," and *metron*, for "measure."

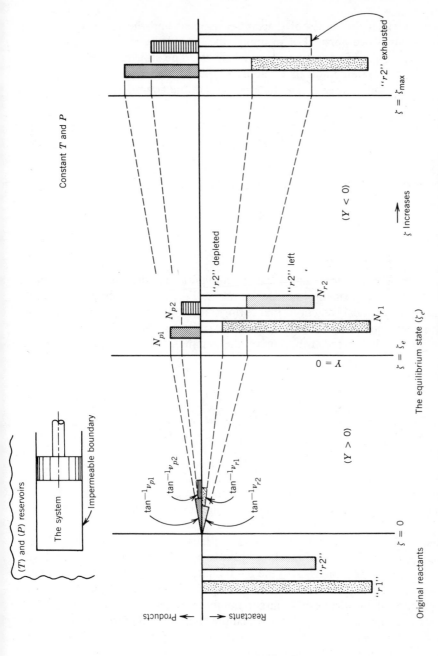

Figure 7.1 The advancement of the reaction (ζ) at constant temperature and pressure and the formation of an equilibrium mixture of reactants and products.

tions in which the moles of reactants disappear and the moles of products appear, and vice versa. For example, the stoichiometric coefficients of H_2 and O_2 in eq. (7.2) are 1 and $\frac{1}{2}$, respectively. The stoichiometric coefficient of a chemical component that belongs to the system but does not participate in the reaction is zero. A shorthand version of the general chemical equation (7.3) is

$$0 \rightleftarrows \sum_{i=1}^{m+n} \nu_i A_i \tag{7.4}$$

where the ν_i coefficients of the reactants are negative:

$$\nu_i = -\nu_{ri} \quad \text{and} \quad A_i = A_{ri} \quad (\text{if } i = 1, 2, \ldots, m)$$

$$\tag{7.4'}$$

$$\nu_i = \nu_{p(i-m)} \quad \text{and} \quad A_i = A_{p(i-m)} \quad (\text{if } i = m+1, m+2, \ldots, m+n)$$

The magnitude and sign of the stoichiometric coefficients are accounted for by the slopes of the lines drawn across Fig. 7.1. We see geometrically that the chemical reaction that brings the "original reactants" to the particular equilibrium state shown in the center of the figure is responsible for the following redistribution of the numbers of moles present in the mixture:

$$
\begin{array}{lll}
(r1) & N_{r1} \rightarrow N_{r1} - \nu_{r1}\zeta \\
(r2) & N_{r2} \rightarrow N_{r2} - \nu_{r2}\zeta \\
(p1) & 0 \rightarrow 0 + \nu_{p1}\zeta \\
(p2) & 0 \rightarrow 0 + \nu_{p2}\zeta
\end{array}
\tag{7.5}
$$

In this summary, ζ is the number of moles with which all the changes experienced by the mole numbers of reactants and products are proportional, the proportionality factors being the stoichiometric coefficients themselves. The number ζ represents the "degree of advancement" of the reaction or the "extent of reaction" [3].[†] In Fig. 7.1, the number of moles ζ is measured on the abscissa. In the $\zeta = 0$ limit, the reactants do not react and products are not produced. In the opposite limit, the equilibrium would be characterized by increasingly larger numbers of moles of products. There exists a ceiling value $\zeta = \zeta_{max}$ for which the changes in the original N_rs are so great that at least one of the reactants is no longer present in the mixture. In the example of Fig. 7.1, the reactant that would be exhausted first happens to be "r2."

[†]A more frequently used symbol for the extent of reaction is ξ. In the present treatment, the ξ symbol is reserved for the specific nonflow exergy (p. 223).

Affinity

At the particular equilibrium state represented by $\zeta = \zeta_e$, the composition of the mixture is represented by the $(m + n)$ mole numbers N_1, $N_2, \ldots, N_i, \ldots, N_{m+n}$. The composition at a sufficiently close neighboring equilibrium state $(\zeta_e + d\zeta)$ is represented by a new set of mole numbers $N_i + dN_i$, where, reasoning as in eq. (7.5),

$$dN_i = \nu_i \, d\zeta \qquad (i = 1, 2, \ldots, m + n) \tag{7.6}$$

Recalling that the mixture system under consideration is single-phase, the outlook of the fundamental relation of the system becomes apparent if we invoke eq. (4.10):

$$dU = T \, dS - P \, dV + \sum_{i=1}^{m+n} \mu_i \, dN_i \tag{7.7}$$

and combine it with eq. (7.6):

$$dU = T \, dS - P \, dV + \left(\sum_{i=1}^{m+n} \mu_i \nu_i \right) d\zeta \tag{7.8}$$

The sum in the parentheses serves as the driving force for the reaction in the same sense that T and P can drive heat transfer and work transfer. The negative of this sum is De Donder's *affinity* function [4]:

$$Y = - \sum_{i=1}^{m+n} \mu_i \nu_i \tag{7.9}$$

which is a function of state. Since it is a linear combination of chemical potentials, the affinity is itself an intensive property.

The internal-energy differential can be written then as

$$dU = T \, dS - P \, dV - Y \, d\zeta \tag{7.10u}$$

which suggests a fundamental relation of the type

$$U = U(S, V, \zeta) \tag{7.11u}$$

Table 4.3 can be consulted in order to see that the H, F, and G representations of the fundamental relation begin with the differentials

$$dH = T \, dS + V \, dP - Y \, d\zeta \tag{7.10h}$$

$$dF = - S \, dT - P \, dV - Y \, d\zeta \tag{7.10f}$$

$$dG = - S \, dT + V \, dP - Y \, d\zeta \tag{7.10g}$$

in other words, that the H, F, and G Legendre transforms of the fundamental relation have the following forms:

$$H = H(S, P, \zeta) \tag{7.11h}$$

$$F = F(T, V, \zeta) \tag{7.11f}$$

$$G = G(T, P, \zeta) \tag{7.11g}$$

Equations (7.10u)–(7.10g) justify the following alternative definitions for affinity:

$$Y = -\left(\frac{\partial U}{\partial \zeta}\right)_{S,V} \tag{7.12u}$$

$$Y = -\left(\frac{\partial H}{\partial \zeta}\right)_{S,P} \tag{7.12h}$$

$$Y = -\left(\frac{\partial F}{\partial \zeta}\right)_{T,V} \tag{7.12f}$$

$$Y = -\left(\frac{\partial G}{\partial \zeta}\right)_{T,P} \tag{7.12g}$$

Maxwell-type relations that connect the partial derivatives of Y to the partial derivatives of other properties can be written by following the example of eqs. (4.56) and Table 4.4. For example, from eqs. (7.10g) and (7.10f), we deduce two particularly useful temperature partial derivatives of the affinity function:

$$\left(\frac{\partial Y}{\partial T}\right)_{P,\zeta} = \left(\frac{\partial S}{\partial \zeta}\right)_{T,P} \tag{7.13}$$

$$\left(\frac{\partial Y}{\partial T}\right)_{V,\zeta} = \left(\frac{\partial S}{\partial \zeta}\right)_{T,V} \tag{7.14}$$

Consider now the special case where the temperature and pressure of the system discussed until now are maintained constant. In the opening section to chapter 6, we learned that at constant T and P, the *equilibrium state* of the system is characterized by the Gibbs free energy minimum, eq. (6.20). The G minimum conclusion consists of two analytical statements, the first of which is

$$dG = 0 \quad \text{(constant } T \text{ and } P\text{)} \tag{7.15}$$

The second statement forms the subject of the next subsection. Comparing eq. (7.15) with eq. (7.10g), we conclude that at equilibrium, the affinity is zero; therefore,

$$Y = -\sum_{i=1}^{m+n} \mu_i \nu_i = 0 \qquad \text{(equilibrium)} \qquad (7.16)$$

Note now that if the state represented by (ζ_e) in the center of Fig. 7.1 is an equilibrium state, then the value of Y at that state is zero.

Just like the G function of eq. (7.11g), the first derivative called "Y" in eq. (7.12g) is generally a function of T, P, and ζ also. The manner in which the affinity Y depends on temperature can be illustrated by holding T and P constant in eq. (7.10h):

$$\left(\frac{\partial H}{\partial \zeta}\right)_{T,P} = T\left(\frac{\partial S}{\partial \zeta}\right)_{T,P} - Y \qquad (7.17)$$

and invoking then eq. (7.13):

$$\left(\frac{\partial H}{\partial \zeta}\right)_{T,P} = T\left(\frac{\partial Y}{\partial T}\right)_{P,\zeta} - Y \qquad (7.18)$$

The quantity on the left side of these last two equations is the *heat of reaction*, that is, the heat transfer administered to the system as the reaction advances from (ζ_e) to $(\zeta_e + d\zeta)$ at constant temperature and pressure. If $(\partial H/\partial \zeta)_{T,P}$ is positive, the reaction is said to be *endothermic*, meaning that during the shift from (ζ_e) to $(\zeta_e + d\zeta)$, the system absorbs heat.[†] Conversely, negative $(\partial H/\partial \zeta)_{T,P}$ values indicate *exothermic* reactions that liberate the heat transfer received ultimately by the temperature reservoir (T). In order to avoid confusing the quantity $(\partial H/\partial \zeta)_{T,P}$ with other heat transfer interactions that can be experienced by a reactive system (e.g., Problem 7.1), it may make more sense to call it "heat of reaction at constant pressure" or "enthalpy of reaction" [5].

Dividing eq. (7.18) by T^2 leads to

$$\frac{1}{T^2}\left(\frac{\partial H}{\partial \zeta}\right)_{T,P} = \frac{\partial}{\partial T}\left(\frac{Y}{T}\right)_{P,\zeta} \qquad (7.19)$$

which can be integrated at constant P and ζ in order to chart the variation of affinity with temperature [3]:

$$\frac{Y(T_2, P, \zeta)}{T_2} - \frac{Y(T_1, P, \zeta)}{T_1} = \int_{T_1}^{T_2} \frac{1}{T^2}\left(\frac{\partial H}{\partial \zeta}\right)_{T,P} dT \qquad (7.20)$$

Conversely, if the dependence between Y and T at constant P and ζ is known sufficiently accurately in the vicinity of the equilibrium state (T, P, ζ), then we can calculate the temperature slope $(\partial Y/\partial T)_{P,\zeta}$ and

---

[†]The prefixes "endo" and "exo" come from the Greek words *endon* (within) and *exo* (outside).

substitute it (and $Y = 0$ for "equilibrium") in eq. (7.18) in order to deduce the heat of reaction. In this way, the heat of reaction can be calculated without having to perform a calorimetric measurement.

The Le Chatelier–Braun Principle

The second part of the analytical statement that at constant T and P, the Gibbs free energy $G(T, P, \zeta)$ is minimum is

$$\left(\frac{\partial^2 G}{\partial \zeta^2}\right)_{T,P} > 0 \tag{7.21}$$

which, according to eq. (7.12g), means

$$\left(\frac{\partial Y}{\partial \zeta}\right)_{T,P} < 0 \tag{7.22}$$

in other words,

$$\left(\frac{\partial}{\partial \zeta} \sum_{i=1}^{m+n} \mu_i \nu_i\right)_{T,P} > 0 \tag{7.23}$$

This strict inequality adds itself to the class of thermodynamic *stability* conditions identified in eqs. (6.28), (6.36), and (6.43). It allows the thermodynamicist to predict the direction in which the equilibrium state (ζ) shifts on the abscissa of Fig. 7.1 if, for example, the temperature of the system changes isobarically from T to $T + dT$. Such a change occurs also at constant Y because, as noted in eq. (7.16), the affinity of all equilibrium states is zero. In order to see this more clearly, note that Y is in general a function of the type[†]

$$Y = Y(T, P, \zeta) \tag{7.24}$$

This means that the equilibrium states of the system reside on the three-dimensional surface:

$$Y(T, P, \zeta) = 0 \quad \text{(equilibrium)} \tag{7.25}$$

or that the extent of the reaction at equilibrium depends on T and P, while the affinity Y remains constant (zero), Fig. 7.2. The inequality (7.22) states that the affinity decreases as ζ increases at constant T and P: this means that immediately above the equilibrium surface, the Y values are negative, and

---

[†]The equation $Y = Y(T, P, \zeta)$ is one of the three equations of state associated with the fundamental relation $G = G(T, P, \zeta)$.

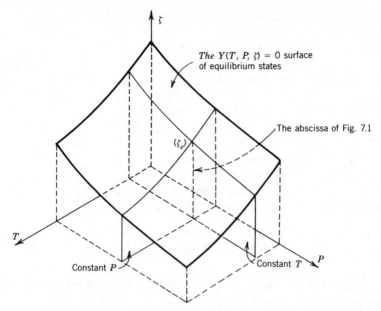

Figure 7.2 The locus of equilibrium states in the (T, P, ζ) space for a reactive system that has two features: (1) its heat of reaction is negative, and (2) its volume increases as ζ increases at constant T and P.

that immediately below the surface, Y is positive. The same point is made in Fig. 7.1, which shows that to the left of the equilibrium state (ζ_e), the affinity is positive, and that to the right of (ζ_e), it is negative.

The effect of a temperature increase on the extent of the reaction is measured by the partial derivative $(\partial \zeta / \partial T)_{P,Y}$. Invoking the cyclical relation among ζ, T, and Y [6],

$$\left(\frac{\partial \zeta}{\partial T} \right)_{P,Y} = - \left(\frac{\partial Y}{\partial T} \right)_{P,\zeta} \bigg/ \left(\frac{\partial Y}{\partial \zeta} \right)_{P,T} \tag{7.26}$$

and noting that at equilibrium $(Y = 0)$, eq. (7.18) reduces to

$$\left(\frac{\partial Y}{\partial T} \right)_{P,\zeta} = \frac{1}{T} \left(\frac{\partial H}{\partial \zeta} \right)_{T,P} \tag{7.27}$$

we combine eqs. (7.26) and (7.27) to conclude that

$$\left(\frac{\partial \zeta}{\partial T} \right)_{P,Y} = \frac{1}{T} \left(\frac{\partial H}{\partial \zeta} \right)_{T,P} \left[- \left(\frac{\partial Y}{\partial \zeta} \right)_{T,P} \right]^{-1} \tag{7.28}$$

The quantity in brackets is always positive because of the stability condition

(7.22). It follows that the sign of derivative of interest, $(\partial\zeta/\partial T)_{P,Y}$, is always the same as the sign of the heat of reaction $(\partial H/\partial\zeta)_{T,P}$. If the reaction is endothermic, the shift $d\zeta$ triggered by a positive dT (at constant P and Y) is positive also. In such a case, the equilibrium state (ζ_e) singled out in Fig. 7.1 would be sliding to the right. In the case of an exothermic reaction, the changes in ζ and T would have opposite signs: raising the temperature of the reactive system of Fig. 7.1 would force the equilibrium state to shift to the left.

The same ideas come into play as we examine the response of the equilibrium to an isothermal pressure change dP. Looking at eq. (7.25), we note that both T and Y remain constant as the pressure and the extent of the reaction change. We are interested in the sign of $(\partial\zeta/\partial P)_{T,Y}$, which, according to the cyclical relation, is also equal to [6]

$$\left(\frac{\partial\zeta}{\partial P}\right)_{T,Y} = -\left(\frac{\partial Y}{\partial P}\right)_{T,\zeta}\bigg/\left(\frac{\partial Y}{\partial\zeta}\right)_{T,P} \qquad (7.29)$$

The derivative $(\partial Y/\partial P)_{T,\zeta}$ can be replaced by noting the Maxwell relation that corresponds to $dT = 0$ in eq. (7.10g):

$$-\left(\frac{\partial Y}{\partial P}\right)_{T,\zeta} = \left(\frac{\partial V}{\partial\zeta}\right)_{T,P} \qquad (7.30)$$

Equation (7.29) assumes the new form

$$\left(\frac{\partial\zeta}{\partial P}\right)_{T,Y} = \left(\frac{\partial V}{\partial\zeta}\right)_{T,P}\left[\left(\frac{\partial Y}{\partial\zeta}\right)_{T,P}\right]^{-1} \qquad (7.31)$$

where the factor in brackets is always negative, eq. (7.22).

In conclusion, the sign of the derivative that characterizes the response of the reaction, $(\partial\zeta/\partial P)_{T,Y}$, is always the opposite of the sign of $(\partial V/\partial\zeta)_{T,P}$. For example, when the pressure increases, the change in ζ is such that the eventual change in the system volume is negative. Conversely, a slight depressurization of the reactive system shifts the extent of the reaction ζ in the direction of larger volumes.

The two trends discussed in this subsection are reviewed graphically in Fig. 7.2. The figure shows first the surface of equilibrium states, eq. (7.25), and, in it, the particular equilibrium state that was labeled (ζ_e) in Fig. 7.1. The temperature and pressure that were held constant in Fig. 7.1 are represented now by the two vertical planes whose line of intersection[†] pierces the $Y = 0$ surface at the point (ζ_e). In view of eq. (7.28) and the discussion that followed it, we note that the zero-affinity surface of Fig. 7.2 corresponds to an exothermic reaction [note the negative sign of

---

[†]The story told in connection with Fig. 7.1 takes place precisely along this line of intersection.

$(\partial \zeta / \partial T)_{P,Y}$]. On the other hand, eq. (7.31) guarantees that the $Y = 0$ surface in Fig. 7.2 corresponds to a particular reactive system whose volume increases as the reaction advances at constant temperature and pressure, $(\partial V / \partial \zeta)_{T,P} > 0$.

The behavior anticipated on the basis of the sign of derivatives such as $(\partial \zeta / \partial T)_{P,Y}$ and $(\partial \zeta / \partial P)_{T,Y}$ is part of an important "stability" chapter in chemical thermodynamics, which in honor of its pioneers is recognized widely as the *Le Chatelier–Braun principle*. Additional reading on this subject can be found in advanced chemical thermodynamics treatises (e.g., Callen [6] and Modell and Reid [7]). I used the opportunity afforded by the present treatment to construct something that is usually not found in the physicochemical treatments, namely, a chronological bibliography of this principle [8–15] and the controversy that it fueled during the first three decades of this century [16].

Ideal Gas Mixtures

The main conclusion of the preceding two subsections is that at constant T and P, the system settles in an equilibrium state characterized by a special value of the extent of the reaction, ζ_e. Figure 7.1 shows that the job of pinpointing this equilibrium position is the same as figuring out the equilibrium chemical composition of the mixture of reactants and products. In this subsection, we illustrate in concrete terms the calculation of the equilibrium chemical composition by focusing on reactive mixtures that fit the ideal gas description given on pages 198–202 of chapter 4. This focus is demanded also by the important role played by the ideal gas model in the analysis of the combustion processes considered later in this chapter.

In the case of an ideal gas mixture, the zero-affinity condition for equilibrium (7.16) can be transformed into a relation between T, P, and the mole fractions x_i expected at equilibrium. In order to do this, we must first determine the chemical potentials μ_i: from eq. (4.139), we have

$$\mu_i(T, P, x_i) = \mu_i^{(x_i=1)}(T, P) + \bar{R}T \ln x_i \qquad (7.32)$$

in which an expression for $\mu_i^{(x_i=1)}$ can be obtained by integrating the single-component eq. (4.141) at constant temperature:

$$\mu_i^{(x_i=1)}(T, P) = \mu_i^{(x_i=1)}(T, P_0) + \bar{R}T \ln \frac{P}{P_0} \qquad (7.33)$$

In this expression, $\mu_i^{(x_i=1)}$ was called "μ" on the left side of eq. (4.141), whereas P_0 is a reference pressure level that is usually set at $P_0 = 1$ atm. The superscript $(x_i = 1)$ stresses the fact that eq. (7.33) refers to the ith species as a single-component system. This superscript will be abandoned shortly.

Taken together, eqs. (7.32) and (7.33) state that

$$\mu_i(T, P, x_i) = \mu_i^{(x_i=1)}(T, P_0) + \bar{R}T \ln \frac{P}{P_0} + \bar{R}T \ln x_i \qquad (7.34)$$

where $\mu_i^{(x_i=1)}(T, P_0)$ is a function of temperature only. The equilibrium condition (7.16) is now converted into

$$-\sum_{i=1}^{m+n} \nu_i \mu_i^{(x_i=1)}(T, P_0) - \bar{R}T \ln \frac{P}{P_0} \sum_{i=1}^{m+n} \nu_i - \bar{R}T \sum_{i=1}^{m+n} \nu_i \ln x_i = 0 \qquad (7.35)$$

where, again, the first sum on the left side is purely a function of temperature. This sum can be written more succinctly as

$$-\sum_{i=1}^{m+n} \nu_i \mu_i^{(x_i=1)}(T, P_0) = \bar{R}T \ln K_P(T) \qquad (7.36)$$

that is, in terms of the new temperature function $K_P(T)$, called the *equilibrium constant* of the reactive system. This quantity is truly a "constant" only if the temperature T is maintained fixed. Using the equilibrium constant notation, we can finally write eq. (7.35) as

$$K_P(T) = \left(\frac{P}{P_0}\right)^b \prod_{i=1}^{m+n} x_i^{\nu_i} \qquad (7.37)$$

or as

$$K_P(T) = P_0^{-b} \prod_{i=1}^{m+n} P_i^{\nu_i} \qquad (7.38)$$

where the dimensionless exponent b is the algebraic sum of all the stoichiometric coefficients:

$$b = \sum_{i=1}^{m+n} \nu_i \qquad (7.39)$$

Despite formulas such as eq. (7.37), whose right side might suggest a dependence on pressure, the equilibrium constant K_P depends solely on temperature, as indicated in the definition (7.36). According to the same definition, K_P is dimensionless: this feature is also evident in eq. (7.37). Figure 7.3 shows that as a good engineering approximation, the logarithm of K_P is a linear function of the inverse absolute temperature:

$$\ln K_P \cong c_1 + \frac{c_2}{T} \qquad (7.40)$$

This relationship and the special meaning of the constant slope c_2 emerge from eq. (7.19), which can be written sequentially as

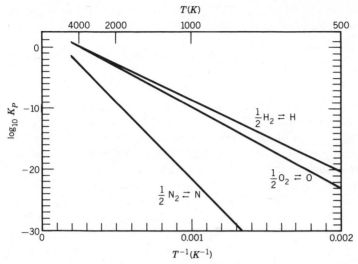

Figure 7.3 The approximately linear relationship between $\log_{10} K_P$ and T^{-1} for some of the equilibria documented in Table 7.1.

$$\frac{1}{T^2}\left(\frac{\partial H}{\partial \zeta}\right)_{T,P} = \frac{\partial}{\partial T}\left(\frac{Y}{T}\right)_{P,\zeta}$$

$$= \frac{\partial}{\partial T}\left(-\frac{1}{T}\sum_{i=1}^{m+n}\nu_i\mu_i\right)_{P,\zeta}$$

$$= \frac{\partial}{\partial T}\left\{-\frac{1}{T}\left[\sum_{i=1}^{m+n}\nu_i\mu_i^{(x_i=1)} + \bar{R}T\ln\left(\frac{P}{P_0}\right)^b + \bar{R}T\sum_{i=1}^{m+n}\nu_i\ln x_i\right]\right\}_{P,\zeta}$$

$$= \frac{\partial}{\partial T}\left[\bar{R}\ln K_P(T) - \bar{R}\ln\left(\frac{P}{P_0}\right)^b - \bar{R}\sum_{i=1}^{m+n}\nu_i\ln x_i\right]_{P,\zeta} \qquad (7.41)$$

The first and second steps outlined above consist of invoking eqs. (7.9) and (7.35), respectively. Note further that the second and third terms in the last set of brackets are constant because the $\partial/\partial T$ derivative is taken at constant P and ζ; therefore,

$$\frac{1}{T^2}\left(\frac{\partial H}{\partial \zeta}\right)_{T,P} = \bar{R}\frac{d}{dT}(\ln K_P) \qquad (7.42)$$

Equation (7.42) is known as the *van't Hoff*[†] *relation*. It shows that the

---

[†]Jacobus Hendricus van't Hoff (1852–1911), Dutch professor of physical chemistry at the Universities of Amsterdam (1878–1896) and Berlin (1896–1911), received the first Nobel Prize awarded in the field of chemistry (1901).

slope of the $\ln K_P$ versus $1/T$ curve is a dimensionless measure of the negative of the heat of reaction:

$$-\frac{1}{\bar{R}}\left(\frac{\partial H}{\partial \zeta}\right)_{T,P} = \frac{d(\ln K_P)}{d(T^{-1})} \tag{7.43}$$

The analytical form of this relation is analogous to the Clapeyron relation developed in Problem 6.2 for a two-phase single-component mixture in which the gaseous phase behaves as an ideal gas. Since the heat of reaction is essentially constant over the temperature intervals that are skipped in tabulations such as Table 7.1, the van't Hoff relation is a precious tool in the effort of finding the needed $\ln K_P$ value by interpolating between the two closest values that are available in the table (Problem 7.3).

The chief message of the preceding discussion and Table 7.1 is that there exists a one-to-one relationship between the value of the equilibrium

TABLE 7.1 The Values of $\text{Log}_{10} K_P(T)$ for Eight Examples of Reactive Ideal Gas

T[K]	$\frac{1}{2}H_2 \rightleftarrows H$	$\frac{1}{2}O_2 \rightleftarrows O$	$\frac{1}{2}N_2 \rightleftarrows N$	$H_2O \rightleftarrows H_2 + \frac{1}{2}O_2$
100	−110.954	−126.73	−243.583	−123.6
298	−35.612	−40.604	−79.8	−40.048
500	−20.158	−22.94	−46.336	−22.886
1000	−8.646	−9.807	−21.528	−10.062
1200	−6.707	−7.604	−17.377	−7.899
1400	−5.315	−6.027	−14.406	−6.347
1600	−4.266	−4.842	−12.175	−5.18
1800	−3.448	−3.918	−10.437	−4.27
2000	−2.790	−3.178	−9.046	−3.54
2200	−2.251	−2.571	−7.905	−2.942
2400	−1.800	−2.065	−6.954	−2.443
2600	−1.417	−1.636	−6.149	−2.021
2800	−1.089	−1.268	−5.457	−1.658
3000	−0.803	−0.949	−4.858	−1.343
3500	−0.231	−0.31	−3.656	−0.712
4000	+0.201	+0.17	−2.752	−0.238
4500	+0.537	+0.543	−2.047	+0.133
5000	+0.806	+0.843	−1.481	+0.43
5500	+1.027	+1.088	−1.016	+0.675
6000	+1.211	+1.292	−0.625	+0.88

[a]The equilibrium constant is defined as

$$K_P(T) = \frac{x_3^{\nu_3}x_4^{\nu_4}}{x_1^{\nu_1}x_2^{\nu_2}}\left(\frac{P}{P_0}\right)^{-\nu_1-\nu_2+\nu_3+\nu_4}$$

in which $P_0 = 1$ atm. Each symbol is identified by comparing the chemical equation:

constant and the composition of the reactive mixture at equilibrium (known T and P). That the knowledge of the equilibrium composition pinpoints also the K_P value is indicated by eq. (7.37). The reverse process—the determination of the equilibrium composition when K_P is known—is illustrated numerically in Example 7.1.

The reading of the $\log_{10} K_P(T)$ information [17] assembled in Table 7.1 should be done with care, because similar information is available in somewhat different forms in other thermodynamics texts and handbooks. To begin with, the numerical values refer to $\log_{10} K_P$ and not to the natural logarithm that is part of the original definition of the equilibrium constant, eq. (7.36). Secondly, the listed numerical values are tied intimately to the way in which the respective chemical equation appears at the head of the table. If, for example, the equation $\frac{1}{2}H_2 \rightleftarrows H$ is rewritten as $H_2 \rightleftarrows 2H$, then the values listed in the first column of Table 7.1 would have to be multiplied by 2. On the other hand, if we choose to write the equation as $H \rightleftarrows \frac{1}{2}H_2$

Mixtures in Equilibrium [17][a]

$H_2O \rightleftarrows OH + \frac{1}{2}H_2$	$\frac{1}{2}O_2 + \frac{1}{2}N_2 \rightleftarrows NO$	$CO_2 \rightleftarrows CO + \frac{1}{2}O_2$	$CO_2 + H_2 \rightleftarrows CO + H_2O$
-143.8	-46.453	-143.2	-19.6
-46.137	-15.171	-45.066	-5.018
-26.182	-8.783	-25.025	-2.139
-11.309	-4.062	-10.221	-0.159
-8.811	-3.275	-7.764	$+0.135$
-7.021	-2.712	-6.014	$+0.333$
-5.677	-2.290	-4.706	$+0.474$
-4.631	-1.962	-3.693	$+0.577$
-3.793	-1.699	-2.884	$+0.656$
-3.107	-1.484	-2.226	$+0.716$
-2.535	-1.305	-1.679	$+0.764$
-2.052	-1.154	-1.219	$+0.802$
-1.637	-1.025	-0.825	$+0.833$
-1.278	-0.913	-0.485	$+0.858$
-0.559	-0.69	$+0.19$	$+0.902$
-0.022	-0.524	$+0.692$	$+0.930$
$+0.397$	-0.397	$+1.079$	$+0.946$
$+0.731$	-0.296	$+1.386$	$+0.956$
$+1.004$	-0.214	$+1.635$	$+0.960$
$+1.232$	-0.147	$+1.841$	$+0.961$

$$\nu_1 A_1 + \nu_2 A_2 \rightleftarrows \nu_3 A_3 + \nu_4 A_4$$

with the particular equation listed at the head of each column of numerical values.

instead of $\frac{1}{2}H_2 \rightleftarrows H$, the sign of the numbers listed in the first column would have to be reversed.

The reactive mixture equilibria covered in Table 7.1 refer to some of the simplest chemical reactions that are encountered in combustion-engineering calculations. The equilibrium constant of systems with considerably more complicated chemical reactions can be deduced from Table 7.1 by recognizing a special property of the logarithm of the equilibrium constant, eq. (7.36). Suppose that the more complicated chemical equation whose $\log_{10} K_P$s are not listed in Table 7.1 has the form given in eq. (7.4), namely,

$$0 \rightleftarrows \sum_{i=1}^{m+n} \nu_i A_i \tag{7.44}$$

Assume further that

$$0 \rightleftarrows \sum_{i=1}^{m_1+n_1} \nu_i^{(1)} A_i \tag{7.45}$$

$$0 \rightleftarrows \sum_{i=1}^{m_2+n_2} \nu_i^{(2)} A_i \tag{7.46}$$

are two simpler equations whose $\log_{10} K_P^{(1)}$ and $\log_{10} K_P^{(2)}$ values are presented in the table. Then, if the original equation (7.44) can be reconstructed as a linear combination of eqs. (7.45) and (7.46), i.e., if

$$\sum_{i=1}^{m+n} \nu_i A_i \equiv c^{(1)} \sum_{i=1}^{m_1+n_1} \nu_i^{(1)} A_i + c^{(2)} \sum_{i=1}^{m_2+n_2} \nu_i^{(2)} A_i \tag{7.47}$$

where $c^{(1)}$ and $c^{(2)}$ are two numbers, then the analytical form of the definition (7.36) assures us that the same linear relationship exists between the logarithms of the respective equilibrium constants:

$$\ln K_P(T) = c^{(1)} \ln K_P^{(1)}(T) + c^{(2)} \ln K_P^{(2)}(T) \tag{7.48}$$

$$\log_{10} K_P(T) = c^{(1)} \log_{10} K_P^{(1)}(T) + c^{(2)} \log_{10} K_P^{(2)}(T) \tag{7.48'}$$

The problem reduces to finding the constants $c^{(1)}$ and $c^{(2)}$ that allow the two-reaction superposition shown in eq. (7.47). Of course, more complicated relations may require more than two reactions to effect the superposition (7.47), in which case the $\ln K_P(T)$ expression (7.48) would have more than two terms on the right side. The procedure is illustrated in Example 7.2.

Example 7.1. Consider an equilibrium mixture containing H_2, O_2, and their oxide (H_2O), and let us determine the composition of this mixture at 3000 K and atmospheric pressure. The equilibrium (reversible) chemical equation is the fourth listed in Table 7.1:

$$H_2O \rightleftharpoons H_2 + \tfrac{1}{2}O_2 \tag{a}$$

At $T = 3000$ K, we read $\log_{10} K_P = -1.343$, which means that

$$K_P = 0.0454 \tag{b}$$

From eq. (7.37) or the more explicit alternative listed under Table 7.1, we learn that the equilibrium mole fractions are related through

$$0.0454 = \frac{x_{H_2}^{\nu_{H_2}} x_{O_2}^{\nu_{O_2}}}{x_{H_2O}^{\nu_{H_2O}}} \tag{c}$$

where $\nu_{H_2O} = 1$, $\nu_{H_2} = 1$, and $\nu_{O_2} = \tfrac{1}{2}$.

The question, next, is how can this single relation—eq. (c)—determine uniquely the three unknowns $(x_{H_2O}, x_{H_2}, x_{O_2})$? In order to proceed, we seek help from Fig. 7.1, which shows that the proportions in which the reactants and products coexist at equilibrium depend solely on the degree of the reaction, ζ. The reasoning goes as follows: if we start with 1 mole of H_2O and no hydrogen and oxygen, then at equilibrium, the H_2O inventory decreases to $(1 - \zeta)$ while the H_2 and O_2 inventories reach ζ and $\tfrac{1}{2}\zeta$, respectively. The total number of moles that coexist at equilibrium is $(1 - \zeta) + \zeta + \tfrac{1}{2}\zeta = 1 + \tfrac{1}{2}\zeta$; therefore, the sought mole fractions can be written as

$$x_{H_2O} = \frac{1 - \zeta}{1 + \zeta/2} \qquad x_{H_2} = \frac{\zeta}{1 + \zeta/2} \qquad x_{O_2} = \frac{\zeta/2}{1 + \zeta/2} \tag{d}$$

The equilibrium condition (c) becomes

$$0.0454 = \frac{\zeta}{1 - \zeta}\left(\frac{\zeta}{2 + \zeta}\right)^{1/2} \tag{e}$$

whose solution is $\zeta = 0.1476$ and, finally,

$$x_{H_2O} = 0.794 \qquad x_{H_2} = 0.137 \qquad x_{O_2} = 0.069 \tag{f}$$

Example 7.2. Consider the problem of determining the equilibrium constant for the mixture of CO_2, H_2, CO, and H_2O, whose reversible chemical equation reads

$$CO_2 + H_2 \rightleftharpoons CO + H_2O \tag{a}$$

Assuming that this case is not covered by Table 7.1 (it is), let us determine its equilibrium constant based on the simpler cases that are very likely to be present in the table. We note that the constituents in question—CO_2, H_2, CO, and H_2O—appear also in two other equations of Table 7.1:

$$CO_2 \rightleftharpoons CO + \tfrac{1}{2}O_2 \tag{b}$$

$$H_2O \rightleftharpoons H_2 + \tfrac{1}{2}O_2 \tag{c}$$

It is not difficult to see that the original equation can be regenerated by subtracting eq. (c) from eq. (b). In terms of the notation used in eq. (7.47), this amounts to the observation that

$$\sum_{i=1}^{m+n} \nu_i A_i = (-CO_2 - H_2 + CO + H_2O)_{(a)}$$

$$= (+1)(-CO_2 + CO + \tfrac{1}{2}O_2)_{(b)} + (-1)(-H_2O + H_2 + \tfrac{1}{2}O_2)_{(c)} \qquad (d)$$

in other words, that $c^{(1)} = 1$, $c^{(2)} = -1$, and

$$[\ln K_P(T)]_{(a)} = [\ln K_P(T)]_{(b)} - [\ln K_P(T)]_{(c)} \qquad (e)$$

The equivalent relation between the logarithms in the base 10 was used to construct the last column of Table 7.1.

IRREVERSIBLE REACTIONS

The focus of the first section of this chapter has been on the equilibrium of a reactive mixture and, in particular, on the chemical composition that prevails at equilibrium. The word "equilibrium" does not mean that the chemical reaction stops, rather, it means that the reaction proceeds in both directions at the same rate. The two-way character of the equilibrium reaction is indicated by the double arrows used in eqs. (7.3) and (7.4). Proceeding from left to right, reactants are depleted in exactly the same proportions and at exactly the same rate at which they are being replenished by the reverse reaction. This is one reason for regarding the equilibrium reactants discussed until now as *reversible* reactions. The more basic reason for using this terminology is that at equilibrium, the two-way reaction generates no entropy: to this aspect of chemically reactive systems, we return in the closing segment of this chapter.

The main problem that concerns us in engineering-thermodynamics practice and throughout much of the present treatment is the issue of thermodynamic irreversibility and, more precisely, the business of identifying irreversibilities and designing the means by which they might be circumvented. It turns out that in actual engineering installations, the processes we call "chemical reactions" are highly irreversible. Those reactions must obviously be different than the equilibrium reactions discussed until now. In order to see the difference as nothing more than a special manifestation of the general principles developed in chapters 3 to 6, consider again the isothermal and isobaric system defined in Fig. 7.1. Recall that at constant T and P, the system settles in the equilibrium state $\zeta = \zeta_e$ because the system's Gibbs-free-energy inventory at that state is the lowest. We can imagine a process $(0) \to (\zeta_e)$ in which the role of initial state is played by the position labeled $\zeta = 0$ and where the final state is the equilibrium position $\zeta = \zeta_e$. In state (0), the system contains only the original reactants, which are being separated by the required impermeable internal partition. If, when the partition is removed, the system shifts from (0) to (ζ_e), it means that the initial G inventory must have been larger than in the final (equilibrium) state; in other words,

$$G_{(0)} > G_{(\zeta_e)} \qquad \text{(constant } T \text{ and } P) \qquad (7.49)$$

Likewise, if the initial state is represented by the combination of mole numbers in line with $\zeta = \zeta_{\max}$ in Fig. 7.1, then, in accordance with the minimum-G principle, the shift from the initial state† $(\zeta_{\max})$ to the equilibrium state (ζ_e) takes place because this change points in the direction of lower Gibbs free energies:

$$G_{(\zeta_e)} < G_{(\zeta_{\max})} \qquad \text{(constant } T \text{ and } P) \qquad (7.50)$$

The constant-T and -P picture that emerges is one where chemical reactions occur "one way" and always in the direction of lower values for the total G inventory of the system. Consider next the process in Fig. 7.4. In the initial state, the system is made up of two parts, (r) and (e), which are separated by an impermeable (otherwise, flexible and diathermal) membrane. The (r) part contains only reactants $(A_{r1}, A_{r2}, \ldots)$: the respective numbers of moles in this conglomerate are the same as the stoichiometric coefficients shown on the left side of eq. (7.3), namely, $\nu_{r1}, \nu_{r2}, \ldots$. The larger part of the system—the subsystem (e)—contains an equilibrium mixture of all the reactants and the products that can coexist at T and P.

In the final state, the system contains again two subsystems, first, an equilibrium mixture of reactants and products that is identical to the

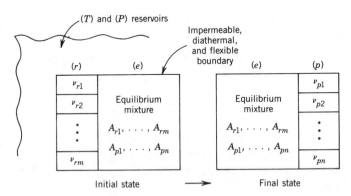

Figure 7.4 Isothermal and isobaric process for the conversion of stoichiometric quantities of reactants into products.

---

†In this state, too, the constituents would have to be separated by impermeable internal partitions.

subsystem (e) described above, and, second, a part (p) made up of only products $(A_{p1}, A_{p2}, \ldots)$. The numbers of moles that are found in (p) are precisely the numbers $\nu_{p1}, \nu_{p2}, \ldots$ that are recommended by stoichiometric calculations, eq. (7.3). This means that during the process described in Fig. 7.4, the boundary of the greater system—the aggregate system drawn as $(r) + (e)$ or $(e) + (p)$—is impermeable to the flow of matter. The change that occurs is that the subsystem of reactants, (r), is replaced entirely by an equivalent batch of products, (p), in the manner indicated by the left-to-right arrow in eq. (7.3):

$$\nu_{r1}A_{r1} + \nu_{r2}A_{r2} + \cdots + \nu_{rm}A_{rm} \rightarrow \nu_{p1}A_{p1} + \nu_{p2}A_{p2} + \cdots + \nu_{pn}A_{pn} \quad (7.51)$$

The process pictured in Fig. 7.4 is possible only if the G inventory of the system decreases in the direction of the arrow:

$$G_r + G_e > G_e + G_p \quad (7.52)$$

which finally means that

$$G_r > G_p \quad \text{(constant } T \text{ and } P) \quad (7.53)$$

The one-way reaction (7.51) evolves in the direction of the arrow because the Gibbs free energy summed over the left side of the equation exceeds the amount calculated for the right side.

The next step in this argument is to assume that the scenario of Fig. 7.4 repeats itself steadily in time. During each unit time interval, the now "open" system admits certain quantities of reactants, in such proportions as are indicated by their stoichiometric coefficients:

$$\dot{n}_{r1} = \nu_{r1}\dot{n} \qquad \dot{n}_{r2} = \nu_{r2}\dot{n} , \qquad \ldots \quad (7.54)$$

At the same time, the isobaric and isothermal flow system (Fig. 7.5) releases a corresponding mixture of products:

$$\dot{n}_{p1} = \nu_{p1}\dot{n} \qquad \dot{n}_{p2} = \nu_{p2}\dot{n} , \qquad \ldots \quad (7.55)$$

where $\dot{n}$ is a number of moles per second, whose job is to account for the overall size and speed of the flow installation. The inequality that in eq. (7.53) referred to a one-shot process continues to apply here, this time on a per-unit-time basis:

$$\dot{G}_r > \dot{G}_p \quad \text{(constant } T \text{ and } P) \quad (7.55')$$

Taking into account the composition of the two streams, we can rewrite this conclusion as

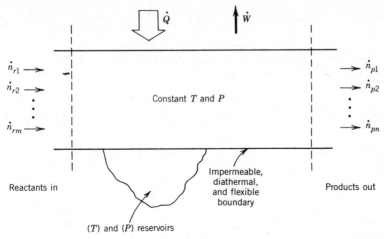

Figure 7.5 Steady-flow apparatus for carrying out a reaction at constant temperature and pressure.

$$\sum_{i=1}^{m} \dot{n}_{ri}\bar{g}_{ri} > \sum_{i=1}^{n} \dot{n}_{pi}\bar{g}_{pi} \tag{7.56}$$

and, in view of eqs. (7.54) and (7.55),

$$\sum_{i=1}^{m} \nu_{ri}\bar{g}_{ri} > \sum_{i=1}^{n} \nu_{pi}\bar{g}_{pi} \quad \text{(constant } T \text{ and } P\text{)} \tag{7.57}$$

There is an important connection between the Gibbs free-energy decrease signaled by eq. (7.57) and the concept of equilibrium constant, eq. (7.36). In terms of the shorter notation defined in eqs. (7.4′), the inequality (7.57) reads

$$0 > \sum_{i=1}^{m+n} \nu_i\bar{g}_i(T, P) \tag{7.58}$$

or

$$0 > \sum_{i=1}^{m+n} \nu_i\mu_i(T, P) \tag{7.59}$$

where both $\bar{g}_i(T, P)$ and $\mu_i(T, P)$ refer to the ith component as a single-component substance [this feature was stressed earlier by the superscript $(x_i = 1)$, which is now abandoned]. Note further that in the special case when P is set at the reference level P_0, the sum that appears on the right side of eq. (7.59) is the same as the sum seen in eq. (7.36). In conclusion, we can write

$$0 < - \sum_{i=1}^{m+n} \nu_i \bar{g}_i(T, P_0) = \bar{R}T \ln K_P(T) \tag{7.60}$$

Or, introducing the notation ΔG for the net *increase* in G when proceeding in the direction of the reaction,

$$\Delta G(T, P_0) = \sum_{i=1}^{n} \nu_{pi} \bar{g}_{pi}(T, P_0) - \sum_{i=1}^{m} \nu_{ri} \bar{g}_{ri}(T, P_0) \tag{7.61}$$

eq. (7.60) indicates that if the reaction takes place, ΔG is negative and

$$\ln K_P(T) = - \frac{\Delta G(T, P_0)}{\bar{R}T} \tag{7.62}$$

These conclusions become a bit clearer as we consider the concrete case of Fig. 7.5 in conjunction with the reaction by which carbon dioxide is formed from its elements:

$$C + O_2 \rightarrow CO_2 \tag{7.63}$$

In this example, both the temperature and the pressure are maintained at the reference levels:

$$T_0 = 25°C \ (298.15 \ K) \quad \text{and} \quad P_0 = 1 \ atm \ (0.101325 \ MPa) \tag{7.64}$$

The inequality (7.57) becomes immediately

$$\bar{g}_C(T_0, P_0) + \bar{g}_{O_2}(T_0, P_0) > \bar{g}_{CO_2}(T_0, P_0) \tag{7.65}$$

This inequality is a warning that reference values for energies (in this case, $\bar{g}$s) cannot be chosen arbitrarily for *all* chemical species. This observation will be reinforced in the next subsection, where we focus on the enthalpy inventories of the "r" and "p" streams. In the case of the $\bar{g}(T_0, P_0)$ values that are being compared in eq. (7.65), it is on the basis of an old convention [18] that we set

$$\bar{g}(T_0, P_0) = 0 \quad \text{(for elements}^\dagger \text{ only)} \tag{7.66}$$

In the present example, $\bar{g}_{O_2} = 0$ and, if the carbon is graphite, $\bar{g}_C = 0$.

---

†The substances that have been chosen as "elements" and appear most often in combustion analyses are C (graphite), H_2 (gas), N_2 (gas), and O_2 (gas). Note, for example, the distinction made between O_2 and O, i.e., that (Table 7.2)

$$\bar{g}_{O_2} = 0 \quad \text{and} \quad \bar{g}_O < 0$$

The $\bar{g}$ value of a chemical compound at 25°C and 1 atm, which cannot be set by convention, is the *Gibbs free energy of formation* $\bar{g}_f^\circ$ of the compound:

$$\bar{g}(T_0, P_0) = \bar{g}_f^\circ \qquad \text{(for compounds)} \tag{7.67}$$

The inequality (7.61) indicates that

$$0 > \bar{g}_{f,\mathrm{CO}_2}^\circ \tag{7.68}$$

which is confirmed by the negative sign of the $\bar{g}_{f,\mathrm{CO}_2}$ value listed in Table 7.2. In summary, there is a net decrease in the Gibbs free energy of the mixture as the reaction proceeds in the direction assumed in eq. (7.63):

$$\Delta G(T_0, P_0) = \bar{g}_{\mathrm{CO}_2}(T_0, P_0) - \bar{g}_{\mathrm{C}}(T_0, P_0) - \bar{g}_{\mathrm{O}_2}(T_0, P_0)$$

$$= \bar{g}_{f,\mathrm{CO}_2}^\circ = -394.39 \,\mathrm{kJ/mol} \tag{7.69}$$

It pays to look a little more closely at the $\bar{g}_f^\circ$ values assembled in Table

ABLE 7.2   The Enthalpy of Formation, Gibbs Free Energy of Formation, and bsolute Entropy of Some of the Most Common Substances Encountered in the nalysis of Combustion Processes [17] ($T = 25°C$, $P = 1$ atm)

ubstance (phase)[a]		M	$\bar{h}_f^\circ$ (kJ/mol)	$\bar{g}_f^\circ$ (kJ/mol)	$\bar{s}^\circ$ (J/mol K)
	Carbon (graphite) (s)	12.011	0	0	5.686
O	Carbon monoxide (g)	28.011	−110.54	−137.16	197.53
O$_2$	Carbon dioxide (g)	44.01	−393.52	−394.39	213.68
H$_4$	Methane (g)	16.043	−74.873	−50.81	186.1
$_2$H$_2$	Acetylene (g)	26.038	+226.73	+209.16	200.85
$_2$H$_4$	Ethylene (g)	28.054	+52.47	+68.35	219.22
$_2$H$_6$	Ethane (g)	30.07	−84.735	−32.91	229.6
$_3$H$_8$	Propane (g)	44.097	−103.92	−23.52	270.1
$_4$H$_{10}$	Butane (g)	58.124	−126.15	−17.04	310.2
$_8$H$_{18}$	n-Octane (g)	114.23	−208.5	+16.6	466.8
$_8$H$_{18}$	n-Octane (l)	114.23	−250.0	+6.71	360.9
	Monatomic hydrogen (g)	1.008	+217.99	+203.27	114.61
$_2$	Hydrogen (g)	2.016	0	0	130.58
$_2$O	Water (g)	18.016	−241.83	−228.59	188.72
$_2$O	Water (l)	18.016	−285.84	−237.15	69.94
	Monatomic nitrogen (g)	14.008	+472.79	+455.49	153.18
O	Nitric oxide (g)	30.008	+90.29	+86.595	210.65
O$_2$	Nitrogen dioxide (g)	46.008	+33.1	+51.24	239.91
$_2$	Nitrogen (g)	28.013	0	0	191.5
	Monatomic oxygen (g)	16.0	+101.7	+91.412	157.79
$_2$	Oxygen (g)	31.999	0	0	205.02

$_=$ gas, l = liquid, and s = solid.

7.2 and to question the correctness of positive numbers such as $\bar{g}^{\circ}_{f,C_2H_2} = +209.16\,\text{kJ/mol}$. After all, did we not learn in the preceding example that during the reaction of "formation" of a compound from its elements, ΔG (or $\bar{g}^{\circ}_{f,\text{compound}}$) must be negative? In the present case, the positive value of $\bar{g}^{\circ}_{f,C_2H_2}$ indicates that the assumed reaction by which acetylene (C_2H_2) is formed from C and H_2 is *impossible*:

$$2C + H_2 \rightarrow C_2H_2 \quad \text{(impossible)}$$
$$\Delta G(T_0, P_0) = +209.16\,\text{kJ/mol} > 0 \tag{7.70}$$

What the positive ΔG indicates is that the *reverse* reaction is possible, because the G inventory decreases in this new direction:

$$C_2H_2 \rightarrow 2C + H_2 \quad \text{(possible)}$$
$$\Delta G(T_0, P_0) = -209.16\,\text{kJ/mol} < 0 \tag{7.71}$$

This observation leads back to the ΔG definition (7.61), which shows that ΔG changes sign when the reactants and products trade places.

Another point is that a negative $\bar{g}^{\circ}_f$ values does not guarantee that the reaction of forming the particular compound will be the one that will actually take place [17]. Consider the case of forming carbon monoxide from C and O_2:

$$C + O_2 \rightarrow CO + \tfrac{1}{2}O_2$$
$$\Delta G(T_0, P_0) = \bar{g}^{\circ}_{f,CO} = -137.16\,\text{kJ/mol} \tag{7.72}$$

Since ΔG is negative, this reaction is not ruled out by the G minimum principle. However, since carbon and oxygen can react along a "steeper descent" on the G scale,

$$C + O_2 \rightarrow CO_2$$
$$\Delta G(T_0, P_0) = \bar{g}^{\circ}_{f,CO_2} = -394.39\,\text{kJ/mol} \tag{7.73}$$

it is this second reaction that is the most likely. We can think of CO_2 as being a "more stable" product (of combining C and O_2) than the products CO and $\tfrac{1}{2}O_2$.

A final observation is that the Gibbs free-energy change of the reaction, ΔG, has the same additive property as the equilibrium constant logarithm, eq. (7.48). This analogy stems from the definition of ΔG and, in particular, from eq. (7.62). Consider the following one-way equations and their respective ΔG values, all at the same temperature and pressure (say, T_0 and P_0):

$$0 \to \sum_{i=1}^{m+n} \nu_i A_i \qquad \Delta G(T_0, P_0) \tag{7.74}$$

$$0 \to \sum_{i=1}^{m_1+n_1} \nu_i^{(1)} A_i \qquad \Delta G^{(1)}(T_0, P_0) \tag{7.75}$$

$$0 \to \sum_{i=1}^{m_2+n_2} \nu_i^{(2)} A_i \qquad \Delta G^{(2)}(T_0, P_0) \tag{7.76}$$

If the chemical equation of the first reaction can be written as a linear combination of the second and third reactions, as shown in eq. (7.47), then

$$\Delta G(T_0, P_0) = c^{(1)} \Delta G^{(1)}(T_0, P_0) + c^{(2)} \Delta G^{(2)}(T_0, P_0) \tag{7.77}$$

One requirement that should not be overlooked is that the two chemical equations used in order to reconstruct the first must *both* be possible. In other words, $\Delta G^{(1)}$ and $\Delta G^{(2)}$ must both be negative for the direction in which the respective chemical equations are written. Of course, if the target reaction is to be possible in the direction in which it is written in eq. (7.74), the ΔG value calculated with eq. (7.77) must also be negative.

In conclusion, the Gibbs free-energy inventory of the stream decreases as the inflow of "reactants" changes into an outflow of "products." This decrease, or the sign that the reaction can take place in the direction assumed in eq. (7.57), is also a sign that the stream can flow only from left to right in Fig. 7.5. Shortly, we shall demonstrate that under certain conditions, the G decrease is also a sign of irreversibility. For example, if the isothermal and isobaric flow of Fig. 7.5 occurs at the reference conditions T_0 and P_0 (i.e., at the "restricted dead state," p. 126), then the rate of Gibbs free-energy decrease across the apparatus becomes a measure of lost mechanical power. This point is discussed further in connection with eq. (7.133).

Example 7.3. Consider the problem of estimating the Gibbs free-energy change during the oxidation of acetylene at reference temperature and pressure:

$$C_2H_2 + O_2 \to 2CO + H_2 \qquad \Delta G = ? \tag{a}$$

This chemical equation can be reconstructed on the basis of two simpler equations [17]:

$$C_2H_2 \to 2C + H_2 \qquad \Delta G^{(1)} = -209.16 \, \text{kJ/mol} \tag{b}$$

$$C + \tfrac{1}{2}O_2 \to CO \qquad \Delta G^{(2)} = -137.16 \, \text{kJ/mol} \tag{c}$$

which are both possible (note the negative ΔGs). The reconstruction formula is

$$(a) = (b) + 2(c)$$

in other words, $c^{(1)} = 1$ and $c^{(2)} = 2$. The ΔG value for the original reaction is

$$\Delta G = \Delta G^{(1)} + 2\Delta G^{(2)} = -483.5 \text{ kJ/mol}$$

Example 7.4.   Suppose that the object to calculate the decrease in G during the reaction [17]

$$4C + H_2 + O_2 \rightarrow C_2H_2 + 2CO \qquad \Delta G = ? \qquad\qquad (a)$$

Reasoning as in the preceding example leads us to

$$\Delta G = \Delta G^{(1)} + 2\Delta G^{(2)} \qquad\qquad (b)$$

where $\Delta G^{(1)}$ and $\Delta G^{(2)}$ refer to the simpler reactions:

$$2C + H_2 \rightarrow C_2H_2 \qquad \Delta G^{(1)} = 209.16 \text{ kJ/mol} \qquad\qquad (c)$$

$$C + \tfrac{1}{2}O_2 \rightarrow CO \qquad \Delta G^{(2)} = -137.16 \text{ kJ/mol} \qquad\qquad (d)$$

Since the calculated ΔG for the chemical equation (a) is negative,

$$\Delta G = [209.16 + 2(-137.16)] \text{ kJ/mol} = -65.16 \text{ kJ/mol}$$

the immediate conclusion drawn by the problem solver might be that reaction (a) is possible as assumed. This conclusion is wrong because one of the pieces of the argument—reaction (c)—is impossible in the direction in which it was assumed (note the positive value of $\Delta G^{(1)}$).

STEADY-FLOW COMBUSTION

The heat transfer and work transfer rate interactions ($\dot{Q}, \dot{W}$) in Fig. 7.5 hint at the engineering relevance of the class of chemically reactive flows depicted in the figure. In this section, we study the limitations that the first law and the second law place on steady-flow chemical reactions that occur in open-system apparatuses similar to the one considered in Fig. 7.5. This time, however, we consider the more general setup in which the temperature and pressure of the products are not necessarily equal to the T and P values of the reactants or of neighboring reservoirs. This more general system is shown in Fig. 7.6: the control surface continues to be impermeable to mass transfer, however, it is not necessarily diathermal and flexible.

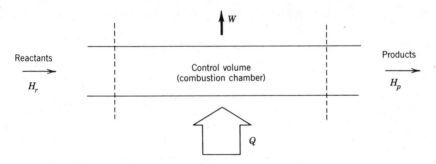

Figure 7.6 Steady flow through a combustion chamber, showing the four terms that enter in the first-law analysis of the system.

Combustion Stoichiometry

As chemical reactions, we consider the burning or rapid oxidation of hydrocarbon fuels. These processes are recognized generally under the more common name of *combustion*. Looking at the left column of Table 7.2, we can write $C_\alpha H_\beta$ as a generic chemical formula for hydrocarbon fuels. In the case of methane, for example, $\alpha = 1$ and $\beta = 4$. The most common oxidant is atmospheric air, in which at 25°C and 1 atm, the diatomic oxygen occupies a volume or mole fraction of 20.35 percent, eq. (7.129). Combustion-energy calculations are usually based on the much simpler model in which the air is viewed as a two-component ideal gas mixture of 21 percent O_2 and 79 percent N_2 on a volume basis. This means that in a chemical reaction in which a fuel is burnt with air, the oxygen and the nitrogen participate in a characteristic proportion:

$$0.21O_2 + 0.79N_2 \quad \text{[for each unit (mole) of air]} \quad (7.78)$$

or, noting that $0.79/0.21 = 3.76$,

$$O_2 + 3.76N_2 \quad \text{[for each unit (mole) of oxygen]} \quad (7.78')$$

The combustion of $C_\alpha H_\beta$ is considered *complete* if in the mixture of products all the carbon appears as CO_2 and all the hydrogen appears as H_2O. We shall see that when the combustion occurs in an insufficient amount of air, some of the fuel burns incompletely. For example, the carbon can appear as both CO_2 and CO in the products. It takes a certain amount of air in order to burn a fuel completely. The *minimum* amount of air that accomplishes this task is the *theoretical air* required by the specified fuel. We can easily calculate the theoretical number of moles of O_2 per mole of $C_\alpha H_\beta$ from the stoichiometry of the chemical equation:

$$C_\alpha H_\beta + \gamma(O_2 + 3.76N_2) \rightarrow \delta CO_2 + \epsilon H_2O + \eta N_2 \qquad (7.79)$$

The unknown stoichiometric coefficients are, in order,

$$\delta = \alpha \qquad \epsilon = \beta/2 \qquad \gamma = \alpha + \beta/4 \qquad \eta = 3.76(\alpha + \beta/4) \quad (7.80)$$

That $(\alpha + \beta/4)$ represents the minimum number of moles of O_2 per mole of $C_\alpha H_\beta$ effecting the complete combustion of $C_\alpha H_\beta$ is indicated by the absence of pure O_2 from the products mixture. During the theoretical combustion (7.79)–(7.80), all the oxygen admitted into the control volume (the combustion chamber) is used up. When the air stream exceeds the theoretical amount calculated above, some of the oxygen turns up "unused" among the products. Let λ be the greater-than-one number representing the ratio between the actual air flowrate and the theoretical flowrate. The chemical equation reads, in this case,

$$C_\alpha H_\beta + \lambda\left(\alpha + \frac{\beta}{4}\right)(O_2 + 3.76N_2) \rightarrow \alpha CO_2 + \frac{\beta}{2} H_2O + 3.76\lambda\left(\alpha + \frac{\beta}{4}\right)N_2$$

$$+ (\lambda - 1)\left(\alpha + \frac{\beta}{4}\right)O_2 \qquad (7.81)$$

If, for example, $\lambda = 4$, the above reaction is said to be carried out with 400 percent theoretical air, or with 300 percent "excess air." The excess air is measured by $(\lambda - 1)$, which also appears as a factor in the amount of O_2 that survives unused on the right side of eq. (7.81).

When the amount of air is less than the theoretical amount, $\lambda < 1$, the combustion is incomplete. The mixture of products contains CO in addition to CO_2, H_2O, N_2, and, possibly other components. The actual proportions in which these components appear in the products depend on the products' temperature and pressure. This aspect is illustrated by the numerical examples and proposed problems that follow. Note that unlike the case in eq. (7.81), one component that is notoriously absent from the products of combustion with less-than-theoretical air is the unused diatomic oxygen, O_2.

The foregoing nomenclature applies almost unchanged to combustion processes based on oxidants other than air. If the role of oxidant is played by pure O_2, the chemical equation for the combustion of $C_\alpha H_\beta$ with $(\lambda - 1)$ excess oxygen reads

$$C_\alpha H_\beta + \lambda\left(\alpha + \frac{\beta}{4}\right)O_2 \rightarrow \alpha CO_2 + \frac{\beta}{2} H_2O + (\lambda - 1)\left(\alpha + \frac{\beta}{4}\right)O_2 \quad (7.82)$$

The First Law

As a direct application of eq. (1.22) to the steady-flow system defined in Fig. 7.6, the First Law of Thermodynamics requires

$$0 = \dot{Q} - \dot{W} + \sum_{i=1}^{m} \dot{n}_{ri}\bar{h}_{ri} - \sum_{i=1}^{n} \dot{n}_{pi}\bar{h}_{pi} \tag{7.83}$$

Let $\dot{n}_{r1} = \dot{n}_{fuel}$ represent the molal flowrate of fuel as one of the inlet streams that flow into the combustion chamber. On a per-mole-of-fuel basis, then, the first law (7.83) assumes the more convenient form

$$0 = \underbrace{\frac{\dot{Q}}{\dot{n}_{fuel}}}_{Q} - \underbrace{\frac{\dot{W}}{\dot{n}_{fuel}}}_{W} + \underbrace{\sum_{i=1}^{m} \nu_{ri}\bar{h}_{ri}}_{H_r} - \underbrace{\sum_{i=1}^{n} \nu_{pi}\bar{h}_{pi}}_{H_p} \tag{7.84}$$

in other words,

$$0 = Q - W + H_r - H_p \tag{7.84'}$$

The shorthand notations H_r and H_p represent the enthalpy inventories of the "reactants" and "products" streams, expressed in units of energy per mole of fuel. The heat transfer and work transfer terms Q and W are expressed in the same units. Note further that since $\dot{n}_{r1} = \dot{n}_{fuel}$, the first of the stoichiometric coefficients in the sum called H_r is equal to 1, $\nu_{r1} = 1$. In other words, the coefficients ν_{ri} and ν_{pi} listed in eq. (7.84) correspond to chemical equations such as eqs. (7.81) and (7.82), which are written on the basis of 1 mole of fuel.

Consider first the case where there is no work transfer, $W = 0$, and where the temperature and pressure of the reactants and the products are at the reference levels T_0 and P_0. The first law (7.84') reduces to

$$Q = H_p - H_r \tag{7.85}$$

where, in accordance with the usual convention, Q is defined positive when it points toward the system (i.e., as in Fig. 7.6). In the present combustion process, in which the reactants are brought in at the restricted dead state (T_0, P_0) and where the products are rejected in the same condition, Q is expectedly negative because combustion reactions are exothermic. Experience suggests that the reactive mixture must be cooled if it is to be discharged at the restricted dead state.

More important than the sign of Q is the observation that the absolute value of Q or the difference $(H_p - H_r)$ is in all likelihood *finite* and measurable. This observation plays a role in the calculation of the H_r and H_p inventories:

$$H_r = (1)\bar{h}_{r1}(T_0, P_0) + \nu_{r2}\bar{h}_{r2}(T_0, P_0) + \cdots \tag{7.86}$$

$$H_p = \nu_{p1}\bar{h}_{p1}(T_0, P_0) + \nu_{p2}\bar{h}_{p2}(T_0, P_0) + \cdots \tag{7.87}$$

in which all the components' enthalpies are evaluated at the reference temperature and pressure. Since the difference $(H_p - H_r)$ is finite and since its value depends on the actual combination of reactants in the "reactants" stream, *not all* the $\bar{h}_{ri}(T_0, P_0)$ and $\bar{h}_{pi}(T_0, P_0)$ enthalpies may be set equal to zero at the reference temperature and pressure. The reference enthalpy convention that mimics eqs. (7.66) and (7.67) is

$$\bar{h}(T_0, P_0) = 0 \quad \text{(for elementary substances only)} \quad (7.88)$$

$$\bar{h}(T_0, P_0) = \bar{h}_f^\circ \quad \text{(for compounds)} \quad (7.89)$$

The quantity $\bar{h}_f^\circ$ is the *enthalpy of formation* of the particular compound: numerical values of this important constant have been compiled in Table 7.2 for some of the most frequently encountered compounds in combustion analyses. As an alternative to eq. (7.88), we can write that the enthalpy of formation of all elementary substances is zero:

$$\bar{h}_f^\circ = 0 \quad \text{(for elementary substances only)} \quad (7.88')$$

An example of a steady-flow, zero-work combustion process of the kind described in the preceding two paragraphs can be constructed using the chemical equation for the combustion of $C_\alpha H_\beta$ with $(\lambda - 1)$ excess air, eq. (7.81). Keeping in mind that all the reactants and products are at T_0 and P_0, we have

$$H_r = \bar{h}_{f,C_\alpha H_\beta}^\circ + 0 + 0 \quad (7.90)$$

$$H_p = \alpha\bar{h}_{r,CO_2}^\circ + \frac{\beta}{2}\,\bar{h}_{f,H_2O}^\circ + 0 + 0 \quad (7.91)$$

The zeros represent the reference enthalpies of the elements, in this case, O_2 and N_2. The heat transfer interaction per mole of fuel is, therefore,

$$Q = \alpha\bar{h}_{f,CO_2}^\circ + \frac{\beta}{2}\,\bar{h}_{f,H_2O}^\circ - \bar{h}_{f,C_\alpha H_\beta}^\circ \quad (7.92)$$

The above heat transfer is recognized also as the *enthalpy of combustion* of $C_\alpha H_\beta$ if, as in eq. (7.81), the combustion is *complete*. An additional requirement that must be met if the "enthalpy-of-combustion" terminology is to apply is that the reactants and the products must be at the same temperature and pressure. As commented already in connection with the sign of $Q = H_p - H_r$, the enthalpy of combustion is negative. The positive or absolute value of the same quantity represents the *heating value* of the particular fuel for the combustion of which the first law (7.85) has been written, in other words,

$$\text{heating value} = H_r - H_p \qquad \text{(complete combustion; } W = 0;$$

$$\text{reactants and products both}$$

$$\text{at } T_0 \text{ and } P_0) \qquad (7.93)$$

Two "heating values" are usually recorded in fuel property tables such as Table 7.3. The *lower heating value* (LHV) refers to the complete combustion process whose products contain H_2O in the vapor phase, that is, as an ideal gas in the mixture of products. The water vapor can be condensed out of the products mixture by appropriately transferring heat out of the stream, without changing the final T_0 and P_0. The heating value calculated with eq. (7.93) for the case when H_2O is a liquid in the (T_0, P_0) products represents the *higher heating value* (HHV) of the fuel. This terminology reminds us that the higher heating value exceeds the lower heating value by an amount corresponding to the heat transfer received by the external entity that effects the condensation of the water vapor contained by the products.

Consider now the more general case where the reactants and the products are not necessarily in the reference condition (T_0, P_0). The first law for the steady state continues to be represented by eq. (7.84); however, this time the constituent enthalpies $(\bar{h}_{ri}, \bar{h}_{pi})$ are not necessarily the reference enthalpies listed in eqs. (7.88) and (7.89). Let T and P, respectively, represent the temperature and pressure of one particular constituent in the stream of reactants or products, and let $\bar{h}(T, P)$ be its enthalpy. The latter can be estimated by writing

$$\bar{h}(T, P) = \bar{h}(T_0, P_0) + \underbrace{\bar{h}(T, P) - \bar{h}(T_0, P_0)}_{\Delta\bar{h}} \qquad (7.94)$$

or, recognizing the enthalpy-of-formation convention,

$$\bar{h}(T, P) = \bar{h}_f^\circ + \Delta\bar{h} \qquad (7.95)$$

The enthalpy change $\Delta\bar{h}$ is in general a function of both T and P. At sufficiently low pressures, i.e., when the constituent in question obeys the ideal gas model, $\Delta\bar{h}$ is only a function of temperature whose reference value is by definition zero, $\Delta\bar{h}(T_0) = 0$. If the constituent exists as liquid, the $\Delta\bar{h}$ part of eq. (7.95) can be evaluated invoking the incompressible-substance model for moderately subcooled liquid states (see the Appendix). The same model can be used for $\Delta\bar{h}$ in cases when the constituent is a solid. In conclusion, the per-mole-of-fuel enthalpy inventories appearing in the first law (7.84') are given by

$$H_r = \sum_{i=1}^{m} \nu_{ri}(\bar{h}_f^\circ + \Delta\bar{h})_{ri} \quad \text{and} \quad H_p = \sum_{i=1}^{n} \nu_{pi}(\bar{h}_f^\circ + \Delta\bar{h})_{pi} \quad (7.96)$$

Numerical tabulations of the $\Delta\bar{h}(T)$ functions of the ideal gases found among the reactants and products of combustion are available in most thermodynamics texts and handbooks [17]. I use this opportunity to present this information in graphical form, Fig. 7.7, partly because the accuracy of the tabulated information is not nearly as high as is suggested by the six-significant-digit reporting chosen by some textbooks. Another purpose of constructing Fig. 7.7 is to show that in sufficiently narrow ranges of

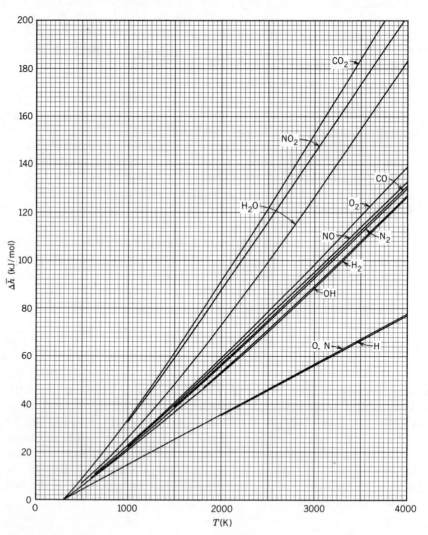

Figure 7.7 The enthalpy change function $\Delta\bar{h}(T)$ for several of the ideal gases encountered in the analyses of combustion processes.

temperature, the $\Delta\bar{h}(T)$ function is practically linear in temperature. Note further that

$$\Delta\bar{h}(T) = \int_{T_0}^{T} \bar{c}_P(T') \, dT' \quad \text{(ideal gas)} \quad (7.97)$$

and that the deviation of the $\Delta\bar{h}(T)$ function from a perfectly straight line is a sign that the specific heat increases with temperature. The $\bar{c}_P(T)$ functions of some of the ideal gases covered by Fig. 7.7 are displayed in Fig. 4.4.

A special limit in the operation of the steady-flow system discussed until now is the *zero-work and adiabatic* limit represented by $Q = W = 0$. In this case, the enthalpy inventory brought in by the reactants is conserved across the combustion chamber, $H_r = H_p$, eq. (7.84'). Typical in this limit is the extremely high temperature reached by the products. Consider, for example, the case of Fig. 7.8, in which the mixture of reactants is admitted at T_0 and P_0. If the products of combustion were to be discharged at T_0 and P_0, then, according to eq. (7.93), the system external to the combustion chamber would be receiving a certain number of joules per mole of fuel. That number would be LHV if the products contain water vapor as opposed to liquid water. Now, when the system is perfectly insulated, it cannot dispose of the LHV: Fig. 7.8 shows that the only option left for LHV is to heat the mixture of products to a new temperature, T_{af}, which is usually much higher than T_0. The special temperature T_{af} represents the *adiabatic flame* temperature associated with the combustion of the given mixture of reactants (fuel + oxidant). In the example in Fig. 7.8, the products' temperature T_{af} is determined by solving the first-law equation:

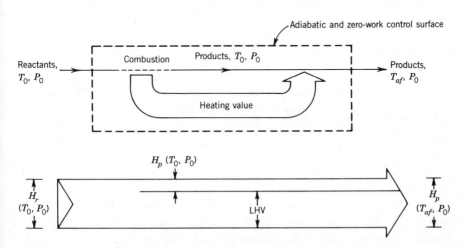

Figure 7.8 The conservation of enthalpy through an adiabatic and zero-work combustion process leading to products at the adiabatic flame temperature T_{af}.

$$H_r(T_0, P_0) = H_p(T_{af}, P_0) \qquad (7.98)$$

Other zero-work combustion systems that deliver at least part of the heat transfer LHV to an external user show correspondingly lower temperatures in the stream of combustion products. The adiabatic flame temperature can be viewed as a *theoretical* ceiling for the temperatures that can be recorded in the steady-flow combustion system. This ceiling value is "theoretical," first, because no combustion chamber is truly perfectly insulated, and, second, because at temperatures of the order of T_{af}, one or more of the original products of combustion can undergo a process of chemical *dissociation*. For example, if T_{af} is sufficiently high, the carbon dioxide contained in the original products can dissociate into carbon monoxide and oxygen:

$$CO_2 \rightarrow CO + \tfrac{1}{2}O_2 \qquad (7.99)$$

Written in this direction, the reaction is endothermic: it proceeds at the expense of the stream of products, which experiences a cooling effect (Fig. 7.9). The dissociation ends when the temperature of the products has become low enough for CO_2 to coexist in equilibrium with CO, O_2, and the other products of combustion. At this new adiabatic flame temperature *with dissociation*, T_{afd}, the dissociation reaction (7.99) proceeds with equal speeds in both directions.

The engineering impact of the dissociation process described above is that it prohibits the establishment of the theoretical adiabatic flame temperature calculated with eq. (7.98), that is, based on the assumption of no dissociation. The actual adiabatic flame temperature reached after partial dissociation, T_{afd}, is less than the theoretical value for combustion without dissociation, T_{af}. Additional examples of dissociation reactions are

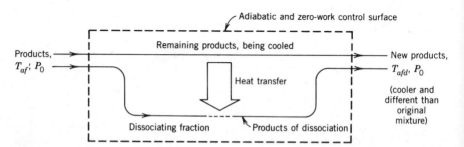

Figure 7.9 The lowering of the adiabatic flame temperature in the wake of the partial dissociation of one of the products of combustion (this figure is a continuation of Fig. 7.8).

$$H_2 \rightarrow 2H$$

$$O_2 \rightarrow 2O$$

$$N_2 \rightarrow 2N \qquad (7.100)$$

$$H_2O \rightarrow H_2 + \tfrac{1}{2}O_2$$

$$H_2O \rightarrow OH + \tfrac{1}{2}H_2$$

The equilibrium constants of the two-way (reversible) limit of these reactions can be deduced from Table 7.1, while paying close attention to the stoichiometric coefficients that have been used in the writing of each chemical equation at the head of the table.

The Second Law

Relative to the first-law analysis centered around Fig. 7.6, the second law places emphasis on those interactions that act as carriers of entropy. For this reason, in Fig. 7.6, we would have to specify unambiguously the temperature of the boundary penetrated by the heat transfer interaction Q. Consider the more specific system defined in Fig. 7.10, in which the heat transfer per mole of fuel Q_i crosses the boundary segment whose temperature is T_i. The work transfer W is not shown because it does not appear in the second-law statement (work transfer is the type of energy transfer interaction that is not accompanied by entropy transfer). As a special application of eq. (2.46) of open systems, the Second Law of Thermodynamics for the steady-flow system of Fig. 7.10 reads

$$\dot{S}_{\text{gen}} = -\sum_i \frac{\dot{Q}_i}{T_i} - \sum_{i=1}^{m} \dot{n}_{ri}\bar{s}_{ri} + \sum_{i=1}^{n} \dot{n}_{pi}\bar{s}_{pi} \geq 0 \qquad (7.101)$$

Dividing by the molal flowrate of the fuel, $\dot{n}_{\text{fuel}}$,

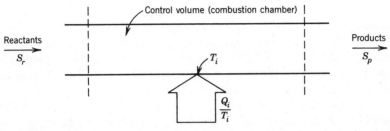

Figure 7.10 The entropy interactions associated with steady flow through a combustion chamber.

$$\frac{1}{\dot{n}_{\text{fuel}}} \underbrace{\dot{S}_{\text{gen}}}_{S_{\text{gen}}} = - \underbrace{\sum_i \frac{\dot{Q}_i/\dot{n}_{\text{fuel}}}{T_i}}_{Q_i/T_i} - \underbrace{\sum_{i=1}^{m} \nu_{ri}\bar{s}_{ri}}_{S_r} + \underbrace{\sum_{i=1}^{n} \nu_{pi}\bar{s}_{pi}}_{S_p} \geq 0 \qquad (7.102)$$

we obtain the per-mole-of-fuel statement:

$$S_{\text{gen}} = - \sum_i \frac{Q_i}{T_i} - S_r + S_p \geq 0 \qquad (7.103)$$

The second law states that the combustion chamber acts as a producer of entropy. For example, in the special case of adiabatic combustion (Fig. 7.8), the second law reduces to the statement that the entropy inventory of the products (S_p) generally exceeds the entropy inventory of the reactants (S_r). These two inventories are evaluated according to the two sums involved in the definition (7.102). Unlike the $\bar{g}_i$ and $\bar{h}_i$ values, which could not be all set equal to zero at one reference state, the entropies $\bar{s}_i$ do have a common state where they are all zero. The basis for this assertion is the empirical evidence summarized under the heading of the *Third Law of Thermodynamics* (pp. 582–583), which suggests that the entropy of any substance in equilibrium approaches a unique value as T approaches absolute zero. By convention, this unique value was set equal to zero, eq. (10.109); therefore, for all the *absolute entropies* $\bar{s}_i$ appearing in S_r and S_p, we recognize that

$$\lim_{T \to 0} \bar{s} = 0 \qquad (7.104)$$

Equation (7.104) suggests that in the absolute-zero limit, the entropy of the constituent depends only on the temperature. Away from absolute zero and, especially, in the high-temperature range spanned by combustion processes, $\bar{s}$ is a function of both T and P. The most readily available entropy information is in the form of numerical tables for $\bar{s}$ versus T at atmospheric pressure, $P = P_0$ [17]. The shorthand notation for these reference-pressure values is

$$\bar{s}(T, P_0) = \bar{s}°(T) \qquad (7.105)$$

The last column in Table 7.2 exhibits the special values of the absolute entropy of various reactants and products at reference pressure and temperature, $\bar{s}°(25°C)$. The $\bar{s}°(T)$ values corresponding to other temperatures can be retrieved from the tables [17] or from Fig. 7.11, which is a novel problem-solving aid in the present treatment.

In order to calculate the entropy at any T and P (where P is the partial pressure of the constituent in the products, for example), we write

$$\bar{s}(T, P) = \bar{s}°(T) + \underbrace{\bar{s}(T, P) - \bar{s}(T, P_0)}_{\Delta\bar{s}} \qquad (7.106)$$

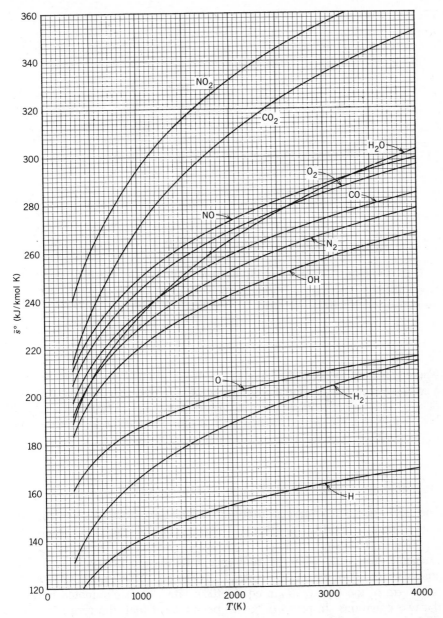

Figure 7.11 The absolute entropy at atmospheric pressure, $\bar{s}°(T)$, for several of the ideal gases encountered in the analyses of combustion processes.

The entropy change $\Delta \bar{s}$ refers to a pressure change executed at constant temperature. Fortunately, in many cases, the temperatures of these constituents are high and the partial pressures sufficiently low to justify the use of the ideal gas model (Table 4.7):

$$\Delta \bar{s} = -\bar{R} \ln \frac{P}{P_0} \tag{7.107}$$

Assuming that all the constituents are ideal gases, the entropy inventories of the reactants and products streams are

$$S_r = \sum_{i=1}^{m} \nu_{ri}\left(\bar{s}^{\circ} - \bar{R} \ln \frac{P}{P_0}\right)_{ri}$$

$$S_p = \sum_{i=1}^{n} \nu_{pi}\left(\bar{s}^{\circ} - \bar{R} \ln \frac{P}{P_0}\right)_{pi} \tag{7.108}$$

If the constituent appears in liquid form, then we can use the incompressible-substance model for moderately compressed liquid states, so that, instead of eq. (7.106), we have

$$\bar{s}(T) \cong \bar{s}^{\circ}(25°C)_{\text{liquid, } \atop \text{Table 7.2}} + [\bar{s}_f(T) - \bar{s}_f(25°C)] \tag{7.109}$$

where this time subscript "f" means "saturated liquid." If the constituent is a solid, then its $\bar{s}(T)$ formula is similar to eq. (7.109), except that—by virtue of the same model—the quantity in brackets is replaced further by $\bar{c} \ln (T/298.15 \text{ K})$. This last move is permissible if the lone specific heat of the solid ($\bar{c}$) is constant over the temperature interval $298.15 \text{ K} - T$.

Maximum Work and "Effective Flame Temperature"

The chief energy-engineering question is to what extent can a given combustion process be used for producing useful mechanical power. Consider this question in the context of Fig. 7.12, in which the steady-flow apparatus can exchange heat only with the reference temperature reservoir (T_0). The analysis consists of writing Q_0 instead of Q in eq. (7.84') and Q_0/T_0 as the only term of the type Q_i/T_i in eq. (7.103): later, Q_0 is eliminated between the two equations. In terms of joules per mole of fuel, the result is

$$W = H_r - H_p - T_0(S_r - S_p) - T_0 S_{\text{gen}}$$

$$= \sum_{i=1}^{m} \nu_{ri}\underbrace{(\bar{h} - T_0\bar{s})_{ri}}_{\bar{b}_{ri}} - \sum_{i=1}^{n} \nu_{pi}\underbrace{(\bar{h} - T_0\bar{s})_{pi}}_{\bar{b}_{pi}} - T_0 S_{\text{gen}} \tag{7.110}$$

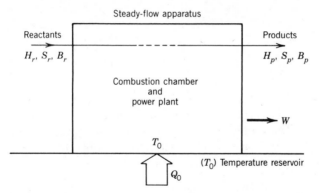

Figure 7.12 Steady-flow apparatus for the production of useful mechanical work from a combustion process in communication with the reference temperature reservoir.

The same conclusion follows directly from eq. (3.41), of course, dividing it first by $\dot{n}_{\text{fuel}}$ and writing $W = \dot{E}_W / \dot{n}_{\text{fuel}}$:

$$W = \underbrace{\sum_{i=1}^{m} \nu_{ri} \bar{b}_{ri}}_{B_r} - \underbrace{\sum_{i=1}^{n} \nu_{pi} \bar{b}_{pi}}_{B_p} - T_0 S_{\text{gen}} \qquad (7.111)$$

This alternative brings out the *flow availability* notation $\bar{b}$ for each of the $(\bar{h} - T_0 \bar{s})$ groups appearing in the last of eq. (7.110). Clearly, the flow availability inventories of the streams of reactants and products are related to their respective enthalpy and entropy inventories:

$$B_r = H_r - T_0 S_r \quad \text{and} \quad B_p = H_p - T_0 S_p \qquad (7.112)$$

If we are able to estimate the enthalpy and entropy inventories of the two streams (Figs. 7.7 and 7.11), then we should also be able to evaluate their flow availability. The resulting chart is presented as Fig. A2 in the Appendix.

One immediate conclusion that can be drawn based on the values of B_r and B_p alone is that the work output per mole of fuel cannot exceed $(B_r - B_p)$. This conclusion follows from setting $S_{\text{gen}} = 0$ for the reversible limit of eq. (7.111):

$$W_{\text{rev}} = B_r - B_p \geq W \qquad (7.113)$$

In the special set of circumstances where both the inlet and the outlet mixtures are at the reference (restricted dead-state) conditions, T_0 and P_0, the reversible work per mole of fuel (7.113) becomes

$$W_{\text{rev}} = \sum_{i=1}^{m} \nu_{ri}(\bar{h} - T_0 \bar{s})_{0,ri} - \sum_{i=1}^{n} \nu_{pi}(\bar{h} - T_0 \bar{s})_{0,pi}$$

$$= \sum_{i=1}^{m} \nu_{ri}\mu_{0,ri} - \sum_{i=1}^{n} \nu_{pi}\mu_{0,pi} \qquad (7.114)$$

Each factor of the type $(\bar{h} - T_0 \bar{s})_0$ represents the partial molal Gibbs free energy or the chemical potential of the particular constituent in the T_0 and P_0 mixture. Therefore, the two sums on the right side of eq. (7.114) represent the per-mole-of-fuel inventories of Gibbs free energy of the inlet mixture (G_r) and the outlet mixture (G_p):

$$W_{\text{rev}} = G_r - G_p \qquad (7.115)$$

The inequality between the reversible work and the actual work, $W_{\text{rev}} \geq W$, eq. (7.113), is correct in an algebraic sense, that is, regardless of the sign of the numerical values taken by W_{rev} and W. The irreversibility $T_0 S_{\text{gen}}$ that separates the values of W_{rev} and W can be caused not only by the combustion process itself, but also by the power plant sandwiched between the combustion chamber and the ambient (T_0) inside the steady-flow apparatus of Fig. 7.12.

One way to distinguish between these two sources of irreversibility is shown in Fig. 7.13: in this figure, the power plant is replaced by a Carnot engine; therefore, the irreversibility $(W_{\text{rev}} - W)$ is due to the combustion chamber alone. The Carnot engine receives the heat input $(H_r - H_p)$, produces the "actual" work discussed above, W, and rejects heat to the reservoir (T_0). In view of the power generation chapter that follows, it is convenient to define the *effective flame temperature* of the combustion chamber as an exergy source, T_f, by writing

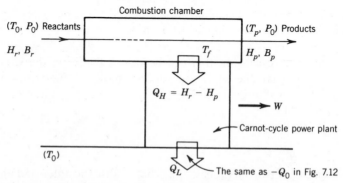

Figure 7.13 Device for defining the effective flame temperature of the combustion chamber as a source of exergy, T_f.

$$W = (H_r - H_p)(1 - T_0/T_f) \qquad (7.116)$$

This definition is written with eq. (3.14) in mind: W is the exergy (useful work) associated with the heat transfer $(H_r - H_p)$ released from the temperature level T_f by the combustion chamber. If T_f and the steady-flow combustion process are known, then W and W_{rev} can be evaluated using eqs. (7.113) and (7.116). The ratio of the two can be viewed as the second-law efficiency of the combustion chamber alone:

$$\eta_{II} = \frac{W}{W_{rev}} = \frac{H_r - H_p}{B_r - B_p}\left(1 - \frac{T_0}{T_f}\right) \qquad (7.117)$$

that is, as an indicator of the degree of irreversibility of the combustion process.

The concept of "effective flame temperature" as the temperature of a source of heat transfer exergy originates from the point of view of the individual who seeks to design and position a power plant between the combustion chamber and the environmental temperature reservoir (T_0). For this reason, the actual T_f value chosen for use in eqs. (7.116) and (7.117) depends on the design of the heat exchanger that couples the hot end of the power cycle to the combustion chamber. In what follows, we consider two possible designs.

A simple model for the temperature distribution inside the combustion chamber consists of the assumption that the gaseous contents of the chamber that come in contact with the hot-end heat exchanger of the power plant are *well mixed*, so that their temperature is uniform and equal to that of the exiting products, T_p. According to this model, the absolute temperature of the "heat source" seen by the power cycle can only be T_p; therefore,

$$T_f = T_p \qquad (7.118)$$

There exists an optimum combustion-chamber temperature T_p (or T_f) that maximizes the potential work output of the Carnot-cycle power plant, W. Although it is not necessary for making this point, superior illustration is made possible by the additional assumption that in the range $(T_{af} - T_p)$, the mole fractions and the specific heats of all the ideal gas products of combustion are constant. In this case, since H_r equals the H_p evaluated at the adiabatic flame temperature, eq. (7.98), the heat transfer drawn by the power plant from the combustion chamber is

$$H_r(T_0, P_0) - H_p(T_p, P_0) = H_p(T_{af}, P_0) - H_p(T_p, P_0)$$

$$= \sum_{i=1}^{n} (\nu \bar{c}_P)_{pi}(T_{af} - T_p)$$

$$= C_p(T_{af} - T_p) \qquad (7.119)$$

where the definition of the "products" constant[†] C_p is evident. Combining eqs. (7.116), (7.118), and (7.119), we see that W reaches a maximum as T_p varies between T_{af} and T_0:

$$W = C_p(T_{af} - T_p)(1 - T_0/T_p) \tag{7.120}$$

namely,

$$W_{\substack{\max \\ \text{Fig. 7.13}}} = C_p T_{af}[1 - (T_0/T_{af})^{1/2}]^2 \tag{7.121}$$

at an optimum effective flame temperature of

$$T_{\substack{f,\text{opt} \\ \text{Fig. 7.13}}} = T_{p,\text{opt}} = (T_0 T_{af})^{1/2} \tag{7.122}$$

At flame temperatures higher than this optimum, the work production decreases because the heat input to the power plane $(H_r - H_p)$ decreases. When the flame temperature decreases below this optimum level, the work output W decreases because of the decreasing Carnot efficiency factor $(1 - T_0/T_f)$ in eq. (7.116). The optimum temperature discovered in eq. (7.122) is a "high" temperature relative to the temperature range spanned by the power cycle itself. For example, if $T_0 = 298$ K and $T_{af} = 3000$ K, eq. (7.122) yields $T_{f,\text{opt}} = 946$ K. If the combustion reaction is followed by dissociation, Fig. 7.9, then the "T_{af}" constant of this theory must be replaced by the corresponding temperature constant for adiabatic and zero-work steady-flow combustion followed by dissociation, T_{afd}.

A second combustion chamber model is shown in Fig. 7.14. It stems from the question of what might be done to improve the performance of the single-T_f scheme of Fig. 7.13 beyond the optimum conditions identified above. One thing that can be done is to use the still hot exhaust (H_p, S_p, B_p) of Fig. 7.13 as an additional heat source in the power cycle. Note that until now, the products were, presumably, used for a different purpose before being released into the environment (to this observation, we return in the last section of this chapter). In the new scheme of Fig. 7.14, the temperature of the products varies inside the combustion chamber. The temperature decrease from T_{af} to T_p is caused by the cooling effect provided by the reversible power plant positioned underneath. In hardware terms, the arrangement can be realized by building a counterflow heat exchanger in which the products of combustion flow to the right while the working fluid of the hot end of the power cycle flows to the left. Anywhere along this counterflow, the temperature difference between the gaseous products and the working fluid is zero.

The work extracted by the reversible power plant from the stream of

[†]The C_p constant should not be confused with specific heat at constant pressure, c_P.

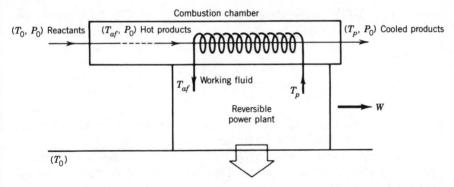

Figure 7.14 Power plant driven by the continuous counterflow cooling of the products of combustion.

products is equal to the drop in the flow availability inventory B_p along the stream:

$$W = B_p(T_{af}, P_0) - B_p(T_p, P_0) \qquad (7.123)$$

Noting that the pressure remains constant and that $B_p = H_p - T_0 S_p$, we write further that

$$W = C_p(T_{af} - T_p) - T_0 C_p \ln \frac{T_{af}}{T_p} \qquad (7.124)$$

in which we have assumed once again that in the range $T_{af} - T_p$, the products are all ideal gases with temperature-independent specific heats and mole fractions. The constant of the products stream, C_p, was defined in the course of writing eq. (7.119). The important feature of this new W estimate is that it increases monotonically as T_p decreases. The maximum W is reached when $T_p = T_0$, that is, when the stream of products reaches the restricted dead state:

$$W_{\substack{\max \\ \text{Fig. 7.14}}} = C_p T_0\left(\frac{T_{af}}{T_0} - 1 - \ln \frac{T_{af}}{T_0}\right) \qquad (7.125)$$

This quantity can now be compared with that of eq. (7.121) in order to show that the counterflow heat exchanger scheme of Fig. 7.14 is indeed superior to the best of the single-temperature products scheme of Fig. 7.13 (see Problem 7.5).

Finally, a unique (average) effective flame temperature can also be estimated for this latest design by viewing the maximum work of eq. (7.125) as the output of the Carnot engine referred to in eq. (7.116). In the present case, $T_p = T_0$, which means that instead of $(H_r - H_p)$ in eq. (7.116), we write $C_p(T_{af} - T_0)$. In the same equation, $T_{f,\text{opt}}$ replaces T_f. Equating the right sides of eqs. (7.116) and (7.125) leads to the optimum, average effective flame temperature of the counterflow scheme of Fig. 7.14:

$$T_{f,\text{opt} \atop \text{Fig. 7.14}} = \frac{T_{af} - T_0}{\ln(T_{af}/T_0)} \qquad (7.126)$$

For example, setting $T_0 = 298$ K and $T_{af} = 3000$ K in the above, we obtain $T_{f,\text{opt}} = 1170$ K. This effective flame temperature is greater than the corresponding temperature given by eq. (7.122). This is another way of stating that the optimized design based on Fig. 7.14 is better than the optimized design based on Fig. 7.13.

Example 7.5. Consider the steady-flow combustion of graphite with theoretical oxygen:

$$C + O_2 \rightarrow CO_2 \qquad (a)$$

and calculate the adiabatic flame temperature, first, by neglecting the partial dissociation of CO_2, and, second, by accounting for the effect of dissociation. The pressure throughout the combustion chamber is 1 atm. The reactants enter the chamber at $T_0 = 25°C$.

For the first part of the problem, we recognize that

$$H_r(T_0, P_0) = H_p(T_{af}, P_0) \qquad (b)$$

where

$$H_r(T_0, P_0) = \bar{h}^{\circ}_{f,C} + \bar{h}^{\circ}_{f,O_2} = 0 \qquad (c)$$

$$H_p(T_0, P_0) = \bar{h}^{\circ}_{f,CO_2} + \Delta\bar{h}_{CO_2}(T_{af}) \qquad (d)$$

According to Table 7.2, the enthalpy of formation of carbon dioxide is -393.52 kJ/mol; therefore, eqs. (b)–(c) yield one relationship for T_{af}:

$$0 = -393.52 \text{ kJ/mol} + \Delta\bar{h}_{CO_2}(T_{af}) \qquad (e)$$

Extrapolating linearly along the CO_2 curve of Fig. 7.7, we find, approximately, of course, that $T_{af} \cong 6770$ K.

The second part of the problem refers to the calculation of T_{afd}, which is expectedly lower than the T_{af} calculated above. The dissociation reaction for CO_2 is

$$CO_2 \rightarrow CO + \tfrac{1}{2}O_2 \qquad (f)$$

Not all the CO_2 formed during the combustion reaction (a) splits into CO and O_2 according to eq. (f). Let ζ be the extent of the reaction of dissociation. From each mole of CO_2 formed at (a), a certain fraction ζ breaks up into ζ moles of CO and

$\frac{1}{2}\zeta$ moles of O_2, as required by the stoichiometry of eq. (f). In the place once occupied by 1 mole of CO_2 before dissociation, in the final mixture of products, we find now $(1 - \zeta)$ moles of CO_2, ζ moles of CO and $\frac{1}{2}\zeta$ moles of O_2. Since the total number of moles is $1 - \zeta + \zeta + \frac{1}{2}\zeta = 1 + \frac{1}{2}\zeta$, the respective mole fractions of the final products are

$$x_{CO_2} = \frac{1 - \zeta}{1 + \zeta/2} \qquad x_{CO} = \frac{\zeta}{1 + \zeta/2} \qquad x_{O_2} = \frac{\zeta/2}{1 + \zeta/2} \tag{g}$$

Assuming that after dissociation, these products leave the combustion chamber in equilibrium, the above mole fractions are related through the equilibrium constant:

$$K_p(T_{afd}) = \frac{x_{CO} x_{O_2}^{1/2}}{x_{CO_2}} = \frac{\zeta}{1 - \zeta} \left(\frac{\zeta}{2 + \zeta} \right)^{1/2} \tag{h}$$

The temperature function $K_p(T_{afd})$ is provided in the seventh column of Table 7.1: combined with eq. (h), it constitutes a one-to-one relationship between T_{afd} and ζ. Here are three points calculated along this curve:

$$\begin{array}{lccc} T_{afd} & 3000\ \text{K} & 3500\ \text{K} & 4000\ \text{K} \\ \zeta & 0.436 & 0.748 & 0.8984 \end{array} \tag{i}$$

The additional information needed for pinpointing ζ (and T_{afd}) is provided by the first-law statement for the adiabatic and zero-work steady-flow process:

$$H_r(T_0, P_0) = H_p(T_{afd}, P_0) \tag{j}$$

where, again, $H_r = 0$. Recognizing the ζ-dependent composition of the products after dissociation, eq. (j) states that

$$0 = (1 - \zeta)[\bar{h}_{f,CO_2}^\circ + \Delta\bar{h}_{CO_2}(T_{afd})] + \zeta[\bar{h}_{f,CO}^\circ + \Delta\bar{h}_{CO}(T_{afd})]$$
$$+ \tfrac{1}{2}\zeta[\bar{h}_{f,O_2}^\circ + \Delta\bar{h}_{O_2}(T_{afd})] \tag{k}$$

where $\bar{h}_{f,O_2}^\circ = 0$. This is another one-to-one relationship between T_{afd} and ζ, yielding, for example,

$$\begin{array}{lccc} T_{afd} & 3000\ \text{K} & 3500\ \text{K} & 4000\ \text{K} \\ \zeta & 0.882 & 0.774 & 0.664 \end{array} \tag{l}$$

The intersection of the curves represented by (i) and (l) is the wanted result:

$$T_{afd} \cong 3540\ \text{K at } \zeta \cong 0.765$$

Example 7.6. Liquid octane is burned with theoretical air in a steady-flow apparatus that houses also the boiler and superheater of a steam-turbine power plant. The temperature and pressure conditions at the four ports of the control volume are indicated directly on Fig. 7.15. Noteworthy is that the size of the boiler and superheater heat exchanger is such that although the superheated-

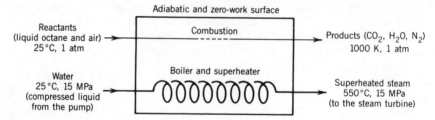

Figure 7.15 Boiler and superheater design for a steam-turbine power plant (Example 7.6).

steam temperature approaches the temperature of the products of combustion, there is a finite temperature difference between the two exiting streams, 1000 K − 823 K = 177 K.

We are interested in a complete first-law, second-law, and combined-law analysis of the installation of Fig. 7.15. Particularly important to the designer is (1) the needed flowrate of fuel per unit water flowrate, and (2) the degree of irreversibility of the installation.

The first step in any of these problems is occupied by stoichiometry, in this case by

$$C_8H_{18} + 12.5(O_2 + 3.76N_2) \rightarrow 8CO_2 + 9H_2O + (12.5)(3.76)N_2 \qquad (a)$$

It is assumed that the products' temperature is sufficiently low and that the mole fraction of the nondissociating component (N_2) is sufficiently high so that the effect of dissociation can be left out of the discussion.

The First Law of Thermodynamics requires

$$0 = \dot{n}_{octane}(H_r - H_p) + \dot{m}_{water}(h_{in} - h_{out}) \qquad (b)$$

where, from the steam tables, h_{in} (25°C, 15 MPa) = 118.7 kJ/kg water and h_{out} (550°C, 15 MPa) = 3448.6 kJ/kg water. The per-mole-of-fuel enthalpies of the reactants and the products are

$$H_r = \bar{h}_{f,C_8H_{18},liquid}^\circ = -250 \text{ kJ/mol octane} \qquad (c)$$

$$H_p = 8(\bar{h}_f^\circ + \Delta\bar{h})_{CO_2} + 9(\bar{h}_f^\circ + \Delta\bar{h})_{H_2O} + 47(\bar{h}_f^\circ + \Delta h)_{N_2}$$
$$= -3793 \text{ kJ/mol octane} \qquad (d)$$

The first law (b) delivers the required fuel-to-water ratio, either as

$$\frac{\dot{n}_{octane}}{\dot{m}_{water}} = 0.94 \ \frac{\text{mol octane}}{\text{kg water}} \qquad (e)$$

or, since $M_{octane} = 114.23$ kg/kmol,

$$\frac{\dot{m}_{octane}}{\dot{m}_{water}} = 0.107 \; \frac{kg \; octane}{kg \; water} \tag{f}$$

The Second Law of Thermodynamics requires

$$S_{gen} = \dot{n}_{octane}(S_p - S_r) + \dot{m}_{water}(s_{out} - s_{in}) > 0 \tag{g}$$

where, from the steam tables, s_{in} (25°C, 15 MPa) = 0.3611 kJ/kg K and s_{out} (550°C, 15 MPa) = 6.5199 kJ/kg K. Since the combustion flow takes place at atmospheric pressure, the entropy inventories are (Table 7.2)

$$S_r = \bar{s}^{\circ}_{C_8H_{18}, liquid} + 12.5\bar{s}^{\circ}_{O_2} + 47\bar{s}^{\circ}_{N_2} = 11{,}924 \; J/K/mol \; octane \tag{h}$$

$$S_p = (8\bar{s}_{CO_2} + 9\bar{s}_{H_2O} + 47\bar{s}_{N_2})_{1000K} = 14{,}964 \; J/K/mol \; octane \tag{i}$$

Putting all the numbers together (including the fuel-to-water ratio calculated earlier), we conclude that the installation of Fig. 7.15 does not violate the second law:

$$\frac{\dot{S}_{gen}}{\dot{m}_{water}} = 9.016 \; \frac{kJ/K}{kg \; water} > 0 \tag{j}$$

Finally, the combined first- and second-law analysis consists of evaluating

$$B_r = H_r - T_0 S_r = -3806 \; kJ/mol \; octane \tag{k}$$

$$B_p = H_p - T_0 S_p = -8256 \; kJ/mol \; octane \tag{l}$$

and the flow-availability (or flow-exergy) increase along the water stream:

$$b_{out} - b_{in} = h_{out} - h_{in} - T_0(s_{out} - s_{in}) = 1493 \; kJ/kg \; water \tag{m}$$

If the steam-turbine power plant (not sketched in Fig. 7.15) is irreversibility free, then the maximum work per mole of octane that could theoretically be delivered by the plant is

$$W_{plant} = (b_{out} - b_{in}) \frac{\dot{m}_{water}}{\dot{n}_{octane}} = 1591 \; kJ/mol \; octane \tag{n}$$

On the other hand, the reversible work that could theoretically be extracted from the combustion process is

$$W_{rev} = B_r - B_p = 4451 \; kJ/mol \; octane \tag{o}$$

We discover in this way a maximum second-law efficiency of only 36 percent:

$$\frac{W_{plant}}{W_{rev}} = \frac{1591}{4451} = 0.36 \tag{p}$$

The lost work $(W_{\text{rev}} - W_{\text{plant}})$, or $T_0 S_{\text{gen}}$, is only partially due to the irreversibility of the combustion reaction and flow. The other part is due to the finite-ΔT heat transfer $(H_r - H_p)$ from the flame to the water stream: the temperature difference is the smallest at the right edge of Fig. 7.15 (namely, 177 K). If we model the combustion products as "well mixed" at the temperature $T_f = T_p = 1000$ K, then the work lost in the nonuniform temperature gap formed between the T_f ceiling and the water stream is

$$W_{\text{lost},\Delta T} = (H_r - H_p)\left(1 - \frac{T_0}{T_f}\right) - (b_{\text{out}} - b_{\text{in}})\frac{\dot{m}_{\text{water}}}{\dot{n}_{\text{octane}}} \tag{q}$$

The first term on the right side of eq. (q) is the exergy associated with the heat input $(H_r - H_p)$ coming down from the temperature level T_f. Numerically, we obtain

$$W_{\text{lost},\Delta T} = 897 \text{ kJ/mol octane} \cong 0.2 W_{\text{rev}} \tag{r}$$

In summary, out of the theoretical work ceiling W_{rev}, eq. (p), 36 percent could in principle be produced by the steam-turbine power plant, 20 percent is destroyed in the finite-ΔT heat transfer process, while the remaining 44 percent is destroyed by the rapid oxidation of the fuel in the combustion chamber.

THE CHEMICAL EXERGY OF FUELS

As a special application of the maximum-work considerations outlined in the preceding section, consider the question of the maximum work that could be extracted per mole of fuel if the reactants and the products are in thermomechanical *and* chemical equilibrium with the ambient. This topic has been treated in detail by Moran [19], after whom the present subsection is patterned. The same topic has been treated by Riekert [20], Szargut [21], Sussman [22], and Wepfer and Gaggioli [23], among others. We considered a similar problem already in the discussion leading to eq. (7.115), where both the mixture of reactants and the mixture of products were only in thermal and mechanical equilibrium with the ambient (at T_0 and P_0, or the restricted dead state). The new feature of the analysis presented below is that each product and reactant (other than the fuel itself) is at the temperature, pressure, and chemical potential at which it exists in the environment. Employing the language of chapter 5, we can say that the products and the reactants other than the fuel are at the unrestricted dead state or unrestricted environmental state.

For the sake of concreteness, consider the complete combustion of one mole of the hydrocarbon $C_\alpha H_\beta$ with diatomic oxygen, that is, without the atmospheric N_2 listed in eq. (7.79):

$$C_\alpha H_\beta + \left(\alpha + \frac{\beta}{4}\right)O_2 \rightarrow \alpha CO_2 + \frac{\beta}{2} H_2O_{\text{(vapor)}} \tag{7.127}$$

The water that is formed during this reaction is in its vapor phase. The fuel enters the steady-flow apparatus as a single-component stream at the restricted dead state (T_0, P_0). The remaining reactant (O_2) and the products (CO_2, H_2O) enter and exit the control volume by passing through appropriately semipermeable membranes, Fig. 7.16. Each of these components is in the condition in which it is found in the environment, i.e., in chemical equilibrium with its natural counterpart in the environment.

For example, the oxygen that crosses the control surface has the same chemical potential as the oxygen found as a component in the atmospherical ideal gas mixture, eq. (4.139),

$$\mu_{O_2}(T_0, P_0, x_{O_2}) = \mu_{O_2}^{(x_{O_2}=1)}(T_0, P_0) + \bar{R}T_0 \ln x_{O_2}$$
$$= \bar{g}_{O_2}(T_0, P_0) + \bar{R}T_0 \ln x_{O_2} \qquad (7.128)$$

The mole fraction x_{O_2} represents the mole fraction of oxygen in the atmosphere. This and the other mole fractions that are needed later are part of a more detailed[†] model for the composition of atmospheric air [19]:

$$x_{N_2} = 0.7567$$
$$x_{O_2} = 0.2035$$
$$x_{H_2O} = 0.0303 \qquad (7.129)$$
$$x_{CO_2} = 0.0003$$
$$x_{other} = 0.0092$$

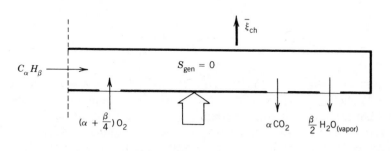

Figure 7.16 Reversible combustion process for calculating the chemical exergy of 1 mole of hydrocarbon $C_\alpha H_\beta$.

[†]More accurate than the two-component model shown on page 369.

The Gibbs free energy $\bar{g}_{O_2}(T_0, P_0)$ appearing in the last of eq. (7.128) refers to oxygen as a single-component batch at reference temperature and pressure. The chemical potentials of the two combustion products of eq. (7.127) can be evaluated in the same manner as in eq. (7.128):

$$\mu_{CO_2}(T_0, P_0, x_{CO_2}) = \bar{g}_{CO_2}(T_0, P_0) + \bar{R}T_0 \ln x_{CO_2} \tag{7.130}$$

$$\mu_{H_2O}(T_0, P_0, x_{H_2O}) = \bar{g}_{H_2O}(T_0, P_0) + \bar{R}T_0 \ln x_{H_2O} \tag{7.131}$$

The maximum (reversible) work per mole of fuel follows from eq. (7.114), in which the "$r1$" reactant is the fuel (note that $\nu_{r_1} = 1$):

$$W_{rev} = \mu_{C_\alpha H_\beta} + \left(\alpha + \frac{\beta}{4}\right)\mu_{O_2} - \alpha\mu_{CO_2} - \frac{\beta}{2}\mu_{H_2O} \tag{7.132}$$

Since the fuel enters the combustion chamber as a single component at T_0 and P_0, its chemical potential is the same as $\bar{g}_{C_\alpha H_\beta}(T_0, P_0)$. Taken together, eqs. (7.128)–(7.132) yield

$$W_{rev} = \bar{\xi}_{ch} = -\Delta G(T_0, P_0) + \bar{R}T_0 \ln\left(\frac{x_{O_2}^{\alpha+\beta/4}}{x_{CO_2}^{\alpha} x_{H_2O}^{\beta/2}}\right) \tag{7.133}$$

where ΔG is the net change in the Gibbs free energy in the direction of the reaction:

$$\Delta G(T_0, P_0) = \underbrace{\alpha\bar{g}_{CO_2}(T_0, P_0) + \frac{\beta}{2}\bar{g}_{H_2O}(T_0, P_0)}_{\text{Products}}$$

$$\underbrace{-\left[\bar{g}_{C_\alpha H_\beta}(T_0, P_0) + \left(\alpha + \frac{\beta}{4}\right)\bar{g}_{O_2}(T_0, P_0)\right]}_{\text{Reactants}} \tag{7.134}$$

Beginning with eq. (7.133), we use $\bar{\xi}_{ch}$ instead of W_{rev} as symbol for the per-mole chemical exergy of the $C_\alpha H_\beta$ fuel. This terminology is consistent with what we spoke in chapter 5 and with the function of the apparatus in Fig. 7.16. That function is to transform reversibly 1 mole of fuel (already at the restricted dead state) into products that are at the unrestricted dead state or environmental state. Or, by changing the direction of all the arrows in Fig. 7.16, $\bar{\xi}_{ch}$ represents the minimum work that is required in order to constitute 1 mole of fuel from substances found naturally in the environment.

All the $\bar{g}$s appearing in eq. (7.134) are being evaluated at T_0 and P_0; therefore, their respective values are the Gibbs free energies of formation listed in Table 7.2. The $(-\Delta G)$ and $\bar{\xi}_{ch}$ values calculated by Moran [19],

TABLE 7.3. Lower Heating Values (LHV), Higher Heating Values (HHV), Gibbs Free-Energy Decrease $(-\Delta G)$, and Chemical Exergy $(\bar{\xi}_{ch})$ of Various Fuels at $T_0 = 25°C$ and $P_0 = 1$ atm [19, 24][a]

Fuel (phase)[b]	LHV (kJ/mol)	HHV (kJ/mol)	$-\Delta G(T_0, P_0)$ (kJ/mol)	$\bar{\xi}_{ch}$ (kJ/mol)
Hydrogen (g), H_2	241.8	285.9	237.2	235.2
Carbon (s), C	393.5	393.5	394.4	410.5
Paraffin Family, C_nH_{2n+2}				
Methane (g), CH_4	802.3	890.4	818	830.2
Ethane (g), C_2H_6	1427.9	1559.9	1467.5	1493.9
Propane (g), C_3H_8	2044	2220	2108.4	2149
Butane (g), C_4H_{10}	2658.5	2878.5	2747.8	2802.5
Pentane (g), C_5H_{12}	3272.1	3536.1	3386.9	3455.8
Pentane (l), C_5H_{12}	3245.5	3509.5	3385.8	3454.8
Hexane (g), C_6H_{14}	3886.7	4194.8	4026.8	4110
Hexane (l), C_6H_{14}	3855.1	4163.1	4022.8	4106
Heptane (g), C_7H_{16}	4501.4	4853.5	4667	4764.3
Heptane (l), C_7H_{16}	4464.9	4816.9	4660	4757.3
Octane (g), C_8H_{18}	5116.2	5512.2	5307.1	5418.6
Octane (l), C_8H_{18}	5074.6	5470.7	5297.2	5408.7
Olefin Family, C_nH_{2n}				
Ethylene (g), C_2H_4	1323	1411	1331.3	1359.6
Propylene (g), C_3H_6	1926.5	2058.5	1957.3	1999.9
Butene (g), C_4H_8	2542.6	2718.6	2598.3	2655.1
Pentene (g), C_5H_{10}	3155.8	3375.9	3236.5	3307.4
Napthene Family, C_nH_{2n}				
Cyclopentane (g), C_5H_{10}	3099.5	3319.5	3196.5	3267.4
Cyclopentane (l), C_5H_{10}	3053.6	3273.6	3189.6	3260.5
Cycloxehane (g), C_6H_{12}	3688.9	3953	3821.2	3906.3
Cycloxehane (l) C_6H_{12}	3655.9	3919.9	3816.2	3901.3
Aromatic Family, C_nH_{2n-6}				
Benzene (g), C_6H_6	3169.5	3301.6	3207.5	3298.5
Toluene (g), C_7H_8	3772	3948	3831.7	3936.9
Toluene (l), C_7H_8	3771	3947	3834	3939.2
Ethylbenzene (g), C_8H_{10}	4387.1	4607.1	4471.6	4591

With permission from Professor Michael J. Moran, the Ohio State University, after Table 7-2 in his book *Availability Analysis: A Guide to Efficient Energy Use*, Prentice-Hall, Englewood Cliffs, NJ, 1982.

[a] g = gas, l = liquid, and s = solid.

based on eqs. (7.133) and (7.134), are reported in Table 7.3. A general characteristic of the tabulated values is, first, that the fuel availability does not differ substantially from the Gibbs free-energy drop $(-\Delta G)$. In other words, the actual proportions in which the gaseous products of combustion find themselves in the atmosphere [or the last term in eq. (7.133)] have only a minor impact on the calculated chemical availability. From the point of view of being able to make engineering calculations, this is fortunate, because the atmospheric air proportions assumed in eqs. (7.129) are by no means universal. The water-vapor mole fraction, for example, varies daily with the relative humidity.

The second characteristic is that both $(-\Delta G)$ and $\bar{\xi}_{ch}$ are comparable with the heating values, especially with the lower heating value (LHV) because the present calculation is based on the assumption that H_2O is gaseous in the products, eq. (7.127). Moran [19] reports the following approximate relations between $\bar{\xi}_{ch}$ and LHV:

$$\frac{\bar{\xi}_{ch}}{\text{LHV}} \cong 1.033 + 0.0169\frac{\beta}{\alpha} - \frac{0.0698}{\alpha} \qquad \text{(for gaseous } C_\alpha H_\beta) \qquad (7.135)$$

$$\frac{\bar{\xi}_{ch}}{\text{LHV}} \cong 1.04224 + 0.011925\frac{\beta}{\alpha} - \frac{0.042}{\alpha} \qquad \text{(for liquid } C_\alpha H_\beta) \qquad (7.136)$$

It is worth noting, finally, that $(-\Delta G)$ represents the reversible work of the reaction in which each reactant and product participates as a single component at T_0 and P_0. In this case, the μs and $\bar{g}$s of eqs. (7.128)–(7.131) are identical and eq. (7.133) reduces to $W_{rev} = -\Delta G$. An example of this kind was discussed in connection with eqs. (7.60)–(7.61). We see now that the decrease registered in the Gibbs free-energy inventory in the direction of the constant-T and -P reaction represents the reversible work, which, if lost, is a measure of the irreversibility of the reaction.

CONSTANT-VOLUME COMBUSTION

In this section, we consider an entirely different class of combustion processes, one in which the *closed system* that houses the reactive mixture is surrounded by a constant-volume enclosure. Because of this feature, the combustion process evolves without work transfer. In the initial state, the system contains the reactants $n_{r1}, n_{r2}, \ldots, n_{rm}$, the first of which represents the fuel, $n_{r1} = n_{fuel}$. In the final state, the constant-volume enclosure houses the products of combustion $(n_{p1}, n_{p2}, \ldots, n_{pn})$. The main features of this type of system and process are illustrated in Fig. 7.17.

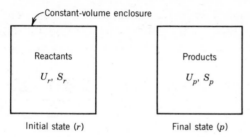

Figure 7.17   The end states of a combustion process confined by a constant volume and impermeable boundary.

The First Law

Writing Q_{r-p} for the net heat transfer received by the system during the zero-work process described above, the first law can be written as

$$Q_{r-p} = \sum_{i=1}^{n} n_{pi}\bar{u}_{pi} - \sum_{i=1}^{m} n_{ri}\bar{u}_{ri} \tag{7.137}$$

The corresponding per-mole-of-fuel statement is

$$Q = \frac{1}{n_{\text{fuel}}} Q_{r-p} = \sum_{i=1}^{n} \nu_{pi}\bar{u}_{pi} - \sum_{i=1}^{m} \nu_{ri}\bar{u}_{ri}$$

$$= \underbrace{\sum_{i=1}^{n} \nu_{pi}(\bar{h}_f^{\circ} + \Delta\bar{h} - P\bar{v})_{pi}}_{U_p} - \underbrace{\sum_{i=1}^{m} \nu_{ri}(\bar{h}_f^{\circ} + \Delta\bar{h} - P\bar{v})_{ri}}_{U_r} \tag{7.138}$$

where U_p and U_r are the internal energy inventories of the products and the reactants, respectively, expressed in joules per mole of fuel.

If the combustion is complete and, additionally, if the temperature and pressure of the reactants (T, P) equal the temperature and pressure of the products, then $(U_p - U_r)$ is recognized as the *internal energy of combustion* associated with the particular fuel that is being considered. Since the combustion is an exothermic reaction, we expect the difference $(U_p - U_r)$ to be negative. The actual heat transfer that is released to the environment during this process, $(U_r - U_p)$, represents the *constant-volume heating value* of the fuel. This is a positive number of joules per mole of fuel. And, just as in the definition of the LHV and HHV quantities of the preceding section, we can distinguish again between a *lower* constant-volume heating value $(\text{LHV})_v$ and a *higher* constant-volume heating value, $(\text{HHV})_v$, depending on whether the H_2O in the products is in vapor or liquid form.

The symmetry between the new quantities $(\text{LHV})_v$ and $(\text{HHV})_v$ and their steady-flow counterparts becomes even more evident if we rewrite the last of eq. (7.138) by noting first that the $P\bar{v}$ term of any gaseous component (reactant or product) is in fact equal to the same $\bar{R}T$, where T is the common temperature of the initial reactants and the final products. The constant-volume heating value $(U_r - U_p)$ can be replaced by

$$U_r - U_p = \underbrace{\sum_{i=1}^{m} \nu_{ri}(\bar{h}_f^\circ + \Delta\bar{h})_{ri}}_{H_r} - \underbrace{\sum \nu_{ri}\bar{R}T}_{\substack{\text{Gaseous} \\ \text{reactants} \\ \text{only}}} - \underbrace{\sum \nu_{ri}P\bar{v}_{ri}}_{\substack{\text{Solid and} \\ \text{liquid} \\ \text{reactants}}}$$

$$\underbrace{-\sum_{i=1}^{n} \nu_{pi}(\bar{h}_f^\circ + \Delta\bar{h})_{pi}}_{H_p} + \underbrace{\sum \nu_{pi}\bar{R}T}_{\substack{\text{Gaseous} \\ \text{products} \\ \text{only}}} + \underbrace{\sum \nu_{pi}P\bar{v}_{pi}}_{\substack{\text{Solid and} \\ \text{liquid} \\ \text{reactants}}}$$

(7.139)

In many cases, the volume occupied by the nongaseous components in the reactants and the products mixtures is negligible when compared with the volume occupied by the gaseous components. This means that the third and sixth terms can be neglected in favor of the second and fifth terms in eq. (7.139); therefore,

$$U_r - U_p \cong H_r - H_p - \underbrace{\bar{R}T\left(\sum \nu_{ri} - \sum \nu_{pi}\right)}_{\substack{\text{Gaseous} \\ \text{components} \\ \text{only}}}$$

(7.140)

What emerges on the right side of eq. (7.140) is the steady-flow heating value $(H_r - H_p)$. In the special case where the number of moles of gaseous products is exactly the same as the original number of moles of gaseous reactants, the last group in eq. (7.140) vanishes and the constant-volume heating value becomes practically the same as the steady-flow heating value. In the same special case then, $(\text{LHV})_v \cong \text{LHV}$ and $(\text{HHV})_v \cong \text{HHV}$.

The Second Law

Returning to the constant-volume process defined in Fig. 7.17, the Second Law of Thermodynamics requires

$$S_{\text{gen},(r)-(p)} = \sum_{i=1}^{n} n_{pi}\bar{s}_{pi} - \sum_{i=1}^{m} n_{ri}\bar{s}_{ri} - \int_{(r)}^{(p)} \frac{\delta Q}{T} \geq 0$$

(7.141)

where δQ is taken as positive when entering the system, and where T is the absolute temperature of the boundary crossed by δQ. On a per-mole-of-fuel basis, the same statement reads

$$S_{\text{gen}} = \frac{1}{n_{\text{fuel}}} S_{\text{gen},(r)-(p)} = S_p - S_r - \int_{(r)}^{(p)} \frac{\delta Q/n_{\text{fuel}}}{T} \geq 0 \qquad (7.142)$$

where the entropy inventories of the reactants (S_r) and products (S_p) are calculated as indicated in eq. (7.102).

Maximum Work

Consider now the question of the maximum work that could be derived from the process of Fig. 7.17 in the presence of the reference temperature reservoir (T_0). The aggregate thermodynamic system under consideration consists of the constant-volume system of Fig. 7.17 and the cyclical device sandwiched between it and the reservoir (T_0). The total per-mole-of-fuel work transfer delivered by the cyclical device as the constant-volume system proceeds from (r) to (p) is W. During the same process, the total per-mole-of-fuel heat transfer interaction that passes from (T_0) to the cyclical device is Q_0. In view of eqs. (7.138) and (7.142), the first-law and the second-law statements for the aggregate system defined above are

$$Q_0 - W = U_p - U_r \qquad (7.143)$$

$$S_{\text{gen}} = S_p - S_r - \frac{Q_0}{T_0} \geq 0 \qquad (7.144)$$

Eliminating Q_0 between these two equations yields

$$W = U_r - T_0 S_r - (U_p - T_0 S_p) - T_0 S_{\text{gen}} \qquad (7.145)$$

The maximum work transfer per mole of fuel is

$$W_{\text{rev},v} = (U - T_0 S)_r - (U - T_0 S)_p \qquad (7.146)$$

i.e., the difference between the values of the new quantity $(U - T_0 S)$ when summed up over the original reactants and, later, over the products.

Considering the nonflow nature of the $(r)-(p)$ process of Fig. 7.17, in place of eq. (7.146), we might have expected to obtain the difference in the respective nonflow availability inventories, $A_r - A_p$, where, for example, $A_r = U_r - T_0 S_r + P_0 V_r$, eq. (3.30). In the present case, the $P_0 V_r$ and $P_0 V_p$ terms do not appear because the reactive mixture evolves at constant volume and does not do any work against the pressure reservoir (P_0). Conversely, we can regard the right side of eq. (7.146) as having been written originally $(A_r - A_p)$, before the terms $P_0 V_r$ and $P_0 V_p$ cancelled one another.

There are cases in which the per-mole-of-fuel maximum work transfer developed for constant-volume combustion, eq. (7.146), is the same as the corresponding result for the same combustion reaction in steady flow, eq.

(7.113). When the assumptions listed above eq. (7.140) are valid, the constant-volume maximum work is also equal to

$$W_{\text{rev},v} \cong B_r - B_p - \bar{R}T\underbrace{\left(\sum \nu_{ri} - \sum \nu_{pi}\right)}_{\substack{\text{Gaseous} \\ \text{components} \\ \text{only}}} \qquad (7.147)$$

The $(B_r - B_p)$ difference appearing on the right side is the reversible work limit encountered earlier in the realm of steady-flow combustion, eq. (7.113). The two reversible work estimates are equal in the case of those reactions that are not accompanied by changes in the total number of moles of all the gaseous components.

SYMBOLS

A_i	chemical symbols [eq. (7.4′)]
b	exponent [eq. (7.39)]
$\bar{b}_i$	partial molal flow availability [J/mol]
B_p	flow availability inventory of the products, per mole of fuel
B_r	flow availability inventory of the reactants, per mole of fuel
c_1, c_2	constants [eq. (7.40)]
$c^{(1)}, c^{(2)}$	constants [eq. (7.47)]
$\bar{c}_P$	specific heat at constant pressure
C_p	constant associated with the products mixture [eq. (7.119)]
F	Helmholtz free energy
$\bar{g}_i$	partial molal Gibbs free energy [J/mol]
$\bar{g}_f^{\,\circ}$	Gibbs free energy of formation [eq. (7.67)]
G	Gibbs free energy
G_p	Gibbs free-energy inventory of the products mixture, per mole of fuel
G_r	Gibbs free-energy inventory of the reactants mixture, per mole of fuel
ΔG	Gibbs free-energy increase in the direction of the reaction [eqs. (7.61) and (7.134)]
$\bar{h}_i$	partial molal enthalpy [J/mol]
$\bar{h}_f^{\,\circ}$	enthalpy of formation [eq. (7.89)]
$\Delta \bar{h}$	enthalpy temperature-correction function [eq. (7.95) and Fig. 7.7]
H	enthalpy
H_p	enthalpy inventory of the products mixture, per mole of fuel
H_r	enthalpy inventory of the reactants mixture, per mole of fuel
HHV	higher heating value

$(HHV)_v$	higher heating value at constant volume
K_P	equilibrium constant [eq. (7.36) and Table 7.1]
LHV	lower heating value
$(LHV)_v$	lower heating value at constant volume
m	number of reactants
$\dot{m}$	mass flowrate
n	number of products
$\dot{n}$	molal flowrate
N_i	number of moles of species i
P	pressure
P_i	partial pressure
P_0	reference (atmospheric) pressure
$\dot{Q}$	heat transfer rate [W]
Q	heat transfer per mole of fuel [J/mol fuel]
Q_H	heat transfer interaction with a high temperature reservoir
Q_0	heat transfer interaction with the reference temperature reservoir, per mole of fuel
$\bar{R}$	universal gas constant
$\bar{s}_i$	partial molal entropy [J/K/mol]
$\bar{s}^\circ$	absolute entropy at reference pressure [eq. (7.105) and Fig. 7.11]
$\Delta\bar{s}$	pressure-correction function for absolute entropy [eq. (7.106)]
S	entropy [J/K]
$\dot{S}_{gen}$	entropy generation rate [W/K]
S_{gen}	entropy generation per mole of fuel [J/K/mol]
S_p	entropy inventory of the products mixture, per mole of fuel
S_r	entropy inventory of the reactants mixture, per mole of fuel
T	temperature
T_{af}	adiabatic flame temperature
T_{afd}	adiabatic flame temperature after dissociation
T_f	effective flame temperature of the combustion chamber when viewed as a source of exergy [eq. (7.116)]
T_p	temperature of the products mixture
T_0	reference temperature (25°C)
$\bar{u}_i$	partial molal internal energy [J/mol]
U	internal energy [J]
U_p	internal-energy inventory of the products mixture, per mole of fuel
U_r	internal-energy inventory of the reactants mixture, per mole of fuel
$\bar{v}_i$	partial molal volume [m³/mol]
V	volume [m³]
W	work transfer per mole of fuel [J/mol]
$\dot{W}$	work transfer rate [W]
W_{rev}	reversible work transfer per mole of fuel [eq. (7.113)]

$W_{rev,v}$ reversible work transfer at constant volume, per mole of fuel [eq. (7.147)]

x_i mole fraction

Y affinity function [eq. (7.9)]

α number of atoms of carbon in $C_\alpha H_\beta$

β number of atoms of hydrogen in $C_\alpha H_\beta$

ζ extent of the reaction, degree of advancement

ζ_e equilibrium extent of the reaction

η_{II} second-law efficiency [eq. (7.117)]

ν_i stoichiometric coefficients [eq. (7.4′)]

$\bar{\xi}_{ch}$ chemical exergy of 1 mole of fuel [eq. (7.133)]

τ_{af} dimensionless adiabatic flame temperature (T_{af}/T_0)

$(\ )_{opt}$ optimum

$(\ )_p$ associated with the products; also the final state of a constant-volume reaction

$(\ )_r$ associated with the reactants; also the initial state of a constant-volume reaction

$(\ )_v$ associated with a constant-volume reaction

REFERENCES

1. J. W. Gibbs, On the equilibrium of heterogeneous substances, *The Collected Works of J. Willard Gibbs*, Vol. I, Longmans, Green, New York, 1928.

2. G. N. Lewis and M. Randall, *Thermodynamics and the Free Energy of Chemical Substances*, McGraw-Hill, New York, 1923.

3. I. Prigogine and R. Defay, *Chemical Thermodynamics*, translated by D. H. Everett, Longmans, Green, London, 1954.

4. Th. De Donder, L'affinité, applications aux gaz parfaits, *Bull. Cl. Sci., Acad. R. Belg.*, Vol. 7(5), 1922, pp. 197–205.

5. E. A. Guggenheim, *Thermodynamics*, 7th ed., North-Holland Publ., Amsterdam, 1985, p. 241.

6. H. B. Callen, *Thermodynamics*, Wiley, New York, 1960, pp. 205–206.

7. M. Modell and R. C. Reid, *Thermodynamics and Its Applications*, Prentice-Hall, Englewood Cliffs, NJ, 1974, pp. 417–427.

8. H. Le Chatelier, Sur un énoncé général des lois des équilibres chimiques, *C. R. Hebd. Seances Acad. Sci.*, Vol. 99, 1884, pp. 786–789; Vol. 104, 1887, p. 679.

9. F. Braun, *Z. Phys. Chem.*, Vol. 1, 1887, p. 259.

10. F. Braun, Ueber einen allgemeinen qualitativen Satz für Zustandsänderungen nebst einigen sich schliessenden Bemerkungen, insbesondere über nicht eindeutige Systeme, *Ann. Phys. Chem.*, Vol. 33, 1888, pp. 337–353.

11. C. Raveau, Le lois du déplacement de l'équilibre et le principe de Le Chatelier, *Journal de Physique*, Vol. 8, 1909, pp. 572–579.

12. P. Ehrenfest, Das Prinzip von Le Chatelier–Braun und die Reziprozitätssätze der Thermodynamik, *Z. Phys. Chem.*, Vol. 77, 1911, pp. 227–244.

13. J. W. Strutt (Baron Rayleigh), The Le Chatelier–Braun principle, *Trans. Chem. Soc. London*, Vol. 111, 1917, pp. 250–252; also in Rayleigh's *Scientific Papers*, Vol. 6, Cambridge University Press, London and New York, 1920, pp. 475–577.

14. T. Ehrenfest-Afanassjewa and G. L. De Haas-Lorentz, Über Intensitätsparameter und über stabiles thermodynamisches Gleichgewicht, *Physica (The Hague)*, Vol. 2, 1935, pp. 743–752.

15. M. Planck, Bemerkungen über Qualitätsparameter, Intensitätsparameter und stabiles Gleichgewicht, *Physica (The Hague)*, Vol. 2, 1935, pp. 1029–1032.

16. M. J. Klein, *Paul Ehrenfest, The Making of a Theoretical Physicist*, 3rd ed., Vol. 1, North-Holland Publ., Amsterdam, 1985, pp. 156–161.

17. D. R. Stull and H. Prophet, project directors, *JANAF Thermochemical Tables*, 2nd ed., NSRDS-NBS 37, National Bureau of Standards, Washington, DC, 1971.

18. G. N. Lewis, The free energy of chemical substances, *J. Am. Chem. Soc.*, Vol. 35, No. 1, 1913, pp. 1–30 (see p. 15).

19. M. J. Moran, *Availability Analysis: A Guide to Efficient Energy Use*, Prentice-Hall, Englewood Cliffs, NJ, 1982, pp. 152–156.

20. L. Riekert, The efficiency of energy-utilization in chemical processes, *Chem. Eng. Sci.*, Vol. 29, 1974, pp. 1613–1620.

21. J. Szargut, International progress in second law analysis, *Energy*, Vol. 5, 1980, pp. 709–718.

22. M. V. Sussman, *Availability (Exergy) Analysis*, Mulliken House, Lexington, MA, 1981.

23. W. J. Wepfer and R. A. Gaggioli, Reference datums for available energy, *ACS Symp. Ser.*, No. 122, 1980, pp. 77–92.

24. F. D. Rossini, D. D. Wagman, W. H. Evans, S. Levine, and I. Jaffe, *Selected Values of Chemical Thermodynamic Properties*, NBS Circ. 500, National Bureau of Standards, Washington, DC, 1952.

PROBLEMS

7.1 The "heat of reaction at constant volume" is the heat transfer interaction experienced by a reactive system as the reaction advances from ζ to $(\zeta + d\zeta)$ at constant temperature and volume. This quantity is represented by the derivative

$$\left(\frac{\partial U}{\partial \zeta}\right)_{T,V}$$

Prove that the relationship between affinity and temperature at constant V and ζ is dictated by the heat of reaction at constant volume in the following manner:

$$\frac{Y(T_2, V, \zeta)}{T_2} - \frac{Y(T_1, V, \zeta)}{T_1} = \int_{T_1}^{T_2} \frac{1}{T^2}\left(\frac{\partial U}{\partial \zeta}\right)_{T,V} dT$$

7.2 Demonstrate that the total differential of Y/T, in which $Y = Y(T, P, \zeta)$, is

$$d\left(\frac{Y}{T}\right) = \frac{1}{T^2}\left(\frac{\partial H}{\partial \zeta}\right)_{T,P} dT - \frac{1}{T}\left(\frac{\partial V}{\partial \zeta}\right)_{T,P} dP + \frac{1}{T}\left(\frac{\partial Y}{\partial \zeta}\right)_{T,P} d\zeta$$

7.3 Consider the equilibrium of the reaction $\frac{1}{2}H_2 \rightleftarrows H$, and assume that the tabulation of $\log_{10} K_P$ values (Table 7.1) is so coarse that the value corresponding to $T = 500$ K is missing. Evaluate $\log_{10} K_P$ (500 K) by interpolating linearly between the two closest values, by assuming that $\log_{10} K_P$ is first a linear function of T and, second, a linear function of T^{-1}. Compare your estimates with the correct value (Table 7.1) and show that the second method is considerably more accurate.

7.4 Consider an equilibrium mixture of H_2, O_2, and H_2O at $T = 3000$ K and $P = 0.1$ atm. Determine the mole fractions of the three components and comment on the direction in which they would vary if P were to increase at constant temperature.

7.5 Consider the optimized power-plant designs based in the text on Figs. 7.13 and 7.14. Show that the relative goodness of the second design is measured by the ratio

$$\frac{W_{\text{max,Fig. 7.14}}}{W_{\text{max,Fig. 7.13}}} = \frac{\tau_{af} - 1 - \ln \tau_{af}}{(\tau_{af}^{1/2} - 1)^2}$$

where τ_{af} is the dimensionless adiabatic flame temperature T_{af}/T_0. Plot this ratio as a function of τ_{af} in the range $1 < \tau_{af} < 10$ and comment on the superiority of the second design.

7.6 A stream of gaseous methane is burned in adiabatic and zero-work steady flow with theoretical air. The reactants enter the combustion chamber at 25°C and 1 atm. The pressure of the products is also 1 atm. Calculate the adiabatic flame temperature T_{af} by neglecting the effect of dissociation.

Repeat the T_{af} calculation two more times, for the case when methane is burned with 200-percent theoretical air and, later, when it is burned with 400-percent theoretical air. Show graphically how the mole fractions of the products vary with the percent excess air brought into the reaction.

7.7 The boiler and superheater of a steam-turbine power plant are heated by the combustion chamber shown in Fig. 7.15. The reactants, products, and the working fluid of the power cycle enter and leave at the conditions indicated on the drawing. Assuming that the combustion of liquid octane is with 100-percent excess air, calculate in order (see also Example 7.6)

(a) the fuel to water ratio, $\dot{n}_{octane}/\dot{m}_{water}$

(b) the entropy generation per mole of octane, S_{gen}

(c) the reversible work of combustion, W_{rev}

(d) the maximum work that could be produced by the power plant, W_{plant}

(e) the work lost due to the heat transfer across the nonuniform temperature difference between the flame ($T_f = T_p = 1000$ K) and the water stream, $W_{lost,\Delta T}$.

Repeat the above calculations by assuming combustion with 200-percent excess air. Use the numerical conclusions of this problem [part (e)] and the conclusion of Example 7.6 in order to discuss the effect of excess air on the pattern of destruction of availability in the installation of Fig. 7.15.

7.8 One mole of carbon monoxide burns completely in the presence of 3 moles of diatomic oxygen in a constant-volume enclosure. The original temperature and pressure of the reactants are 25°C and 1 atm.

(a) The combustion takes place adiabatically. Neglecting the effect of dissociation, calculate the final pressure and temperature of the products.

(b) The combustion does not occur adiabatically. Calculate the heat transfer released per mole of CO when the products are cooled to a final temperature of 500 K.

8

Power Generation

The generation of mechanical power is the original *raison d'être* of the science of engineering thermodynamics. It would be easy to argue that the world has come a long way from the early steam engines and reciprocating machinery that started the field. One could also argue that the major post-Watt technological breakthroughs in power generation triggered their own new brands of engineering thermodynamics. There is, of course, truth in this argument, as even a quick glance at the contents of Fig. 8.1 [1, 2] invites the reader to think of the beginnings of more specialized modern fields, for example, internal combustion engines, turbomachinery, jet propulsion, etc. Yet I will argue that of more fundamental importance is what links all these developments rather than what sets them apart. The technological developments that led to today's explosive growth in electrical power production are all manifestations of a common philosophy.

It is a real challenge to try to bring under the same theoretical roof so many advances; in fact, it is impossible to do justice to all the developments that deserve to be mentioned in a course such as this. Nevertheless, in this chapter, I will argue that the left-to-right reading of Fig. 8.1 is a review of successive entropy-generation minimization attempts in the realm of heat-engine development. This common and by now familiar story (chapter 3) has a special modern twist. The modern view is that the heat-engine efficiency increments recorded by history are secondary results of the more practical engineering effort of maximizing the *per-unit-time* production of work in power plants whose goodness (e.g., size, heat transfer area) is constrained by economic considerations.

MAXIMUM POWER CONDITIONS

The simplest way of illustrating the conceptual difference between efficiency maximization and power maximization is to analyze a Carnot engine that functions between the temperature reservoirs (T_H) and (T_L), as shown in Fig. 8.2. In earlier discussions of the Carnot-engine concept, it was tacitly

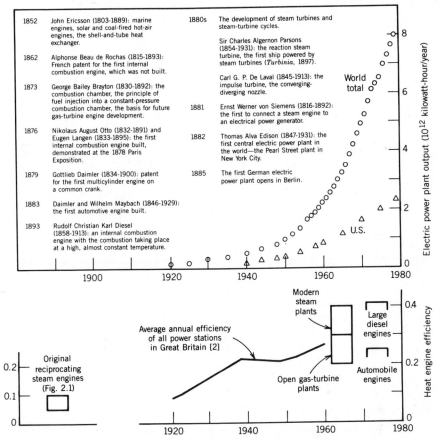

Figure 8.1 Highlights of the early developments that led to today's heat-engine efficiencies and electric-power generation level (the power production data are from Loftness [1]: note the kink in world production during the energy crisis of the mid-1970s; the average efficiency curve is from Haywood [2]).

assumed that the heat engine is in thermal equilibrium with each temperature reservoir during the respective heat transfer interactions. Such equilibria would require either an infinitely slow cycle or an infinitely large engine–reservoir contact surface, in order to allow the working fluid to reside indefinitely in the vicinity of the engine–reservoir interface. In the real engineering world, in which not only heat transfer area but also time are in limited supply, it makes more sense to recognize the energy interaction Q_H and Q_L on a per-unit-time basis, i.e., as heat transfer rates:

$$\dot{Q}_H = (\bar{h}A)_H(T_H - T_{HC}) \tag{8.1}$$

$$\dot{Q}_L = (\bar{h}A)_L(T_{LC} - T_L) \tag{8.2}$$

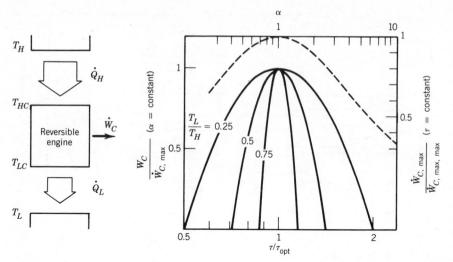

Figure 8.2   The maximization of power output in a power plant of finite size.

The product of type $(\bar{h}A)$ represents the respective heat transfer conductance, or, in heat-exchanger design terms, the engine–reservoir thermal contact surface times the overall heat transfer coefficient based on that surface. Since each conductance $(\bar{h}A)$ is, in general, finite, each heat transfer rate occurs across a finite temperature difference. In Fig. 8.2, the temperature gap $(T_H - T_L)$ that in the past was occupied only by the Carnot engine, is filled now by a three-layer sandwich: the high-temperature gap $(T_H - T_{HC})$, the Carnot engine itself, and, finally, the low-temperature gap $(T_{LC} - T_L)$. The engine is "endoreversible" in the sense that no entropy is being generated between the extreme temperatures seen by the working fluid:

$$\frac{\dot{Q}_H}{T_{HC}} = \frac{\dot{Q}_L}{T_{LC}} \tag{8.3}$$

Note finally that since we are analyzing specifically a heat engine, the heat transfer rates are represented in Fig. 8.2 by arrows that point in the direction of the energy transfer.

Optimum Temperature Ratio

Interesting conclusions are reached in the process of maximizing the instantaneous power output:

$$\dot{W}_C = \dot{Q}_H (1 - T_{LC}/T_{HC}) \tag{8.4}$$

Reviewing the content of eqs. (8.1)–(8.4), we see that the power output depends on $(\bar{h}A)_H$, $(\bar{h}A)_L$, T_{HC}, and T_{LC}. The second law (8.3) constrains one of these four parameters. At this early stage, let us assume also that the installation is built, i.e., that $(\bar{h}A)_H$ and $(\bar{h}A)_L$ are two known constants. The Carnot power $\dot{W}_C$ is then a function of only one parameter, namely, T_{HC} or T_{LC}. Or, we can express $\dot{W}_C$ as a function of the ratio of the temperatures that actually border the Carnot cycle:

$$\tau = T_{LC}/T_{HC} \tag{8.5}$$

Therefore, by eliminating T_{HC} and T_{LC} using eqs. (8.1)–(8.3) and (8.5), the instantaneous Carnot power can be written as

$$\frac{\dot{W}_C}{(\bar{h}A)_H T_H} = \frac{(\tau - T_L/T_H)(1 - \tau)}{\tau[1 + (\bar{h}A)_H/(\bar{h}A)_L]} \tag{8.6}$$

This expression shows clearly the trade-off in τ, and that the optimum Carnot-engine temperature ratio depends only on the overall (extreme, imposed) temperature ratio T_L/T_H:

$$\tau_{\text{opt}} = (T_L/T_H)^{1/2} \tag{8.7}$$

The maximum Carnot power level that corresponds to τ_{opt} is

$$\dot{W}_{C,\text{max}} = \frac{(\bar{h}A)_H T_H}{1 + (\bar{h}A)_H/(\bar{h}A)_L}\left[1 - \left(\frac{T_L}{T_H}\right)^{1/2}\right]^2 \tag{8.8}$$

That $\dot{W}_C$ reaches a well-defined maximum at $\tau = \tau_{\text{opt}}$ is illustrated in Fig. 8.2

Heat-Engine Efficiency for Maximum Power

A complementary result of the preceding analysis is the heat-engine efficiency that corresponds to the power maximum:

$$\eta' = (\dot{W}_C/\dot{Q}_H)_{\tau=\tau_{\text{opt}}} = 1 - (T_L/T_H)^{1/2} \tag{8.9}$$

This expression deserves to be called the *Curzon and Ahlborn efficiency* after the University of British Columbia thermodynamicists who determined it first [3]. This result is representative of a new emphasis in fundamental engineering thermodynamics [e.g., Refs. 4–6], which I believe is closely related to my own effort of bridging the gap between engineering thermodynamics and heat transfer engineering [7–10]. The derivation presented here is a modified version of the simpler approach offered by De Vos [11].

The Curzon and Ahlborn efficiency is not the "maximum" efficiency of the heat engine, in other words, it should not be confused with the Carnot efficiency $\eta_C = 1 - T_L/T_H$. The η' value is simply the efficiency for which—

in the presence of finite $(\bar{h}A)_H$ and $(\bar{h}A)_L$—the instantaneous power output reaches its maximum. Significant is the fact that η' does not depend on the size of the power plant $[(\bar{h}A)_H + (\bar{h}A)_L]$ and on the relative size of its two heat exchangers $[(\bar{h}A)_H/(\bar{h}A)_L]$. This means that the heat-engine efficiency of power plants designed for maximum power output should approach the value given by eq. (8.9) regardless of the many design parameters (fuel, working fluid, cooling medium, size) that enter the design process. The power plant examples gathered in Table 8.1 justify this expectation.

TABLE 8.1   The Observed Efficiency of Ten Power Plants, as an Illustration of Curzon and Ahlborn's Efficiency for Maximum Power Output

Symbol on Fig. 8.3	Power Plant	T_L (°C)	T_H (°C)	η_C	η'	η (observed)
①	West Thurrock (UK) 1962 conventional coal-fired steam plant [12]	25	565	0.64	0.4	0.36
②	CANDU (Canada), PHW nuclear reactor [13]	25	300	0.48	0.28	0.3
③	Larderello (Italy), geothermal steam plant [14]	80	250	0.32	0.18	0.16
④	1936–1940 central steam-power stations in the UK [12]	25	425	0.57	0.35	0.28
⑤	Calder Hall (UK) 1956 nuclear reactor [12]	25	310	0.49	0.28	0.19
⑥	Dungeness "A" (UK), 1965 nuclear reactor [12]	25	390	0.55	0.33	0.33
⑦	1956 steam-power plant in the U.S. [12]	25	650	0.68	0.43	0.40
⑧	1949 combined-cycle (steam and mercury) plant in the U.S. [12]	25	510	0.62	0.38	0.34
⑨	1944 closed-cycle gas turbine in Switzerland [12]	25	690	0.69	0.44	0.32
⑩	1950 closed-cycle gas turbine in France [12]	25	680	0.69	0.44	0.34

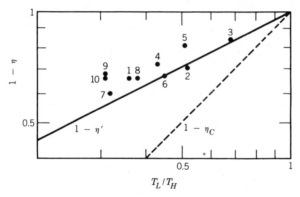

Figure 8.3 Compilation of power-plant efficiencies, showing the emergence of a correlation of the type suggested by eq. (8.9) (the numbers refer to the power-plant examples listed in Table 8.1).

I use this opportunity to correlate the data of Table 8.1 using the double-logarithmic plot of Fig. 8.3. On this plot, the observed efficiencies increase in the direction of the lower-left corner. The efficiencies calculated with Curzon and Ahlborn's formula form a straight line of slope $\frac{1}{2}$. The Carnot efficiencies, on the other hand, form a line of slope 1. It is quite clear that the actual power-plant efficiencies compiled in Table 8.1 [12–14] follow the trend suggested by the theory of instantaneous power maximization. Furthermore, the Curzon and Ahlborn efficiency η' appears to be an upper bound for the observed efficiencies. The examples that deviate the most from the maximum-power trend $(1 - \eta') = (T_L/T_H)^{1/2}$ are the oldest of the power plants listed in Table 8.1 [see examples (4), (5), (9), and (10)].

Optimum Distribution of Heat-Exchanger Equipment

It is possible to go beyond the power-maximization step executed by Curzon and Ahlborn, and to question whether $\dot{W}_{C,\max}$ can be maximized further by the proper distribution of heat-exchanger area between the two ends of the power cycle. Let the new constant $(\bar{h}A)$ symbolize the total (combined) size of the heat-exchanger equipment invested in the power plant:

$$(\bar{h}A) = (\bar{h}A)_H + (\bar{h}A)_L \tag{8.10}$$

We can then nondimensionalize the once-maximized power output (8.8) by dividing it by the constant $(\bar{h}A)T_L$:

$$\frac{\dot{W}_{C,\max}}{(\bar{h}A)T_L} = \frac{\alpha}{(1+\alpha)^2} \frac{T_H}{T_L} \left[1 - \left(\frac{T_L}{T_H}\right)^{1/2}\right]^2 \tag{8.11}$$

where α is the dimensionless hot-end/cold-end heat-exchanger size ratio:

$$\alpha = \frac{(\bar{h}A)_H}{(\bar{h}A)_L} \tag{8.12}$$

The upper graph that has been superimposed on Fig. 8.2 shows the effect of α on power output: optimum allocation of heat-exchanger equipment occurs when the overall $(\bar{h}A)$ is divided equally between the two temperature ends of the power plant:

$$\alpha_{\text{opt}} = 1 \tag{8.13}$$

which means that the ultimate power maximum is

$$\frac{\dot{W}_{C,\text{max},\text{max}}}{(\bar{h}A)T_L} = \frac{T_H}{4T_L}\left[1 - \left(\frac{T_L}{T_H}\right)^{1/2}\right]^2 \tag{8.14}$$

In conclusion, the maximum power that can be produced by a power plant (whose size is dominated by the two heat exchangers) increases proportionally with the overall size of the plant, $(\bar{h}A)$. The maximum power also increases with the imposed temperature ratio T_H/T_L or, as shown in Fig. 8.4, with the dimensionless temperature difference spanned by the three-part system of Fig. 8.2, $(T_H - T_L)/T_L$. In the limit of small temperature differences or low-temperature heat sources, the maximum power is given by

$$\frac{\dot{W}_{C,\text{max},\text{max}}}{(\bar{h}A)T_L} \rightarrow \left(\frac{T_H - T_L}{4T_L}\right)^2 \qquad (\text{as } T_H \rightarrow T_L) \tag{8.15}$$

This limiting result and Fig. 8.4 show that throughout the $(T_H - T_L)$ range

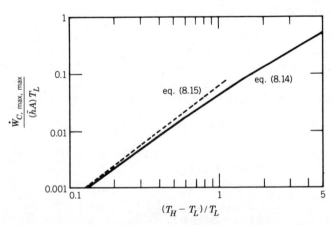

Figure 8.4 The maximum power level of a power plant of finite size, showing that it increases proportionally with the overall size $(\bar{h}A)$ and, roughly, with $(T_H - T_L)^2$.

spanned by today's power plants (Table 8.1), the maximum power level increases roughly as $(T_H - T_L)^2$. This theoretical conclusion is particularly interesting: it reinforces the message of Fig. 8.3, which is that the pursuit of higher and higher T_H/T_L ratios is demanded by the maximization of instantaneous power in plants of finite size, not by the maximization of the forever elusive Carnot efficiency. Other aspects of this design are presented in Problem 8.20.

Application to an Ideal Brayton Cycle

The preceding conclusions were made to a large extent possible by the simplicity of the power-plant model adopted in Fig. 8.2. More realistic power-plant models—even the highly idealized ones—require more complex analytical and numerical work that tends to obscure the message conveyed here by Figs. 8.3 and Fig. 8.4. The purpose of this subsection is to demonstrate that the trade-offs illustrated by Fig. 8.2 govern the maximization of power in more realistic, textbook-type, power-plant descriptions.

Consider for this purpose the ideal Brayton cycle in Fig. 8.5. The working fluid (ideal gas with constant c_P) is heated at constant pressure P_H as it flows from the compressor outlet (1) to the turbine inlet (2). The compressor and the turbine process the fluid reversibly and adiabatically, i.e., isentropically. Between the turbine outlet (3) and the compressor inlet (4), the working fluid is cooled at constant pressure P_L. The power plant is sandwiched

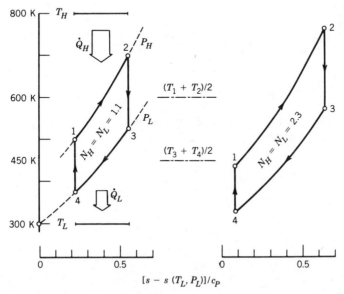

Figure 8.5 The Brayton cycle with $(\bar{h}A)$ divided equally between the heater and the cooler (scale drawing for $T_H = 800$ K, $T_L = 300$ K, and $a = 4/3$).

between two temperature reservoirs, (T_H) and (T_L), meaning that the maximum working fluid temperature can never exceed T_H, and that the minimum working fluid temperature can never fall below T_L. The Brayton cycle is said to be "ideal" (or endoreversible) because the only irreversibilities accounted for by this model are external to the space occupied by the working fluid itself. These irreversibilities are associated with the heat transfer from (T_H) to the heater tube $(1) \rightarrow (2)$, and with the heat transfer from the cooler $(3) \rightarrow (4)$ to the low-temperature reservoir (T_L).

To maximize the instantaneous power output W is to minimize the entropy-generation rate for the entire installation:

$$\dot{S}_{gen} = \dot{m}c_P\left(\frac{T_3 - T_4}{T_L} - \frac{T_2 - T_1}{T_H}\right) \tag{8.16}$$

Note that this expression comes from writing

$$\dot{S}_{gen} = \frac{\dot{Q}_L}{T_L} - \frac{\dot{Q}_H}{T_H} \tag{8.17}$$

where

$$\dot{Q}_L = \dot{m}c_P(T_3 - T_4) \qquad \text{and} \qquad \dot{Q}_H = \dot{m}c_P(T_2 - T_1)$$

Alternatively, the overall entropy-generation rate (8.16) can be derived by adding the contributions made by the two components that operate irreversibly:

$$\dot{S}_{gen,(T_H)-(heater)} = \dot{m}c_P\left(\ln\frac{T_2}{T_1} - \frac{T_2 - T_1}{T_H}\right) \tag{8.18}$$

$$\dot{S}_{gen,(cooler)-(T_L)} = \dot{m}c_P\left(\ln\frac{T_4}{T_3} - \frac{T_4 - T_3}{T_L}\right) \tag{8.19}$$

The variation of the four corner temperatures, T_1 to T_4, is constrained by four additional relations:

$$\frac{T_H - T_1}{T_H - T_2} = e^{N_H} \tag{8.20}$$

$$\frac{T_3 - T_L}{T_4 - T_L} = e^{N_L} \tag{8.21}$$

$$\frac{T_1}{T_4} = \frac{T_2}{T_3} = a \text{ (constant)} \tag{8.22}$$

where the Ns represent the "number of heat transfer units" of each heat exchanger [Ref. 7, p. 139]:

$$N_H = \frac{(\bar{h}A)_H}{\dot{m}c_P} \qquad N_L = \frac{(\bar{h}A)_L}{\dot{m}c_P} \qquad (8.23)$$

The temperature ratio (8.22) depends only on the pressure ratio, $a = (P_H/P_L)^{(k-1)/k}$. Finally, as a measure of the overall size of the heat-exchanger equipment, we assume again that the total $(\bar{h}A)$ is fixed, eq. (8.10). In the present terminology, the $(\bar{h}A)$ constraint reads

$$N_H + N_L = \frac{(\bar{h}A)}{\dot{m}c_P} \qquad (8.24)$$

where the flowrate $\dot{m}$ is not necessarily constant.

The overall entropy-generation rate can be written as (Problem 8.1)

$$\frac{\dot{S}_{gen}}{(\bar{h}A)} = \frac{\dot{m}c_P}{(\bar{h}A)} \left[\left(\frac{T_H}{aT_L} \right)^{1/2} - \left(\frac{T_H}{aT_L} \right)^{-1/2} \right]^2 \frac{(e^{N_H} - 1)(e^{N_L} - 1)}{e^{N_H}e^{N_L} - 1} \qquad (8.25)$$

This expression shows that $\dot{S}_{gen}$ decreases monotonically as the pressure ratio (a) increases, i.e., as the cycle expands vertically and eliminates the irreversibility plagued temperature gaps (note also that $\dot{S}_{gen}$ becomes zero in the limit $a = T_H/T_L$). Next, we see that $\dot{S}_{gen}$ depends on both N_H and N_L; however, only one of these parameters constitutes a degree of freedom [see eq. (8.24)]. The minimization of $\dot{S}_{gen}$ with respect to this single degree of freedom yields $N_H = N_L$, in other words (Problem 8.1),

$$(\bar{h}A)_{H,opt} = (\bar{h}A)_{L,opt} = \tfrac{1}{2}(\bar{h}A) \qquad (8.26)$$

$$\frac{\dot{S}_{gen,min}}{(\bar{h}A)} = \left[\left(\frac{T_H}{aT_L} \right)^{1/2} - \left(\frac{T_H}{aT_L} \right)^{-1/2} \right]^2 M \tanh \left(\frac{1}{4M} \right) \qquad (8.27)$$

where M is dimensionless shorthand for the mass flowrate, $M = \dot{m}c_P/(\bar{h}A)$. The same group can be viewed as the inverse of the number of heat transfer units of the two heat exchangers combined. At large flowrates $(M \gg 1)$, the entropy-generation-rate minimum (8.27) is practically independent of M, as the product $M \tanh (1/4M)$ approaches $\tfrac{1}{4}$. On the other hand, $\dot{S}_{gen,min}$ decreases proportionally with M when $M \ll 1$. In conclusion, there is a thermodynamic incentive for slowing down the working fluid as it bathes the two heat-exchanger surfaces. In the limit $M = 0$, the extreme corners of the Brayton cycle match the imposed fuel and ambient temperatures, $T_2 = T_H$ and $T_4 = T_L$, respectively [see Fig. 8.5 and eqs. (8.20) and (8.21)].

Either in terms of $(\bar{h}A)$ or the number of heat transfer units, the optimum heat-exchanger size distribution (8.26) verifies the conclusion reached earlier on the basis of a much simpler power-plant model, eq. (8.10). The meaning of this design rule is illustrated in the remainder of Fig. 8.5. At the optimum, the arithmetic average temperatures of the heater and the cooler depend solely on the imposed temperatures and pressures:

$$(T_1 + T_2)_{\text{opt}} = T_H + aT_L \tag{8.28}$$

$$(T_3 + T_4)_{\text{opt}} = \frac{1}{a} T_H + T_L \tag{8.29}$$

This conclusion matches in spirit the earlier result that the power cycle must occupy a certain position on the temperature scale between T_H and T_L, eq. (8.7). Regardless of whether eq. (8.26) applies, the heat-engine efficiency is a function of P_H/P_L only:

$$\eta = \frac{\dot{W}_{\text{net}}}{\dot{Q}_H} = 1 - \left(\frac{P_L}{P_H}\right)^{(k-1)/k} < 1 - \frac{T_L}{T_H} \tag{8.30}$$

Therefore, although the cycle efficiency is independent of how $(\bar{h}A)$ is distributed among the two heat exchangers, the entropy-generation rate per unit $(\bar{h}A)$ is minimized when the cycle is positioned between T_H and T_L in the manner indicated by eqs. (8.28) and (8.29). The Brayton cycle occupies an increasingly taller area on the T–s diagram as M decreases, i.e., as the numbers of heat transfer units of both heat exchangers increase.

EXTERNAL IRREVERSIBILITIES

The irreversibility that plagued the performance of the examples treated in the preceding section was due entirely to the heat transfer across the finite temperature gaps that occur between the actual power cycle and the extreme temperature reservoirs (T_H) and (T_L) (see Figs. 8.2 and 8.5). In this section, we take a closer look at the origin of this "external" irreversibility. The objective is to first define the exergy-based terminology that is being used more and more in the analysis of power cycles, and, second, to bring into view the common reason for many of the design changes that account for the "advanced" cycles of today's power industry.

It is sufficient to consider the ideal Rankine cycle with superheat, as shown in Fig. 8.6. The cycle is said to be "ideal" because the heat exchangers $(1) \rightarrow (2)$ and $(3) \rightarrow (4)$ are modeled as having zero pressure drop, and because the compressor $(4) \rightarrow (1)$ and the turbine $(2) \rightarrow (3)$ are modeled as reversible and adiabatic (isentropic). Three temperature levels are relevant to the discussion that follows, namely, the ambient temperature T_0, the maximum fluid temperature $T_{\text{max}} = T_2$, and the effective flame temperature T_f, which was defined on p. 383. The maximum fluid temperature T_{max} is constrained by the degradation of the mechanical properties of the turbine-blade material at very high temperatures. The flame tempera-

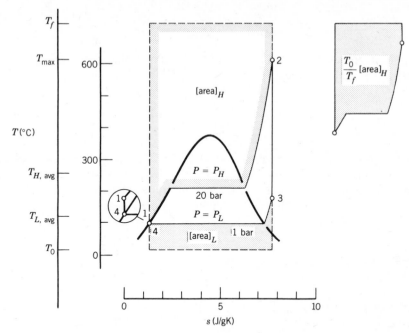

Figure 8.6 External irreversibilities in an ideal Rankine cycle with superheat (here and in subsequent Rankine-cycle diagrams, the T–s diagram is a correct scale drawing of the T–s diagram for water above 0°C; one feature of these drawings is that at subcritical pressures, the compressed-liquid states that reside on the P_H isobar are graphically indistinguishable from the neighboring saturated-liquid states).

ture T_f is the theoretical "ceiling" temperature in the entire installation. Fixing T_f is another way of saying that on the back of the heat transfer rate to the power cycle $(\dot{Q}_H)$ rides the maximum power-producing potential (exergy) of the spent fuel:

$$\dot{E}_{Q_H} = \dot{Q}_H(1 - T_0/T_f) \qquad (8.31)$$

and that the ratio $\dot{E}_{Q_H}/\dot{Q}_H$ is fixed. Equation (8.31) is the definition of T_f: as such, it constitutes a compact summary of the chemical exergy and combustion calculations that lead to $\dot{E}_{Q_H}$. For this reason, T_f is an involved function of many design parameters, such as the type of fuel, the fuel/air ratio, the combustion-chamber design (e.g., the direct heat leak to T_0), etc. Of importance here is only that the T_f value that might be calculated with eq. (8.31) is always higher than $T_{\max}$, and that it is *fixed*. The T_f temperature is the effective temperature of the combustion chamber as an exergy source.

The heat-engine efficiency of the ideal Rankine cycle can be written in a way that shows the impact of external irreversibilities. By definition, we have

$$\eta = \frac{\dot{W}_{net}}{\dot{Q}_H} = 1 - \frac{\dot{Q}_L}{\dot{Q}_H} = 1 - \frac{h_3 - h_4}{h_2 - h_1} \tag{8.32}$$

and since processes $(1) \rightarrow (2)$ and $(3) \rightarrow (4)$ are isobaric, and $dh = T\,ds + v\,dP$, eq. (8.32) becomes

$$\eta = 1 - \frac{(\int_4^3 T\,ds)_{P=P_L}}{(\int_1^2 T\,ds)_{P=P_H}} \tag{8.33}$$

The second term on the right side of this last expression is the ratio of the T–s area under the cooling curve $(3) \rightarrow (4)$ divided by the area measured under the heating curve $(1) \rightarrow (2)$. Since both areas have the same horizontal dimension, $s_2 - s_1 = s_3 - s_4$, it makes sense to define *the average high and low temperatures* of the cycle (see also p. 454):

$$T_{H,avg} = \frac{(\int_1^2 T\,ds)_{P=P_H}}{s_2 - s_1} \qquad T_{L,avg} = \frac{(\int_4^3 T\,ds)_{P=P_L}}{s_3 - s_4} \tag{8.34}$$

so that the heat-engine efficiency becomes

$$\eta = 1 - \frac{T_{L,avg}}{T_{H,avg}} \tag{8.35}$$

The average temperatures $T_{L,avg}$ and $T_{H,avg}$ are at the left of Fig. 8.6: the figure and the Carnot-type efficiency (8.35) show that the ideal Rankine cycle is equivalent to a Carnot cycle operating between $T_{L,avg}$ and $T_{H,avg}$. With $T_f > T_{H,avg}$ and $T_0 < T_{L,avg}$, it is clear that $\eta < 1 - T_0/T_f$ and that the irreversibilities are restricted to the temperature gaps above $T_{H,avg}$ and below $T_{L,avg}$.

The relative size of the two external irreversibilities can be illustrated graphically by first calculating the entropy-generation rates associated with the two temperature gaps:

$$\dot{S}_{gen,H} = \dot{m}(s_2 - s_1) - \frac{\dot{Q}_H}{T_f}$$

$$= \frac{\dot{m}}{T_f} \left[T_f(s_2 - s_1) - \left(\int_1^2 T\,ds \right)_{P=P_H} \right] \tag{8.36}$$

$$\dot{S}_{gen,L} = \frac{\dot{Q}_L}{T_0} + \dot{m}(s_4 - s_3)$$

$$= \frac{\dot{m}}{T_0} \left[\left(\int_4^3 T\,ds \right)_{P=P_L} - T_0(s_3 - s_4) \right] \tag{8.37}$$

Note that the quantities in brackets represent the T–s areas trapped between $T = T_f$ and $P = P_H$, and between $P = P_L$ and $T = T_0$, respectively. Let $[\text{area}]_H$ and $[\text{area}]_L$ denote the two T–s areas listed in eqs. (8.36) and (8.37). In accordance with the Gouy–Stodola theorem (3.7), each entropy-generation rate $\dot{S}_{\text{gen}}$ contributes a $T_0 \dot{S}_{\text{gen}}$ share to the destruction of available work:

$$\dot{W}_{\text{lost}} = T_0 \dot{S}_{\text{gen},H} + T_0 \dot{S}_{\text{gen},L} \tag{8.38}$$

in other words,

$$\frac{\dot{W}_{\text{lost}}}{\dot{m}} = \frac{T_0}{T_f} [\text{area}]_H + [\text{area}]_L \tag{8.39}$$

This shows that although both T–s temperature gap areas are related to the destruction of available work, the contribution of the high-temperature area is, relatively speaking, diminished in proportion with the small number T_0/T_f. This point is made graphically in Fig. 8.6 by the shaded areas: the low-temperature area is shaded as it is, whereas the upper area had to be reduced by the factor T_0/T_f before being shaded. [Note that the horizontal and vertical linear dimensions of $[\text{area}]_H$ were both reduced by the factor $(T_0/T_f)^{1/2}$.]

These graphic conclusions are relevant to the conceptual development of advanced Rankine cycles, in which the positions occupied by the constant-pressure heating and cooling processes are readjusted in an attempt to reduce as much as possible the two temperature-gap areas, $[\text{area}]_H$ and $[\text{area}]_L$. We see here that one unit of T–s area reduction yields a greater return in useful power savings when applied to the low-temperature portion of the cycle.

Another way of looking at the irreversibility of the power plant is to consider the fate of the exergy invested per unit time, eq. (8.31). The power plant is a closed system in communication with (T_f) and (T_0); therefore, according to eq. (3.18),

$$\dot{E}_{Q_H} = \dot{W}_{\text{lost}} + \dot{W}_{\text{net}} \tag{8.40}$$

In this special (ideal Rankine) case, the $\dot{W}_{\text{net}}/\dot{m}$ ratio equals the T–s area delineated by the cycle itself:

$$\frac{\dot{W}_{\text{net}}}{\dot{m}} = h_2 - h_3 - (h_1 - h_4)$$

$$= \left(\int_1^2 T\, ds \right)_{P=P_H} - \left(\int_4^3 T\, ds \right)_{P=P_L} \tag{8.41}$$

Note that the sides of this T–s area are perfectly vertical, i.e., that the equality between $\dot{W}_{\text{net}}/\dot{m}$ and cyclical integral $\oint T\, ds$ holds only when the compressor and the turbine operate isentropically. In conclusion, eq. (8.40) shows that the two shaded areas of Fig. 8.6 and the area enclosed by the cycle $(1) \rightarrow (2) \rightarrow (3) \rightarrow (4) \rightarrow (1)$ represent the specific exergy input $\dot{E}_{Q_H}/\dot{m}$. These three areas have been redrawn stacked together in the left half of Fig. 8.7: the meaning of each component in the stack is made clear by its geometrical shape, which has been cut out directly from Fig. 8.6. To improve the efficiency of the cycle, then, means to enlarge the area of the uppermost component in the stack $(\dot{W}_{net}/\dot{m})$ at the expense of the lower components, which together represent $\dot{W}_{lost}/\dot{m}$.

A more direct way of constructing the T–s area that represents the specific-exergy input $\dot{E}_{Q_H}/\dot{m}$ is to rely on eq. (8.31) and note that

$$\frac{\dot{Q}_H}{\dot{m}} = \left(\int_1^2 T \, ds \right)_{P = P_H} \tag{8.42}$$

The ratio $\dot{Q}_H/\dot{m}$ emerges as the area trapped between the high-pressure heating line $(1) \rightarrow (2)$ and absolute zero. This area occupies now the right side of Fig. 8.7: reducing it by the factor $(T_f - T_0)/T_f$, as indicated by eq.

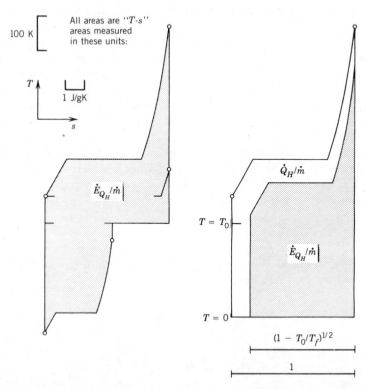

Figure 8.7 Two graphic constructions of the T–s area that represents the specific-exergy intake $\dot{E}_{Q_H}/\dot{m}$.

(8.31), yields a $T–s$ area that equals $\dot{E}_{Q_H}/\dot{m}$. Note that this geometric construction consists of reducing each linear dimension by the factor $[(T_f - T_0)/T_f]^{1/2}$. Of the two graphic constructions presented in Fig. 8.7, the first is more instructive because it shows the partitioning of $\dot{E}_{Q_H}$ into $\dot{W}_{net}$ and $\dot{W}_{lost}$, as well as the extent to which the two external irreversibilities contribute to $\dot{W}_{lost}$.

As an example of how the preceding conclusions can be blended in the conceptual design of a Rankine power cycle, consider again the ideal cycle shown in Fig. 8.6. This time, however, the low pressure P_L is fixed (say, $P_L = 1$ atm), while the maximum temperature T_{max} is constrained by structural design considerations. We are interested in how the selection of the high pressure P_H affects the partitioning of the exergy intake into net (delivered) power and lost power. Indeed, is there an optimum high pressure that maximizes the cycle efficiency when both P_L and T_{max} are fixed?

Figure 8.8 shows the relationship between η and P_H, where all the calculations are based on the Mollier chart of Keenan et al. [15]. At

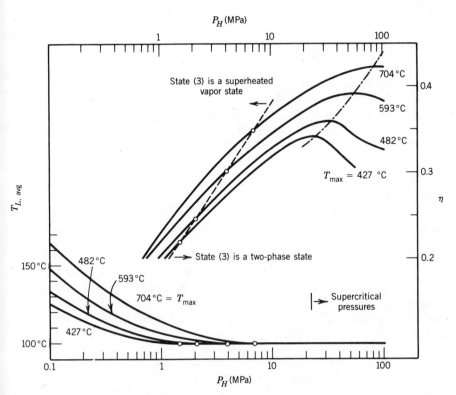

Figure 8.8 The effect of high pressure on the cycle efficiency and the average low temperature of the ideal Rankine cycle of Fig. 8.6 at constant T_{max} and $P_L (=1$ atm).

subcritical pressures, the cycle efficiency increases as P_H increases. During this process, the state of the expander exhaust [state (3)] shifts from the dry steam region into the wet steam dome. Each $\eta(P_H)$ curve shows eventually a maximum at supercritical pressures. The maximum becomes less pronounced as the temperature ceiling T_{max} increases. Correlating the efficiency maxima, we note that the optimum P_H increases with T_{max}; in other words, maximum efficiency means operating at higher pressures when the material temperature limit T_{max} increases. Unfortunately, the quest for higher pressures P_H (also called initial turbine pressures) runs into the turbine-blade erosion problem, which is aggravated as the moisture content of the turbine exhaust increases, [note the location of state (3)].

Figure 8.8 shows also the evolution of the average low temperature $T_{L,avg}$ as P_H increases. This temperature decreases as state (3) moves on the $P_L = 1$ atm line toward the two-phase region, and remains constant once state (3) enters the dome. Since $T_{L,avg}$ is relatively insensitive to P_H, the $\eta(P_H)$ curves of Fig. 8.8 and eq. (8.35) suggest that the average high temperature of the cycle $(T_{H,avg})$ reaches a maximum in the same range of supercritical pressures. A plot of $T_{H,avg}$ versus P_H curves was constructed for a related steam-plant calculation by Baehr [16].

INTERNAL IRREVERSIBILITIES

Each of the four components of the Rankine-cycle power plant generates entropy. This feature was assumed away in the definition of the "ideal" cycle of Fig. 8.6, where the focus was on the remaining "external" irreversibilities associated with the heat transfer between the working fluid and the (T_f) and (T_0) reservoirs. In this section, we analyze these additional, "internal," irreversibilities, and show how each of them works toward enlarging the fraction occupied by lost exergy in the exergy-area maps of Figs. 8.6 and 8.7.

Heater

In general, the flow of the high-pressure working fluid through the heater (or, in this case, boiler and superheater) is accompanied by a finite pressure drop (ΔP_H). The exit state $(2')$ is now represented by the maximum temperature $T_{max} = T_{2'}$ and the pressure $P_H - \Delta P_H$, as shown in Fig. 8.9. The entropy-generation rate contributed by the heater, while in communication with the (T_f) reservoir, is

$$\dot{S}_{gen,heater} = \dot{m}(s_{2'} - s_1) - \frac{\dot{Q}_H}{T_f} \tag{8.43}$$

where the heat transfer rate is given by

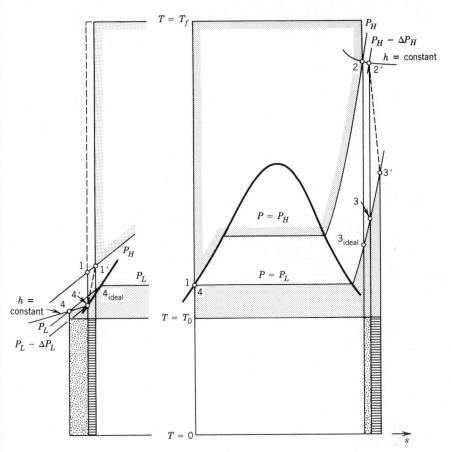

Figure 8.9 The consistent expansion of the lost-exergy area as a result of internal irreversibilities in the four components of the ideal Rankine cycle of Fig. 8.6.

$$\dot{Q}_H = \dot{m}\left(\int_1^2 T\, ds\right)_{P=P_H} \tag{8.44}$$

Note here the relationship between the present exit state (2′) and the neighboring state (2), namely, $h_{2'} = h_2$. This means that state (2) would assume the role of exit state as ΔP_H tends to zero in the presence of the same (constant) $\dot{Q}_H/\dot{m}$.

By converting the entropy generation (8.43) into lost exergy, $\dot{W}_{lost} = T_0 \dot{S}_{gen}$, it is possible to arrive at an expression that spells out the additive contribution of the internal pressure-drop irreversibility:

$$\frac{1}{\dot{m}}\, \dot{W}_{lost,heater} = \frac{T_0}{T_f}\, [\text{area}]_H + T_0(s_{2'} - s_2) \tag{8.45}$$

The first term on the right side accounts for the external heat transfer irreversibility encountered already in the discussion of Fig. 8.6. The second term represents the specific lost exergy associated with the heater pressure drop: this particular contribution occupies the rectangular area covered with dots in the lower-right corner of Fig. 8.9. It is clear from the geometric construction of this area that any pressure drop or "throttling" effect is accompanied by an increase in the lost-exergy fraction of the original exergy intake of the cycle. In the closing segment of this section, we see that the dotted rectangular area is only a part of the lost-exergy area that must be attributed to the heater pressure drop.

Expander

We move on to the inherent irreversibility of the flow process through the expander or turbine. Continuing to model this process as adiabatic, we write that, in general, the expander generates entropy, $s_{3'} \geq s_{2'}$, or that

$$\dot{S}_{\text{gen,expander}} = \dot{m}(s_{3'} - s_{2'}) \qquad (8.46)$$

In the isentropic expansion limit, the expander outlet state occupies the position labeled (3); in other words, $s_3 = s_{2'}$. The exergy destroyed directly by the expander is

$$\frac{1}{\dot{m}} \dot{W}_{\text{lost,expander}} = T_0(s_{3'} - s_3) \qquad (8.47)$$

which measures up to the rectangular area covered with horizontal lines in the lower-right corner of Fig. 8.9. This area increases in width as the expander isentropic efficiency (η_t) decreases. There is also an indirect exergy-destruction effect that is due to the expander inefficiency, namely, the increase in the heat-rejection rate from the cooler to the ambient and in the associated heat transfer irreversibility. This indirect effect is examined next.

Cooler

Taking into account the finite pressure drop across the flow through the cooler or condenser means that if the pressure at the inlet (3′) is P_L, the final pressure of the condensate is $P_L - \Delta P_L$. Let state (4′) represent the saturated-liquid state at the condenser outlet. The entropy-generation rate of the system composed of the stream that flows from (3′) to (4′), and the stream–ambient temperature gap is

$$\dot{S}_{\text{gen,cooler}} = \frac{\dot{Q}_L}{T_0} - \dot{m}(s_{3'} - s_{4'}) \qquad (8.48)$$

The cooling rate $\dot{Q}_L$ is equal to $\dot{m}(h_{3'} - h_{4'})$ or, according to the drawing enlarged to the left of the two-phase dome,

$$\dot{Q}_L = \dot{m}\left(\int_4^{3'} T\, ds\right)_{P=P_L} \tag{8.49}$$

State (4) is situated inside the subcooled liquid region, on the same constant-enthalpy line as the saturated-liquid state (4'). Note the increase in temperature as the pressure drops at constant h from state (4) to state (4'): this effect is the fingerprint of incompressible-liquid behavior known also as the *negative* Joule–Thomson coefficient (Table 4.5).

By combining eqs. (8.48) and (8.49) into the usual "specific-lost-exergy" expression, we have

$$\frac{1}{\dot{m}}\dot{W}_{\text{lost,cooler}} = \left(\int_4^{3'} T\, ds\right)_{P=P_L} - T_0(s_{3'} - s_4) + T_0(s_{4'} - s_4) \tag{8.50}$$

Here, the last term on the right side represents the rectangular area covered with dots in the lower-left corner of Fig. 8.9. The first two terms account for the shaded area trapped between the P_L isobar and the T_0 isotherm. Following this shaded area from $s = s_{3'}$, all the way through the enlarged drawing, to $s = s_4$, we note that it is greater than the area associated with the low-temperature external irreversibility of ideal cycles ([area]$_L$ of Fig. 8.6):

$$
\begin{aligned}
\text{shaded area of Fig. 8.9} = &\left(\int_4^{4_{\text{ideal}}} T\, ds\right)_{P=P_L} - T_0(s_{4_{\text{ideal}}} - s_4) \\
&+ \left(\int_{4_{\text{ideal}}}^{3_{\text{ideal}}} T\, ds\right)_{P=P_L} - T_0(s_{3_{\text{ideal}}} - s_{4_{\text{ideal}}}) \\
&+ \left(\int_{3_{\text{ideal}}}^{3} T\, ds\right)_{P=P_L} - T_0(s_3 - s_{3_{\text{ideal}}}) \\
&+ \left(\int_3^{3'} T\, ds\right)_{P=P_L} - T_0(s_{3'} - s_3) \tag{8.51}
\end{aligned}
$$

The second line in this decomposition represents the specific lost exergy in the ideal Rankine cycle, which was labeled [area]$_L$ in Fig. 8.6. The first line represents the augmentation of the cooler heat transfer irreversibility because of the pressure drop experienced by the stream. This is to say that the dotted rectangular area of width $(s_{4'} - s_4)$ and height T_0, i.e., the last term in eq. (8.50), does not represent *all* the exergy lost by the power plant because of the pressure drop ΔP_L. Exactly the same observation can be made with regard to the meaning of the third line in eq. (8.51): an added effect of the heater pressure drop ΔP_H is to increase the cooler heat-rejection rate and, with it, the cooler–ambient heat transfer irreversibility.

Finally, the last line in eq. (8.51) shows the amount by which the irreversible operation of the expander contributes to the heat transfer irreversibility of the cooler. Overall, the expander irreversibility is responsible for the entire trapezoidal area contained between the low-pressure segment (3)–(3') and absolute zero.

Pump

The entropy-generation rate contributed by the pump is

$$\dot{S}_{gen,pump} = \dot{m}(s_{1'} - s_{4'}) \tag{8.52}$$

where the heat leak from the pump to the ambient has been neglected. The corresponding contribution to the exergy destroyed by the cycle per unit time is

$$\frac{1}{\dot{m}} \dot{W}_{lost,pump} = T_0(s_{1'} - s_{4'}) \tag{8.53}$$

which is represented by the slender rectangular area ruled with horizontal lines in the lower-left corner of Fig. 8.9. The size of this area increases as the isentropic efficiency (η_p) of the pump decreases, i.e., as state (1') climbs up on the high-pressure isobar. At the same time, however, the upper area labeled [area]$_H$ is being diminished from the left (note the vertical dashed line at the left edge of Fig. 8.9). This effect is hardly visible in a steam-turbine power cycle, which is why we postpone its discussion until Fig. 8.20 and the irreversibility of gas-turbine power plants.

The Relative Importance of Internal Irreversibilities

The graphic blowup of the left side of the main $T–s$ diagram of Fig. 8.9 is intended to stress the observation that not all the internal irreversibilities identified in this section are "crucial" from the point of view of maximizing the overall efficiency of the cycle. Some irreversibilities have to be magnified to be seen. Consider, as a first example, the "lost-exergy" effect of pump irreversibility relative to that of the expander irreversibility:

$$\frac{\dot{W}_{lost,pump}}{total\ \dot{W}_{lost,expander}} = \frac{T_0(s_{1'} - s_{4'})}{\left(\int_3^{3'} T\,ds\right)_{P=P_L}} \tag{8.54}$$

Note that in the denominator, we have included the entire trapezoidal $T–s$ area that is borne as a result of expander inefficiency, that is, the rectangular area of eq. (8.47) and the shaded trapezoidal area situated immediately above it. The pump/expander irreversibility ratio can be written more explicitly as (see the necessary modeling, Problem 8.5)

$$\frac{\dot{W}_{\text{lost,pump}}}{\text{total } \dot{W}_{\text{lost,expander}}} = \left(\frac{v_1 P_L}{c_{P,3}\bar{T}_1}\right) \frac{T_0(r-1)(\eta_p^{-1}-1)}{\tilde{T}_3[r^{(k-1)/k}-1](1-\eta_t)} \qquad (8.55)$$

where r is the pressure ratio P_H/P_L, η_p is the isentropic efficiency of the pump, $(h_1 - h_{4'})/(h_{1'} - h_{4'})$, and where η_t is the isentropic efficiency of the expander or turbine, $(h_{2'} - h_3)/(h_{2'} - h_3)$. Noteworthy in this last expression is that the dimensionless group that occupies the first pair of parentheses on the right side is a number much smaller than 1. For example, if $P_L = 1$ atm, this leading group is equal to 0.00014. Since the remainder of the pump/expander irreversibility ratio takes values that are not significantly greater than 1, the message of eq. (8.55) is that in a Rankine-cycle steam-power plant, the exergy destroyed as a result of turbine inefficiency is much greater than the exergy lost because of pump inefficiency.

Two internal irreversibilities that are not necessarily negligible with respect to one another are the irreversibilities associated with the pressure drops ΔP_H and ΔP_L. The relative importance of these exergy-destruction mechanisms can be expressed quantitatively as (Problem 8.6)

$$\frac{\text{total } \dot{W}_{\text{lost,}\Delta P_L}}{\text{total } \dot{W}_{\text{lost,}\Delta P_H}} = \frac{\left(\displaystyle\int_4^{4_{\text{ideal}}} T ds\right)_{P=P_L} - T_0(s_{4_{\text{ideal}}} - s_{4'})}{\left(\displaystyle\int_{3_{\text{ideal}}}^3 T\, ds\right)_{P=P_L}}$$

$$= C\, \frac{(\Delta P/P)_L}{(\Delta P/P)_H} \qquad (8.56)$$

where the numerical coefficient C is shorthand for

$$C = \frac{P_L}{R_2 \bar{T}_3}\left(\frac{db_f}{dP}\right)_{\substack{\text{near}\\ P=P_L}} \qquad (8.57)$$

The derivative db_f/dP concerns the variation of flow availability (or exergy) with pressure along the saturated-liquid-states curve, $dh_f/dP - T_0\, ds_f/dP$, while R_2 represents the ideal gas constant of steam near state (2). Note that the irreversibility ratio defined on the left side of eq. (8.56) is based on accounting for *all* the exergy destroyed as a result of ΔP_H and ΔP_L. The places occupied by the numerator and the denominator of this ratio can easily be identified on Fig. 8.9. Other features of the analytical model that leads to eq. (8.56) are evident from the statement of Problem 8.6.

One interesting aspect of this result is that the value of C is a constant if the condenser pressure P_L is fixed. In the case of water, we have $C \cong 0.15$ if $P_L = 1$ atm, and $C \cong 0.036$ if $P_L = 0.1$ atm. Therefore, the relative importance of the two pressure drops as sources of exergy destruction is established by first calculating C and comparing the $\Delta P/P$ numbers that characterize the heater and the cooler. It turns out that the dimensionless pressure drop $\Delta P/P$ is a very important "number" in the thermodynamic analysis of any heat exchanger (chapter 11; see also Problem 8.7).

Summarizing this section, we see that the combined effect of all the internal irreversibilities is to decrease the rate of exergy delivery from the cycle. Writing $\dot{W}_{net} = \dot{Q}_H - \dot{Q}_L$, we obtain in order

$$\frac{1}{\dot{m}}\,\dot{W}_{net} = (h_{2'} - h_{1'}) - (h_{3'} - h_{4'})$$

$$= \left(\int_{1'}^{2} T\,ds\right)_{P=P_H} - \left(\int_{4}^{3'} T\,ds\right)_{P=P_L}$$

$$= \int_{s=s_{1'}}^{s=s_2} (T_{P=P_H} - T_{P=P_L})\,ds - \left(\int_{s=s_4}^{s=s_{1'}} T\,ds\right)_{P=P_L} - \left(\int_{s=s_2}^{s=s_{3'}} T\,ds\right)_{P=P_L}$$

$$(8.58)$$

The first integral on the right side of the last line is the value of $\dot{W}_{net}/\dot{m}$ in the ideal limit, where all the internal irreversibilities are absent. This area was encountered earlier in Fig. 8.6 as the "cycle" area bordered from above by $[area]_H$ and from below by $[area]_L$. The second integral represents the Fig. 8.9 area (dotted, ruled, and shaded) that has grown to the left of $s = s_{1'}$ as a result of pump inefficiency and cooler pressure drop. Finally, the third integral totals the area that expanded to the right of $s = s_2$ in the wake of the heater pressure drop, expander inefficiency, and the additional cooler irreversibility caused by these two effects. The result delivered by eq. (8.58) is that any internal irreversibility, no matter how small, detracts from the exergy output of the cycle, which is already limited by external irreversibilities.

ADVANCED STEAM-TURBINE POWER PLANTS

In view of the preceding two sections, a consistent strategy for improving the efficiency of power cycles is to minimize the entropy generated both externally and internally. It is important, of course, to attack most aggressively those irreversibilities that *dominate* the overall rate of exergy destruction that plagues the power plant. Three examples of the comparative evaluation of various internal irreversibilities were illustrated in the preceding section and in Problems 8.5–8.7. In this section, we take another look at a sequence of classical design modifications whose common objective is to decrease the external irreversibilities of the cycle.

Superheater, Reheater, and Partial Condenser Vacuum

Figure 8.10 shows the raising of the cycle-average high temperature ($T_{H,\mathrm{avg}}$) by first installing a superheater, Fig. 8.10b, and then splitting the expansion

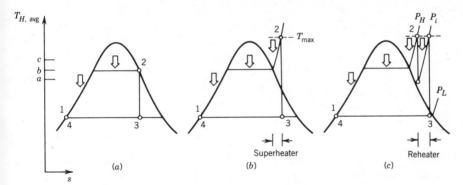

Figure 8.10 The successive augmentation of the average high temperature of a simple ideal Rankine cycle (a) through the insertion of a superheater (b) and a reheater (c).

process between a high-pressure turbine and a low-pressure turbine, and by inserting a reheater between these two stages, Fig. 8.10c. In each design, the cycle segments in which the working fluid is exposed to heating from the (T_f) reservoir are indicated by wide arrows that point downward. In the case of a cycle with superheater and reheater (or with "resuperheat"), there exists an optimum reheating pressure (P_i) that maximizes the cycle efficiency when T_{max}, P_H, and P_L are fixed (Problem 8.8).

Geometrically, it is clear that in moving from Fig. 8.10a to 8.10c, the average temperature $T_{H,avg}$ increases. The price paid for these gains, however, is measured in terms of the additional hardware and the development of heat-exchanger and turbine-blade materials that can withstand increased temperatures. Steam-turbine plants with reheat (or "resuperheat") appeared around 1925, at a time when turbine-blade materials limited T_{max} to 400°C and when the blade-erosion problem limited the turbine-exhaust moisture content to no more than 12 percent. Since in a steam-turbine plant the moisture content at state (3) increases as the initial turbine pressure P_H increases, Fig. 8.10b, the blade-erosion problem constrained P_H to pressures below 30 atm [17]. Therefore, another advantage of the reheating method—an advantage in addition to the increase in the cycle efficiency—is that it reduces the turbine-exhaust moisture content when T_{max} and P_H are fixed. Worth noting is that these upper temperature and pressure limits have climbed steadily through the years, reaching the maximum temperatures listed as T_H in Table 8.1, and initial turbine pressures of the order of 150 atm in subcritical-pressure cycles and 340 atm in supercritical-pressure cycles in the 1950s and 1960s [2].

Another classical route to higher steam-cycle efficiencies is to expand to a "partial vacuum," as shown in Fig. 8.11. The lowering of P_L to subatmospheric levels means lower condenser temperatures and, consequently, lower cycle-averaged $T_{L,avg}$ values. The essential effect is that the shaded area that

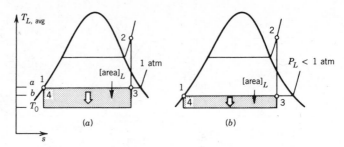

Figure 8.11 The minimization of condenser external irreversibility by lowering the condenser pressure.

represents the exergy lost through condenser–ambient heat transfer shrinks as $T_{L,\text{avg}}$ decreases.

Although the use of subatmospheric condenser pressures is desirable, one cannot think of eliminating the condenser external irreversibility $[\text{area}]_L$ entirely. Assuming for the sake of simplicity that the turbine exhaust is in the two-phase region, the specific lost exergy $[\text{area}]_L$ decreases proportionally with the condenser–ambient temperature difference. In heat transfer engineering terms, the $T_{L,\text{avg}} \to T_0$ limit demands condensers of increasingly larger sizes $(\bar{h}A)$. The same limit is accompanied by increasingly larger condenser pressure drops (ΔP_L); in other words, hand in hand with the shrinking of $[\text{area}]_L$ comes the expansion of the lost-exergy area that is due to ΔP_L, Fig. 8.9. So, at least from the point of view of selecting the right "partial vacuum" for maximum irreversibility reduction, one should minimize not $[\text{area}]_L$ alone, but the sum of $[\text{area}]_L$ and the total specific lost exergy that can be associated with ΔP_L. This trade-off is another manifestation of the competition between heat transfer and fluid-flow irreversibilities in heat transfer devices (chapter 11), which in the context of two-phase-flow heat exchangers is discussed in Ref. 10.

Regenerative Feed Heating

Another approach to maximizing $T_{H,\text{avg}}$ is to remove the low-temperature beginning of the $P = P_H$ heating curve. Conceptually, this can be done by taking the compressed liquid (the feed water) from the pump and leading it in counterflow through the turbine expansion stages, Fig. 8.12. The counterflow heat transfer between the two streams raises the temperature of the compressed-liquid stream close to its boiling point. It is worth keeping in mind that the states (1), (c), (d), . . . represent compressed-liquid states on the P_H isobar, despite their graphic proximity to the saturated-liquid states line. The intended outcome of the feed-heating arrangement is to avoid the degradation of the heat transfer from the (T_f) reservoir in the course of heating the cold compressed-liquid stream produced by the pump.

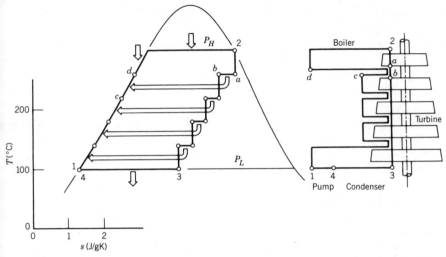

Figure 8.12 Simple Rankine cycle with four stages of regenerative feed heating.

The thermodynamic concept of regenerator (or economizer) is the 1816 invention of the Reverend Robert Stirling, the earliest construction of the device having appeared in his engine and later in James Stirling's and Captain Ericsson's. A theory of the regenerative heating method was presented in 1854 by Rankine [18], who viewed the method as a means of approaching the Carnot efficiency in a machine whose expansion and compression cover a temperature range that is narrower than the overall (extreme) temperature gap of the cycle, i.e., in a physically *smaller* machine that would be required by the theoretical Carnot-engine design. Indeed, as the number of turbine expansion stages of Fig. 8.12 becomes infinite, the cycle efficiency approaches the Carnot efficiency dictated by the saturation temperatures associated with P_H and P_L (see also Problem 8.9 later in this chapter).

The feeding of the compressed-liquid stream through the turbine expansion stages is illustrated as a sequence of four steps in Fig. 8.12. The figure was drawn to scale assuming that the expansion through each turbine stage is isentropic and that enough contact area exists between the liquid stream and the expanding stream so that the liquid temperature is brought up to the temperature of the condensing stream (i.e., $T_d = T_b$). Next, the lateral move from state (a) to state (b) on the T–s diagram is dictated by the first-law analysis of the counterflow heat exchanger, $h_a - h_b = h_d - h_c$. When the number of feed-heating stages if finite, the regenerative heating scheme is irreversible because of the heat transfer across finite temperature gaps of type $(T_b - T_c)$. In the limit of infinitely many feed heating stages, this sort of "distributed" internal irreversibility disappears and the regenerative feed-heating arrangement becomes reversible. In that limit, the expanding stream $(2) \rightarrow (3)$ traces a well-defined curve $s(T)$ inside the two-phase dome of Fig. 8.12.

A practical version of these considerations is encountered in the process of improving the performance of a steam-turbine plant *with superheat*, where, unlike in Fig. 8.12, a good stretch of the stream that expands through the turbine is single-phase. Figure 8.13 shows a simple way of effecting the regenerative heating of the feed-water stream. The new engineering components in this power plant are the *contact heaters*, in which the feed-water stream is heated step by step and augmented by mixing with steam (wet at first, and later superheated) bled from the turbine. Each contact heater has its own pump, whose function is to raise the pressure of the feed to the saturation pressure of the steam drawn from the turbine. Note that the distribution of the pump effect among the three contact heaters exhibited in Fig. 8.13 has been exaggerated on the left side of the two-phase dome.

The feed-heating scheme of the superheat cycle derives its irreversibility from two main sources, first, the mixing of subcooled liquid with a saturated mixture of higher temperature, and, in the case of the later feed-heating stages, the mixing of superheated steam with a saturated mixture of lower temperature. With the irreversibility of each feed-heating stage depending on the temperature gaps that separate it from adjacent stages and from its own turbine-pressure tap, it is clear that the effectiveness of the entire string of contact feed heaters depends on their distribution along the turbine, i.e., on the sequence of turbine-pressure taps P_i.

The problem of determining the optimum distribution of a finite number

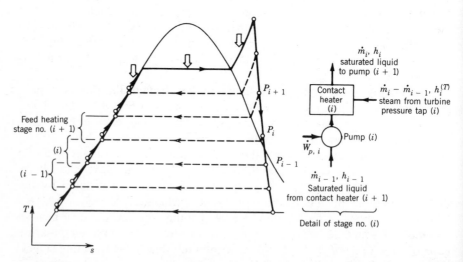

Figure 8.13 Steam cycle with superheat and a train of contact feed heaters and feed pumps.

of feed-heating stages is one of the most fundamental topics in steam-turbine power-plant design. It was first studied by Chambadal [19], Schäff [20], and Ricard [21], and on the basis of a concise analytical model by Haywood [22; also Ref. 2, pp. 130–140]. In what follows, we discuss Haywood's formulation of this fundamental problem because, although approximate, it is an effective way of illustrating the essence of the optimization procedure and the power of concise analytical models in engineering thermodynamics in general.

Consider a superheat steam cycle in which the peripheral parameters are fixed $(P_H, P_L, T_{max}, \eta_t, \eta_p)$ and the only design decision that is to be made is the positioning of the n contact feed heaters along the turbine. The object is to maximize the cycle efficiency:

$$\eta = 1 - \frac{\dot{m}_0(h_3 - h_4)}{\dot{m}_n(h_2 - h_n)} \qquad (8.59)$$

in which the corner enthalpies h_2, h_3, h_4, and h_B are fixed (Fig. 8.14). For now, we assume that the enthalpy of the inlet to the boiler (h_n) is fixed, even though its value is the direct result of the position occupied by the nth feed heater. We reconsider this assumption at the end of this subsection. The flowrates $\dot{m}_0$ and $\dot{m}_n$ are the condenser and boiler flowrates, respectively. The boiler flowrate $\dot{m}_n$ is always greater than $\dot{m}_0$, because it is the compounded result of n incidents of feed-stream augmentation by mixing with steam bled from the turbine. In conclusion, to maximize η is the same as maximizing the boiler/condenser flowrate ratio $\dot{m}_n/\dot{m}_0$.

An analytical expression for $\dot{m}_n/\dot{m}_0$ is obtained from the first-law analysis of an individual contact heater, Fig. 8.13,

$$\dot{m}_{i-1}h_{i-1} + (\dot{m}_i - \dot{m}_{i-1})h_i^{(T)} + \dot{W}_{p,i} = \dot{m}_i h_i \qquad (8.60)$$

where $h_i^{(T)}$ is the enthalpy of the turbine steam bled through the pressure tap P_i. The enthalpy of the feed water, h_i, is the enthalpy of saturated liquid. One important feature of this analytical model is that the feed-pump power $\dot{W}_{p,i}$ is regarded as negligible in eq. (8.60): this is why, in Fig. 8.14, the train of feed heaters is represented only by the sequence of saturated-liquid states $(1), \ldots, (i), \ldots, (n)$. The mass flow ratio that shows the augmentation of the feed stream through one contact heater is then

$$\frac{\dot{m}_i}{\dot{m}_{i-1}} = 1 + \frac{r_i}{\beta_i} > 1 \qquad (8.61)$$

where, according to Haywood's notation, r_i is the enthalpy rise experienced by the feed water through the contact heater (i):

$$r_i = h_i - h_{i-1} \qquad (8.62)$$

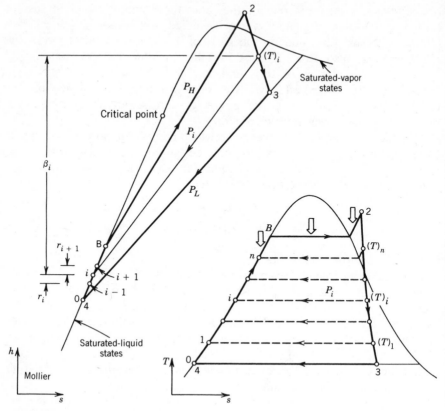

Figure 8.14 *T–s* diagram and Mollier chart (*h–s*) showing the distribution of *n* contact feed heaters.

and β_i is the enthalpy drop experienced by the steam bled by heater (*i*) from the turbine:

$$\beta_i = h_i^{(T)} - h_i \tag{8.63}$$

The always positive quantities r_i and β_i correspond to the vertical segments shown in the Mollier chart of Fig. 8.14.

Reexamining the makeup of $\dot{m}_n/\dot{m}_0$, we rely on the recurrence relation (8.61) to conclude that the optimization process boils down to maximizing the product

$$\frac{\dot{m}_n}{\dot{m}_0} = \prod_{i=1}^{n} \left(1 + \frac{r_i}{\beta_i}\right) \tag{8.64}$$

That the selection of each tap pressure has a maximization effect on this product is illustrated by considering the case where only P_i is free to vary, i.e., the case in which the conditions into and out of the remaining $n - 1$ feed heaters $(1), \ldots, (i - 1)$ and $(i + 1), \ldots, (n)$ are held fixed. In geometrical terms, the variation of P_i amounts to the travel of point (i) between the fixed points $(i - 1)$ and $(i + 1)$ on the saturated-liquid line, Fig. 8.14, i.e., to the variation of r_i and r_{i+1} subject to the constraint

$$r_i + r_{i+1} = r \quad \text{(fixed)} \tag{8.65}$$

Of the total n factors in the product of eq. (8.64), only two vary as a result of this travel:

$$\frac{\dot{m}_{i+1}}{\dot{m}_{i-1}} = \left(1 + \frac{r_i}{\beta_i}\right)\left(1 + \frac{r_{i+1}}{\beta_{i+1}}\right)$$
$$= \left(1 + \frac{r_i}{\beta_i}\right)\left(1 + \frac{r - r_i}{\beta_{i+1}}\right) \tag{8.66}$$

The maximization of this expression is greatly simplified by the observation that the position of the cycle is such that β_i is practically independent of r_i (notice how β_i shifts unchanged as r_i varies on the Mollier chart, Fig. 8.14). Writing, therefore,

$$\beta_i \cong \beta_{i+1} = \beta \quad \text{(fixed)} \tag{8.67}$$

the maximization of eq. (8.66) occurs when [22]

$$r_i = r_{i+1} = r/2 \tag{8.68}$$

In conclusion, the optimum position of the contact heater (i) is the one that splits the feed-enthalpy rise evenly between heaters (i) and $(i + 1)$. The reasoning that led to this conclusion can be repeated for any other pair of adjacent feed heaters. In the end, we learn that the entire product (8.64) reaches its maximum when the n feed heaters are distributed evenly between states (0) and (n) on the saturated-liquid line of the Mollier chart, Fig. 8.14, i.e., when the total enthalpy rise experienced by the feed $(h_n - h_0)$ is divided evenly between the n feed heaters.

It remains to determine the optimum location of the last feed heater (n) relative to the boiling onset (B). This is the same as deciding how much preheating of the single-phase fluid goes on in the boiler before boiling actually takes over. The heat transfer rate in question, $\dot{Q}_{n-B} = \dot{m}_n(h_B - h_n)$, originates from the high-temperature ceiling (T_f). Therefore, thermodynamically, it makes no difference if $\dot{Q}_{n-B}$ is first absorbed by an intermediate entity and only later used for the purpose of heating the feed water between (n) and (B). Let this intermediate entity be the more voluminous boiler stream $\dot{m}_{n+1}$ shown on the right side of Fig. 8.15. Note that $\dot{m}_{n+1} > \dot{m}_n$ because $\dot{m}_n(h_2 - h_n) = \dot{m}_{n+1}(h_2 - h_B)$, where both quantities in the equality are equal to $\dot{Q}_{n-B} + \dot{Q}_{B-2}$.

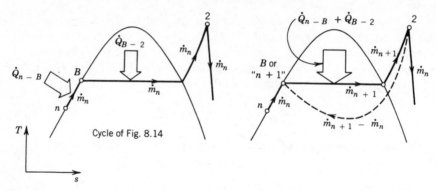

Figure 8.15   The equivalence between the cycle of Fig. 8.14 and a cycle with $(n + 1)$ feed-heating stages and a saturated liquid of state (B) as inlet to the boiler (the right side).

The left side of Fig. 8.15 shows the upper part of the cycle seen on the T–s diagram of Fig. 8.14. We conclude that equivalent to the cycle of Fig. 8.14 is the one where the feed is heated from (n) to (B) in an additional contact heater that draws superheated steam directly from state (2). The feed flowrate into this $(n + 1)$th contact heater is $\dot{m}_n$, the bled superheated steam flowrate is $(\dot{m}_{n+1} - \dot{m}_n)$, and the flowrate through the boiler between sections (B) and (2) is $\dot{m}_{n+1}$. The cycle efficiency in this case can be expressed as

$$\eta = 1 - \frac{\dot{m}_0(h_3 - h_4)}{\dot{m}_{n+1}(h_2 - h_B)} \tag{8.69}$$

where η is exactly the same η as in eq. (8.59). In this last expression, all the hs are fixed, and the efficiency-maximization problem reduces to the maximization of $\dot{m}_{n+1}/\dot{m}_0$. According to the analysis outlined between eqs. (8.61) and (8.68), the maximum η occurs when the $n + 1$ contact heaters contribute evenly to the overall enthalpy rise experienced by the feed-water stream, $(h_B - h_0)$. This means that the optimum location of the inlet to the boiler in the original configuration of Fig. 8.14 is given by

$$\frac{h_B - h_n}{h_B - h_0} \cong \frac{1}{n + 1} \tag{8.70}$$

where, again, n is the actual number of contact heaters that are being contemplated. The approximately equal sign of eq. (8.70) is a reminder of the Mollier-chart approximation that was adopted as eq. (8.67).

The existence of an optimum ultimate feed-water temperature (or h_n) is illustrated in Fig. 8.16, where each curve corresponds to a design with a finite number of feed heaters (n). Plotted on the ordinate is the "theoretical

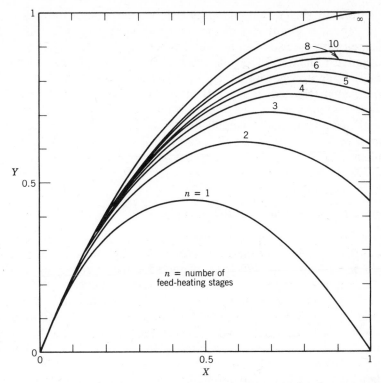

Figure 8.16 The relative reduction in the heat transfer input $\dot{Q}_H$, due to the use of n contact feed heaters (drawn after Salisbury [23]; $P_H = 82.6$ atm, $P_L = 0.0334$ atm, $T_{max} = 482°C$).

gain percentage" [23], or the relative reduction in the heat input $\dot{Q}_H$ that is attributed to the use of n contact heaters:

$$Y = \frac{\dot{Q}_H(0) - \dot{Q}_H(n)}{\dot{Q}_H(0) - \dot{Q}_H(\infty)} \qquad (8.71)$$

In this dimensionless group, $\dot{Q}_H(0)$ represents the heat transfer input received from (T_f) when the cycle does not employ any regenerative feed heating. $\dot{Q}_H(\infty)$ is the heat input in the case of an infinite number of feed heaters $(h_n = h_B)$, for a cycle delineated by a fixed set of peripheral conditions, $P_H = 82.6$ atm, $P_L = 0.0334$ atm, and $T_{max} = 482°C$. The abscissa parameter is the relative enthalpy rise experienced by the feed water:

$$X = \frac{h_n - h_0}{h_B - h_0} \qquad (8.72)$$

Figure 8.16 shows that, if optimized, the installation of a single feed heater generates about half of the theoretically maximum savings that are associated with the use of an infinite number of feed heaters. As n increases, the relative reduction in $\dot{Q}_H$ decreases, while the optimum ultimate feed enthalpy h_n approaches h_B, cf. eq. (8.70). The curves of Fig. 8.16 are based on the actual (charted, tabulated) properties of steam, not on simplifying assumptions such as eq. (8.67), which appeared almost 10 years later. Nevertheless, Haywood's simplified analysis [22] anticipates very well the position of the maximum of each constant-n curve in the X–Y plane (see Problem 8.10).

Combined Feed Heating and Reheating

Looking back at the Fig. 8.10, we get the idea that if regenerative feed heating can improve the performance of the superheat cycle (b), then it should also upgrade the efficiency of the superheat + reheat cycle (c). This expectation is based on the geometric similarity between the first (low-temperature) portions of the constant-P_H heating curves: in both Fig. 8.10b and Fig. 8.10c, the use of regenerative feed heating promises to eliminate the low-temperature heat transfer process needed to heat the feed water to the saturation temperature associated with P_H. Hand in hand with the implementation of feed-water heating in superreheat cycles comes the fundamental problem of determining the optimum distribution of feed heaters, or the optimum distribution of enthalpy rise among the few feed heaters. This problem was considered by Weir [24] in a study that included also the application of feed heating to supercritical pressure cycles. For the sake of continuity with the preceding subsection, we discuss the main conclusion of this work in terms of Haywood's simplified analysis, which in 1960 was extended to resuperheat cycles [25].

The new element in the analysis is the reheater that imparts the heat transfer rate $\dot{Q}_R$ to the partially expanded stream of pressure P_R, Fig. 8.17. Let $(i+1)$ and (i) represent the pressure taps (or contact heaters) that at the end of the optimization process are positioned upstream and downstream, respectively, of the reheater. It is clear from the analysis of the preceding section and from the Mollier charts of Figs. 8.14 and 8.17 that the optimization of contact-heater distribution upstream $(i+1, \ldots, n)$ and downstream of the reheater $(1, \ldots, i)$ leads to the compact conclusion that the feed-enthalpy rise must be divided evenly between adjacent heaters, eqs. (8.68)–(8.70). This conclusion is illustrated in Fig. 8.17 by means of two strings of equidistant points placed on the line of saturated-liquid states. The spacing of the two strings is different in order to emphasize the only unknown left in the present problem, namely, the relative size of the two feed-enthalpy rises r_{i+1} and r_i.

The efficiency of the power cycle is

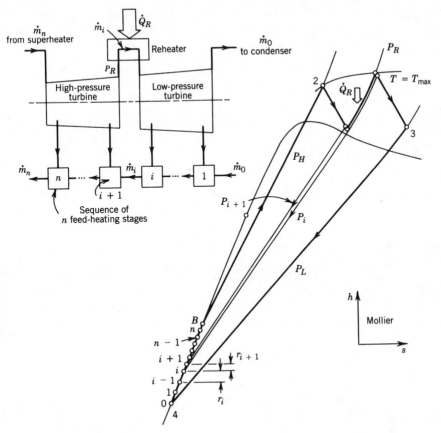

Figure 8.17 The optimum positioning of contact feed-water heaters around the reheater of a steam cycle.

$$\eta = 1 - \frac{\dot{Q}_C}{\dot{Q}_B + \dot{Q}_R} \qquad (8.73)$$

where $\dot{Q}_C$ and $\dot{Q}_B$ are the condenser heat-rejection rate and the (boiler + superheater) heat input that precedes $\dot{Q}_R$ along the heating side of the cycle. Writing $\Delta h_C = h_3 - h_4$ for the fixed enthalpy drop through the condenser, and $\dot{m}_0$ for the condenser flowrate, the efficiency becomes

$$\eta = 1 - \frac{\Delta h_C}{\dot{Q}_B/\dot{m}_0 + \dot{Q}_R/\dot{m}_0} \qquad (8.74)$$

This shows that for maximum η, we must pursue the maximization of the denominator of the second term on the right side. The maximization consists of selecting P_i (or r_i) while holding all the other parameters fixed, including the positions of the $i-1$ contact heaters situated below heater (i), and the positions of the $n-i$ heaters situated above heater (i). This means that the following flowrate ratios are fixed:

$$\frac{\dot{m}_{i-1}}{\dot{m}_0} = k_0 \qquad \frac{\dot{m}_n}{\dot{m}_{i+1}} = k_n \tag{8.75}$$

where the constants k_0 and k_n depend on the number of optimally spaced contact heaters that fit below and above (i). By using the recurrence relation (8.61), it is easy to show that

$$\frac{\dot{Q}_B}{\dot{m}_0} = k_0 k_n \, \Delta h_B \, \frac{\alpha_i \alpha_{i+1}}{\beta_i \beta_{i+1}} \tag{8.76}$$

$$\frac{\dot{Q}_R}{\dot{m}_0} = k_0 \, \Delta h_R \, \frac{\alpha_i}{\beta_i} \tag{8.77}$$

where $\alpha_i = \beta_i + r_i$ and $\alpha_{i+1} = \beta_{i+1} + r_{i+1}$. Note further that the enthalpy excursion from $(i-1)$ to $(i+1)$ on the saturated-liquid line of the Mollier chart, Fig. 8.17, is fixed, and that the constraint (8.65) applies here unchanged. Of interest is the sum of the two derivatives:

$$\frac{d(\dot{Q}_B/\dot{m}_0)}{dr_i} = \frac{\dot{Q}_B}{\dot{m}_0} \left(\frac{1}{\alpha_i} - \frac{1}{\alpha_{i+1}} \right) \tag{8.78}$$

$$\frac{d(\dot{Q}_R/\dot{m}_0)}{dr_i} = \frac{\dot{Q}_R}{\dot{m}_0 \alpha_i} \tag{8.79}$$

Setting the sum equal to zero, yields the wanted result [25]:

$$\frac{r_i + \beta_i}{r_{i+1} + \beta_{i+1}} = 1 + \frac{\dot{Q}_R}{\dot{Q}_B} \tag{8.80}$$

The preceding analysis is based on the approximation that both β_i and β_{i+1} are insensitive to changes in r_i. This makes eq. (8.68) a particular case of eq. (8.80), namely, the case of a cycle with zero reheat ($\dot{Q}_R/\dot{Q}_B = 0$ and $\beta_i \cong \beta_{i+1}$). A more general version of this analysis—one in which the possible effect of r_i on β_i and β_{i+1} is taken into account—can be found in Haywood's original communication [25].

Even though in eq. (8.80) the bled-steam enthalpy drops β_i and β_{i+1} are treated as constant, they are, in general, different[†]:

$$\beta_i - \beta_{i+1} = \Delta\beta > 0 \tag{8.81}$$

[†]The relative size of β_i and β_{i+1} can be measured from Fig. 8.17, keeping in mind the graphic definition of β_i given in Fig. 8.14.

The optimum-design rule (8.80) can be rewritten in a way that shows the relative size of r_i and r_{i+1}:

$$r_i - r_{i+1} = \frac{\dot{Q}_R}{\dot{Q}_B} (\beta_{i+1} + r_{i+1}) - \Delta\beta \qquad (8.82)$$

Because in most superheat cycles the ratio $\dot{Q}_R/\dot{Q}_B$ is not negligibly small when compared with unity, eq. (8.82) and the dimensions of the Mollier-chart cycle indicate that r_i must exceed r_{i+1} by a certain, optimum amount. In conclusion, the outlet states of the n contact heaters must be arranged equidistantly in two strings along the saturated-liquid line of the Mollier chart, in such a way that the density of the upper string exceeds the density of the lower string in the manner indicated by eq. (8.82).

It is fitting to end this section with the remark that although both "reheating" and "feed heating" lead to improvements in the efficiency of the steam-turbine cycle, the η change caused by one method is greater when the method is implemented alone. Note the distinction that is being made between efficiency (η) and the efficiency increment ($\Delta\eta$) associated with the implementation of one or more design changes. The efficiency is certainly the highest when both reheating and feed heating are used. However, the impact of a single method depends on whether the other method has been applied already. This point is made graphically in Fig. 8.18, in which the relative efficiency increase $\Delta\eta/\eta$ is plotted in the vertical direction.

A more concrete illustration of the same point is presented in Fig. 8.19, which shows the percent increase in the efficiency of a steam-turbine plant that uses one reheater and a varying number of feed-heating stages. Note

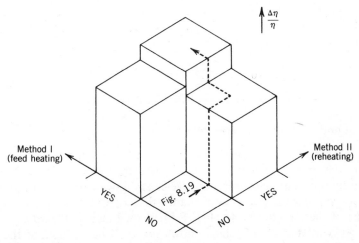

Figure 8.18 The interaction between reheating and feed heating en route to increasing the cycle efficiency.

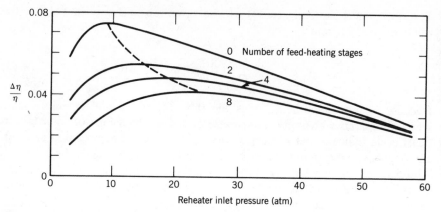

Figure 8.19 The effect of regenerative feed heating on the η increase caused by reheating (drawn after Reynolds [17]; $P_H = 100$ atm, $P_L = 0.05$ atm, $T_{max} = 538°C$, reheater pressure drop $\Delta P_R = 0.1 P_R$).

first the existence of the optimum reheater pressure addressed in Problem 8.8, and that this pressure increases as the number of feed heaters increases. The beneficial effect of using one reheater decreases steadily as the number of feed-heating stages increases. This trend can be detected also by comparing the $T_{H,avg}$ increase registered while proceeding from Fig. 8.10b to Fig. 8.10c, first, without any regenerative feed heaters, and, second, with an infinite number of feed heaters. In the second case, the low-temperature beginnings of the $(1) \rightarrow (2)$ isobaric heating processes (the first vertical arrows) disappear: this modification makes the $T_{H,avg}$ values of Figs. 8.10b and 8.10c very similar, as all the vertical heat transfer arrows enter the working fluid at about the same temperature level. Therefore, in the second case, the exhaustive use of feed heating robs the reheater of most of its opportunity to improve the cycle.

ADVANCED GAS-TURBINE POWER PLANTS

External and Internal Irreversibilities

Following the extensive coverage of steam-turbine-cycle power plants provided in the preceding section, it is now easier to analyze in the same light an entirely different class of power cycles and power plants. In this section, we focus on gas-turbine power plants that operate based on improved versions of the Brayton cycle (Fig. 8.5), known also as the *Joule cycle* [2]. We begin with Fig. 8.20, which shows the realistic version of the simple Brayton cycle encountered in Fig. 8.5. This time the flow through the heater $(1') \rightarrow (2')$ is accompanied by the pressure drop ΔP_H. Similarly, the flow through the cooler $(3') \rightarrow (4')$ is driven by the pressure difference ΔP_L. Finally, the gas turbine and the compressor are plagued by irreversibilities lumped in the respective isentropic efficiencies (η_t, η_c), which are both less than 1. The turbine and the compressor are both modeled as adiabatic, that is, their respective heat leaks to the ambient (T_0) are assumed negligible.

Figure 8.20 invites the analogy between the irreversibilities of the present power plant and that of the steam-turbine power plant disected in Fig. 8.9. An irreversibility source or mechanism that is called by the same name is indicated in the same way on both figures (for example, by using the same shading). Starting from the top of the figure, we recognize the external irreversibility due to heating the working fluid (flowrate $\dot{m}$) across the temperature gap between the combustion chamber (T_f) and the bulk temperature of the heater stream $(1') \rightarrow (2')$. In units of lost exergy per unit flowrate, this external heat transfer irreversibility amounts to

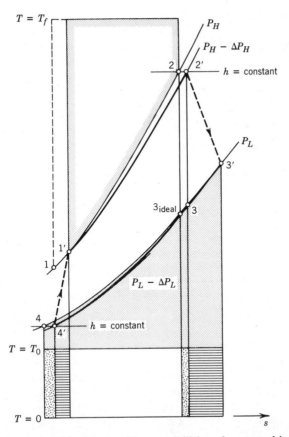

Figure 8.20 The external and internal irreversibilities of a gas-turbine power plant.

$$\frac{T_0}{\dot{m}} \dot{S}_{\substack{\text{gen,heater,} \\ \text{external heat transfer}}} = \frac{T_0}{T_f} \left[T_f(s_2 - s_{1'}) - \left(\int_{1'}^{2} T\, ds \right)_{P=P_H} \right] \quad (8.83)$$

The quantity in brackets represents the almost-trapezoidal area whose border has been shaded in Fig. 8.20. This area is analogous to the $[\text{area}]_H$ quantity studied in connection with Fig. 8.6. Once again, this area must first be reduced by the factor T_0/T_f if it is to represent lost exergy per unit flowrate. In the case of eq. (8.83), the quantity between brackets can be written explicitly in terms of T_2, $T_{1'}$, and c_P if the working fluid is modeled as an ideal gas with constant specific heat.

The internal irreversibility of the heater is due to the pressure drop ΔP_H. The size of this contribution is

$$\frac{T_0}{\dot{m}} \dot{S}_{\substack{\text{gen,heater} \\ \text{pressure drop}}} = T_0(s_{2'} - s_2) \cong T_0 R \frac{\Delta P_H}{P_H} \quad (8.84)$$

where the last expression is based on the ideal gas model and the assumption that $\Delta P_H \ll P_H$. The heater pressure-drop irreversibility is represented by the dotted rectangular area at the right of the base of Fig. 8.20.

Proceeding clockwise around the power cycle, we encounter next the exergy destroyed by the turbine. In the $T \times s$ area units of Fig. 8.20, the turbine irreversibility is represented by the rectangle with horizontal lines in the lower-right corner of the figure (Problem 8.13):

$$\frac{T_0}{\dot{m}} \dot{S}_{\text{gen,turbine}} = T_0(s_{3'} - s_2)$$

$$= T_0 c_P \ln \left\{ 1 + (1 - \eta_t) \left[\left(\frac{P_H - \Delta P_H}{P_L} \right)^{R/c_P} - 1 \right] \right\}$$

$$(8.85)$$

The closed-form expression that concludes eq. (8.85) is based on modeling the working fluid as an ideal gas with constant c_P.

The irreversibility associated with the cooler $(3') \rightarrow (4')$ can be divided into a heat transfer contribution stemming from the stream–ambient temperature difference and a fluid-flow contribution due to the pressure drop ΔP_L. In order to see how this separation of the two effects takes place analytically, in Fig. 8.20, I show that the flow through the cooler $(3') \rightarrow (4')$ is equivalent to the flow through a slightly warmer cooler with zero pressure drop, $(3') \rightarrow (4)$, followed by the flow through a throttle that drops the pressure from P_L to $(P_L - \Delta P_L)$, that is, from state (4) to $(4')$. The per-unit-$\dot{m}$ destruction of exergy during the cooler–ambient interaction is

$$\frac{T_0}{\dot{m}} \dot{S}_{\text{gen,cooler}} = (h_{3'} - h_{4'}) - T_0(s_{3'} - s_{4'})$$

$$= \left(\int_4^{3'} T \, ds\right)_{P=P_L} - T_0(s_{3'} - s_{4'})$$

$$= \left(\int_4^{3'} T \, ds\right)_{P=P_L} - T_0(s_{3'} - s_4) + T_0(s_{4'} - s_4) \quad (8.86)$$

$$\underbrace{\phantom{= \left(\int_4^{3'} T \, ds\right)_{P=P_L} - T_0(s_{3'} - s_4)}}_{\substack{\text{Caused by the heat transfer} \\ \text{from the stream } (3') \to (4) \\ \text{to the ambient } (T_0)}} \quad \underbrace{\phantom{+ T_0(s_{4'} - s_4)}}_{\substack{\text{Caused by the} \\ \text{pressure drop} \\ \Delta P_L}}$$

The heat transfer part noted on the right side is represented by the almost-trapezoidal area shaded in full in Fig. 8.20. Relative to this area, the size of the pressure-drop contribution to eq. (8.86) is measured by the dotted rectangular area shown in the lower-left corner of Fig. 8.20.

Finally, the compressor irreversibility can be estimated using the same approach as in Problem 8.13:

$$\frac{T_0}{\dot{m}} \dot{S}_{\text{gen,compressor}} = T_0(s_{1'} - s_{4'})$$

$$= T_0 c_P \ln \left\{ 1 + \frac{1}{\eta_c} \left[\left(\frac{P_H}{P_L - \Delta P_L}\right)^{R/c_P} - 1 \right] \right\} \quad (8.87)$$

where the last expression is again the result of modeling the fluid as a constant-c_P ideal gas. The top expression on the right side of eq. (8.87) shows that the compressor irreversibility is well represented by the rectangular area with horizontal lines in the lower-left corner of Fig. 8.20.

Summing up the quantities expressed in eqs. (8.83)–(8.87), we obtain the total exergy-destruction rate of the entire power plant. One general observation to stress at this point is that the four components of the power plant interact with one another as they contribute to the overall irreversibility of the plant. The fact that in the preceding analysis we have assigned one $T_0 \dot{S}_{\text{gen}}/\dot{m}$ quantity and one $T \times s$ area in Fig. 8.20 to each of the four components of the power plant does not mean that these components contribute "independently" to degrading the thermodynamic performance of the power plant. In an example discussed earlier in connection with Fig. 8.9, we saw that both the boiler (ΔP_H) irreversibility and the turbine (η_t) irreversibility enhance the irreversibility of the condenser. If we are interested in minimizing the irreversibility caused by a single component, we must carefully evaluate the *total* irreversibility introduced by that component in the greater system to which it belongs. This theme is expanded upon in chapter 11, which is on thermodynamic design.

Another example of the close interaction between components can be constructed by asking the question: "To what extent does the inefficiency (nonideality) of the compressor *alone* affect the irreversibility figure of the entire power plant?" Figure 8.20 shows that if the compressor isentropic efficiency η_c decreases, the outlet state (1') moves steadily to the right along the P_H isobar. Two irreversibility "areas" of Fig. 8.20 are affected by this movement. First, the rectangular area of height T_0 and width $(s_{1'} - s_{4'})$ increases in accordance with the last part of eq. (8.87). But, at the same time, the external (heat transfer) irreversibility area of the heater is being cut off from the left, as indicated by the dotted frame corresponding to the case $\eta_c = 1$. This second effect of decreasing η_c is beneficial. The *net* irreversibility increase associated with the deterioration of η_c *alone*,[†] from $\eta_c = 1$ to $\eta_c < 1$, is

$$\frac{T_0}{\dot{m}} \dot{S}_{\text{gen},\eta_c \text{ alone}} = T_0(s_{1'} - s_{4'}) - \frac{T_0}{T_f}(T_f - T_{1,\text{avg}})(s_{1'} - s_1) \quad (8.88)$$

where $T_{1,\text{avg}}$ is the average temperature between (1) and (1'):

$$T_{1,\text{avg}} = \frac{1}{s_{1'} - s_1}\left(\int_1^{1'} T\,ds\right)_{P=P_H} \quad (8.89)$$

The first group on the right side of eq. (8.88) represents the rectangular-area quantity seen already in eq. (8.87). The second group is negative and represents the area cut off from the left in the upper portion of Fig. 8.20, namely, $(T_f - T_{1,\text{avg}})(s_{1'} - s_1)$ times the area-reduction factor T_0/T_f justified earlier in eq. (8.83). Noting that $(s_{1'} - s_{4'})$ is the same as $(s_{1'} - s_1)$, we conclude that the irreversibility increase associated with the decrease of η_c while keeping everything else fixed is

$$\frac{T_0}{\dot{m}} \dot{S}_{\text{gen},\eta_c \text{ alone}} = \frac{T_{1,\text{avg}}}{T_f} T_0(s_{1'} - s_{4'}) \quad (8.90)$$

This quantity is $(T_{1,\text{avg}}/T_f)$ times smaller than the compressor exergy loss identified in eq. (8.87).

Regenerative Heat Exchanger, Reheaters, and Intercoolers

The approach to improving the performance of the gas-turbine cycle of Fig. 8.20 consists of eliminating wherever possible the shaded irreversibility areas in the figure. Three important steps that have been made in this direction are illustrated in Figs. 8.21 to 8.23. It is assumed in these figures that the gas turbines and compressors function reversibly (isentropically) and that the pressure drops measured across all the heat exchangers are negligible. In brief, the overall irreversibility of each power plant is due entirely to the

---

[†]The deterioration of η_c occurs while keeping all the other features of the cycle intact.

external heat transfer irreversibilities illustrated by means of $[area]_L$ and $(T_0/T_f)[area]_H$ in Fig. 8.6.

The two areas that play the analogous roles in the case of a closed gas-turbine power cycle are shown in Fig. 8.21. The lack of perfection of that cycle can be measured in terms of the size of the two shaded areas or in terms of the closeness of the average heater and cooler temperatures, $T_{H,\text{avg}}$ and $T_{L,\text{avg}}$, respectively. Note that the η, $T_{H,\text{avg}}$, and $T_{L,\text{avg}}$ definitions [(8.32)–(8.35)] apply here unchanged. Close $T_{H,\text{avg}}$ and $T_{L,\text{avg}}$ mean small heat-engine cycle efficiencies, η.

Figure 8.22 shows the effect of installing a counterflow heat exchanger between the high-temperature and low-temperature ends of the cycle. One job of the regenerative heat exchanger is to use the hot exhaust from the turbine, state (3), in order to preheat the compressed stream that eventually enters the heater. Because of this preheating effect, the heater operates at temperatures between states (1') and (2), which are in an average sense much higher than the heater temperatures in the simplest design (Fig. 8.21). Note the much higher position occupied by $T_{H,\text{avg}}$ in Fig. 8.22. The irreversibility $(T_0/T_f)[area]_H$ shrinks considerably in going from Fig. 8.21 to 8.22: what is eliminated is the heating of the coldest stretch of the compressed stream, that is, the heat transfer irreversibility associated with

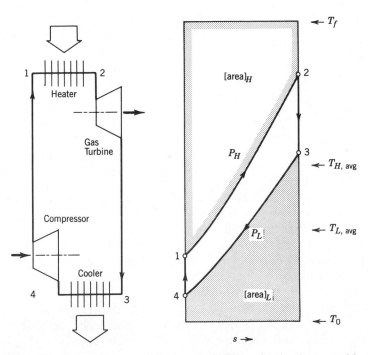

Figure 8.21   The external heat transfer irreversibilities of a simple gas-turbine power plant.

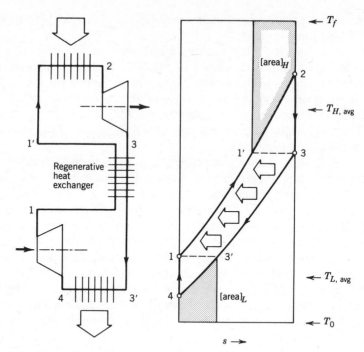

Figure 8.22 The effect of the regenerative heat exchanger on the heat transfer external irreversibilities of the gas-turbine power plant of Fig. 8.21.

the biggest temperature differences across the original heater of Fig. 8.21. For this reason, the use of a regenerative heat exchanger in a closed gas-turbine cycle is conceptually analogous to the use of regenerative feed heating in advanced steam-turbine power plants (Fig. 8.12).

The other job of the regenerative heat exchanger of Fig. 8.22 is to prevent the cooler from rejecting heat to the ambient across excessive temperature differences. This second job is quite clear if we compare the $[area]_L$ sections of Figs. 8.21 and 8.22. Another way of describing this job is to say that the regenerative heat exchanger lowers the average temperature experienced by the low-pressure stream in the cooler, $T_{L,avg}$.

In keeping up with the neglect of all internal irreversibilities, the counter-flow heat exchanger $(1) \rightarrow (1') \rightarrow (3) \rightarrow (3')$ has been modeled as having not only zero pressure drops, but also zero temperature differences between its two streams. Note that states such as $(1')$ and (3), which face each other across the heat exchanger surface, are positioned at the same temperature level in Fig. 8.22. In the present treatment, the interplay between pressure drops and stream-to-stream temperature differences in dictating the internal irreversibility of the regenerative heat exchanger forms the subject of the first part of the thermodynamic design chapter (chapter 11).

Figure 8.23 shows two improvements that can be contemplated in relation to the power plant of Fig. 8.22. The first is the idea of using a reheater to break up the expansion from (2) to (3) into two expansion stages, $(2) \rightarrow (2')$ and $(2'') \rightarrow (3)$. The reheating concept is the same as that in Fig. 8.10c. The beneficial effect of installing a reheater is that it raises the average tempera-ture $T_{H,\text{avg}}$ to a level higher than in Fig. 8.22.

The second idea illustrated in Fig. 8.23 is the use of an intercooler between the two compression stages $(4) \rightarrow (4')$ and $(4'') \rightarrow (1)$. Like the cooler, the intercooler is a heat exchanger in which the stream is cooled by thermal contact with the ambient. The pressure in the intercooler, P_i, is an intermediate pressure, $P_L < P_i < P_H$. The same observation holds for the reheater pressure (labeled also P_i in Fig. 8.10c). Relative to a cycle without intercooling (Fig. 8.22), the effect of the intercooler is to lower the average temperature of the working fluid while in thermal communication with the ambient (note the position of $T_{L,\text{avg}}$ in Fig. 8.23). The use of multiple compression stages that alternate with intercoolers (or "aftercoolers") is a standard feature of efficient gas compressors. This feature is encountered again in the study of refrigeration methods (chapter 10). The optimum selection of the intermediate pressure of each intercooler is an important design problem addressed in Problem 10.3.

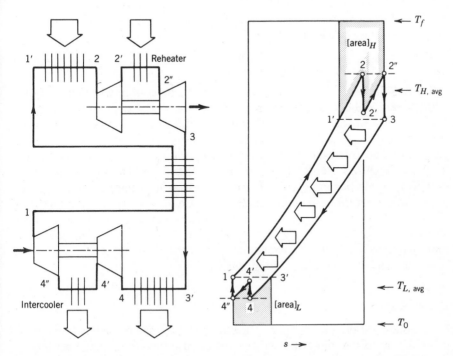

Figure 8.23 The effect of reheating and intercooling on the heat transfer external irreversibilities of the power plant of Fig. 8.22.

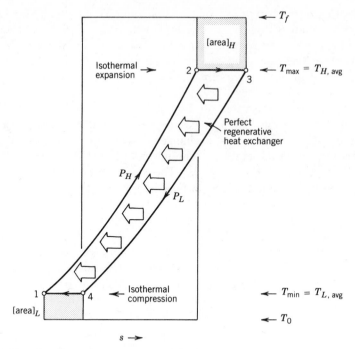

Figure 8.24 The theoretical limit of a gas-turbine cycle with perfect regenerator and infinitely many reheating and intercooling stages.

Figures 8.21–8.23 form a sequence that can be continued indefinitely by adding more reheating and intercooling stages to the expansion and compression ends of the cycle. The theoretical limit of this process of optimization is shown in Fig. 8.24. The high-temperature end of the cycle becomes effectively an isothermal gas turbine of temperature T_{max}. This temperature level had been represented by T_2 in Figs. 8.21 to 8.23. The low-temperature end of the cycle becomes an isothermal compressor of temperature T_{min} (T_4 in the preceding three figures). Either based on eq. (8.35) or other methods (Problem 8.14), it is easy to see that the efficiency reached in this theoretical limit is the Carnot efficiency $(1 - T_{min}/T_{max})$. This limiting result and idea that one needs a perfect regenerator in order to achieve it are due to Rankine [see Ref. 18, Proposition XI].

Cooled Turbines

Figure 8.24 and the $(1 - T_{min}/T_{max})$ ceiling to the cycle efficiency are a good introduction to the modern trend toward the use of turbine-blade cooling in the development of gas-turbine cycles for both power plants and propulsion systems. In the quest for increasingly higher efficiencies, the power-plant designer seeks to decrease T_{min} and to increase T_{max}. The former is limited by the size of the heat-exchanger surface built into the cooler and by the temperature of the atmosphere itself (T_0). For example, in an "open" gas-turbine cycle, where the cooler is missing and the compressor draws air directly from the ambient, T_{min} is clearly equal to T_0. The progress toward higher efficiencies hinges then on the designer's ability to augment T_{max}. Here, one runs into the temperature limitation placed by the strength of the turbine-blade and vane material.

One way to circumvent this problem is to allow the temperature of the working fluid at the turbine inlet to rise above T_{max}—as demanded by the augmentation of η—and, at the same time, to hold the blade and vane material temperature below T_{max} by providing sufficient cooling to the metal parts [see, for example, Refs. 26–30]. The most accessible coolant in a gas-turbine power plant is the compressed gas of state (1), Fig. 8.25. A fraction of the stream delivered by the compressor is diverted to the turbine for the purpose of cooling those early expansion stages that operate above T_{max}. In one possible flow circuit, the compressed gas (1) flows through each blade and is discharged into the main stream that expands through the turbine.

The coolant flowrate demanded by the safe operation of each blade and vane increases monotonically with the temperature difference by which the local main stream exceeds the metal temperature, T_{max}. We note here the emergence of an important design trade-off, for higher turbine-stream inlet temperatures mean higher efficiencies, whereas larger fractions bled from the compressor stream and not processed through the combustion chamber and the earliest turbine stages mean lower efficiencies. A peak efficiency is registered at an optimum coolant-flowrate fraction drawn from the compressor outlet, in other words, at an optimum temperature difference between

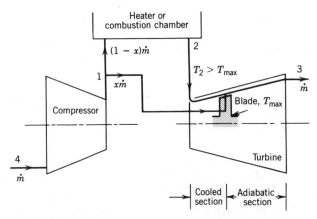

Figure 8.25 The high-temperature part of a gas-turbine cycle, showing the cooled section of the turbine and the acceptance of higher turbine inlet temperatures.

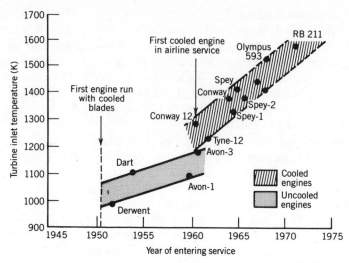

Figure 8.26 The evolution of the gas-turbine inlet temperature in Rolls Royce engines (courtesy of Prof. Bernard W. Martin and Alec Brown, University of Wales, Cardiff).

the inlet temperature of the main turbine stream and the temperature level tolerated by the mechanical strength and corrosion resistance of the blade material.

For example, a look at the 1980-level design of cooled gas-turbine blades for aircraft at General Electric Co. reveals coolant flowrates of the order of 3.5 percent of the main turbine flowrate and turbine inlet temperatures of the order of 1700 K [31]. Figure 8.26 presents a bird's-eye view of the 1950–1975 history of the turbine inlet temperature in Rolls Royce aircraft engines, showing also the dramatic shift associated with the use of cooled gas turbines. The steady progress that has been registered over the years in increasing the turbine inlet temperature (and η) is related closely to the introduction of new alloys and blade manufacturing processes.

COMBINED STEAM-TURBINE AND GAS-TURBINE CYCLES

Most of the material covered in the preceding two sections deals with the step-by-step improvement of the thermodynamic performance of, first, steam-turbine power plants and, second, gas-turbine power plants. Now is the time to review the contents of these two sections and to recognize the common design philosophy that stimulates each series of step-by-step improvements. That philosophy consists of avoiding or, at least, minimizing the irreversibility that accompanies the various components of a power plant. In graphic terms, it is useful to retain the conclusion that the effect of gradually minimizing the irreversibility of the power cycle is to "stretch" the cycle vertically on the T–s diagram so that the thermal contact between the working fluid and the T_f and T_0 temperature reservoirs is more and more intimate.

The graphic manifestation of each chain of cycle improvements is reducing the irreversibility areas $(T_0/T_f)[\text{area}]_H$ and $[\text{area}]_L$ *relative* to the $T \times s$ area enclosed by the cycle. Furthermore, in the ideal limit, the cycle must become "flat" along the top and bottom portions; in other words, the heating and cooling processes experienced by the working fluid must approach the "isothermal" description. This brings us to an interesting observation concerning the relative willingness of steam-turbine cycles and gas-turbine cycles to submit to this process of thermodynamic upgrading.

Compare, for example, the shape of the gas-turbine cycle of Fig. 8.23 with that of a steam-turbine cycle with one superheater and reheater and with a large number of feed-heating stages. The latter can be visualized by piecing together the cycle on the T–s diagram of Fig. 8.14 and the cycle with superheater and reheater shown in Fig. 8.10c. These two cycles—the gas-turbine cycle of Fig. 8.23 and the steam-turbine cycle of the combined Figs. 8.14 and 8.10c—are similar with regard to their complexity. Each has only one reheating stage. Furthermore, the regenerative heat exchanger of the gas-turbine cycle is conceptually equivalent to the train of feed heaters used in the steam-turbine cycle. Most striking in this comparison of the two cycles is the difference between the traces of the respective heating and cooling processes. By looking first at the cooling process, the steam-turbine cycle (Fig. 8.10c) hugs the (T_0) reservoir considerably more smoothly than the gas-turbine cycle of Fig. 8.23, even when the gas-turbine cycle employs one intercooler. Of course, the relative goodness (maleability) of the steam-turbine cycle has to do with the constant-temperature condensation process that accompanies the cooling between states (3) and (4).

The roles are reversed as we look at the hot end of each cycle. The trace of the heating process of the steam-turbine cycle is considerably more rugged than its gas-turbine-cycle counterpart. At the same time, the temperature gap between (T_f) and the trace of the heated-water stream is considerably wider and more difficult to decrease than the $[\text{area}]_H$ shaded above the cycle of Fig. 8.23. The difference is again due to the phase change experienced by water during its run between the nth feed-heating stage and the inlet to the low-pressure steam turbine. This difference persists even if we do away with the phase change and take P_H to be a supercritical pressure. The shape of a supercritical steam-turbine cycle on the T–s diagram resembles a right triangle whose hypotenuse is the constant-P_H heating process. And, even though the water flows as a single phase between the last feed heater and the steam turbine, it is unlike the single-phase fluid heated in the gas-turbine cycle because it experiences large variations in specific heat as it rounds the critical-point region of the dome.

From this comparison, we learn that gas-turbine cycles are better suited for efficient operation at high temperatures than steam-turbine cycles. On the other hand, the steam-turbine cycle is more attractive from the point of view of minimizing the temperature gap between the cold end of the cycle and the low-temperature reservoir. It seems natural then to *combine* these two cycles in such a way that on the temperature scale, the gas-turbine cycle is situated *above* the steam-turbine cycle. In this way, the gas-turbine cycle reaches up to the (T_f) ceiling in the advantageous manner shown in the upper part of Fig. 8.23. In its turn, the steam-turbine cycle coats the (T_0) reservoir in the smooth manner indicated in Fig. 8.10c. The engineering challenge that remains is to optimally mesh the two cycles along that seam of intermediate temperatures where the upper (warmer) cycle must act as a heat source for the lower one.

One possibility is illustrated in Fig. 8.27. In this arrangement, the gas-turbine cycle is an "open" circuit, in the sense that the expanded stream (air and products of combustion) exhausts directly into the atmosphere. Otherwise, the gas-turbine part of the combined-cycle power plant of Fig. 8.27 is the same as in Fig. 8.22, where the isobaric heating process $(1') \rightarrow (2)$ is one way of modeling the operation of the combustion chamber of Fig.

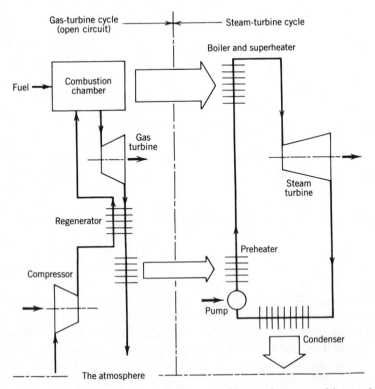

Figure 8.27   Example of a combined gas-turbine and steam-turbine cycle.

8.27. The steam-turbine part of the combined cycle resembles most closely the steam cycle with superheat shown in Fig. 8.10b.

The interface between the two parts of the combined cycle is crossed by two heat transfer interactions. During one of these interactions, the compressed-water stream produced by the pump is preheated en route to the boiler. The preheating is done at the expense of the low-pressure gas stream that is about to be discharged into the atmosphere (the temperature of this stream immediately downstream of the regenerator is higher than atmospheric: see $T_{3'}$ in Fig. 8.22). The second heat transfer interaction occurs between the reacting (air + fuel) mixture of the combustion chamber and the boiler tubes of the steam cycle. In terms of the language employed in the preceding two sections, this upper heat transfer interaction means that both streams of the combined-cycle power plant are heated from the same (T_f) temperature reservoir. The "meshing" of the two cycles is accomplished primarily by the first of the heat transfer interactions discussed above. Through this interaction, the gas-turbine part of the combined cycle does the job that in a *solo* steam-turbine cycle would have been done by a train of feed-heating stages. In return, the steam-turbine part of the cycle does a job that would be necessary if one is to minimize the $[area]_L$ patch shaded in Fig. 8.22.

The thermodynamic design work of optimizing the "fit" between the two cycles of the combined-cycle power plant and between the power plant as a whole and the (T_f) and (T_0) reservoirs is notoriously complicated. This feature is due to the many design parameters[†] of each cycle taken separately, plus the parameters of the components that constitute the interface between the two cycles. Examples of the second-law analysis and optimization of combined gas-turbine and steam-turbine cycles have been published by Sciubba et al. [32] and El-Masri [33]. Considerable work is presently being devoted to the design of combined cycles whose two parts are individually superior to the relatively simple gas-turbine and steam-turbine parts in Fig. 8.27. Noteworthy in this respect is the combined cycle incorporating a reheat gas-turbine cycle, which is described by Rice and Jenkins [34, 35].

On the steam-turbine side of the combined-cycle interface, worthy of mention is the current work on the *Kalina cycle* [36–39], in which the working fluid is a mixture of water and ammonia. The shape of the Kalina cycle on the $T–s$ diagram approaches a right triangle in which the hypotenuse is the high-pressure heating process. For this reason, the Kalina cycle fits much better than a classical Rankine cycle in the trapezoidal gap $[area]_L$ left uncovered under a gas-turbine cycle with regenerator (Fig. 8.22).

Another direction of combined-cycle optimization has to do with the selection of fuel and the matching of its combustion to the users of the ensuing heating effect. An interesting study is described by Tsujikawa and

[†]These parameters include turbine and compressor isentropic efficiencies, heat-exchanger pressure drops, heat-exchanger areas (numbers of heat transfer units, chapter 11), etc.

Sawada [40, 41], who considered a combined gas-turbine and steam-turbine cycle powered by cryogenic (liquefied) hydrogen. An additional exergy that this type of fuel brings into the exergy-flow diagram of the combined cycle is the exergy of liquefaction alone, that is, the exergy stored in the hydrogen stream during its cryogenic liquefaction. An extreme example of the work-producing potential of a cryogenic liquid is Dooley and Hammond's [42] automotive engine, which is powered solely by the exergy of a cryogen. In Dooley and Hammond's case, that cryogen is liquid nitrogen.

The trend toward improved second-law efficiencies is reflected finally by the interest in optimally combining not only two power cycles, but also a power cycle and another process requiring heat transfer from a relatively low temperature. The general name for this research arena is *cogeneration* or, literally, the generation of two or more useful things at the same time. For example, if the low-pressure steam leaving the turbine of a Rankine-cycle power plant is superheated [e.g., state (3') in Fig. 8.9], then before condensation, that stream can be used to provide heat transfer to an industrial or domestic process whose temperature range is compatible with the temperature of the superheated steam. The combined second-law and economic ("thermoeconomic") optimization of such cogenerating plants has been considered by Garceau and Wepfer [43] and Sciubba et al. [44]. Cogeneration, of course, can be contemplated in connection with other power cycles as well. For example, an open-circuit gas-turbine power cycle in which the high-pressure stream is heated without mixing with fuel (in a heater, as in Fig. 8.12) can be designed to produce not only shaft power, but also a stream of clean warm air for a separate industrial process. The thermodynamic analysis of this class of cogenerating plants was presented by Huang and Egolfopoulos [45].

CONCLUDING REMARKS

The second-law-oriented coverage of power plants outlined in this chapter can be extended to more specialized subfields of power engineering. Among the applications that have received renewed attention during the past decade worth mentioning are those associated with the harvesting and conversion of geothermal energy [46, 47], ocean thermal energy [48], and solar energy. An introduction to this relatively new and active arena is found in Shepard et al. [49]. In the next chapter, we zero in on one of these applications—solar power—not because this application might be more promising than others, but simply because it is a topic about which I have thought and read more.

Before closing, I have an opportunity to convey an observation made to me by Prof. Zilberberg. The "average" temperatures $T_{H,\text{avg}}$ and $T_{L,\text{avg}}$, which have been so helpful in the present treatment, beginning with their definition (8.34), have been used for many years by Martynovskiy and the Odessa school of refrigeration [50, 51]. With reference to the T–s area measurement that serves as a basis for defining $T_{H,\text{avg}}$ and $T_{L,\text{avg}}$, Martynovskiy called these temperatures "planimetric."

SYMBOLS

a	isentropic temperature ratio [eq. (8.22)]
A	heat transfer area
b_f	flow availability of a saturated liquid
c_P	specific heat at constant pressure
C	coefficient [eq. (8.57)]
$\dot{E}_{Q_H}$	exergy-rate input [eq. (8.31)]
h	specific enthalpy
$h_i^{(T)}$	specific enthalpy of the steam bled from the turbine
$\bar{h}$	heat transfer coefficient
k	ratio of specific heats, c_P/c_v
k_0, k_n	mass flowrate ratios [eqs. (8.75)]
$\dot{m}$	mass flowrate
M	dimensionless flowrate [eq. (8.27)]
n	number of feed-heating stages
N_H, N_L	number of heat transfer units [eqs. (8.23)]
P	pressure
P_i	intermediate pressure
$\dot{Q}$	heat transfer rate
r	pressure ratio, P_H/P_L; also, constant in eq. (8.65)
r_i	feed-water enthalpy rise [eq. (8.62)]
R	ideal gas constant
s	specific entropy
$\dot{S}_{\text{gen}}$	entropy-generation rate
T	absolute temperature
$\bar{T}$	average temperature (Problem 8.5)
T_f	equivalent fuel temperature, or heat transfer source temperature
$T_{H,\text{avg}}, T_{L,\text{avg}}$	average temperatures [eqs. (8.34)]
T_0	ambient temperature
v	specific volume
$\dot{W}$	work transfer rate (power output)
X	relative enthalpy rise experienced by the feed water [eq. (8.72)]
Y	theoretical gain [eq. (8.71)]
α	heat-exchanger-area ratio [eq. (8.12)]
β	constant [eq. (8.67)]
β_i	enthalpy drop experienced by the steam bled from the turbine [eq. (8.63)]
ΔP	pressure drop
η	heat-engine (cycle) efficiency, or first-law efficiency [eq. (8.32)]
η_c	compressor isentropic efficiency
η_p	pump isentropic efficiency
η_t	turbine isentropic efficiency

η' Curzon and Ahlborn's efficiency [eq. (8.9)]

η'_{II} second-law efficiency corresponding to maximum-power operating conditions

τ temperature ratio [eq. (8.5)]

$(\ )_{avg}$ average

$(\ )_{B}$ boiler

$(\ )_{C}$ condenser; also Carnot

$(\ )_{H}$ high temperature

$(\ )_{i}$ pertaining to the ith feed heater

$(\ )_{L}$ low temperature

$(\ )_{max}$ maximum

$(\ )_{opt}$ optimum

$(\ )_{R}$ reheater

REFERENCES

1. R. L. Loftness, *Energy Handbook*, 2nd ed., Van Nostrand-Reinhold, New York, 1984, p. 121; source: *EEI Pocketbook of Electric Utility Industry Statistics*, 27th ed., Edison Electric Institute, Washington, DC.

2. R. W. Haywood, *Analysis of Engineering Cycles*, Pergamon, Oxford, 1967, p. 261.

3. F. L. Curzon and B. Ahlborn, Efficiency of a Carnot engine at maximum power output, *Am. J. Phys.*, Vol. 43, 1975, pp. 22–24.

4. P. Salamon, A. Nitzan, B. Andresen, and R. S. Berry, Minimum entropy production and the optimization of heat engines, *Phys. Rev. A*, Vol. 21, 1980, pp. 2115–2129.

5. P. Salamon and A. Nitzan, Finite time optimization of a Newton's law Carnot cycle, *J. Chem. Phys.*, Vol. 74, 1981, pp. 3546–3560.

6. B. Andresen, P. Salamon, and R. S. Berry, Thermodynamics in finite time, *Phys. Today*, September 1984, pp. 62–70.

7. A. Bejan, *Entropy Generation Through Heat and Fluid Flow*, Wiley, New York, 1982.

8. A. Bejan, Second law analysis in heat transfer and thermal design, *Adv. Heat Transfer*, Vol. 15, 1982, pp. 1–58.

9. A. Bejan, Second law analysis in heat transfer, *Energy*, Vol. 5, 1980, pp. 721–732.

10. A. Bejan, Second law aspects of heat transfer engineering, in T. N. Veziroglu and A. E. Bergles, eds., *Multi-Phase Flow and Heat Transfer III*, Vol. 1A, Elsevier, Amsterdam, 1984, pp. 1–22 (the Keynote Address to the Third Multi-Phase Flow and Heat Transfer Symposium/Workshop, Miami Beach, Florida, April 18–20, 1983).

11. A. De Vos, Efficiency of some heat engines at maximum-power conditions, *Am. J. Phys.*, Vol. 53, 1985, pp. 570–573.

12. D. B. Spalding and E. H. Cole, *Engineering Thermodynamics*, 3rd ed., Edward Arnold, London, 1973, p. 209.

13. G. M. Griffiths, CANDU—A Canadian success story, *Physics in Canada*, Vol. 30, 1974, pp. 2–6.

14. A. Chierici, *Planning of a Geothermal Power Plant: Technical and Economic Principles*, Vol. 3, UN Conference on New Sources of Energy, New York, 1964, pp. 299–311.

15. J. H. Keenan, F. G. Keyes, P. G. Hill, and J. G. Moore, *Steam Tables*, Wiley, New York, 1969.

16. H. D. Baehr, *Thermodynamik*, 3rd ed., Springer-Verlag, Berlin, 1973, p. 378.

17. R. L. Reynolds, Reheating in steam turbines, *Trans. ASME*, Vol. 71, 1949, pp. 701–706.

18. W. J. M. Rankine, On the geometrical representation of the expansive action of heat, and the theory of thermo-dynamic engines, *Philos. Trans. R. Soc. London*, Vol. 144, 1854, pp. 115–175.

19. P. Chambadal, Le fractionnement du réchauffage de l'eau d'alimentation des chaudières, *Chal. Ind.*, Vol. 18, 1937, pp. 279–372.

20. K. Schäff, Die Theorie der Speisewasservorwärmung, *AEG Mitt.*, No. 1, 1938, p. 14.

21. J. Richard, *Equipement Thermique des Usines Génératrices d'Energie Electrique*, Dunod, Paris, 1942.

22. R. W. Haywood, A generalized analysis of the regenerative steam cycle for a finite number of heaters, *Proc.—Inst. Mech. Eng.*, Vol. 161, 1949, pp. 157–162.

23. J. K. Salisbury, The steam-turbine regenerative cycle—an analytical approach, *Trans. ASME*, Vol. 64, 1942, pp. 231–245.

24. C. D. Weir, Optimization of heater enthalpy rises in feed-heating trains, *Proc.—Inst. Mech. Eng.*, Vol. 174, 1960, pp. 769–783.

25. R. W. Haywood, Communications, *Proc.—Inst. Mech. Eng.*, Vol. 174, 1960, pp. 784–787.

26. W. R. Hawthorne, The thermodynamics of cooled turbines. Part 1. The turbine stage, *Trans. ASME*, Vol. 78, 1956, pp. 1765–1779.

27. W. R. Hawthorne, The thermodynamics of cooled turbines. Part 2. The multi-stage turbine, *Trans. ASME*, Vol. 78, 1956, pp. 1781–1786.

28. W. M. Rohsenow, Effect of turbine-blade cooling on efficiency of a simple gas-turbine power plant, *Trans. ASME*, Vol. 78, 1956, pp. 1787–1794.

29. J. C. Burke, B. L. Buteau, and W. M. Rohsenow, Analysis of the effect of blade cooling on gas-turbine performance, *Trans. ASME*, Vol. 78, 1956, pp. 1795–1806.

30. M. A. El-Masri, On thermodynamics of gas-turbine cycles. Part 2. A model for expansion in cooled turbines, *J. Eng. Gas Turbines Power*, Vol. 108, 1986, pp. 151–170.

31. R. E. Allen and J. E. Sidenstick, Aircraft gas turbine blades—present and future technology, *Mech. Eng.*, Vol. 104, April 1982, pp. 58–63.

32. E. Sciubba, W. J. Kelnhofer, and H. Esmaili, Second law analysis of a combined gas turbine–steam turbine cycle powerplant, in A. Bejan and R. L. Reid, eds., *Second Law Aspects of Thermal Design*, HTD-Vol. 33, Am. Soc. Mech. Eng., New York, 1984, pp. 55–68.

33. M. A. El-Masri, On the thermodynamics of gas turbine cycles. Part 1. Second

law analysis of combined cycles, *J. Eng. Gas Turbines Power*, Vol. 107, 1985, pp. 881–889.

34. I. G. Rice and P. E. Jenkins, Comparison of the HTTT reheat-gas-turbine combined cycle with the HTTT nonreheat gas-turbine combined cycle, *J. Eng. Power*, Vol. 104, 1982, pp. 129–142.

35. I. G. Rice, The reheat-gas-turbine combined cycle, *Mech. Eng.*, Vol. 104, April 1982, pp. 46–57.

36. A. I. Kalina, Generation of energy by means of a working fluid, and regeneration of a working fluid, U.S. Patent 4,346,561, August 31, 1982.

37. A. I. Kalina, Combined cycle system with novel bottoming cycle, *Am. Soc. Mech. Eng. (Pap.)* No. 84-GT-173, 1984.

38. A. I. Kalina, Combined cycle and waste heat recovery power systems based on a novel thermodynamic energy cycle utilizing low-temperature heat for power generation, *Am. Soc. Mech. Eng. (Pap.)* No. 83-JPGC-GT-3, 1983.

39. Y. M. El-Sayed and M. Tribus, A theoretical comparison of the Rankine and Kalina cycles, in R. A. Gaggioli, ed., *Analysis of Energy Systems—Design and Operation*, AES-Vol. 1, Am. Soc. Mech. Eng., New York, 1985.

40. Y. Tsujikawa and T. Sawada, On the utilization of hydrogen as a fuel for gas turbine (1st report, on the utilization of low temperature exergy of liquid hydrogen), *Bull. JSME*, Vol. 23, No. 183, September 1980, pp. 1506–1513.

41. Y. Tsujikawa and T. Sawada, Analysis of a gas turbine and steam turbine combined cycle with liquefied hydrogen as fuel, *Int. J. Hydrogen Energy*, Vol. 7, No. 6, 1982, pp. 499–505.

42. J. L. Dooley and R. P. Hammond, The cryogenic nitrogen automotive engine, *Mech. Eng.*, Vol. 106, October 1984, pp. 66–73.

43. R. M. Garceau and W. J. Wepfer, Thermoeconomic optimization of a Rankine cycle cogeneration system, *ACS Symp. Ser.*, No. 235, 1983.

44. E. Sciubba, W. J. Kelnhofer, and P. L. della Vida, On the optimization of cogenerating plants, in A. Bejan and R. L. Reid, eds., *Second Law Aspects of Thermal Design*, HTD-Vol. 33, Am. Soc. Mech. Eng., New York, 1984, pp. 45–54.

45. F. F. Huang and E. Egolfopoulos, Performance analysis of an indirect fired air turbine cogeneration system, *Am. Soc. Mech. Eng. [Pap.]*, No. 85-IGT-3, 1985.

46. J. Kestin, R. DiPippo, H. E. Khalifa, and D. J. Ryley, *Sourcebook on the Production of Electricity from Geothermal Energy*, DOE/RA/28320-2, U.S. Gov. Printing Office, Washington, DC, 1980.

47. D. H. Freeston and R. McKibbin, *Proceedings of the 6th New Zealand Geothermal Workshop*, University of Auckland Geothermal Institute, Auckland, New Zealand, 1984.

48. D. Bharathan, F. Kreith, D. Schlepp, and W. L. Owens, Heat and mass transfer in open-cycle OTEC systems, *Heat Transfer Eng.*, Vol. 5, No. 1–2, 1984, pp. 17–30.

49. M. L. Shepard, J. B. Chaddock, F. H. Cocks, and C. M. Harman, *Introduction to Energy Engineering*, Ann Arbor Science, Ann Arbor, MI, 1976.

50. V. S. Martynovskiy, *Termodinamitcheskiye Kharakteristiki Tsiklov Tyeplovykh i Kholodil'nykh Mashin (Thermodynamic Characteristics of Power and Refrigeration Cycles)*, Gosenergoizdat, Moscow and Leningrad, 1952, pp. 19–24.

51. V. S. Martynovskiy, *Analiz Deystvitel'nykh Termodinamitcheskikh Tsiklov* (*Analysis of Real Thermodynamic Cycles*), Energiya, Moscow, 1972, pp. 51–53.
52. A. Bejan, Theory of heat transfer-irreversible power plants, *Int. J. Heat Mass Transfer*, Vol. 31, 1988, to appear. (See also *Mech. Eng.*, Vol. 110, May 1988, p. 64).

PROBLEMS

8.1 Show that the entropy-generation rate of the Brayton-cycle power plant of Fig. 8.5 is given by eq. (8.25). Find the optimum distribution of heat-exchanger area (8.26) by minimizing $\dot{S}_{gen}$ with respect to N_H and N_L subject to the total $(\bar{h}A)$ constraint (8.24). Note that this step is equivalent to minimizing the aggregate expression

$$\frac{\dot{S}_{gen}}{(\bar{h}A)} + \lambda\left(N_H + N_L - \frac{(\bar{h}A)}{\dot{m}c_p}\right)$$

subject to no constraints, where λ is a Lagrange multiplier. Verify the validity of the average heater and cooler temperatures given by eqs. (8.28) and (8.29).

8.2 With reference to the ideal Rankine cycle in Fig. 8.6, show that the lost mechanical power (the irreversibility rate) is equal to

$$\dot{W}_{lost} = \dot{m}T_0(s_2 - s_1)\left[\frac{T_{L,avg}}{T_0} - \frac{T_{H,avg}}{T_f}\right]$$

and that $\dot{W}_{lost} \geq 0$. Verify that this result is the same as eq. (8.38).

8.3 Demonstrate analytically that the two areas labeled $\dot{E}_{Q_H}/\dot{m}$ in Fig. 8.7 must be equal.

8.4 Consider the ideal Rankine cycle of Fig. 8.6 in the special case where state (3) is always a two-phase state, and the condenser temperature is practically the same as the ambient temperature T_0. Show that the heat-engine efficiency of this class of cycles is [2]

$$\eta = \frac{b_2 - b_1}{h_2 - h_1}$$

8.5 Modeling the subcooled liquid states of Fig. 8.9 as "incompressible liquid" states, show that the entropy increase through the pump equals

$$s_{1'} - s_1 = \frac{v_1}{T_1}(P_H - P_L)(\eta_p^{-1} - 1)$$

where η_p is the pump isentropic efficiency, v_1 is the specific volume of the incompressible liquid, and $\bar{T}_1$ is an average temperature defined by the relation

$$\bar{T}_1(s_{1'} - s_1) = \left(\int_1^{1'} T\,ds\right)_{P=P_H}$$

Next, modeling the turbine exit states as "ideal gas" states, show that

$$h_{3'} - h_3 = c_{P,3}T_3\left[\left(\frac{P_H}{P_L}\right)^{(k-1)/k} - 1\right](1 - \eta_t)$$

where $c_{P,3}$ is the specific heat at constant pressure, k is the c_P/c_v ratio of near-state-(3) steam as an ideal gas, and where η_t is the turbine isentropic efficiency. Rely on these two findings in order to derive eq. (8.55).

8.6   Modeling the behavior of the steam near state (2) (Fig. 8.9) as that of an ideal gas whose constant is R_2, show that

$$\left(\int_{3_{\text{ideal}}}^3 T\,ds\right)_{P=P_L} = \bar{T}_3 R_2 \frac{\Delta P_H}{P_H}$$

where $\bar{T}_3$ is an appropriately defined average temperature of the states that reside on the P_L isobar between state (3_{ideal}) and state (3). Rely on this result in proving the validity of eqs. (8.56) and (8.57).

8.7   Examine the relative importance of the total irreversibility due to ΔP_H and the total irreversibility due to expander inefficiency ($\eta_t < 1$), Fig. 8.9. Show that when the pressure ratio P_H/P_L is fixed, the ratio of the total ΔP_H and expander irreversibilities varies as $(\Delta P_H/P_H)/(1 - \eta_t)$.

8.8   Consider the ideal Rankine cycle with superheat and reheat in Fig. 8.10c, and assume $P_H = 10\,\text{MPa}$, $P_L = 0.1\,\text{MPa}$, and $T_{\text{max}} = 500°C$. Show numerically that there exists an optimum reheater pressure P_i for which the heat-engine efficiency of the cycle reaches a maximum.

8.9   In the limit of infinitely many regenerative feed-heating stages in Fig. 8.12, the path of the expanding stream $(2) \rightarrow (3)$ assumes the shape of a smooth curve inside the two-phase dome. Determine analytically this smooth curve $s(T)$ as a function of T and $h_f(T)$, where the saturated-liquid enthalpy h_f can be curve-fitted from tabulated data [e.g. Ref. 15]. Report the closed-form expression of the $s(T)$ curve that holds when the $(T_2 - T_3)$ temperature range is narrow enough so that h_f increases linearly with T.

8.10 Evaluate the relative reduction in $\dot{Q}_H$ caused by regenerative feed heating (parameter Y, Fig. 8.16) using Haywood's compact design rule for h_n. Show that the maxima of the constant-n curves fall on the $Y \cong X$ diagonal, and that the maxima are approximately

$$Y_{\max}(n) \cong \frac{n}{n+1}$$

8.11 Figure 8.16 shows only the "relative" effect of increasing the number of contact feed heaters. Develop a quantitative feel for the actual effect of feed heating on the cycle efficiency η by comparing the efficiency of a cycle without any feed heaters with the efficiency of the same cycle in the limit of an infinite number of feed heaters. For this calculation, use Salisbury's P_H, P_L, and $T_{\max}$ values (Fig. 8.16) coupled with the assumptions that the expansion through the turbine is isentropic and that the pump power is negligible. In terms of maximum cycle efficiency, then, what is the effect of increasing the number of feed-heating stages? In other words, combine the result of your calculation with the $Y_{\max}(n)$ values read off Fig. 8.16 and construct the corresponding table $\eta_{\max}(n)$.

8.12 Show that the second-law efficiency of a power plant designed for maximum power according to Curzon and Ahlborn's rule (8.7) is

$$\eta'_{\mathrm{II}} = \frac{1}{1 + (T_L/T_H)^{1/2}}$$

8.13 Evaluate the irreversibility of the turbine-expansion portion of the cycle of Fig. 8.20. Show that if the fluid is an ideal gas with constant specific heat and if $\Delta P_H \ll P_H$ and $(1 - \eta_t) \ll 1$, then the turbine irreversibility is approximately equal to

$$\dot{S}_{\mathrm{gen,turbine}} \cong \dot{m}c_P(1 - \eta_t)[(P_H/P_L)^{R/c_P} - 1]$$

8.14 Consider the cycle of Fig. 8.24 and assume that throughout the cycle the working fluid can be modeled as an ideal gas with constant specific heat. Demonstrate that the heat-engine efficiency of this cycle is

$$\eta = 1 - T_{\min}/T_{\max}$$

8.15 The condenser of a Rankine-cycle steam engine receives a stream of 1 kg/s of dry saturated steam at 30 kPa, and produces a stream of saturated liquid at the same pressure. In the condenser, the stream is cooled by the atmospheric temperature reservoir $T_0 = 25°C$. Calculate the external irreversibility associated with the condenser–atmosphere heat transfer interaction and represented by $[\mathrm{area}]_L$ in Fig. 8.6.

8.16   An advanced gas-turbine power plant is designed so that the ideal gas stream $\dot{m}$ delivered by the heater (high pressure P_H, high temperature T_H) is expanded first in a "high-pressure" turbine to the intermediate pressure P_i. The stream is later reheated at constant pressure until its temperature again reaches T_H. Finally, the (P_i, T_H) stream is allowed to expand through a "low-pressure" turbine to the low-pressure level P_L.

Determine the optimum intermediate pressure P_i for which the total power delivered by the two turbines is maximum. In the analysis, assume that $\dot{m}$, P_H, P_L, and T_H are fixed and that the isentropic efficiencies of the high-pressure and low-pressure turbines are $\eta_{t,H}$ and $\eta_{t,L}$, respectively.

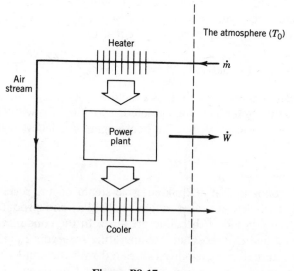

Figure P8.16

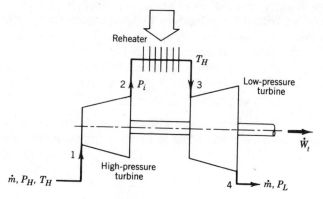

Figure P8.17

8.17 A method for the production of mechanical power ($\dot{W}$) from the atmosphere, T_0, is shown in the figure. The air stream ($\dot{m}$) is used first as heat source in the heater of the actual power plant. The temperature of the air stream drops as it passes through the heater; therefore, the stream is used next as heat sink while flowing through the cooler at the power plant. The spent air is later discharged into the atmosphere. Investigate the feasibility of this design.

8.18 The steam turbine and condenser of a simple Rankine cycle are connected as shown in the figure. The turbine inlet state is characterized by $T_1 = 200°C$ and $P_1 = 0.4\,MPa$. The operation of the turbine can be modeled as reversible and adiabatic. The condenser pressure is $P_2 = P_3 = 0.1\,MPa$ (note: the condenser pressure drop is neglected). The water stream exiting the condenser is saturated liquid.

  (a) Calculate the specific power output of the turbine ($\dot{W}_t/\dot{m}$) and the specific-heat rejection rate of the condenser ($\dot{Q}_C/\dot{m}$).

  (b) Write the second law for the "aggregate" system defined by the dashed line in the figure (note that the temperature of the boundary crossed by $\dot{Q}_C$ is in this case the saturation temperature corresponding to the condenser pressure). Rely on this second-law statement to prove that the specific-heat rejection rate calculated in part (a) is a minimum. In other words, prove that the $\dot{Q}_C/\dot{m}$ value of an aggregate system that operates irreversibly is greater than the $\dot{Q}_C/\dot{m}$ value calculated in part (a).

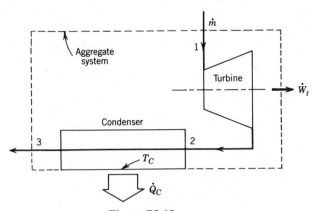

Figure P8.18

8.19 The ideal Brayton cycle in the figure employs one reheater, which is positioned between states (2′) and (2″). The compressor (4)–(1), the high-pressure turbine (2)–(2′), and the low-pressure turbine (2″)–(3) function reversibly and adiabatically. The heater (1)–(2) and the reheater and the cooler (3)–(4) cause negligible pressure drops along the fluid circuit. The heater, reheater, and cooler pressures are known (P_H, P_R, P_C, respectively). The working fluid is an ideal gas with constant specific heat.

(a) Determine the heat-engine efficiency of the cycle and express your result as

$$\eta_R = \text{function}\,(\tau, \tau_{\max}, x)$$

where τ is the isentropic temperature ratio (T_1/T_4), $\tau_{\max}$ is the overall temperature ratio spanned by the cycle (T_2/T_4), and x is the inverse of the isentropic temperature ratio across the high-pressure turbine, i.e.,

$$x = T_{2'}/T_2$$

(b) From the η_R formula derived above, deduce the heat-engine efficiency expression for a simple Brayton cycle in which the reheater is absent. Label this new result η_B.

(c) Demonstrate analytically that $\eta_R < \eta_B$, in other words, that the addition of one reheater to a closed Brayton cycle always diminishes the efficiency. This conclusion seems paradoxical in view of the treatment received by "reheating" in this chapter. Do you accept this conclusion?

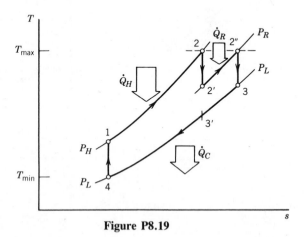

Figure P8.19

8.20 Review the power plant design in which the power output has been maximized twice, eq. (8.14). Show that in this design the optimum temperature differences across the two heat exchangers obey the proportionality

$$\frac{T_H - T_{HC}}{T_{LC} - T_L} = \left(\frac{T_H}{T_L}\right)^{1/2}$$

Next, construct the absolute temperature scale $0-T_H$, that is, draw a straight line and mark on it the point of absolute zero, 0, and the temperatures T_L and T_H. The segments $\overline{0T_L}$ and $\overline{0T_H}$ should be measured so that they are proportional to the known temperatures T_L and T_H, which are expressed in degrees Kelvin.

On the absolute temperature scale, pinpoint *graphically* the position of the optimum temperature levels T_{HC} and T_{LC} that correspond to the design (8.14). Execute the graphic construction using no more than a straight edge (unmarked ruler) and a compass. The solution to this problem is given in Ref. 52.

9

Solar Power

Solar-energy technology has become one of the most talked about subjects in applied thermodynamics and heat transfer research today. This development has understandably been stimulated by the global energy "problem," that is, by the quest for new energy resources and for new ideas of how to exploit energy resources. Although the body of solar-energy-engineering research is voluminous, the thinking of how this research fits in the greater engineering thermodynamics picture is underdeveloped. The true mission of the harvesting of solar energy is only rarely addressed in the field of thermal engineering [e.g., Ref. 1]; furthermore, it is almost never discussed in the engineering-thermodynamics textbooks and, I assume, in the classrooms.

The mission of solar-energy technology is to place "work" (exergy, availability) at our disposal, for consumption. In this chapter, I construct a thermodynamic framework for the debate of controversial topics such as the maximum exergy content of solar radiation and the efficient use of collected solar energy. The chapter begins with an outline of the equilibrium thermodynamics of thermal radiation, and continues with a review of theoretical advances that demonstrate that the energy irradiated by the sun is rich in exergy. The focus shifts eventually to the thermodynamically irreversible operation of solar collectors, and to the issue of optimum operating conditions from the point of view of maximum exergy extraction. More complex features of solar collector and power-plant operation, as well as the current literature devoted to this thermodynamics sector are discussed in the closing sections of this chapter.

THERMODYNAMIC PROPERTIES OF THERMAL RADIATION

Any discussion of the "efficient" or "economic" use of solar energy must be based on an understanding of the maximum work-producing potential that is associated with the solar thermal radiation. This potential—the exergy of thermal radiation—has been the subject of an intense research activity, which is reviewed partly in my 1982 monograph [1]. One striking feature of

466

this research is the controversy that is generated by each new answer to the basic question of how much exergy can be derived from sunlight. Although qualitatively the competing answers that have been given all say that the thermal radiation received from the sun is rich in exergy, they disagree in quantitative terms and with regard to methodology. Recent examples of this disagreement are Gribik and Osterle's paper [2], Jeter's [3], and De Vos and Pauwels' [4] comments on it, and Gribik and Osterle's response [5].

Controversy is valuable, especially in a young field like solar-energy engineering. However, in order to understand the ideas that are being debated, and in order to clarify the terminology that will be used as this debate continues, it is important to revisit the classical heat transfer subject of thermal radiation, this time from the point of view of engineering thermodynamics. The shift from the heat transfer view to the thermo-dynamics view involves two conceptual changes. The first is that in heat transfer, we are used to focusing on the net energy interaction (heat transfer) between two or more surfaces with different temperatures. The analytical framework of classical thermodynamics on the other hand is based on the concept of thermodynamic equilibrium; therefore, at least in the first part of this presentation, we consider surfaces that are in equilibrium with each other and with the radiation contained between them.

The second conceptual change consists of abandoning the electromag-netic-wave description of thermal radiation (which is the traditional descrip-tion in heat transfer) in favor of the discrete-particle (photon) description. This change allows us to define the thermodynamic properties of thermal radiation and to condense the thermodynamic behavior of the thermal radiation "system" into a Gibbsian fundamental relation of type $U = U(S, V)$. The analytical similarities between the description of a space filled by a collection of photons and the description of the same space filled by an ideal gas are the reason for the *photon gas* name that is usually given to the thermodynamic photon model. Worth keeping in mind is that the photon gas—ideal gas analogy is a superficial one, as stressed in the discussion of eqs. (9.26) and (9.32) later in this section.

Photons

As a relativistic particle, the photon is characterized by zero rest mass, by the energy

$$\epsilon = h\nu \tag{9.1}$$

and the momentum

$$p = \frac{h\nu}{c} \tag{9.2}$$

where $h = (6.626)10^{-34}$ J s is Planck's constant and $c = (2.998)10^{8}$ m/s is the photon's speed of propagation in an empty space (the speed of light). The

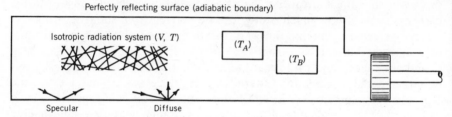

Figure 9.1 Enclosure with perfectly reflecting internal surfaces, for the containment of isotropic radiation.

photon is also characterized by a direction of propagation and a frequency (ν), or, conversely, a wavelength

$$\lambda = \frac{c}{\nu} \tag{9.3}$$

In what follows, we restrict the discussion to *unpolarized* radiation.

The steady emission of energy as radiation from a surface can be viewed as the departure of a stream of photons, in which the emission of each photon is associated with a drop in the energy level of an atom in the solid. On the other hand, absorption of photons is accompanied by energy transfer to the surface, as the atoms of the solid material experience transitions to higher energy states [6].

Consider now a space of size V that is delineated by a "perfectly reflecting" surface (Fig. 9.1). Recall that the reflectivity of a surface is defined as that fraction (percentage) of the incident-energy stream that is neither transmitted through nor absorbed by the surface. A perfectly reflecting surface reflects the incident radiation stream 100 percent: this means that for each photon of frequency ν_i absorbed, the surface has the property of emitting one photon of the same frequency and energy. From the point of view of transferring energy as heat across the boundary of the V space, the perfectly reflecting surface is *adiabatic*.

Physically, the name "perfectly reflecting" invites us to think first of a surface that resembles a highly polished (mirror-type) metallic surface, i.e., of a *specular* reflector in which the reflected and incident beams are positioned symmetrically about the normal to the surface (Fig. 9.1). The present treatment, however, is not restricted to specular reflectors. The surface of the enclosed space V can reflect the incident radiation in any direction (for example, it can be a *diffuse* reflector, Fig. 9.1), as long as it reflects the incident radiation entirely.

Temperature

Assuming that initially the V space is completely evacuated, consider introducing into this space a body (A), whose special property is that its surface can emit and absorb photons of a certain frequency, ν_A. If, as in the example addressed in the next paragraph, photons of many other frequencies travel through the enclosure, then the (A) body is assumed to be completely transparent to these photons. In time, the V space fills with monochromatic radiation of frequency ν_A, while the (A) body reaches the equilibrium temperature T_A. At this stage, we can speak of the monochromatic radiation as having the temperature T_A, because it is in equilibrium with a body whose measurable temperature is T_A. The act by which we assign a temperature to the collection of photons that account for the monochromatic radiation is, therefore, another application of the Zeroth Law of Thermodynamics. The temperature of this thermal radiation can only be determined indirectly, i.e., by measuring the temperature of the body with which the radiation is in equilibrium.

The above argument can be repeated by replacing the (A) body with a *blackbody* (B). The defining feature of the blackbody is that its surface absorbs entirely the incident radiation, regardless of frequency. This means that the monochromatic (spectral) and total absorptivities of the blackbody surface are equal to 1, or that the corresponding reflectivities and transmissivities are all equal to 0. We are particularly interested in the type of radiation with which the blackbody is in equilibrium, because at least above the atmosphere, the solar thermal radiation mimics the radiation that in Fig. 9.1 would be in equilibrium with a (B) body of temperature 5762 K [7, 8].

With the (B) body placed inside it, the V space will be filled by photons of all frequencies, assuming, of course, that the linear dimension of the V space is large enough to accommodate the radiation with the longest wavelength (namely, the low-frequency photons whose aggregate contribution to the energy inventory of the V space is still meaningful; the relationship between photon density and frequency is discussed in the next subsection). Since the V space is completely isolated with respect to its environment, after a sufficiently long time, the system housed by it reaches internal equilibrium. The eventual temperature T_B that is reached by the blackbody can also be assigned to the radiation or instantaneous collection of photons with which the blackbody is in equilibrium.

Summing up these first two examples, we can think of the case where both bodies (A) and (B) reside inside the enclosure. At equilibrium, we will have $T_A = T_B$, which means that (A) and (B) will be in equilibrium and that the temperature of the monochromatic radiation with which (A) is in equilibrium will be the same as the temperature of the blackbody radiation sustained by the blackbody (B). The monochromatic radiation is, therefore, the same as that sample of photons that occupy the narrow frequency band $(\nu_A) - (\nu_A + d\nu_A)$ in the frequency-diverse population of photons that make up the blackbody radiation of the same temperature. And since the ν_A frequency is arbitrary, we can vary it and regenerate line by line the frequency dependence of blackbody radiation. The point of this summarizing discussion is that although two batches of thermal radiation can contain photon populations of different numbers and frequencies, at equilibrium, they are characterized by the same temperature.

Energy

Consider from now on the case where the V space is filled only by blackbody radiation of temperature T, which is in equilibrium with an infinitesimally small blackbody (e.g., a grain of soot). The energy inventory of the blackbody radiation as a thermodynamic system can be calculated from Planck's distribution of photon volumetric density:

$$n_\nu = \frac{8\pi\nu^2 c^{-3}}{\exp(h\nu/kT) - 1} \qquad (9.4)$$

where $k = (1.38)10^{-23}$ J/K is Boltzmann's constant, and n_ν represents the number of photons per unit volume and unit frequency interval, i.e., the units of n_ν are the number of photons/m^3/s^{-1}. The refractive index of the medium is being assumed equal to 1, as a representative value for vacuum and common gases [Ref. 6, p. 700].

The energy per unit volume associated with the density n_ν is then $u_\nu = n_\nu h\nu$; hence,

$$u_\nu = \frac{8\pi h\nu^3 c^{-3}}{\exp(h\nu/kT) - 1} \qquad (9.5)$$

Obtained first by Planck [9], eq. (9.5) constituted the starting point in the development of quantum theory. Finally, integrating eq. (9.5) over the entire frequency domain, we obtain the volumetric-specific energy of blackbody radiation:

$$u = \int_0^\infty u_\nu \, d\nu = aT^4 \qquad (9.6)$$

where

$$a = \frac{8\pi^5}{15}\frac{k^4}{h^3 c^3} = (7.565)10^{-6}\,\text{J}\,\text{m}^{-3}\,\text{K}^{-4} \qquad (9.7)$$

The total energy inventory of the blackbody radiation that fills the space of size V is, therefore,

$$U = uV = aVT^4 \qquad (9.8)$$

Heat transfer engineers can recognize at this stage the connection between eqs. (9.4)–(9.8) and equivalent formulas that serve as the backbone of most radiation heat transfer calculations. With reference to eq. (9.4) and the "pencil of rays" of unit solid angle shown in the detail of Fig. 9.2, we can evaluate the energy that arrives per unit time, unit area (dA), unit solid angle ($d\Omega$), and unit frequency interval ($d\nu$) from a particular direction that serves as axis for the pencil:

$$i'_{\nu b} = u_\nu c/4\pi = \frac{2h\nu^3 c^{-2}}{\exp{(h\nu/kT)} - 1} \tag{9.9}$$

Equation (9.9) represents the spectral intensity of blackbody radiation, where, in accordance with the notation rules of Siegel and Howell [6], the subscripts "b" and "ν" stand for "blackbody" and "per-unit frequency," respectively, and where the superscript ( )' is meant to suggest that $i'_{\nu b}$ is a per-unit-solid-angle quantity associated with a certain direction. More popular in heat transfer is the equivalent expression for the spectral intensity on a per-unit-wavelength basis:

$$i'_{\lambda b} = \frac{\nu^2}{c} i'_{\nu b} = \frac{2hc^2 \lambda^{-5}}{\exp{(hc/k\lambda T)} - 1} \tag{9.10}$$

as well as the corresponding spectral hemispherical emissive power, $e_{\lambda b} = \pi i'_{\lambda b}$.

Equations (9.9) and (9.10) reveal a very important one-to-one relationship that exists between the spectral intensity of blackbody radiation and the

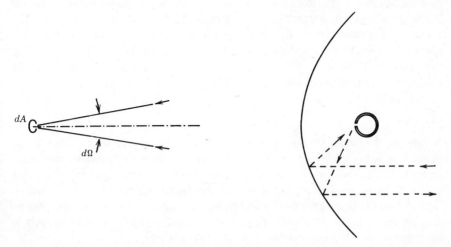

Figure 9.2 Pencil of rays of unit solid angle (left side), and the concentration of a parallel beam of radiation into the "pupil" of an enclosure filled with isotropic radiation (right side).

temperature of radiation, T. This relationship is even more evident in the makeup of the total intensity:

$$i'_b = \int_0^\infty i'_{\nu b}\, d\nu = \int_0^\infty i'_{\lambda b}\, d\lambda = \frac{c}{4\pi}\, aT^4 \tag{9.11}$$

Through this relationship, we can attach a temperature (T) not only to the directionally random radiation that is trapped inside the perfectly reflecting enclosure of Fig. 9.1, but also to *beamed* radiation [10].

This conceptual step is important in the realm of solar-energy utilization, because the incident extraterrestrial solar radiation is confined to a very narrow solid angle. The beamed radiation is also a good working model for the solar radiation that reaches the earth on a clear day. The implication of the intensity–temperature relationship is that in a concentrating device like the one in Fig. 9.2, the maximum theoretical temperature that might be realized at the focal point of the parabolic mirror is that of the radiation beam itself. As the incident radiation hits the mirror, it experiences a compression (concentration), with the result that the energy flux per unit area normal to one ray increases. However, the intensity or the energy flux per unit normal area and *per unit solid angle* is the same anywhere along the incident and reflected ray, and so is the temperature of beamed radiation. This observation is based on the assumption that the mirror is a specular reflector; we shall see that a diffusely reflecting surface decreases the temperature and increases the entropy of the incident beam.

In the analysis that follows, we will continue to exploit the *Stefan–Boltzmann law*, i.e., the conclusion that the volumetric-specific total energy of blackbody radiation is a function of temperature only, eq. (9.6). This law was established experimentally in 1879 by Stefan [11] and derived on the basis of statistical arguments five years later by Boltzmann [12]. The exclusive dependence of u_ν on temperature is one feature that fuels the photon gas–ideal gas analogy. In this sense, eq. (9.6) can be regarded as that equation of state whose equivalent in the ideal gas model is the calorimetric equation of state $u = u(T)$.

Pressure

If one of the walls of the perfectly reflecting enclosure of Fig. 9.1 could move like a piston inside a frictionless sleeve, then the environment must restrain this movement by applying a net pressure on the piston. The same piston is being pushed outward by the averaged impact of all the photon-wall collisions: the "pressure" exerted by the trapped radiation (the photon gas) on the wall of the enclosure can be estimated from a classical result of the kinetic theory of monatomic gases [13], namely,

$$P = \frac{1}{3}\frac{N}{V} mV_{\text{avg}}^2 \tag{9.12}$$

where N is the total number of particles that occupy the volume V, and m and V_{avg} are the mass and average speed of one particle, respectively. If the enclosure is occupied by monochromatic radiation of frequency ν, then in the present notation, we have $N/V = n_\nu$, $m = \epsilon/c^2$, and $V_{avg} = c$, so that eq. (9.12) yields the partial photon-gas pressure associated with one narrow frequency band:

$$P_\nu = \frac{1}{3} n_\nu h\nu = \frac{1}{3} u_\nu \qquad (9.13)$$

Note that the units of P_ν are Pa/s^{-1}. In the case of blackbody radiation, the same reasoning leads to the total pressure:

$$P = \frac{1}{3} nh\nu = \frac{1}{3} u \qquad (9.14)$$

Comparing eqs. (9.13) and (9.14), we learn that the pressure exerted by blackbody radiation is the sum of all partial (spectral) pressures P_ν:

$$P = \int_0^\infty P_\nu \, d\nu \qquad (9.15)$$

much in the way that the pressure of an ideal gas is the sum of the partial pressures of the ideal gas components that, one at a time, would fill the same volume at the same temperature. One feature that distinguishes the blackbody photon gas from an ideal gas is the combined message of eqs. (9.14) and (9.6), namely, $P = (a/3)T^4$. This means that if the temperature of blackbody radiation is maintained constant, then constant is not only its volumetric-specific energy, but also its pressure. As a thermodynamic system, the blackbody radiation is completely specified by V and T, or V and P, or V and U.

Entropy

There are at least two ways in which to determine the entropy of the photon-gas system. First, we can consider an infinitesimal reversible process that proceeds from (V, U) to $(V + dV, U + dU)$. The First Law of Thermodynamics requires

$$\delta Q_{rev} - P \, dV = dU \qquad (9.16)$$

where $P \, dV$ represents the infinitesimal work done reversibly by the system on the environment, and δQ_{rev} is the reversible heat transfer from the environment to the system. The latter is accompanied by a decrease in the entropy inventory of the environment:

$$dS_{env} = -\frac{\delta Q_{rev}}{T} \tag{9.17}$$

where T is the absolute temperature shared by both the system and the environment during the reversible heat transfer interaction. Invoking next the Second Law of Thermodynamics for the aggregate system composed of the environment and the photon-gas system, we write that for this reversible process, the entropy generation is zero:

$$dS_{env} + dS = 0 \tag{9.18}$$

where dS accounts for the entropy change experienced by the photon-gas system. Combining eqs. (9.16)–(9.18), we obtain the "combined law" for the photon-gas system:

$$T\,dS - P\,dV = dU \tag{9.19}$$

in other words,

$$dS = \frac{P}{T}\,dV + \frac{1}{T}\,dU \tag{9.20}$$

For blackbody radiation, eq. (9.20) yields

$$dS = \frac{u}{3T}\,dV + \frac{1}{T}\,(u\,dV + V\,du)$$

$$= \frac{4}{3}\,aT^3\,dV + 4aT^2 V\,dT$$

$$= \frac{4}{3}\,a\,d(VT^3) \tag{9.21}$$

Setting $S = 0$ at absolute zero, we obtain finally

$$S = \frac{4}{3}\,aVT^3 = \frac{4U}{3T} \tag{9.22}$$

or per-unit volume:

$$s = \frac{S}{V} = \frac{4}{3}\,aT^3 = \frac{4u}{3T} \tag{9.23}$$

We conclude that, like the volumetric-specific energy (u), the volumetric-specific entropy of blackbody radiation (s) is only a function of absolute temperature.

A more direct alternative to determining S is by invoking the Euler equation in entropy representation:

$$S = \frac{1}{T} U + \frac{P}{T} V \tag{9.24}$$

and using again eqs. (9.8) and (9.14). The resulting expression is then the *fundamental relation*, either in entropy representation:

$$S(U, V) = \frac{4}{3} a^{1/4} U^{3/4} V^{1/4} \tag{9.25}$$

or in energy representation:

$$U(S, V) = \left(\frac{3}{4}\right)^{4/3} a^{-1/3} S^{4/3} V^{-1/3} \tag{9.26}$$

It is easy to show that eq. (9.25) is the same as eq. (9.22), if V is eliminated via eq. (9.8).

I use this opportunity to construct graphically the three-dimensional surface dictated by the fundamental relation, eqs. (9.25–9.26). In the triple-logarithmic dimensionless space of Fig. 9.3, the fundamental surface is

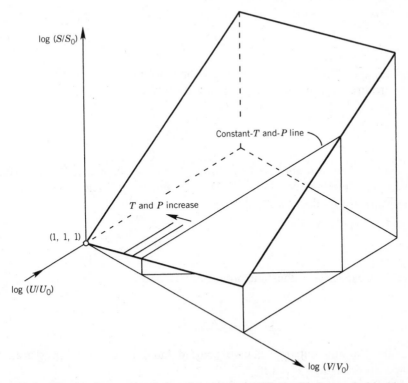

Figure 9.3 The fundamental relation for a volume filled with isotropic blackbody radiation, showing the location of isotherms (or isobars) on the $U(S, V)$ surface.

a plane whose trace has a slope of $\frac{3}{4}$ in any $V=$ constant plane, and a slope of only $\frac{1}{4}$ in a $U=$ constant plane. Figure 9.3 shows also the manner in which the fundamental surface is covered by isotherms and isobars, stressing again the combined message of eqs. (9.14) and (9.6): in equilibrium blackbody radiation, constant temperature also means constant pressure. The isotherms (isobars) result from the intersection of the fundamental surface with vertical planes (constant-T and -P planes) whose traces show a slope equal to 1 in each horizontal ($S=$ constant) plane.

Similar entropy expressions can be developed for monochromatic radiation as well as for any multifrequency combination of photons other than the blackbody radiation distribution of eq. (9.4). For equilbrium monochromatic radiation of frequency ν, the Euler relation (9.24) is replaced by

$$S_\nu = \frac{1}{T} U_\nu + \frac{P_\nu}{T} V \tag{9.27}$$

where S_ν is the entropy-per-unit-frequency interval, and $U_\nu = u_\nu V$. Recalling the simple relationship between spectral pressure and volumetric-specific spectral energy, eq. (9.13), we find that S_ν is related to U_ν and T exactly the same way as (S, U, T) in eq. (9.22):

$$S_\nu = \frac{4U_\nu}{3T} \tag{9.28}$$

Expressed in $J/K/s^{-1}$, the entropy of monochromatic radiation of volume V, temperature T, and frequency ν is, therefore,

$$S_\nu = \frac{32\pi V h \nu^3 c^{-3}}{3T[\exp(h\nu/kT) - 1]} \tag{9.29}$$

Turning our attention to the concept of heat capacity, for blackbody radiation heated at constant volume, we find a coefficient that increases rapidly with the absolute temperature:

$$C_V = T\left(\frac{\partial S}{\partial T}\right)_V = 4aVT^3 \tag{9.30}$$

The heat capacity at constant pressure is infinite:

$$C_P = T\left(\frac{\partial S}{\partial T}\right)_P \to \infty \tag{9.31}$$

because for any entropy change contemplated at constant pressure, the temperature change is zero.

Finally, a peculiar property of the photon gas (either monochromatic or blackbody) is that the Gibbs free energy is identically zero:

$$G = U + PV - TS = 0$$
$$G_\nu = U_\nu + P_\nu V - TS_\nu = 0$$

(9.32)

This means that the "chemical potential" of the photon gas is also zero:

$$\mu = \left(\frac{\partial G}{\partial N}\right)_{T,P} = 0$$

(9.33)

where N is the total number of photons in the system. Equation (9.33) draws attention to the fact that G (as well as U and S, Fig. 9.3) do not depend on the total number of photons that instantaneously inhabit volume V. The same feature is evident in the makeup of the fundamental relation (9.26), where U depends on only two extensive properties (S, V) and is independent on the total number of particles [recall that the classical form of the fundamental relation for a single-component system is $U = U(S, V, N)$]. Even if the photon-gas system is completely isolated, the total number of photons is not conserved [14].

REVERSIBLE PROCESSES

Relevant to the conceptual design of thermodynamically efficient operations for the utilization of solar energy are a number of benchmark reversible processes that can be executed by a blackbody radiation system contained in the deformable enclosure of Fig. 9.1. The processes that are to be described are all characterized by zero entropy generation.

Reversible and Adiabatic Expansion/Compression

Consider the reversible volume change from V_1 to V_2: in the absence of heat transfer, the entropy S remains constant; therefore, according to eq. (9.22), the path of the process is

$$VT^3 = \text{constant}$$

(9.34)

or, using eqs. (9.14) and (9.6),

$$VP^{3/4} = \text{constant}$$

(9.35)

The reversible work transfer interaction $\int_1^2 P \, dV$ is

$$W_{1-2,\text{rev}} = 3P_1 V_1 [1 - (V_1/V_2)^{1/3}]$$

(9.36)

We learn that the work delivered during an isentropic expansion is influenced greatly by the initial pressure (temperature) of blackbody radiation.

Reversible and Isothermal Expansion/Compression

During this process, the pressure (P) also remains constant, which means that the work transfer interaction is

$$W_{1-2,\text{rev}} = P(V_2 - V_1) = \frac{a}{3} T^4(V_2 - V_1) \qquad (9.37)$$

The work output of a reversible and isothermal expansion again depends very strongly on the temperature of the process (T). The heat transfer interaction experienced by the blackbody radiation system during the same process is

$$Q_{1-2,\text{rev}} = W_{1-2,\text{rev}} + U_2 - U_1 = \frac{4}{3} aT^4(V_2 - V_1) \qquad (9.38)$$

During an expansion, the heat input is exactly four times greater than the work output.

Carnot cycle

Now we can alternate the preceding processes in the reconstruction of the classical four-process Carnot cycle, this time using the trapped blackbody radiation as a working substance. To place this cycle in the "heat-engine" mode, we position the reversible and isothermal expansion process at the high-temperature end of the cycle, T_H. The reversible and isothermal compression process is executed while in communication with the low-temperature reservoir (T_L). On a T–V plane, the cycle is sandwiched between the temperature levels T_H and T_L. Since the reversible and isothermal processes are also isobaric, on a P–V plane, the same cycle is bounded from above by $P_H = (a/3)T_H^4$ and from below by $P_L = (a/3)T_L^4$.

Finally, in the T–S plane, the Carnot cycle is a rectangle with horizontal sides represented by $T = T_H$ and $T = T_L$ and vertical sides by $S = S_1 = (4/3)aV_1T_H^3$ and $S = S_2 = (4/3)aV_2T_H^3$, where the subscripts "1" and "2" denote the initial and final states, respectively, of the high-temperature reversible and isothermal expansion. The net work output of the cycle is clearly

$$\oint \delta W_{\text{rev}} = \oint \delta Q_{\text{rev}} = \oint T \, dS = \frac{4}{3} aT_H^3(V_2 - V_1)(T_H - T_L) \quad (9.39)$$

which, compared with the high-temperature heat transfer interaction $Q_H = (4/3)aT_H^4(V_2 - V_1)$, reveals the traditional Carnot heat-engine efficiency

$$\eta_C = \frac{1}{Q_H} \oint \delta W_{\text{rev}} = 1 - \frac{T_L}{T_H} \qquad (9.40)$$

This result is purely of academic interest. It reminds us of Sadi Carnot's argument that the heat-engine efficiency of a reversible cycle should be independent of the nature of the working fluid. This is a useful reminder even in solar-energy engineering, because it highlights the engineer's freedom to employ traditional working fluids in the design of the actual power cycle. It also highlights a basic conceptual difference between the postulated reversible cycle that led to eq. (9.40) and an actual cycle that might be powered by the concentrating collector in Fig. 9.2. The issue that has not been addressed yet is how *efficiently* can solar thermal radiation be collected in order to account for the heat input Q_H that is delivered to the high-temperature end of the power cycle. It turns out that the process of "collecting" solar energy is inherently irreversible, which is why we now turn to the issues assembled in the next section.

IRREVERSIBLE PROCESSES

In this section, we focus on some of the irreversibilities that can accompany the transformation of solar radiation into mechanical power. The discussion begins with the illustration of an entropy-generation mechanism in the photon-gas system of Fig. 9.1, and culminates with the entropy-generation analysis of the purely radiative heat transfer process. Of special interest in all these examples is the quantitative description of entropy generation as a measure of lost work, as well as the physical description of those solar-system design features that function as mechanisms (sources) of entropy generation.

Adiabatic Free Expansion

Let us assume that the perfectly reflecting enclosure of Fig. 9.1 contains at some stage equilibrium blackbody radiation of volume V_1 and temperature T_1. The inner surface of the piston is subjected to the photon-gas pressure $(a/3)T_1^4$. If the external mechanism that restrains the piston is removed, the photon gas expands without delivering work to its environment. Let V_2 be the larger volume occupied by the photon-gas system at the end of this "free-expansion" process. The energy and entropy inventories before and after this process are

$$U_1 = aT_1^4 V_1 \qquad U_2 = aT_2^4 V_2$$

$$S_1 = \frac{4a}{3} T_1^3 V_1 \qquad S_2 = \frac{4a}{3} T_2^3 V_2 \tag{9.41}$$

Since the system was completely isolated during this process, Q_{1-2} and W_{1-2} are both zero; the first law then allows us to estimate the final equilibrium temperature:

$$U_2 = U_1; \text{ hence, } T_2 = T_1(V_1/V_2)^{1/4} \tag{9.42}$$

We note also that since $U = \text{constant}$ and $S = 4U/3T$, the drop in temperature from T_1 to T_2 is accompanied by an increase in entropy:

$$S_{\text{gen},1-2} = S_1[(V_2/V_1)^{1/4} - 1] > 0$$

$$= S_1\left(\frac{T_1}{T_2} - 1\right) > 0 \tag{9.43}$$

The entropy generation increases with the volume of the final space, reaching infinity in the case of adiabatic free expansion into an infinite space.

The Transformation of Monochromatic Radiation into Blackbody Radiation

If the volume V of Fig. 9.1 contains monochromatic radiation of frequency ν and temperature T, then the presence of a small grain of soot will act as "catalyst" and trigger a redistribution of the original energy inventory. The result will be a blackbody radiation of a considerably lower temperature. For the original energy inventory of the isolated system, we write

$$U = \frac{8\pi h\nu^3 c^{-3}}{\exp(h\nu/kT) - 1} V \,\Delta\nu \tag{9.44}$$

where $\Delta\nu$ is the width of the narrow band of monochromatic radiation (no "monochromatic" radiation has mathematically a unique frequency, rather, the frequencies span a narrow domain centered around ν). The resulting equilibrium temperature of blackbody radiation, T_b, follows from the condition that the energy of the isolated system remains constant (see detail of Fig. 9.4)

$$U = aVT_b^4 \tag{9.45}$$

which yields

$$\frac{kT_b}{h\nu} = \frac{15^{1/4}}{\pi}\left[\frac{\Delta\nu/\nu}{\exp(h\nu/kT) - 1}\right]^{1/4} \tag{9.46}$$

The irreversibility of the monochromatic $\rightarrow$ blackbody radiation transition is measured as the entropy generation $S_{\text{gen}} = S_b - S > 0$, where S_b is the final entropy of blackbody radiation:

$$\frac{S_b}{S} = \frac{\pi}{15^{1/4}}\left(\frac{\Delta\nu}{\nu}\right)^{-1/4}\frac{kT}{h\nu}[\exp(h\nu/kT) - 1]^{1/4} \tag{9.47}$$

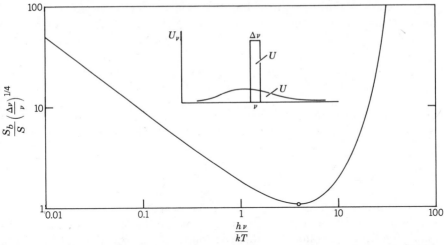

Figure 9.4   The entropy increase associated with the constant-energy transformation of monochromatic radiation into blackbody radiation.

and where S is the original entropy inventory of the system, $S = 4U/3T$. In dimensionless terms, the entropy generation increases as the slenderness ratio of the frequency band $(\Delta v/v)$ decreases. Figure 9.4 shows that relative to the dimensionless frequency hv/kT, the entropy generation exhibits a minimum at $hv/kT = 3.921$, which in terms of wavelength means $\lambda T = 0.367$ cm K. This minimum is close to the maximum of the spectral energy distribution (9.5), which is located at $\lambda T = 0.29$ cm K.

In conclusion, the entropy generated during the transformation is minimized if the original frequency falls "in the middle" of the frequency range, i.e., in that part of the spectrum that accounts for most of the energy of blackbody radiation.

The monochromatic $\rightarrow$ blackbody radiation transformation just analyzed is the simplest in a class of similar irreversible processes. Any sample of equilibrium radiation whose spectral energy does not fit a blackbody distribution can undergo an irreversible constant-energy transition to blackbody radiation of *lower* temperature. An example of radiation whose spectral energy distribution does not conform to the Planck distribution (9.5) is the solar radiation filtered by the atmosphere: the energy spectrum of this radiation shows sizeable gaps (absorption bands) caused primarily by the presence of H_2O and CO_2 in the atmosphere [Ref. 8, p. 7].

Scattering

Consider next the dispersion of a certain radiation beam over a solid angle Ω_2 that is greater than the original angle Ω_1. With reference of the detail displayed in the lower right of Fig. 9.5, think of the isotropic scattering of solar radiation at a point. Writing $i'_{vb} = i_{vb}/\Omega$, where i_{vb} represents the number of watts arriving per unit area, for the incoming radiation, we have

$$\frac{i_{vb}}{\Omega_1} = \frac{2h\nu^3 c^{-2}}{\exp{(h\nu/kT_1)} - 1} \tag{9.48}$$

The solid angle subtended by the solar disc seen from earth is $\Omega_1 = (6.8)10^{-5}$ steradians. Similarly, for the outgoing radiation, we write

$$\frac{i_{vb}}{\Omega_2} = \frac{2h\nu^3 c^{-2}}{\exp{(h\nu/kT_2)} - 1} \tag{9.49}$$

where, for example, $\Omega_2 = 4\pi$ if the incident beam is scattered isotropically in all directions. The numerator i_{vb} is the same on the left side of eqs. (9.48) and (9.49), to account for the continuity of energy flow through the scattering site. Dividing eqs. (9.48) and (9.49) side by side, we learn that scattering decreases the temperature of the original radiation:

$$\frac{T_1}{T_2} = \frac{1}{x} \ln\left[\frac{\Omega_2}{\Omega_1}(e^x - 1) + 1\right], \text{ where } x = \frac{h\nu}{kT_1} \tag{9.50}$$

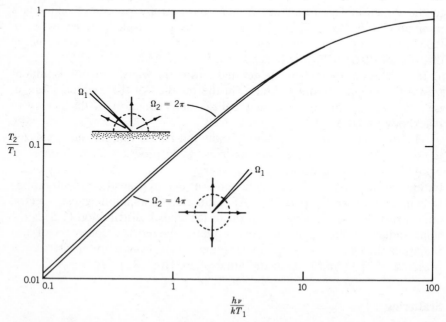

Figure 9.5   The temperature decrease (entropy increase) induced by the scattering of monochromatic solar radiation; $\Omega_1 = (6.8)10^{-5}$ steradians.

This effect is illustrated in Fig. 9.5 for solar radiation scattered over 4π and over 2π (diffuse reflection by an opaque nonabsorbing surface).

Associated with the temperature drop induced by scattering is a rise in the total entropy current of the beam (i.e., the number of watts per unit area and degrees Kelvin)

$$L_{\nu b} = L'_{\nu b}\Omega = \frac{4}{3T} i_{\nu b} \qquad (9.51)$$

where $L'_{\nu b}$ is the corresponding per-unit-solid-angle quantity:

$$L'_{\nu b} = \frac{4}{3T} i'_{\nu b} \qquad (9.52)$$

Scattering, or the "dilution" of beam radiation, is an entropy-generation mechanism:

$$S_{\text{gen}} = \frac{4}{3} i_{\nu b}\left(\frac{1}{T_2} - \frac{1}{T_1}\right) \geq 0 \qquad (9.53)$$

where T_1 and T_2 represent the monochromatic radiation temperatures before and after scattering, respectively.

Figure 9.5 shows also how the temperature drop and the entropy generation depend on wavelength: these effects become increasingly more pronounced as the frequency of the monochromatic beam decreases. This means that if the incident beam contains blackbody radiation of one temperature T_1, the scattered beam contains an entire spectrum of temperatures lower than T_1, i.e., the scattered radiation is not "blackbody." According to the constant-energy scenario constructed in Fig. 9.4, if this nonblackbody scattered radiation were to be trapped in a reflecting enclosure and seeded with soot particles, then its conversion into a new blackbody radiation would be accompanied by an additional generation of entropy.

Net Radiative Heat Transfer

The simplest configuration for pure radiation heat transfer is shown in Fig. 9.6. A blackbody of temperature T and total surface area A is surrounded from all angles by an evacuated enclosure whose wall has a different temperature T_e. The enclosure is itself a blackbody. The two temperatures T and T_e are maintained constant by placing the inner body and the enclosure in thermal communication with appropriate temperature (heat) reservoirs. Given the temperature difference $T - T_e$, we expect from the outset the occurrence of net heat transfer *and* entropy generation [cf. Ref. 1, p. 99]. These two features are illustrated by the relative size of the heat transfer and entropy transfer arrows shown in Fig. 9.6, where it was assumed that $T > T_e$.

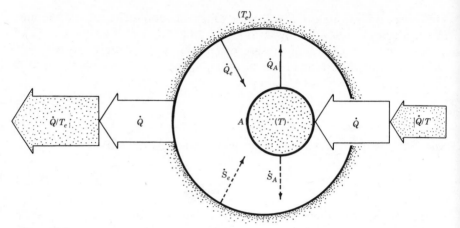

Figure 9.6 Energy and entropy currents between a blackbody (A, T) and a surrounding black surface of a different temperature (T_e).

The following analysis contains two parts. First, we account for the flow of energy and entropy through the surface of the inner body and conclude that in the presence of net heat transfer, this surface is the locus of irreversibility. Second, we account for similar flows and reach a similar conclusion in connection with the surface of the outer enclosure.

The energy emitted per unit time by the inner body is calculated by integrating $i_b'(T)$ over the solid angle 2π and area A

$$\dot{Q}_A = \int_0^A \int_0^{2\pi} i_b'(T) \cos\beta \, d\Omega \, dA \tag{9.54}$$

where β is the sharp angle formed between the direction of i_b' and the normal to the area element dA [Ref. 6, p. 17]. Recognizing $d\Omega = \sin\beta \, d\beta \, d\theta$, in which β varies from 0 to $\pi/2$, and where θ increases from 0 to 2π in the plane of dA, eq. (9.54) yields

$$\dot{Q}_A = A\sigma T^4 \tag{9.55}$$

Note here the emergence of the new coefficient σ, or the Stefan–Boltzmann constant:

$$\sigma = ac/4 = (5.669)10^{-8} \, \text{W m}^{-2} \, \text{K}^{-4} \tag{9.56}$$

The photon stream that carries $\dot{Q}_A$ is responsible also for the entropy outflow

$$\dot{S}_A = \frac{4}{3T} \dot{Q}_A = \frac{4}{3} A\sigma T^3 \tag{9.57}$$

Next, we consider the energy and entropy streams that *arrive* at the surface of the inner blackbody. The energy current $\dot{Q}_e$ emitted by the enclosure and absorbed by the inner surface A follows from a double integral similar to eq. (9.54), using this time $i_b'(T_e)$ instead of $i_b'(T)$ in the integrand:

$$\dot{Q}_e = A\sigma T_e^4 \tag{9.58}$$

Similarly, the entropy current emitted by the enclosure wall and absorbed by the A surface is

$$\dot{S}_e = \frac{4}{3T_e}\,\dot{Q}_e = \frac{4}{3}\,A\sigma T_e^3 \tag{9.59}$$

Also arriving at the A surface are the heat current $\dot{Q}$ and the entropy current $\dot{Q}/T$ furnished by the heat reservoir of temperature T. The conservation of energy through the A surface requires

$$\dot{Q} = \dot{Q}_A - \dot{Q}_e = A\sigma(T^4 - T_e^4) \tag{9.60}$$

which means also that the entropy current delivered by the (T) reservoir is $A\sigma(T^4 - T_e^4)/T$.

The irreversibility of the absorption and emission process that takes place on surface A is demonstrated by the entropy-generation rate calculated for the inner blackbody as a system:

$$\dot{S}_{\text{gen},A} = \dot{S}_A - \dot{S}_e - \dot{Q}/T$$

$$= \frac{\sigma A}{3T}\,(T - T_e)^2(T^2 + 3T_e^2 + 2TT_e) \geq 0 \tag{9.61}$$

This important result, which was obtained first by Planck [see Ref. 9, p. 98], was reactivated in an informative 1983 article by De Vos and Pauwels [15]. The message of eq. (9.61) is that the A surface is the locus of irreversibility as soon as a temperature difference $|T - T_e|$ exists. Said another way, the interaction between a blackbody (T) with blackbody radiation of a different temperature (T_e) is intrinsically irreversible. This conclusion is essential from the point of view of solar-energy utilization, because, as a first-cut model, a solar collector can be viewed as a blackbody that is exposed to blackbody radiation of a higher temperature.

In the extreme situation where the return streams $(\dot{Q}_e, \dot{S}_e)$ are both negligible, the entropy generated at the A surface is definitely positive, namely, $\sigma A T^3/3 > 0$. This extreme—the case of emission without simultaneous absorption—might occur when the inner blackbody is surrounded by a background (enclosure) whose temperature approaches absolute zero. In conclusion, emission in the absence of simultaneous absorption acts as a definite source of irreversibility.

The reverse extreme—the case of absorption in the absence of simultaneous emission—is especially relevant to deciding how much work can be extracted from solar radiation. This extreme is relevant because in the hope of maximizing the production of exergy per unit of collector area, the engineer may wish to "design" a system that absorbs all the solar radiation that lands on it, without returning any radiation back to the sun, i.e., without simultaneously emitting radiation. For this, the engineer chooses a collector surface that makes the collector act like a blackbody, of course. And, driven by the same exergy-maximization objective, the engineer may also wish to design his collector so that the collector temperature is the highest that it can be, namely, the same as the sun's temperature as a blackbody. With reference to Fig. 9.6, these wishes amount to assuming $\dot{Q}_A = 0$ and $T_e \rightarrow T$. In this extreme, the first law (9.60) simply states that the absorbed heat current ($\dot{Q}_e$) must be rejected to the (T) heat reservoir, $\dot{Q} = -\dot{Q}_e < 0$. There is nothing illegal and controversial about this "first-law" conclusion, which is why the second law is an integral part of engineering thermodynamics. For in the same extreme, the entropy-generation expression (9.61) would read

$$\dot{S}_{\text{gen},A} = -\dot{S}_e - \dot{Q}/T$$

$$= -\frac{4}{3T}\dot{Q}_e + \frac{\dot{Q}_e}{T} < 0 \qquad (9.62)$$

which would be a violation of the Second Law of Thermodynamics. In conclusion, it is impossible for a blackbody to absorb radiation of the same temperature, without simultaneously emitting radiation (see Kirchhoff's law in the next subsection).

The radiation heat transfer arrangement of Fig. 9.6 has not one but two black surfaces that emit and absorb. Repeating the preceding analysis for the enclosure body of temperature T_e, in place of eq. (9.60), we find [15]

$$\dot{S}_{\text{gen},e} = \frac{\sigma A}{3T_e}(T - T_e)^2(T_e^2 + 3T + 2T_eT) \geq 0 \qquad (9.63)$$

Note that the entropy generated by emission and absorption at the enclosure surface becomes infinite as T_e approaches zero.

Finally, adding eqs. (9.61) and (9.63), we arrive at the aggregate entropy-generation rate of the radiative heat transfer arrangement:

$$\dot{S}_{\text{gen}} = \dot{S}_{\text{gen},A} + \dot{S}_{\text{gen},e}$$

$$= \frac{\sigma A}{TT_e}(T - T_e)^2(T + T_e)(T^2 + T_e^2) \geq 0 \qquad (9.64)$$

Regardless of the relative size of T and T_e, i.e., regardless of the physical direction of $\dot{Q}$, the occurrence of net heat transfer is accompanied by the generation of entropy. The same result can be obtained directly by writing the second law for a control volume that includes the inner blackbody and the enclosure wall, $\dot{S}_{gen} = \dot{Q}/T_e - \dot{Q}/T$, where $\dot{Q}$ is given by eq. (9.60), as done in the thermodynamic optimization of low-temperature radiation shields [Ref. 1, pp. 187 and 197; also Ref. 16].

Kirchhoff's Law

No coverage of the thermodynamics of radiative heat transfer would be complete without mentioning one of the earliest invocations of the second law in this field [17, 18]. Consider the more general case where the A surface of Fig. 9.6 is not "black," meaning that the outgoing energy and entropy currents are now

$$\dot{Q}_A = \int_{A=0}^{A} \int_{\Omega=0}^{2\pi} \int_{\nu=0}^{\nu} \epsilon(\nu, T, \beta, \theta)\, i'_{\nu b}(\nu, T) \cos \beta \, d\nu \, d\Omega \, dA \quad (9.65)$$

$$\dot{S}_A = \int_{A=0}^{A} \int_{\Omega=0}^{2\pi} \int_{\nu=0}^{\infty} \epsilon(\nu, T, \beta, \theta)\, L'_{\nu b}(\nu, T) \cos \beta \, d\nu \, d\Omega \, dA \quad (9.66)$$

where $d\Omega = \sin \beta \, d\beta \, d\theta$, $0 \le \theta \le 2\pi$ and $0 \le \beta \le \pi/2$. The spectral intensity of the blackbody entropy current per unit solid angle $(L'_{\nu b})$ is defined in eq. (9.52). The spectral emissivity $\epsilon \le 1$ is, in general, a direction-dependent quantity. Note that when ϵ is definitely less than 1, the emission of radiation from a surface of temperature T is the same as having diluted the blackbody monochromatic radiation that originated from a black surface of the same temperature.

Assuming next that the nonblack body of area A is placed in an enclosure filled with diffuse blackbody radiation of temperature T_e (as in Fig. 9.6), the energy and entropy currents absorbed by this body are

$$\dot{Q}_e = \int_{A=0}^{A} \int_{\Omega=0}^{2\pi} \int_{\nu=0}^{\infty} \alpha(\nu, T, \beta, \theta)\, i'_{\nu b}(\nu, T_e) \cos \beta \, d\nu \, d\Omega \, dA \quad (9.67)$$

$$\dot{S}_e = \int_{A=0}^{A} \int_{\Omega=0}^{2\pi} \int_{\nu=0}^{\infty} \alpha(\nu, T, \beta, \theta)\, L'_{\nu b}(\nu, T_e) \cos \beta \, d\nu \, d\Omega \, dA \quad (9.68)$$

In both integrands, we distinguish now the direction-dependent spectral absorptivity $\alpha(\nu, T, \beta, \theta) \le 1$, which is a property of the absorbing surface of temperature T.

Finally, if the entire system (the black enclosure with the nonblack absorber–emitter in it) is isolated, the First Law of Thermodynamics (9.60) requires $\dot{Q}_A = \dot{Q}_e$, which in view of eqs. (9.65) and (9.67) means

$$\epsilon(\nu, T, \beta, \theta)\, i'_{\nu b}(\nu, T) = \alpha(\nu, T, \beta, \theta)\, i'_{\nu b}(\nu, T_e) \quad (9.69)$$

If, in addition, the isolated system is abandoned for a sufficiently long time so that it reaches internal *equilibrium*, then according to the second law (9.64), the entropy generation of the entire system or any of its subsystems is zero. In particular, $\dot{S}_{\text{gen},A} = 0$ means $\dot{S}_A = \dot{S}_e$, or

$$\epsilon(\nu, T, \beta, \theta) \frac{4}{3T} i'_{\nu b}(\nu, T) = \alpha(\nu, T, \beta, \theta) \frac{4}{3T_e} i'_{\nu b}(\nu, T_e) \qquad (9.70)$$

Dividing eqs. (9.69) and (9.70) side by side, we conclude that the equilibrium of the isolated system is characterized by $T = T_e$. Substituting this conclusion into eq. (9.69), we arrive at the most frequent analytical statement of *Kirchhoff's law*:

$$\epsilon(\nu, T, \beta, \theta) = \alpha(\nu, T, \beta, \theta) \qquad (9.71)$$

The spectral directional emissivity of a certain surface of temperature T is always exactly equal to the spectral directional absorptivity—this, of course, when the surface is in equilibrium with the radiation that surrounds it. Which brings us back to the observation made in connection with eq. (9.62). The same analysis can be generalized to the realm of nonblack surfaces, to conclude that it is impossible for a body of temperature T to absorb radiation of the same temperature, without simultaneously emitting radiation.

THE IDEAL CONVERSION OF ENCLOSED BLACKBODY RADIATION

The controversy that accompanies the issue of how much work can be extracted from thermal radiation is largely due to the fact that historically this question has been stated in two different settings. One group of analyses considers as a "thermodynamic system" a certain volume that is initially filled with equilibrium blackbody radiation of a certain temperature, and asks the question of how much work could be extracted as the system reaches the "dead state" defined by a specified ambient. The second group of analyses considers the steady emission of solar radiation into the surrounding cold universe, and asks the practical question of how much mechanical power can be produced on earth with that portion of the solar emission that can be intercepted with a collector. Among these alternatives, the second represents more closely the solar-energy-engineering point of view. Both types of questions are fundamentally important, especially since not everybody agrees with the answer given to each question taken separately. In this section, we consider the first question type—namely, the exergy content of trapped blackbody radiation—and reserve the more practical issue of collector-efficiency maximization for the remaining sections of this chapter.

Petela's Theory

In a 1964 seminal paper, Petela [19] reported the exergy of isotropic blackbody radiation by considering the deformable reflecting enclosure of Fig. 9.7, in which the initial state of the system is represented by (V_1, T_1). He argued that the system delivers maximum work as it settles into the "dead state" defined by the ambient temperature T_0. We denote the end state as (V_2, T_2), Fig. 9.7, and note that since $T_2 = T_0$, the dead-state pressure is fixed, $P_2 = (a/3)T_0^4$. The exergy calculation for trapped black-body radiation differs from the usual exergy calculations in this very important respect: to define the environmental (dead-state) conditions for blackbody radiation means to specify only one intensive property (T_0). This peculiarity is another manifestation of the special two-argument appearance of the fundamental relation $U(S, V)$, which was discussed immediately after eq. (9.33). Since the chemical potential of the photon-gas system is zero, the Gibbs–Duhem relation reduces to $S \, dT = V \, dP$, which means that only one intensive property can be varied independently.

Imagine then a batch of blackbody radiation (V_1, T_1) immersed in an environment of blackbody radiation of temperature T_0. For the extraction of

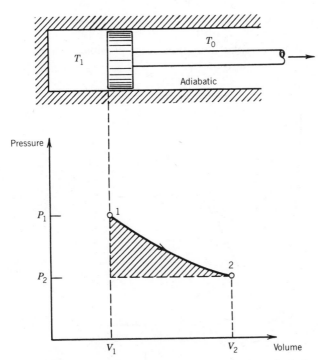

Figure 9.7 Reversible and adiabatic expansion process, for calculating the nonflow exergy of enclosed blackbody radiation (Bejan [1], p. 207).

maximum work, Petela proposed the reversible and adiabatic expansion to the end state (V_2, T_2), as shown in the lower half of Fig. 9.7. The net work transferred through the piston rod is

$$W_{1-2} = \int_1^2 P\,dV - P_2(V_2 - V_1) \tag{9.72}$$

where, according to eq. (9.35), the path of the process is $P = (\text{constant})V^{-4/3}$. The net work (or exergy) W_{1-2} resulting from eq. (9.72) can be arranged as

$$W_{1-2} = U_1\left[1 - \frac{4}{3}\frac{T_2}{T_1} + \frac{1}{3}\left(\frac{T_2}{T_1}\right)^4\right] \tag{9.73}$$

where U_1 is the original energy inventory of the system ($U_1 = aT_1^4 V_1$). It has become fashionable to remember this result as an "efficiency" of converting U_1 into work:

$$\eta_P = \frac{W_{1-2}}{U_1} = 1 - \frac{4}{3}\frac{T_2}{T_1} + \frac{1}{3}\left(\frac{T_2^*}{T_1}\right)^4 \tag{9.74}$$

where the subscript "P" stands for Petela. The same result was determined independently 12 years later by Press [20] and Landsberg and Mallinson [21]. The dependence of η_P on the temperature ratio T_1/T_2 is illustrated in Fig. 9.8. Thinking of solar radiation as a possible initial charge for the

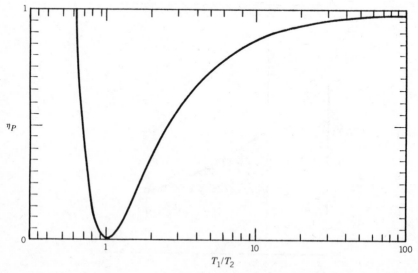

Figure 9.8 Petela's efficiency as a function of the absolute temperature ratio T_1/T_2 (after Bejan [1], p. 209).

work-producing stroke of Fig. 9.7, we set $T_1 = 5762 \, \text{K}$, $T_2 = 300 \, \text{K}$, and find a remarkably high conversion efficiency, $\eta_P = 0.9306$. Similar plots of η_P vs. T_1/T_2 that have appeared in the literature [e.g., Refs. 4, 15] show only the $T_1 > T_2$ abscissa range, because of its relevance to solar and power-production applications in general. In this particular range, the eye sees $\eta_P \leq 1$, which looks familiar in view of the engineer's earlier encounters with the thermodynamics of energy-conversion devices.

Before questioning this widely quoted and accepted result, it is instructive to review two alternative derivations of eq. (9.74) that have appeared subsequently in the literature. The first is based on the straightforward application of the concept of nonflow availability [Ref. 1, p. 34] to the process $(1) \rightarrow (2)$ of Fig. 9.7; this amounts to writing

$$
\begin{aligned}
W_{1-2} &= A_1 - A_2 \\
&= (U_1 - U_2) + P_0(V_1 - V_2) - T_0(S_1 - S_2)
\end{aligned}
\tag{9.75}
$$

where P_0 and T_0 are the traditional environmental conditions. As noted already, in the present problem $P_0 = (a/3)T_2^4$ and $T_0 = T_2$, i.e., P_0 is not atmospheric pressure [for example, note that if $T_0 = 300 \, \text{K}$, the corresponding photon-gas pressure is $P_0 = (a/3)T_0^4 = (2.02)10^{-11} \, \text{atm}$]. Recalling the formulas that were developed earlier for energy (9.8) and entropy (9.22), it is easy to show that eq. (9.75) yields the same result as eq. (9.73). The availability-based derivation of Petela's efficiency is discussed in Refs. 2 and 22–24.

Another way of deriving eq. (9.73) is to note that the (V_1, T_1) system of Fig. 9.7 can be brought to its dead-state temperature T_2 [or dead-state pressure $(a/3)T_2^4$] during a reversible constant-volume cooling process, in which the temperature gap between the system and the ambient (T_2) reservoir is bridged by a Carnot engine [23]. During the process from (V_1, T_1) to (V_1, T_2), the energy of the system U drops from $aT_1^4V_1$ to $aT_2^4V_1$. Any infinitesimal energy drop $(-dU)$ acts as a positive heat input to the high-temperature end of the Carnot cycle, whose instantaneous high temperature is T (this temperature drops gradually from T_1 to T_2). The total work extracted during the entire process is, therefore,

$$
W_{1-2} = \int_{U=aT_1^4V_1}^{aT_2^4V_1} \left(1 - \frac{T_2}{T}\right)(-dU)
\tag{9.76}
$$

where $(1 - T_2/T)$ is the instantaneous efficiency of the Carnot engine, and $dU = d(aV_1T^4)$. Performing the above integral leads back to eq. (9.73).

The Controversy

Now we turn our attention to the controversy generated by the fact that at least two other "efficiency" expressions have been proposed in place of eq.

(9.74). The first alternative is a 1964 formula due to Spanner [10] and championed forcefully by Gribik and Osterle [2, 5]:

$$\eta_S = 1 - 4T_2/3T_1 \tag{9.77}$$

and the second alternative was proposed in 1981 by Jeter [23]:

$$\eta_J = 1 - T_2/T_1 \tag{9.78}$$

In these expressions, T_1 and T_2 have exactly the same meaning as until now; however, contrary to the impression emanating from the current literature, not all the ηs fit the definition given in eq. (9.74). The purpose of the following analysis is not to determine "who was right," and certainly not to take sides. The objective is to establish the relationship between these seemingly competing results. Indeed, it will be shown that all these results do not "compete," rather, they complement one another.

A Unifying Theory

My own questioning of the physical relevance of Petela's result consists of asking [25]:

(i) What is the origin (the source) of the equilibrium blackbody radiation system (V_1, T_1) postulated in the beginning of the process $(1) \rightarrow (2)$? A "supply" of such radiation does not exist, and even if it does, the job of filling the V_1 space with T_1 radiation must not be overlooked.

(ii) What is the ultimate fate of the blackbody radiation of temperature T_2 left when the system reaches its dead state? Clearly, there is no such thing as an "environment" of isotropic blackbody radiation (and pressure), as is assumed most visibly in the availability-type derivation of eq. (9.75).

Questions (i) and (ii) are particularly appropriate in the field of solar-energy engineering, because inserting $T_1 = 5762$ K and $T_2 = 300$ K into eq. (9.74) is considerably easier than picturing the (V_1, T_1) enclosure as actually containing radiation of *solar* origin.

Answers to these questions follow from the analysis of the reversible three-part process shown in Fig. 9.9. The middle process $(1) \rightarrow (2)$ is the same as in Petela's argument or in the two alternative arguments reviewed after eq. (9.74): it is a reversible process executed in communication with the temperature reservoir (T_2); during this process, the net work transfer interaction with an external mechanism is the quantity W_{1-2} represented by eq. (9.73).

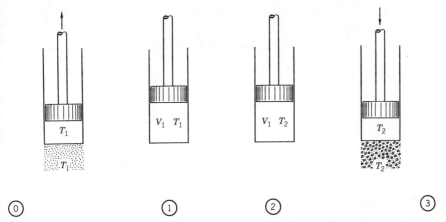

Figure 9.9 Reversible cycle executed by an enclosed radiation system while in communication with two heat reservoirs.

The first process $(0) \rightarrow (1)$ accounts for the reversible manufacture of the (V_1, T_1) radiation system, while in thermal contact with the temperature reservoir (T_1). The system volume V increases from zero to V_1 as the (T_1) reservoir heats the system isothermally and as the system performs quasi-static work of type $\delta W_{rev} = P_1 \, dV$, where P_1 is constant because T_1 is constant. It is not difficult to show that the net energy interactions during this process are

$$Q_{0-1} = \frac{4}{3} U_1 \tag{9.79}$$

$$W_{0-1} = \frac{1}{3} U_1 \tag{9.80}$$

where U_1 is the energy inventory of the system in state (1) [i.e., the same U_1 as in eq. (9.73)]. The entropy generation is, of course, zero:

$$S_{gen,0-1} = -\frac{Q_{0-1}}{T_1} + S_1 = 0 \tag{9.81}$$

In conclusion, in order to produce the assumed state (1) reversibly, it is necessary to extract heat from the (T_1) reservoir while delivering work to a user. This brings up the observation that there are infinitely many irreversible processes whose end state is (V_1, T_1): in each of these processes, both energy interactions are algebraically smaller than the respective values listed in eqs. (9.79) and (9.80). An example of such an alternative (irreversible) process is given at the end of this subsection.

The $(0) \to (1)$ scenario parades in the reverse direction in the last process $(2) \to (3)$ of Fig. 9.9. This time, the system loses heat to the (T_2) reservoir and, since the process is reversible, the net interactions are

$$Q_{2-3} = -\frac{4}{3} U_2 \qquad (9.82)$$

$$W_{2-3} = -\frac{1}{3} U_2 \qquad (9.83)$$

$$S_{\text{gen},2-3} = -\frac{Q_{2-3}}{T_2} - S_2 = 0 \qquad (9.84)$$

In other words, it takes work to eliminate the dead-state batch of radiation left at the end of Petela's process. In any irreversible process $(2) \to (3)$, the required work input (i.e., the absolute value of W_{2-3}) will be greater than the reversible-limit result (9.83).

Putting these three-part results together, we note that $(0) \to (1) \to (2) \to (3)$ represents actually a cycle whose net work transfer interaction is maximum in the reversible limit discussed here:

$$\oint \delta W = W_{0-1} + W_{1-2} + W_{2-3} = \frac{4}{3} U_1 (1 - T_2/T_1) \qquad (9.85)$$

What "maximum efficiency" corresponds to this result depends, of course, on how the efficiency ratio is defined. In this case—a cycle executed in communication with two heat reservoirs—the most reasonable definition is to divide the net work output (9.85) by the heat input supplied by the high-temperature reservoir (Q_{0-1}). The result is then the expected Carnot efficiency:

$$\eta_C = \frac{\oint \delta W}{Q_{0-1}} = 1 - \frac{T_2}{T_1} \qquad (9.86)$$

This result should be "expected" because the efficiency of a reversible cycle executed by a closed system while in communication with two heat reservoirs is independent of the nature of the working fluid employed by the engine. In the $(0) \to (1) \to (2) \to (3)$ scenario of Fig. 9.9, the presence of blackbody radiation is purely intermediary. It is important to note at this stage that the right side of eq. (9.86) is the same as in Jeter's efficiency (9.78), even though Jeter's analysis addresses an entirely different kind of process, system, and surroundings. The equivalence of eqs. (9.78) and (9.86) is discussed in the closing part of this section.

Educated by the Petela terminology, we may be tempted to define "efficiency" by dividing the net work output of the cycle by U_1, as in eq. (9.74). If we do this, the efficiency of the cycle is exactly $\frac{4}{3}$ times the Carnot efficiency, which should be enough of a hint that the η_P ratio is not really a

"heat-engine efficiency" in the usual engineering-thermodynamics interpretation. I believe that η_P is no more than a convenient albeit artificial way of
nondimensionalizing the calculated work output W_{1-2}.

A slightly modified version of the $(0)\rightarrow(1)\rightarrow(2)\rightarrow(3)$ cycle of Fig. 9.9
is relevant to understanding the position of the Spanner–Gribik–Osterle
efficiency (9.77) relative to Jeter's (9.78) and Petela's (9.74). The modification consists of replacing the reversible expansion process $(0)\rightarrow(1)$ with a
spontaneous (irreversible) process in which the system does not deliver any
work to the assumed external mechanism (the "user"). We label this new
starting process $[(0)\rightarrow(1)]_{\text{zero-work}}$. Physically, this new process begins with
a completely evacuated enclosure of volume V_1, which is placed suddenly in
thermal communication with the (T_1) reservoir. Seeded with a soot particle,
the (V_1) enclosure fills eventually with equilibrium blackbody radiation of
temperature T_1. Repeating the first- and second-law analyses that earlier led
to eqs. (9.79)–(9.81), we find that the new process $[(0)\rightarrow(1)]_{\text{zero-work}}$ is
characterized by the following net interactions:

$$Q_{0-1,\text{zero-work}} = U_1 \tag{9.87}$$

$$W_{0-1,\text{zero-work}} = 0 \tag{9.88}$$

$$S_{\text{gen},0-1,\text{zero-work}} = -\frac{1}{T_1} Q_{0-1,\text{zero-work}} + S_1 > 0 \qquad \text{(irreversible)} \tag{9.89}$$

The obvious reason for considering this modification of the first leg of the
$(0)\rightarrow(1)\rightarrow(2)\rightarrow(3)$ reversible cycle is that the spontaneous filling process
$[(0)\rightarrow(1)]_{\text{zero-work}}$ requires considerably less hardware than the original
reversible version (Fig. 9.9). However, the real reason for drawing attention
to the $[(0)\rightarrow(1)]_{\text{zero-work}}$ alternative is to show that the aggregate work
output of the modified (now labeled "irreversible") cycle is

$$\left(\oint \delta W\right)_{\text{irrev}} = W_{1-2} + W_{2-3}$$

$$= aT_1^4 V_1(1 - 4T_2/3T_1) \tag{9.90}$$

and that the proper *heat-engine* efficiency for this new cycle is identical to
Spanner's efficiency (9.77):

$$\frac{(\oint \delta W)_{\text{irrev}}}{Q_{0-1,\text{zero-work}}} = 1 - \frac{4T_2}{3T_1} \tag{9.91}$$

In conclusion, eqs. (9.90) and (9.91) represent the maximum work output
and the maximum heat-engine efficiency for that class of cycles where the
filling of the cylinder and piston apparatus occurs spontaneously, i.e.,
without the possibility of delivering useful work.

The equality between the right sides of eqs. (9.91) and (9.77) is co-incidental, as η_S was historically defined by dividing $(W_{1-2} + W_{2-3})$ by U_1, in an analysis of a reversible process that begins with state (1), not in the analysis of a heat-engine cycle. The coincidence is due to eq. (9.87), which shows that the energy inventory of the gas system at state (1) happens to be numerically equal to the net heat input delivered by the (T_1) reservoir to the modified (irreversible) version of the $(0) \rightarrow (1) \rightarrow (2) \rightarrow (3)$ cycle.

One very interesting characteristic of Spanner's efficiency is that it becomes negative for sufficiently small temperature ratios, namely, for $T_1/T_2 < 4/3$. This feature, which is certainly unexpected in the "maximum-efficiency" context in which η_S was originally proposed, prompted some of Gribik and Osterle's critics to dismiss the η_S result as being obviously suspicious and in error [3, 4]. Yet, in view of the scenario of Fig. 9.9, the Spanner-type results, eqs. (9.90) and (9.91), are correct: they both *must* become negative below a certain T_1/T_2 ratio, because, as T_1/T_2 decreases, the value of Petela's work output (W_{1-2}) drops below the value of the work input required to eliminate the T_2 radiation (W_{2-3}). Note, on the other hand, that the maximum heat-engine efficiency of the entire cycle, eq. (9.86), is positive in the entire $T_1/T_2 > 1$ domain, because of the positive contribution made by the work produced during the reversible charging phase (W_{0-1}).

Relative to questions (i) and (ii), which in combination with Petela's problem led to the reversible cycle analysis of Fig. 9.9, it is worth noting that Spanner [10] and Gribik and Osterle [2, 5] properly and openly raised the question (ii), in that they insisted on the "destruction" of the T_2 radiation that is left after Petela's process $(1) \rightarrow (2)$. Gribik and Osterle's way of emphasizing this subtlety is to visualize the lowering of the tempera-ture of the remanent T_2 radiation down all the way to absolute zero (note that the photon population decreases to zero in the $T \rightarrow 0$ limit). In my opinion, the reason why this subtle point has not attracted the attention that it deserves is because it relies on the invocation of a third temperature (absolute zero) in the maximization of a conversion efficiency, i.e., in the discussion of a concept that has been traditionally associated with the presence of only two temperature reservoirs.

Figure 9.10 presents a graphic summary of the three distinct points of view that are succinctly represented by the "efficiency" ratios of Petela (9.74), Spanner (9.77), and Jeter (9.78). The narrowest domain is occupied by the Petela–Press–Landsberg and Mallinson line of inquiry, which de-scribes the maximum work that could be extracted during the transforma-tion of isotropic T_2 radiation into isotropic T_1 radiation. While the W_{1-2} result is correct in the narrow context in which the Petela problem was formulated, the η_P ratio defined in eq. (9.74) is not a "conversion efficien-cy" in the usual sense. Spanner's point of view (and especially Gribik and Osterle's reformulation) covers a territory that extends beyond Petela's, because it considers also the reversible removal of the remanent T_2 radia-

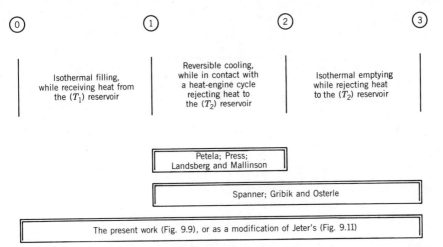

Figure 9.10 The relative position of the three different theories on the domain defined by the cycle $(0) \to (1) \to (2) \to (3)$.

tion. The broadest territory of Fig. 9.10 is covered by the analysis of the three-part cycle of Fig. 9.9. The cycle analysis consisted of complementing Petela's problem with answers to questions (i) and (ii). Assuming all the way that T_1 is greater than T_2, we learned that W_{0-1} is positive and W_{2-3} negative, which means that the following inequalities must hold if all the Ws are referenced to (produced with) the same heat input Q_{0-1}:

$$W_{0-3} \left(\text{written also} \oint \delta W\right) > W_{1-2} \text{ (Petela)} > W_{1-3} \text{ (Spanner)} \quad (9.92)$$

The same inequalities would exist between the corresponding heat-engine efficiencies, again, provided that they are all defined in the same way.

Closed-System Reformulation of Jeter's Analysis

Finally, we seek an explanation for the coincidence that the heat-engine efficiency of the reversible cycle $(0) \to (1) \to (2) \to (3)$ is the same as Jeter's efficiency (9.78). Jeter [23] described the "steady flow" of radiant energy through the apparatus that here is reproduced intentionally as the solid-line centerpiece of Fig. 9.11. The words "steady flow" are in quotation marks because they are not the most appropriate name for an apparatus that actually processes its contents in a "one-shot" fashion. Jeter's apparatus is the familiar cylinder and piston enclosure with perfectly reflecting surfaces. The environment that acts on the back side of the piston is assumed to be blackbody radiation of temperature T_2 (hence, pressure P_2). In the beginning of the one-shot process, the piston is in its "start" position (i.e., the swept volume is zero), while the left end of the cylinder is suddenly opened

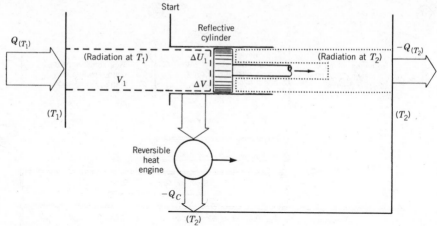

Figure 9.11   An extension of Jeter's analysis, showing that the operation of the "flow" apparatus is analogous to the cyclical operation of the apparatus shown in Fig. 9.9.

to a cavity filled with isotropic blackbody radiation of temperature T_1. The "one-shot" process amounts to the following sequence of events:

(a) The piston moves slightly to the right, sweeping the volume ΔV and delivering work to two other systems, namely,

$$P_2 \, \Delta V \qquad \text{(to the environment)}$$

and

$$(P_1 - P_2) \, \Delta V \qquad \text{(to an external mechanism, the "user")}$$

During this movement, the swept volume ΔV fills with (T_1) radiation: the internal energy inventory of the ΔV volume is $\Delta U_1 = u_1 \, \Delta V$, or since $P_1 = u_1/3$,

$$\Delta U_1 = 3P_1 \, \Delta V \tag{9.93}$$

(b) The second step consists of using the internal energy of the admitted radiation (ΔU_1) for producing work while in contact with the (T_2) radiation environment. In order to execute this step, the open side of the cylinder must first be closed and replaced with another perfectly reflecting wall. The maximum work produced as the trapped radiation $(\Delta V, \Delta U_1)$ has its temperature lowered from T_1 to T_2 is the one calculated with Petela's formula (9.73), namely,

$$\Delta U_1 \left[1 - \frac{4T_2}{3T_1} + \frac{1}{3} \left(\frac{T_2}{T_1} \right)^4 \right]$$

Adding this work contribution to the part delivered earlier in step (a), Jeter arrived at the total maximum work delivered during the one-shot process:

$$W = 4P_1 \, \Delta V (1 - T_2/T_1) \qquad (9.94)$$

While the Carnot efficiency appears now explicitly on the right side of eq. (9.94), the proper definition of conversion efficiency is not as obvious [the automatic use of Petela's and Spanner's definition would lead to $W/\Delta U_1 = (4/3)(1 - T_2/T_1)$]. Jeter departed from tradition and defined his efficiency as

$$\eta_J = \frac{W}{4P_1 \, \Delta V} = 1 - \frac{T_2}{T_1} \qquad (9.95)$$

by arguing that $4P_1 \, \Delta V$ is the total "quantity of radiation" that entered the cylinder to occupy the volume ΔV (this name for the "$4P_1 \, \Delta V$" denominator comes from the argument that the "quantity of radiation" admitted from the left side of the apparatus is responsible for the internal energy of the trapped radiation, $\Delta U_1 = 3P_1 \, \Delta V$, *and* the total work performed during the filling stroke, $P_1 \, \Delta V$). He used the same argument and terminology in his rebuttal [3] to Gribik and Osterle's criticism [2]. The disagreement persists (in fact it appears to have deepened [5]), because the notion of "quantity of radiation" has not been explained. A definition of this concept follows.

My own analysis of Jeter's argument consists of asking the same questions as before, namely, questions (i) and (ii). First, what is the origin of the radiation from which Jeter's one-shot process bleeds a finite amount? The isotropic (T_1) radiation must always be enclosed, even during the charging step (a). Consider then all the enclosed (T_1) radiation as a thermodynamic system whose volume increases quasistatically from V_1 to $(V_1 + \Delta V)$. The boundary of this system is indicated by the dashed line in Fig. 9.11. Now, if the system is surrounded by an adiabatic boundary, the expansion is accompanied by a drop in the temperature of the radiation system (recall the path of a reversible and adiabatic expansion, $T^3 V = \text{constant}$). This temperature drop is not allowed by the definition of Jeter's apparatus (the left, upstream, side), especially if the one-shot process is to be repeated indefinitely for the "steady" production of work. Therefore, the boundary cannot be adiabatic, and that leaves only one possibility—thermal communication with a heat reservoir of temperature T_1.

Step (a) is, therefore, a reversible and isothermal (isobaric) expansion for which the first law requires

$$Q_{(T_1)} - P_1 \, \Delta V = \Delta U_1 \qquad (9.96)$$

and since $\Delta U_1 = 3P_1 \, \Delta V$, it follows that the heat transfer received by the enclosed radiation system from the (T_1) temperature reservoir is $Q_{(T_1)} = 4P_1 \Delta V$. In conclusion, what Jeter has called the "quantity of radiation" is in fact the heat input that must be provided by the temperature reservoir (T_1) in order to *regenerate* the portion of the isotropic (T_1) radiation system that is consumed by the one-shot apparatus.

Question (ii) relates to the task of maintaining the (T_2) environment isothermal and isobaric despite the compression stroke provided by the back side of the piston during the execution of step (a). With reference to the dotted line system of (T_2) radiation sketched on the right side of Fig. 9.11, the first law for step (a) requires

$$Q_{(T_2)} + P_2 \Delta V = \Delta U_2 \tag{9.97}$$

where $P_2 \Delta V$ is the work transfer received by the system, and $\Delta U_2 = -3P_2 \Delta V$ is the chunk of (T_2) radiation that occupied the ΔV space in the beginning of step (a). The heat transfer interaction with the (T_2) temperature reservoir is negative, $Q_{(T_2)} = -4P_2 \Delta V$.

Reviewing now the greater picture that emerged in Fig. 9.11—an aggregate system that executes a reversible cycle[†] while in communication with two temperature reservoirs—we see that Jeter's work (9.94) can be recalculated as

$$W = Q_{(T_1)} + Q_{(T_2)} + Q_C \tag{9.98}$$

where Q_C is the heat transfer interaction between the Carnot engine and the (T_2) reservoir.[‡] The value of Q_C can be calculated by applying the second law to the step (b) executed by the reversible heat engine and the trapped radiation volume ΔV, as one system,

$$Q_C / T_2 + \Delta S_2 - \Delta S_1 = 0 \tag{9.99}$$

where $\Delta S_{1,2}$ is the entropy inventory of the ΔV space before and after the reversible cooling process:

$$\Delta S_1 = \frac{4 \Delta U_1}{3 T_1} \qquad \Delta S_2 = \frac{4 \Delta U_2}{3 T_2} \tag{9.100}$$

Combining eqs. (9.98)–(9.100) leads back to Jeter's work (9.94), proving that Jeter's efficiency, eqs. (9.78) and (9.95), is the proper heat-engine

---

[†]The one-shot process can be repeated if, following step (b), an open window allows the two sides of the piston face to communicate as the piston is moved back to its "start" position.
[‡]Note that all heat transfer interactions are defined positive when entering the system; in accordance with the same "heat-engine" sign convention, the work transfer interactions are defined positive when exiting the system.

efficiency for the cycle referred to in the footnote, $\eta_J = W/Q_{(T_1)}$. The analogy between the aggregate system of Fig. 9.11 and the $(0) \rightarrow (1) \rightarrow (2) \rightarrow (3)$ cycle of Fig. 9.9 is now evident.

The present modification of Jeter's analysis also clarifies the source of an important misunderstanding in the current literature. Referring to the "Carnot" appearance of his efficiency formula, Jeter wrote that it "supports the very widely held apprehension that thermal radiation is heat" [Ref. 3, p. 80]. Interpreted literally, this statement invites the criticism offered by Gribik and Osterle [5] on the grounds that in the current context "radiation" is a thermodynamic system that must not be confused with the energy interaction called "heat" (or, better, heat transfer). However, the analysis of Fig. 9.11 spells out in concrete terms the intuitive content and, perhaps, the real intention of Jeter's statement: what the appearance of eq. (9.95) suggests is not that radiation is heat, but that the denominator $4P_1 \Delta V$ (which was arbitrarily named "quantity of radiation") represents the magnitude of a precious heat transfer interaction.

To summarize, this section brought under the same roof three divergent theories concerning the ideal conversion of thermal radiation into work. Contrary to some of the claims made during the ongoing controversy, it showed that all these theories are correct, and that each occupies a well-deserved place in the theoretical domain outlined in Fig. 9.10. What makes each theoretical result different is, first, each individual's understanding of what constitutes an appropriate description for the "investment" made in the production of work from thermal radiation, and, second, what is an appropriate model for the fully spent radiation. Whether or not these descriptions are relevant to quantizing the ideal conversion of solar radiation into mechanical power is an entirely different matter that forms the subject of the remainder of this chapter.

THE MAXIMIZATION OF MECHANICAL POWER PER UNIT COLLECTOR AREA

The closed-system reformulation of Jeter's analysis showed that the maximum conversion efficiency η_C refers to situations where a reversible heat-engine cycle uses the sun as a *heat reservoir* of high temperature, $T_1 = T_s$ (e.g., Figs. 9.9 and 9.11). The role of ambient is correspondingly played by a heat reservoir of a lower temperature, $T_2 = T_0$. The presence of radiation in the reversible cycles invoked in the development of eq. (9.86) is purely intermediary: this conclusion parallels the comments made after eq. (9.40). Nevertheless, the Carnot efficiency is a useful concept in the field of solar-energy utilization, because it emphasizes very well the need for a better engineering "figure of merit."

Ideal Concentrators

In order to appreciate this need, try to conceptualize a reversible heat-engine cycle that uses the sun not as a source of thermal radiation but as a heat reservoir, that is, as a supply of heat transfer at a very high, constant temperature. At least in principle, an arrangement of this kind can be visualized as was done already on the right side of Fig. 9.2, where the aperture of the enclosure occupies the focal point of the parabolic concentrating mirrors (for an alternative design, see Jeter [23] and Winston [26]). The geometry of this arrangement is such that the aperture sees only the sun, and, in turn, the solar beam intercepted by the mirror is totally directed into the aperture (i.e., over the hemispherical solid angle 2π). This "focal enclosure" is filled by isotropic blackbody radiation of temperature T, where T is also the temperature of the enclosure wall $(T_s \geq T \geq T_0)$. If the focal enclosure is insulated perfectly, the total energy current that leaves through the aperture $(\sigma A T^4)$ must match the solar-energy stream that arrives through the same solid angle (2π) into the enclosure $(\sigma A T_s^4)$, the A area being the aperture cross-section. In this limiting case then, the focal enclosure reaches thermal equilibrium with the solar radiation channeled into it, $T = T_s$. Note that this case is the same as setting $\dot{Q} = 0$ in the analysis of the two-body arrangement shown in Fig. 9.6.

Although the maximization of the enclosure wall temperature T for the purpose of creating a body whose temperature approaches T_s is appealing, the focal enclosure imagined here will never be a "heat reservoir" of temperature T_s, because in the $T = T_s$ limit, there can be no heat transfer between this enclosure and its environment. If the enclosure serves steadily as "heater" for a Carnot engine between temperatures T and T_0, then the Carnot power output of the engine is

$$\dot{W}_C = \dot{Q}(1 - T_0/T) \qquad (9.101)$$

where $\dot{Q}$ is the heat transfer rate into the pupil of the focal enclosure:

$$\dot{Q} = \sigma A(T_s^4 - T^4) \qquad (9.102)$$

We see immediately that although in the limit $T = T_s$, the Carnot efficiency of the engine $(\dot{W}_C/\dot{Q})$ reaches its maximum value, in the same limit, both $\dot{W}_C$ and $\dot{Q}$ tend to zero. The only way to visualize finite $\dot{W}_C$ and $\dot{Q}$ in the $T = T_s$ limit is by invoking the totally unrealistic limit $A \to \infty$.

This brings us to the more practical proposition of maximizing not $\dot{W}_C/\dot{Q}$ but $\dot{W}_C/A$. To maximize the Carnot power output subject to constant A means to maximize the dimensionless expression[†]

[†]The η_A notation is simply a way of nondimensionalizing the target quantity $\dot{W}_C/A$; this notation is not intended to suggest that $\dot{W}_C$ might ever become equal to $\sigma A T_s^4$ in a certain "ideal" set of circumstances.

$$\eta_A = \frac{\dot{W}_C}{\sigma A T_s^4} = \left[1 - \left(\frac{\theta}{\theta_s}\right)^4\right]\left(1 - \frac{1}{\theta}\right) \tag{9.103}$$

where $\theta = T/T_0$, and $\theta_s = T_s/T_0$. There is only one degree of freedom in the design of the collector and engine installation, namely, the "collector temperature" T, or, conversely, the heat transfer rate put into the Carnot engine, $\dot{Q}$. The optimization problem reduces to solving $d\eta_A/d\theta = 0$, namely,

$$4\theta^5 - 3\theta^4 - \theta_s^4 = 0 \tag{9.104}$$

and substituting the θ_{opt} solution to this equation back into the η_A expression (9.103). The complete function $\eta_{A,max} = \eta_A(\theta_{opt})$ can be generated numerically, however, most relevant to solar-energy utilization is the case $\theta_s \gg 1$ for which the following closed-form results hold [15, 27]:

$$\theta_{opt} \cong 4^{-1/5}\theta_s^{4/5} \quad \text{and} \quad \eta_{A,max} \cong 1 - 5(4\theta_s)^{-4/5} \tag{9.105}$$

For example, if $T_0 = 300\,\text{K}$ and $T_s = 5762\,\text{K}$, i.e., if $\theta_s = 19.21$, the exact numerical solution to eq. (9.104) is $\theta_{opt} = 8.216$; hence, $\eta_{A,max} = 0.849$. In the same case, the approximate expressions (9.105) yield $\theta_{opt} \cong 8.06$ and $\eta_{A,max} \cong 0.845$.

The analytical development that ended with eq. (9.104), and, especially the point of view that a ratio of type $\dot{W}_C/A$ is an important figure of merit in the field of solar-energy conversion, were presented independently by Müser [27], Castans [28, 29], Jeter [23, 30], and De Vos and Pauwels [15, 31, 32]. Although the collector geometry and collector heat-loss mechanism differ from the case discussed until now, the same point of view was articulated independently by Howell and Bannerot [33], Bejan et al. [34], and Haught [35].

One important observation concerns the meaning of the Carnot power whose maximization led to the ratio that was labeled $\eta_{A,max}$. Despite the "Carnot" terminology, the operation of the collector and engine arrangement is not reversible, because the collector surface (A, T) is not in equilibrium with the incoming radiation. The reversible compartment of the arrangement is only the Carnot engine, which bridges the gap between T and T_0. Indeed, the lone degree of freedom that permits the optimization of this power-producing installation amounts to varying the irreversibility associated with the "collection" of the net heat input delivered to the Carnot engine, $\dot{Q}$. With the Gouy–Stodola theorem in mind, we know to expect that the optimum design represented by eq. (9.104) can also be accomplished by minimizing the total entropy-generation rate per unit area, $\dot{S}_{gen}/A$. If we pursue this alternative derivation, however, we reach a very interesting paradox. Basing the analysis on a two-body arrangement like the one shown in Fig. 9.6, one might easily argue that the entropy-generation rate associated with the collection of a net $\dot{Q}$ is [see the discussion under eq. (9.64)]:

$$\dot{S}_{\text{gen}} = \frac{\dot{Q}}{T} - \frac{\dot{Q}}{T_s} \qquad (9.106)$$

Minimizing this expression with respect to T yields

$$4\left(\frac{\theta}{\theta_s}\right)^5 - 3\left(\frac{\theta}{\theta_s}\right)^4 - 1 = 0 \qquad (9.107)$$

which is clearly in disaccord with eq. (9.104). What then is the proper entropy-generation rate associated with the collector and engine arrangement?

An answer to the above question can be formulated in two steps (Fig. 9.12). First, the Gouy–Stodola idea of minimizing entropy generation for maximizing the production of mechanical power requires that we include in the picture the ambient heat reservoir (T_0). Second, if the net heat transfer rate into the collector $\dot{Q}$ is to float freely for the purpose of choosing the optimum operating conditions, then the actual heat input from the sun must be greater than any $\dot{Q}$ value that might be required in the course of collector and engine optimization. Let $\dot{Q}_+$ be this sufficiently large (and fixed) heat transfer rate. It is easy to see that the $\dot{Q}_+$ constant must be at least as large as $\sigma A T_s^4$. A portion of $\dot{Q}_+$ is intercepted by the concentrating collector ($\dot{Q}$),

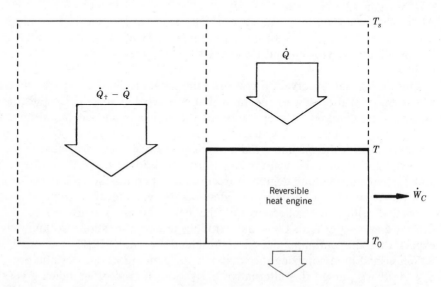

Figure 9.12 Heat transfer across two temperature gaps, $(T_s - T)$ and $(T_s - T_0)$, as source of entropy generation in a fixed-area collector and engine installation.

while the remainder $(\dot{Q}_+ - \dot{Q})$ must necessarily be rejected to the ambient reservoir (T_0): both portions, $\dot{Q}$ and $(\dot{Q}_+ - \dot{Q})$ vary with the only degree of freedom of the system $(T$ or $\dot{Q})$.

The correct entropy-generation rate of the collector and engine installation is, therefore, (Fig. 9.12)

$$\dot{S}_{\text{gen}} = \frac{\dot{Q}}{T} + \frac{\dot{Q}_+ - \dot{Q}}{T_0} - \frac{\dot{Q}_+}{T_s} \tag{9.108}$$

The minimization of this expression with respect to T and subject to constant A, T_0, T_s, and $\dot{Q}_+$ leads to the same conclusion as the Carnot power-maximization procedure outlined earlier. The paradox that has just been clarified is similar to the one identified in Problem 2.2 of Ref. 1. As far as the entropy-generation analysis of solar collector systems is concerned, the error that was illustrated via eq. (9.106) can be avoided by recognizing the following theorem [1, 34]:

> The task of maximizing the work (exergy) delivered per unit of collector cross-section A is equivalent to minimizing the entropy generated in the "column" of cross-section A extending from T_0 all the way up to T_s.

Omnicolor Series of Ideal Concentrators

The essential point of the preceding subsection is that there exists a unique collector temperature for which the Carnot power produced per unit collector area reaches a maximum value. The same $\dot{W}_C/A$-maximization analysis can be repeated for the case where the intercepted T_s radiation is monochromatic (ν) instead of blackbody. In place of eq. (9.104), then we reach the conclusion that to each frequency ν corresponds an optimum collector temperature $T_{\text{opt}}(\nu)$. In physical terms, the transition from the ideal blackbody concentrator of Fig. 9.2 to the single-color scheme discussed here can be effected by first decomposing the incoming flux $\sigma A T_s^4$ into an infinite number of monochromatic components, $\pi A i'_{\nu b}$, where $i'_{\nu b}$ is a function of ν and T_s. Second, the heat current $\dot{Q}_\nu$ drawn from each single-color collector is used to drive a Carnot engine whose power output is

$$\dot{W}_{C,\nu}(\nu, T) = \dot{Q}_\nu (1 - T_0/T) \tag{9.109}$$

where

$$\dot{Q}_\nu = \pi A [i'_{\nu b}(\nu, T_s) - i'_{\nu b}(\nu, T)] \tag{9.110}$$

By continuing to regard A as fixed, the power $\dot{W}_{C,\nu}$ can be maximized by finding the optimum temperature of the single-color collector, $T_{\text{opt}}(\nu)$. A related calculation that is discussed later in connection with Fig. 9.15 shows that $T_{\mathrm{opt}}(\nu)$ increases monotonically with ν. Substituting $T_{\mathrm{opt}}(\nu)$ into eq. (9.109), we obtain the maximum power output of the collector of frequency ν:

$$\dot{W}_{C,\nu,\max} = \dot{W}_{C,\nu}(\nu, T_{\mathrm{opt}}) \qquad (9.111)$$

and the total maximum power of this "omnicolor" series of single-color collectors:

$$\dot{W}_{C,\max,\mathrm{omnicolor}} = \int_0^\infty \dot{W}_{C,\nu,\max}(\nu)\, d\nu \qquad (9.112)$$

This analytical development was reported by De Vos [36] and, independently by Haught [35] (see the next subsection). The maximum power calculated with eq. (9.112) can be summarized in dimensionless form as

$$\eta_{A,\max,\mathrm{omnicolor}} = \frac{\dot{W}_{C,\max,\mathrm{omnicolor}}}{\sigma A T_s^4}$$

$$= \frac{15}{\pi^4}\,\theta_s^{-4}\int_0^\infty \frac{u^2 x^2 e^{u-x}}{(e^{u-x}-1)^2}\, du \qquad (9.113)$$

where $x(u, \theta_s)$ is the solution of the following transcendental equation [37]:

$$\frac{(1+x)e^{u-x}-1}{(e^{u-x}-1)^2} = (e^{u/\theta_s}-1)^{-1} \qquad (9.114)$$

When θ_s is considerably greater than 1, the dimensionless power per unit area (9.113) can be expressed in closed form as [37]

$$\eta_{A,\max,\mathrm{omnicolor}} \cong 1 - (1.567 + 0.37\ln\theta_s)\theta_s^{-1} \qquad (\theta_s \gg 1) \quad (9.115)$$

Note also that this limit was reprinted incorrectly in Ref. 4. Considering again the numerical example $\theta_s = 5762\,\mathrm{K}/300\,\mathrm{K}$, eq. (9.115) yields $\eta_{A,\max,\mathrm{omnicolor}} \cong 0.861$, which is only 2 percent greater than the one calculated with eq. (9.105). Therefore, the refined optimization procedure that gives us an omnicolor series of individually optimized collectors improves only slightly on the maximum $\dot{W}_C/A$ ratio of ideal concentrators that have a single temperature, eqs. (9.103)–(9.105). Equation (9.113) represents also the theoretically maximum power that could be drawn from a photovoltaic energy-conversion device [31, 32, 37].

Unconcentrated Solar Radiation

The same power-maximization ideas were advanced by Haught [35] in the context of a Carnot engine driven by a solar collector exposed to unconcentrated solar radiation. In other words, unlike in the ideal concentrators discussed until now, any point on Haught's collector sees the sun only over the narrow solid angle $\Omega_1 = (6.8)10^{-5}$ steradians. The collector receives radiation from both the sun and the ambient modeled as a black surface of temperature T_0. The T_0 radiation arrives at the collector surface through the hemispherical solid angle 2π. The radiation emitted back by the collector leaves also through the solid angle 2π.

The Carnot power output is given by eq. (9.101), where the heat input $\dot{Q}$ this time is

$$\dot{Q} = \pi A \int_0^\infty [i'_{vb}(v, T_s)\Omega_1/\pi + i'_{vb}(v, T_0) - i'_{vb}(v, T)]\alpha(v, T)\, dv \quad (9.116)$$

The radiation input from the sun and the ambient, and the departing radiation are listed sequentially in the brackets of the integrand. One additional feature of Haught's collector model is the nonuniform absorptivity $\alpha(v, T)$, which allows us to consider *selective* absorber surfaces. The simplest model of such a surface is the step-absorptivity function shown against the right ordinate of Fig. 9.13. The left ordinate of the same figure shows qualitatively the size of the three terms that appear inside the integral (9.116). A surface that absorbs and emits above a certain "cutoff" frequency v_0 has the potential of absorbing most of the incoming solar energy, while avoiding most of the radiative heat loss that would occur if α were equal to 1 for all vs. The visual subtraction of the superimposed shaded areas of Fig.

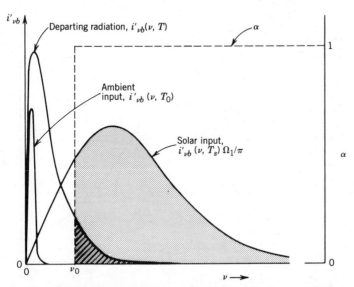

Figure 9.13 The relative size and frequency position of the three radiative contributions to $\dot{Q}$, eq. (9.116), and the opportunity for maximizing $\dot{Q}$ by varying the cutoff frequency v_0 (after Haught [35]).

9.13 suggests that there exists an optimum cutoff frequency for which the net heat transfer input $\dot{Q}$ associated with the intensity spectra of Fig. 9.13 is a maximum.

The optimization of the collector and engine arrangement just described consists of selecting not only the collector temperature but also the cutoff frequency. By holding ν_0 fixed, the $\dot{W}_C/A$ ratio reaches its maximum for the $T_{\mathrm{opt}}(\nu_0)$ values plotted in Fig. 9.14. The dimensionless $\eta_{A,\mathrm{max}}(\nu_0)$ ratio that corresponds to each $T_{\mathrm{opt}}(\nu_0)$ is shown on the same graph [the η_A ratio is defined as in eq. (9.103)]. The additional effect of varying ν_0 is clear: although T_{opt} increases with ν_0 monotonically, there exists an optimum cutoff frequency for which $\eta_{A,\mathrm{max}}(\nu_0)$ is itself a maximum [35]:

$$\eta_{A,\mathrm{max},\mathrm{max}} = 0.54$$
$$\nu_{0,\mathrm{opt}} = (2.22)10^{14}\ \mathrm{s}^{-1} \tag{9.117}$$
$$T_{\mathrm{opt}} = 863\ \mathrm{K}$$

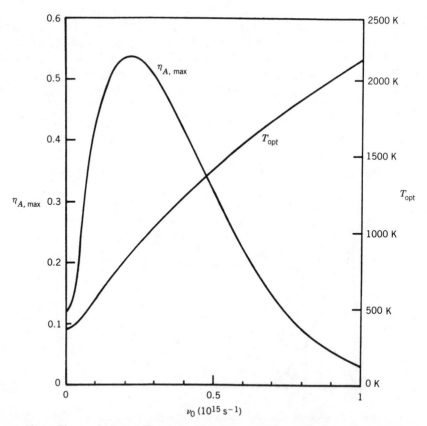

Figure 9.14 The optimum collector temperature and the maximum dimensionless $\dot{W}_C/A$ ratio as a function of cutoff frequency (after Haught [35]).

Figure 9.14 and the numerical values listed above are all based on $T_s =$ 6000 K and $T_0 = 300$ K. The peak η_A value of 0.54 is lower than the value calculated for ideal concentrators (namely, 0.849), because of the considerably lower energy flux (W/m^2) that is associated with unconcentrated solar radiation [note the smallness of the factor Ω_1/π that multiplies the solar term in the $\dot{Q}$ integral (9.116)].

Another interesting result is the maximum power delivered by an infinite series of single-frequency collectors that are optimized individually. In an analysis that parallels the one discussed between eqs. (9.109) and (9.115), Haught found that the optimum temperature of a single-frequency collector increases with ν in the manner shown in Fig. 9.15. This analysis consists of maximizing eq. (9.109) for a fixed ν, where $\dot{Q}$ is the one obtained by setting $\alpha(\nu, T) = 1$ in eq. (9.116). Again, the numerical values assumed for T_s and T_0 are 6000 K and 300 K, respectively. To each $T_{opt}(\nu)$ value plotted in Fig. 9.15 corresponds a maximum single-frequency Carnot power output, eq. (9.111): integrating this result over frequency yields the following result for the maximum Carnot power for a series of single-frequency collectors:

$$\eta_{A,max,omnicolor} = 0.683 \qquad (9.118)$$

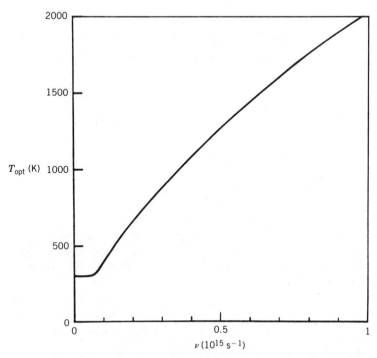

Figure 9.15 The optimum temperature of a single-frequency collector that receives unconcentrated solar radiation (after Haught [35]).

Comparing this result with the $\eta_{A,\text{max,max}}$ of eqs. (9.117), we find that the optimized series of single-frequency collectors is 26 percent more productive than the optimized single-temperature collector. On the other hand, in view of the numerical example listed under eq. (9.115), the "omnicolor" series of collectors that receive unconcentrated solar radiation are about 20 percent less productive than the corresponding series of ideal concentrators.

CONVECTIVELY COOLED COLLECTORS

One common feature of the collector models discussed until now is the presence of radiation as the only heat transfer mechanism for the leakage of heat from the collector surface. Another common feature that is intimately linked to the first is the high collector-temperature domain that is identified by each process of power maximization. The recommended high operating temperatures come in serious conflict with material worthiness and structural consideration that most certainly must enter the solar-power-plant design. The fundamental probem that is borne out of this conflict is that of determining the maximum power per unit collector area in "low-temperature" installations where, by default, the dominant collector heat-loss mechanism is convection [34].

Linear Convective-Heat-Loss Model

Consider for the sake of analytical clarity the simple linear model:

$$\dot{Q}_0 = UA(T - T_0) \tag{9.119}$$

where U is the overall convective heat transfer coefficient based on A, and $\dot{Q}_0$ is the convective heat loss. The heat transfer rate that would drive a Carnot engine positioned between T and T_0 is given by the expression

$$\dot{Q} = \dot{Q}_s - UA(T - T_0) \tag{9.120}$$

in which $\dot{Q}_s$ represents the net heat transfer rate of solar origin that is absorbed by the collector (note that $\dot{Q}_s$ is proportional to A). As shown in Fig. 9.16, if U is assumed to be independent of temperature, this simple convective cooling model translates into a linear collector-"efficiency" curve [38]:

$$\eta_{\text{collector}} = \frac{\dot{Q}}{\dot{Q}_s} = 1 - \frac{\theta - 1}{\theta_{\text{max}} - 1} \tag{9.121}$$

where $\theta = T/T_0$. The dimensionless maximum (or "stagnation") temperature θ_{max} is defined as

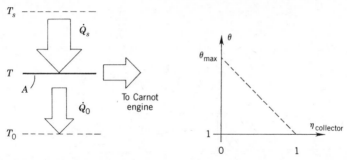

Figure 9.16 Linear convective-cooling model for low-temperature solar collectors (after Bejan et al. [34]).

$$\theta_{max} = 1 + \frac{\dot{Q}_s}{UAT_0} \tag{9.122}$$

The optimum collector temperature for maximum power per unit area can be determined in three ways: by maximizing $\dot{W}_C$; by minimizing $\dot{S}_{gen}$, as indicated in the theorem (p. 505); or by maximizing the exergy streaming out of the column of cross-section A and temperature height $T_s - T_0$ [1, 34]. In all cases, the thermodynamic optimum occurs when

$$\theta_{opt} = \theta_{max}^{1/2} \quad \text{or} \quad \eta_{collector,opt} = \frac{\theta_{max}^{1/2}}{1 + \theta_{max}^{1/2}} < 1 \tag{9.123}$$

which means the following dimensionless $\dot{W}_C/A$ ratio:

$$\frac{\dot{W}_{C,max}}{\dot{Q}_s} = \frac{\theta_{max}^{1/2} - 1}{\theta_{max}^{1/2} + 1} \tag{9.124}$$

In the low-temperature limit $\theta_{max} \to 1$, the $\dot{W}_{C,max}/\dot{Q}_s$ ratio approaches $(\theta_{max} - 1)/4$; in other words, $\dot{Q}_s/4UAT_0$. This limiting result shows most visibly how the power production per unit area increases with the absorbed flux $(\dot{Q}_s/A)$, and how it decreases when the convective-loss coefficient U increases.

The Effect of Collector–Engine Heat-Exchanger Irreversibility

As a step in the direction of a more realistic model of the collector and engine installation, it is time to recognize that the net heat transfer rate $\dot{Q}$ requires a finite temperature difference in order to flow from the collector surface into the high-temperature end of the power cycle. Let $(UA)_i$ represent the internal heat transfer conductance of the heat exchanger that places the engine in thermal contact with the collector, such that (Fig. 9.17)

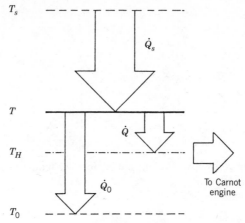

Figure 9.17 The position of the internal collector–engine heat-exchanger irreversibility in a collector cooled by convection (after Bejan et al. [34]).

$$\dot{Q} = (UA)_i (T - T_H) \qquad (9.125)$$

where T_H is the high-temperature level of the Carnot cycle that uses $\dot{Q}$. Equation (9.125) introduces one more variable (T_H) and one more constraint; therefore, there is still only one degree of freedom left in the maximization of the Carnot power $\dot{Q}(1 - T_0/T_H)$. Thermodynamic optimum operation is achieved when

$$\theta_{opt} = \frac{\theta_{max}^{1/2} + R\theta_{max}}{1 + R} \qquad (9.126)$$

where R is the external/internal conductance ratio:

$$R = \frac{UA}{(UA)_i} \qquad (9.127)$$

Obviously, the infinite internal-conductance limit $R = 0$ represents the case treated in the preceding subsection. The maximum power produced per unit of collector area now assumes the more general form:

$$\frac{\dot{W}_{C,max}}{\dot{Q}_s} = \frac{\theta_{max}^{1/2} - 1}{(R+1)(\theta_{max}^{1/2} + 1)} \qquad (9.128)$$

which shows explicitly how the work productivity of the collector decreases as the thermal contact between the collector and the power cycle worsens (i.e., as R increases).

The model-refining procedure illustrated here by the shift from Fig. 9.16 to Fig. 9.17 can be continued. One interesting direction might be to explore the effect of imperfect thermal contact (or finite heat transfer area) between the low-temperature end of the power cycle and the ambient. This effect is also relevant to the performance of the radiation-cooled collector models considered earlier. An illustration of this idea is presented in the section on extraterrestrial solar power plants later in this chapter.

Combined Convective and Radiative Heat Loss

A general collector-heat-loss model that bridges the gap between the purely radiative and purely convective models discussed until now was studied by Howell and Bannerot [33]. For the net heat transfer rate into the high-temperature end of the endoreversible power cycle, these authors wrote

$$\dot{Q} = \tau \alpha'_s q_s A_c - \epsilon_b \sigma A_b T^4 - (UA)_b (T - T_0) \tag{9.129}$$

with the following notation:

τ = transmittance of the cover-plate assembly

α'_s = effective solar absorptance of the base of the collector

q_s = direct solar flux, $q_s = \dot{Q}_s / A_c$

A_c = collector projected area

ϵ_b = effective infrared emittance of the absorber plate

A_b = base-plate area (projected area of the absorber plate)

U_b = overall convective heat transfer coefficient referenced to the base-plate area A_b

T = base-plate temperature (also the temperature of the hot end of the Carnot power cycle)

The maximization of the Carnot power (9.101) based on the above $\dot{Q}$ as high-temperature heat input yields an equation for the optimum collector temperature $\theta = T/T_0$:

$$\theta^5 - \frac{3}{4} \theta^4 + \frac{b}{4a} \theta^2 = \frac{1+b}{4a} \tag{9.130}$$

where

$$a = \frac{\epsilon_b A_b \sigma T_0^4}{\tau \alpha'_s q_s A_c} \qquad \text{(radiative heat-loss parameter)} \tag{9.131}$$

$$b = \frac{(UA)_b T_0}{\tau \alpha'_s q_s A_c} \qquad \text{(convective heat-loss parameter)} \tag{9.132}$$

Figure 9.18 shows the values taken by the optimum collector temperature in the domain $10^{-3} < a < 1$ and $0 < b < 2$. It is worth looking back at the models considered earlier in this chapter and recognizing Howell and

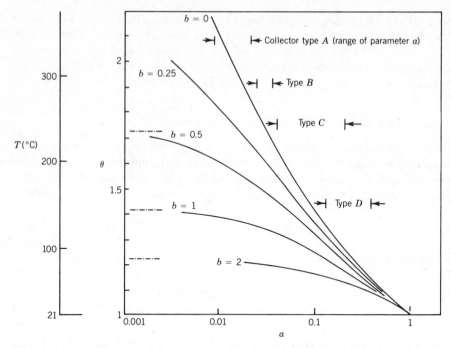

Figure 9.18 Optimum collector temperatures for maximum Carnot power, showing the combined effect of radiative and convective heat losses (after Howell and Bannerot [33]).

Bannerot's line $b = 0$ as the radiation cooling limit, and the $a \to 0$ limit as the domain of relatively cold collectors whose heat loss is dominated by convection. Shown also on Fig. 9.18 are the (a) ranges that characterize four classes of collector designs, labeled [33]:

(A) Spectrally selective absorbers, concentrating collectors ($A_b = A_c/3$, $6 < \alpha_s'/\epsilon_b < 15$, which means $0.009 < a < 0.022$; tracking concentrators attain A_c/A_b ratios of order 10^3, meaning $a < 10^{-3}$).

(B) Spectrally selective flat plate collectors ($A_b = A_c$, $10 < \alpha_s'/\epsilon_b < 15$, which means $0.026 < a < 0.039$).

(C) Directionally selective, nontracking black absorbers ($A_b \approx A_c/3$, $0.6 < \alpha_s'/\epsilon_b < 3$, or $0.04 < a < 0.21$).

(D) Flat plate collectors using black absorbers ($A_b = A_c$, $1 < \alpha_s'/\epsilon_b < 3$, or $0.13 < a < 0.39$).

From (A) to (D), all these calculations have been based on $q_s = 1 \text{ kW/m}^2$ and $T_0 = 294 \text{ K}$ (21°C). Howell and Bannerot produced similar graphs for the optimum thermodynamic condition of other ideal power cycles (Stirling, Ericsson, Brayton), heat pumps, and absorption refrigeration systems.

SINGLE-STREAM NONISOTHERMAL COLLECTORS

One of the simplest heat-exchanger designs for the removal of heat from a collector is the single-stream arrangement shown on the left side of Fig. 9.19. In this installation, the user circulates a stream ($\dot{m}c_P$) of single-phase fluid whose temperature increases with the longitudinal position x [34]. At any position along the collector (or along the stream), the insolation per unit area ($\dot{Q}_s/A$) is fixed. The heat loss to the ambient can be assumed proportional to the local collector–ambient temperature difference. Assuming also that the stream and the collector surface are locally in thermal equilibrium, the First Law of Thermodynamics requires at any x:

$$\frac{\dot{Q}_s}{A} \, dA = U(T - T_0) \, dA + \dot{m}c_P \, dT \qquad (9.133)$$

where the collector area increases from 0 to A as the stream flows from $x = 0$ to $x = L$.

The nonisothermal collector–stream system of Fig. 9.19 can be viewed as a succession of many isothermal collectors, each with a temperature $T(x)$. The preceding examples taught us to expect the existence of an optimum temperature level (or heat transfer delivery rate) in cases where the collector

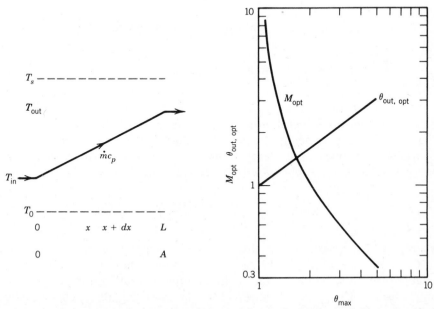

Figure 9.19 Nonisothermal collector cooled by a single-phase fluid stream (left side), and the optimum flowrate and outlet temperature for minimum entropy generation or maximum exergy-delivery rate (right side) (after Bejan et al. [34]).

is used for the production of power. Reasoning that in the present installation the heat transfer delivered per unit area (dA) is proportional to $(\dot{m}c_P)$, we seek to determine the optimum flowrate for which the power produced by the entire collector–stream installation is maximum. The same result is found by minimizing the overall entropy-generation rate

$$\dot{S}_{\text{gen}} = \dot{m}c_P \ln \frac{T_{\text{out}}}{T_{\text{in}}} - \frac{\dot{Q}_s}{T_s} + \frac{\dot{Q}_0}{T_0} \qquad (9.134)$$

where T_{in} and T_{out} are the inlet and outlet temperatures of the stream, respectively, and where the overall convective heat loss obeys the first law for the single-stream heat exchanger:

$$\dot{Q}_0 = \dot{Q}_s - \dot{m}c_P(T_{\text{out}} - T_{\text{in}}) \qquad (9.135)$$

Combining eqs. (9.134) and (9.135), we can express the subject of our minimization effort as

$$N_S = M\left(\ln \frac{\theta_{\text{out}}}{\theta_{\text{in}}} - \theta_{\text{out}} + \theta_{\text{in}}\right) - \frac{1}{\theta_s} + 1 \qquad (9.136)$$

where, in addition to $\theta = T/T_0$, we adopt the following dimensionless terminology:

$$N_S = \frac{\dot{S}_{\text{gen}}}{\dot{Q}_s/T_0} \qquad \text{(entropy-generation number [1])}$$

$$M = \frac{\dot{m}c_P}{\dot{Q}_s/T_0} \qquad \text{(mass flow number)}$$

When the fluid inlet temperature is fixed (for example, $T_{\text{in}} = T_0$, or $\theta_{\text{in}} = 1$), the entropy-generation number is a function of two parameters, M and θ_{out}. These parameters, however, are not independent, since higher flowrates yield lower outlet temperatures and vice versa. The relationship between M and θ_{out} is obtained by integrating the local first law (9.133):

$$M = \left[(\theta_{\text{max}} - 1) \ln \left(\frac{\theta_{\text{max}} - \theta_{\text{in}}}{\theta_{\text{max}} - \theta_{\text{out}}}\right)\right]^{-1} \qquad (9.137)$$

where, as in the past, θ_{max} is the stagnation temperature $(1 + \dot{Q}/UAT_0)$.

The right side of Fig. 9.19 shows the optimum flowrate number M_{opt} resulting from the minimization of the entropy-generation number (9.136) subject to the $M(\theta_{\text{out}})$ constraint (9.137). The curve $M_{\text{opt}}(\theta_{\text{max}})$ corresponds to the case $\theta_{\text{in}} = 1$. Since θ_{max} varies inversely with the collector–ambient convective-heat-loss coefficient, Fig. 9.19 suggests that for minimum $\dot{S}_{\text{gen}}/A$, the collector fluid should be circulated faster through high-heat-loss collectors. The outlet temperature corresponding to M_{opt} is also plotted on Fig. 9.19: note that the optimum θ_{out} increases with the stagnation temperature roughly as $\theta_{max}^{0.7}$. This outlet temperature is higher than the optimum isothermal-collector temperature determined earlier, eq. (9.123).

Figure 9.20 shows the variation of the entropy-generation rate with the mass flow number and the stagnation temperature. The quantity plotted on the ordinate $(N_s + \theta_s^{-1})$ approaches unity as M tends to zero or infinity. In the $M = 0$ limit, there is no collector–user (power-plant) interaction, and the isolation $\dot{Q}_s$ is lost entirely to the ambient. Likewise, when M tends to infinity, the collector temperature is depressed to the inlet (ambient-level) temperature T_0: the irreversibility rate is again maximum since in this case no collector exists between T_s and T_0. For mass flow numbers of order 1, and for stagnation temperatures significantly greater than 1, the entropy-generation rate is considerably smaller than what it would be if no collector were present.

Another way of looking at the thermodynamic optimum conditions identified above is by evaluating the rate of exergy extraction from the collector:

$$\dot{E}_x = \dot{m}\{[h - h_0 - T_0(s - s_0)]_{out} - [h - h_0 - T_0(s - s_0)]_{in}\} \quad (9.138)$$

or, in dimensionless form,

$$\frac{\dot{E}_x}{\dot{Q}_s} = M\left(\theta_{out} - \theta_{in} - \ln\frac{\theta_{out}}{\theta_{in}}\right) \quad (9.139)$$

Comparing eqs. (9.139) and (9.136), we see that the dimensionless exergy-delivery rate is the complement of the entropy-generation number plotted on the original ordinate:

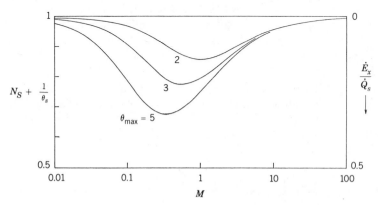

Figure 9.20 The effect of the mass flowrate on the entropy-generation rate and the rate of exergy extraction from a single-stream nonisothermal collector (after Bejan et al. [34]).

$$\frac{\dot{E}_x}{\dot{Q}_s} = 1 - (N_S + \theta_s^{-1}) \tag{9.140}$$

In other words, the area above each well-shaped curve represents the potential for exergy extraction from the single-stream collector. This is another illustration of the message embodied in the theorem of page 505: the maximization of the exergy output (Carnot power equivalent) from a collector system is equivalent to the minimization of the entropy-generation rate.

The thermodynamic analysis of solar collectors has been reconsidered in a number of more recent fundamental studies, which are listed chronologically as Refs. 39–46. For example, the irreversibility due to the pressure drop across the flow passage, which was neglected in the foregoing analysis, was taken into consideration by Fujiwara [42]. The same effect and the competition between it and the heat transfer irreversibility of the collector heat exchanger are also discussed in Ref. 34. Second-law optimization procedures based on more practical first-law models of collector operation have been proposed by Bošnjaković [39], Manfrida [43], and Zarea and Mayer [45]. The ideal collection and conversion of blackbody radiation was considered further by Landsberg [41, 44] and Badescu [46].

UNSTEADY OPERATION

In all the models discussed until now, the operation was assumed to be time-independent. One step in the direction of a more realistic description is to consider the time-varying conditions prompted by the daily insolation cycle [47]. The focus of the analysis becomes the trade-off between storing and not storing exergy, with the objective of maximizing the exergy harvested over a finite time period.

Consider the convection-cooled model in Fig. 9.21, where a new feature is the thermal inertia of the liquid mass M_0 that resides inside the collector. The insolation flux $q_s = \dot{Q}_s/A$ is time-dependent. The collector is cooled by two independent effects, the heat loss to the ambient, $UA(T - T_0)$, and the user (power cycle) that circulates the stream $\dot{m}$ through the collector. Assuming that the inlet temperature of this stream is T_0 and that the stream and the collector fluid mix well inside the collector enclosure, the user bleeds enthalpy at a rate equal to $\dot{m}c_P(T - T_0)$. The pressure drop across the collector is assumed negligible. The first-law statement for the system of Fig. 9.21 is

$$M_0 c_P \frac{dT}{dt} = q_s A - UA(T - T_0) - \dot{m}c_P(T - T_0) \tag{9.141}$$

where T, q_s, and $\dot{m}$ are functions of time. The exergy extracted over a finite period of time t_0 is

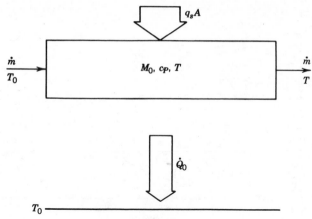

Figure 9.21 Model of solar collector with thermal inertia and single-stream exergy delivery to a power cycle [47].

$$E_x = \int_0^{t_0} \dot{m} c_P T_0 (\theta - 1 - \ln \theta) \, dt \qquad (9.142)$$

where, as in the past, $\theta = T/T_0$. We seek to maximize the above integral, while keeping in mind that $\dot{m}(t)$ and $\theta(t)$ are related through the first law (9.141). In dimensionless terms, the problem reduces to finding the optimum function $\theta(t^*)$ that maximizes the integral:

$$\frac{E_x}{M_0 c_P T_0} = \int_0^{t_0^*} \left(\frac{i - d\theta/dt^*}{\theta - 1} - K_0 \right) (\theta - 1 - \ln \theta) \, dt^* \qquad (9.143)$$

with the notation

$$t^* = \frac{q_{s,\text{ref}} A}{M_0 c_P T_0} t \qquad K_0 = \frac{U T_0}{q_{s,\text{ref}}} \qquad i(t^*) = \frac{q_s}{q_{s,\text{ref}}} \qquad (9.144)$$

where $q_{s,\text{ref}}$ is a reference value of the absorbed solar flux (e.g., the peak insolation for that day and location).

The variational calculus solution $\theta_{\text{opt}}(t^*)$ is given implicitly by [1, 47]

$$\frac{(\theta_{\text{opt}} - 1)^3}{\theta_{\text{opt}} \ln \theta_{\text{opt}} - \theta_{\text{opt}} + 1} = \frac{i(t^*)}{K_0} \qquad (9.145)$$

This result is general since the insolation function $i(t^*)$ is yet to be specified. Figure 9.22 shows that optimum collector temperature increases monotonically as the insolation increases and/or the convective-heat-loss parameter K_0 decreases. In conclusion, the optimum collector temperature must vary "in step" with the insolation function.

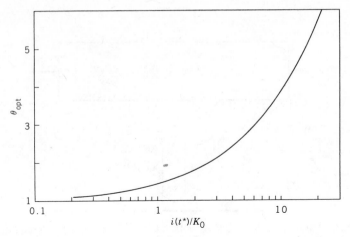

Figure 9.22 The optimum collector temperature versus the level of time-dependent insolation [47].

A more general version of the collector model of Fig. 9.21 formed the subject of a recent study [48] in which the mass inventory of the collector (M_0) was allowed to vary with time. The analysis showed that the relative timing of the filling and discharging processes has a significant effect on the total exergy delivered by the installation. The main conclusion of the study reinforces the one highlighted in this section, namely, that the daily regime of a collector with storage capability can be selected by design in order to maximize the harvesting of solar exergy per unit of collector area.

EXTRATERRESTRIAL POWER PLANTS

An interesting extension of the power-maximization point of view exhibited in the second half of this chapter is the problem of maximizing the production of power per unit area in an extraterrestrial solar power-plant application. As a first cut, we can model the actual power cycle as a Carnot cycle that operates between a high-temperature collector of area A_H and temperature T_H, and a low-temperature radiator (A_L, T_L), Fig. 9.23. The high-temperature collector is an ideal concentrator in the sense that the net heat input delivered to the heat-engine cycle is

$$\dot{Q}_H = \sigma A_H (T_s^4 - T_H^4) \tag{9.146}$$

Likewise, the heat-rejection rate managed by the radiator is

$$\dot{Q}_L = \sigma A_L (T_L^4 - T_\infty^4) \tag{9.147}$$

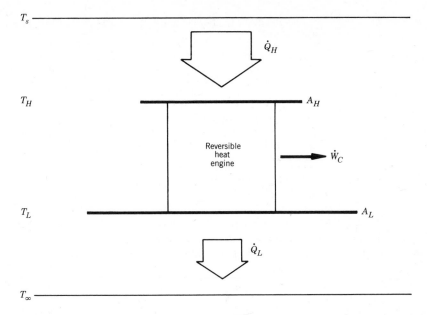

Figure 9.23 Schematic of an extraterrestrial power plant rejecting heat radiatively to the universe.

where T_∞ is the absolute temperature of the cold background. In what follows, T_∞^4 is considered negligible with respect to T_L^4.

The object of what follows is to maximize the mechanical power output of the endoreversible engine:

$$\dot{W}_C = \dot{Q}_H(1 - T_L/T_H) \qquad (9.148)$$

Out of the four parameters that affect $\dot{W}_C$ (namely, A_H, T_H, A_L, and T_L), only two are veritable degrees of freedom, because of the condition that the engine cycle is internally reversible:

$$\frac{\dot{Q}_H}{T_H} + \frac{\dot{Q}_L}{T_L} = 0 \qquad (9.149)$$

and because of an upper limit on the overall size of the collector–engine–radiator installation:

$$A = A_H + A_L = \text{constant} \qquad (9.150)$$

First, eliminating T_L between eqs. (9.147)–(9.149), the maximization of $\dot{W}_C$ with respect to T_H yields

$$\dot{W}_{C,\max} = \sigma A_H T_s^4 \frac{\xi - 1}{\xi + 3} \qquad (9.151)$$

where the optimum high-temperature parameter $\xi = (T_s/T_{H,\text{opt}})^4$ is given implicitly by

$$(\xi - 1)\left(\frac{\xi}{3} + 1\right)^3 = \frac{A_L}{A_H} \qquad (9.152)$$

In view of the total area constraint (9.150), the once maximized power output emerges as a function of the area ratio A_H/A_L: maximizing the power output for the second time, we obtain the ultimate maximum:

$$\dot{W}_{C,\max,\max} = 0.0414\sigma A T_s^4 \qquad (9.153)$$

for which the optimum area ratio is

$$(A_H/A_L)_{\text{opt}} = 0.538 \qquad (9.154)$$

Assuming as in the previous numerical examples that $T_s = 5762\,\text{K}$, this theoretical upper limit means $T_{H,\text{opt}} = 5174\,\text{K}$ and $T_{L,\text{opt}} = 3423\,\text{K}$. Although these temperatures are too high to be considered in the design of an actual power plant, it is worth noting that the ideal design is one with roughly twice as much heat transfer area at the cold end of the power cycle than at the warm end.

The maximum power output (9.153) is considerably lower than, for example, eq. (9.105), because in the present analysis the cold-end radiative area A_L was assumed finite. In the earlier analyses, on the other hand, the assumption that the low temperature of the power cycle is *fixed* at T_0 is equivalent to assuming an infinite heat transfer area between the cycle and the ambient temperature reservoir.

The present maximization of $\dot{W}_C$ with respect to T_H is a special case of a larger class of power-maximization problems treated by De Vos [49] and De Vos and Vyncke [50] as a generalization of a seminal paper by Curzon and Ahlborn [51] (see also Boehm [52] and Kamiuto [53]). The eventual maximization with respect to the constant-A constraint, eqs. (9.153)–(9.154), is the contribution of the present treatment. Although unreachable in a practical design, the optimum features revealed by the analysis are relevant concepts in the design of extraterrestrial power stations. A review of current power-plant concepts has been published recently by Hendricks et al. [54].

SYMBOLS

a	radiation constant $[(7.565)10^{-16}\,\text{J/m}^3/\text{K}^4]$
a	radiative-loss parameter [eq. (9.131)]
A	area $[\text{m}^2]$
A	nonflow availability [J]
b	convective-loss parameter [eq. (9.132)]
c	speed of light $[(2.998)10^8\,\text{m/s}]$
c_P	specific heat at constant pressure [J/kg/K]
C_P	isobaric heat capacity
C_V	isochoric heat capacity
$\dot{E}_x$	exergy-delivery rate [W]
G	Gibbs free energy
h	Planck's constant $[(6.626)10^{-34}\,\text{J s}]$
h	specific enthalpy [J/kg]
i'_b	total intensity of blackbody radiation
$i'_{\nu b}$	spectral intensity of blackbody radiation $[\text{W/m}^2/\text{s}^{-1}/(\text{unit solid angle})]$
k	Boltzmann's constant $[(1.38)10^{-23}\,\text{J/K}]$
K_0	convective-heat-loss parameter [eq. (9.144)]
L	length [m]
$L'_{\nu b}$	spectral entropy flux of blackbody radiation $[\text{W/K/m}^2/\text{s}^{-1}/(\text{unit solid angle})]$
m	mass of one particle
$\dot{m}$	mass flowrate [kg/s]
M	mass flow number
M_0	liquid mass inventory
n	number of photons per unit volume $[\text{m}^{-3}]$
N	total number of particles
N_S	entropy-generation number
p	photon momentum
P	pressure [Pa]
P_ν	partial pressure of photon gas of frequency ν
q	heat flux $[\text{W/m}^2]$
Q	heat transfer interaction [J]
$Q_\dagger$	heat transfer intercepted by collector
R	external/internal conductance ratio
s	specific entropy [eq. (9.138), J/K/kg]
S	entropy [J/K]
S_{gen}	entropy generation [J/K]
t	time [s]
t^*	dimensionless time [eq. (9.144)]
T	absolute temperature [K]
u	volumetric-specific internal energy $[\text{J/m}^3]$
U	internal energy [J]

UA	heat transfer conductance [W/K]
$(UA)_i$	internal conductance [W/K]
V	volume [m^3]
V_{avg}	average speed of one particle
W	work transfer interaction [J]
x	longitudinal coordinate [m]
α	absorptivity
ϵ	photon energy
ϵ	emissivity
η_A	dimensionless Carnot power per unit area [eq. (9.103)]
$\eta_{collector}$	collector efficiency [eq. (9.121)]
η_C	Carnot efficiency
η_J	Jeter's efficiency
η_P	Petela's efficiency
η_S	Spanner's efficiency
θ	absolute temperature ratio (T/T_0)
λ	wavelength [m]
μ	chemical potential
ν	frequency [s^{-1}]
ν_0	cutoff frequency [s^{-1}]
σ	Stefan–Boltzmann constant [eq. (9.56)]
Ω	solid angle
$(\ )_A$	associated with the surface A
$(\ )_b$	blackbody radiation
$(\ )_C$	Carnot
$(\ )_e$	associated with the enclosure
$(\ )_{env}$	environment
$(\ )_H$	high
$(\ )_{in}$	inlet
$(\ )_{irrev}$	irreversible
$(\ )_L$	low
$(\ )_{max}$	maximum
$(\ )_{opt}$	optimum
$(\ )_{out}$	outlet
$(\ )_{rev}$	reversible
$(\ )_s$	sun
$(\ )_0$	ambient
$(\ )_\nu$	per-unit-frequency interval
$(\dot{\ })$	per unit time

REFERENCES

1. A. Bejan, *Entropy Generation Through Heat and Fluid Flow*, Wiley, New York, 1982, Chapter 11.

2. J. A. Gribik and J. F. Osterle, The second law efficiency of solar energy conversion, *J. Sol. Energy Eng.*, Vol. 106, 1984, pp. 16–21.

3. S. M. Jeter, Discussion of "The second law efficiency of solar energy conversion," *J. Sol. Energy Eng.*, Vol. 108, 1986, pp. 78–80.

4. A. De Vos and H. Pauwels, Discussion of "The second law efficiency of solar energy conversion," *J. Sol. Energy Eng.*, Vol. 108, 1986, pp. 80–83.

5. J. A. Gribik and J. F. Osterle, Authors' closure, *J. Sol. Energy Eng.*, Vol. 108, 1986, pp. 83, 84.

6. R. Siegel and J. R. Howell, *Thermal Radiation Heat Transfer*, McGraw-Hill, New York, 1972, pp. 405–409.

7. M. P. Thekaekara, ed., *The Energy Crisis and Energy from the Sun*, Institute of Environment Sciences, Mt. Prospect, IL, 1974.

8. J. R. Howell, R. B. Bannerot, and G. C. Vliet, *Solar-Thermal Energy Systems*, McGraw-Hill, New York, 1982, p. 5.

9. M. Planck, *The Theory of Heat Radiation*, 2nd ed., translated by M. Masius, Dover, New York, 1959, p. 176 (original: *Waermestrahlung*, 1913); also, M. Planck, Ueber eine Verbesserung der Wien'schen Spectralgleichung, *Verh. Dtsch. Phys. Ges.*, Vol. 2, 1900, pp. 202–204.

10. D. C. Spanner, *Introduction to Thermodynamics*, Academic Press, London, 1964, p. 218.

11. J. Stefan, Über die Beziehung zwischen der Wärmestrahlung und der Temperatur, *Sitzber. Akad. Wiss. Wien*, Vol. 79, 1879, pp. 391–428.

12. L. Boltzmann, Ableitung des Stefan'schen Gesetzes, betreffend die Abhängigkeit der Wärmestrahlung von der Temperatur aus der electromagnetischen Lichtteorie, *Ann. Phys. (Leipzig)*, Ser. 3, Vol. 22, 1884, pp. 291–294.

13. W. C. Reynolds, *Thermodynamics*, 2nd ed., McGraw-Hill, New York, 1968, p. 227.

14. H. B. Callen, *Thermodynamics and an Introduction to Thermostatics*, 2nd ed., Wiley, New York, 1985, p. 412.

15. A. De Vos and H. Pauwels, Comment on a thermodynamic paradox presented by P. Würfel, *J. Phys. C: Solid State Phys.*, Vol. 16, 1983, pp. 6897–6909.

16. A. Bejan, *Solutions Manual for Entropy Generation Through Heat and Fluid Flow*, Wiley, 1984, p. 46.

17. G. Kirchhoff, *Gesammelte Abhandlungen*, Johann Ambrosius Barth, Leipzig, 1882, p. 574.

18. E. Pringsheim, Einfache Herleitung des Kirchhoff'schen Gesetzes, *Verh. Dtsch. Phys. Ges.*, Vol. 3, 1901, pp. 81–84.

19. R. Petela, Exergy of heat radiation, *J. Heat Transfer*, Vol. 86, 1964, pp. 187–192.

20. W. H. Press, Theoretical maximum for energy from direct and diffuse sunlight, *Nature (London)*, Vol. 264, 1976, pp. 734–735.

21. P. T. Landsberg and J. R. Mallinson, *Thermodynamic Constraints, Effective Temperatures and Solar Cells*, CNES, Toulouse, 1976, pp. 27–46.

22. P. T. Landsberg and G. Tonge, Thermodynamics of the conversion of diluted radiation, *J. Phys. A: Math. Gen.*, Vol. 12, 1979, pp. 551–562.

23. S. J. Jeter, Maximum conversion efficiency for the utilization of direct solar radiation, *Sol. Energy*, Vol. 26, 1981, pp. 231–236.

24. P. I. Moynihan, *Second-law Efficiency of Solar-Thermal Cavity Receivers*, JPL Publ. 83-97, Jet Propulsion Laboratory, Pasadena, CA, 1983; published in *J. Sol. Energy Eng.*, Vol. 108, 1986, pp. 67–84.

25. A.Bejan, Unification of three different theories concerning the ideal conversion of enclosed radiation, *J. Sol. Energy Eng.*, Vol. 109, 1987, pp. 46–51.

26. R. Winston, Principles of solar concentrators of a novel design, *Sol. Energy*, Vol. 16, 1974, pp. 89–95.

27. H. Müser, Thermodynamische Behandlung von Electronenprozessen in Halbleiter-Randschichten, *Z. Phys.*, Vol. 148, 1957, pp. 380–390.

28. M. Castans, Bases fisicas del aprovechamiento de la energia solar, *Rev. Geofis.*, Vol. 35, 1976, pp. 227–239.

29. M. Castans, Comments on "Maximum conversion efficiency for the utilization of direct solar radiation," *Sol. Energy*, Vol. 30, 1983, p. 293.

30. S. M. Jeter, Response to Dr. M. Castans comments, *Sol. Energy*, Vol. 30, 1983, p. 293.

31. A. De Vos and H. Pauwels, On the thermodynamic limit of photovoltaic energy conversion, *Appl. Phys.*, Vol. 25, 1981, pp. 119–125.

32. H. Pauwels and A. De Vos, Determination and thermodynamics of maximum efficiency photovoltaic device, *Proc. 15th IEEE Photovoltaic Spec. Conf.*, Orlando, FL, 1981, pp. 377–382.

33. J. R. Howell and R. B. Bannerot, Optimum solar collector operation for maximizing cycle work output, *Sol. Energy*, Vol. 19, 1977, pp. 149–153.

34. A. Bejan, D. W. Kearney, and F. Kreith, Second law analysis and synthesis of solar collector systems, *J. Sol. Energy Eng.*, Vol. 103, 1981, pp. 23–30.

35. A. F. Haught, Physics considerations of solar energy conversion, *J. Sol. Energy Eng.*, Vol. 106, 1984, pp. 3–15.

36. A. De Vos, Detailed balance limit of the efficiency of tandem solar cells, *J. Phys. D: Appl. Phys.*, Vol. 13, 1980, pp. 839–846.,

37. A. De Vos, C. Grosjean, and H. Pauwels, On the formula for the upper limit of photovoltaic solar energy conversion efficiency, *J. Phys. D: Appl. Phys.*, Vol. 15, 1982, pp. 2003–2015.

38. F. Kreith, D. Kearney, and A. Bejan, End-use matching of solar energy systems, *Energy*, Vol. 5, 1980, pp. 875–890.

39. F. Bošnjaković, Solar collectors as energy converters, in J. P. Hartnett, T. F. Irvine, Jr., E. Pfender, and E. M. Sparrow, eds., *Studies in Heat Transfer: A Festschrift for E. R. G. Eckert*, Hemisphere, Washington, DC, 1979, pp. 331–381.

40. J. Harris, Thermodynamic availability and effectiveness with particular reference to solar heaters, *Luso, J. Sci. Technol. (Malaŵi)*, Vol. 3(2), 1982, pp. 75–88.

41. P. T. Landsberg, Some maximal thermodynamic efficiencies for the conversion of blackbody radiation, *J. Appl. Phys.*, Vol. 54(5), 1983, pp. 2841–2843.

42. M. Fujiwara, Exergy analysis for the performance of solar collectors, *J. Sol. Energy Eng.*, Vol. 105, 1983, pp. 163–167.

43. G. Manfrida, The choice of an optimal working point for solar collectors, *Sol. Energy*, Vol. 34, 1985, pp. 513–515.

44. P. T. Landsberg, An introduction to nonequilibrium problems involving electromagnetic radiation, in J. Casas-Vázquez et al., eds., *Recent Developments in Nonequilibrium Thermodynamics*, Springer-Verlag, Berlin, 1986, pp. 224–267.

45. M. Zarea and E. Mayer, Second law optimization procedure for concentrating collectors, in H. Yüncü, E. Paykoc, and Y. Yener, eds., *Solar Energy Utilization: Fundamentals and Applications*, Martinus Nijhoff, The Netherlands, 1987, pp. 255–270.

46. V. Badescu, L'éxergie de la radiation solaire directe et diffuse sur la surface de la Terre, *Entropie* (to be published).

47. A. Bejan, Extraction of exergy from solar collectors under time-varying conditions, *Int. J. Heat Fluid Flow*, Vol. 3, 1982, pp. 67–72.

48. D. E. Chelghoum and A. Bejan, Second-law analysis of solar collectors with energy storage capability, *J. Sol. Energy Eng.*, Vol. 107, 1985, pp. 244–251.

49. A. De Vos, Efficiency of some heat engines at maximum-power conditions, *Am. J. Phys.*, Vol. 53, 1985, pp. 570–573.

50. A. De Vos and D. Vyncke, Solar energy conversion: photovoltaic versus photothermal conversion, *Proc. 5th E. C. Photovoltaic Solar Energy Conf.*, Athens, October 17–21, 1983, pp. 186–190.

51. F. L. Curzon and B. Ahlborn, Efficiency of a Carnot engine at maximum power output, *Am. J. Phys.*, Vol. 43, 1975, pp. 22–24.

52. R. F. Boehm, Maximum performance of solar heat engines, *Appl. Energy*, Vol. 23, 1986, pp. 281–296.

53. K. Kamiuto, Determination of the optimum pond temperature for maximizing power production of a convecting solar pond thermal-energy conversion system, *Appl. Energy*, Vol. 28, 1987, pp. 47–57.

54. R. C. Hendricks, R. J. Simoneau, and J. W. Dunning, Jr., Heat transfer in space power and propulsion systems, *Mech. Eng.*, Vol. 108, February 1986, pp. 41–52.

10

Refrigeration

The engineering-thermodynamics growth documented in the preceding two chapters is equally evident in the field of refrigeration. Unlike power engineering, which needed the energy crisis jolt of the 1970s to rejuvenate its fundamental research component, the field of refrigeration engineering benefited from a steady stimulus throughout its 150-year development. Originally, that stimulus was provided by a combination of the quest for absolute zero and the need of less-expensive refrigeration methods in general. The second component of this stimulus dominates today, as the great power-engineering projects of the future (e.g., controlled fusion, superconductivity) require enormous amounts of refrigeration at low temperatures.

Refrigeration has come a long way from the late 1800s when center stage belonged to the engine developers, and when refrigeration engineers devoted themselves primarily to ice manufacturing and the transoceanic transport of frozen meat. In the meantime, the field of refrigeration produced not only the tested techniques that are being used today, but also much of the current exergy and entropy-generation methodology of basic engineering thermodynamics. Nowhere on the absolute temperature scale is the issue of entropy generation more critical than at low temperatures. And, proving again that Rankine was right (p. 27 footnote), refrigeration engineering continues to be a showcase of engineering creativity in a perennial "supporting role" to a long list of seemingly more philosophical and noble pursuits that catch the headlines, for example, physics (superconductivity, fusion), medicine (preservation of living tissues, surgery), and aerospace science (rocket motors, insulation systems).

For all these reasons, this chapter presents an outline of modern refrigeration thermodynamics, with an emphasis on low-temperature refrigeration, or cryogenics. The presentation is structured to show the steady progress made toward lower temperatures, more efficient and advanced refrigeration installations, and, overall, toward a better understanding of engineering-thermodynamic fundamentals.

JOULE–THOMSON EXPANSION

In terms of hardware—i.e., the number of components and, especially, components with moving parts—the simplest installation is the vapor-compression refrigeration cycle shown in Fig. 10.1. The refrigeration effect, or the removal of the heat transfer rate $\dot{Q}_L$ from a system whose temperature (T_L) is lower than the ambient temperature (T_0), is the result of placing the system in thermal contact with an evaporating stream. The evaporation temperature T_1 is in general lower than the refrigeration temperature T_L.

The purpose of the refrigeration machine then is to manufacture the stream of low-temperature fluid whose complete evaporation accounts for the refrigeration effect, $\dot{Q}_L = \dot{m}(h_1 - h_4)$. This is done in three steps, first, by compressing the vapor produced by the absorption of $\dot{Q}_L$ (in Fig. 10.1, this step is shown as $(1) \to (2)$, suggesting also that in general the compressor generates entropy); second, by cooling the compressed fluid through an *aftercooler* that rejects heat to the ambient, $\dot{Q}_H = \dot{m}(h_2 - h_3)$; and, third, by expanding the stream through a valve. The last step $(3) \to (4)$ is clearly the constant-enthalpy Joule–Thomson expansion process that was studied in detail in chapter 4.

For this scheme to be successful, the Joule–Thomson coefficient μ_J at state (3) must be positive, which means that the constant-h line that passes through state (3) must have a negative slope on the T–s diagram. Another important requirement is that, given the evaporation pressure P_L, the working fluid must be such that its evaporation temperature (T_1) falls below the refrigeration temperature (T_L). Table 10.1 shows a partial listing of the evaporation temperatures of various working fluids used in refrigeration installations, although only the upper entries can be used in the scheme of Fig. 10.1. Moving down through the table, the condition that the Joule–Thomson coefficient must be positive at room temperature is gradually being placed in jeopardy and violated most notoriously by helium. This is why the simple vapor-compression cycle of Fig. 10.1 is far from being the answer to the quest for low-temperature refrigeration.

One feature that will stay with us throughout this chapter is the physical sense chosen for the arrows that indicate energy transfer in figures such as Fig. 10.1. This decision is equivalent to the one made in the illustration of power systems (e.g., Fig. 8.2). The energy transfer sign convention of Fig. 1.3 is appropriate for abstract arguments and analyses in which the actual direction of the interaction cannot be anticipated (e.g., chapter 2). To use it blindly in the analysis of a "power" plant or a "refrigeration" machine, i.e., for a system whose name spells out the physical direction of energy transfer, is to sacrifice clarity and good engineering in the name of mathematical consistency.

The physical components of the machine are organized vertically in a way that distinguishes between the cold region of the machine and those

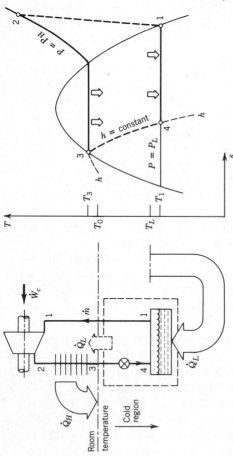

Figure 10.1 Simple vapor-compression refrigeration cycle with one Joule–Thomson expansion stage.

TABLE 10.1 The Main Temperature and Pressure Characteristics of Some of the Most Common Refrigerants

Refrigerant	Evaporation Temperature (K) $P_L = 1$ atm	Evaporation Temperature (K) $P_H = 10$ atm	Critical Point Temperature (K)	Critical Point Pressure	Triple Point Temperature (K)	Triple Point Pressure
Ammonia (NH_3)	234.7	301.4	405.5	111.3 atm (11.28 MPa)	—	—
Sulphur dioxide (SO_2)	263	330.6	430.7	77.8 atm (7.88 MPa)	—	—
Dichlorodifluormethane, Freon-12, or Refrigerant 12 (CCl_2F_2)	243	317.6	384.7	39.6 atm (4.01 MPa)	—	—
Carbon Dioxide (CO_2)			304	73 atm (7.4 MPa)	216.6	5.11 atm (0.518 MPa)
Ethylene (C_2H_4)	170.5	222	282.4	50.5 atm (5.12 MPa)	—	—
Methane (CH_4)	120	149	191.1	45.8 atm (4.64 MPa)	—	—
Oxygen (O_2)	90.2	120	154.6	49.8 atm (5.04 MPa)	~54	0.0015 atm (0.15 kPa)
Nitrogen (N_2)	77.4	105	126.1	33.6 atm (3.4 MPa)	63.14	0.124 atm (12.5 kPa)
Hydrogen, normal (H_2)	20.3	31.3	33.3	12.83 atm (1.3 MPa)	13.96	0.071 atm (7.21 kPa)
Helium-4 (He)	4.214	—	5.2	2.25 atm (0.228 MPa)	2.172	0.05 atm (5.05 kPa)
Helium-3	3.195	—	3.32	1.17 atm (0.119 MPa)		

components that operate above room temperature, Fig. 10.1. The warm components are positioned above the cold components, the demarcation line being the room temperature level T_0. This vertical arrangement makes sense for two reasons: first, because from his earliest encounter with thermometric scales, the engineer has been taught to associate "up" with high temperatures and "down" with low temperatures; and, second, because in many designs, the minimization of the convection heat leak from room temperature to the evaporator demands that colder components be placed at progressively lower levels. This vertical stacking shows how the machine "pulls" the refrigeration load $\dot{Q}_L$ up and through the room-temperature demarcation line.

Invoking the first law for the cold region of the machine, i.e., for the dashed-line subsystem on the left side of Fig. 10.1, we discover that

$$\dot{Q}_L = \dot{m}(h_1 - h_3) \tag{10.1}$$

In other words, the function of the counterflow that pierces the room-temperature demarcation line is to convect to room temperature the heat transfer that was extracted from the cold system (T_L). The "continuity" of $\dot{Q}_L$ vertically through the cold region of the machine is a common feature of systems that rely exclusively on the Joule–Thomson process for the purpose of expanding the refrigerant. (See also the dashed-line system on the left side of Fig. 10.2.)

It is possible to focus on each of the four components of the vapor-compression and throttling installation and to disect the entropy generated by the whole plant into "external" and "internal" irreversibilities, and into irreversibility contributions that can be blamed specifically on one component (e.g., the compressor) or on one functional feature of a component (e.g., the pressure drop through the aftercooler). This methodology formed already the subject of the first part of chapter 8; therefore, it can be practiced at least mentally in conjunction with the refrigeration systems of this chapter. In order to avoid repetition, we reserve the discussion of irreversibility minimization for critical occasions, and continue the description of refrigeration machines that reach toward progressively lower temperatures.

Figure 10.2 shows an evolved version of the vapor-compression and throttling cycle. The new component is the regenerative (counterflow) heat exchanger that "insulates" the Joule–Thomson expansion stage and the evaporator from the warm zone of the machine. The longitudinal thermal-insulation function served by the counterflow heat exchanger is the subject of a special section later in this chapter (pp. 545–547). In the field of refrigeration, the regenerative counterflow heat exchanger was first used by John Gorrie [1, 2] and by Charles Siemens [3].

The reason for introducing a counterflow heat exchanger in the cycle of Fig. 10.2 is the pursuit of lower refrigeration temperatures. For example, in

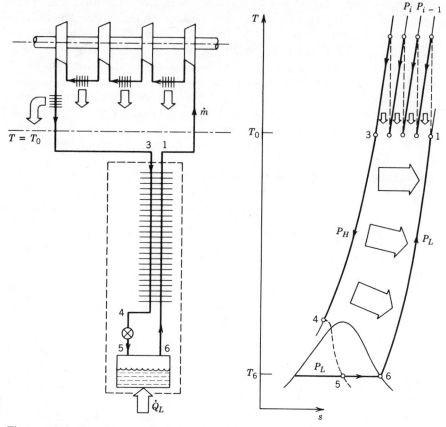

Figure 10.2 Reaching toward lower refrigeration temperatures by inserting a counterflow heat exchanger in a cycle that relies exclusively on the Joule–Thomson expansion effect.

the absence of a regenerative heat exchanger, the Joule–Thomson expansion would proceed from state (3) along the local constant-enthalpy line, yielding temperatures that are far above the refrigeration temperature. The refrigeration temperature T_L is slightly greater than the evaporation temperature T_6: these temperatures constrain the selection of the refrigerant and the evaporation pressure P_L.

Inside the counterflow heat exchanger, the compressed stream is cooled down from state (3) to state (4) using the sensible heat of the cold gas (6) that leaves the evaporator. The Joule–Thomson expansion takes the refrigerant from state (4) to state (5), highlighting again the requirement that μ_J must be positive in the area inhabited by these states.

The limited reach into low temperatures that is provided by the regenerative scheme of Fig. 10.2 becomes evident if we analyze the meaning of eq.

(10.1). That equation applies here as well. "Refrigeration" means $\dot{Q}_L > 0$ and, at room temperature, $h_1 > h_3$. Without defeating the conclusion that we are about to reach, we assume that the heat transfer area of the counterflow heat exchanger (3)–(4)–(6)–(1) is infinite, in other words, that thermal equilibrium rules the top end of the heat exchanger, $T_3 = T_1$. These observations can be summarized as

$$\text{state (3)} \quad \text{state (1)}$$

$$h_3 < h_1$$

$$T_3 = T_1$$

$$P_H > P_L$$

which means that "refrigeration" at T_L demands a refrigerant that at room temperature shows

$$\left(\frac{\partial h}{\partial P}\right)_T < 0 \tag{10.2}$$

Seen through the cyclical relation

$$\left(\frac{\partial h}{\partial T}\right)_P \left(\frac{\partial T}{\partial P}\right)_h \left(\frac{\partial P}{\partial h}\right)_T = -1 \tag{10.3}$$

the refrigeration condition (10.2) is the same as requiring a refrigerant with positive Joule–Thomson coefficient *at room temperature*:

$$\mu_J = -\frac{1}{c_P}\left(\frac{\partial h}{\partial P}\right)_T > 0 \tag{10.4}$$

The same requirement was identified earlier in the discussion of the throttling process $(3) \rightarrow (4)$ of Fig. 10.1. We reach the important conclusion that the refrigerants that approach ideal gas behavior at room temperature ($\mu_J \cong 0$) cannot be used in the arrangement of Fig. 10.2, despite the obvious attractiveness of their low boiling points. Even more unfortunate is the fact that the inequality (10.4) represents the optimistic side of a more stringent criterion that holds when a finite temperature difference exists across the top end of the counterflow heat exchanger, $T_3 > T_1$. In the general case, the refrigeration effect $\dot{Q}_L$ is possible only when μ_J exceeds a positive critical value. This threshold value increases as the effectiveness of the heat exchanger decreases (Problem 10.1). It is this limitation that created the need for the more advanced schemes discussed in the next section.

One last observation concerns the geometry of the low-temperature end of the cycle plotted on the T–s diagram of Fig. 10.2. In heat transfer engineering, we learn to expect vanishing temperature differences as the heat-exchanger area becomes infinite: on this basis, we wrote $T_1 = T_3$ and derived eq. (10.4). Yet, a look at the fixed position of the P_H and P_L isobars and their intersection with constant-h lines of type $(4) \rightarrow (5)$ in Fig. 10.2 tells us that T_4 must always be greater than T_5. In other words, a temperature difference appears to persist across the bottom end of the counterflow heat exchanger despite the infinite heat transfer area assumption.

An important diagram that clarifies this apparent paradox is the plot of counterflow stream-to-stream temperature difference (ΔT) versus the longitudinal temperature variation of one of the two streams (Fig. 10.3). This diagram can be drawn to scale for each counterflow heat-exchanger design by writing the first law for the dashed-line control volume:

$$h_{4'}(T_{4'}, P_H) = h_3 - h_1 + h_{6'}(T_{6'}, P_L) \qquad (10.5)$$

The calculation begins with assuming the top-end temperature difference $(T_3 - T_1)$, which fixes the value of $(h_3 - h_1)$ on the right side of eq. (10.5). Next, selecting a temperature for the low-pressure stream $(T_{6'})$ and determining $h_{6'}(T_{6'}, P_L)$ from property tables, we use eq. (10.5) to calculate the enthalpy of the high-pressure stream at the point that resides vis-à-vis with point $(6')$. Using the same tables, we rely on the calculated $h_{4'}$ to deduce the temperature $T_{4'}$. The path of this calculation is indicated by a dashed line in Fig. 10.3. Repeating the calculation for a new starting value $T_{6'}$ corresponds to moving up or down the bottom cut made by the control volume across the heat exchanger.

In this manner, we can trace the entire curve T_{P_H} vs. T_{P_L} or, alternative-

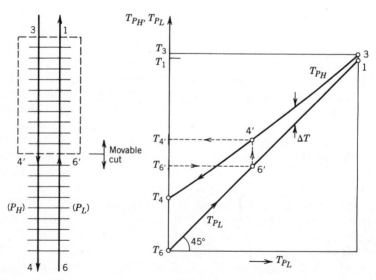

Figure 10.3 Graphic construction of the temperature distribution along the two sides of a regenerative heat exchanger.

ly, ΔT vs. T_{P_L}. In this discussion, the "T_{P_L}" notation means the temperature of the low-pressure stream. The shape of the $T_{P_H}(T_{P_L})$ curve in Fig. 10.3 is similar to what one discovers if this diagram is constructed for the regenerative heat exchanger (3)–(4)–(6)–(1) of Fig. 10.2. The stream-to-stream temperature difference increases gradually toward the cold end of the heat exchanger. This feature is due to the different c_P values of the two streams, the greater value being on the high-pressure side.

As the heat-exchanger area increases, the T_{P_H} curve approaches T_{P_L} *until* the two curves touch in at least one point. In the present example, the shape of the T_{P_H} curve is such that the "pinch" point will occur at the top end, assuming, of course, that the area can be infinite.

One danger to guard against is the calculation of a $T_{P_H}(T_{P_L})$ line that intersects the diagonal $T_{P_L}(T_{P_L})$, i.e., a heat exchanger in which the two temperature distributions cross. In the example of Fig. 10.3, the T_{P_H} and T_{P_L} lines would cross if the selected $(h_1 - h_3)$ value that starts the calculation corresponds to a temperature difference that points in the wrong direction, $T_1 > T_3$. Physical situations of this kind are ruled out by the second law, because the crossing implies the existence of negative ΔTs or spontaneous heat transfer against the temperature gradient. The test that deserves to accompany every heat-exchanger calculation is the Second Law of Thermodynamics, which in the present example requires

$$\frac{1}{\dot{m}} \dot{S}_{\text{gen}} = s_1 + s_{4'} - s_3 - s_{6'} > 0 \qquad (10.6)$$

If the sign in this calculation is negative, we can be sure that the temperature distribution inside the dash-line control volume of Fig. 10.3 is physically impossible.

WORK-PRODUCING EXPANSION

Consider the cold region of the refrigerators discussed until now, and recall that the refrigeration load $\dot{Q}_L$ must escape through the room-temperature roof of this region. Two room-temperature features conspire to prevent this escape, the smallness of the Joule–Thomson expansion coefficient and the finite ΔT across the warm end of the regenerator. With reference to the left side of Fig. 10.4, the refrigeration that eventually is convected through the warm end can be viewed as the difference between the maximum value that corresponds to infinite heat-exchanger area, $\dot{Q}_{L,\text{max}}$, and a downward convective heat leak that is due to the finiteness of the actual area, $\dot{Q}_{\text{leak}}$,

$$\dot{Q}_L = \dot{m}(h_1 - h_3)$$
$$= \underbrace{\dot{m}[(h_1)_{T_1 = T_3} - h_3]}_{\uparrow\ \dot{Q}_{L,\text{max}}} - \underbrace{\dot{m}[(h_1)_{T_1 = T_3} - h_1]}_{\downarrow\ \dot{Q}_{\text{leak}}} \qquad (10.7)$$

The first term is the one threatened by the smallness of μ_J, whereas the second term increases proportionally with the ΔT across the warm end.

The issue boils down to providing a different escape route for $\dot{Q}_L$. This is precisely the function served by the intermediate temperature expander on the right side of Fig. 10.4. The expander bleeds a fraction ($\dot{m}_e$) from the high-pressure stream, expands it, lowers its temperature, and returns it to a colder position on the low-pressure side of the counterflow heat exchanger. Modeling the expanding fluid as an ideal gas, and assuming that the expander functions adiabatically and reversibly, the mechanical power extracted by the expander is

$$\frac{\dot{W}_e}{\dot{m}_e} = \frac{k}{k-1} RT_e \left[1 - \left(\frac{P_f}{P_e} \right)^{(k-1)/k} \right] \tag{10.8}$$

Noting that $P_f/P_e = P_L/P_H$ and that in the nitrogen and helium refrigeration cycles that use such expanders, $P_L/P_H \ll 1$, the specific expander power is approximately $kRT_e/(k-1)$.

The positive and dominating effect of $\dot{W}_e$ on the refrigeration rate $\dot{Q}_L$ can be seen by writing the first law for the entire system on the right side of Fig. 10.4:

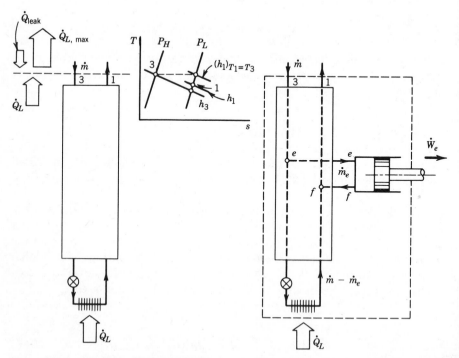

Figure 10.4 Work-producing expansion as another escape route for $\dot{Q}_L$ out of the cold region of the refrigeration machine.

$$\frac{\dot{Q}_L}{\dot{m}} = \frac{\dot{W}_e}{\dot{m}} + h_1 - h_3$$

$$\cong \underbrace{\frac{\dot{m}_e kRT_e}{\dot{m}(k-1)}\left[1 - \left(\frac{P_L}{P_H}\right)^{(k-1)/k}\right]}_{\uparrow\,\left(\frac{\dot{Q}_L}{\dot{m}}\right)_{\max}} + \overbrace{\overline{\mu_J c_P}(P_H - P_L) - c_P(T_3 - T_1)}^{h_1 - h_3}$$

$$\downarrow\,\left(\frac{\dot{Q}_{\text{leak}}}{\dot{m}}\right)$$

$$(10.9)$$

The approximation of $(h_1 - h_3)$ in terms of the linear two-term expression above is the focus of Problem 10.1. Note that $\overline{\mu_J c_P}$ is an average value of all $(\mu_J c_P)$ values found on the $T = T_3$ isotherm between $P = P_L$ and $P = P_H$, while the lone c_P factor is the c_P value found on the P_L isobar between T_1 and T_3. Numerically, one can show that the expansion work term dominates the remaining two terms on the right side of eq. (10.9), even when $\mu_J = 0$ (Problem 10.2).

A practical dimensionless figure of merit of the installation is the ratio $\dot{Q}_L/\dot{W}_c$, where $\dot{W}_c$ represents the compressor power input. The inverse of this ratio is also in use [4]. Worth noting is that the $\dot{Q}_L/\dot{W}_c$ ratio is identical to the classical coefficient of performance (COP) only in installations that rely exclusively on the Joule–Thomson expansion effect (e.g., Figs. 10.1 and 10.2), and in the special case where the expander power $\dot{W}_e$ of Fig. 10.4 is dissipated through a brake. Since $\dot{W}_e$ is in general greater than zero, the $\dot{Q}_L/\dot{W}_c$ ratio is slightly smaller than the corresponding COP.

The actual compressor power $\dot{W}_c$ depends on the design of the compressor, for example, on the isentropic efficiency of each compression stage, the number of compression stages, the heat-exchanger effectiveness of each aftercooler, etc. When the focus is on the design of the cold region for the purpose of maximizing $\dot{Q}_L/\dot{m}$, it is convenient to use as standard $\dot{W}_c$ value the minimum compressor power that can theoretically pressurize the $\dot{m}$ stream from P_L to P_H at room temperature. That power is the rate of flow-exergy increase from (T_0, P_L) to (T_0, P_H), or the power input required by a reversible and isothermal compressor that is in thermal equilibrium with the ambient:

$$\dot{W}_{c,\min} = \dot{m}RT_0 \ln \frac{P_H}{P_L} \qquad (10.10)$$

Therefore, to the extent that the last two terms of eq. (10.9) can be neglected, the $\dot{Q}_L/\dot{W}_c$ ratio of the system shown on the right side of Fig. 10.4 is

$$\frac{\dot{Q}_L}{\dot{W}_{c,\min}} \cong \frac{\dot{m}_e kT_e}{\dot{m}(k-1)T_0 \ln(P_H/P_L)}\left[1 - \left(\frac{P_L}{P_H}\right)^{(k-1)/k}\right] \qquad (10.11)$$

Two classical examples of refrigeration cycles that combine the Joule–

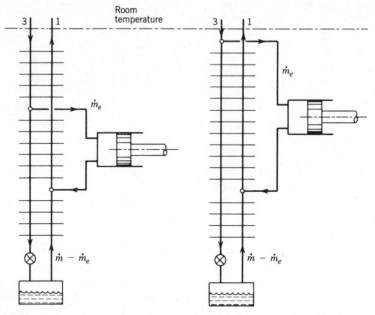

Figure 10.5 Two cycles that combine bottom-end Joule–Thomson expansion with work-producing expansion at intermediate temperatures: Claude (left) and Heylandt (right).

Thomson effect with expansion in a work-producing device (originally cylinder and piston, lately, turbines; also called expander, expansion engine, or "engine") are the Claude [5] and Heylandt [6] machines shown in Fig. 10.5. Claude's design and especially his development of a sufficiently efficient and reliable expander proved instrumental in the growth of an industry for the liquefaction of air and the manufacture of nitrogen, oxygen, and other gases. In 1920, Claude used an expansion engine in the industrial liquefaction of hydrogen and the production of refrigeration at and around the normal boiling point of hydrogen (Table 10.1).

The Heylandt cycle differs from Claude's in that the expander receives room-temperature compressed gas. With inlet pressures as high as 200 atm and outlet pressures of about 5 atm, the temperature at the expander outlet is almost as low as the temperature of the cold end of the counterflow heat exchanger. This cycle was used extensively in the production of nitrogen and oxygen.

THE BRAYTON CYCLE

Another way of looking at the technological breakthrough represented by the insertion of work-producing expanders in the cold zone is to visualize the expander stream $\dot{m}_e$ as it flows through the remainder of the installation. Shown on the left side of Fig. 10.6 is the usual drawing, in which the compressor processes the total flowrate $\dot{m}$, while the refrigeration load $\dot{Q}_L$ is absorbed by the smaller stream $(\dot{m} - \dot{m}_e)$. On the right side of Fig. 10.6, the circuits completed by the expander stream $(\dot{m}_e)$ and the refrigeration load stream $(\dot{m} - \dot{m}_e)$ are shown separately. The load stream $(\dot{m} - \dot{m}_e)$ completes a circuit "similar" to the one of Fig. 10.2, in the sense that its only expansion occurs through the Joule–Thomson valve. The crucial difference between Fig. 10.2 and Fig. 10.6 is that the counterflow heat exchanger traveled by the $(\dot{m} - \dot{m}_e)$ stream on the right side of Fig. 10.6 is *cooled* along the segment embraced by the expander. This feature—the cooling of the cold zone of the installation at intermediate temperatures between T_0 and T_L—is fundamentally important, as we shall see.

The intermediate cooling effect is provided by the refrigeration cycle executed by the expander stream $\dot{m}_e$, namely, expander – heater (P_L) – regenerator (P_L) – compressor – aftercooler (P_H) – regenerator (P_H). Since this sequence is exactly the reverse of what in the field of power engineering is called the regenerative Brayton cycle, in memory of George Brayton's 1873 heat engine, a good name for it here is the "refrigeration Brayton cycle" with a regenerative counterflow heat exchanger.

As a self-standing refrigeration cycle (i.e., not as in Fig. 10.6), the Brayton cycle has been studied in detail [4, 7] because of its extensive use in

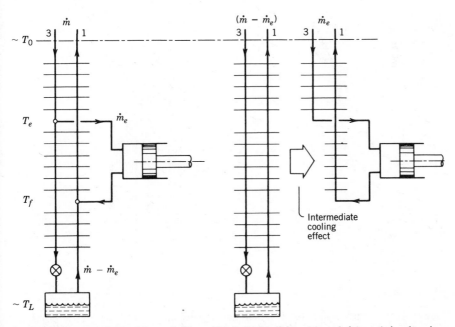

Figure 10.6 Decomposition of the cycles executed by $\dot{m}_e$ and $(\dot{m} - \dot{m}_e)$, showing that the $\dot{m}_e$ cycle cools the midsection of the long counterflow heat exchanger.

aircraft cooling and in the conceptual design of operations in space. My reason for focusing on the Brayton cycle at this stage is twofold. First, the Brayton cycle is hidden in any refrigeration scheme that uses an expander (Fig. 10.6). The second reason is that in combination with the ideal gas model, the Brayton cycle makes possible a highly compact and instructive bit of analysis that shows the direct adverse effect of various irreversibilities on the refrigeration rate managed by the machine, $\dot{Q}_L$.

Consider the T–s outlook of the regenerative Brayton cycle, Fig. 10.7, and continue to by-pass the discussion of room-temperature compression by adopting eq. (10.10) as the minimum compressor power that would take the stream from state (1) to state (3). In the first phase of this analysis, let us

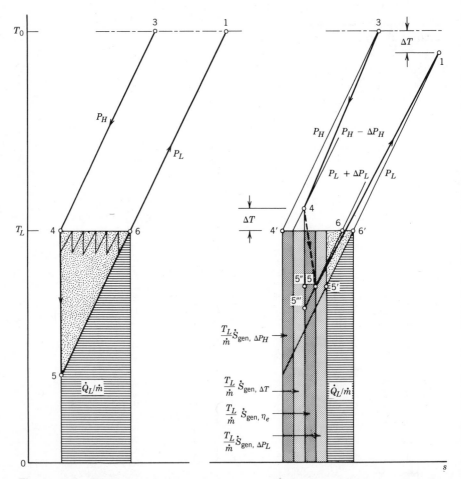

Figure 10.7 The irreversibilities that accompany the cold zone of the Brayton refrigeration cycle: the ideal cycle (left side) and the added effect of internal irreversibilities (right side).

assume that the refrigeration cycle is ideal, i.e., there are no stream-to-stream temperature differences in the regenerator ($T_3 = T_1$, $T_4 = T_6$), no pressure drops ($P_3 = P_4 = P_H$, $P_6 = P_1 = P_L$), and the expander works reversibly and adiabatically ($s_5 = s_4$). This ideal cycle is shown on the left side of Fig. 10.7.

In general, the ideal Brayton cycle can collect the refrigeration load $\dot{Q}_L$ from the entire temperature range $T_5 - T_6$: note that an example of this kind of "distributed" refrigeration effect is provided by the $\dot{m}_e$ cycle of Fig. 10.6. For simplicity, we assume that the refrigeration load is extracted from a unique temperature T_L, which is equal to T_6. This implies a cold-end heat exchanger with infinite heat transfer area, since, in general, T_L must be greater than T_6.

The refrigeration rate extracted by the ideal Brayton cycle is

$$\frac{\dot{Q}_L}{\dot{m}} = h_6 - h_5 = \left(\int_5^6 T\,ds \right)_{P=P_L} \tag{10.12}$$

in other words, the specific[†] refrigeration $\dot{Q}_L/\dot{m}$ is represented by the lined trapezoidal area trapped between the P_L isobar and absolute zero. The expansion from P_H to P_L can take place in one stage, e.g., $(4) \rightarrow (5)$ in Fig. 10.7, or in more than one stage. The multiparameter optimization procedure required in Problem 10.3 can be used to determine the optimum pressures of the isobaric heating processes that would alternate with the n expanders, as the stream would follow a horizontal zig-zag course from state (4) to state (6). According to eq. (10.12), the refrigeration rate $\dot{Q}_L$ increases with the number of expansion stages, because the specific refrigeration area on the T–s diagram rises from $T = 0$ all the way into the teeth of the zig-zag path. In the limit of infinitely many expansion stages, the expansion proceeds reversibly and isothermally from state (4) to state (6), while the refrigeration rate reaches the maximum represented by the rectangular area of height T_L:

$$\frac{\dot{Q}_{L,\max}}{\dot{m}} = T_L(s_6 - s_4) = T_L R \ln \frac{P_H}{P_L} \tag{10.13}$$

Graphically as well as analytically, the difference between the refrigeration rate (10.12) and the ceiling value (10.13) is due to the irreversibility of heating the stream from state (5) to state (6) across the finite temperature gap that extends downward from (T_L). The entropy generated in the cold-end heat exchanger is

$$\dot{S}_{\text{gen},L} = \dot{m}(s_6 - s_5) - \dot{Q}_L/T_L \tag{10.14}$$

[†]The specific refrigeration is the refrigeration rate per unit of mass flowrate.

Multiplied by T_L, this entropy generation accounts for the dotted triangular area of Fig. 10.7 (left), that is, for the difference between the refrigeration rate and its theoretical maximum value:

$$\dot{Q}_L = \dot{Q}_{L,\max} - T_L \dot{S}_{\text{gen},L} \qquad (10.15)$$

The refrigeration effect $(\dot{Q}_L)$ produced by the ideal Brayton cycle shown on the left side of Fig. 10.7 is reduced further by a conglomerate of internal irreversibilities. In the second phase of this analysis, we consider the more general situation shown on the right side of Fig. 10.7, where, proceeding counterclockwise around the bottom end of the cycle, the following sources of entropy generation are present:

(i) pressure drop along the high-pressure side of the counterflow heat exchanger, $P_3 - P_4 = \Delta P_H$
(ii) finite stream-to-stream temperature difference in the counterflow heat exchanger, $T_4 - T_6 = \Delta T$ (note that since the same ideal gas and mass flowrate flows through the two sides of the counterflow, $T_{P_H} - T_{P_L} = \Delta T$, constant, i.e., $T_3 - T_1 = \Delta T$)
(iii) nonisentropic expansion through the expander, $\eta_e < 1$, where, for ideal gases, $\eta_e = (T_4 - T_{5''})/(T_4 - T_{5'''})$
(iv) pressure drop along the low-pressure side of the counterflow heat exchanger, $P_5 - P_1 = \Delta P_L$

Equations (10.12) and (10.14) continue to hold; however, for the sake of a much better drawing, we note that $h_5 = h_{5'}$ and $h_6 = h_{6'}$. This allows us to translate the T–s areas for $\dot{Q}_L/\dot{m}$ and $\dot{S}_{\text{gen},L}/\dot{m}$ slightly to the right of their true position on the s axis. The specific refrigeration area $\dot{Q}_L/\dot{m}$ is now smaller than in the ideal case treated on the left side of Fig. 10.7, because each of the above irreversibilities takes a rectangular slice out of the maximum refrigeration area $\dot{Q}_{L,\max}/\dot{m} = T_L(s_{6'} - s_{4'})$. Geometrically, it is clear that

$$\dot{Q}_L = \dot{Q}_{L,\max} - T_L(\dot{S}_{\text{gen},L} + \dot{S}_{\text{gen},\Delta P_H} + \dot{S}_{\text{gen},\Delta T} + \dot{S}_{\text{gen},\eta_e} + \dot{S}_{\text{gen},\Delta P_L}) \quad (10.16)$$

where the last four entropy-generation contributions are listed in the same order as their sources (i)–(iv). In the case of negligibly small pressure drops and temperature differences $(\Delta P_H \ll P_H, \Delta P_L \ll P_L, \Delta T \ll T_L)$, these four terms reduce to

$$\dot{S}_{\text{gen},\Delta P_H} \cong \dot{m}R \frac{\Delta P_H}{P_H} \qquad (10.17)$$

$$\dot{S}_{\text{gen},\Delta T} \cong \dot{m}c_P \frac{\Delta T}{T_L} \qquad (10.18)$$

$$\dot{S}_{\text{gen},\eta_e} \cong \dot{m}c_P(1 - \eta_e)\left[\left(\frac{P_H}{P_L}\right)^{(k-1)/k} - 1\right] \tag{10.19}$$

$$\dot{S}_{\text{gen},\Delta P_L} \cong \dot{m}R \frac{\Delta P_L}{P_L} \tag{10.20}$$

Their combined effect, or the sum of the first four shaded strips in Fig. 10.7 (right), can be compared with (i.e. divided by) the maximum refrigeration potential offered by the cycle:

$$\frac{T_L}{\dot{Q}_{L,\text{max}}} (\dot{S}_{\text{gen},\Delta P_H} + \dot{S}_{\text{gen},\Delta P_L} + \dot{S}_{\text{gen},\Delta T} + \dot{S}_{\text{gen},\eta_e})$$

$$= \left(\ln \frac{P_H}{P_L}\right)^{-1}\left\{\frac{\Delta P_H}{P_H} + \frac{\Delta P_L}{P_L} + \frac{k}{k-1}\left(\frac{\Delta T}{T_L}\right)\right.$$

$$\left. + \frac{k}{k-1}(1 - \eta_e)\left[\left(\frac{P_H}{P_L}\right)^{(k-1)/k} - 1\right]\right\} \tag{10.21}$$

This dimensionless expression serves two functions. First, its numerical value begins[†] to tell the story of how the internal irreversibilities (i)–(iv) threaten the overall performance of the cycle. The complete picture emerges only on the basis of a numerical analysis that avoids the small-ΔP and -ΔT assumption that preceded eqs. (10.17)–(10.20). The second function of eq. (10.21) is to show the *relative* effect of each internal irreversibility. The first three terms contained between the braces are due to the counterflow heat exchanger: the manner in which they influence one another is a fundamental thermodynamics problem in heat transfer engineering [8, 9] (see also chapter 11 in this book). The last term represents the irreversibility of the expander. For example, in a typical liquid-nitrogen-temperature application, we have $T_L \sim 80$ K, $P_H/P_L \sim 200$, and $k = 1.4$, which means that the expander and the counterflow heat-exchanger ΔT contribute equally when $\Delta T \cong 280(1 - \eta_e)$ K. Therefore, the irreversibility of an expander whose efficiency is $\eta_e = 0.9$ rivals that of a counterflow heat exchanger in which the stream-to-stream temperature difference is 28 K.

The above conclusions can be compared with the numerical example in Fig. 10.8, which is based on a set of calculations reported by Thirumal-eshwar [10]. With reference to the T–s diagram of Fig. 10.7, the overall coordinates of the helium Brayton cycle considered by Thirumaleshwar are $P_H = 10$ atm, $P_L = 1$ atm, $T_0 = 300$ K, and $T_L = 30$ K. The compression from state (1) to state (3) was assumed to be carried out in a 60-percent-efficient isothermal compressor. The destruction of exergy is dominated by the compressor and the low-temperature expansion process. The loss due to

---

[†]In view of the small-ΔP and -ΔT assumption, the analytical expression (10.21) holds only when its numerical value is at least one order of magnitude smaller than unity.

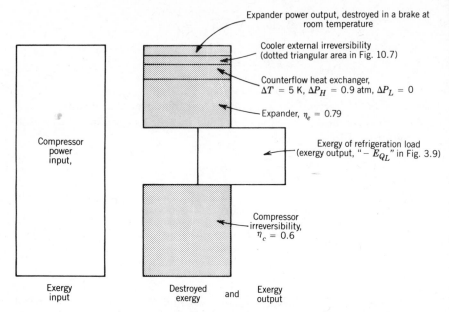

Figure 10.8   Example of the distribution of exergy destruction among the components of a helium-gas Brayton cycle with a regenerative heat exchanger.

counterflow heat-exchanger irreversibility is noticeably small because of the favorable (small) ΔT and ΔPs assumed in the model. The utilization factor or second-law efficiency of this particular cycle is roughly 25 percent; the utilization factor η_{II} is defined in eq. (3.24).

OPTIMUM INTERMEDIATE COOLING

Returning to the "intermediate cooling effect" provided by the expansion of the $\dot{m}_e$ stream on the right side of Fig. 10.6, we must face the questions of exactly "how much" intermediate cooling is advantageous, and at "what location" between T_0 and T_L should this cooling effect be installed? In the single-expander configuration of Fig. 10.6 (right), these questions result in determining the optimum expander mass fraction $(\dot{m}_e/\dot{m})$ and the position of the expansion process on the temperature scale (T_e)—under the assumption that the pressure ratio (P_H/P_L), the expander efficiency (η_e) and the size of the counterflow heat exchanger $(\bar{h}A)$ are fixed. The two-parameter optimization problem identified here can be tackled numerically. However, since the final conclusions of such a study are part of a much more fundamental design principle in refrigeration engineering, in this section, we consider the general problem of finding the thermodynamically optimum "distribution" of intermediate refrigeration along the cold zone (column) of a refrigeration machine.

The Main Counterflow Heat Exchanger

With reference to Fig. 10.9, we seek to determine the optimum distribution of the intermediate cooling effect $d\dot{Q}$ that minimizes the entropy-generation rate contributed by the counterflow heat exchanger. Note that the minimization of entropy generation leads directly to a reduction in the mechanical-power input demanded by the refrigerator when the load $\dot{Q}_L$ is fixed (Problem 10.5), or to the augmentation of the refrigeration effect $\dot{Q}_L$ when the room-temperature compressor power is fixed [Fig. 10.7 and eq. (10.21)]. The flowrate $\dot{m}_L$ is the same on both sides of the heat exchanger: the subscript "L" draws attention to the fact that like the $(\dot{m} - \dot{m}_e)$ flowrate of the counterflow of Fig. 10.6 (right), the present flowrate is constant throughout the temperature range $T_0 - T_L$ spanned by the heat exchanger.

Neglecting the pressure-drop irreversibility and writing ΔT for the local stream-to-stream temperature difference, we find that the entropy-generation rate contributed by the segment of height dT is

$$dS_{\text{gen}} = \dot{m}_L c_P \ln \frac{T + dT}{T} + \dot{m}_L c_P \ln \frac{T + \Delta T}{T + \Delta T + d(T + \Delta T)}$$

$$\cong \dot{m}_L c_P \left(\frac{dT}{T} - \frac{dT + d(\Delta T)}{T + \Delta T} \right)$$

$$\cong \frac{\dot{m}_L c_P \, \Delta T}{T^2} \, dT \tag{10.22}$$

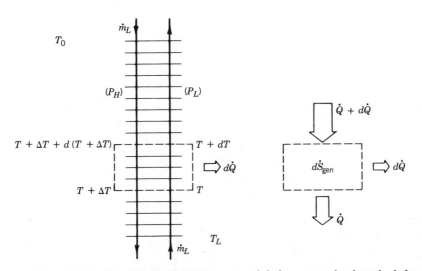

Figure 10.9 Counterflow heat exchanger as a conduit for convective heat leak from T_0 to T_L, and the calculation of entropy generation.

The compact expression deduced above holds if $\Delta T \ll T$. The group $\dot{m}_L c_P \Delta T$ plays the role of longitudinal convective heat leak in the same sense as the term $\dot{Q}_{leak}$ identified in eq. (10.7) and Fig. 10.4 (left). By writing

$$\dot{Q} = \dot{m}_L c_P \Delta T \qquad (10.23)$$

the subject of the minimization effort is the integral

$$\dot{S}_{gen} = \int_{T_L}^{T_0} \frac{\dot{Q}}{T^2} dT \qquad (10.24)$$

To determine the optimum distribution of intermediate cooling $d\dot{Q}$ is the same as finding the optimum heat-leak distribution $\dot{Q}(T)$, or the optimum stream-to-stream temperature difference ΔT as a function of T. Worth noting is that ΔT, $\dot{Q}$, and $\dot{S}_{gen}$ all tend to zero as the contact area between the two streams becomes sufficiently large. The present challenge is to minimize the integral (10.24) for a certain heat exchanger whose size is fixed.

The enthalpy increase along any of the two branches of the counterflow is

$$\dot{m}_L c_P \, dT = \bar{h} \, \Delta T \, dA \qquad (10.25)$$

where dA is the stream-to-stream contact area contained in the slice of height dT, Fig. 10.9, and $\bar{h}$ is the overall heat transfer coefficient based on dA. The fixed area of the counterflow heat exchanger and eqs. (10.23) and (10.25) account for an integral constraint that acts on the function $\dot{Q}(T)$ that forms the subject of this analysis:

$$A = \int_{T_L}^{T_0} \frac{(\dot{m}_L c_P)^2}{\bar{h} \dot{Q}} \, dT \qquad (10.26)$$

The variational calculus solution to the problem of minimizing the $\dot{S}_{gen}$ integral (10.24) subject to the A constraint (10.26) is [11, 12]

$$\dot{Q}_{opt} = \left[\frac{\dot{m}_L c_P}{\bar{h} A} \ln \frac{T_0}{T_L} \right] \dot{m}_L c_P T \qquad (10.27)$$

$$\dot{S}_{gen,min} = \left[\frac{\dot{m}_L c_P}{\bar{h} A} \ln \frac{T_0}{T_L} \right]^2 \bar{h} A \qquad (10.28)$$

where it has been assumed that both $\bar{h}$ and c_P are constant throughout the temperature interval $T_0 - T_L$. The special physical meaning of the dimensionless constant placed between the brackets of eqs. (10.27) and (10.28) is made evident by the convective heat-leak expression (10.23):

$$\left[\frac{\dot{m}_L c_P}{\bar{h} A} \ln \frac{T_0}{T_L} \right] = \left(\frac{\Delta T}{T} \right)_{opt} \qquad \text{(constant)} \qquad (10.29)$$

The optimum dimensionless temperature difference $(\Delta T / T)$ decreases as the number of heat transfer units of the heat exchanger $(\bar{h} A / \dot{m}_L c_P)$ increases.

In conclusion, the optimum $\dot{Q}$ and stream-to-stream ΔT must be proportional to the absolute temperature T, i.e., the optimum intermediate cooling effect must be distributed *evenly* along the temperature interval spanned by the counterflow heat exchanger, $(d\dot{Q}/dT)_{opt} = \text{constant}$. The preceding analysis and results were reported in 1979 [11] in order to illustrate the general applicability of a thermodynamic optimization procedure developed in the design of low-temperature mechanical supports [13–15]. The importance of the "$\Delta T / T = \text{constant}$" design principle is well known in cryogenic engineering, especially, in the German language literature [16–18].

A related thermodynamic optimization problem is that of determining the heat-exchanger design for which A is minimum while the overall irreversibility of the apparatus is fixed. The problem consists of minimizing the A integral (10.26) subject to the $\dot{S}_{gen}$ integral constraint (10.24). The result is again the "constant-$\Delta T / T$" rule:

$$\frac{\Delta T}{T} = \frac{\dot{S}_{gen}}{\dot{m}_L c_P \ln (T_H / T_L)} \qquad \text{(constant)} \qquad (10.30)$$

where $\dot{S}_{gen}$ is the fixed entropy-generation rate of the heat exchanger.

The Distribution of Expanders

If the optimum intermediate cooling effect is to be provided by a separate stream of cold gas like the $\dot{m}_e$ stream of Fig. 10.6 (right), then in the present case, $(d\dot{Q}/dT)_{opt} = \dot{m}_e c_P$ and

$$\left(\frac{\dot{m}_e}{\dot{m}_L} \right)_{opt} = \left(\frac{\dot{m}_L c_P}{\bar{h} A} \ln \frac{T_0}{T_L} \right) \qquad (10.31)$$

The flowrate ratio that expresses the "imbalance" between the warm and cold branches of the three-stream arrangement shown in Fig. 10.10 is exactly the same as the $(\Delta T / T)$ constant of the counterflow heat exchanger. The $\dot{m}_e$ stream could in principle be produced through the same method as in Fig. 10.6 (right), provided that the P_H / P_L ratio is large enough so that the expander can embrace the entire heat exchanger, $T_e = T_0$ and $T_f = T_L$. The expander flowrate and the counterflow imbalance decrease as the number of heat transfer units $\bar{h} A / \dot{m}_L c_P$ increases.

When the temperature span of one expander is too narrow relative to $T_0 - T_L$, the tapering of the ΔT versus T distribution can be achieved by installing two or more expanders along the main heat exchanger. This

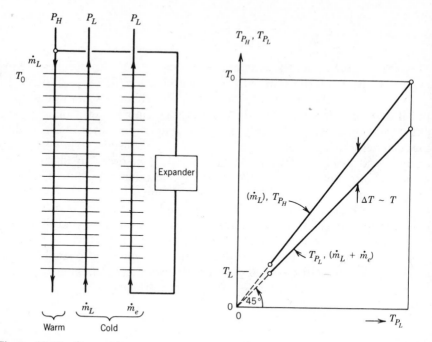

Figure 10.10 The optimum temperature distribution in an ideal gas counterflow heat exchanger.

technique is illustrated in Fig. 10.11 for a sequence of one, two, and, finally, three expanders. Given the fixed-temperature span of one expander, there exists an optimum position for inserting each expander along the $T_0 - T_L$ scale. One effect of the string of expanders is that the flowrate handled by the counterflow heat exchanger decreases in the direction of lower temperatures. In other words, the cold-end flowrate ($\dot{m}_L$) that removes the refrigeration load is smaller than the flowrate processed by the room-temperature compressor ($\dot{m}_0$).

The relationship between the distribution of expanders and the overall performance of the cycle can be illustrated by considering the continuous distribution shown in Fig. 10.12. In each temperature interval of height dT, the high-pressure stream loses a fraction of its flowrate, $d\dot{m}$. This fraction is expanded through a small isothermal expander, as shown on the right side of Fig. 10.12. The expander power output $d\dot{W}$ is less than the output of a reversible and isothermal expander working between P_H and P_L; therefore, we model $d\dot{W}$ as

$$dW = \eta_e \, d\dot{W}_{\substack{\text{max} \\ \text{(reversible and} \\ \text{isothermal)}}} = \eta_e (d\dot{m}) RT \ln \frac{P_H}{P_L} \qquad (10.32)$$

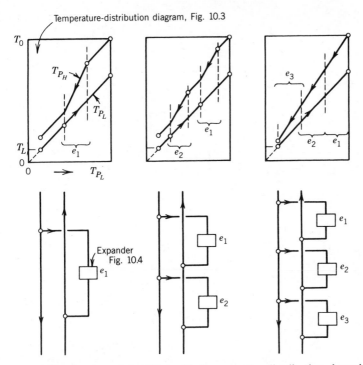

Figure 10.11   The tapering of the $(T_{P_H} - T_{P_L})$ versus T_{P_L} distribution, by using one, two, or three intermediate expanders.

where the efficiency η_e is less than 1. The expander producing $d\dot{W}$ can be viewed as a reversible and isothermal expander working between P_H and the intermediate pressure P_i, followed by a throttle between P_i and P_L (Fig. 10.12). The "internal" effect of the expander is the cooling of the counter-flow heat-exchanger segment (dT): Note the white "heat-transfer" arrow absorbed by the isothermal expander.

We are interested in the flowrate distribution $\dot{m}(T)$ and how this impacts the ratio of refrigeration load divided by room-temperature compressor power. The first-law statement for the (dT) segment is

$$(\eta_e bT - \Delta T)\, d\dot{m} = \dot{m}\, d(\Delta T) \tag{10.33}$$

where b is shorthand for

$$b = \frac{R}{c_P} \ln \frac{P_H}{P_L} \tag{10.34}$$

The product $\eta_e b$ is a number of order 1, whereas in most designs $\Delta T/T$ is less than 0.1. Therefore, neglecting the ΔT term on the left side of eq.

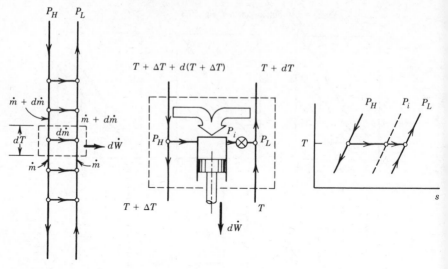

Figure 10.12   Counterflow heat exchanger with continuously distributed isothermal expanders.

(10.33), assuming that η_e is constant, and that the "constant-$\Delta T/T$" rule (10.29) holds, the first-law statement can be integrated to yield the flowrate distribution:

$$\frac{\dot{m}}{\dot{m}_0} = \left(\frac{T}{T_0}\right)^{\nu} \qquad (10.35)$$

The exponent ν is a constant smaller than 1:

$$\nu = \frac{\Delta T/T}{\eta_e b} \qquad (10.36)$$

The actual value of this exponent (or of $\Delta T/T$) can be calculated from the heat-exchanger area constraint (10.25)–(10.26), which in this case reads

$$A = \int_{T_L}^{T_0} \frac{\dot{m} c_P}{\bar{h}\,\Delta T}\, dT \qquad (10.37)$$

Keeping in mind that $\dot{m}$ is a function of temperature, the result of combining eqs. (10.35)–(10.37) is

$$N_{\text{tu},0}\,\eta_e b = \frac{1}{\nu^2}\left[1 - \left(\frac{T_L}{T_0}\right)^{\nu}\right] \qquad (10.38)$$

which for $\nu \ll 1$ means

$$\nu \cong \frac{\ln\left(T_0/T_L\right)}{N_{\mathrm{tu},0}\,\eta_e b} \ll 1 \qquad (10.39)$$

The number of heat transfer units $N_{\mathrm{tu},0}$ is based on the largest flowrate:

$$N_{\mathrm{tu},0} = \frac{\bar{h}A}{\dot{m}_0 c_P} \qquad (10.40)$$

Examining now eq. (10.35), we see that the flowrate $\dot{m}(T)$ decreases toward lower temperatures, the decrease becoming more pronounced as ν increases, i.e., as the expanders and the heat exchanger become less efficient. This trend is even more visible in the construction of the $\dot{Q}_L/\dot{W}_c$ ratio. If we model the room-temperature compressor as an isothermal compressor with the efficiency η_0, then

$$\dot{W}_c = \frac{1}{\eta_0}\,\dot{m}_0 R T_0 \ln \frac{P_H}{P_L} \qquad (10.41)$$

Likewise, if the low-temperature expansion and refrigeration-load removal are effected by means of an isothermal expander of efficiency η_L, then

$$\dot{Q}_L = \eta_L \dot{m}_L R T_L \ln \frac{P_H}{P_L} \qquad (10.42)$$

and, finally,

$$\frac{\dot{Q}_L}{\dot{W}_c} \cong \eta_L \eta_0 \left(\frac{T_L}{T_0}\right)^{1+\nu} \qquad (\nu \ll 1) \qquad (10.43)$$

This figure of merit decreases as soon as one of the components of the installation becomes less efficient: note the monotonic relationships between $\dot{Q}_L/\dot{W}_c$ and η_L, η_0, $N_{\mathrm{tu},0}$, or η_e.

One final observation concerns the mass flowrate function $\dot{m}(T)$, eq. (10.35). The flowrate becomes temperature-independent as ν approaches zero. On the other hand, it is worth noting that the "constant-$\Delta T/T$" design rule used to derive eq. (10.35) applies when the flowrate of the balanced counterflow heat exchanger is constant, $\dot{m} = \dot{m}_L$, eqs. (10.23)–(10.29). These observations suggest that the conclusions drawn based on the analysis presented above are valid only for vanishingly small νs. The more general problem of optimizing the installation of Fig. 10.12 when $\Delta T/T$ is not necessarily negligible remains to be studied. The chief unknown in this new problem is the thermodynamically optimum $\Delta T/T$ function that must replace the simple constant-$\Delta T/T$ rule invoked until now. One additional source of irreversibility in the more general problem identified here is the inefficient operation of the continuously distributed isothermal expanders, i.e., the effect of $\eta_e < 1$. One step in the direction of solving this new problem is outlined in Problem 10.6.

The Thermal-Insulation Superstructure

The intermediate cooling technique encountered in the optimization of the counterflow heat exchanger has general applicability in the design of the many features that constitute the "thermal insulation" of any low-temperature apparatus [11]. These applications and the fundamental research generated by them have been reviewed in detail in chapters 9 and 10 of Ref. 9.

Consider as a first example the heat leak made possible by a one-dimensional *mechanical support* stretching from T_0 all the way down to T_L (Fig. 10.13a):

$$\dot{Q} = kA_c \frac{dT}{dx} \tag{10.44}$$

where $k(T)$ is the thermal conductivity of the structural material. Parameters A_c and x represent the cross-sectional area pierced by $\dot{Q}$ and the longitudinal coordinate measured in the direction $T_L \rightarrow T_0$ (note that $\dot{Q}$ flows toward lower temperatures). The Fourier law (10.44) places an integral constraint on the heat-leak function $\dot{Q}(T)$:

$$\frac{L}{A_c} = \int_{T_L}^{T_0} \frac{k}{\dot{Q}} \, dT \tag{10.45}$$

where L is the length of the support (the length of the heat-leak path). The entropy-generation rate contributed by the entire support is given by the same integral as in eq. (10.24). It is easy to see that the problem of selecting the optimum function $\dot{Q}(T)$ or the optimum intermediate cooling effect $d\dot{Q}/dT$ is the same as the heat-exchanger problem that ended with eqs. (10.27)–(10.29). The optimum design of a mechanical support is represented by [13]

$$\dot{Q}_{\text{opt}} = \left(\frac{A_c}{L} \int_{T_L}^{T_0} \frac{k^{1/2}}{T} \, dT \right) k^{1/2} T \tag{10.46}$$

$$\dot{S}_{\text{gen,min}} = \frac{A_c}{L} \left(\int_{T_L}^{T_0} \frac{k^{1/2}}{T} \, dT \right)^2 \tag{10.47}$$

The desirability of this design relative to the simple "no-cooling" design in which $\dot{Q}$ is constant is best illustrated by considering the special case of a constant-conductivity structural material. Setting $k = k_0$ in eqs. (10.46) and (10.47), we obtain

$$\dot{Q}_{\text{opt}} = \left(\frac{k_0 A_c}{L} \ln \frac{T_0}{T_L} \right) T \tag{10.48}$$

$$\dot{S}_{\text{gen,min}} = \frac{k_0 A_c}{L} \left(\ln \frac{T_0}{T_L} \right)^2 \tag{10.49}$$

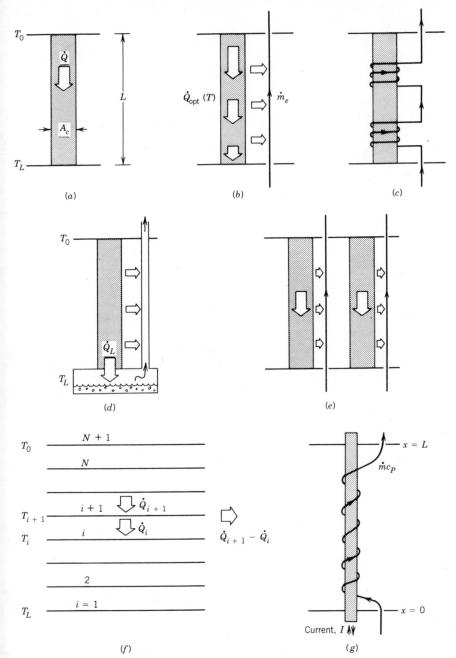

Figure 10.13 Elements of thermal insulation: (*a*) mechanical support without intermediate cooling; (*b*) continuous cooling provided by a single stream of cold gas; (*c*) discrete cooling concentrated in two cooling stations; (*d*) continuous cooling provided by boil-off gas; (*e*) continuous cooling for two parallel insulations (e.g., two mechanical supports); (*f*) discrete cooling of a stack of $(N-1)$ radiation shields; and (*g*) continuous, single-stream cooling of an electrical power cable.

In this case, the optimum heat-leak function is proportional to the absolute temperature, which means that the intermediate cooling effect must be distributed evenly along the support. If the intermediate cooling is to be provided by a cold stream of constant-c_p gas like the $\dot{m}_e$ stream in Fig. 10.13b or on the left side of Fig. 10.10, then the optimum flowrate of this stream can be calculated by writing $\dot{m}_e c_P = d\dot{Q}_{opt}/dT$. The entropy-generation reduction due to the use of intermediate cooling is illustrated dividing $\dot{S}_{gen,min}$ of eq. (10.49) by the entropy-generation rate of the same support in the absence of any intermediate cooling:

$$\dot{S}_{gen,no\ cooling} = k_0 \frac{A_c}{L} (T_0 - T_L)\left(\frac{1}{T_L} - \frac{1}{T_0}\right) \tag{10.50}$$

The resulting ratio is the entropy-generation number N_S of Fig. 10.14:

$$N_S = \frac{\dot{S}_{gen,min}}{\dot{S}_{gen,no\ cooling}} = \frac{T_0}{T_L}\left[\frac{\ln\ (T_0/T_L)}{T_0/T_L - 1}\right]^2 \tag{10.51}$$

The irreversibility reduction is sizeable in cryogenic applications, where T_0/T_L is of the order of 10 or larger.

The thermal conductivity of most structural materials varies strongly with the temperature, particularly at low temperatures. Consequently, the very popular [19] one-stream intermediate cooling design discussed in the preceding paragraph can only approximately approach the irreversibility minimum listed in eq. (10.47). In such cases, one still faces the question of determining the optimum coolant flowrate of the single-stream arrangement. Answers to this question are presented in Ref. 13 for a number of common low-temperature structural materials.

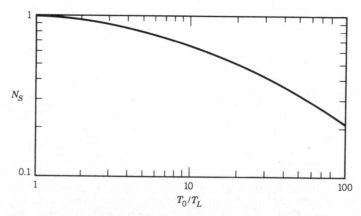

Figure 10.14   Entropy-generation reduction due to the use of optimum intermediate cooling in the design of a constant-conductivity mechanical support [9].

Another aspect of the optimum intermediate cooling of low-temperature mechanical supports is the design and fabrication of the stream-to-solid heat exchanger that effects the intermediate cooling demanded by the heat-leak function $\dot{Q}_{opt}(T)$. This is a difficult task, when we realize that to bring a gaseous stream in contact with a solid structural member means to weaken the solid, by machining holes and channels in it. The thermodynamic optimization of this class of heat exchangers is discussed in Ref. 15. The fabrication of the heat exchanger is less difficult and more economical if instead of building a continuous heat exchanger from T_L to T_0 between the gas stream and the mechanical support, one builds only a sequence of *discrete* points of thermal contact (discrete cooling stations, Fig. 10.13c). The optimum heat-leak function demanded by eq. (10.46) is approximated in this case by a stepwise-varying function. This problem was considered by Hilal and Boom [20], who reported a procedure for calculating the optimum spatial distribution of discrete cooling stations, and also the optimum temperature of each cooling station. Taken together, Hilal and Boom's results instruct the refrigeration-system designer how much intermediate cooling to invest in the mechanical support structure, and where to position it [see also Ref. 9, pp. 190–192].

A simpler discrete-cooling problem related to the one solved by Hilal and Boom is the problem of finding the optimum location of cooling stations along a mechanical support (or any other insulation element) that is cooled by a stream of boil-off gas [21]. In many applications, the cold space (T_L) is not cooled steadily by a refrigerator but by a pool of liquefied refrigerant (e.g., liquid helium, Fig. 10.13d). The liquid-helium pool generates naturally a stream of vapor ("boil-off") that is proportional to the total heat leak reaching the cold space, $\dot{Q}_L$. In this problem, one is faced with determining only the spatial distribution of discrete-cooling stations, because the gas flowrate no longer constitutes a degree of freedom in the design. The geometric layout of the problem can be seen by combining Fig. 10.13c and Fig. 10.13d. Specific rules for selecting the position of discrete stations along a conducting support cooled by boil-off gas are reported in Ref. 21.

The insulation "structure" of any low-temperature installation contains more than one heat-leak path. This state of affairs is shown symbolically in Fig. 10.13e by means of two mechanical supports in parallel. Another example of two parallel insulation elements would be the long counterflow heat exchanger of Fig. 10.9 and the mechanical support structure that connects the (T_L) zone with the room-temperature casing. The intermediate cooling regime of parallel insulation elements can be optimized similarly [22, 23; also Ref. 9, pp. 182–185]. One conclusion that emerges from such studies is that parallel insulations must be fitted with optimum continuous distributions of intermediate cooling along both legs of the insulation [23].

Another important strategy in the field of thermal insulation is the use of *radiation shields*. The thermodynamic optimization of a stack of radiation shields follows the same entropy-generation minimization route as the optimization of counterflow heat exchangers and one-dimensional mechanical supports. First, if the number of shields is sufficiently large, the variational results (10.46) and (10.47) continue to apply, provided we make an appropriate change in notation. Let $(N-1)$ be the number of parallel radiation shields suspended in the vacuum between T_0 and T_L (Fig. 10.13f). The heat transfer rate between two adjacent shields is

$$\dot{Q}_i = \sigma A_c F(T_{i+1}^4 - T_i^4) \tag{10.52}$$

where σ is the Stefan–Boltzmann constant, A_c is the shield area (the cross-section pierced by $\dot{Q}_i$), and F is the effective view factor that accounts for the emissivities of the two surfaces. In the limit $N \to \infty$, the shield-to-shield heat leak approaches [9]

$$\dot{Q}_i \to 4\sigma A_c F T_i^3 \frac{\Delta T_i}{\Delta i} \tag{10.53}$$

where $\Delta T_i = T_{i+1} - T_i$ and $\Delta i = (i+1) - i$. In the "many-shields" limit, eq. (10.53) plays the same role as the Fourier law (10.44) in the optimization of mechanical supports. Rearranging eq. (10.53), writing $\dot{Q}(T)$ instead of $\dot{Q}_i(T_i)$, and integrating across the N-thick stack of shields, we obtain the integral constraint:

$$\frac{N}{A_c} = \int_{T_L}^{T_0} \frac{4\sigma F T^3}{\dot{Q}} \, dT \tag{10.54}$$

Comparing this constraint with eq. (10.45), we see that before applying the optimum results (10.46) and (10.47) in the realm of radiation shielding, we first effect the transformation $L \to N$ and $k \to 4\sigma F T^3$. As a first approximation then, the effective thermal conductivity of a stack of many shields increases with the temperature as T^3. Finally, the optimum heat-leak function derived from eq. (10.46) can later be combined with eq. (10.53) to determine the optimum distribution of shield temperatures and the cooling effect that must be provided to each shield.

When the number of shields is small, the variational results exploited above do not apply. Instead of the heat-leak integral (10.24), now we have to minimize the sum of the entropy-generation rates associated with the N shield-to-shield temperature gaps, Fig. 10.13f:

$$\dot{S}_{\text{gen}} = \sum_{i=1}^{N} \sigma A_c F(T_{i+1}^4 - T_i^4)\left(\frac{1}{T_i} - \frac{1}{T_{i+1}}\right) \tag{10.55}$$

Solving the system of $(N-1)$ equations:

$$\frac{\partial \dot{S}_{\text{gen}}}{\partial T_i} = 0 \qquad (i = 1, 2, \ldots, N-1) \tag{10.56}$$

is sufficient for determining the optimum set of shield temperatures T_i, while A_c and N are assumed fixed. Once the optimum T_is are known, one can use eq. (10.52) to calculate the optimum stepwise distribution of heat leak across the stack. Finally, the difference between two adjacent heat-leak values in this stepwise distribution represents the optimum cooling effect that must be applied to the shield that separates these two values, for example (Fig. 10.13f),

$$\dot{Q}_{i+1} - \dot{Q}_i = \text{the cooling rate of the } i\text{th shield} \qquad (10.57)$$

The thermodynamic optimization of radiation shielding is covered extensively in the literature [24–31]. For example, Martynovskii et al. [24, 25] used the entropy-generation minimization scheme, eqs. (10.55)–(10.57), to determine the optimum shield temperatures for insulations with up to three shields. A most complete study that combines the optimization of conducting supports with that of radiation shields was reported by Chato and Khodadadi [31]. These authors considered a one-dimensional insulation whose effective thermal conductivity varies as

$$k = k_1 T^m + k_2 T^n \qquad (10.58)$$

where $m \cong 1$ accounts for the contribution of pure conduction (at low temperatures) and where $n \cong 3$ accounts for the presence of radiation heat transfer. This effective thermal-conductivity construction is a good model for thermal-insulation structures consisting of cooled shields and structural material built into the gaps between shields. Chato and Khodadadi's problem is a most general one, because it asks not only for the optimum distribution of shield temperatures, but also for the optimum spacing (conduction path) between two adjacent shields. Note that since the shields are cooled, they play the role of discrete cooling stations for the conducting structure that immobilizes and supports the shields between T_0 and T_L.

A sample of Chato and Khodadadi's thermodynamic-optimization conclusions is presented in Figs. 10.15 and 10.16. The sample refers to the design of a stack of three shields (cooling stations) in the limit where the conduction term $k_1 T$ dominates the effective conductivity function (10.58). According to the general nomenclature of Fig. 10.13f, the three shield temperatures are T_2, T_3, and T_4. At the optimum, these temperatures decrease as the overall temperature ratio T_L/T_0 decreases (Fig. 10.15). The spacing (Δx) between the shields is relatively insensitive to the ratio T_L/T_0; however, the three shields migrate together toward the cold surface T_L as the ratio T_L/T_0 decreases (Fig. 10.16).

As a last example of a heat-leak path that can be "obstructed" by means of intermediate cooling, consider the functioning of an *electrical power cable* that crosses the temperature gap $(T_0 - T_L)$. The thermodynamic optimization of this class of systems leads to the by now familiar conclusion that an

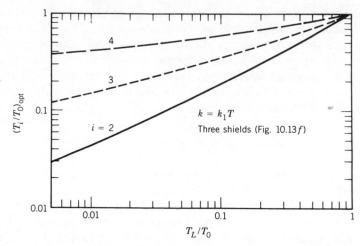

Figure 10.15 The optimum shield temperatures for a three-shield insulation with $k = k_1 T$ (after Chato and Khodadadi [31]).

optimum distribution of lateral cooling must be provided to the cable, if the cable is to perform its task (the transfer of the electrical current I) with minimum irreversibility. The new feature in the thermodynamic optimization of a power cable is that the entropy generation is due to *two* effects, the heat conduction along the cable and the dissipation of electrical power in the resistance posed by the cable. Because of this new feature, an additional

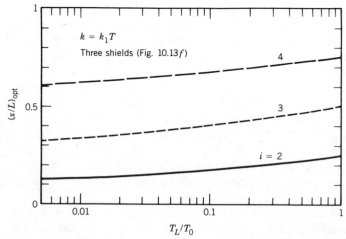

Figure 10.16 The optimum location of the three shields of Fig. 10.15 (after Chato and Khodadadi [31]).

function of the lateral-cooling effect recommended by thermodynamic optimization is to prevent the cable from overheating and burning up [32].

The simplest model for illustrating the applicability of the design methodology of this section to power cables is to consider a cable of length L, cross-sectional area A_c, thermal conductivity k_0 (constant), and electrical resistivity that varies as T:

$$\rho_e = \frac{L_0}{k_0} T \tag{10.59}$$

In writing eq. (10.59), we are assuming that the cable material obeys the Widemann–Franz law [33], in which $L_0 = (2.45)10^{-8}\,(\text{W/A K})^2$. It is not difficult to show that based on the present model, the total entropy-generation rate of the cable assumes the form [Ref. 9, p. 199]:

$$\dot{S}_{\text{gen}} = \int_{T_L}^{T_0} \frac{\dot{Q}}{T^2}\, dT + \frac{L_0 I^2 L}{k_0 A_c} \tag{10.60}$$

where $\dot{Q}(T)$ is the longitudinal heat leak. The critical observation is that the second term on the right side is a constant, and that the fixed-geometry constraint (10.45) applies here as well. It means that the optimum conduction heat leak that minimizes eq. (10.60) is exactly the same as in eq. (10.48): combining that $\dot{Q}_{\text{opt}}(T)$ expression with the Fourier law (10.44) with $k = k_0$ yields the following optimum temperature distribution along the cable [34]:

$$T_{\text{opt}}(x) = T_L \exp\left(\frac{x}{L} \ln \frac{T_0}{T_L}\right) \tag{10.61}$$

In conclusion, the optimum cable-temperature distribution is independent of the electrical current I. To maintain this distribution, i.e., to apply the appropriate distribution of intermediate cooling, one must take into account the electrical power dissipation. Note that the optimum intermediate-cooling regime demanded by eq. (10.61) can be achieved by cooling the cable with a single stream of cold gas (Fig. 10.13g). The optimum flowrate of this stream is obtained by combining eq. (10.61) with the first law for a cable slice of thickness dx:

$$k_0 A_c \frac{d^2 T}{dx^2} - \dot{m}c_P \frac{dT}{dx} + \rho_e \frac{I^2}{A_c} = 0 \tag{10.62}$$

The result is

$$\frac{\dot{m}c_P L}{k_0 A_c} = \ln \frac{T_0}{T_L} + \left(\ln \frac{T_0}{T_L}\right)^{-1}\left(\frac{IL_0^{1/2} L}{k_0 A_c}\right)^2 \tag{10.63}$$

in other words, the optimum coolant flowrate generally increases as I increases. The optimum flowrate–current operating curve (10.63) is shown in the dimensionless plane of Fig. 10.17, for a liquid-helium temperature application, $T_L = 4.2$ K, $T_0 = 300$ K [14]. The zero-current point on this operating curve indicates the optimum coolant flowrate for a one-dimensional conducting support with constant thermal conductivity, as was observed immediately below eq. (10.49). The optimum operating curve is situated close to the curve of "self-sufficient" operation, when the coolant flowrate is the boil-off generated by the cable cold-end heat leak $\dot{Q}_L$ that reaches the liquid-helium space (as in Fig. 10.13d). Figure 10.17 shows also the "burn-up" domain in which a steady-state temperature distribution cannot exist [32]. This last feature is due to the thermal-runaway instability caused by the fact that the resistivity ρ_e increases with the temperature, eq. (10.59).

Recalling the entropy-generation minimization (10.49), we see that in the present case, the minimum of eq. (10.60) must be

$$\dot{S}_{\text{gen,min}} = \frac{k_0 A_c}{L}\left(\ln \frac{T_0}{T_L}\right)^2 + \frac{L_0 I^2 L}{k_0 A_c} \tag{10.64}$$

This expression shows the competition between heat-conduction and electrical-resistance irreversibilities: to see this competition better, imagine varying the cable cross-sectional area. We reach the interesting conclusion that in addition to the optimum intermediate-cooling regime (10.63), there exists an optimum cable size for minimum irreversibility [Ref. 9, p. 201]:

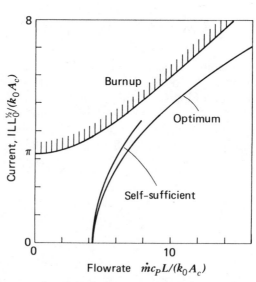

Figure 10.17   The operating domain of a current cable stretching from 300 to 4.2 K (after Ref. 14).

$$A_{c,\text{opt}} = \frac{IL_0^{1/2}L}{k_0 \ln(T_0/T_L)} \tag{10.65}$$

This conclusion follows from minimizing eq. (10.64) with respect to A_c. The optimum cable cross-sectional area increases linearly with the current and the cable length.

LIQUEFACTION

The main ideas responsible for the technological developments reviewed until now are expressed succinctly by the refrigeration load formula (10.9). Successful and efficient refrigeration requires a combination of two design features:

(1) one or more paths by which the refrigeration load can exit the cold zone [two such paths account for the two-term expression of $(\dot{Q}_L/\dot{m})_{\text{max}}$ under eq. (10.9)], and

(2) the minimization of the total heat leak into the cold zone.

Starting with Fig. 10.4 and eq. (10.7), we recognized the top end of the counterflow heat exchanger as the entrance of the convective heat leak $\dot{Q}_{\text{leak}}$, which is due to the finite size of the heat exchanger. In the preceding section, we learned that the possible heat-leak paths are quite numerous (Fig. 10.13), and that, fortunately, they can all be obstructed by applying the general design principle of irreversibility minimization via intermediate cooling.

Liquefiers vs. Refrigerators

In this section, we examine the result of putting together the above features in the design of large-scale refrigeration installations. To be efficient, I focus on a special class of such installations, namely, on liquefaction systems or "liquefiers." The feature that distinguishes the liquefier of Fig. 10.18 from the refrigerator on the right side of Fig. 10.4 is the fact that a fraction (x_l) of the cold-end flowrate $(\dot{m})$ is separated as liquid and stored for later use. The steady-state operation of the liquefier is preserved by replenishing the low-pressure stream with a stream of room-temperature makeup gas $(x_l\dot{m})$ right before entering the compressor.

From the point of view of the operator who supplies the makeup gas (T_0, P_L) and collects an equal mass flowrate in liquid form (T_L, P_L), the stream $x_l\dot{m}$ simply "descends" through the cold zone of the closed-loop refrigerator whose cold-end flowrate is $\dot{m}$. This different point of view is illustrated on the right side of Fig. 10.18. What in the preceding discussion of refrigeration cycles was labeled $\dot{Q}_L$ is now used to deliver the final blow

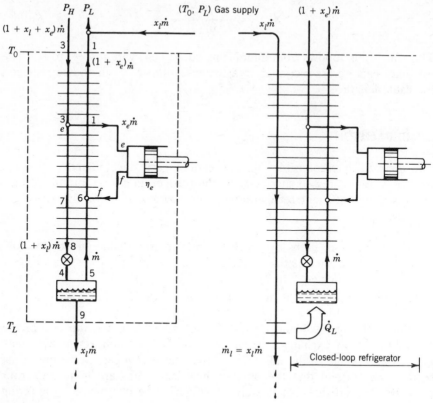

Figure 10.18 Liquefier with a single expander, and its relationship to a closed-loop refrigerator.

to the $x_l\dot{m}$ side stream and liquefy it. The connection between refrigerators and liquefiers is now evident.

The number of expanders that must be used along the main counterflow heat exchanger depends on the overall ratio T_0/T_L and the pressure ratio P_H/P_L. One expander is sufficient in liquid-nitrogen liquefiers, as we shall later see in the numerical example of Fig. 10.20. Helium liquefiers require three expanders (Fig. 10.19); when liquid nitrogen is readily available at the helium-liquefaction site, the uppermost expander can be replaced by a liquid-nitrogen "precooling" stage [35]. Keeping in mind the intrinsic irreversibility of any throttling process (p. 725), it is easy to see that the performance of refrigerators and liquefiers improves if the Joule–Thomson valve is replaced by an additional expander. This design change is shown in the lower-right section of Fig. 10.19.

Although we devoted considerable space to the desirability of work-producing expanders, we can rediscover the same idea in the context of liquefiers by considering the one-expander installation of Fig. 10.18. If

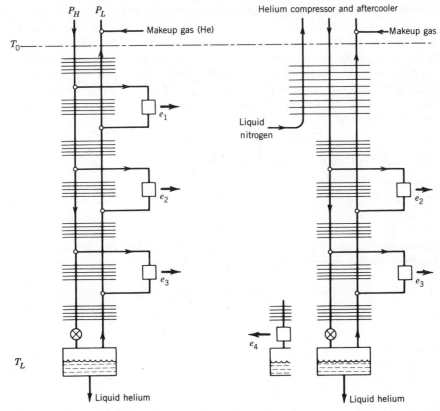

Figure 10.19 Helium liquefier with three expanders (left side) and with liquid-nitrogen precooling (right side).

$(x_e \dot{m})$ is the flowrate processed by the expander, then the total flowrate handled by the room-temperature compressor is $(1 + x_l + x_e)\dot{m}$. The second-law statement for the cold zone of the left side of Fig. 10.18 is

$$\dot{S}_{\text{gen}} = x_l \dot{m} s_f + (1 + x_e)\dot{m} s_1 - (1 + x_l + x_e)\dot{m} s_3 \geq 0 \qquad (10.66)$$

from which we deduce

$$x_l < (1 + x_e) \frac{s_3 - s_1}{s_3 - s_f} \qquad (10.67)$$

The second law prescribes an upper bound for the liquid fraction (the "yield") x_l. Worth noting is that the denominator $(s_3 - s_f)$ is fixed by the high pressure and low temperature of the process, $s_3(T_0, P_H)$ and $s_f(P_L)$. In the limit of negligible top heat-exchanger temperature difference ($T_3 \cong T_1$), the numerator $(s_3 - s_1)$ approaches the constant $R \ln P_L/P_H$. Therefore, the message of eq. (10.67) is that the ceiling value of x_l increases as the expander fraction x_e increases. The actual position of the expander on the temperature scale of the heat exchanger does not affect the ceiling value of x_l. However, we know from the study of the constant-$\Delta T/T$ rule in the preceding section that the expander location will most certainly influence the degree to which the actual x_l approaches the ceiling value listed on the right side of eq. (10.67).

A Heylandt Nitrogen-Liquefier Example

Insight into how the various design parameters and irreversibilities of the process influence the overall figure of merit of a liquifier is provided by the numerical calculations summarized in Fig. 10.20. The calculations refer to a single-expander Heylandt nitrogen liquefier whose flow circuit and numbering convention are shown on the left side of Fig. 10.18. The parameters that are fixed in this example are

$$T_0 = 300 \text{ K} \qquad P_H = 200 \text{ atm} \qquad \Delta P_H/P_H = 0 \qquad \eta_e = 0.8$$
$$\tag{10.68}$$
$$T_L = 77.4 \text{ K} \qquad P_L = 1 \text{ atm} \qquad \Delta P_L/P_L = 0$$

The pressure ratio is high enough for the expander to embrace most of the counterflow heat exchanger. For this reason, the expander intake is placed

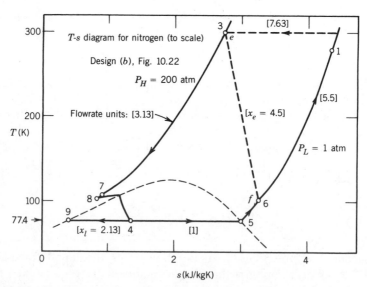

Figure 10.20 T–s diagram for a Heylandt-type nitrogen liquefaction scheme, Fig. 10.18, left side, also eqs. (10.68), $T_3 - T_1 = 20 \text{ K}$, and $T_7 - T_6 = 6 \text{ K}$.

at room temperature (Fig. 10.5, right side), which means that in this example the uppermost section of the heat exchanger of Fig. 10.18 is absent. To indicate the absence of the uppermost section, the top-end states (1) and (3) are written again at the level of the expander inlet.

In general, the cold zone of this liquefier has four sources of irreversibility: the finite-size heat exchanger, the throttling from state (8) to state (4), the expansion through the less than perfect expander, and, finally, the constant-pressure mixing between the exhaust from the expander $(x_e\dot{m})$ and the low-pressure stream $\dot{m}$. For simplicity, the irreversibility of mixing the $x_e\dot{m}$ and $\dot{m}$ streams was neglected; in other words, the states (f) and (6) were assumed to fall at the same temperature on the P_L isobar. This assumption is shown clearly on the T–s diagram of Fig. 10.20.

Our final objective in analyzing this liquefaction scheme is to determine the ratio between the liquefied-stream flowrate $x_f\dot{m}$ and the flowrate handled by the compressor, so that we might calculate the number of kilowatts of compressor power needed to produce one unit (1 kilogram, or 1 liter) of liquid nitrogen at atmospheric pressure. The analysis proceeds the usual way, by identifying all the states (corners) visited by the working fluid on the T–s diagram. For example, state (6) is determined easily from the notion that the isentropic efficiency η_e is 80 percent.

In order to pinpoint the heat exchanger state (7), which is positioned vis-à-vis with state (6), we have to make additional assumptions concerning the temperature differences that exist along the counterflow heat exchanger. For example, Fig. 10.20 was drawn assuming $\Delta T_{\text{top}} = 20\,\text{K}$ and $\Delta T_{\text{bottom}} = 6\,\text{K}$, where $\Delta T_{\text{top}} = T_3 - T_1$ and $\Delta T_{\text{bottom}} = T_7 - T_6$. The act of assuming these temperature differences is not sufficient, because we have no assurance that the temperature distribution inside the black box (3)–(1)–(6)–(7) is physically possible. We can use the second law to make sure that the black box is a source of entropy, not a sink, or we can draw the T_{P_H} vs. T_{P_L} lines as learned in Fig. 10.3. The latter course leads to Fig. 10.21, which shows the temperature distribution inside the counterflow heat exchanger of Fig. 10.20, that is, when the top and bottom ΔTs are 20 K and 6 K, respectively.

The chief message of Fig. 10.21 is that the assumed top and bottom ΔTs are just large enough so that the intersection of the T_{P_H} and T_{P_L} curves is avoided (recall that according to the second law, an intersection is physically impossible, p. 136). It follows that the assumption of a ΔT_{bottom} value smaller than 6 K while holding ΔT_{top} fixed at 20 K will yield a crossed-over distribution of T_{P_H} and T_{P_L} lines, hence, an erroneous set of calculations. The same thing would happen if ΔT_{top} is assumed smaller than 20 K while holding $\Delta T_{\text{bottom}} = 6\,\text{K}$. Proceeding in the opposite direction (i.e., away from the origin in Fig. 10.22) means assuming realistic heat exchangers with progressively smaller N_{tu} values. The critical $(\Delta T_{\text{top}}, \Delta T_{\text{bottom}})$ values for which the T_{P_H} and T_{P_L} lines are tangent correspond to an infinitely large heat exchanger: Figure 10.21 shows that the values $(\Delta T_{\text{top}} = 20\,\text{K}, \Delta T_{\text{bottom}} = 6\,\text{K})$ come close to representing one such critical design.

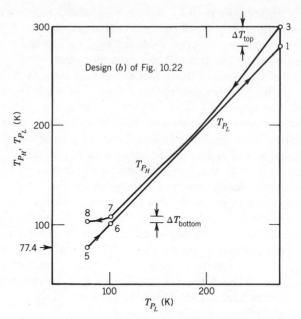

Figure 10.21 The temperature distribution inside the imbalanced counterflow heat exchanger (3)–(1)–(6)–(7) of Fig. 10.20.

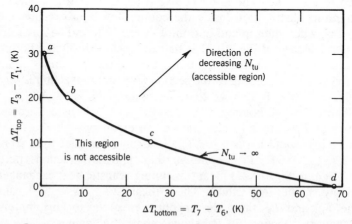

Figure 10.22 The domain of permissible pairs $(\Delta T_{top}, \Delta T_{bottom})$ for the design of the (3)–(1)–(6)–(7) heat exchanger of Fig. 10.18, left side.

For each finite ΔT_{bottom} value there exists a critical ΔT_{top} value below which the heat exchanger cannot exist. The frontier between allowed and not-allowed $(\Delta T_{\text{top}}, \Delta T_{\text{bottom}})$ pairs was shown roughly in Fig. 10.22 on the basis of four separate plots like Fig. 10.21. The frontier is also the locus of infinitely large counterflow heat exchangers, which, despite their infinite size and zero pressure drops, continue to contribute a remanent ("imbalance") irreversibility to the greater system. This aspect is treated in detail in the next chapter.

The final phase of these calculations consists of determining the necessary imbalance between the two streams of the counterflow and, later, all the flowrates listed in relative terms in brackets around the T–s diagram of Fig. 10.20. This phase of the analysis is based on invoking the first law. Taking as reference unit the flowrate of saturated vapor that leaves the two-phase separator, we obtain the results plotted in Fig. 10.23, namely the compressor flowrate $(1 + x_l + x_e)$ and the corresponding yield (x_l) and expander fraction (x_e). Letters (a)–(d) in this figure indicate the four heat-exchanger designs that helped trace the $N_{\text{tu}} \to \infty$ frontier in Fig. 10.22. Instructive in Fig. 10.23 is the "figure-of-merit" curve

$$\frac{\text{compressor power}}{\text{mass flowrate of liquefied gas}} = \frac{\dot{W}_c}{\dot{m}_l} \sim \frac{1 + x_l + x_e}{x_l} \qquad (10.69)$$

where $\dot{m}_l = x_l \dot{m}$. This curve has a minimum in the vicinity of design (b): it means that the brute-force approach of investing an infinite N_{tu} in the heat exchanger does not represent the end of the effort of minimizing heat-exchanger irreversibility. Moving from design (a) to design (d), one changes

a–d: heat-exchanger designs shown in Fig. 10.22

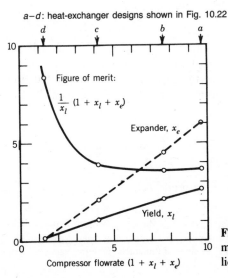

Figure 10.23 The existence of a minimum compressor flowrate per unit of liquefied nitrogen, $(1 + x_l + x_e)/x_l$.

a number of parameters[†] including the degree of parallelism (the mean angle) between the T_{P_H} and T_{P_L} lines of Fig. 10.21. That design (b) is the best among the four options (a)–(d) is a reflection of the fact that when N_{tu} is fixed, there exists an optimum orientation of the T_{P_H} line relative to the T_{P_L} line. This optimum was the basis for the constant-$\Delta T/T$ rule encountered earlier in this chapter. The same rule would have become visible again in Fig. 10.21 had the nitrogen behaved like a constant-c_P ideal gas all along the high-pressure track (3)–(7).

The Efficiency of Liquefiers and Refrigerators

How does the minimum $\dot{W}_c/\dot{m}_l$ ratio (10.69) compare with the ultimate minimum that would characterize a perfectly reversible process? First, we can write the $\dot{W}_c/\dot{m}_l$ ratio of eq. (10.69) in complete form by modeling the battery of room-temperature compressors and aftercoolers as an isothermal (T_0) ideal gas compressor with the efficiency η_c (Problem 10.3):

$$\dot{W}_c = \eta_c^{-1}\dot{m}_c RT_0 \ln \frac{P_H}{P_L} \tag{10.70}$$

Here $\dot{m}_c$ is the compressor flowrate $(1 + x_l + x_e)\dot{m}$; therefore, eq. (10.69) becomes

$$\frac{\dot{W}_c}{\dot{m}_l} = \frac{1 + x_l + x_e}{x_l} \eta_c^{-1} RT_0 \ln \frac{P_H}{P_L} \tag{10.71}$$

In the case of design (b), this ratio equals

$$\left(\frac{\dot{W}_c}{\dot{m}_l}\right)_{\text{design } (b)} = \frac{1695}{\eta_c} \text{ kJ/kg liquid N}_2 \tag{10.72}$$

Obviously, the price paid for 1 kilogram of liquefied gas increases if the compressor efficiency decreases.

The thermodynamic "floor" value of the $\dot{W}_c/\dot{m}_l$ ratio is simply the flow exergy of saturated liquid nitrogen at atmospheric pressure (Problem 10.10):

$$\left(\frac{\dot{W}_c}{\dot{m}_l}\right)_{\text{rev}} = e_{x,f}$$
$$= b_f(P_L) - b_0(T_0, P_L)$$
$$= 770 \text{ kJ/kg liquid N}_2 \tag{10.73}$$

[†]Related changes occur in the irreversibilities due to the throttling process (8)→(4), the intermediate expansion (3)→(6), and the low-temperature counterflow heat exchanger (7)–(6)–(5)–(8). The last is plagued by a gaping ΔT, because of the much greater c_P value of N_2 on the high-pressure side.

This theoretical value can be expressed also as the minimum mechanical power required per *volumetric* flowrate of liquid cryogen; in this case,

$$\left(\frac{\dot{W}_c}{v_f \dot{m}_l}\right)_{rev} = 173 \text{ W hr/liter of liquid } N_2 \qquad (10.74)$$

where v_f is the specific volume of saturated liquid nitrogen at atmospheric pressure, $0.001237 \text{ m}^3/\text{kg}$. The same limit can be put in dimensionless form by dividing the minimum power requirement by the refrigeration capacity provided by the vaporization of the $\dot{m}_l$ stream at temperature T_L:

$$\dot{Q}_L = \dot{m}_l h_{fg} \qquad (10.75)$$

namely,

$$\left(\frac{\dot{W}_c}{\dot{Q}_L}\right)_{rev} = 3.87 \text{ W/W} \qquad (10.76)$$

Table 10.2 [36] shows the theoretically minimum power required by the liquefaction of eight cryogenic fluids.

The second-law efficiency or utilization factor of liquefier design (*b*) is obtained by combining eqs. (10.72) and (10.73):

$$\eta_{II} = \frac{(\dot{W}_c/\dot{m}_l)_{rev}}{(\dot{W}_c/\dot{m}_l)_{\text{design }(b)}} = 0.45\eta_c \qquad (10.77)$$

The product $(100\eta_{II})$ is recognized also as the "percent Carnot" of the liquefaction process [37]. Taking $\eta_c \cong 0.7$ as representative of the efficiency of room-temperature compressors, we find that the highest η_{II} value among the $N_{tu} \to \infty$ designs is $\eta_{II} \cong 0.3$. This conclusion agrees with the efficiency of large-scale liquefiers, as indicated by the surveys of Strobridge and Chelton [36–38] and Fig. 10.24. In the case of liquefiers, the refrigeration capacity $\dot{Q}_L$ used on the abscissa of Fig. 10.24 is the load removed by the vaporiza-

TABLE 10.2 The Minimum Mechanical Power Required by the Liquefaction of Cryogenic Fluids[a]

Fluid	T_L [K]	$(\dot{W}_c/v_f \dot{m}_l)_{rev}$ [W hr/liter of liquid]	$(\dot{W}_c/\dot{Q}_L)_{rev}$ [W/W]
Helium	4.2	236	326
Hydrogen	20.4	278	31.7
Neon	27.1	447	15.5
Nitrogen	77.4	173	3.87
Fluorine	85	238	3.26
Argon	87.3	185	2.95
Oxygen	90.2	195	2.89
Methane	111.5	129	2.15

[a]After Strobridge [36].

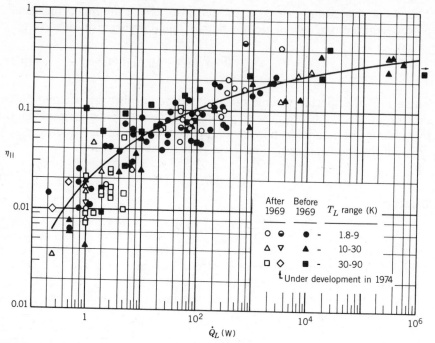

Figure 10.24 Compilation of second-law efficiencies of refrigerators and liquefiers (after Strobridge [36]).

tion of the $\dot{m}_l$ stream, eq. (10.75). On the other hand, in the case of refrigerators, $\dot{Q}_L$ is simply the load that the machine can steadily remove from T_L.

The trend revealed by the data assembled in Fig. 10.24 is one where the performance of refrigerators and liquefiers deteriorates as they become smaller. This trend is due to the greater role played by heat leaks as the machine decreases in size. For example, remembering the counterflow heat exchanger as one heat-leak avenue, we shall see in the next chapter that the minimum irreversibility of a heat exchanger varies as the inverse of the linear dimension of the heat exchanger (p. 621).

Now we are in a position to bring back the controversy that was effectively put to rest by the publication of Strobridge's very important display, shown here as η_{II} vs. $\dot{Q}_L$. About Fig. 10.24, Strobridge wrote [36, 38]:

> Historically the contention has been that higher temperature refrigerators (or liquefiers) are more efficient. The data for refrigeration temperatures between 10 and 30 K (and 30 to 90 K) refute that notion at least when presented on this common basis. To be sure, less input power is required to produce the same number of watts of cooling at higher temperatures, but the losses relative to ideal are proportionally the same.

Although not wrong, Strobridge's conclusion regarding the lack of effect of T_L on η_{II} is inappropriate, because if there is one effect that has been removed intentionally from Fig. 10.24, it is the effect of T_L (note the wide temperature ranges listed for the actual T_L values of the surveyed liquefiers and refrigerators). Whether T_L has an effect on η_{II} can only be decided by allowing T_L to vary while comparing machines of roughly the same size. Lacking better data, I used the η_{II} values that I could read from Fig. 10.24 in the vertical column of width $10^2 \le \dot{Q}_L \le 10^5$ W. I sorted these values according to the temperature ranges 1.8–9 K, 10–30 K, and 30–90 K, and obtained an average η_{II} for each range. The three average values calculated in this manner are shown as horizontal bars in Fig. 10.25: in each case, the extremities of the bar indicate the width of the T_0/T_L range to which the calculated η_{II} value refers. The figure shows a definite trend of higher η_{II} values toward lower T_0/T_Ls. Higher-temperature refrigerators have higher second-law efficiencies indeed.

The challenge now is to predict the downward trend of Fig. 10.25 theoretically. Consider the steady-state refrigeration-plant model in Fig. 10.26. According to this model, the plant owes its irreversibility solely to the "internal" heat leak $\dot{Q}_i$ that flows through the machine "vertically," i.e., from the ambient temperature level T_0 to the cold-end temperature T_L. In an actual refrigeration plant, $\dot{Q}_i$ represents the combined effect of vertical heat leaks through components such as the main counterflow heat exchanger, the supports that tie the cold end (T_L) mechanically to the room-temperature frame of the machine, and the thermal-insulation system sandwiched between T_L and T_0.

In a machine of fixed size, the simplest model for the combined internal heat-leak effect $\dot{Q}_i$ amounts to writing:

$$\dot{Q}_L = C_i(T_0 - T_L) \tag{10.78}$$

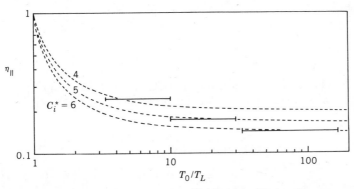

Figure 10.25 Empirical data (horizontal bars) showing that η_{II} decreases as T_0/T_L increases, and the theoretical curves based on the model of Fig. 10.26 (the dashed lines).

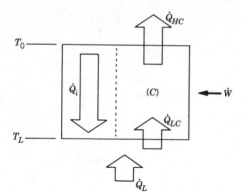

Figure 10.26 Model of refrigeration plant with internal heat transfer irreversibility.

where the constant C_i is the internal conductance of the refrigeration plant. One could, of course, refine equation (10.78) by raising the temperature difference $(T_0 - T_L)$ to a certain exponent, if radiation heat leaks and temperature-dependent conductivities are dominant in the evaluation of $\dot{Q}_i$. It turns out that the simple linear model (10.78) captures enough of the heat transfer reality of the refrigeration plant in order to explain the downward trend revealed by Fig. 10.25.

The model of Fig. 10.26 consists of a Carnot refrigerator (C) working in parallel with the internal conductance discussed above. Applied only to the Carnot refrigerator, the Second Law of Thermodynamics states that

$$\frac{\dot{Q}_{HC}}{T_0} = \frac{\dot{Q}_{LC}}{T_L} \tag{10.79}$$

in which $\dot{Q}_{LC}$ is to be distinguished from the refrigeration capacity (load) $\dot{Q}_L$ which is fixed, $\dot{Q}_{LC} = \dot{Q}_L + \dot{Q}_i$. Using the second law (10.79) and noting also that the refrigerator power input $\dot{W}$ is equal to $(\dot{Q}_{HC} - \dot{Q}_{LC})$, the coefficient of performance of the overall plant is simply

$$\text{COP} = \frac{\dot{Q}_L}{\dot{W}} = \frac{1 - \dot{Q}_i/\dot{Q}_{LC}}{T_0/T_L - 1} \tag{10.80}$$

The Carnot limit of this coefficient is $(\text{COP})_C = (T_0/T_L - 1)^{-1}$; therefore, the second-law efficiency of the actual plant modeled in Fig. 10.26 is

$$\eta_{II} = \frac{\text{COP}}{(\text{COP})_C} = 1 - \frac{\dot{Q}_i}{\dot{Q}_{LC}}$$

$$= [1 + C_i^*(1 - T_L/T_0)]^{-1} \tag{10.81}$$

In this final result, C_i^* is a dimensionless version of the internal conductance introduced first in equation (10.78), namely,

$$C_i^* = C_i \frac{T_0}{\dot{Q}_L} \qquad (10.81')$$

It is already evident from the analytical form of equation (10.81) that η_{II} decreases monotonically as the refrigeration temperature level T_L decreases. The same point is made graphically in Fig. 10.25: we see that a constant-C_i^* curve whose heat-leak number C_i^* is of order 5 passes right through the band of η_{II} values calculated based on the data of Fig. 10.24. The second-law efficiency decreases monotonically also as C_i^* increases (this trend should be expected, as leakier refrigerated enclosures make less-efficient refrigeration plants).

One remarkable discovery made possible by Fig. 10.25 is that the statement

$$C_i^* = \text{constant of order 5}$$

recommends itself as an empirical scaling law of the large population of built refrigeration plants compiled by Strobridge and Chelton [36–38]. This law states that the built-in conductance C_i scales with the refrigeration capacity $\dot{Q}_L$ and that the combined internal leak $\dot{Q}_i$ is consistently of the order of five times the refrigeration load $\dot{Q}_L$.

MAGNETIC REFRIGERATION

The structure of the preceding coverage of refrigeration concepts and methods is one that highlights the progress toward lower temperatures. The same structure can be interpreted in at least two other ways. From a practical engineering standpoint, it shows persistence in the face of increasingly greater obstacles posed by the thermodynamic irreversibilities as the refrigeration temperature decreases. Note, for example, the monotonic relationship between η_{II} and T_L in Fig. 10.25. The same structure can also be appreciated from a historical point of view because, as is illustrated in Fig. 10.27, the refrigeration history of the past century is also a history of steady progress toward absolute zero.

The technological breakthrough that is responsible for the modern work at temperatures below 1 K is Debye and Giauque's independent invention of "magnetic refrigeration" or, more specifically, the process of adiabatic demagnetization [39, 40]. This development is in a way analogous to Claude's use of an intermediate-temperature expander in order to provide to the refrigeration load another exit out of the cold zone of the machine. In the process of adiabatic demagnetization, the "working substance" that occupies the coldest region of the machine is allowed to do work on its environment. The work output is of the magnetic kind, while the working substance is not a fluid but a solid (a paramagnetic salt). Accounts of the

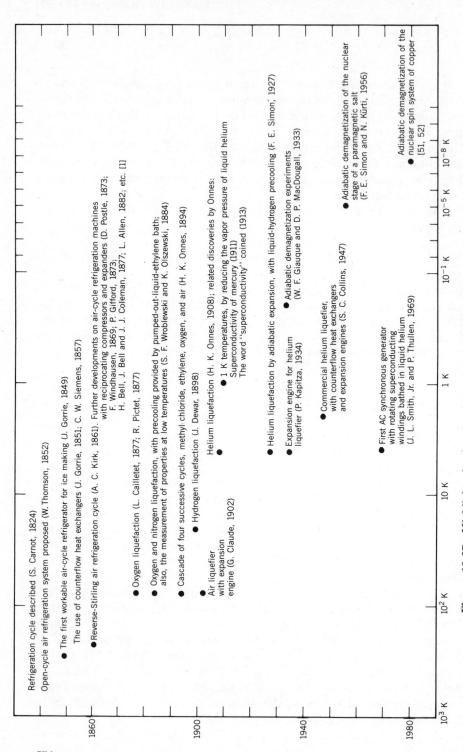

Figure 10.27 Highlights in the development of refrigeration technology and the consequent progress toward absolute zero.

basic process and the theory behind it are available in the low-temperature physics literature (e.g., Casimir [41] and Zemansky [42]).

Although intended originally for small-scale physics experiments at extremely low temperatures, the possible applicability of the magnetic refrigeration method to the design of large-scale systems draws increasing attention from the engineering community [43–45]. In this section, we study the thermodynamic basis for a cooling effect via adiabatic demagnetization. An additional incentive for undertaking this study is that it constitutes a compact review of some of the analytical steps described in chapter 4.

Fundamental Relations

Consider a certain amount of solid paramagnetic material, that is, a substance (or body) that in the absence of a magnetic field is not a magnet. If this substance is immersed in a magnetic field, its permeability does not depart appreciably from unity. From the theory of electromagnetism, we know that during a process of magnetization, the microscopic magnetic dipole moments of the material oppose changes made in their orientation, hence, the infinitesimal magnetic work transfer interaction:

$$\delta W^* = - \mathcal{H} \, d\mathcal{M} \tag{10.82}$$

where $\mathcal{H}$ is the intensity of the applied magnetic field, and where $d\mathcal{M}$ is the increase in the magnetization (i.e., total magnetic moment) of the material. The minus sign in eq. (10.82) is contrary to the convention used in low-temperature physics: it is however consistent with the heat-engine sign convention of engineering thermodynamics, and it states that in order to increase the magnetization of our system ($d\mathcal{M} > 0$), work must be done on the system ($\delta W^* < 0$). The work is done by the entity that creates the magnetic field of intensity $\mathcal{H}$. If that entity is an electrical coil powered by a battery, then, as our system experiences the magnetization increase $d\mathcal{M}$, the coil witnesses an electromotive force that acts toward diminishing the work dissipated by Joule heating in the coil, while the total electrical work done by the battery does not change. A $d\mathcal{M}$-size portion of the battery work that would have been dissipated by the coil is now transferred to the paramagnetic-solid system.

The system possesses three modes of reversible energy transfer interactions, heat transfer ($T \, dS$), mechanical work transfer ($P \, dV$), and magnetic work transfer ($- \mathcal{H} \, d\mathcal{M}$). In the immediate vicinity of an equilibrium state, we can write

$$dU^* = T \, dS - P \, dV + \mathcal{H} \, d\mathcal{M} \tag{10.83}$$

where U^* is the internal energy of the paramagnetic solid. The processes that are described next occur at atmospheric pressure; they are also accompanied by negligible changes in the volume of our solid; therefore, we retain the simpler form

$$dU^* = T \, dS + \mathcal{H} \, d\mathcal{M} \qquad (10.84u)$$

Recognizing a fundamental relation of type $U^*(S, \mathcal{M})$ behind eq. (10.84u), it is a simple matter to repeat the string of Legendre transforms outlined in Table 4.3 and obtain the corresponding differential forms for the *magnetic enthalpy*, Helmholtz free-energy, and Gibbs free-energy functions:

$$H^*(S, \mathcal{H}) = U^* - \mathcal{H}\mathcal{M} \qquad dH^* = T \, dS - \mathcal{M} \, d\mathcal{H} \qquad (10.84h)$$

$$F^*(T, \mathcal{M}) = U^* - TS \qquad dF^* = -S \, dT + \mathcal{H} \, d\mathcal{M} \qquad (10.84f)$$

$$G^*(T, \mathcal{H}) = U^* - TS - \mathcal{H}\mathcal{M} \qquad dG^* = -S \, dT - \mathcal{M} \, d\mathcal{H} \qquad (10.84g)$$

Finally, examining the partial-differential coefficients in the differential forms (10.84u)–(10.84g), we deduce the corresponding "Maxwell's relations" of the system:

$$\left(\frac{\partial T}{\partial \mathcal{M}}\right)_S = \left(\frac{\partial \mathcal{H}}{\partial S}\right)_{\mathcal{M}} \qquad (10.85u)$$

$$\left(\frac{\partial T}{\partial \mathcal{H}}\right)_S = -\left(\frac{\partial \mathcal{M}}{\partial S}\right)_{\mathcal{H}} \qquad (10.85h)$$

$$-\left(\frac{\partial S}{\partial \mathcal{M}}\right)_T = \left(\frac{\partial \mathcal{H}}{\partial T}\right)_{\mathcal{M}} \qquad (10.85f)$$

$$\left(\frac{\partial S}{\partial \mathcal{H}}\right)_T = \left(\frac{\partial \mathcal{M}}{\partial T}\right)_{\mathcal{H}} \qquad (10.85g)$$

One purpose of this analytical apparatus is to show that during a process of isentropic demagnetization, the system can be expected to experience a drop in temperature. Two additional measurable quantities are useful at this juncture, the heat capacity at constant magnetization ($C_{\mathcal{M}}$) and the heat capacity at constant intensity ($C_{\mathcal{H}}$). The former is defined as the incremental heat transfer $(\Delta Q)_{\mathcal{M}}$ responsible for the temperature rise ΔT while $\mathcal{M}$ remains fixed [i.e., during a zero-work process, eq. (10.84u)]; hence, $(\Delta Q)_{\mathcal{M}} = (\Delta U^*)_{\mathcal{M}}$ and the partial-differential definition:

$$C_{\mathcal{M}} = \left(\frac{\Delta Q}{\Delta T}\right)_{\mathcal{M}} = \left(\frac{\partial U^*}{\partial T}\right)_{\mathcal{M}} \qquad (10.86)$$

Alternatively, since $(\Delta Q)_{\mathcal{M}} = T(\Delta S)_{\mathcal{M}}$, the heat capacity at constant magnetization is defined also as

$$C_{\mathcal{M}} = T\left(\frac{\partial S}{\partial T}\right)_{\mathcal{M}} \qquad (10.87)$$

The heat capacity at constant intensity $C_{\mathscr{H}}$ is defined as the heat transfer increment needed to raise the system's temperature by an amount ΔT while $\mathscr{H}$ is constant. The heating is accompanied by magnetic work transfer $(\Delta Q)_{\mathscr{H}} = (\Delta U^*)_{\mathscr{H}} - \mathscr{H}(\Delta M)_{\mathscr{H}} = \Delta(H^*)_{\mathscr{H}}$; therefore,

$$C_{\mathscr{H}} = \left(\frac{\Delta Q}{\Delta T}\right)_{\mathscr{H}} = \left(\frac{\partial H^*}{\partial T}\right)_{\mathscr{H}} \tag{10.88}$$

More useful is the second definition resulting from $(\Delta Q)_{\mathscr{H}} = T(\Delta S)_{\mathscr{H}}$, namely,

$$C_{\mathscr{H}} = T\left(\frac{\partial S}{\partial T}\right)_{\mathscr{H}} \tag{10.89}$$

It is now possible to calculate and plot the changes in U^* and S based on heat-capacity information and the measured magnetization curves at known temperatures, $M = M(\mathscr{H}, T)$. For the magnetic internal energy U^*, eq. (10.84u) yields directly

$$\left(\frac{\partial U^*}{\partial \mathscr{H}}\right)_T = T\left(\frac{\partial S}{\partial \mathscr{H}}\right)_T + \mathscr{H}\left(\frac{\partial M}{\partial \mathscr{H}}\right)_T \tag{10.90}$$

and, using the fourth of Maxwell's relations, eq. (10.85g),

$$U^*(\mathscr{H}, T) - U^*(0, T) = \int_0^{\mathscr{H}} \left[T\left(\frac{\partial M}{\partial T}\right)_{\mathscr{H}} + \mathscr{H}\left(\frac{\partial M}{\partial \mathscr{H}}\right)_T \right] d\mathscr{H} \tag{10.91}$$

As a generalization of the heat transfer expressions mentioned in the derivation of eqs. (10.86) and (10.88), the heat transfer during an infinitesimal change of state is (Problem 10.11)

$$\delta Q = C_{\mathscr{H}}\, dT + T\left(\frac{\partial M}{\partial T}\right)_{\mathscr{H}} d\mathscr{H} \tag{10.92}$$

By writing $\delta Q = T\, dS$, the corresponding formula for the entropy change is

$$dS = C_{\mathscr{H}} \frac{dT}{T} + \left(\frac{\partial M}{\partial T}\right)_{\mathscr{H}} d\mathscr{H} \tag{10.93}$$

An alternate dS formula is obtained by considering S as a function of T and M (Problem 10.11):

$$dS = C_M \frac{dT}{T} - \left(\frac{\partial \mathscr{H}}{\partial T}\right)_M dM \tag{10.94}$$

Figure 10.28 shows the shape of the calculated $S(T, \mathscr{H})$ curves for a paramagnetic salt like gadolinium sulphate or gadolinium gallium garnet, at temperatures below 1 K. An extensive comparison of the various properties of paramagnetic materials contemplated for magnetic refrigeration systems has been provided by Barclay and Steyert [43].

Adiabatic Demagnetization

We are interested in the behavior of the system's temperature during an isentropic deintensification of the imposed field, that is, when $dS = 0$ and $d\mathcal{H} < 0$. For such a process, eq. (10.91) recommends

$$dT = \frac{T}{C_{\mathcal{H}}} \left(-\frac{\partial \mathcal{M}}{\partial T} \right)_{\mathcal{H}} d\mathcal{H} \qquad (10.95)$$

It is found experimentally that the magnetization decreases as T increases at constant $\mathcal{H}$:

$$\left(\frac{\partial \mathcal{M}}{\partial T} \right)_{\mathcal{H}} < 0 \qquad (10.96)$$

therefore, the message of eq. (10.95) is that the sign of dT is always the same as the sign of $d\mathcal{H}$. When $\mathcal{H}$ drops rapidly enough so that the system does not experience a significant heat transfer interaction, the temperature T drops also. Thinking again of the Claude-expander analogy referred to at the start of this section, it is worth noting that during a process with $dS = 0$ and $d\mathcal{H} < 0$, the system *delivers* work to its environment (Problem 10.12).

A refrigeration system with a low-temperature magnetic work-producing stage functions according to the process $(0) \rightarrow (1) \rightarrow (2)$ shown in Fig. 10.28. During the first part of the process, the system is magnetized isothermally, as the intensity $\mathcal{H}$ increases from zero to $\mathcal{H}_1$. The system releases heat:

$$Q_{0-1} = T_1 \int_0^{\mathcal{H}_1} \left(\frac{\partial \mathcal{M}}{\partial T} \right)_{\mathcal{H}} d\mathcal{H} < 0 \qquad (10.97)$$

which is removed by contact with the cold extremity of a conventional refrigerator residing above T_1. Between states (1) and (2), the magnetic

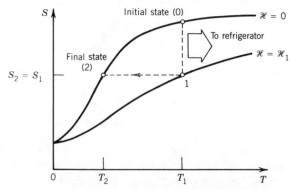

Figure 10.28   The orientation of the constant-$\mathcal{H}$ lines of the T–S plane of a paramagnetic salt, and the position of a process of adiabatic demagnetization.

field is switched off and the temperature decreases isentropically to the new value

$$T_2 = T_1 - \int_2^1 \frac{T}{C_{\mathscr{H}}} \left(-\frac{\partial \mathscr{M}}{\partial T} \right)_{\mathscr{H}} d\mathscr{H} , \quad \text{i.e.,} \quad T_2 < T_1 \qquad (10.98)$$

Beyond state (2), the mass of paramagnetic salt can be used as heat sink in a variety of applications that require refrigeration at temperatures below 1 K, especially in physics experiments concerning the properties of matter at extremely low temperatures. Even when not used, the system acts as heat sink for the heat leak that eventually reaches down to T_2.

Paramagnetic Thermometry

A very interesting question concerns the temperature of the newly established state (2). Indeed, the original mission of the adiabatic demagnetization scheme was to extend man's reach toward absolute zero, in a low-temperature domain that had not been accessible to those using liquid-helium refrigeration alone. Prior to Giauque and MacDougall's first demag-netization experiments in Berkeley [46, 47], the lowest temperatures that could be reached with normal liquid helium ("Helium-4") were of the order of 1 K. The standard technique for achieving such temperatures is to pump out the vapor from the space above a pool of liquid helium, in other words, to lower the vapor pressure and, with it, the saturation temperature. Nowadays, it is possible to obtain even lower temperatures (~ 0.3 K) by lowering the vapor pressure of the light-helium isotope "Helium-3." Compared at the same pressure, the boiling point of Helium-3 is substantially lower than that of Helium-4 (Fig. 10.29): for example, at a vapor pressure of 1 torr, these temperatures are $T(\text{He}^4) = 1.27$ K and $T(\text{He}^3) = 0.668$ K [48]. Figure 10.29 makes the additional point that under certain conditions (Problem 6.2), the relationship between $\ln(P_{sat})$ and $1/T$ is practically linear.

If the T_1 temperature in Fig. 10.28 is a boiling liquid-helium temperature of order 1 K, then the experimentalist faces the challenge of deducing the absolute temperature T_2 that is being reached at the end of the adiabatic demagnetization. This ultimate temperature measurement is made possible by a special property of paramagnetic substances, a property that makes them thermometric materials at low temperatures in the same sense as the ideal gas behavior makes helium gas a thermometric fluid at higher temperatures. An *ideal paramagnetic substance* is one whose properties obey the equation of state:

$$\mathscr{M} = f\left(\frac{\mathscr{H}}{T} \right) \qquad (10.99)$$

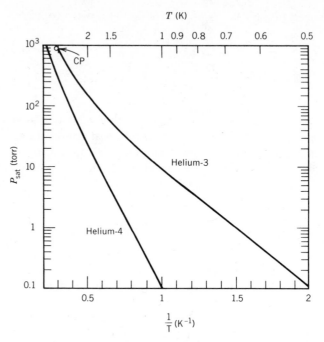

Figure 10.29 The saturation pressure of liquid Helium-3 and Helium-4.

For a number of paramagnetic salts at temperatures not much different than 1 K, this equation takes an even simpler form known as *Curie's law*:

$$\mathcal{M} = C_C \frac{\mathcal{H}}{T} \tag{10.100}$$

where C_C is Curie's constant. Curie's law states that the susceptibility of the material $(\mathcal{M}/\mathcal{H})$ varies as the inverse of the absolute temperature.

With reference to the final state (2) of Fig. 10.28, it is clear that if the unknown T_2 is sufficiently high so that Curie's law still applies, we can simply invoke eq. (10.100) and calculate the final temperature:

$$T_2 = \frac{C_C}{(\mathcal{M}/\mathcal{H})_2} \quad \text{(if Curie's law applies)} \tag{10.101}$$

In general, Curie's law breaks down as we progress toward lower temperatures. Even though in such cases we cannot use it to calculate T_2 directly, we can still use it to define an intermediate quantity (T^*) called *magnetic temperature*, in this case,

$$T_2^* = \frac{C_C}{(\mathcal{M}/\mathcal{H})_2} \quad \text{(if Curie's law does not apply)} \tag{10.102}$$

As far as the actual thermodynamic properties of state (2) are concerned, we note that we can calculate its entropy by using eq. (10.97):

$$S_2 = S_1 = S_0 - \int_0^{\mathcal{H}_1} \left(-\frac{\partial \mathcal{M}}{\partial T} \right)_{\mathcal{H}} d\mathcal{H} \qquad (10.103)$$

We can also evaluate its energy relative to that of state (0), by using calorimetry and the first-law statement for the zero-work process $(2) \rightarrow (0)$ along the $\mathcal{H} = 0$ curve of Fig. 10.28:

$$U_2^* = U_0^* - Q_{2-0} \qquad (10.104)$$

Summing up, for the state (2), we can calculate T_2^*, S_2, and U_2^*. By repeating the adiabatic demagnetization experiment for different values of maximum field intensity $\mathcal{H}_1$, it is possible to travel up and down the $\mathcal{H} = 0$ curve and calculate (T^*, S, U^*) for each state of type (2). Based on this information, at a state of type (2), we can also calculate the slopes $\partial U^*/\partial T^*$ and $\partial S/\partial T^*$. Finally, we recall eq. (10.84u) and note that for any state (2), the true absolute temperature is

$$T = \frac{\partial U^*}{\partial S} \qquad (\mathcal{H} = 0) \qquad (10.105)$$

therefore, in terms of the information supplied by experiment,

$$T = \frac{\partial U^*/\partial T^*}{\partial S/\partial T^*} \qquad (\mathcal{H} = 0) \qquad (10.106)$$

The method outlined between eqs. (10.102) and (10.106) produces a relationship between the provisional temperature T^*, which is easier to calculate, and the desired absolute temperature T. This relationship is not unique, because the susceptibility $(\mathcal{M}/\mathcal{H})_2$ of eq. (10.102) depends on the geometrical shape of the paramagnetic salt sample. A unique relationship does exist between T and the T^* values of a spherical sample of a certain (fixed type of) paramagnetic salt. The T^* values calculated using spherical "paramagnetic thermometers" are also labeled as $T^{\circledR}$. When the paramagnetic thermometer employs an ellipsoidal salt sample, the calculated T^* value can readily be converted into a corresponding $T^{\circledR}$ value that would have been measured using a perfectly spherical sample [48].

Giauque and MacDougall's [46, 47] first adiabatic demagnetization experiments produced temperatures of roughly 0.2 K. The work in this area sprung simultaneously in Holland [49] and England [50]. It has evolved considerably en route to record low temperatures of about $(5)10^{-8}$ K [51, 52]. These milestones are listed also in Fig. 10.27.

The Third Law of Thermodynamics

The geometry of the two constant-$\mathcal{H}$ lines shown in Fig. 10.28 is an excellent opportunity for reviewing the physical evidence that supports the "Third Law" of Thermodynamics. One question that arises after the study of the magnetic refrigeration scheme $(0) \rightarrow (1) \rightarrow (2)$ is whether such processes can be used to reach absolute zero. We are certainly free to theorize that processes of type $(0) \rightarrow (1) \rightarrow (2)$ can be repeated in the direction of lower temperatures, as shown by the steps nested between the $\mathcal{H} = 0$ and $\mathcal{H} = \mathcal{H}_1$ lines in Fig. 10.30. The assumption made here is that during each subsequent isothermal magnetization process [e.g., $(2) \rightarrow (1')$], the heat rejected by the "working" paramagnetic salt is absorbed by a sufficiently large thermal mass of temperature T_2—the latter being the result of the first magnetic refrigeration stage $(0) \rightarrow (1) \rightarrow (2)$.

The answer that emerges from the geometric construction of Fig. 10.30 is that the temperature $T = 0$ cannot be reached through a finite sequence of operations. This answer is "geometric" because it stems from the fact that the constant-$\mathcal{H}$ lines intersect at $T = 0$. The same geometric feature can be expressed as

$$\lim_{T \to 0} (\Delta S)_T = 0 \qquad (10.107)$$

where $(\Delta S)_T$ is shorthand for the magnitude of the isothermal entropy change experienced by the substance, for example,

$$(\Delta S)_T = |S_0 - S_1| \qquad (10.108)$$

The statement that summarizes the observation made above is [53]

It is impossible by any procedure, no matter how idealized, to reduce any system to the absolute zero in a finite number of operations.

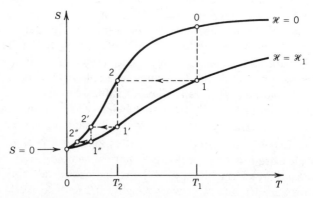

Figure 10.30 The unattainability of absolute zero by means of a finite sequence of isothermal magnetization and adiabatic demagnetization processes of the type illustrated in Fig. 10.28.

This statement is largely based on the work of Simon [53–55], who did much to clarify the limitations of the 1906 theoretical development known as *Nernst's Theorem* [56] (see also Planck [57]). With direct reference to processes of type $(1) \rightarrow (2)$ in Fig. 10.30, the principle of the unattainability of absolute zero can also be stated as [58]

No reversible adiabatic process starting at nonzero temperature can possibly bring a system to zero temperature.

A more analytical third-law alternative is to begin with eq. (10.107) and the intersection of the constant-$\mathcal{H}$ lines at absolute zero in Fig. 10.30, and to recognize the uniqueness of the entropy value at the point of intersection on the $T = 0$ isotherm. Since Planck [57], the entropy at absolute zero temperature was assigned the value zero; therefore, an alternative third-law statement is

$$S = 0 \text{ at } T = 0 \tag{10.109}$$

It means that for any system at equilibrium, the $T = 0$ isotherm coincides with the $S = 0$ adiabat. Related to Planck's statement (10.109) is Callen's postulate IV in his axiomatic construction of equilibrium thermodynamics [Ref. 58, p. 30]; in the case of the paramagnetic system of Figs. 10.28 and 10.30, the fourth postulate states that

$$S = 0 \text{ when } \left(\frac{\partial U^*}{\partial S} \right)_{\mathcal{M}} = 0 \tag{10.110}$$

in other words, when $T = 0$. Relative to the class of simple systems analyzed in chapter 4, this postulate becomes (Fig. 2.7)

$$S = 0 \text{ when } \left(\frac{\partial U}{\partial S} \right)_{V, N_1, \ldots, N_n} = 0 \tag{10.111}$$

where $N_1, \ldots, N_n$ represent the numbers of moles in the system. The reflection of the Third Law of Thermodynamics in the behavior of certain thermodynamic properties in the $T \rightarrow 0$ limit is illustrated in Problem 10.14.

SYMBOLS

A	heat-exchanger (heat transfer) area
A_c	cross-sectional area
b	constant [eq. (10.34)]
b	flow availability [eq. (10.73)]
c_v	specific heat at constant volume
c_P	specific heat at constant pressure

C_C	Curie's constant
$C_{\mathscr{H}}$	heat capacity at constant $\mathscr{H}$
C_i	internal conductance
C_i^*	dimensionless internal conductance [eq. (10.81')]
$C_{\mathscr{M}}$	heat capacity at constant $\mathscr{M}$
COP	coefficient of performance of a refrigerator
$(\mathrm{COP})_C$	coefficient of performance of a Carnot refrigerator
e_x	flow exergy [eq. (10.73)]
F	radiation view factor [eq. (10.52)]
F^*	magnetic Helmholtz free energy
G^*	magnetic Gibbs free energy
h	specific enthalpy
$\bar{h}$	heat transfer coefficient
h_{fg}	latent heat of vaporization
H^*	magnetic enthalpy
$\mathscr{H}$	magnetic field intensity
I	electric current
k	specific-heat ratio c_P/c_v
k	thermal conductivity [eq. (10.44)]
k_0	constant thermal conductivity
L	length
L_0	constant [eq. (10.59)]
$\dot{m}$	mass flowrate
$\mathscr{M}$	magnetization (total magnetic moment)
N	number of shield-to-shield gaps (Fig. 10.13f)
$N_1, \ldots, N_n$	numbers of moles [eq. (10.111)]
N_S	entropy-generation number [eq. (10.51)]
N_{tu}	number of heat transfer units
P	pressure
ΔP	pressure drop
P_H	high pressure
P_L	low pressure
Q	heat transfer interaction
$\dot{Q}$	heat transfer rate
$\dot{Q}_{HC}$	heat rejection rate from the Carnot compartment of Fig. 10.26
$\dot{Q}_i$	internal heat leak
$\dot{Q}_{\mathrm{leak}}$	convective heat leak [eqs. (10.7) and (10.23)]
$\dot{Q}_L$	refrigeration load
$\dot{Q}_{LC}$	refrigeration load felt by the Carnot compartment of Fig. 10.26
R	ideal gas constant
s	specific entropy
S	entropy
$(\Delta S)_T$	entropy change at constant temperature

$\dot{S}_{gen}$	entropy-generation rate
T	absolute temperature
ΔT	temperature difference
T^*	magnetic temperature
$T^{\circledast}$	magnetic temperature measured with spherical paramagnetic salt sample
T_L	refrigeration load temperature
T_{P_L}	temperature distribution along the low-pressure side
T_{P_H}	temperature distribution along the high-pressure side
T_0	ambient (room) temperature
U	internal energy
U^*	magnetic internal energy
v	specific volume
V	volume
$\dot{W}$	mechanical power
W^*	magnetic work transfer interaction
x	longitudinal coordinate
x_e	mass flowrate fraction passing through the expander
x_l	mass flowrate fraction transformed into liquid
β	coefficient of thermal expansion
η_c	efficiency of compressor
η_e	expander isentropic efficiency
η_L	efficiency of cold-end isothermal expander
η_0	efficiency of room-temperature isothermal compressor
η_{II}	second-law efficiency [eq. (10.77)]
μ_J	Joule–Thomson expansion coefficient
ν	function [eq. (10.36)]
ρ_e	electrical resistivity
σ	Stefan–Boltzmann constant
χ_m	isothermal magnetic susceptibility [Problem 10.12]
$(\ )_c$	compressor
$(\ )_e$	expander
$(\ )_i$	radiation shield index
$(\ )_L$	refrigeration load, cold-end heat exchanger
$(\ )_s$	solid, shield
$(\ )_0$	warm-end, room-temperature quantity
$(\ )_{max}$	maximum
$(\ )_{min}$	minimum
$(\ )_{opt}$	optimum

REFERENCES

1. S. C. Collins and R. L. Cannaday, *Expansion Machines for Low Temperature Processes*, Oxford University Press, London and New York, 1958.

2. J. Gorrie, Apparatus for the artificial production of ice, in tropical climates, U.S. Patent 8080, 1851.

3. C. W. Siemens, Improvements in refrigerating and producing ice, and in apparatus or machinery for that purpose, British Patent 2064, 1857.

4. T. R. Strobridge, Refrigeration, in R. H. Kropschot, B. W. Birmingham, and D. B. Mann, eds., *Technology of Liquid Helium*, Monogr. No. 111, National Bureau of Standards, Washington, DC, 1968, Chapter 4.

5. G. Claude, Sur la liquéfaction de l'air par détente avec travail extérieur récupérable, *C. R. Hebd. Seances Acad. Sci.*, Vol. 134, 1902, pp. 1568–1571.

6. C. W. P. Heylandt, Expansion-engine for producing low temperature, U.S. Patent 1,019,791, 1912.

7. R. C. Muhlenhaupt and T. R. Strobridge, An analysis of the Brayton cycle as a cryogenic refrigerator, *NBS Tech. Note (U.S.)*, No. 366, 1968.

8. A. Bejan, The concept of irreversibility in heat exchanger design: Counterflow heat exchangers for gas-to-gas applications, *J. Heat Transfer*, Vol. 99, 1977, pp. 374–380.

9. A. Bejan, *Entropy Generation Through Heat and Fluid Flow*, Wiley, New York, 1982, pp. 138–141.

10. M. Thirumaleshwar, Exergy method of analysis and its application to a helium cryorefrigerator, *Cryogenics*, Vol. 19, 1979, pp. 355–361.

11. A. Bejan, A general variational principle for thermal insulation system design, *Int. J. Heat Mass Transfer*, Vol. 22, 1979, pp. 219–228.

12. A. Bejan, Second law analysis in heat transfer and thermal design, *Adv. Heat Transfer*, Vol. 15, 1982, pp. 1–58.

13. A. Bejan and J. L. Smith, Jr., Thermodynamic optimization of mechanical supports for cryogenic apparatus, *Cryogenics*, Vol. 14, 1974, pp. 158–163.

14. A. Bejan, Improved thermal design of the cryogenic cooling system for the rotor of a superconducting generator, Ph.D. thesis, Department of Mechanical Engineering, Massachusetts Institute of Technology, Cambridge, MA, 1975.

15. A. Bejan and J. L. Smith, Jr., Heat exchangers for vapor-cooled conducting supports of cryostats, *Adv. Cryog. Eng.*, Vol. 21, 1976, pp. 247–256.

16. P. Grassmann and J. Kopp, Zur günstingsten Wahl der Temperaturdifferenz und der Wärmeübergangszahl in Wärmeaustauschern, *Kaeltetechnik*, Vol. 9, 1957, pp. 306–308.

17. C. Trepp, Refrigeration systems for temperatures below 25°K with turboexpanders, *Adv. Cryog. Eng.*, Vol. 7, 1962, pp. 251–261.

18. J. Szargut, International progress in second law analysis, *Energy*, Vol. 5, 1980, pp. 709–718.

19. R. B. Scott, *Cryogenic Engineering*, Van Nostrand, Princeton, NJ, 1959.

20. M. A. Hilal and R. W. Boom, Optimization of mechanical supports for large superconductive magnets, *Adv. Cryog. Eng.*, Vol. 22, 1977, pp. 224–232.

21. A. Bejan, Discrete cooling of low heat leak supports to 4.2 K, *Cryogenics*, Vol. 15, 1975, pp. 290–292.

22. M. A. Hilal and Y. M. Eyssa, Minimization of refrigeration power for large cryogenic systems, *Adv. Cryog. Eng.*, Vol. 25, 1980, pp. 350–357.

23. W. Schultz and A. Bejan, Exergy conservation in parallel thermal insulation systems, *Int. J. Heat Mass Transfer*, Vol. 26, 1983, pp. 335–340.

24. V. S. Martynovskii, V. T. Cheilyakh, and T. N. Shnaid, Thermodynamic effectiveness of a cooled shield in vacuum low-temperature insulation, *Energ. Transp.*, Vol. 2, 1971.

25. J. E. Ahern, *The Exergy Method of Energy Systems Analysis*, Wiley, New York, 1980, pp. 263–270.

26. Y. M. Eyssa and O. Okasha, Thermodynamic optimization of thermal radiation shields for cryogenic apparatus, *Cryogenics*, Vol. 18, 1978, pp. 305–307.

27. P. J. Murto, A gas-shielded storage and transport vessel for liquid helium, *Adv. Cryog. Eng.*, Vol. 7, 1962, pp. 291–295.

28. J. A. Paivanas, A. W. Francis, and D. I.-J. Wang, Insulation construction, U.S. Patent 3,133,422, May 19, 1964.

29. J. A. Paivanas, O. P. Roberts, and D. I.-J. Wang, Multishielding—an advanced superinsulation technique, *Adv. Cryog. Eng.*, Vol. 10, 1965, pp. 197–207.

30. G. R. Cunnington, Thermodynamic optimization of a cryogenic storage system for minimum boiloff, *AIAA Pap.*, No. AIAA-82-0075, 1982.

31. J. C. Chato and J. M. Khodadadi, Optimization of cooled shields in insulations, *J. Heat Transfer*, Vol. 106, 1984, pp. 871–875.

32. A. Bejan and E. M. Cluss, Jr., Criterion for burn-up conditions in gas-cooled cryogenic current leads, *Cryogenics*, Vol. 16, 1976, pp. 515–518.

33. R. G. Scurlock, *Low Temperature Behavior of Solids: An Introduction*, Dover, New York, 1966, p. 62.

34. R. Agsten, Thermodynamic optimization of current leads into low temperature regions, *Cryogenics*, Vol. 13, 1973, pp. 141–146.

35. S. C. Collins, Liquefaction techniques, in R. H. Kropschot, B. W. Birmingham, and D. B. Mann, eds., *Technology of Liquid Helium*, Monogr. No. 111, National Bureau of Standards, Washington, DC, 1968, Chapter 3.

36. T. R. Strobridge, Cryogenic refrigerators—an updated survey, *NBS Tech. Note (U.S.)*, No. 655, June 1974.

37. T. R. Strobridge and D. B. Chelton, Size and power requirements of 4.2°K refrigerators, *Adv. Cryog. Eng.*, Vol. 12, 1967, pp. 576–584.

38. T. R. Strobridge, Refrigeration for superconducting and cryogenic systems, *IEEE Trans. Nucl. Sci.*, Vol. NS–16, No. 3, Part 1, 1969, pp. 1104–1108.

39. P. Debye, Einige Bemerkungen zur Magnetisierung bei tiefer Temperatur, *Ann. Phys. (Leipzig)*, Ser. 4, Vol. 81, 1926, pp. 1154–1160.

40. W. F. Giauque, A thermodynamic treatment of certain magnetic effects. A proposed method of producing temperatures considerably below 1° absolute, *J. Am. Chem. Soc.*, Vol. 49, 1927, pp. 1864–1870, 1870–1877.

41. H. B. G. Casimir, *Magnetism and Very Low Temperatures*, Dover, New York, 1961 (originally published in 1940 by Cambridge University Press).

42. M. W. Zemansky, *Heat and Thermodynamics*, 5th ed., McGraw-Hill, New York, 1968, Chapter 14.

43. J. A. Barclay and W. A. Steyert, Materials for magnetic refrigeration between 2 K and 20 K, *Cryogenics*, Vol. 22, 1982, pp. 73–79.

44. J. A. Barclay, A comparison of the efficiency of gas and magnetic refrigerators, in A. Bejan and R. L. Reid, eds., *Second Law Aspects of Thermal Design*, HTD-Vol. 33, Am. Soc. Mech. Eng., New York, 1984, pp. 69–76.

45. J. A. Barclay, *Magnetic Refrigeration: A Review of a Developing Technology*, Paper CA–1, Presented at the Cryogenic Engineering Conference, June 14–18, 1987.

46. W. F. Giauque and D. P. MacDougall, Attainment of temperatures below 1° absolute by demagnetization of $Gd_2(SO_4)_3 \cdot 8H_2O$, *Phys. Rev.*, Vol. 43, 1933, p. 768.

47. W. F. Giauque and D. P. MacDougall, The heat capacity of gadolinium sulfate octahydrate below 1° absolute, *Phys. Rev.*, Vol. 44, 1933, pp. 235–236.

48. A. C. Rose-Innes, *Low Temperature Techniques*, English Universities Press, London, 1964.

49. W. J. de Haas, E. C. Wiersma, and H. A. Kramers, Temperature below 0.27°K reached in Holland, *Nature (London)*, Vol. 131, 1933, p. 719.

50. N. Kürti and F. Simon, Production of very low temperatures by the magnetic method: Supraconductivity of cadmium, *Nature (London)*, Vol. 133, 1934, pp. 907–908.

51. G. J. Ehnholm, J. P. Ekström, J. F. Jacquinot, M. T. Laponen, O. V. Lounasmaa, and J. K. Soini, NMR studies on nuclear ordering in metallic copper below 1 μK, *J. Low Temp. Phys.*, Vol. 39, 1980, pp. 417–450.

52. O. V. Lounasmaa, Low temperature laboratory at Helsinki University of Technology, *Cryogenics*, Vol. 22, 1982, pp. 213–223.

53. R. Fowler and E. A. Guggenheim, *Statistical Thermodynamics*, Cambridge University Press, London and New York, 1939, p. 224.

54. F. Simon, Fünfundzwanzig Jahre Nernstscher Wärmesatz, *Ergeb. Exakten Naturwiss.*, Vol. 9, 1930, pp. 222–274.

55. E. A. Guggenheim, *Thermodynamics*, 7th ed., North-Holland Publ., Amsterdam, 1985, pp. 157–159 (this and Ref. 53 are based on F. Simon, *Sci. Mus. Handb.*, Book 3, 1937, p. 61).

56. W. Nernst, *The New Heat Theorem*, 2nd ed., translated by G. Barr, E. P. Dutton, New York, 1926 (Nernst's idea first appeared in *Nacht. Ges. Wiss. Goettingen, Math. Phys. Kl.*, Vol. 1, 1906, and in *Sitzungsber. Preuss. Akad. Wiss.*, Vol. 20, December 1906).

57. M. Planck, *Treatise on Thermodynamics*, 3rd Engl. ed., translated by A. Ogg, Longmans, Green, London, 1927, pp. 272–292.

58. H. B. Callen, *Thermodynamics and an Introduction to Thermostatistics*, 2nd ed., Wiley, New York, 1985, p. 281.

PROBLEMS

10.1 The derivation of the positive μ_J criterion (10.4) was based on the assumption that the heat-exchanger area is infinite so that $T_1 = T_3$ in Fig. 10.2. In general, the low-pressure stream is colder than the high-pressure stream even at the top of the heat exchanger, $T_1 < T_3$, which means that $h_1(T_1, P_L)$ is lower than in the ideal limit treated in the text, $h_1(T_3, P_L)$. Demonstrate that in this general case, the refrigeration effect is possible only if the room temperature μ_J exceeds a critical positive value, and that when μ_J and c_P are both constant, this criterion is approximately

$$\mu_J \gtrsim \frac{T_3 - T_1}{P_H - P_L}$$

10.2 Consider a nitrogen refrigeration cycle ($k \cong 1.4$) in which the pressure ratio $P_H/P_L = 100$ and the expander flowrate fraction $\dot{m}_e/\dot{m} = 0.6$. With reference to the right side of Fig. 10.4, note further that $T_e = 250$ K, $T_3 - T_1 = 25$ K, $P_L = 1$ atm, and $\overline{\mu_J c_P} \cong 0.0075$ m^3/kg. Show that under these circumstances, the expander work term of eq. (10.9) dominates the remaining two terms in the constitution of the specific refrigeration rate $\dot{Q}_L/\dot{m}$.

10.3 The refrigerant of Fig. 10.2 is to be compressed from (T_0, P_L) to (T_0, P_H) in n stages, that is, with n compressors alternating with n aftercoolers. The right side of Fig. 10.2 shows the special case $n = 4$. The refrigerant can be modeled as an ideal gas (k, R). Each aftercooler performs well enough so that the temperature of the inlet to each compressor is approximately equal to T_0. The isentropic efficiency of each compressor is η_c. Also known are the low and high pressures of the cycle, P_L and P_H, respectively.

There are $(n-1)$ degrees of freedom left in the design of this multistage compressor, namely, the pressures of the first $(n-1)$ aftercoolers. Let these unknowns be $P_1, P_2, \ldots, P_{n-1}$, so that P_i/P_{i-1} is the pressure ratio seen by the ith compressor. Show that the total compressor power required by the n compressors is minimized when the $(n-1)$ aftercooler pressures are chosen such that the pressure ratio P_i/P_{i-1} does not vary from one compression stage to the next. Determine this optimum per-stage pressure ratio. Show that the minimum total compressor power is

$$\dot{W}_c = \frac{n}{\eta_c} \dot{m} c_P T_0 \left[\left(\frac{P_H}{P_L} \right)^{(k-1)/(nk)} - 1 \right]$$

and that in the limit $(n \to \infty, \eta_c \to 1)$, this power matches the minimum claimed in eq. (10.10).

10.4 Consider the refrigeration cycle of Fig. 10.2 in the limit of infinite counterflow heat-exchanger area. Modeling the refrigerant as an ideal gas at room temperature, show that the $\dot{Q}_L/\dot{W}_{c,min}$ ratio increases with the pressure ratio as $(\pi - 1)/\ln \pi$, where $\pi = P_H/P_L$.

10.5   A certain cold region is maintained at a temperature T_L below the environment temperature T_0 by the steady operation of an irreversible refrigeration machine. The refrigeration load $\dot{Q}_L$ is attributed to the leakage of heat from T_0 to T_L. Prove that the total mechanical power demanded by this machine ($\dot{W}$) is equal to T_0 times the *total* rate of entropy generation ($\dot{S}_{gen}$), where $\dot{S}_{gen}$ equals the entropy generated by the irreversible refrigerator and by the leakage of $\dot{Q}_L$ from T_0 to T_L.

10.6   Consider the counterflow heat exchanger with continuously distributed isothermal expanders shown in Fig. 10.12. Assume that the working fluid is an ideal gas with constant specific heat and that the expander efficiency is a function of temperature, $\eta_e(T)$. Verify that the First Law of Thermodynamics is represented by eq. (10.33) in the text. Assuming that ΔT is an unspecified function of T, and that $\Delta T \ll T$, show that the total entropy generated by the apparatus (heat exchanger + expanders) is

$$\dot{S}_{gen} = \int_{T_L}^{T_0} \dot{m} c_P \left[\left(\frac{1}{\eta_e} - 1 \right) \frac{d\tau}{dT} + \frac{\tau}{\eta_e T} \right] dT$$

where $\tau = \Delta T/T$ is, in general, a function of temperature.

10.7   A vessel contains a slowly vanishing amount of T_L-temperature liquid cryogen (e.g., liquid helium) boiling at constant pressure under the influence of the heat leak $\dot{Q}_L$ that penetrates the insulation and reaches the pool of liquid. The boil-off flowrate $\dot{m}$ escapes through a top vent port into the atmosphere.

(a) Demonstrate that as long as there is liquid inside the vessel, the boil-off rate $\dot{m}$ is proportional to the heat leak $\dot{Q}_L$ and that the proportionality is

$$\dot{Q}_L = \dot{m} \left(h_{fg} - \frac{v_g}{v_{fg}} u_f + \frac{v_f}{v_{fg}} u_g \right)$$

The quantities in the parentheses represent the saturation properties of the cryogen at atmospheric pressure.

(b) Assuming that the insulation that surrounds the vessel is not cooled at intermediate temperatures between T_L and T_0 (i.e., that $\dot{Q}_L$ originates from room temperature T_0), demonstrate that the boil-off rate $\dot{m}$ is proportional to the entropy-generation rate of the aggregate system that is colder than T_0 (the T_L cryogen and its insulation).

10.8   Consider the one-dimensional insulation shown in Fig. 10.13a. The thermal conductivity is constant. Assume that the insulation is cooled at only one intermediate point (location x, temperature T_m).

Determine the optimum x and T_m so that the entropy-generation rate of the entire insulation is minimum.

10.9 Consider the case of a single shield separating T_0 and T_L in Fig. 10.13f, i.e., the case $N = 2$. Determine the optimum shield temperature T_s so that the entropy generated in the space between T_0 and T_L is minimum. In this analysis, model the radiation heat transfer in accordance to eq. (10.52) and assume that F is constant.

10.10 The $\dot{m}_l$ stream drawn vertically on the right side of Fig. 10.18 enters the cold zone at atmospheric temperature and pressure (T_0, P_L). Assume that the gas flows through a thermodynamically reversible installation, and exits as saturated liquid of pressure P_L. The installation uses the ambient (T_0) as its only heat reservoir. Invoking the first and the second laws for an irreversibility-free open system with one inlet (T_0, P_L) and one outlet (saturated liquid, P_L), determine the mechanical power required by the liquefaction process. Compare your result with the general conclusions reached already in chapter 3.

10.11 Starting with $\delta Q = dU^* - \mathcal{H}\, d\mathcal{M}$, prove that the infinitesimal heat transfer into a paramagnetic solid is given by

$$\delta Q = C_{\mathcal{H}}\, dT + T\left(\frac{\partial \mathcal{M}}{\partial T}\right)_{\mathcal{H}}\, d\mathcal{H}$$

Derive the dS formula (10.94) in order to show that an alternate expression for the heat transfer is

$$\delta Q = C_{\mathcal{M}}\, dT - T\left(\frac{\partial \mathcal{H}}{\partial T}\right)_{\mathcal{M}}\, d\mathcal{M}$$

10.12 Consider an infinitesimal process of adiabatic demagnetization $(dS = 0,\ d\mathcal{H} < 0)$, and show that the magnetic work transfer δW^* is positive. Note that in addition to eq. (10.96), experiments show that the isothermal magnetic susceptibility χ_m is positive:

$$\chi_m = \left(\frac{\partial \mathcal{M}}{\partial \mathcal{H}}\right)_T > 0$$

10.13 An ideal paramagnetic substance is defined by the equation of state (10.99). Prove that the properties of such a substance also obey the relations

$$\left(\frac{\partial U^*}{\partial \mathcal{H}}\right)_T = 0$$

$$dS = -\frac{\mathcal{H}}{T}\, d\mathcal{M} \qquad \text{(at } T = \text{constant)}$$

10.14 Invoke the third-law statement (10.111) in order to prove that the following quantities of a closed system must approach zero in the limit $T \to 0$:

(a) the coefficient of thermal expansion β,

(b) the specific heats c_v and c_P,

(c) the dP/dT derivative of the $P(T)$ curve for an equilibrium mixture of liquid and solid of the same pure substance.

10.15 It is contemplated to solve simultaneously the heating requirements of an office building and the refrigeration requirements of an ice-storage facility by installing the refrigerator/heat pump system shown. The temperature of the ice-storage facility (T_1) and the temperature of the office building interior (T_2) are fixed by design. Environmental and heat transfer engineering experts advise that the steady-state heat leak from the ambient to the ice-storage facility $(\dot{Q}_1)$ and the heat leak from the building interior to the ambient $(\dot{Q}_2)$ are both proportional to the respective temperature differences that drive the leak, i.e.,

$$\dot{Q}_1 = C_1(T_0 - T_1) \quad \text{and} \quad \dot{Q}_2 = C_2(T_2 - T_0)$$

where T_0 is the ambient temperature, and heat conductances C_1 and C_2 are known constants. The refrigerator/heat pump system is also in thermal contact with the ambient and experiences the heat transfer input $\dot{Q}_0$ (note the physically positive direction of the energy interaction arrows and the vertical alignment of temperatures on the figure). Finally, it is assumed that the refrigerator/heat pump system operates cyclically and reversibly.

(a) Determine the mechanical power requirement $\dot{W}$ as a function of T_0, T_1, T_2, C_1, and C_2.

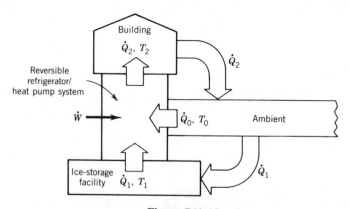

Figure P10.15

(b) Recognizing that the ambient temperature T_0 can vary daily or seasonally while T_1 and T_2 remain fixed, determine the ambient temperature $T_{0,\text{opt}}$ that would minimize the power requirement $\dot{W}$.

(c) Show that the optimum condition described above in part **(b)** corresponds to the special case where the refrigerator/heat pump system does not exchange heat with the ambient ($\dot{Q}_0 = 0$).

(d) What is the entropy-generation rate $\dot{S}_{\text{gen}}$ of the aggregate system that includes the refrigerator/heat pump apparatus, the ice-storage facility, and the office building?

(e) Show that regardless of weather conditions (T_0), the power requirement $\dot{W}$ is directly proportional to $\dot{S}_{\text{gen}}$.

11

Thermodynamic Design

The irreversibility-minimization philosophy displayed in the treatment of large power and refrigeration systems, chapters 8 to 10, can be employed also at the system-component level. The fundamental idea that justifies the work of irreversibility minimization at the system-component level is that the overall entropy-generation rate of the system—the right side of eq. (3.7)—is in fact the sum of the entropy-generation contributions made by all the system's components. If the irreversibility of one component is minimized while leaving the other components untouched[†] then the irreversibility reduction registered at the component level shows up also at the overall "system" level.

The objective of this chapter is to outline the most basic steps of the procedure of entropy-generation minimization (thermodynamic design) at the system-component level. As such, this chapter is a continuation of the review work I attempted on two earlier occasions [1, 2]; therefore, an additional objective of this chapter is to review the fundamental work that was published in this area in the 1980s.

Blending basic thermodynamics with the core problems of thermal design is an important bridge that must be constructed between the two fields. In view of the rebirth of "design" in engineering education, this chapter serves also as an introduction to the more applied questions that are being asked in heat transfer engineering and thermal design.

THE TRADE-OFF BETWEEN ·COMPETING IRREVERSIBILITIES

The basic design problem we face in this chapter is the problem of determining the thermodynamically optimum size or operating regime of a

[†]The principle of thermodynamically "isolating" the component that is being optimized is absolutely essential. Violations of this principle can lead to "paradoxes" of the type illustrated later in this chapter, in the section on "Heat Exchangers with Negligible Pressure-Drop Irreversibility."

594

certain engineering system, where by "optimum" we mean the condition in which the system destroys the least exergy while still performing its fundamental engineering function. It turns out that in many systems, the various mechanisms and design features that account for irreversibility *compete* with one another. For this reason, the thermodynamic optimum that concerns us here is the condition of the most desirable trade-off between two or more competing irreversibilities.

Internal Flow and Heat Transfer

In view of the diversity of thermodynamic optimization problems—at least one problem of this type is contained in each power- and refrigeration-system design—we begin with some of the simplest illustrations of the basic design principle. Later, we focus on more complex systems as we review the expanding literature of this field. Two very simple elementary features (subcomponents) in the constitution of most power and refrigeration installations are the "heat-exchanger passage" and the "fin." Both features account for most of the heat exchanger.

Consider first a heat-exchanger passage, which is a duct of arbitrary cross-section A and arbitrary wetted perimeter p. The engineering function of the passage is specified in terms of the heat transfer rate per unit length q' that is to be transmitted to the stream $\dot{m}$. In other words, both q' and $\dot{m}$ are specified. In the steady state, the heat transfer q' crosses the temperature gap ΔT formed between the wall temperature $(T + \Delta T)$ and the bulk temperature of the stream (T). The stream flows with friction in the x direction, hence, the pressure gradient $(-dP/dx) > 0$.

Taking as a thermodynamic system a passage of length dx, the first law and the second law state:

$$\dot{m}\,dh = q'\,dx \qquad (11.1)$$

$$\dot{S}'_{\text{gen}} = \dot{m}\,\frac{ds}{dx} - \frac{q'}{T + \Delta T} \geq 0 \qquad (11.2)$$

where $\dot{S}'_{\text{gen}}$ is the entropy-generation rate per unit length. Combining these statements with $dh = T\,ds + v\,dP$, the design-important quantity $\dot{S}'_{\text{gen}}$ becomes [3]:

$$\dot{S}'_{\text{gen}} = \frac{q'\,\Delta T}{T^2(1 + \Delta T/T)} + \frac{\dot{m}}{\rho T}\left(-\frac{dP}{dx}\right)$$

$$\cong \frac{q'\,\Delta T}{T^2} + \frac{\dot{m}}{\rho T}\left(-\frac{dP}{dx}\right) \geq 0 \qquad (11.3)$$

Note here the use of density (ρ) instead of the inverse of specific volume $(1/v)$; in addition, the denominator of the first term on the right side has been simplified by assuming that the local temperature difference ΔT is negligible compared with the local absolute temperature T.

The heat-exchanger passage is a site for both flow with friction and heat transfer across a finite ΔT—this is why the $\dot{S}'_{\text{gen}}$ expression has two terms, each term accounting for one irreversibility mechanism. We record this observation by rewriting eq. (11.3) as

$$\dot{S}'_{\text{gen}} = \dot{S}'_{\text{gen},\Delta T} + \dot{S}'_{\text{gen},\Delta P} \tag{11.4}$$

in other words, the first term on the right side of eq. (11.3) represents the entropy generation contributed by heat transfer. The relative importance of the two irreversibility mechanisms is described by the *irreversibility distribution ratio* ϕ, which is defined as

$$\phi = \frac{\text{fluid-flow irreversibility}}{\text{heat transfer irreversibility}} \tag{11.5}$$

Equation (11.4) can then be rewritten as

$$\dot{S}'_{\text{gen}} = (1 + \phi)\dot{S}'_{\text{gen},\Delta T} \tag{11.6}$$

A remarkable feature of the $\dot{S}'_{\text{gen}}$ expression (11.3), and of many like it for other simple devices, is that a proposed design change (e.g., making the passage narrower) induces *changes of opposite signs* in the two terms of the expression. There exists then an *optimum trade-off* between the two irreversibility contributions, an optimum "design" for which the overall measure of exergy destruction $(\dot{S}'_{\text{gen}})$ is minimum, while the system continues to serve its specified function $(q', \dot{m})$.

The trade-off between heat transfer and fluid-flow irreversibilities becomes clearer if we convert eq. (11.3) into the language of heat transfer engineering, in which the "heat-exchange passage" subject is usually discussed. For this purpose, we recall the definitions of friction factor (f), Stanton number (St), mass velocity (G), Reynolds number (Re), and hydraulic diameter (D_h):

$$f = \frac{\rho D_h}{2G^2}\left(-\frac{dP}{dx}\right) \tag{11.7}$$

$$\text{St} = \frac{q'/p\,\Delta T}{c_P G} \tag{11.8}$$

$$G = \frac{\dot{m}}{A} \tag{11.9}$$

$$\text{Re} = \frac{GD_h}{\mu} \tag{11.10}$$

$$D_h = \frac{4A}{p} \tag{11.11}$$

where $q'/p \, \Delta T$ of eq. (11.8) is the average heat transfer coefficient, whose symbol $\bar{h}$ should not be confused with that of specific enthalpy (h). The entropy-generation rate formula (11.3) becomes

$$\dot{S}'_{gen} = \frac{(q')^2 D_h}{4T^2 \dot{m} c_p \, \text{St}} + \frac{2\dot{m}^3 f}{\rho^2 T D_h A^2} \tag{11.12}$$

Considering that both q' and $\dot{m}$ are fixed, we note that the thermodynamic design of the heat-exchanger passage has two degrees of freedom, the wetted perimeter p and the cross-sectional area A, or any other pair of independent parameters, such as (Re, D_h) or (G, D_h).

The competition between heat transfer and fluid-flow irreversibilities is hinted at by the positions occupied by St and f on the right side of eq. (11.12). The Reynolds and Colburn analogies regarding turbulent momentum and heat transfer teach us that St and f usually increase simultaneously [4] as the designer seeks to improve the thermal contact between wall and fluid. Consequently, what is good for reducing the heat transfer irreversibility is bad for the fluid-flow irreversibility, and vice versa.

The trade-off between the two irreversibilities can be illustrated by assuming a special case of passage geometry, namely, the straight tube with circular cross-section. In this case, p and A are related through the pipe inner diameter D, which is the only degree of freedom left in the design process. Writing

$$D_h = D \qquad A = \pi D^2 / 4 \qquad \text{and} \qquad p = \pi D \tag{11.13}$$

eq. (11.12) becomes

$$\dot{S}'_{gen} = \frac{(q')^2}{\pi T^2 k \, \text{Nu}} + \frac{32 \dot{m}^3 f}{\pi^2 \rho^2 T D^5} \tag{4}$$

where $\text{Re} = 4\dot{m}/\pi \mu D$. The Nusselt number definition and the rela' between Nu, St, and the Prandtl number ($\text{Pr} = \nu/\alpha$) are

$$\text{Nu} = \frac{\bar{h} D_h}{k} = \text{St Re Pr} \tag{11.15}$$

Invoking two reliable correlations for Nu and f in fully developed turbulent pipe flow

$$\text{Nu} \cong 0.023 \, \text{Re}^{0.8} \, \text{Pr}^{0.4} \qquad (0.7 < \text{Pr} < 160; \text{Re} > 10^4) \tag{11.16}$$

$$f \cong 0.046 \, \text{Re}^{-0.2} \qquad (10^4 < \text{Re} < 10^6) \tag{11.17}$$

and combining them with eq. (11.14), yields an expression for $\dot{S}'_{gen}$ that depends only on Re. Solving $d\dot{S}'_{gen}/d(\text{Re}) = 0$, we find that the entropy-generation rate is minimized when the Reynolds number (or pipe diameter) reaches the optimum value [5]:

$$Re_{opt} = 2.023 \, Pr^{-0.071} B_0^{0.358} \tag{11.18}$$

This compact formula allows the designer to select the optimum tube size for minimum irreversibility. Parameter B_0 is fixed as soon as q', $\dot{m}$, and the working fluid are specified:

$$B_0 = \dot{m}q' \frac{\rho}{\mu^{5/2}(kT)^{1/2}} \tag{11.19}$$

The effect of Re on $\dot{S}'_{gen}$ can be expressed in relative terms as

$$\frac{\dot{S}'_{gen}}{\dot{S}'_{gen,min}} = 0.856 \left(\frac{Re}{Re_{opt}} \right)^{-0.8} + 0.144 \left(\frac{Re}{Re_{opt}} \right)^{4.8} \tag{11.20}$$

where $\dot{S}'_{gen,min} = \dot{S}'_{gen}(Re_{opt})$.

Figure 11.1 shows that the entropy-generation rate of the tube increases sharply on either side of the optimum. The irreversibility distribution ratio varies along the V-shaped curve, increasing in the direction of small Ds (large Re's, because $\dot{m}$ = constant) in which the overall entropy-generation

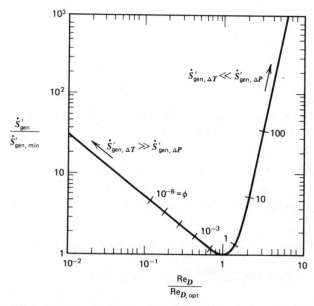

Figure 11.1 The relative entropy-generation rate for forced-convection heat transfer through a smooth tube [5].

rate is dominated by fluid-friction effects. At the optimum, the irreversibility distribution ratio assumes the value $\phi_{opt} = 0.168$. This means that the optimum trade-off between $\dot{S}'_{gen,\Delta T}$ and $\dot{S}'_{gen,\Delta P}$ does not coincide with the design where the two irreversibility mechanisms are in perfect balance, even though setting $\phi = 1$ is a fairly good (back-of-the-envelope) way of locating the optimum.

An important conclusion that follows from Fig. 11.1 is that the dependence between irreversibility and design parameters can be *nonmonotonic*. For this reason, it is difficult to predict in advance (say, on the basis of a "rule of thumb") the change induced in the overall irreversibility figure by a certain design modification. A more general example of how rules of thumb can fail us in the realm of thermodynamic design is illustrated in Fig. 11.2. Plotted in the horizontal plane is the dimensionless temperature difference

$$\tau = \Delta T / T \qquad (11.21)$$

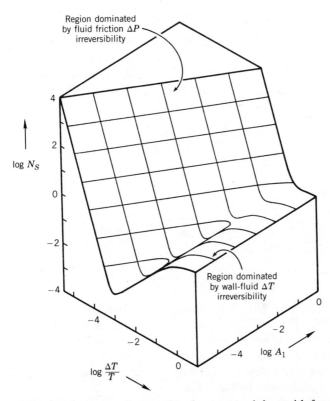

Figure 11.2 The entropy-generation number for a general duct with forced-convection heat transfer, as a function of the wall–fluid temperature difference (τ) and the additional parameter A_1 [3].

while the intent of the figure is to show that the pure-heat-transfer rule that a small wall–fluid ΔT might be desirable (Table 3.1, the first problem) is not necessarily correct. In the case of steady forced convection through a most general heat-exchanger passage, we start with the first of eq. (11.3) and write it in dimensionless form as [3]

$$N_S = \frac{\tau}{1 + \tau} + \frac{J^2 f \, \text{Re}^2}{32 \, \text{St}^3 \, \tau^3} \tag{11.22}$$

where

$$J = \frac{\mu q'}{\rho \dot{m} (c_P T)^{3/2}} \tag{11.23}$$

The left side of eq. (11.22) shows the *entropy-generation number* defined in this case as

$$N_S = \frac{\dot{S}'_{\text{gen}}}{q'/T} \tag{11.24}$$

It is apparent that the heat transfer and fluid-flow contributions to N_S, eq. (11.22), are coupled through the temperature-difference number τ. The minimization of N_S requires the optimum selection of τ, and, as shown in Fig. 11.2, there exists a well-defined optimum τ especially at low values of the A_1 parameter

$$A_1 = J \left(\frac{3f}{32 \, \text{St}} \right)^{1/2} \frac{\text{Re}}{\text{St}} \tag{11.25}$$

In the limit $A_1 \rightarrow 0$, that is, in the limit of small ΔTs, the optimum is described by

$$\tau_{\text{opt}} = A_1^{1/2} \tag{11.26}$$

$$N_{S,\text{min}} = \frac{4}{3} A_1^{1/2} \tag{11.27}$$

When $\tau < \tau_{\text{opt}}$, the ΔT irreversibility is small relative to the fluid-friction irreversibility. Conversely, when $\tau > \tau_{\text{opt}}$, the entropy-generation rate is dominated by the effect of inadequate thermal contact. The lack of monotoneity between N_S and τ means that decreasing the wall–fluid ΔT is a good idea only for those designs that fall on the right side (high-τ side) of the N_S–τ valley illustrated in Fig. 11.2.

Another example of a rule of thumb that clashes with the thermodynamic objective of irreversibility minimization is summarized in Fig. 11.3. In many instances, designers seek to maximize the heat transfer coefficient to pump-

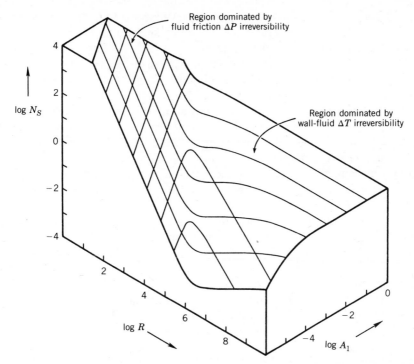

Figure 11.3   The entropy-generation number for a general duct with forced-convection heat transfer, as a function of the ratio of heat transfer coefficient divided by pumping power (R) and the additional parameter A_1 [3].

ing power ratio. A dimensionless group that is proportional to that ratio is

$$R = \frac{\bar{h}\rho p T}{\dot{m}(-dP/dx)} \qquad (11.28)$$

The entropy-generation-number expression (11.22) can be put now in terms of R and A_1:

$$N_S = \frac{A_1(R/3)^{1/2}}{1 + A_1(R/3)^{1/2}} + \frac{3^{1/2}}{R^{2/3}} \qquad (11.29)$$

and, as shown in Fig. 11.3, the features of the surface $N_S(R, A_1)$ are very similar to the features of the $N_S(\tau, A_1)$ surface of Fig. 11.2. If A_1 is constant, there exists an optimum ratio R for which the irreversibility is minimized, for example,

$$R_{\text{opt}} = 3/A_1 \qquad (A_1 \to 0) \qquad (11.30)$$

In conclusion, the decision to increase the ratio of heat transfer coefficient divided by the pumping power is not sufficient if the objective is to improve thermodynamic performance. Since N_S depends on more than just R, the true effect of a proposed design change can only be evaluated by estimating first the changes induced in both R and A_1, and eventually in N_S.

This most fundamental issue of thermodynamic irreversibility at the heat-exchanger-passage level was followed up by Kotas and Shakir [6]. They took into account the temperature dependence of transport properties and showed that the operating temperature of the heat-exchanger passage has a profound effect on the thermodynamic optimum. For example, the optimum Reynolds number increases as the absolute temperature T decreases. The minimum irreversibility that corresponds to this optimum design also increases as T decreases.

A related research direction has been explored by Nag and Mukherjee [7], who minimized the entropy-generation rate of a finite-length section of duct along which the bulk temperature of the fluid decreases exponentially. They showed that the trade-off between heat transfer and fluid-friction irreversibilities results in an optimum value for the inlet temperature difference, that is, for the difference between the inlet temperature of the fluid and the constant temperature of the duct wall. This result is equivalent to the discovery of an optimum distribution of wall–fluid ΔT along the duct. The second-law aspects of the entrance region of a duct with swirling flow have been analyzed similarly by Mukherjee et al. [8].

Heat Transfer Augmentation

Another example of the competition between different irreversibility mechanisms occurs in connection with the general problem of heat transfer augmentation, in which the main objective is to devise a technique that increases the wall–fluid heat transfer coefficient relative to the coefficient of the unaugmented (i.e., untouched) surface. A parallel objective, however, is to register this improvement without causing a damaging increase in the pumping power demanded by the forced-convection arrangement. These two objectives reveal the conflict that accompanies the application of any augmentation technique: a design modification that improves the thermal contact (e.g., roughening the heat transfer surface) is likely to "augment" also the mechanical pumping-power requirement.

The true effect of a proposed augmentation technique on thermodynamic performance can be evaluated by comparing the irreversibility of the heat-exchange apparatus before and after the implementation of the augmentation technique [9]. Consider again the general heat-exchanger passage referred to in eqs. (11.1)–(11.3) and in Figs. 11.2 and 11.3, and let $\dot{S}'_{gen,0}$ represent the degree of irreversibility in the reference (unaugmented, untouched) passage. Writing $\dot{S}'_{gen,a}$ for the heat transfer-augmented version of the same device, we can evaluate the *augmentation entropy-generation number*:

$$N_{S,a} = \frac{\dot{S}'_{gen,a}}{\dot{S}'_{gen,0}} \tag{11.31}$$

Augmentation techniques whose $N_{S,a}$ values are less than 1 are thermo-dynamically advantageous.

If the function of the heat-exchanger passage is fixed, i.e., if $\dot{m}$ and q' are given, the augmentation entropy-generation number can be put in the more explicit form:

$$N_{S,a} = \frac{1}{1 + \phi_0} N_{S,\Delta T} + \frac{\phi_0}{1 + \phi_0} N_{S,\Delta P} \tag{11.32}$$

In this form, ϕ_0 is the irreversibility distribution ratio of the reference design, whereas $N_{S,\Delta T}$ and $N_{S,\Delta P}$ represent the values of $N_{S,a}$ in the limits of pure heat transfer irreversibility and pure fluid-flow irreversibility, respectively. It is not difficult to show that these limiting values are

$$N_{S,\Delta T} = \frac{St_0 \, D_{h,a}}{St_a \, D_{h,0}} \tag{11.33}$$

$$N_{S,\Delta P} = \frac{f_a D_{h,0} A_0^2}{f_0 D_{h,a} A_a^2} \tag{11.34}$$

The geometric parameters (A, D_h) before and after augmentation are linked through the $\dot{m} = $ constant constraint, which reads

$$Re_a \frac{A_a}{D_{h,a}} = Re_0 \frac{A_0}{D_{h,0}} \tag{11.35}$$

Equations (11.32)–(11.35) show that $N_{S,a}$ is, in general, a function of both the heat transfer coefficient ratio St_a/St_0 and the friction-factor ratio f_a/f_0. The relative importance of the friction-factor ratio is dictated by the numerical value of ϕ_0: this value is known because the reference design is known. Note that ϕ_0 describes the thermodynamic regime of operation of the heat-exchanger passage (ΔT losses vs. ΔP losses), much in the way that Re_0 indicates the fluid-mechanics regime (laminar vs. turbulent).

The $N_{S,a}$ calculation outlined above was used to evaluate a large number of heat transfer augmentation techniques, ranging from surface roughening to the use of inserts that promote swirl flow [9, 10; see also Ref. 2, chapter 6]. As an illustration, consider the effect of sand-grain roughness on the irreversibility of forced-convection heat transfer through a straight pipe. Since this technique does not change the hydraulic diameter and the cross-sectional area appreciably $(D_{h,a} \cong D_{h,0}, A_a \cong A_0)$, the augmentation entropy-generation number assumes the particularly simple form

$$N_{S,a} = \frac{1}{1 + \phi_0} \frac{St_0}{St_a} + \frac{\phi_0}{1 + \phi_0} \frac{f_a}{f_0} \tag{11.36}$$

This relationship is presented in Fig. 11.4 for three values of ϕ_0 and four values of the relative roughness height e/D. The figure was constructed using Nikuradse's correlation for sand-grain roughness friction factor f_a [11], Dipprey and Sabersky's correlation for heat transfer coefficient St_a [12], the Karman–Nikuradse relation for friction factor in a smooth tube f_0 [13], and, finally, Petukhov's heat transfer correlation for smooth tubes [14].

For a fixed Reynolds number Re_0 and grain size e/D, there exists a critical irreversibility distribution ratio, where $N_{S,a} = 1$, i.e., where the augmentation technique has absolutely no effect on irreversibility. Figure 11.5 shows this critical value of ϕ_0 as a function of Re_0 and e/D. If, in a certain design, the actual ϕ_0 exceeds the critical ϕ_0 value of Fig. 11.5, the use of sand-grain roughening will not reduce the rate of entropy generation in the tube. This will always be the case, despite the fact that sand-grain roughening always increases the heat transfer coefficient [15].

Figure 11.5 reinforces the observation made in connection with the ϕ_{opt} value suggested by Fig. 11.1. The irreversibility distribution ratio can be smaller than unity in that critical design region, where the fluid-friction irreversibility begins to have an effect on the overall irreversibility of the heat-exchanger passage.

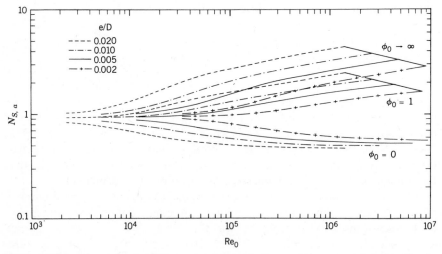

Figure 11.4 The augmentation entropy-generation number associated with the application of sand-grain roughness to a pipe [9].

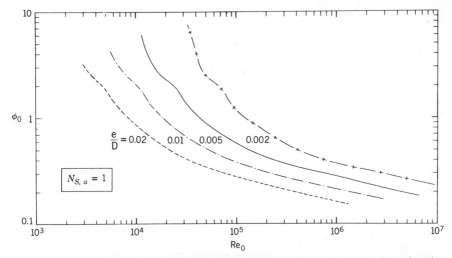

Figure 11.5 The critical irreversibility distribution ratio for the use of sand-grain roughness in a straight pipe [9]: the actual ϕ_0 must be smaller than the critical ϕ_0 if $N_{S,a}$ is to be less than 1.

The impact of heat transfer augmentation on entropy generation was reconsidered more recently by Perez-Blanco [16]. In place of a passage of length dx, Perez-Blanco took as the system a single-stream heat-exchanger tube of finite length L. For simplicity, he assumed that the tube wall temperature is uniform, and developed analytical means of calculating the overall entropy-generation rate of the finite-size system in terms of potential design variables. Particularly interesting are the results showing the maximum friction-factor range that can be tolerated during heat transfer enhancement, in order to maintain the overall entropy-generation rate unchanged. This point of view is related to the one used to construct Fig. 11.5, where the critical ϕ_0s mark the boundary of the domain in which heat transfer enhancement is thermodynamically acceptable.

External Flow and Heat Transfer

The competition between flow and heat transfer irreversibilities rules also the thermodynamic design of "external" convection heat transfer arrangements, in which the flow engulfs the solid body (walls) with which it exchanges heat transfer. In order to determine the overall entropy-generation rate associated with an external convection configuration, consider the heat transfer generalization of the flow-irreversibility analysis given in chapter 3 (p. 140). We ride on the flow and select as a closed system the entire fluid reservoir through which the solid body is being dragged. The speed of the body relative to the fluid reservoir is U_∞, and the total drag force that an external mechanism must apply on the body is F_D. At the same time, the instantaneous heat transfer rate between the body and the fluid reservoir is $\dot{Q}_B$. The temperature of the system boundary crossed by $\dot{Q}_B$ is the solid-wall temperature T_B (recall the definition of the "system"). Except for the interface that it shares with the solid body, the system is surrounded by an adiabatic and rigid (zero-work) boundary.

The instantaneous form of the first law and the second law for the closed system defined above is

$$\dot{Q}_B + F_D U_\infty = dU/dt \qquad (11.37)$$

$$\dot{S}_{gen} = \frac{dS}{dt} - \frac{\dot{Q}_B}{T_B} \geq 0 \qquad (11.38)$$

where U and S are the instantaneous internal energy and entropy inventories of the fluid reservoir (U should not be confused with the relative speed U_∞). If T_∞ is the temperature of the fluid reservoir, then the constant-volume version of eq. (3.66):

$$dU = T_\infty \, dS \qquad (11.39)$$

allows us to combine eqs. (11.37) and (11.38) into

$$\dot{S}_{gen} = \frac{\dot{Q}_B(T_B - T_\infty)}{T_\infty T_B} + \frac{F_D U_\infty}{T_\infty} \qquad (11.40)$$

This is a remarkably simple result that proves again that inadequate thermal contact (the first term) and fluid friction (the second term) contribute hand in hand to degrading the thermodynamic performance of the external convection arrangement. The derivation presented above was first reported in the study of entropy generation from fins with forced-convection heat transfer [17]. A lengthier derivation in which the role of the "system" is played by the stream tube that sweeps the body is given in chapter 5 of Ref. 2. The pure heat transfer half of this problem—the half whose answer is the first term on the right side of eq. (11.40)—was considered most recently by Moody [18], also in a study of the irreversibility contributed by heat transfer from fins.

One area in which eq. (11.40) has found application is the problem of selecting the size and number (density) of fins for the design of extended surfaces. The thermodynamic optimization of fins and fin arrays of various geometries is described in Refs. 17 and 19. As an example, consider the thermodynamic optimization of a *cylindrical pin fin*, whose geometry is described by only two dimensions, the length L and the diameter D. The relationship between the overall heat transfer rate $\dot{Q}_B$ and the base–fluid temperature difference $T_B - T_\infty$ is a classical result in conduction heat transfer [20]:

$$\frac{\dot{Q}_B}{T_B - T_\infty} = \frac{\pi}{4} k D^2 m \tanh(mL) \qquad m = \frac{4\bar{h}}{kD} \qquad (11.41)$$

where k is the thermal conductivity of the fin material, and $\bar{h}$ is the fin–stream heat transfer coefficient (assumed constant). Eliminating the difference $T_B - T_\infty$ between eqs. (11.40) and (11.41), and assuming a sufficiently small base–stream temperature difference, $T_B - T_\infty \ll T_\infty$, the entropy-generation rate (11.40) can be put in the following dimensionless form:

$$N_s = \frac{2(k/\lambda \, \mathrm{Nu})^{1/2}}{\pi \, \mathrm{Re}_D \tanh[2(\lambda \, \mathrm{Nu}/k)^{1/2} \, \mathrm{Re}_L/\mathrm{Re}_D]} + \frac{1}{2} B_1 C_D \, \mathrm{Re}_L \, \mathrm{Re}_D \quad (11.42)$$

where

$$N_S = \dot{S}_\mathrm{gen} \frac{k\nu T_\infty^2}{\dot{Q}_B^2 U_\infty} \qquad \text{(entropy-generation number)}$$

$$\mathrm{Re}_L = L \frac{U_\infty}{\nu} \qquad \text{(dimensionless fin length)}$$

$$\mathrm{Re}_D = D \frac{U_\infty}{\nu} \qquad \text{(dimensionless fin diameter)}$$

$$B_1 = \frac{\rho \nu^3 k T_\infty}{\dot{Q}_B^2} \qquad \text{(duty parameter)}$$

$$\mathrm{Nu} = \frac{\bar{h} D}{\lambda} \qquad \text{(average Nusselt number)}$$

$$C_D = \frac{F_D/LD}{\frac{1}{2}\rho U_\infty^2} \qquad \text{(drag coefficient)}$$

The optimization problem consists of finding the optimum fin size for minimum entropy generation, i.e., the $(\mathrm{Re}_L, \mathrm{Re}_D)$ solution to the system:

$$\frac{\partial N_S}{\partial \mathrm{Re}_L} = 0 \qquad \text{and} \qquad \frac{\partial N_S}{\partial \mathrm{Re}_D} = 0 \qquad (11.43)$$

For this, we must rely on heat transfer and drag-coefficient correlations. For example, if the fin is sufficiently slender, we can use the correlations developed for a single cylinder in cross-flow [21], in the range $40 < \mathrm{Re}_D < 10^3$:

$$\mathrm{Nu} \cong 0.68 \, \mathrm{Re}_D^{0.466} \, \mathrm{Pr}^{0.33} \qquad (11.44)$$

$$C_D \cong 5.48 \, \mathrm{Re}_D^{-0.246} \qquad (11.45)$$

From eqs. (11.42), (11.44), and (11.45), the entropy-generation number emerges as a function of five dimensionless groups:

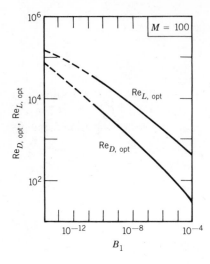

Figure 11.6 Optimum pin-fin length and diameter for minimum entropy generation ($M = 100$) [17].

$$N_S = N_S(\mathrm{Re}_L, \mathrm{Re}_D, \mathrm{Pr}, k/\lambda, B_1) \tag{11.42'}$$

the first two governing the fin geometry, and the last three accounting for the choice of fin material, working fluid, and operating conditions such as $\dot{Q}_B$ and U_∞. Solving the first of eqs. (11.43) yields directly:

$$\mathrm{Re}_{L,\mathrm{opt}} = \frac{1}{2}\,\mathrm{Re}_D\left(\frac{k}{\lambda\,\mathrm{Nu}}\right)^{1/2} \sinh^{-1}\left[\left(\frac{8}{\pi\,\mathrm{Re}_D^3\,C_D B_1}\right)^{1/2}\right] \tag{11.46}$$

which means that the optimum pin length can be calculated immediately if the other dimension (D) is known. There exists also an optimum D, which is obtained numerically by combining eqs. (11.42) and (11.46) and solving the second of eqs. (11.43).

Figure 11.6 shows a sample of $(\mathrm{Re}_L, \mathrm{Re}_D)_{\mathrm{opt}}$ results obtained for $M = 100$, where M is shorthand for $(k/\lambda)^{1/2}\,\mathrm{Pr}^{-1/6}$. The optimum fin length and diameter increase as the heat transfer duty $\dot{Q}_B$ increases (i.e., as B_1 decreases). The optimum slenderness ratio L/D is of the order of 10 throughout the B_1 range depicted in Fig. 11.6: this result validates the slenderness assumption made before using the single-cylinder correlations (11.44) and (11.45). Other examples of thermodynamic sizing procedures for fins can be found in Refs. 2, 17, and 19.

Convective Heat Transfer in General

What all the preceding examples have in common is a two-term expression for the entropy-generation rate, or two distinct mechanisms of thermodynamic irreversibility—heat transfer and flow with friction. These two mechanisms are at work at any point in a convective field, as can be seen from the point-size control-volume formulation of the mass-conservation principle (1.29), the first law (1.26) with "u" in place of "e," and the second law (2.48):

$$\frac{\partial \rho}{\partial t} = -\rho \nabla \cdot \mathbf{v} \qquad (1.29)$$

$$\rho \frac{\partial u}{\partial t} = -\nabla \cdot \mathbf{q} - P\nabla \cdot \mathbf{v} - w''' \qquad (1.26)$$

$$s'''_{\text{gen}} = \rho \frac{\partial s}{\partial t} + \nabla \cdot \left(\frac{\mathbf{q}}{T}\right) \geq 0 \qquad (2.48)$$

The vectors $\mathbf{v}$ and $\mathbf{q}$ represent the velocity and the heat flux, respectively, at the point surrounded by the infinitesimally small control volume. Recall also that w''' represents the work done by the system, per unit time and per unit volume. Eliminating u and s between these laws and the per-unit-time version of $du = T\,ds - P\,dv$, namely,

$$\frac{\partial u}{\partial t} = T \frac{\partial s}{\partial t} + \frac{P}{\rho^2} \frac{\partial \rho}{\partial t} \qquad (11.47)$$

yields the suspected two-term expression for the volumetric rate of entropy generation:

$$s'''_{\text{gen}} = -\frac{1}{T^2}\, \mathbf{q} \cdot \nabla T - \frac{w'''}{T} \geq 0 \qquad (11.48)$$

Now, it is known that in the case of incompressible flow, the place of $(-w''')$ in the first law (1.26) is occupied by $\mu \Phi$, where Φ is the viscous dissipation function [Ref. 4, p. 11]. Furthermore, for almost 200 years, the heat-flux vector and the local temperature gradient have been related through the *Fourier law* of thermal diffusion:

$$\mathbf{q} = -k\nabla T \qquad (11.49)$$

which means that on the right side of eq. (11.48), both terms are positive:

$$s'''_{\text{gen}} = \frac{k}{T^2} (\nabla T)^2 + \frac{\mu}{T} \Phi \geq 0 \qquad (11.50)$$

This aspect becomes even more evident if we write the two-dimensional version of eq. (11.50) for a flow field (x, y) in which the local velocity components are (v_x, v_y):

$$s'''_{\text{gen}} = \frac{k}{T^2} \left[\left(\frac{\partial T}{\partial x}\right)^2 + \left(\frac{\partial T}{\partial y}\right)^2 \right] + \frac{\mu}{T} \left\{ 2\left[\left(\frac{\partial v_x}{\partial x}\right)^2 + \left(\frac{\partial v_y}{\partial y}\right)^2 \right] + \left(\frac{\partial v_x}{\partial y} + \frac{\partial v_y}{\partial x}\right)^2 \right\}$$
$$(11.51)$$

The first term on the right side is clearly the contribution due to finite heat transfer down finite temperature gradients, whereas the second term represents the irreversibility due to frictional dissipation. This is why the finite-control-volume analyses that were highlighted in the preceding subsections led to similar two-term expressions for the entropy-generation rate. The entropy-generation rate of the finite-size control volume is the volume integral of the volumetric entropy-generation rate s_{gen}''' [see also the statement under eq. (2.48)].

How a convective flow field generates entropy at any point in the flow is illustrated in Fig. 11.7. In the horizontal plane, we see the familiar shape of

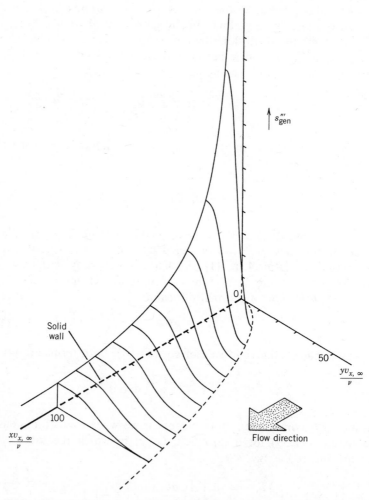

Figure 11.7   The distribution of entropy-generation rate through a laminar boundary layer flow over a flat wall with heat transfer [22].

a laminar boundary layer forming in the x direction. The free-stream velocity of the flow, $v_{x,\infty}$, was used to nondimensionalize the (x, y) coordinates. Measured on a linear scale (any scale) in the vertical direction is the size of s'''_{gen}, which is due to both heat transfer and near-wall friction [22]. The entropy-generation phenomenon is concentrated near the wall; proceeding along the wall, the entropy-generation rate blows up as $1/x$ at the tip of the boundary layer. Entropy-generation "maps" for other convective heat transfer flow fields can be examined in Refs. 1, 2, and 22. The spatial distribution of entropy generation in flames has been illustrated by Arpaci and Selamet [23].

Thermal "Damping"

On the pedagogical side of the challenge that we face in the engineering-thermodynamics profession, one contribution of the irreversibility trade-offs illustrated in this section is that they give physical meaning to relatively ellusive concepts such as the thermodynamic "loss" due to heat transfer across a finite ΔT. The two-term expressions seen previously demonstrate that the heat transfer thermodynamic "loss" is directly comparable with a scaled version of the easier-to-grasp mechanical concept of frictional dissipation. The same lesson is taught by an ingenious exercise conceived by Moody [18] and summarized here in Fig. 11.8.

Figure 11.8a shows a spring–mass system (K, M_p) whose mass can serve as piston for compressing a batch of ideal gas (M_g, R, c_v). Leaving the gas out of the discussion for a moment, we assume that the piston slides in the cylindrical sleeve *with friction*, in such a way that the friction force is proportional to the piston velocity. The motion of piston is governed by the well-known equation:

$$\frac{d^2Z}{dt^2} + b\,\frac{dZ}{dt} + \omega^2 Z = 0 \tag{11.52}$$

where b is the "damping coefficient" and $\omega^2 = K/M_p$. The motion is an undamped sinusoidal oscillation if $b = 0$. In the general case $b > 0$, the oscillation is damped and the amplitude decreases exponentially in time:

$$Z = c_1 \exp\left(-\tfrac{1}{2}bt\right) \sin\left\{\omega[1 - (b/2\omega)^2]^{1/2}(t + c_2)\right\} \tag{11.53}$$

We shall see that the same kind of "damped" oscillation is possible if $b = 0$ and irreversibility is provided by a mechanism other than "sliding with friction." This time we assume the leak-proof seal between piston and cylinder is frictionless, and that the ideal gas compresses and expands in response to the motion $Z(t)$. The entire apparatus is surrounded by the atmospheric temperature and pressure reservoir (T_0, P_0). Since the temperature of the ideal gas (T) is expected to vary, we model the instantaneous heat transfer rate between it and the atmosphere as

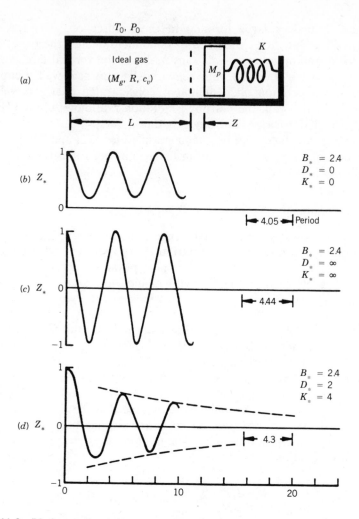

Figure 11.8 Undamped and damped oscillations of a spring–mass–ideal gas system: "thermal damping" regime as the fingerprint of entropy generation due to heat transfer [18].

$$\dot{Q}_0 = \bar{h} A_g (T - T_0) \qquad (11.54)$$

Here A_g is the gas–atmosphere contact area, and $\bar{h}$ the corresponding heat transfer coefficient. Using Moody's notation, the differential equation for the piston motion can be written as [18]

$$\frac{d^3 Z_*}{dt_*^3} + D_* \frac{d^2 Z_*}{dt_*^2} + B_* \frac{dZ_*}{dt_*} + K_* Z_* = 0 \qquad (11.55)$$

where the dimensionless variables are

$$t_* = t\left(\frac{AP_0}{M_p L}\right)^{1/2} \qquad B_* = k + \frac{KL}{AP_0} \qquad K_* = D_*\left(1 + \frac{KL}{AP_0}\right)$$

$$D_* = (k-1)\frac{\bar{h}A_g T_0}{LAP_0}\left(\frac{M_p L}{AP_0}\right)^{1/2} \qquad Z_* = Z/Z_0 \tag{11.56}$$

Other parameters used in this formulation are L, the rest length of the gas column; A, the cylinder cross-sectional area; and k, the ratio c_p/c_v of the gas, which should not be confused with the thermal conductivity. We focus next on three distinct regimes.

(i) The reversible and adiabatic limit corresponds to $\bar{h} = 0$, which means that $D_* = K_* = 0$ and eq. (11.55) reduces to

$$\frac{d^2 Z_*}{dt_*^2} + B_* Z_* = \text{constant} \tag{11.57}$$

In this case, the motion is a pure (undamped) sinusoid, as shown in Fig. 11.8b. The ideal gas expands and contracts reversibly and adiabatically, and it is the lack of irreversibility in the entire system that is responsible for the absence of damping.

(ii) The reversible and isothermal limit is represented by $\bar{h} \to \infty$; hence, $(D_*, K_*) \to \infty$. In this limit, the equation of motion prescribes again an undamped oscillation:

$$\frac{d^2 Z_*}{dt_*^2} + \left(1 + \frac{KL}{AP_0}\right) Z_* = 0 \tag{11.58}$$

One example of this motion is given in Fig. 11.8c. The oscillation is again undamped because of the lack of irreversibility in the system.

(iii) The general class of systems where $\bar{h}$ is finite can be called "thermally damped" because the oscillation has all the features that were revealed by the damped mechanical oscillator of eq. (11.52). Figure 11.8d shows one numerical solution to the complete eq. (11.55). The damping effect in this case is provided by exactly what was missing in the limits (i) and (ii) above, namely, irreversibility or entropy generation. Since frictional effects have been ruled out, the only irreversibility mechanism that is present in regime (iii) is the heat transfer $\dot{Q}_0$ across the finite temperature gap $T - T_0$.

The message of Moody's example is that from an overall "system" perspective, the effect of heat transfer irreversibility is qualitatively the same as that of friction-based irreversibility. In Moody's example, that effect was "damping." The message, however, is of general importance in thermodynamic design, and the effect of irreversibilities of all origins is always the same—lost exergy.

BALANCED COUNTERFLOW HEAT EXCHANGERS

In this section, we increase the degree of complexity of the system component and address the problem of irreversibility minimization in heat exchangers. The classical approach to heat-exchanger design suffers from a traditional bias towards first-law analysis and against second-law consideration of any kind. The very name "heat exchanger" suggests that the function of the apparatus might be to transfer a certain amount of heat between two or more entities (streams, most often) at different temperatures. This is not generally true. For example, in power and refrigeration cycles, the function of the heat-exchange equipment is to allow various components of the cycle to communicate with one another in the least irreversible way possible. In order to see this, review the mission of the counterflow heat exchangers of the refrigeration systems of the preceding chapter.

In connection with the above observation, consider the Brayton-cycle heat engine with a regenerative heat exchanger, which is shown in Fig. 11.9. The high-temperature end of the cycle (heater + expander) *must* communicate with the low-temperature end (cooler + compressor) in order to exchange low-pressure fluid for high-pressure fluid. The most efficient communication is established when

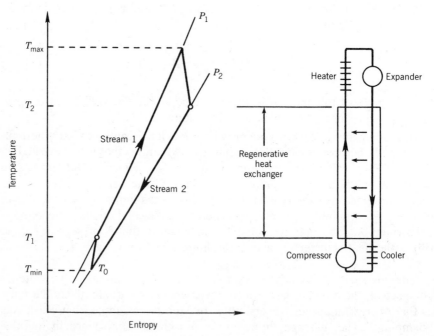

Figure 11.9 Brayton-cycle heat engine with regenerative (counterflow) heat exchanger [24].

(1) there is no pressure drop in the regenerative heat exchanger,
(2) the stream flowing into the heater is already as hot as possible, i.e., as hot as the expander outlet,
(3) the inlet to the cooler is as cold as the compressor outlet.

This limiting regime of operation corresponds to a completely reversible regenerator, one with zero pressure drops and zero stream-to-stream temperature difference. It is clear that the effective stream-to-stream heat exchange is only a part of the true function of the regenerative heat exchanger: its true function is to connect the hot and cold ends of the power cycle in the least irreversible manner.

The Ideal Limit

The trade-off between heat transfer and fluid-flow irreversibilities becomes visible once more if we consider the class of balanced counterflow heat exchangers in the ideal limit of small ΔPs and ΔT. "Balance" means that the capacity flowrates are the same on the two sides of the heat transfer surface:

$$(\dot{m}c_P)_1 = (\dot{m}c_P)_2 = \dot{m}c_P \tag{11.59}$$

The two sides are indicated by the subscripts "1" and "2". With reference to the counterflow heat exchanger in Fig. 11.9, we write T_1 and T_2 for the fixed (given) inlet temperatures of the two streams, and P_1 and P_2 for the respective inlet pressures. The entropy-generation rate of the entire heat exchanger is

$$\dot{S}_{gen} = (\dot{m}c_P)_1 \ln \frac{T_{1,out}}{T_1} + (\dot{m}c_P)_2 \ln \frac{T_{2,out}}{T_2}$$
$$- (\dot{m}R)_1 \ln \frac{P_{1,out}}{P_1} - (\dot{m}R)_2 \ln \frac{P_{2,out}}{P_2} \tag{11.60}$$

where the working fluid has been modeled as an ideal gas with constant specific heat. The outlet temperatures $T_{1,out}$ and $T_{2,out}$ can be eliminated by bringing in the concept of heat exchanger *effectiveness*: in the present example, $(\dot{m}c_P)_1 = (\dot{m}c_P)_2$; therefore, the effectiveness is simple[†]

$$\epsilon = \frac{T_1 - T_{1,out}}{T_1 - T_2} = \frac{T_{2,out} - T_2}{T_1 - T_2} \tag{11.61}$$

If we assume that $(1 - \epsilon) \ll 1$ and that the pressure drops along each stream are sufficiently small relative to the absolute pressure levels, the entropy-generation rate may be nondimensionalized as [2, 24]

[†]The general effectiveness definition is listed in eqs. (11.103).

$$N_S = (1 - \epsilon) \frac{(T_2 - T_1)^2}{T_1 T_2} + \frac{R}{c_P} \left[\left(\frac{\Delta P}{P} \right)_1 + \left(\frac{\Delta P}{P} \right)_2 \right] \qquad (11.62)$$

where N_S is the *entropy-generation number*:

$$N_S = \frac{1}{\dot{m} c_P} \dot{S}_{\text{gen}} \qquad (11.63)$$

In this form, it is clear that the overall entropy generation-rate N_S receives contributions from three sources of irreversibility, namely, from the stream-to-stream heat transfer [regardless of the sign of $(T_2 - T_1)$]; the pressure drop along the first stream, ΔP_1; and the pressure drop along the second stream, ΔP_2. The heat transfer irreversibility term can be split into two terms, each describing the contribution made by one side of the heat transfer surface. We are assuming that the stream-to-stream ΔT is due to the heat transfer across the two convective thermal resistances that sandwich the solid wall separating the two streams, and that the thermal resistance of the wall itself is negligible. In other words, we write

$$\frac{1}{\bar{h} A_1} = \frac{1}{\bar{h}_1 A_1} + \frac{1}{\bar{h}_2 A_2} \qquad (11.64)$$

where A_1 and A_2 are the heat transfer surface areas swept by each stream, and $\bar{h}_1$ and $\bar{h}_2$ the side heat transfer coefficients based on these respective areas (the area A_1 should not be confused with the dimensionless group A_1 used in Figs. 11.2 and 11.3). On the left side of eq. (11.64), we see $\bar{h}$, which is the overall heat transfer coefficient based on A_1.

The thermal-resistance summation (11.64) means also that

$$\frac{1}{N_{\text{tu}}} = \frac{1}{N_{\text{tu},1}} + \frac{1}{N_{\text{tu},2}} \qquad (11.65)$$

where each N_{tu} is a *number of heat transfer units*:

$$N_{\text{tu}} = \frac{\bar{h} A_1}{\dot{m} c_P} \qquad N_{\text{tu},1} = \frac{\bar{h}_1 A_1}{\dot{m} c_P} \qquad N_{\text{tu},2} = \frac{\bar{h}_2 A_2}{\dot{m} c_P} \qquad (11.66)$$

In a balanced counterflow heat exchanger, the $\epsilon(N_{\text{tu}})$ relationship is particularly simple [25]:

$$\epsilon = \frac{N_{\text{tu}}}{1 + N_{\text{tu}}} \qquad (11.67)$$

and even simpler in the limit of vanishingly small stream-to-stream ΔT (i.e., in the limit $\epsilon \to 1$):

$$1 - \epsilon = 1/N_{\text{tu}} \qquad (\epsilon \to 1) \qquad (11.68)$$

Combining eqs. (11.62), (11.65), and (11.68), we find that in the "ideal" heat exchanger limit (small ΔT and ΔPs), the entropy-generation number N_S splits into four distinct terms:

$$N_S = \underbrace{\frac{\tau^2}{N_{tu,1}} + \frac{R}{c_P}\left(\frac{\Delta P}{P}\right)_1}_{N_{S,1}} + \underbrace{\frac{\tau^2}{N_{tu,2}} + \frac{R}{c_P}\left(\frac{\Delta P}{P}\right)_2}_{N_{S,2}} \qquad (11.69)$$

The contribution of the ideal-limit analysis is that it separates N_S into all the pieces that contribute to the irreversibility of the apparatus. The first pair of terms on the right side of eq. (11.69) represents the irreversibility contributed solely by side "1" of the heat transfer surface, $N_{S,1}$. The first term in this first pair is the entropy-generation number due to heat transfer irreversibility on side "1," where τ^2 is shorthand for a parameter fixed by T_1 and T_2:

$$\tau^2 = \frac{(T_2 - T_1)^2}{T_1 T_2} \qquad (11.70)$$

The one-side entropy-generation numbers $N_{S,1}$ and $N_{S,2}$ have the same analytical form; therefore, we can concentrate on the minimization of only one of them (say, $N_{S,1}$) and keep in mind that the minimization analysis can be repeated identically for the other ($N_{S,2}$).

Despite the additive form of $N_{S,1}$, eq. (11.69), the heat transfer and fluid-friction contributions to it are in fact coupled through the geometric parameters of the heat-exchanger duct (passage) that resides on side "1" of the heat-exchanger surface. This coupling is brought to light by rewriting $N_{S,1}$ in terms of the passage slenderness ratio $(4L/D_h)_1$:

$$N_{S,1} = \frac{\tau^2}{St_1}\left(\frac{D_h}{4L}\right)_1 + \frac{R}{c_P}\, g_1^2 f_1\left(\frac{4L}{D_h}\right)_1 \qquad (11.71)$$

where f_1 and St_1 are defined according to eqs. (11.7) and (11.8):

$$f_1(Re_1) = \frac{\rho D_{h,1}}{2G_1^2}\left(\frac{\Delta P_1}{L_1}\right) \qquad St_1(Re_1, Pr) = \frac{\bar{h}_1}{c_P G_1} \qquad (11.72)$$

An important ingredient in the step from the $N_{S,1}$ form (11.69) to eq. (11.71) is the N_{tu}–St relation:

$$N_{tu,1} = \left(\frac{4L}{D_h}\right)_1 St_1 \qquad (11.73)$$

which follows from definitions (11.8), (11.11), and (11.66) in combination with $A_1 = L_1 p_1$, where p_1 is the wetted perimeter of passage "1." Finally, g_1 is a dimensionless mass velocity defined as

$$g_1 = \frac{G_1}{(2\rho P_1)^{1/2}} \tag{11.74}$$

Equation (11.71) shows clearly that the slenderness ratio $(4L/D_h)_1$ has opposite effects on the two terms of $N_{S,1}$. When the mass velocity and Reynolds number are fixed, there is an optimum slenderness ratio for minimum entropy generation:

$$\left(\frac{4L}{D_h}\right)_{1,\text{opt}} = \frac{\tau}{g_1\left(\frac{R}{c_P} f_1 \, \text{St}_1\right)^{1/2}} \tag{11.75}$$

$$N_{S,1,\text{min}} = 2\tau \left(\frac{R}{c_P}\right)^{1/2} g_1\left(\frac{f_1}{\text{St}_1}\right)^{1/2} \tag{11.76}$$

The main features of this minimum are shown qualitatively in Fig. 11.10. The minimum entropy-generation rate is proportional to g_1. Furthermore, in the case of the most common heat-exchanger surfaces, the group $(f_1/\text{St}_1)^{1/2}$ is only a weak function of Re_1. This means that eq. (11.76) resembles a one-to-one correspondence between mass velocity and minimum rate of entropy generation on one side of a heat-exchanger surface.

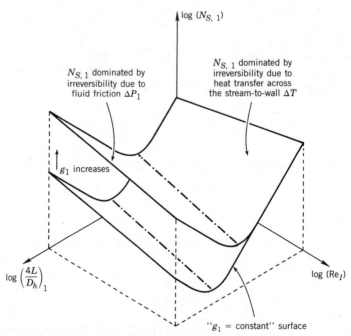

Figure 11.10   Entropy-generation number for one side of the heat-exchanger surface as a function of $(4L/D_h)_1$, g_1, and Re_1 [24].

Area Constraint

As summary to the ideal limit analyzed in the preceding section, we note that the one-side irreversibility depends on two types of parameters:

$$
\begin{array}{ccc}
\tau & R/c_P & \text{Pr} \\
(4L/D_h)_1 & \text{Re}_1 & g_1
\end{array}
\tag{11.77}
$$

The upper row contains the parameters fixed by the selection of working-fluid and inlet conditions. The second row lists the three parameters that depend on the size and geometry of the heat-exchanger passage. How many of these three parameters are true "degrees of freedom" depends on the number of design constraints. One important constraint concerns the heat transfer area A_1. The minimization of irreversibility subject to constant area is important in cases where the cost of constructing the heat-exchanger surface is a major component in the overall cost formula for the heat exchanger.

In dimensionless form, the constant-area condition can be expressed as [2]

$$
a_1 = \frac{A_1}{\dot{m}} (2\rho P_1)^{1/2} \qquad \text{(constant)}
\tag{11.78}
$$

where a_1 is the dimensionless area of side "1" of the surface. It is easy to show that

$$
a_1 g_1 = \left(\frac{4L}{D_h} \right)_1
\tag{11.79}
$$

and that only two degrees of freedom remain for the minimization of $N_{S,1}$:

$$
N_{S,1}(g_1, \text{Re}_1) = \frac{\tau^2}{a_1 g_1 \, \text{St}_1} + \frac{R}{c_P} a_1 f_1 g_1^3
\tag{11.80}
$$

Minimizing the entropy-generation number subject to fixed (known) Reynolds number yields the optimum mass velocity:

$$
g_{1,\text{opt}} = \left[\frac{\tau^2}{(3R/c_P) a_1^2 f_1 \, \text{St}_1} \right]^{1/4}
\tag{11.81}
$$

$$
N_{S,1,\text{min}} = \left[\frac{256 \tau^6 (R/c_P) f_1}{27 a_1^2 \, \text{St}_1^3} \right]^{1/4}
\tag{11.82}
$$

The minimum entropy-generation number varies as $a_1^{-1/2}$; therefore, the thermodynamic goodness of the heat exchanger is enhanced by investing more area in the design of each side.

The above problem can be stated in the reverse direction to find the minimum area subject to fixed $N_{S,1}$ and Re_1, that is, to find the least-expensive design that guarantees a certain (hopefully, low) degree of irreversibility. The answer to this reverse problem turns out the same as in eq. (11.82), in which a_1 is replaced by $a_{1,min}$, and $N_{S,1,min}$ is replaced by $N_{S,1}$ (fixed):

$$a_{1,min} = \frac{1}{N_{S,1}^2} \left[\frac{256\tau^6 (R/c_P) f_1}{27 \, St_1^3} \right]^{1/2} \tag{11.82'}$$

In conclusion, the minimum area varies as $(1/N_{S,1})^2$.

Volume Constraint

The constant-volume constraint is important in the design of heat exchangers for applications where "space" is an expensive commodity (e.g., power plants for naval and airborne applications). The dimensionless constant-volume constraint can be written as

$$v_1 = V_1 \frac{8P_1}{v\dot{m}} \qquad \text{(constant)} \tag{11.83}$$

where V_1 is the volume of the passage (duct) on side "1." Noting that V_1 equals L_1 times the cross-sectional flow area A_1 [labeled A in eqs. (11.9) and (11.11)], we have also

$$v_1 g_1^2 = \left(\frac{4L}{D_h} \right)_1 Re_1 \tag{11.84}$$

This allows us to express $N_{S,1}$ in terms of only g_1 and Re_1 as degrees of freedom (note that Re_1 governs also the variation of St_1 and f_1):

$$N_{S,1} = \frac{\tau^2 \, Re_1}{v_1 g_1^2 \, St_1} + \frac{R}{c_P} \frac{v_1 f_1 g_1^4}{Re_1} \tag{11.85}$$

Regarding Re_1 as given, the optimum mass flowrate and corresponding minimum irreversibility are [2]

$$g_{1,opt} = \left[\frac{\tau^2 \, Re_1^2}{2(R/c_P) v_1^2 f_1 \, St_1} \right]^{1/6} \tag{11.86}$$

$$N_{S,1,min} = \left[\frac{27\tau^4 (R/c_P) \, Re_1 \, f_1}{4 v_1 \, St_1^2} \right]^{1/3} \tag{11.87}$$

The minimum irreversibility decreases as $v_1^{-1/3}$, i.e., as the size of the flow passage increases, and as the stream spends a longer time in residence in the passage. The reverse design problem of finding the minimum volume subject to the constraint of fixed entropy-generation rate has an answer that can be read directly off eq. (11.87), by replacing v_1 with $v_{1,\min}$ and $N_{S,1,\min}$ with $N_{S,1}$ (fixed):

$$v_{1,\min} = \frac{1}{N_{S,1}^3} \frac{27\tau^4 (R/c_P)\, \mathrm{Re}_1\, f_1}{4\, \mathrm{St}_1^2} \tag{11.87'}$$

In other words, one finds that the minimum volume varies as $(1/N_{S,1})^3$.

It is instructive to examine side by side the two conclusions that ended these last two subsections, rewriting them as

$$\text{minimum area} \sim l^2$$
$$\text{minimum volume} \sim l^3 \tag{11.88}$$

where l is another notation for $1/N_{S,1}$ (which was considered given). These two conclusions suggest that l plays the role of a dimensionless linear dimension of the side "1" of the heat-exchanger surface. Therefore, in order to build efficient heat exchangers (low N_Ss on both sides of the surface), one must contemplate sufficiently *large* units (large ls).

Combined Area and Volume Constraint

When the area A_1 and the volume V_1 of the heat exchanger passage are constrained simultaneously, there is only one degree of freedom left for the thermodynamic optimization procedure [see the second row, eq. (11.77)]. This problem was formulated originally as an exercise [26], and, next to the other thermodynamic design problems reviewed in this chapter, illustrates the directness of the irreversibility-minimization method. Combining the area and volume constraints (11.78) and (11.83) with the $N_{S,1}$ form collected from the right side of eq. (11.69), we obtain

$$N_{S,1} = \frac{\tau^2 v_1}{a_1^2\, \mathrm{St}_1\, \mathrm{Re}_1} + \frac{R}{c_P} \frac{a_1^4 f_1\, \mathrm{Re}_1^3}{v_1^3} \tag{11.89}$$

Here the only variable is Re_1, which affects $N_{S,1}$ both directly and through St_1 and f_1. In some designs, St_1 and f_1 are relatively insensitive to changes in Re_1 (e.g., in rough-wall pipes, at sufficiently high Reynolds numbers): in these cases, the thermodynamic optimum corresponds to

$$\mathrm{Re}_{1,\mathrm{opt}} = \frac{v_1}{a_1^{3/2}} \left[\frac{\tau^2}{3(R/c_P)\, \mathrm{St}_1\, f_1} \right]^{1/4} \tag{11.90}$$

It can be shown that the irreversibility distribution ratio ϕ at this optimum [i.e., the second term of eq. (11.89) divided by the first] is equal to $\frac{1}{3}$.

N_{tu} Constraint

Another way of exploiting the N_S-minimization idea was proposed recently by Sekulic and Herman [27]. What they consider fixed is the "operating point" in the heat transfer sense, that is, the overall number of heat transfer units (N_{tu}), the capacity flowrate ratio $[\omega = (\dot{m}c_P)_1/(\dot{m}c_P)_2]$, and the flow arrangement (counterflow, cross flow, etc.). Since, in general, the effectiveness is a function of N_{tu}, ω, and flow configuration, it means that Sekulic and Herman's optimization is carried out also at constant ϵ.

In the case of a general two-stream heat exchanger (finite ΔT and ΔPs, unspecified flow arrangement), the entropy-generation rate $\dot{S}_{gen}$ is given by eq. (11.60), where the first two terms on the right side account together for the heat transfer irreversibility, $\dot{S}_{gen,\Delta T}$. Writing

$$N_S = \frac{\dot{S}_{gen}}{(\dot{m}c_P)_2} \qquad N_{S,\Delta T} = \frac{\dot{S}_{gen,\Delta T}}{(\dot{m}c_P)_2} \qquad (11.91)$$

eq. (11.60) assumes the dimensionless form:

$$N_S = N_{S,\Delta T} + \underbrace{\left[-\omega \left(\frac{R}{c_P} \right)_1 \ln \left(1 - \frac{\Delta P_1}{P_1} \right) \right]}_{N_{S,\Delta P_1}} + \underbrace{\left[-\left(\frac{R}{c_P} \right)_2 \ln \left(1 - \frac{\Delta P_2}{P_2} \right) \right]}_{N_{S,\Delta P_2}}$$

$$(11.92)$$

where $N_{S,\Delta P_1}$ and $N_{S,\Delta P_2}$ are clearly the pressure-drop irreversibilities contributed by the two sides of the heat-exchanger surface. The heat transfer entropy-generation number $N_{S,\Delta T}$ is fixed in this case, because it is in general a function of ϵ (or N_{tu}), ω, and flow arrangement. The special form taken by $N_{S,\Delta T}$ in the case of balanced counterflow heat exchangers in the ideal limit (small ΔT and ΔPs) is listed as the first term on the right side of eq. (11.62), or as the sum of the first and third terms in eq. (11.69). The general $N_{S,\Delta T}$ expression for crossflow heat exchangers is listed in Ref. 27.

In conclusion, the overall irreversibility of the heat exchanger N_S varies on account of the two fluid-flow irreversibilities, $N_{S,\Delta P_1}$ and $N_{S,\Delta P_2}$. These two flow terms are coupled through the constant-N_{tu} constraint (11.65) in such a way that there exists an optimum pair of flow-passage slenderness ratios for which N_S (or $N_{S,\Delta P_1} + N_{S,\Delta P_2}$) is minimum. The presence of this optimum is illustrated qualitatively by Table 11.1, where $\Delta(\ )$ represents the sign of the change in the indicated quantity. The reading of this table proceeds from left to right.

Table 11.1 demonstrates that as the slenderness ratio of one passage increases, the slenderness of the companion passage decreases, and vice versa. The $N_{S,\Delta P}$-type terms of eq. (11.92) increase monotonically with their respective $(4L/D_h)$ ratios, hence, the summary constructed in the right side

TABLE 11.1   The Trade-off Between Pressure-Drop Irreversibilities in a Heat Exchanger with Constant Overall N_{tu}

$\Delta(4L/D_h)_1 \rightarrow$	$\Delta(N_{tu,1}) \rightarrow$ eq. (11.73)	$\Delta(N_{tu,2}) \rightarrow$ eq. (11.65) with N_{tu} = constant	$\Delta(4L/D_h)_2$ eq. (11.73) written for side "2"	$\Delta(N_{S,\Delta P_1}) \rightarrow$	$\Delta(N_{S,\Delta P_2})$ Conclusion
(+)	(+)	(−)	(−)	(+)	(−)
(−)	(−)	(+)	(+)	(−)	(+)

of the table. The search for the thermodynamic optimum was carried out numerically for certain heat-exchanger surfaces in cross flow [27]: a sample of this work is reproduced here in Fig. 11.11 [28]. The overall N_{tu} constraint is listed on each of the V-shaped curves. As N_{tu} increases, the optimum slenderness ratio of one side of the surface increases.

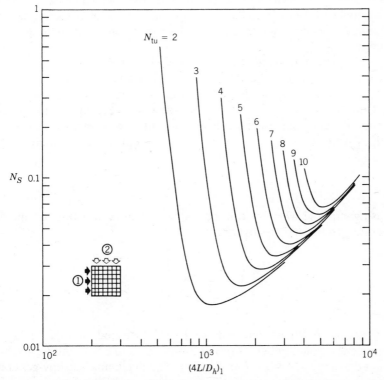

Figure 11.11   The minimization of entropy-generation number subject to constant overall N_{tu} (drawn after Ref. 27; the results refer to a compact crossflow heat exchanger with $\omega = 0.5$, inlet temperatures ratio $T_1/T_2 = 0.8$, and two surfaces out of Ref. 28, namely, "plate fin 11.1" and "louvered fin 3/8–6.06").

HEAT EXCHANGERS WITH NEGLIGIBLE PRESSURE-DROP IRREVERSIBILITY

In this section, we take a closer look at the limit in which the $N_{S,\Delta P}$ terms are negligible in the general expression (11.92):

$$N_S \cong N_{S,\Delta T} \tag{11.93}$$

Much of the attention that is being devoted to this limit in the literature is centered around the seemingly paradoxical conclusion that the irreversibility of such heat exchangers decreases both as $\epsilon \to 1$ [expected; e.g., eq. (11.62)] *and* as $\epsilon \to 0$ (unexpected). I was the first to write about this paradox [5], when in 1980 I published a drawn example used by Professor Tribus in his "Thermoeconomics" class at MIT [29]. Back then I felt that the origin of this paradox was fairly clear, and that the "entropy maximum" associated with it is of little practical consequence [see also Ref. 1, pp. 28–29]. In the meantime, the subject continued to draw attention in settings (configurations) that are considerably more general than the original balanced counterflow example of Ref. 5. In fact, the essential idea of the existence of a maximum N_S versus ϵ was rediscovered independently on three additional continents [30–34].

It is useful to reconsider the maximum-entropy paradox [5], because it constitutes an excellent illustration of the importance of the principle of thermodynamic "isolation" in the optimization of an engineering component. Consider again a balanced counterflow heat exchanger, for which in the limit of negligible pressure drop, the entropy-generation rate (11.60) combines with eqs. (11.61) into the one-term entropy-generation number:

$$N_S = \ln \left\{ \left[1 - \epsilon \left(1 - \frac{T_2}{T_1} \right) \right] \left[1 + \epsilon \left(\frac{T_1}{T_2} - 1 \right) \right] \right\} \tag{11.94}$$

Or, using eq. (11.67), N_S can be expressed as a function of the inlet temperature ratio and the overall N_{tu}:

$$N_S = \ln \frac{\left(1 + \frac{T_1}{T_2} N_{tu} \right) \left(1 + \frac{T_2}{T_1} N_{tu} \right)}{(1 + N_{tu})^2} \tag{11.95}$$

The behavior of N_S at constant T_1/T_2 is illustrated in Fig. 11.12. The entropy-generation number is zero at both $\epsilon = 0$ and $\epsilon = 1$, and its maximum is situated exactly at $\epsilon = \frac{1}{2}$ (or at $N_{tu} = 1$). The maximum entropy-generation number increases as soon as T_1/T_2 goes above or below 1:

$$N_{S,\max} = \ln \left[\frac{1}{2} + \frac{1}{4} \left(\frac{T_1}{T_2} + \frac{T_2}{T_1} \right) \right] \tag{11.96}$$

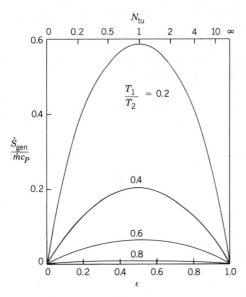

Figure 11.12 Entropy-generation rate in a balanced counterflow heat exchanger with zero pressure-drop irreversibility [1, 5] (drawn after Tribus [29]).

The behavior of N_S in the "good" heat-exchanger limit ($\epsilon \rightarrow 1$) is well understood and expected. The behavior in the $\epsilon \rightarrow 0$ extreme is neither expected nor intuitively obvious, because we expect any heat transfer irreversibility to increase monotonically as the heat-exchanger area (or N_{tu}) decreases. One physical explanation for the anomalous behavior exhibited in Fig. 11.12 is that in the $\epsilon \rightarrow 0$ limit, the stream-to-stream temperature difference ΔT is practically constant and practically equal to $|T_1 - T_2|$. In other words, ΔT is insensitive to the vanishing N_{tu}. Consequently, what decreases as N_{tu} and the heat-exchanger area approach zero is the total heat transfer rate between the two streams, and, along with it, the heat transfer irreversibility (see the second problem in Table 3.1). The vanishing N_S seen in the limit $\epsilon \rightarrow 0$ is first and foremost a sign that the heat exchanger *disappears* as an engineering component. So, if $\epsilon \rightarrow 1$ is the "good" heat-exchanger limit, we can then view $\epsilon \rightarrow 0$ as the "absent" heat-exchanger limit.

Heat exchangers, in general, contribute to the overall irreversibility of the installations that incorporate them. This is why the lower-left corner of Fig. 11.12 is technically correct, because an absent heat exchanger ($\epsilon = 0$) can only contribute zero irreversibility as a heat exchanger ($N_S = 0$). However, if we think of the power and refrigeration applications that over the past 200 years defined the need for inventing heat exchangers, we begin to appreciate the error in associating goodness with the declining-N_S trend observed toward $\epsilon = 0$. Zero irreversibility is certainly good if the heat exchanger exists and does its job; however, a vanishing heat exchanger will most definitely have a negative effect on the overall irreversibility of the mother system.

The absent heat-exchanger limit ($\epsilon \to 0$) is a territory in which the design changes experienced by the vanishing heat exchanger have a profound effect on the irreversibility of the system components with which the heat exchanger is in direct communication. To see how the analysis of the entropy maximum violates the principle of system isolation, consider the greater problem of minimizing the irreversibility of the Brayton-cycle power plant alluded to already in Fig. 11.9. The power plant has five components, which, proceeding clockwise, are

the heater (H)
the expander
the regenerator (R)
the cooler (C)
the compressor

The regenerator (R) is the balanced counterflow heat exchanger for which Fig. 11.12 was drawn. For simplicity, we assume that the expander and compressor function reversibly and adiabatically, so that the entropy generation of the entire Brayton-cycle power plant (B) has only three components:

$$\dot{S}_{gen,B} = \dot{S}_{gen,H} + \dot{S}_{gen,R} + \dot{S}_{gen,C} \qquad (11.97)$$

Dividing by the capacity flowrate $\dot{m}c_P$ (constant), we have also

$$N_{S,B} = N_{S,H} + N_{S,R} + N_{S,C} \qquad (11.98)$$

Note that the $N_{S,R}$ contribution is what in eqs. (11.94) and (11.95) was labeled N_S. Note also that since the expander and compressor function isentropically, the traces left by them on the T–S diagram of Fig. 11.9 would be represented by vertical segments in the Brayton cycle.

By modeling the working fluid as an ideal gas with constant specific heat, it is a simple matter to show that the heat-engine (first-law) efficiency of the power plant (η) depends on three dimensionless parameters, the regenerator effectiveness ϵ, the overall temperature ratio T_{max}/T_{min}, and the absolute temperature ratio across the expander or the compressor. The latter is fixed by the pressure ratio, which is assumed given:

$$\frac{T_{max}}{T_2} = \frac{T_1}{T_{min}} = \left(\frac{P_1}{P_2}\right)^{R/c_P} \qquad (11.99)$$

The first-law efficiency formula turns out to be

$$\eta = \frac{r_{\text{large}}(1 - r_{\text{small}}^{-1}) - r_{\text{small}} + 1}{r_{\text{large}}(1 - \epsilon r_{\text{small}}^{-1}) - (1 - \epsilon)r_{\text{small}}} \qquad (11.100)$$

where $r_{\text{large}} = T_{\max}/T_{\min}$ and $r_{\text{small}} = T_{\max}/T_2$.

Figure 11.13a proves what we expected from the beginning: "more" regenerator (larger ϵ) is always better from the point of view of upgrading the efficiency of the entire power system. Figure 11.13b shows, first, that the irreversibility of the entire system $(N_{S,B})$ decreases monotonically as ϵ

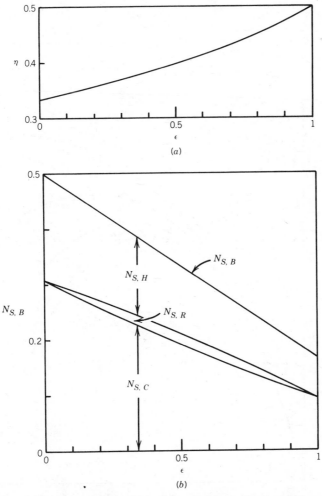

Figure 11.13 (a) The monotonic dependence between Brayton-cycle power-plant efficiency and the size of the regenerative counterflow heat exchanger (drawn for $r_{\text{large}} = 3$ and $r_{\text{small}} = \frac{3}{2}$; hence, $T_1/T_2 = 0.75$). Part (b) shows the distribution of power-plant irreversibility among the three heat exchangers: (H), (R), and (C).

increases (this agrees with Fig. 11.13a and the Gouy–Stodola theorem). Figure 11.13b shows also the manner in which $N_{S,B}$ is distributed among the three heat exchangers: (H), (R), and (C). The heater entropy-generation rate was calculated assuming that the heat input originates from T_{max} and the outlet temperature of the heated stream (i.e., the inlet to the expander) equals T_{max}. In other words, the heater is assumed to have an infinite number of heat transfer units. An identical model was used for the cooler, where the outlet temperature of the cooled stream equals the temperature of the cold side of the cooler, T_{min}.

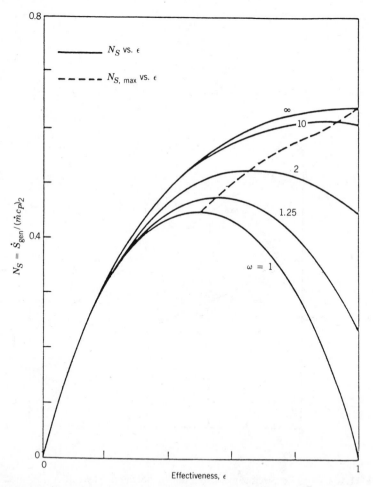

Figure 11.14 The irreversibility of imbalanced counterflow heat exchangers with negligible pressure-drop irreversibility and $T_1/T_2 = 0.25$ (after Sarangi and Chowdhury [30]).

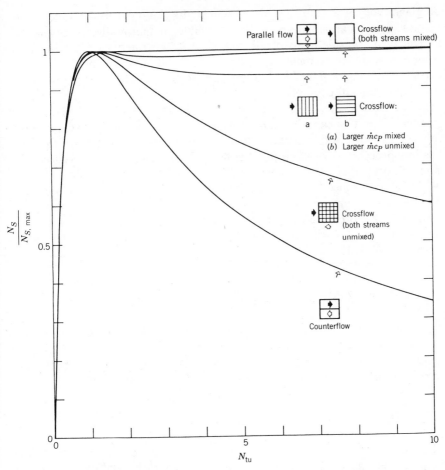

Figure 11.15 The occurrence of a maximum in the irreversibility of various heat-exchanger configurations in which the pressure-drop irreversibility is negligible $(T_1/T_2 = 0.5, \omega = 1;$ after Sekulic [34]).

Sandwiched between $N_{S,H}$ and $N_{S,C}$ is the irreversibility contribution made by the regenerator: this slice has exactly the same features as the $T_1/T_2 = 0.75$ curve that could be drawn near the base of Fig. 11.12, namely, zero height at $\epsilon = 0$ and $\epsilon = 1$, and a maximum at $\epsilon = \frac{1}{2}$. It is now clear that as ϵ decreases below $\frac{1}{2}$, the vanishing of the regenerator has the effect of *augmenting* the heat transfer irreversibilities contributed by the surviving heat exchangers, so that, overall, the irreversibility of the power plant increases monotonically. The only practical significance that I can attach to the maximum thickness exhibited by the $N_{S,R}$ slice is that it marks the order of magnitude of ϵ (or N_{tu}) below which the analysis of the regenerator "alone" is a clear violation[†] of the principle of thermodynamic isolation.

The heat transfer irreversibility maximum illustrated here for balanced counterflow heat exchangers reappears in the analysis of other heat-exchanger configurations. For example, Sarangi and Chowdhury [30] found it in imbalanced counterflow heat exchangers, that is, when $(\dot{m}c_P)_1 \neq (\dot{m}c_P)_2$. Their results are illustrated in Fig. 11.14, where N_S is based on the smaller of the two capacity rates, $(\dot{m}c_P)_2$. Sekulic and Baclic [31] plotted it for counterflow and crossflow heat exchangers, showing also that the maximum occurs at $\epsilon = 1$ in parallel-flow heat exchangers. This last conclusion was drawn also by da Costa and Saboya [32] in a comparative study of $N_{S,\Delta T}$ for imbalanced counterflow and parallel-flow heat exchangers. The presence of the maximum in the irreversibility of crossflow heat exchangers with negligible pressure-drop irreversibility is illustrated in Fig. 11.15, where the N_S value of each configuration has been divided by the respective maximum N_S [34].

REMANENT (FLOW-IMBALANCE) IRREVERSIBILITIES

The study of balanced counterflow heat exchangers led to the conclusion that the overall irreversibility of the device decreases to zero as the design approaches the ideal limit of infinite overall N_{tu} and zero ΔP on both sides of the surface. The focus of this section is strictly on the "perfect" design:

$$N_{tu} = \infty \qquad \Delta P_1 = \Delta P_2 = 0 \qquad (11.101)$$

while its objective is to show that in this limit, the heat-exchanger configurations that are not "balanced counterflow" are characterized by an unavoidable irreversibility that is due solely to the flow arrangement. For historical reasons [24] and for lack of a better name, we shall refer to this irreversibility as the *irreversibility due to flow imbalance* or *remanent irreversibility*.

Consider first an *imbalanced counterflow heat exchanger*, where

$$\omega = \frac{(\dot{m}c_P)_1}{(\dot{m}c_P)_2} > 1 \qquad (11.102)$$

[†]Indeed, the reason why the ideal-limit optimization rules of the preceding section enjoy general validity is that the assumption of vanishingly small ΔT and ΔPs makes the outlet conditions of the two streams practically insensitive to the optimization work performed inside the heat exchanger, i.e., insensitive to the changes in the already small ΔT and ΔPs. This decision effectively isolates the counterflow heat exchanger from the rest of the power or refrigeration installation to which it may belong.

and where the perfect design (11.101) means $P_{1,\text{out}} = P_1$, $P_{2,\text{out}} = P_2$, and $\epsilon = 1$. For imbalanced counterflow heat exchangers, the effectiveness–N_{tu} relations (11.61), (11.66), and (11.67) are replaced by the more general definitions [25]:

$$N_{\text{tu}} = \frac{\bar{h}A_1}{(\dot{m}c_P)_2} \qquad \epsilon = \omega\,\frac{T_1 - T_{1,\text{out}}}{T_1 - T_2} = \frac{T_{2,\text{out}} - T_2}{T_1 - T_2} \qquad (11.103)$$

$$\epsilon = \frac{1 - \exp\left[-N_{\text{tu}}(1 - \omega^{-1})\right]}{1 - \omega^{-1}\exp\left[-N_{\text{tu}}(1 - \omega^{-1})\right]} \qquad (11.104)$$

where $(\dot{m}c_P)_2$ is the *smaller* of the two capacity flowrates. In this case, the overall entropy-generation rate (11.60) has a finite value:

$$N_{S,\text{imbalance}} = \frac{\dot{S}_{\text{gen}}}{(\dot{m}c_P)_2} = \ln\left\{\left[1 - \frac{1}{\omega}\left(1 - \frac{T_2}{T_1}\right)\right]^{\omega}\frac{T_1}{T_2}\right\} \qquad (11.105)$$

As expected, this imbalance irreversibility decreases to zero in the balanced counterflow limit treated earlier, i.e., as $\omega \to 1$ while $\omega > 1$,

$$N_{S,\text{imbalance}} \to (\omega - 1)\left(\frac{T_1}{T_2} - 1 - \ln\frac{T_1}{T_2}\right) \qquad (11.106)$$

The imbalance irreversibility of a two-stream *heat exchanger with phase change on one side* is a special case of eq. (11.105), namely, the limit $\omega \to \infty$, where the stream that bathes side "1" does not experience a temperature variation from inlet to outlet, $T_{1,\text{out}} = T_1$. For this class of heat exchangers, eq. (11.105) reduces to

$$N_{S,\text{imbalance}} = \frac{T_2}{T_1} - 1 - \ln\frac{T_2}{T_1} \qquad (\omega = \infty) \qquad (11.107)$$

The quantity listed above increases if the ratio T_2/T_1 (or T_1/T_2) moves away from the value 1: this behavior is illustrated in Fig. 3.12, which was drawn based on an analytically identical expression, eq. (3.35).

The imbalance irreversibility of two-stream *parallel-flow heat exchangers* is obtained similarly, by combining eq. (11.60) with the perfect-design conditions (11.101) and the $\epsilon(\omega, N_{\text{tu}})$ relation for parallel flow [25]:

$$\epsilon = \frac{1 - \exp\left[-N_{\text{tu}}(1 + \omega^{-1})\right]}{1 + \omega^{-1}} \qquad (11.108)$$

The resulting expression is

$$N_{S,\text{imbalance}} = \frac{\dot{S}_{\text{gen}}}{(\dot{m}c_P)_2} = \ln\left\{\left(\frac{T_2}{T_1}\right)^{\omega}\left[1 + \left(\frac{T_1}{T_2} - 1\right)\frac{\omega}{1 + \omega}\right]^{1+\omega}\right\} \qquad (11.109)$$

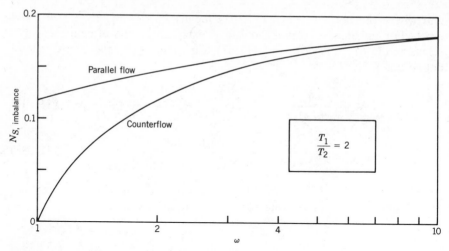

Figure 11.16 The remanent (flow-imbalance) irreversibility in parallel flow is consistently greater than in counterflow; $\omega = (\dot{m}c_P)_1/(\dot{m}c_P)_2 > 1$.

A first observation is that in the limit of extreme imbalance ($\omega \to \infty$), this expression becomes the same as eq. (11.107). In this limit, of course, the side "1" stream is so large that its temperature remains equal to T_1 from inlet to outlet; seen from the outside, it behaves like a stream that condenses or evaporates isobarically.

A second worthwhile observation is that when the two streams and their inlet conditions are given, the imbalance irreversibility of the parallel flow arrangement, eq. (11.109), is consistently greater than the imbalance irreversibility of the counterflow scheme, eq. (11.105). Figure 11.16 shows the behavior of the respective entropy-generation numbers, and how they both approach the value indicated by eq. (11.107), as the flow-imbalance ratio ω increases. Taking the $\omega = 1$ limit of eq. (11.109), it is easy to see that the remanent irreversibility of the parallel-flow arrangement is finite even in the "balanced flow" case (see also $\omega = 1$ in Fig. 11.16).

OVERVIEW: THE STRUCTURE OF HEAT-EXCHANGER IRREVERSIBILITY

There is an important structure to be recognized in the heat-exchanger irreversibility treatment reviewed in the preceding three sections. First, there is the competition between heat transfer and fluid-flow (pressure-drop) irreversibilities, whose various trade-offs were illustrated by considering the analytically simple limit of neary ideal, balanced counterflow heat exchangers. Second, it is essential to keep track of whether the optimization of one heat exchanger causes the thermodynamic degradation of other components to which it might be hooked up in the greater engineering system. We saw the violation of the principle of proper "isolation" by reexamining the maximum-entropy paradox of heat exchangers with zero pressure-drop irreversibility. And, finally, there is the recognition of remanent or flow-imbalance irreversibilities, that is, irreversibilities that persist even in the limit of perfect heat exchangers, eqs. (11.101).

This structure is summarized in Fig. 11.17. The remanent irreversibility deserves to be calculated first in the thermodynamic optimization of any heat exchanger, because it establishes the level (order of magnitude) below which the joint minimization of heat transfer (finite ΔT, or N_{tu}) and fluid-flow (finite ΔP) irreversibilities falls in the realm of diminishing returns. In other words, it would no longer make sense to invest heat-exchanger area and "engineering" into minimizing the sum $(N_{S,\Delta T} + N_{S,\Delta P})$ when this sum[†] is already negligible compared with the remanent irreversibility $N_{S,\text{imbalance}}$.

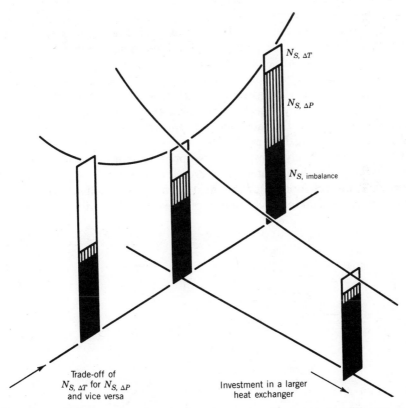

Figure 11.17   The structure of the total entropy-generation rate of a heat exchanger.

[†]In this summary $N_{S,\Delta P}$ is shorthand for the combined effect of pressure-drop irreversibilities [e.g., the sum of the last two terms in eq. (11.62)].

Only in very special cases does the entropy-generation rate of a heat exchanger break up into a sum of three terms such that each term accounts for one of the irreversibilities reviewed above:

$$N_S = N_{S,\text{imbalance}} + N_{S,\Delta T} + N_{S,\Delta P} \tag{11.110}$$

One such case is the balanced counterflow heat exchanger in the nearly balanced and nearly ideal limit ($\omega \to 1, \Delta T \to 0, \Delta Ps \to 0$), which is discussed in detail in Ref. 24. In general, these three irreversibilities contribute in a more complicated way to the eventual size of the overall N_S. Deep down, however, the behavior of the three is the same as in the simple limits that were singled out for discussion until now. This behavior is illustrated qualitatively in Fig. 11.17.

TWO-PHASE-FLOW HEAT EXCHANGERS

The analysis of other classes of heat exchangers reveals the basic structure outlined in Fig. 11.17. One important class are the heat exchangers in which at least one of the streams is a two-phase mixture. We alluded to the remanent irreversibility of some heat exchangers of this type in eq. (11.107). The irreversibility characteristics of this class were discussed first in Ref. 35 and, independently, by London and Shah [36]. Detailed sizing rules for thermodynamic and thermoeconomic optimization were developed by Zubair et al. [37] and Lau et al. [38].

Consider the steady flow of a two-phase mixture through a duct in thermal contact with a heat reservoir of temperature T_0 (Fig. 11.18). To understand the functioning of this heat exchanger, it helps to think of the condenser in a Rankine-cycle power plant, where T_0 is the absolute tem-

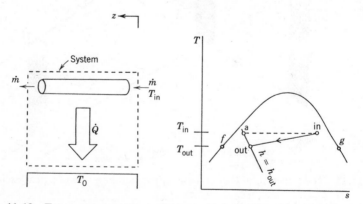

Figure 11.18 Entropy-generation analysis of two-phase flow through a heat-exchanger duct [35].

perature of the atmosphere. The result of the following analysis, however, is quite general. Invoking the first law and the second law for the dashed-line control volume shown on the left side of Fig. 11.18, we find that the total entropy-generation rate is

$$\dot{S}_{gen} = \dot{m}(s_{out} - s_{in}) + \frac{\dot{m}(h_{in} - h_{out})}{T_0} \geq 0 \qquad (11.111)$$

where $\dot{m}(h_{in} - h_{out})$ is the heat-rejection rate to the ambient, $\dot{Q}$. From a design standpoint, we are interested in how the pressure drop ($\Delta P = P_{in} - P_{out}$) and the fluid–ambient temperature difference ($\Delta T = T_{in} - T_0$) affect the overall irreversibility level, $\dot{S}_{gen}$. Assuming that the inlet and outlet states are both in the two-phase domain, Fig. 11.18 (right), we write

$$s_{out} - s_{in} = (s_a - s_{in}) + (s_{out} - s_a) \qquad (11.112)$$

where the auxiliary state (a) is defined by the two properties:

$$T_a = T_{in} \quad \text{and} \quad h_a = h_{out} \qquad (11.113)$$

In other words, state (a) represents the outlet state (out) in the theoretical limit of zero pressure drop. We note further that

$$T_{in}(s_a - s_{in}) = h_a - h_{in}$$
$$= h_{out} - h_{in} \qquad (11.114)$$

Combining eqs. (11.111)–(11.114), we find that the entropy-generation rate separates into two terms:

$$\dot{S}_{gen} = \dot{Q}\left(\frac{1}{T_0} - \frac{1}{T_{in}}\right) + \dot{m}(s_{out} - s_a) \qquad (11.115)$$

where, quite visibly, the first term represents the contribution due to imperfect stream–ambient thermal contact, and the second term accounts for the pressure-drop irreversibility. By keeping in mind the one-to-one relationship between temperature and pressure inside the two-phase dome, the geometric construction of state (a) in Fig. 11.18 (right) makes it clear that ($s_{out} - s_a$) is finite and positive as soon as a pressure drop occurs between inlet and outlet [the flow (in)→(out) does not have to be from right to left, as pictured in Fig. 11.18]. In the limit of sufficiently small ΔP, the relationship between ($s_{out} - s_a$) and ΔP can also be expressed analytically:

$$s_{out} - s_a = [h_f' - T_{in}s_f' + x_{out}(h_{fg}' - T_{in}s_{fg}')]\frac{\Delta P}{T_{in}} \qquad (11.116)$$

where $(\ )'$ means $d(\ )/dP$, and x_{out} is the quality of the outflowing mixture. Properties at saturation, such as h_f, s_f, h_{fg}, and s_{fg}, are known functions of pressure (or temperature): therefore, the quantity in brackets in eq. (11.116) can be calculated once the absolute pressure and the outlet quality are known. The quantity calculated in this manner has the units of specific volume; therefore, we can substitute the following shorthand notation for it:

$$\tilde{v}(P_{in}, x_{out}) = h'_f - T_{in}s'_f + x_{out}(h'_{fg} - T_{in}s'_{fg})$$
$$= h'_f - T_{in}s'_f + x_{out}v_{fg} \tag{11.117}$$

By using this new notation, the entropy-generation rate assumes now a more familiar form:

$$\dot{S}_{gen} = \underbrace{\dot{Q}\frac{\Delta T}{T_{in}T_0}}_{\substack{\text{Heat transfer} \\ \text{irreversibility}}} + \underbrace{\frac{\dot{m}\tilde{v}}{T_{in}}\Delta P}_{\substack{\text{Fluid-flow} \\ \text{irreversibility}}} \tag{11.118}$$

Repeating the analysis for a duct of length dz, where the longitudinal coordinate z is measured in the direction of the flowrate $\dot{m}$, we obtain the per-unit-length result:

$$\frac{d\dot{S}_{gen}}{dz} = \frac{\Delta T}{T_{in}^2}\frac{d\dot{Q}}{dz} + \frac{\dot{m}\tilde{v}}{T_{in}}\left(-\frac{dP}{dz}\right) \tag{11.119}$$

This result contains the additional assumption that $(T_{in} - T_0) \ll T_{in}$. The structure of eq. (11.119) is the same as that of entropy-generation formulas encountered earlier. Therefore, by combining eq. (11.119) with appropriate correlations for heat transfer coefficient and pressure drop in two-phase flow, it is possible to select a design (e.g., duct inner diameter) so that $d\dot{S}_{gen}/dz$ is minimum.

If the duct of Fig. 11.18 is surrounded by still air, and if T_w is the wall temperature, then

$$T_{in} - T_0 = (T_{in} - T_w) + (T_w - T_0) \tag{11.120}$$

and the local rate of entropy generation becomes

$$\frac{d\dot{S}_{gen}}{dz} = \left(\frac{d\dot{Q}/dz}{T_{in}}\right)^2\left[\frac{1}{(p\bar{h})_i} + \frac{1}{(p\bar{h})_o}\right] + \frac{\dot{m}\tilde{v}}{T_{in}}\left(-\frac{dP}{dz}\right) \tag{11.121}$$

In this expression, p and $\bar{h}$ denote the wetted perimeter of the duct cross-section and the heat transfer coefficient, while subscripts "i" and "o"

stand for the inner side and outer side of the duct wall, respectively. In writing only one pressure-drop term in eq. (11.121), we are assuming that the flow on the outside of the duct is driven by buoyancy effects. The entropy-generation rate is due to only three contributors, the imperfect thermal contact between two-phase mixture and wall, the imperfect thermal contact between wall and ambient, and, finally, the flow with friction through the duct.

If the tube of Fig. 11.18 is surrounded not by a stagnant fluid reservoir, but by an evaporating stream at a lower temperature, then the entropy-generation rate of this two-stream heat exchanger is

$$\frac{d\dot{S}_{gen}}{dz} = \left[\frac{1}{T^2 p\bar{h}} \left(\frac{d\dot{Q}}{dz} \right)^2 + \frac{\dot{m}\tilde{v}}{T} \left(-\frac{dP}{dz} \right) \right]_H + \left[\frac{1}{T^2 p\bar{h}} \left(\frac{d\dot{Q}}{dz} \right)^2 + \frac{\dot{m}\tilde{v}}{T} \left(-\frac{dP}{dz} \right) \right]_C$$

$$(11.122)$$

Subscripts "H" and "C" represent the condensing (hot) and evaporating (cold) sides of the heat-exchanger surface, respectively. Equation (11.122) follows directly from eq. (11.119), by first writing eq. (11.119) for each side of the surface (T_0 is replaced by T_w in this case), and then adding the two expressions side by side. Note further that the First Law of Thermodynamics requires $(d\dot{Q}/dz)_H = (d\dot{Q}/dz)_C$. The competition between heat transfer and fluid-flow irreversibilities, or the opportunity for reaching a thermodynamic optimum based on the proper selection of duct geometry is evident in both groups of terms on the right side of eq. (11.122).

Finally, regarding the shorthand $\tilde{v}$ for what is listed in eq. (11.117), it is important to keep in mind the geometrical layout of the T–s diagram of Fig. 11.18, on which the $\tilde{v}$ expression is based. Regardless of whether the process (in) $\rightarrow$ (out) represents condensation or evaporation, it was assumed that the fictitious isenthalpic process (a) $\rightarrow$ (out) is situated fully inside the two-phase dome. If the tube of Fig. 11.18 works as a *condenser*, then the outlet state will be situated on the left-hand frontier of the dome, in other words (out) $\equiv$ (f). In this case, the role of (a) will be played by a slightly subcooled (compressed) liquid state, and the process (a) $\rightarrow$ (out) will be executed by a single-phase stream. Invoking $dh = T\,ds + v\,dP$ and the constancy of enthalpy from (a) to (out), it is easy to prove that when (out) means saturated liquid, the $\tilde{v}$ of eq. (11.117) must be replaced by v_f.

Similarly, if the stream boils so that the outlet state is a single-phase saturated-vapor state, (out) is replaced by a state (g) situated on the right side of the two-phase dome. If the orientation of the constant-enthalpy line that passes through (g) is such that (a) would be situated outside the dome (i.e., to the right), then, based on the argument of the preceding paragraph, the $\tilde{v}$ of eq. (11.117) is replaced by v_g. Conversely, if the orientation of the constant-enthalpy line would make (a) fall inside the dome, the $\tilde{v}$ expression (11.117) holds. Note that in this last case, $x_{out} = 1$.

OTHER HEAT-EXCHANGER CONFIGURATIONS AND WAYS OF MEASURING IRREVERSIBILITY

In the few years that passed since the pioneering papers of the 1970s, the calculation and minimization of irreversibility in heat-exchanger design has become a self-standing topic that continues to gather momentum. For example, this topic recently attracted Professor London, a most creative and influential figure whose long Stanford career had a lot to do with the streamlining and popularization of first-law analysis in heat-exchanger design: in 1982, he published his first paper on the second-law analysis of heat exchangers [39], and, through him, the language of second-law analysis was spoken in the prestigious Max Jakob Award acceptance lecture (Denver, 1985). In this section, we review a number of additional advances in heat-exchanger second-law analysis, placing a special emphasis on alternate approaches that have been proposed for the dimensionless reporting of irreversibility (i.e., in place of N_S).

Extensions to the study of entropy generation in counterflow heat exchangers [24] have been published by Sarangi and Chowdhury [30] and Huang [40]. A study of compact *crossflow heat exchangers* was conducted along similar lines by Baclic and Sekulic [41]. Their study reveals once again the trade-off between heat transfer and fluid-flow irreversibilities, and the remanent (flow-imbalance) irreversibility associated purely with the crossflow arrangement. Basic studies of the thermodynamics of forced-convection heat transfer was undertaken also by Dr. Soumerai in Switzerland [42–44]. The relationship between irreversibility minimization and cost minimization was illustrated by Wepfer et al. in the problem of deciding the optimum size of a steam pipe and its insulation [45].

A new and promising direction has been traced in a sequence of studies by Professor Zilberberg [46–49]. He draws attention to the unsteady (often, periodic) character of the operation of most power and refrigeration plants, and to the irreversibility that is due solely to this unsteadiness. He calls this effect *dynamic irreversibility*. Problems of plant start-up and shutdown fall also in the domain identified by him, and so does the basic thermal-energy storage problem that is reviewed in the next section.

The design principle that works at the component and subcomponent level works also at the overall system level. In the realm of heat-exchanger design, then, it is worth noting the application of second-law concepts to the optimization of entire *heat-exchanger networks*. Studies of this kind have been contributed most recently by Chato and Damianides [50] and Hesselmann [51].

Finally, we turn our attention to the choice of dimensionless reporting of the calculated irreversibility figure. In most of the examples reviewed until now, the entropy-generation rate was nondimensionalized by dividing it through a capacity flowrate, say, $(\dot{m}c_P)_2$ in the case of imbalanced two-stream heat exchangers, eq. (11.105). The *entropy-generation numbers* that can be defined in this manner (N_S, $N_{S,\Delta T}$, $N_{S,\Delta P_1}$, etc.) are second-law "relatives" of the older concept of the number of heat transfer units (N_{tu}), which is used in traditional first-law analyses of heat exchangers. And, just like the N_{tu}, the N_S value can vary from 0 all the way to ∞: whether the calculated N_S represents a "high" or "low" entropy-generation rate depends on the size of the heat-exchanger N_S that can be tolerated economically [see the discussion centered around eq. (11.110)]; on the magnitude of the remanent irreversibility, $N_{S,\text{imbalance}}$ (Fig. 11.17); and, certainly, on the entropy-generation levels shown by the other components that make up the greater system.

In some problems, it is possible to nondimensionalize $\dot{S}_{gen}$ by dividing it through a known entropy-generation rate, which is regarded as reference. An example of this kind is the *augmentation entropy-generation number* $N_{S,a}$, eq. (11.31). It is because of this definition that the value $N_{S,a} = 1$ acquires a special meaning, Fig. 11.5. However, $N_{S,a}$ can be both smaller and greater than 1, depending on whether the augmentation of heat transfer is thermodynamically advantageous.

Another dimensionless measure of heat-exchanger irreversibility is the rational (second-law) effectiveness introduced by Bruges [52] and Reistad [53]:

$$\epsilon_R = \frac{\text{availability (exergy) gained by the cold stream}}{\text{availability (exergy) donated by the warm stream}}$$

$$= \frac{\dot{m}_C(e_{x,\text{out}} - e_{x,\text{in}})_C}{\dot{m}_H(e_{x,\text{in}} - e_{x,\text{out}})_H} \tag{11.123}$$

This quantity varies monotonically with the entropy-generation number N_S:

$$\epsilon_R = 1 - \frac{T_0 \dot{S}_{gen}}{\dot{m}_H(e_{x,\text{in}} - e_{x,\text{out}})_H}$$

$$= 1 - \frac{T_0 c_{P,H}}{(e_{x,\text{in}} - e_{x,\text{out}})_H} N_S \tag{11.124}$$

where, for the sake of the argument, $N_S = \dot{S}_{gen}/(\dot{m}c_P)_H$, and T_0 is the absolute temperature of the ambient. In the limit of reversible heat-exchanger operation (zero ΔT and ΔPs), ϵ_R is equal to 1. The lowest possible value for ϵ_R is zero: this occurs in the limit in which the heat exchanger physically disappears. Assuming that the pressure-drop irreversibility contribution is negligible, Golem and Brzustowski [54] showed that ϵ_R reduces to

$$\epsilon_R = \pm \frac{(\dot{m}c_P)_C[T_{\text{out}} - T_{\text{in}} - T_0 \ln(T_{\text{out}}/T_{\text{in}})]_C}{(\dot{m}c_P)_H[T_{\text{out}} - T_{\text{in}} - T_0 \ln(T_{\text{out}}/T_{\text{in}})]_H} \tag{11.125}$$

where the subscripts "in" and "out" refer to the outlets and the inlets, respectively. The "+" sign applies to counterflow, and the "−" sign to parallel flow. Equation (11.125) holds for ideal gases and for incompressible liquids with negligible pressure drop. The same authors extended the ϵ_R concept to the local level, showing that when the longitudinal temperature distributions $T_C(x)$ and $T_H(x)$ are known, one can evaluate locally the destruction of exergy via heat transfer across the stream-to-stream temperature difference.

The newest proposal for the dimensionless reporting of heat-exchanger irreversibility is due to Professors Witte and Shamsundar of the University of Houston [55]. Their second-law heat-exchanger efficiency is defined as

$$\eta_{W-S} = 1 - \frac{T_0 \dot{S}_{\text{gen}}}{\dot{Q}} \tag{11.126}$$

where $\dot{Q}$ is the total stream-to-stream heat transfer rate:

$$\dot{Q} = \dot{m}_H (h_{\text{in}} - h_{\text{out}})_H = \dot{m}_C (h_{\text{out}} - h_{\text{in}})_C \tag{11.127}$$

The reason for choosing the efficiency expression (11.126) is that while evaluating the thermodynamic imperfection of a real heat exchanger by comparing it with an ideal one that operates reversibly, Witte and Shamsundar regard $\dot{Q}$ as fixed. The η_{W-S} efficiency is related to ϵ_R and N_S in the following ways:

$$\eta_{W-S} = 1 - \frac{T_0 c_{P,H}}{(h_{\text{in}} - h_{\text{out}})_H} N_S \tag{11.128}$$

$$\frac{1 - \eta_{W-S}}{1 - \epsilon_R} = \left(\frac{e_{x,\text{in}} - e_{x,\text{out}}}{h_{\text{in}} - h_{\text{out}}} \right)_H \tag{11.129}$$

where N_S was defined again as $\dot{S}_{\text{gen}}/(\dot{m}c_P)_H$. If pressure drops are neglected, a visually pleasing alternative to writing η_{W-S} is [55]

$$\eta_{W-S} = 1 + \frac{T_0}{\bar{T}_H} - \frac{T_0}{\bar{T}_C} \tag{11.130}$$

where $\bar{T}_H$ and $\bar{T}_C$ are two average temperatures defined by

$$\bar{T}_{H,C} = \left(\frac{\int_{\text{in}}^{\text{out}} dh}{\int_{\text{in}}^{\text{out}} ds} \right)_{H,C} \tag{11.131}$$

It can be demonstrated that the highest value of the Witte–Shamsundar efficiency is $\eta_{W-S} = 1$, and that it occurs in the limit of reversible operation [55]. The same conclusion is reached by substituting $\epsilon_R \leq 1$ into eq. (11.129). However, not noted until now is that η_{W-S} can assume *negative* values, and that its full range is $-\infty < \eta_{W-S} \leq 1$. A negative η_{W-S} value would characterize a counterflow heat exchanger working at cryogenic temperatures (e.g., $T_0 = 300$ K, $\bar{T}_H = 30$ K, $\bar{T}_C = 26$ K; hence, $\eta_{W-S} = -0.54$). The η_{W-S} efficiency was employed further in a study of the second-law optimization of heat exchangers [56].

THERMAL-ENERGY STORAGE

Energy Storage versus Exergy Storage

The growing emphasis placed on energy-conservation measures has renewed the interest in "thermal-energy storage" systems [57], that is, in the type of system whose job is to store[†] temporarily the energy received during a heat transfer interaction. Such a system is capable of providing at a later time a heat transfer interaction of its own. The traditional view in this design area is that a storage unit is efficient when the energy increase experienced during the storage phase approaches the maximum energy increase of which the unit is capable. For example, a batch of incompressible liquid of mass m, constant specific heat c, and initial temperature T_0 can experience a maximum energy increase equal to $mc(T_\infty - T_0)$ if the temperature of the heat source that heats the batch is T_∞. In the traditional sense, the goodness of this unit can be quantized in terms of a first-law efficiency ratio:

$$\eta_I = \frac{\text{actual energy increase}}{\text{maximum energy increase}} = \frac{mc(T - T_0)}{mc(T_\infty - T_0)} \qquad (11.132)$$

where T is the temperature of the batch of liquid at the end of the storage process.

The first-law efficiency η_I can have values greater than zero and less than one. The desirable limit of $\eta_I \to 1$ is approached through a number of design decisions, for example, by increasing the size of the heat exchanger placed between the liquid batch and the heat source, and by increasing the time of thermal communication between heat source and storage material.

The traditional view was challenged on thermodynamic grounds in 1978 [58]. It seemed that if the upgrading of power-system performance depends on the designer's ability to eliminate or, at least, reduce irreversibilities, then the real purpose of using storage systems in the power-system area must also be the reduction of irreversibility. And if the reduction of irreversibility amounts to an exergy flow that is more nearly "conserved" as it descends through the power plant (Fig. 3.7, right side), then, too, the mission of the storage device is to temporarily store *exergy*, not "energy." This new point of view has developed into a distinct subfield in the thermal design of energy-storage systems, as exemplified by the recent work of Krane [59, 60], Mathiprakasam and Beeson [61], and Taylor [62].

[†]Energy storage in the system is as an increase in its *energy* inventory, not as "thermal energy."

The Optimum Duration of the Storage Process

The destruction of exergy in a storage system and the opportunity for minimizing this destruction become apparent if we examine briefly the first phase (the storage phase) in the operation of the system in Fig. 11.19. An extensive treatment of this example is given in Ref. 58 as well as in chapter 8 of Ref. 2. The storage system (the left side of the figure) contains the batch of incompressible liquid (m, c) alluded to earlier. The liquid is held in an insulated vessel. The hot-gas stream $\dot{m}$ enters the system through one port and is gradually cooled as it flows through a heat exchanger immersed in the liquid bath. The spent gas is discharged directly into the atmosphere. As time passes, the bath temperature T and the gas outlet temperature T_{out} approach the hot-gas inlet temperature, T_∞.

If we model the hot gas (steam, products of combustion) as an ideal gas with constant specific heat c_P, the temperature history of the storage system is expressed in closed form by the equations

$$\frac{T(t) - T_0}{T_\infty - T_0} = 1 - \exp(-y\theta) \qquad (11.133)$$

$$\frac{T_{\text{out}}(t) - T_0}{T_\infty - T_0} = 1 - y \exp(-y\theta) \qquad (11.134)$$

where y and the dimensionless time θ are defined as

$$y = 1 - \exp(-N_{\text{tu}}) \qquad N_{\text{tu}} = \frac{\bar{h}_b A_b}{\dot{m} c_P} \qquad (11.135)$$

$$\theta = \frac{\dot{m} c_P}{mc} t \qquad (11.136)$$

In these equations, A_b is the total heat-exchanger surface separating the gaseous stream from the liquid bath, and $\bar{h}_b$ is the overall heat transfer

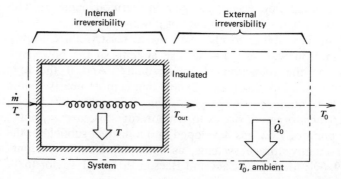

Figure 11.19 Two sources of irreversibility in a batch heating process (after Ref. 58).

coefficient based on A_b. Built into the above model is also the assumption that the liquid bath is well mixed, i.e., that the liquid temperature (T) is a function of the time (t) only. As expected, both T and T_{out} approach T_∞ asymptotically—the higher the N_{tu} value, the faster. The first-law efficiency η_I is the same as the ratio calculated with eq. (11.133): it shows that the ability to store energy increases with increasing charging time (θ) and the size of the heat exchanger (N_{tu}).

Turning our attention to the irreversibility of the energy-storage process, we see in Fig. 11.19 that the irreversibility is divided between two distinct parts of the apparatus. First, there is the finite-ΔT irreversibility associated with the heat transfer between the hot gaseous stream and the cold liquid bath. Second, the stream exhausted into the atmosphere is eventually cooled down to T_0, again by heat transfer across a finite ΔT. Neglected in the present model is the irreversibility due to the pressure drop across the heat exchanger traveled by the stream $\dot{m}$.

The combined effect of the competing irreversibilities noted in Fig. 11.19 is a characteristic of all sensible-heat energy-storage systems. Because of it, only a fraction of the exergy content of the hot stream can ever be stored in the liquid bath. In order to see this, consider the instantaneous rate of entropy generation in the overall system delineated in Fig. 11.19:

$$\dot{S}_{gen} = \dot{m}c_P \ln \frac{T_0}{T_\infty} + \frac{\dot{Q}_0}{T_0} + \frac{d}{dt}(mc \ln T) \qquad (11.137)$$

where $\dot{Q}_0 = \dot{m}c_P(T_{out} - T_0)$. More important than $\dot{S}_{gen}$ is the entropy generated during the entire "charging"-time interval $0-t$, which, using eqs. (11.133)–(11.137) can be put in dimensionless form as

$$\frac{1}{mc} \int_0^t \dot{S}_{gen}\, dt = \theta\left(\ln \frac{T_0}{T_\infty} + \tau\right) + \ln(1 + \tau\eta_I) - \tau\eta_I \qquad (11.138)$$

where η_I is shorthand for the right side of eq. (11.133) [see also eq. (11.132)], and where

$$\tau = \frac{T_\infty - T_0}{T_0} \qquad (11.139)$$

Multiplied by T_0, the entropy-generation integral $\int_0^t \dot{S}_{gen}\, dt$ calculated above represents the bite taken by irreversibilities out of the total exergy supply brought into the system by the hot stream:

$$E_x = t\dot{E}_x = t\dot{m}c_P \ln\left(T_\infty - T_0 - \ln \frac{T_\infty}{T_0}\right) \qquad (11.140)$$

On this basis, we define the entropy-generation number N_S as the ratio of the lost exergy divided by the total exergy invested during the time interval $0-t$:

$$N_S(\theta, \tau, N_{tu}) = \frac{T_0}{E_x} \int_0^t \dot{S}_{gen} \, dt = 1 - \frac{\tau\eta_I - \ln(1 + \tau\eta_I)}{\theta[\tau - \ln(1 + \tau)]} \quad (11.141)$$

This particular entropy-generation number takes values in the range 0–1, the $N_S = 0$ limit representing the elusive case of reversible operation. Worth noting is the relation $N_S = 1 - \eta_{II}$, where η_{II} is the second-law efficiency of the installation during the charging process.

Charts of the $N_S(\theta, \tau, N_{tu})$ surface show [2, 58] that N_S decreases steadily as the heat-exchanger size (N_{tu}) increases. This effect is expected, in fact, it matches the N_{tu}-related conclusion drawn based on first-law arguments, eq. (11.132). Less expected is the fact that N_S goes through a minimum as the dimensionless time θ increases. For example, the optimum time for minimum N_S can be calculated analytically in the limit $\tau \ll 1$, where eq. (11.141) reduces to

$$N_S = 1 - \frac{1}{\theta}[1 - \exp(-y\theta)]^2 \quad (11.142)$$

The solution of the equation $\partial N_S / \partial \theta = 0$ is

$$\theta_{opt} = 1.256[1 - \exp(-N_{tu})]^{-1} \quad (11.143)$$

in other words, for the common range of N_{tu} values (1–10), the optimum

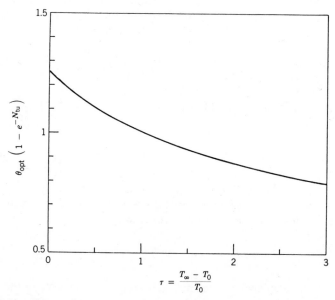

Figure 11.20 The optimum charging time for minimum irreversibility during the energy-storage phase (after Ref. 58).

dimensionless charging time is consistently *a number of order 1*. This conclusion continues to hold as τ takes values greater than 1, Fig. 11.20.

Away from the optimum charging time illustrated above (i.e., when $\theta \to 0$ or $\theta \to \infty$), the entropy-generation number N_S approaches unity. In the short-time limit ($\theta \ll \theta_{opt}$), the entire exergy content of the hot stream is destroyed by heat transfer to the liquid bath, which was initially at atmospheric temperature T_0. In the long-time limit ($\theta \gg \theta_{opt}$), the external irreversibility takes over: in this limit, the used stream exits the heat exchanger as hot as it enters ($T_{out} = T_\infty$) and, because of this, its exergy content is destroyed entirely by the heat transfer (or mixing) with the T_0 atmosphere. The first-law rule of thumb of increasing the time of communication between heat source and storage material, eqs. (11.132)–(11.133), is counterproductive from the point of view of avoiding the destruction of exergy.

The Optimum Size of the Heat Exchanger

Continuing the example constructed based on Fig. 11.19, we inquire into the effect of heat-exchanger N_{tu} on the overall irreversibility of the energy-storage phase. Figure 11.21 shows that the minimum entropy-generation number corresponding to the optimum charging time of Fig. 11.20 decreases steadily as N_{tu} increases. This trend is due to the fact that all the irreversibilities accounted for in the constitution of N_S are ΔT-type irreversibilities.

In the study of heat-exchanger irreversibilities, we learned to expect a trade-off with respect to N_{tu} as a result of the competition between heat

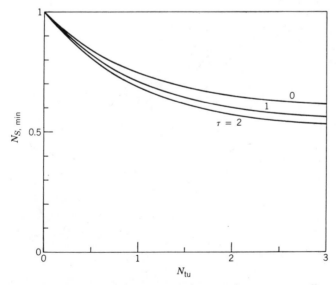

Figure 11.21 The minimum entropy-generation number corresponding to the optimum charging time of Fig. 11.20 (after Ref. 58).

transfer and fluid-flow irreversibilities. The same trade-off appears in the design of the heat exchanger of Fig. 11.19 as soon as we take into account the pressure drop (ΔP) between the stream inlet and outlet. It has been shown that when the pressure-drop entropy generation is not neglected, the N_S expression (11.141) contains an additional term [58], now labeled $N_{S,\Delta P}$:

$$N_S = \underbrace{\frac{(R/c_P)fg^2 N_{\mathrm{tu}}}{[\tau - \ln(1 + \tau)]\,\mathrm{St}}}_{N_{S,\Delta P}} + \underbrace{1 - \frac{\tau\eta_\mathrm{I} - \ln(1 + \tau\eta_\mathrm{I})}{\theta[\tau - \ln(1 + \tau)]}}_{\substack{N_{S,\Delta T}\ \text{or}\ N_S\ \text{of}\\ \text{eq. (11.141)}}} \qquad (11.141')$$

In this additional term, f represents the friction factor and St is the Stanton number on the gas side of the heat exchanger. It is being assumed also that the overall N_{tu}, which was defined in eq. (11.135), is practically equal to the number of heat transfer units for the gas side of the heat exchanger. Finally, the dimensionless mass velocity g is defined according to eq. (11.74).

Recalling that the result of minimizing the $N_{S,\Delta T}$ part with respect to θ forms the subject of Fig. 11.21, it is now clear that N_{tu} has competing effects on the fluid-flow and heat transfer irreversibilities of the energy-storage process. In the case where the $N_{S,\Delta T}$ part has already been minimized with respect to θ, the optimum N_{tu} that minimizes the whole N_S expression (11.141') is

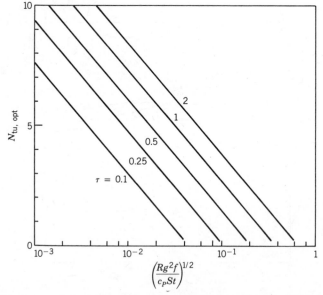

Figure 11.22 The optimum number of heat transfer units for the gas–liquid heat exchanger of the storage process of Fig. 11.19 (after Ref. 58).

$$N_{tu,opt} = \ln\left[\frac{\tau^2\eta_I(1-\eta_I)}{1+\tau\eta_I}\right] - \ln\left(\frac{Rg^2f}{c_P St}\right) \qquad (11.144)$$

In the optimum charging-time regime, Fig. 11.20, the first term on the right side depends only on τ. Therefore, the optimum number of heat transfer units depends only on τ and the group $(Rg^2f/c_P St)$, as is illustrated in Fig. 11.22. Since for most heat-exchanger surface types, the ratio f/St is only a weak function of the Reynolds number, the optimum N_{tu} depends primarily on τ and g.

Storage Followed by Removal of Exergy

The two optima analyzed until now rule the design of more complex processes executed by energy-storage systems. A necessary step in the direction of completing the thermodynamic treatment of such systems was taken by Professor Krane of the University of Tennessee, who considered the cyclical operation of the device [60]. Figure 11.23 shows the schematic evolution of the liquid-bath temperature during the storage phase and

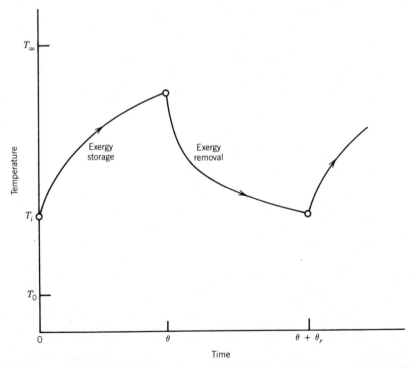

Figure 11.23 The batch-system temperature evolution during a complete exergy-storage and exergy-removal cycle (after Krane [60]).

during the exergy-removal phase that follows immediately. The liquid temperature varies periodically without ever reaching the limiting temperature levels T_0 and T_∞.

Insight into the irreversibility composition of the storage and removal cycle is provided by Fig. 11.24. The only parameter that varies in this example is the duration of the "storage" part of the cycle, θ. The storage part is accompanied by the irreversibilities that were discussed already, namely, the contributions due to heat-exchanger ΔT, heat-exchanger ΔP, and the dumping of the used stream into the atmosphere. The exergy-removal part of the cycle is plagued by irreversibilities due only to heat-exchanger ΔT and ΔP. The gas stream $\dot{m}_r$ heated by the liquid pool during the removal phase—the fruit of the entire scheme—is delivered to a power cycle that can use its exergy content. In Fig. 11.24, the two pressure-drop effects (during storage and removal) are shown added under the same curve.

Krane's Fig. 11.24 begins to show the importance of fine-tuning the timing of the storage and removal phases in order to minimize the cycle-integrated destruction of the original exergy content of the hot stream $(\dot{m}, T_\infty)$. The optimization of the whole cycle and the gas–liquid heat

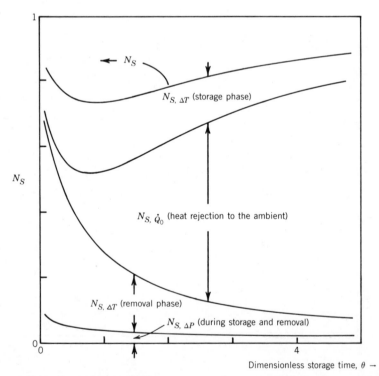

Figure 11.24 The effect of charging time on the irreversibilities of the storage and removal cycle (after Krane [60]).

exchanger can be accomplished numerically by minimizing the total N_S with respect to the charging interval (θ) and the heat-exchanger size (N_{tu}). For the design case illustrated in Fig. 11.24, Krane obtained:

$\theta_{opt} = 0.863$, optimum charging (storage) interval
$N_{tu,opt} = 5.53$, optimum number of heat transfer units
$N_{S,min} = 0.734$, minimum entropy-generation number, i.e., $\eta_{II} = 0.266$
    (under the same conditions, $\eta_I = 0.577$)

The inlet temperature of the stream of cold gas $\dot{m}_r$ was assumed to be the same as T_0. The dimensionless time interval of the exergy-removal part of this cycle, θ_r, was found to be equal to 1.83. Other parameters that were held fixed during this optimization example are

$$\dot{m}_r/\dot{m} = 1 \qquad g = 0.0354 \qquad T_i/T_0 = 1.1 \qquad Pr = 0.71 \qquad R/c_p = 0.286$$

$$(11.145)$$

The dimensionless pressure-drop ratios ($\Delta P/P$) during the storage and removal phases were 0.021 and 0.01, respectively.

The conclusions reached during the study of the storage phase alone are reinforced by Krane's study of the complete storage and removal cycle. The optimum storage-time interval, for example, is such that in dimensionless terms, it emerges once more as a number of order 1. There exists again an optimum number of heat transfer units for the gas–liquid heat exchanger. The minimum N_S values revealed by Krane's study are generally greater than what we see in Fig. 11.21. This effect is due to the irreversibilities contributed by the exergy-removal phase of the cycle.

Figure 11.24 makes the point that the task of perfecting the thermo-dynamic performance of a storage system hinges on the ability to minimize the heat transfer across three temperature gaps, namely, the ΔT between gas and liquid during both storage and removal, and the ΔT between the stream exhausted at the end of the storage phase. These ΔTs can be reduced by bringing the inlet temperature of the stream closer to the temperature of the liquid bath, and by keeping the exhaust temperature T_{out} as close to T_0 as possible. This proposal can be executed in strikingly simple fashion by using a large number of storage units positioned in series. During the storage phase, the stream exhausted by the ith unit becomes the exergy-source stream of the $(i + 1)$th unit, etc. [2]. In other words, the stream exhausted by the ith unit does not reject heat to T_0 but to a higher temperature—the temperature of the unit $(i + 1)$.

In the series arrangement described above, the temperature of the storage units decreases monotonically in the direction of flow. During the exergy-recovery phase, the cold stream is led in the opposite direction, that is, the direction of increasing temperature, or in "counterflow" relative to the stream used during the storage phase. The ΔTs between the stream and the storage material and the exhaust stream and the ambient are considerably smaller in this arrangement. This proposal was investigated in great detail by Taylor [62] based on a solid "distributed-storage-element" model in which the storage-material temperature varied continuously along the stream. Taylor shows among other things that the longitudinal conduction of heat through the storage material during the periodic operation of the heat exchanger can have a major impact on the overall irreversibility of the installation. The overall irreversibility figure N_S is again a strong function of the time interval required by the storage part of the cycle: the identification of the optimum storage-time interval is critical. The overall N_S is affected also by the geometric aspect ratio of the storage material. The numerical examples documented in Taylor's study reveal N_S values that cover the range 0.2–0.8. This range compares favorably with the 0.7–0.9 range covered by the N_S results obtained for a single sensible-heat element during complete storage and removal cycles [60].

The cyclical storage and removal of exergy from a continuous one-dimensional stretch of storage material was studied also by Mathiprakasam and Beeson [61]. One interesting effect illustrated by them is that of the direction of flow during the removal phase. They found that the second-law efficiency $(1 - N_S)$ is always lower if the exergy-removal stream flows in the same direction as the original exergy-supply stream (i.e., in "parallel"), lower than in the "counterflow" arrangement discussed in the preceding paragraph. The relative inferiority of the parallel-flow arrangement was illustrated also by Taylor [62].

Closely related to the continuous one-dimensional storage scheme with periodic counterflow circulation is the class of periodic heat exchangers recognized as "regenerators." The design of this type of heat exchanger was approached on the basis of entropy-generation minimization by San et al. [63]. Their model consists of two-dimensional parallel-plate channels sandwiched between slabs of energy-storage material. The longitudinal conduction of heat through the storage material is neglected. An important difference between this regenerator model and the continuous storage system analyzed by Taylor [62] is that in the case of the regenerator, the stream exhausted during the storage phase is not dumped into the atmosphere. That stream and the exergy still in it are considered usable. Therefore, the total entropy-generation figure of one full cycle in the operation of the regenerator is due to four contributions, namely, the ΔT- and ΔP-inspired irreversibilities of the storage part and the removal part of the cycle. The irreversibility of periodic-flow regenerative heat exchangers was studied also by Hutchinson and Lyke [64].

Worthy of mention in this section is the thermodynamic investigation of another simple device for temporary exergy storage, one where the heating during the storage phase is provided by an electrical resistance (Joulean heating) [59]. The sources of irreversibility identified in the operation of this device are three: the electrical resistive heating itself, the heat transfer across the finite ΔT between the storage material and the stream used during the exergy-removal phase, and the flow with pressure drop through the heat exchanger built between the stream and the storage material. The entropy-generation numbers (N_S) revealed by this study fall in the range 0.6–0.8.

Heating and Cooling Subject to Time Constraint

Related to the lumped-system model in Fig. 11.19 are the basic metallurgical problem of heating an object to a prescribed temperature level [65, 66] and the "cooldown" problem of cryogenics, where large-scale superconducting windings must first be cooled to liquid-helium temperature before they can be operational [67]. For the sake of concreteness, consider the cooldown process by which the lumped system (m, c) is cooled from an original temperature T_i to a lower temperature T_f by a single-phase stream $(\dot{m}c_P)$ whose temperature T_L is lower than T_f. The expensive commodity in this operation is the total amount of cold gas used to do the job:

$$m_{0-t_c} = \int_0^{t_c} \dot{m}(t)\, dt \qquad (11.146)$$

where t_c is the duration of the cooldown process. The total mass m_{0-t_c} is an "expensive" commodity because it is directly proportional to its exergy content and to the actual refrigerator power required to produce it (chapter 10).

When the cooldown time is fixed by logistic and economic considerations, there exists an optimum cold-gas flowrate history $\dot{m}(t)$ that minimizes the overall expenditure of cryogen, m_{0-t_c}. The details of the analysis that produces this result [65, 67] are given in Ref. 2 (pp. 166–169). In short, the instantaneous heat transfer between the object (m, c, T) and the cold stream $(\dot{m}, c_P)$ can be modeled by writing

$$\bar{h}_b A_b (T - T_{\text{out}}) = \dot{m}c_P(T_{\text{out}} - T_L) \qquad (11.147)$$

$$mc\,\frac{dT}{dt} = -\bar{h}_b A_b (T - T_{\text{out}}) \qquad (11.148)$$

According to this model, the heat transfer between the object and the cold stream occurs across the temperature difference $(T - T_{\text{out}})$, where T_{out} is the time-dependent temperature of the stream at the point where it leaves the object. The assumption, therefore, is that at any point in time, the lumped mass m is permeated by a well-mixed cold stream at temperature T_{out} (this, despite the fact that the inlet temperature of the stream is fixed at the low temperature T_L).

An expression for the function $\dot{m}(t)$ is obtained by eliminating T_{out} between eqs. (11.147) and (11.148).

$$\dot{m} = \frac{mc/c_P}{T_L - T - \dfrac{mc}{h_b A_b} \dfrac{dT}{dt}} \frac{dT}{dt} \tag{11.149}$$

Using this expression as integrand in eq. (11.146) yields the temperature integral [2]:

$$m_{0-t_c} = \int_{T_i}^{T_f} \frac{(mc/c_P)\, dT}{T_L - T - \dfrac{mc}{h_b A_b} \dfrac{dT}{dt}} \tag{11.150}$$

where $T_i = T(t = 0)$ and $T_f = T(t = t_c)$. The optimum flowrate history $\dot{m}(t)$ is found indirectly, by first determining the temperature history $T(t)$ that minimizes the integral (11.150) (see the section on variational calculus in the Appendix):

$$\frac{dT_{\text{opt}}}{dt} = \frac{\bar{h}_b A_b}{mc} \frac{T_L - T_{\text{opt}}(t)}{1 + (C^* \bar{h}_b A_b / c_P)^{1/2}} \tag{11.151}$$

The C^* constant depends on the constrained cooldown time t_c: its value is found by integrating eq. (11.149) from $t = 0$ to $t = t_c$, while regarding $\bar{h}_b$ as a known function of temperature. Finally, the optimum flowrate history is obtained by combining eqs. (11.149) and (11.151):

$$\dot{m}_{\text{opt}} = \left[\frac{\bar{h}_b(T) A_b}{C^* c_P(T)} \right]^{1/2} \tag{11.152}$$

In conclusion, the optimum flowrate is time-independent only in those cases where $\bar{h}_b$ and c_P do not vary as the temperature of the ensemble decreases. In general, however, the overall heat transfer coefficient varies with temperature: from the optimum cooldown regime (11.152), we learn that during periods of poor heat transfer (low $\bar{h}_b$), the coolant flowrate must be decreased. If during the same cooldown run, the specific heat of the cold gas (c_P) increases as T decreases (as in N_2 gas at constant P, for example), then the coolant flowrate must again decrease. The optimum flowrate $\dot{m}_{\text{opt}}(t)$ depends on T indirectly, via $\bar{h}_b(T)$ and $c_P(T)$.

The savings in coolant mass m_{0-t_c} associated with employing the optimum flowrate regime (11.152) can be evaluated by calculating the mass ratio [67]:

$$N_m = \frac{(m_{0-t_c})_{\text{optimum-}\dot{m}}}{(m_{0-t_c})_{\text{constant-}\dot{m}}} = \frac{\text{total mass used under regime (11.152)}}{\text{total mass used when } \dot{m} \text{ is kept constant}} \tag{11.153}$$

The numerator and denominator in this ratio refer to the same (fixed) cooldown time t_c. A more revealing N_m expression is [67]

$$N_m = \frac{N_{\text{tu}}(T_L) + (I_2/I_1)^2}{N_{\text{tu}}(T_L) + I_3/I_1} \tag{11.154}$$

where

$$N_{\text{tu}}(T_L) = \frac{\bar{h}_b(T_L)A_b}{\dot{m}_{\text{const.}}c_P} \qquad \begin{array}{l}\text{(cold-end number of heat transfer units;} \\ \dot{m}_{\text{const.}} \text{ is the constant flowrate referred} \\ \text{to in the denominator of } N_m) \end{array} \tag{11.155}$$

$$\bar{h}_b(T) = h_b(T_L)\left(\frac{T}{T_L}\right)^q \qquad \text{(heat transfer coefficient model)} \tag{11.156}$$

$$c(T) = c(T_L)\left(\frac{T}{T_L}\right)^p \qquad \begin{array}{l}\text{(model for the specific heat of the} \\ \text{object being cooled)}\end{array} \tag{11.157}$$

$$I_1 = \int_{\tau_f}^{\tau_i} \frac{\tau^p}{\tau - 1}\, d\tau \qquad \tau = \frac{T}{T_L} \tag{11.158}$$

$$I_2 = \int_{\tau_f}^{\tau_i} \frac{\tau^{p-q/2}}{\tau - 1}\, d\tau \tag{11.159}$$

$$I_3 = \int_{\tau_f}^{\tau_i} \frac{\tau^{p-q}}{\tau - 1}\, d\tau \tag{11.160}$$

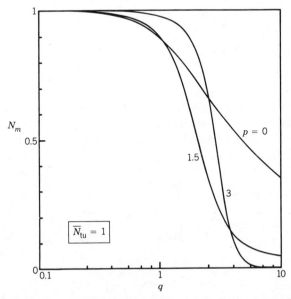

Figure 11.25 The coolant total mass ratio N_m versus the heat transfer coefficient exponent q, showing the effect of the specific-heat exponent p (after Ref. 67).

Figure 11.25 shows that sizeable savings can be made using the $\dot{m}_{opt}(t)$ cooldown technique when q is greater than 1. The curves were drawn for a numerical example concerning the cooldown of a large superconducting structure precooled to liquid-nitrogen temperature ($T_L = 4.2$ K, $T_f = 4.5$ K, $T_i = 80$ K). The average number of heat transfer units held constant in this figure, $\bar{N}_{tu} = 1$, is defined as

$$\bar{N}_{tu} = \frac{1}{\tau_i - \tau_f} \int_{\tau_f}^{\tau_i} N_{tu}(\tau)\, d\tau \tag{11.161}$$

where $N_{tu}(\tau)$, or $N_{tu}(T)$, is the general version of definition (11.155). The specific heat data for Al, Fe, and Cu below 80 K suggest that $p \cong 2.85$ is a representative value for the exponent in the specific-heat model (11.157) [68]. This is why the curves of Fig. 11.25 cover the p range 0–3. Additional numerical examples can be examined in Ref. 67.

MASS EXCHANGERS

The thermodynamic optima identified in the study of heat exchangers and related heat transfer applications surface again in the thermodynamic design of mass transfer devices. The work on the minimization of entropy generation in the mass transfer domain is even newer than what has been accomplished in heat transfer; therefore, one objective of the present section is to place in perspective the mass transfer thermodynamic design subfield that is beginning to emerge.

Convective Mass Transfer

As an analogy to the most basic problem of heat-exchanger irreversibility at the flow-passage level investigated in Refs. 2 and 3, the competition between fluid-flow and mass transfer irreversibilities in a mass exchanger was studied by San et al. [69]. The general expression for the rate of entropy generation in a flow field with both heat and mass transfer and without body force and chemical-reaction effects is [70]

$$s'''_{gen} = \underbrace{-\frac{1}{T}\sum_i \mathbf{j}_i \cdot \nabla \mu_i}_{\text{Mass diffusion}} + \underbrace{\frac{\mu}{T}\Phi}_{\substack{\text{Fluid}\\\text{friction}}} - \underbrace{\frac{1}{T^2}\mathbf{q}\cdot\nabla T}_{\substack{\text{Thermal}\\\text{diffusion}}} - \underbrace{\frac{1}{T}\sum_i \bar{s}_i \mathbf{j}_i \cdot \nabla T}_{\substack{\text{Coupling between}\\\text{thermal diffusion}\\\text{and mass diffusion}}}$$

$$\tag{11.162}$$

where $\mathbf{j}_i$ is the mass-diffusion flux vector of the "i" species and $\bar{s}_i$ the partial molal entropy of the species, eq. (4.119),

$$\bar{s}_i = -\left(\frac{\partial \mu_i}{\partial T}\right)_{P,N_k} \qquad (k \neq i) \qquad (11.163)$$

Worth noting is eq. (11.48) as a special case of eq. (11.162), namely, the limiting case of zero mass transfer. In the *zero heat transfer limit*, on the other hand, we retain only the first two terms on the right side of eq. (11.162) and concentrate on the interplay between fluid-friction and mass-diffusion irreversibilities. Regarding the chemical potential[†] gradient that appears in the mass-diffusion term, we recall from the study of convection mass transfer [Ref. 4, Chapter 9] that the mass transfer part of the problem is usually described in terms of species-concentration distributions, C_i (moles of i/m^3), not in terms of chemical potential. Useful then is the ideal gas mixture model:

$$\mu_i(T, P_i) = \mu_i(T, P_0) + \bar{R}T \ln \frac{P_i}{P_0} \qquad (11.164)$$

where the partial pressure of "i" in the mixture of pressure P_0 and temperature T is

$$P_i = C_i \bar{R} T \qquad (11.165)$$

If, in addition, we invoke *Fick's law* of mass diffusion:

$$\mathbf{j}_i = -\mathcal{D}_i \cdot \nabla C_i \qquad (11.166)$$

the entropy-generation expression (11.162) reduces to [69]

$$s_{\text{gen}}''' = \bar{R} \sum_i \frac{\mathcal{D}_i}{C_i} (\nabla C_i)^2 + \frac{\mu}{T} \Phi \qquad (11.167)$$

This two-term expression is the analog of eq. (11.50) of convective heat transfer. For example, in the case of two-dimensional flow (v_x, v_y) through the space (x, y) in which only the "i" species diffuses, eq. (11.167) reads

$$\begin{aligned} s_{\text{gen}}''' = \bar{R} \frac{\mathcal{D}_i}{C_i} &\left[\left(\frac{\partial C_i}{\partial x}\right)^2 + \left(\frac{\partial C_i}{\partial y}\right)^2\right] \\ &+ \frac{\mu}{T}\left\{2\left[\left(\frac{\partial v_x}{\partial x}\right)^2 + \left(\frac{\partial v_y}{\partial y}\right)^2\right] + \left(\frac{\partial v_x}{\partial y} + \frac{\partial v_y}{\partial x}\right)^2\right\} \end{aligned} \qquad (11.168)$$

Both terms on the right side of eq. (11.168) are positive, indicating the permanent collaboration between fluid friction and mass diffusion in determining the irreversibility rate at each point in the flow field. The similarity

[†]The chemical potentials μ_i should not be confused with the viscosity μ.

between eq. (11.168) and eq. (11.51) of convective heat transfer assures us that the thermodynamic trade-offs discovered in the field of heat transfer and thermal design exist also in the design of mass transfer devices. San et al. [69] demonstrated this fact by considering the design of a two-dimensional mass exchanger in which the fluid mixture flows through a parallel-plate channel with imposed uniform mass flux of "i" normal to the flow. They assumed also the "small diffusion-rate" limit, in which the velocity profile in the channel cross-section is not affected by the species that diffuses in the direction normal to the flow. Minimizing the entropy-generation rate integrated over the channel cross-section, San et al. [69] developed a complete sizing procedure for the plate-to-plate spacing of the channel, for both laminar and turbulent flow. Their paper and results are highly recommended.

Simultaneous Mass and Heat Transfer by Convection

The problem of entropy-generation minimization in combined convective-heat and mass transfer was considered in a subsequent paper by San et al. [71]. As indicated by eq. (11.162), in this problem, the irreversibility is due to four distinct effects: pure mass diffusion, fluid friction, pure thermal diffusion, and the coupling between thermal diffusion and mass diffusion. The first three terms are all positive, as indicated by eqs. (11.50) and (11.167). The fourth term requires special care in view of the partial molal entropy factor $\bar{s}_i$, whose definition is eq. (11.163). It is necessary to bring this factor and the fourth term of eq. (11.162) to a form compatible with the nomenclature of convective heat and mass transfer.

Guided by the definition (11.163), we work first on the chemical potential of the "i" species, eq. (11.164) [71],

$$
\begin{aligned}
\mu_i(T, P_i) &= \mu_i(T, P_0) + \bar{R}T \ln \frac{P_i}{P_0} \\
&= \bar{c}_{P,i}\left(T - T_0 - T \ln \frac{T}{T_0}\right) + \bar{h}_{i,0} - T\bar{s}_{i,0} + \bar{R}T \ln \frac{P_i}{P_0}
\end{aligned} \tag{11.169}
$$

where $\bar{h}_{i,0}$ and $\bar{s}_{i,0}$ are evaluated at T_0 and P_0. The last step in this derivation is based on the assumption that $\bar{c}_{P,i}$ is constant over the interval $T_0 - T$. Applying the definition (11.163) and eq. (11.165), we conclude that

$$
\bar{s}_i = \bar{R} \ln \frac{C_0}{C_i} + \bar{c}_{v,i} \ln \frac{T}{T_0} + \bar{s}_{i,0} \tag{11.170}
$$

where C_0 is the concentration at reference conditions, $C_0 = P_0/\bar{R}T_0$.

After invoking once more the Fick and Fourier laws of diffusion, the local entropy-generation rate (11.162) can finally be written as

$$s_{gen}''' = \bar{R} \sum_i \frac{\mathcal{D}_i}{C_i} (\nabla C_i)^2 + \frac{\mu}{T} \Phi + \frac{k}{T^2} (\nabla T)^2 + \frac{1}{T} \sum_i \bar{s}_i \mathcal{D}_i (\nabla C_i) \cdot (\nabla T)$$

$$(11.171)$$

where $\bar{s}_i$ is given by eq. (11.170). Whether all these terms are important in the final s_{gen}''' figure depends, of course, on the particular convection problem that is being investigated. For example, consider the heat and mass transfer to fully developed laminar flow through a parallel-plate channel of size (spacing) D, and let ΔC_i and ΔT represent the scales of the wall—mean concentration and temperature differences, respectively. Taking $\bar{R}$ as the scale of $\bar{s}_i$ in eq. (11.170), the importance of the fourth term (coupling) relative to the first term (mass diffusion) in eq. (11.171) is measured by the ratio:

$$\frac{\text{heat and mass transfer coupling}}{\text{pure mass diffusion}} \sim \frac{\Delta T/T}{\Delta C_i/C_i} \qquad (11.172)$$

We see here the emergence of another family of dimensionless groups for second-law analysis, namely, the dimensionless ratios $(\Delta C/C)_i$. In the description of mass transfer irreversibility, these ratios play the same role as the $\Delta T/T$ ratios in heat transfer and the $\Delta P/P$ ratios in duct flow with friction.

The complete four-term entropy-generation-rate expression (11.171) was used by San et al. [71] for determining the optimum spacing D of a parallel-plate heat and mass exchanger with uniform heat and mass fluxes. They showed that the thermodynamic optimum is due to the fact that the fluid-friction term varies as D^{-3}, whereas the remaining terms are directly proportional to D.

Of the thermodynamic design work devoted to mass transfer processes, a substantial part deals with the optimization of drying and moistening processes (e.g., Sieniutycz [72–75]), the dehumidification of air [76], and dessicant cooling systems [77, 78]. A general framework for the calculation of mass transfer irreversibility in chemical-separation systems was constructed by Moore and Wepfer [79].

Convective Heat and Mass Transfer Through a Saturated Porous Medium

Considering the step-by-step evolution from convective heat transfer, eq. (11.50), to combined heat, mass, and fluid flow, eq. (11.171), this is an opportunity to generalize the entropy-generation-rate formula for convection through a saturated porous medium [first stated in Ref. 4, p. 355]. Writing term by term the porous-medium equivalent of eq. (11.171):

$$s_{gen}''' = \bar{R} \sum_i \frac{\mathcal{D}_{pm,i}}{C_{pm,i}} (\nabla C_{pm,i})^2 + \frac{\mu}{KT} (v_{pm})^2 + \frac{k_{pm}}{T^2} (\nabla T)^2$$

$$+ \frac{1}{T} \sum_i \bar{s}_{pm,i} \mathcal{D}_{pm,i} (\nabla C_{pm,i}) \cdot (\nabla T) \qquad (11.173)$$

we note the special form of the fluid-friction contribution (the second term), in which $\mathbf{v}_{pm}$ is the volume-averaged velocity vector through the mixture-saturated porous medium. The second term owes its compact form to the Darcy flow model, K being the permeability constant of the medium. The saturated porous medium is further modeled as homogeneous and isotropic with an effective thermal conductivity k_{pm}. The solid matrix is locally in thermal equilibrium with the fluid that seeps through it. The concentration $C_{pm,i}$ is a volume-averaged quantity also, as it represents the number moles of "i" per cubic meter of porous medium saturated with fluid. The partial molal entropy $\bar{s}_{pm,i}$ is also a volume-averaged quantity. The coefficient $\mathscr{D}_{pm,i}$ is the mass diffusivity of the "i" species through the porous medium saturated with the fluid mixture to which "i" belongs.

SYMBOLS

a_1	area constraint [eq. (11.78)]
A	area, cross-sectional area, flow cross-section
A_b	gas–liquid bath contact area
A_g	gas–atmosphere contact area
A_1	dimensionless group [eq. (11.25)]
A_1, A_2	areas of the two sides of the heat-exchanger surface [eq. (11.64)]
b	damping coefficient
B_0	dimensionless group [eq. (11.19)]
B_1	dimensionless group [eq. (11.42)]
c	specific heat of a solid or incompressible liquid
c_P	specific heat at constant pressure [J/kg K]
$\bar{c}_P$	specific heat at constant pressure [J/kgmole K]
$\bar{c}_{v,i}$	specific heat at constant volume [J/kgmole K]
C_D	drag coefficient
C_i	concentration [kgmole/m^3]
C^*	constant [eq. (11.151)]
D	tube diameter, plate-to-plate spacing
D_h	hydraulic diameter
$\mathscr{D}_i$	mass diffusivity
e	sand grain size
e_x	specific flow exergy [J/kg]
E_x	flow exergy [J]
$\dot{E}_x$	flow exergy flowrate [W]
f	friction factor
F_D	drag force
g	dimensionless mass velocity [eq. (11.74)]
G	mass velocity [eq. (11.9)]
h	specific enthalpy [J/kg]
$\bar{h}$	heat transfer coefficient [W/m^2 K]

$\bar{h}_i$	partial molal enthalpy [J/kgmole]
I_1, I_2, I_3	integrals [eqs. (11.158)–(11.160)]
$\mathbf{j}$	mass-flux vector, [kgmole/m^2 s]
J	dimensionless group [eq. (11.23)]
k	thermal conductivity
k	ratio of specific heats, c_P/c_v
K	spring constant (Fig. 11.8)
K	permeability of a porous medium in the Darcy-flow regime [eq. (11.173)]
l	dimensionless linear size, $1/N_{S,1}$
L	length
m	mass [eq. (11.136)]
m	fin conduction parameter [eq. (11.41)]
$\dot{m}$	mass flowrate
$\dot{m}_r$	mass flowrate during the exergy-removal phase
M	dimensionless group $[(k/\lambda)^{1/2}\,\mathrm{Pr}^{-1/6}]$
M_p	piston mass
N_k	number of moles of species "k"
N_m	total coolant mass ratio [eq. (11.153)]
N_S	entropy-generation number
$N_{S,a}$	augmentation entropy-generation number
$N_{S,\mathrm{imbalance}}$	imbalance (remanent) entropy-generation number
$N_{S,1}, N_{S,2}$	entropy-generation numbers of the two sides of a heat-exchanger surface
N_{tu}	number of heat transfer units [eq. (11.66)]
$\bar{N}_{\mathrm{tu}}$	average number of heat transfer units [eq. (11.161)]
Nu	Nusselt number [eq. (11.15)]
p	wetted perimeter [m, eq. (11.11)]
p	exponent [eq. (11.157)]
P	pressure
P_i	partial pressure of species "i"
P_0	reference pressure
Pr	Prandtl number (ν/α)
q	exponent [eq. (11.156)]
$\mathbf{q}$	heat-flux vector [W/m^2]
q'	heat transfer rate per unit length [W/m]
q''	heat flux [W/m^2]
$\dot{Q}$	heat transfer rate [W]
$\dot{Q}_B$	heat transfer rate through the fin base
$\dot{Q}_0$	heat transfer rate interaction with the ambient
$r_{\mathrm{large}}, r_{\mathrm{small}}$	temperature ratios [eq. (11.100)]
R	dimensionless group [eq. (11.28)]
R	ideal gas constant [J/kg K]
$\bar{R}$	universal gas constant [J/kgmole K]
Re	Reynolds number

s	specific entropy [J/kg K]
S	entropy [J/K]
$\dot{S}_{gen}$	entropy-generation rate [W/K]
$\dot{S}'_{gen}$	entropy-generation rate per unit length [W/m K]
s'''_{gen}	volumetric entropy-generation rate [W/m^3 K]
$\bar{s}_i$	partial molal entropy [J/kgmole K]
St	Stanton number
t	time
t_c	cooldown time
T	temperature
T_B	fin base temperature
$\bar{T}_C$	average low temperature [eq. (11.131)]
$\bar{T}_H$	average high temperature [eq. (11.131)]
T_i	initial temperature (Fig. 11.23)
T_L	refrigerant (cold-gas) temperature
T_0	reference or ambient temperature
T_∞	free-stream temperature, gas-supply temperature
u	specific internal energy
U	internal energy
U_∞	free-stream velocity
$\mathbf{v}$	velocity vector
v	specific volume
v_x, v_y	velocity components
$\tilde{v}$	parameter [m^3/kg, eq. (11.117)]
v_1	volume constraint [eq. (11.83)]
V	volume
x	quality
x, y, z	cartesian coordinates
y	dimensionless parameter [eq. (11.135)]
Z	displacement
w'''	volumetric rate of work done [W/m^3]
α	thermal diffusivity
$\Delta(\ )$	difference
ϵ	heat-exchanger effectiveness [eq. (11.103)]
ϵ_R	rational effectiveness [eq. (11.123)]
η_I	first-law efficiency [eq. (11.132)]
η_{II}	second-law efficiency
η_{W-S}	Witte and Shamsundar's efficiency [eq. (11.126)]
θ	dimensionless time [eq. (11.136)]
λ	fluid thermal conductivity in fin analysis
μ	viscosity
μ_i	chemical potential of species "i"
ν	kinematic viscosity
ρ	density

τ	dimensionless temperature difference [eqs. (11.21), (11.70), and (11.139)]
τ	temperature ratio [eq. (11.158)]
ϕ	irreversibility distribution ratio [eq. (11.5)]
Φ	viscous dissipation function [eqs. (11.50)–(11.51)]
ω	capacity rate ratio $[(\dot{m}c_P)_1/(\dot{m}c_P)_2 > 1]$
ω	$(K/M_p)^{1/2}$ (Fig. 11.8)
$(\ )_a$	augmented
$(\ )_b$	batch of liquid or solid storage material
$(\ )_B$	Brayton-cycle power plant
$(\ )_C$	cooler, cold side
$(\ )_f$	saturated liquid, or final state
$(\ )_{fg}$	shorthand for $(\ )_g - (\ )_f$
$(\ )_g$	saturated vapor
$(\ )_H$	heater, hot side
$(\ )_i$	initial state, inner surface, species
$(\ )_{in}$	inlet
$(\ )_{max}$	maximum
$(\ )_{min}$	minimum
$(\ )_o$	outer surface
$(\ )_{opt}$	optimum
$(\ )_{out}$	outlet
$(\ )_{pm}$	fluid-saturated porous medium
$(\ )_{\Delta P}$	due to fluid flow ΔP
$(\ )_r$	removal phase of storage cycle
$(\ )_R$	regenerator
$(\ )_{\Delta T}$	due to heat transfer ΔT
$(\ )_w$	wall
$(\ )_*$	dimensionless variables [eq. (11.56)]
$(\ )_0$	reference, ambient conditions

REFERENCES

1. A. Bejan, Second-law analysis in heat transfer and thermal design, *Adv. Heat Transfer*, Vol. 15, 1982, pp. 1–58.
2. A. Bejan, *Entropy Generation Through Heat and Fluid Flow*, Wiley, New York, 1982, Chapters 5–9.
3. A. Bejan, General criterion for rating heat-exchanger performance, *Int. J. Heat Mass Transfer*, Vol. 21, 1978, pp. 655–658.
4. A. Bejan, *Convection Heat Transfer*, Wiley, New York, 1984, Chapter 7.
5. A. Bejan, Second law analysis in heat transfer, *Energy*, Vol. 5, 1980, pp. 721–732.
6. T. J. Kotas and A. M. Shakir, Exergy analysis of a heat transfer process at sub-environmental temperature, in R. A. Gaggioli, ed., *Computer-Aided En-*

gineering of Energy Systems, AES-Vol. 2–3, Am. Soc. Mech. Eng., New York, 1986, pp. 87–92.

7. P. K. Nag and P. Mukherjee, Thermodynamic optimization of convective heat transfer through a duct with constant wall temperature, *Int. J. Heat Mass Transfer*, Vol. 30, 1987, pp. 401–405.

8. P. Mukherjee, G. Biswas, and P. K. Nag, Second-law analysis of heat transfer in swirling flow through a cylindrical duct, *J. Heat Transfer*, Vol. 109, 1987, pp. 308–313.

9. A. Bejan and P. A. Pfister, Jr., Evaluation of heat transfer augmentation techniques based on their impact on entropy generation, *Lett. Heat Mass Transfer*, Vol. 7, 1980, pp. 97–106.

10. W. R. Ouellette and A. Bejan, Conservation of available work (exergy) by using promoters of swirl flow, *Energy*, Vol. 5, 1980, pp. 587–596.

11. J. Nikuradse, Laws of flow in rough pipes, *NACA Tech. Memo.*, No. 1292, November 1950.

12. D. F. Dipprey and R. H. Sabersky, Heat and momentum transfer in smooth and rough tubes at various Prandtl numbers, *Int. J. Heat Mass Transfer*, Vol. 6, 1963, pp. 329–353.

13. W. M. Kays and H. C. Perkins, Forced convection, internal flow in ducts, in W. M. Rohsenow and J. P. Hartnett, eds., *Handbook of Heat Transfer*, McGraw-Hill, New York, 1973, Chapter 7, p. 4.

14. B. S. Petukhov, Heat transfer and friction in turbulent pipe flow with variable physical properties, *Adv. Heat Transfer*, Vol. 6, 1970, pp. 504–564.

15. R. L. Webb, Toward a common understanding of the performance and selection of roughness for forced convection, *Am. Soc. Mech. Eng.* [*Pap.*], No. 78-WA/HT-61, 1978.

16. H. Perez-Blanco, Irreversibility in heat transfer enhancement, in A. Bejan and R. L. Reid, eds., *Second Law Aspects of Thermal Design*, HTD-Vol. 33, Am. Soc. Mech. Eng., New York, 1984, pp. 19–26.

17. D. Poulikakos and A. Bejan, Fin geometry for minimum entropy generation in forced convection, *J. Heat Transfer*, Vol. 104, 1982, pp. 616–623.

18. F. J. Moody, Second law thinking—example applications in reactor and containment technology, in A. Bejan and R. L. Reid, eds., *Second Law Aspects of Thermal Design*, HTD-Vol. 33, Am. Soc. Mech. Eng., New York, 1984, pp. 1–9.

19. D. Poulikakos, Fin geometry for minimum entropy generation, M.S. thesis, Department of Mechanical Engineering, University of Colorado, Boulder, December 1980.

20. D. Q. Kern and A. D. Kraus, *Extended Surface Heat Transfer*, McGraw-Hill, New York, 1972.

21. B. Gebhart, *Heat Transfer*, 2nd ed., McGraw-Hill, New York, 1971, pp. 212, 270.

22. A. Bejan, A study of entropy generation in fundamental convective heat transfer, *J. Heat Transfer*, Vol. 101, 1979, pp. 718–725.

23. V. S. Arpaci and A. Selamet, Entropy production in flames, *Am. Soc. Mech. Eng.* [*Pap.*], No. 87-HT-55, 1987.

24. A. Bejan, The concept of irreversibility in heat exchanger design: Counterflow

heat exchangers for gas-to-gas applications, *J. Heat Transfer*, Vol. 99, 1977, pp. 374–380.

25. W. M. Rohsenow and H. Y. Choi, *Heat, Mass and Momentum Transfer*, Prentice-Hall, Englewood Cliffs, NJ, 1961, p. 315.

26. A. Bejan, *Solutions Manual for Entropy Generation through Heat and Fluid Flow*, Wiley, New York, 1984, Problem 7.2, pp. 36–37.

27. D. P. Sekulic and C. V. Herman, One approach to irreversibility minimization in compact crossflow heat exchanger design, *Int. Commun. Heat Mass Transfer*, Vol. 13, 1986, pp. 23–32.

28. W. M. Kays and A. L. London, *Compact Heat Exchangers*, McGraw-Hill, New York, 1964.

29. M. Tribus, private communication, course notes "Thermoeconomics," Massachusetts Institute of Technology, Cambridge, MA, September 1978.

30. S. Sarangi and K. Chowdhury, On the generation of entropy in a counterflow heat exchanger, *Cryogenics*, Vol. 22, 1982, pp. 63–65.

31. D. P. Sekulic and B. S. Baclic, Enthalpy exchange irreversibility, *Faculty of Technical Sciences*, No. 15, University of Novi Sad, Yugoslavia, 1984, pp. 113–123.

32. C. E. S. M. da Costa and F. E. M. Saboya, Second law analysis for parallel flow and counterflow heat exchangers, *Proc. 8th Braz. Congr. Mech. Eng.*, São José dos Campos, S.P., Brazil, December 1985, pp. 185–187.

33. D. P. Sekulic, Unequally sized passes in two-pass crossflow heat exchangers: A note on the thermodynamic approach to the analysis, *Faculty of Technical Sciences*, No. 16, University of Novi Sad, Yugoslavia, 1985–1986.

34. D. P. Sekulic, Entropy generation in a heat exchanger, *Heat Transfer Eng.*, Vol. 7, Nos. 1–2, 1986, pp. 83–88.

35. A. Bejan, Second-law aspects of heat transfer engineering, in T. N. Veziroglu and A. E. Bergles, eds., *Multi-Phase Flow and Heat Transfer III*, Vol. 1A, Elsevier, Amsterdam, 1984, pp. 1–22 (Keynote Address to the Third Multi-Phase Flow and Heat Transfer Symposium/Workshop, Miami Beach, FL, April 18–20, 1983).

36. A. L. London and R. K. Shah, Costs of irreversibilities in heat exchanger design, *Heat Transfer Eng.*, Vol. 4, No. 2, 1983, pp. 59–73.

37. S. M. Zubair, P. V. Kadaba, and R. B. Evans, Design and optimization of two-phase heat exchangers, in J. T. Pearson and J. B. Kitto, Jr., eds., *Two-Phase Heat Exchanger Symposium*, HTD-Vol. 44, Am. Soc. Mech. Eng., New York, 1985; published as Second-law-based thermoeconomic optimization of two-phase heat exchangers, *J. Heat Transfer*, Vol. 109, 1987, pp. 287–294.

38. S. C. Lau, K. Annamalai, and S. V. Shelton, Optimization of air-cooled condensers, *J. Energy Resour. Technol.*, Vol. 109, 1987, pp. 90–95.

39. A. L. London, Economics and the second law: An engineering view and methodology, *Int. J. Heat Mass Transfer*, Vol. 25, 1982, pp. 743–751.

40. S. Y. Huang, The heat exchanger for capturing energy from waste heat (ORC Technology), *VDI-Ber.*, No. 539, 1984, pp. 623–629.

41. B. Baclic and D. Sekulic, A crossflow compact heat exchanger of minimum irreversibility (in Serbo-Croatian), *Termotehnika*, Vol. 4, 1978, pp. 34–42.

42. H. P. Soumerai, Second law thermodynamic treatment of heat exchangers, in A. Bejan and R. L. Reid, eds., *Second Law Aspects of Thermal Design*, HTD-Vol. 33, Am. Soc. Mech. Eng., New York, 1984, pp. 11–18.

43. H. P. Soumerai, Thermodynamic considerations on a "mirror image" concept—forced convection single-phase tube flow applications, *Am. Soc. Mech. Eng.*, [*Pap.*], No. 85-WA/HT-18, 1985.

44. H. P. Soumerai, Thermodynamic aspects of adiabatic and diabatic tube flow regime transitions—single-phase fluids, *AIAA Pap.*, No. AIAA-86-1364, 1986.

45. W. J. Wepfer, R. A. Gaggioli, and E. F. Obert, Economic sizing of steam piping and insulation, *J. Eng. Ind.*, Vol. 101, 1979, pp. 427–433.

46. Y. M. Zilberberg, On dynamic irreversibilities, *Energy*, Vol. 9, 1984, pp. 1005–1007.

47. Y. M. Zilberberg, Impact of heat transfer dynamics on the second law analysis of thermodynamic cycles, in A. Bejan and R. L. Reid, eds., *Second Law Aspects of Thermal Design*, HTD-Vol. 33, Am. Soc. Mech. Eng., New York, 1984, pp. 39–43.

48. Y. M. Zilberberg, Dynamic irreversibilities and their impact on energy efficiency of refrigeration/heat-pump thermodynamic cycles, in P. O. Fanger, ed., *CLIMA 2000*, Vol. 6, VVS Kongres-VVS Messe, Copenhagen, 1985, pp. 77–85.

49. Y. M. Zilberberg, Dynamic exergy destruction of statically-reversible heat-transfer processes, *Am. Soc. Mech. Eng.* [*Pap.*], No. 85-WA/HT-21, 1985.

50. J. C. Chato and C. Damianides, Second-law-based optimization of heat exchanger networks using load curves, *Int. J. Heat Mass Transfer*, Vol. 29, 1986, pp. 1079–1086.

51. K. Hesselmann, Optimization of heat exchanger networks, in A. Bejan and R. L. Reid, eds., *Second Law Aspects of Thermal Design*, HTD-Vol. 33, Am. Soc. Mech. Eng., New York, 1984, pp. 95–99.

52. E. A. Bruges, *Available Energy and the Second Law Analysis*, Butterworth, London, 1959, p. 62.

53. G. M. Reistad, Availability: Concepts and applications, Ph.D. thesis, University of Wisconsin, Madison, 1970.

54. P. J. Golem and T. A. Brzustowski, Second-law analysis of energy processes. Part II. The performance of simple heat exchangers, *Trans. Can. Soc. Mech. Eng.*, Vol. 4, 1976–1977, pp. 219–226.

55. L. C. Witte and N. Shamsundar, A thermodynamic efficiency concept for heat exchange devices, *J. Eng. Power*, Vol. 105, 1983, pp. 199–203.

56. L. C. Witte, Second law optimization of heat exchangers, *Am. Soc. Mech. Eng.* [*Pap.*], No. 87-HT-65, 1987.

57. F. W. Schmidt and A. J. Willmott, *Thermal Energy Storage and Regeneration*, Hemisphere, Washington, DC, 1981.

58. A. Bejan, Two thermodynamic optima in the design of sensible heat units for energy storage, *J. Heat Transfer*, Vol. 100, 1978, pp. 708–712.

59. R. J. Krane, A second law analysis of a thermal energy storage system with Joulean heating of the storage element, *Am. Soc. Mech. Eng.* [*Pap.*], No. 85-WA/HT-19, 1985.

60. R. J. Krane, A second law analysis of the optimum design and operation of

thermal energy storage systems, *Int. J. Heat Mass Transfer*, Vol. 30, 1987, pp. 43–57.

61. B. Mathiprakasam and J. Beeson, Second law analysis of thermal energy storage devices, *Proc. AIChE Symp. Ser., Natl. Heat Transfer Conf.*, Seattle, Washington, 1983.

62. M. J. Taylor, Second law optimization of a sensible heat thermal energy storage system with a distributed storage element, M.S. thesis, Department of Mechanical and Aerospace Engineering, University of Tennessee, Knoxville, June 1986.

63. J. Y. San, W. M. Worek, and Z. Lavan, Second-law analysis of a two-dimensional regenerator, *Energy*, Vol. 12, 1987, pp. 485–496.

64. R. Hutchinson and S. Lyke, Microcomputer analysis of regenerative heat exchangers for oscillating flow, *Proc. ASME/JSME Therm. Eng. Jt. Conf.*, Vol. 2, 1987, p. 653.

65. E. S. Geskin, Second law analysis of fuel consumption in furnaces, *Energy*, Vol. 5, 1980, pp. 949–954.

66. E. S. Geskin, Assessment of energy use in material heating, in R. A. Gaggioli, ed., *Computer-Aided Engineering of Energy Systems*, AES-Vol. 2-2, Am. Soc. Mech. Eng., New York, 1986, pp. 107–112.

67. A. Bejan and W. Schultz, Optimum flowrate history for cooldown and energy storage processes, *Int. J. Heat Mass Transfer*, Vol. 25, 1982, pp. 1087–1092.

68. G. G. Haselden, *Cryogenic Fundamentals*, Academic Press, New York, 1971, p. 680.

69. J. Y. San, W. M. Worek, and Z. Lavan, Entropy generation in convective heat transfer and isothermal convective mass transfer, *J. Heat Transfer*, Vol. 109, 1987, pp. 647–652.

70. J. O. Hirschfelder, C. F. Curtiss, and R. B. Bird, *The Molecular Theory of Gases and Liquids*, Wiley, New York, 1964.

71. J. Y. San, W. M. Worek, and Z. Lavan, Entropy generation in combined heat and mass transfer, *Int. J. Heat Mass Transfer*, Vol. 30, 1987, pp. 1359–1369.

72. S. Sieniutycz, The thermodynamic approach to fluidized drying and moistening optimization, *AIChE J.*, Vol. 19, No. 2, 1973, pp. 277–285.

73. S. Sieniutycz, A synthesis of mathematical models and optimization algorithms of invariant imbedding type for a class of adiabatic drying processes with granular solid suspension, *Chem. Eng. Sci.*, Vol. 37, No. 10, 1982, pp. 1557–1568.

74. S. Sieniutycz, Lumped parameter modelling and an introduction to optimization of one-dimensional nonadiabatic drying systems, *Int. J. Heat Mass Transfer*, Vol. 27, No. 11, 1984, pp. 1971–1983.

75. S. Sieniutycz, A development of the relation between drying energy savings and thermodynamic irreversibility, *Chem. Eng. Sci.*, Vol. 39, No. 12, 1984, pp. 1647–1659.

76. E. Van den Bulck, S. A. Klein, and J. W. Mitchell, Second law analysis of solid desiccant rotary dehumidifiers, *Am. Soc. Mech. Eng. [Pap.]*, No. 85-HT-71, 1985.

77. Z. Lavan, J.-B. Monnier, and W. M. Worek, Second law analysis of desiccant cooling systems, *J. Sol. Energy Eng.*, Vol. 104, 1982, pp. 229–236.

78. I. L. Maclaine-cross, High-performance adiabatic desiccant open-cooling cycles, *J. Sol. Energy Eng.*, Vol. 107, 1985, pp. 102–104.

79. B. B. Moore and W. J. Wepfer, Application of second law based design optimization to mass transfer processes, *ACS Symp. Ser.*, Vol. 235, 1983, pp. 289–306.

PROBLEMS

11.1   Consider the classical Hagen–Poiseuille solution for fully developed laminar flow and heat transfer through a parallel-plate channel [Ref. 4, chapter 3]. The plate-to-plate spacing D and the wall heat flux q'' are given. The wall heat flux is uniform; therefore, the longitudinal temperature gradient dT/dx is constant and fixed by design. Derive an expression for the volumetric rate of entropy generation s'''_{gen} and plot the profile of entropy-generation distribution over the channel cross-section.

11.2   The job of the section of a heat exchanger is to transfer the heat transfer rate per unit length q' to an ideal-gas stream $(\dot{m}, R, c_P)$ flowing through a stack of n parallel-plate channels. The total thickness of the stack, D, is fixed. The plate-to-plate spacing of each channel is D/n, the mass flowrate through each channel is $\dot{m}/n$, and the heat transfer rate transmitted to the $\dot{m}/n$ stream is q'/n. Assuming that the flow regime is laminar, determine the optimum number of channels (n_{opt}) so that the overall entropy-generation rate of the stack (S'_{gen}) is minimized.

11.3   The object of this problem is to determine the optimum diameter of a sphere, D, whose job is to transmit in the least irreversible way possible the heat transfer rate $\dot{Q}$ to a uniform external flow (U_∞, T_∞). Assume that the Reynolds number $Re_D = DU_\infty/\nu$ is in the range 10^3-10^5 so that the drag coefficient can be assumed constant:

$$C_D = \frac{F_D/(\pi D^2/4)}{\rho U_\infty^2/2} \cong 0.5$$

11.4   Calculate the entropy-generation rate contributed by a rectangular plate fin whose dimension parallel to the flow (and to the base wall) is L. The boundary layer that coats the length L is turbulent: it is known that the fin-surface shear stress τ_w and the fin-stream heat transfer coefficient $\bar{h}$ are functions of the longitudinal position x measured along the boundary layer:

$$\frac{\tau_w}{\rho U_\infty^2/2} = 0.0576\left(\frac{U_\infty x}{\nu}\right)^{-1/5} = \frac{\bar{h}}{\rho c_P U_\infty}\, Pr^{2/3}$$

such that $x = 0$ represents the leading edge of the plate fin, and $x = L$, the trailing edge. The free-stream velocity and temperature are U_∞ and T_∞, respectively. If q'' (constant) is the local heat flux from the fin surface to the turbulent boundary layer, then the design you are about to execute is one where the heat transfer rate to the entire boundary layer (q') is considered fixed:

$$q' = q''L$$

By minimizing the total entropy-generation rate $\dot{S}'_{gen}$ of the L-wide fin, determine the optimum L that should be in contact with the flow.

11.5 We wish to determine whether it is better (i.e., less irreversible) to drive a car with the window up and the air conditioner on than to drive with the window down and the air conditioner off. The car can be modeled as a blunt object with frontal area $A = 5\,\mathrm{m}^2$ traveling through an infinite air reservoir with the speed U_∞. It is known that in the flow regime of interest, the drag coefficients are

$$C_D \cong 0.47 \quad \text{(with the window up)}$$

$$C_D \cong 0.51 \quad \text{(with the window down)}$$

When turned on, the air conditioner consumes electrical power at a constant rate, $\dot{W}_{AC} = 746\,\mathrm{W}$ (i.e., 1 horsepower). Your answer will depend on the traveling speed U_∞.

11.6 A counterflow heat exchanger consists of two concentric pipes, such that a hot-oil stream flows through the inner pipe and a cold-water stream flows through the annular space. The following data are given:

	Oil stream	Water stream
Mass flowrate, $\dot{m}$	2500 kg/hr	2000 kg/hr
Specific heat, c	1.67 J/g K	4.15 J/g K
Inlet temperature, T_{in}	77°C	32°C
Outlet temperature, T_{out}	55°C	—

It is assumed that the pressure drops along both streams contribute negligibly to the overall irreversibility of the apparatus. Determine in order:

- the water outlet temperature, $T_{out,water}$
- the effectiveness, ϵ
- the number of heat transfer units, N_{tu}
- the entropy-generation number, $N_S = \dot{S}_{gen}/(\dot{m}c)_{oil}$

Consider next the idea of increasing the heat transfer area by 20 percent. You can do this by simply lengthening the concentric-pipe arrangement by 20 percent, while the pipe diameters, the flow rates, the fluid properties, and the two inlet temperatures remain the same. Determine in order:

- the new N_{tu}, ϵ, $T_{out,oil}$, $T_{out,water}$ values
- the new entropy-generation number, N_S

You will find that the new N_S value is greater than the value calculated in the first part of the problem. This is a paradoxical result, considering that the new N_S corresponds to a 20-percent larger heat exchanger. Is this paradox related to the one discussed in connection with Fig. 11.12 and Fig. 11.13?

11.7   The optimum slenderness ratio (11.75) was obtained by regarding the mass velocity g_1 as fixed. Let $v_{1,g}$ and $N_{S,1,min}$ represent the dimensionless volume and entropy-generation number of the heat exchanger side optimized in this manner. Compare $v_{1,g}$ with the minimum volume that the designer can achieve for the same number of entropy-generation units as $N_{S,1,min}$. In other words, compare $v_{1,g}$ with the minimum volume $v_{1,min}$ indicated by eq. (11.87'), in which $N_{S,1}$ is equal to the $N_{S,1,min}$ discussed above. Show that $v_{1,g}$ exceeds $v_{1,min}$ by only 18.5 percent.

11.8   Let $a_{1,g}$ represent the dimensionless area of the counterflow heat-exchanger side optimized at fixed g_1, i.e., in accordance with eq. (11.75). Compare this area with the minimum area discovered in eq. (11.82'), in the case when the number of entropy-generation units is the same in both designs [i.e., when the $N_{S,1}$ of eq. (11.82') is the same as the $N_{S,1,min}$ of eq. (11.76)]. Show that $a_{1,g}$ exceeds $a_{1,min}$ by 30 percent.

11.9   Consider an adiabatic chamber in which two streams carrying the same fluid are being mixed steadily at constant pressure. Show that the maximum entropy-generation rate discovered in a counterflow heat exchanger with negligible pressure drop irreversibility, eq. (11.96), is the same as the entropy-generation rate of the mixing chamber that receives the same two streams as the inlet streams to the counterflow heat exchanger. This problem is a special case of a more general theorem uncovered by Sekulic [33].

11.10   (a) Let $v_{1,a}$ represent the dimensionless volume of side 1 of a nearly ideal balanced counterflow heat exchanger designed for minimum area subject to a fixed $N_{S,1}$. Show that $v_{1,a}/v_{1,min} = 256/243$, where $v_{1,min}$ is the minimum dimensionless volume that could be achieved subject to the same number of entropy-generation units, $N_{S,1}$.

(b) Examine next the area $(a_{1,v})$ of one side of a nearly ideal counterflow heat exchanger that has been designed for minimum volume subject to a fixed number of entropy-generation units. Compare $a_{1,v}$ with $a_{1,min}$, that is, with the minimum dimensionless area that could be achieved at the same number of entropy-generation units. Show that $a_{1,v}/a_{1,min} = 1.033$.

12

Irreversible Thermodynamics

Aside from its obvious engineering-design flavor, one purpose of the preceding chapter has been to bring under the larger umbrella of thermodynamics the beginnings of the fluid-flow and heat and mass transfer work that is an integral part of thermal-engineering practice. Another way of looking at the connection between thermodynamic and transport phenomena is represented by the method of *irreversible thermodynamics*, elements of which are outlined in this chapter. Unlike the Gibbsian formulation of chapter 4, where the focus was on a system in equilibrium internally and vis-à-vis with its environment, in the present chapter, we are concerned exclusively with the *rates* of energy and mass transfer interactions that occur in the space betwen macroscopic systems that are not mutually in equilibrium.

As it turns out, this new "irreversible-thermodynamics" approach does not generate additional physical insight into the relationship between interaction rates and the local thermodynamic properties of the system (space) penetrated by these interactions. The analytical form of these relation is "assumed" in the same manner that Fourier's law of heat conduction was written almost 200 years ago. And, just like Fourier's thermal-conductivity coefficient, the many coefficients of the newly assumed relations can only be determined empirically or based on an entirely different theory. This new approach does make a contribution in the realm of "coupled" transport phenomena, where two or more transport processes coexist and influence one another. This contribution is represented most succinctly by Onsager's reciprocity relations [1, 2], which here are discussed beginning with eq. (12.38). Another important contribution of this approach is the promotion of a unified and analytically compact treatment of all irreversible-flow phenomena, this, despite the fact that many of these phenomena had been figured out and treated as individual cases long before the emergence of irreversible thermodynamics.

670

CONJUGATE FLUXES AND FORCES

In the course of developing the equilibrium conditions of chapter 4, we learned that descriptions such as "equilibrium" and "reversibility" depend on the maintenance of uniform intensive properties throughout the system, that is, uniform fields of temperature, pressure, and chemical potentials. The concepts of irreversibility and entropy generation, on the other hand, are associated with the changes driven by the intensive property nonuniformities that can exist in the system. The preceding chapter illustrated this relationship by means of examples drawn from the practice of heat and mass transfer engineering. The objective of this opening section is to develop formally the relationship between the local entropy-generation rate and the local gradients of the intensive properties. The emphasis is placed on those situations in which the entropy-generation process is associated with changes caused by two or more intensive property gradients.

Consider for this purpose the single-phase system sketched on the left side of Fig. 12.1. This system is in a state of equilibrium characterized by the uniform temperature T, pressure P, and n chemical potentials μ_i. The system contains n chemical species, which are labeled $i = 1, 2, \ldots, n$. Thanks to the Gibbs–Duhem relation (4.32) only $(n + 1)$ of the total $(n + 2)$ intensive properties of the system can be specified independently. In the following discussion, we regard T and the n chemical potentials μ_i as the $(n + 1)$ intensive properties that are specified independently. In other words, the *intensive state* of the left-side system is pinpointed by the numerical values of T, μ_1, μ_2, $\ldots$, μ_{n-1}, and μ_n.

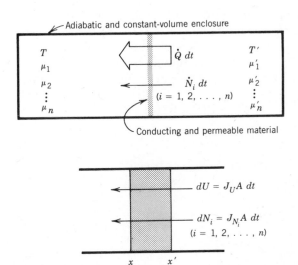

Figure 12.1 Unidirectional flow of heat and mass (top) and internal energy and mass (bottom) through a conducting and permeable material.

The single-phase system described is held inside a constant-volume enclosure (control volume). The right wall of this enclosure is permeable to both heat transfer and mass transfer. On the other side of the right wall of this system, we see another constant-volume enclosure containing another single-phase system. The uniform intensive properties of this second system are indicated with the prime [$( )'$]; its intensive state is determined by the specification of the $(n + 1)$ uniform intensities T', μ'_1, $\mu'_2, \ldots, \mu'_{n-1}$, and μ'_n.

The individual single-phase equilibrium systems described are old news. The new feature on which we begin to focus is the vertical partition across which we now note the temperature and chemical potential discontinuities, respectively, $(T' - T)$, $(\mu'_1 - \mu_1), \ldots, (\mu'_n - \mu_n)$. At a particular point in time, these differences are together responsible for the $(n + 1)$ heat and mass transfer interactions ($\dot{Q}\, dt$, $\dot{N}_1\, dt, \ldots, \dot{N}_n\, dt$, respectively) that cross the partition. These interactions are considered positive when they proceed from right to left: this sign convention is indicated by the arrows in Fig. 12.1. Assuming that the interactions are sufficiently slow so that during the time interval dt, the $(T, \mu_1, \ldots, \mu_n)$ and $(T', \mu'_1, \ldots, \mu'_n)$ distributions remain unafffected (i.e., uniform), for each of the single-phase systems, we can write

$$dS = \frac{1}{T}\, dU - \sum_{i=1}^{n} \frac{\mu_i}{T}\, dN_i \tag{12.1}$$

$$dS' = \frac{1}{T'}\, dU' - \sum_{i=1}^{n} \frac{\mu'_i}{T'}\, dN'_i \tag{12.2}$$

These equations follow from the differential form of the fundamental relation in entropy representation, eq. (4.18), while keeping in mind that each single-phase system evolves at constant volume. Taken together, the two single-phase systems constitute an *isolated* system (note the zero-work, adiabatic, and impermeable boundary, Fig. 12.1); therefore,

$$dU + dU' = 0 \tag{12.3}$$

$$dN_i + dN'_i = 0 \qquad (i = 1, 2, \ldots, n) \tag{12.4}$$

Eliminating dU' and the (dN'_i)s between eqs. (12.1)–(12.4), we obtain the entropy generated inside the isolated (aggregate) system:

$$\Delta\dot{S}_{\text{gen}}\, dt = dS + dS' = \left(\frac{1}{T} - \frac{1}{T'}\right) dU - \sum_{i=1}^{n} \left(\frac{\mu_i}{T} - \frac{\mu'_i}{T'}\right) dN_i \geq 0 \tag{12.5}$$

or, per unit time,

$$\Delta\dot{S}_{\text{gen}} = \frac{dS + dS'}{dt} = \left(\frac{1}{T} - \frac{1}{T'}\right)\frac{dU}{dt} - \sum_{i=1}^{n} \left(\frac{\mu_i}{T} - \frac{\mu'_i}{T'}\right)\frac{dN_i}{dt} \geq 0 \tag{12.6}$$

Since the single-phase systems that sandwich the partition enjoy individually states of internal equilibrium, the locus of the entropy generation $(dS + dS')$ can only be the partition itself. The special region occupied by the partition is expanded for the purpose of a closer examination in the lower part of Fig. 12.1. All the flows that cross the partition are unidirectional; therefore, the analysis can be put on a per-unit-area basis by dividing eq. (12.6) through the partition cross-section A:

$$\frac{\Delta \dot{S}_{gen}}{A} = \left(\frac{1}{T} - \frac{1}{T'}\right) \underbrace{\frac{dU/dt}{A}}_{J_U} - \sum_{i=1}^{n} \left(\frac{\mu_i}{T} - \frac{\mu_i'}{T'}\right) \underbrace{\frac{dN_i/dt}{A}}_{J_{N_i}} \geq 0 \qquad (12.7)$$

In this expression, J_{N_i} represents the *molal flux* of species "i," that is, the number of moles of "i" that cross the partition per unit area and per unit time. Note further that

$$J_{N_i} = \frac{\dot{N}_i}{A} \qquad (i = 1, 2, \ldots, n) \qquad (12.8)$$

where $\dot{N}_i$ represents the n molal flowrates mentioned earlier. Similarly, by adopting Callen's notation [3], we write J_U for the net *energy flux* across the partition. Although the units of J_U are W/m², it is worth keeping in mind that, in general, J_U is not equal to the heat flux q'' that may be crossing the partition. We return to this observation in the course of writing eq. (12.13).

What has been distilled in the form of eq. (12.7) is the entropy generated per unit area by $(n + 1)$ fluxes $(J_U, J_{N_1}, \ldots, J_{N_n})$. These fluxes appear to be associated with the $(n + 1)$ intensive property differences that can exist across the partition. As a final step in this development, consider the fact that the partition has a finite thickness whose linear measure in the direction of flow is

$$\Delta x = x - x' \qquad (12.9)$$

It is permissible to divide eq. (12.7) by Δx:

$$\frac{\Delta \dot{S}_{gen}}{A \, \Delta x} = \frac{T^{-1} - (T')^{-1}}{x - x'} J_U + \sum_{i=1}^{n} \frac{(-\mu_i/T) - (-\mu_i'/T')}{x - x'} J_{N_i} \geq 0 \quad (12.10)$$

and to imagine the limit in which the thickness Δx and the intensive property discontinuities vanish:

$$s'''_{gen} = \lim_{\Delta x \to 0} \frac{\Delta \dot{S}_{gen}}{A \, \Delta x} = J_U \underbrace{\frac{\partial}{\partial x}\left(\frac{1}{T}\right)}_{X_U} + \sum_{i=1}^{n} J_{N_i} \underbrace{\frac{\partial}{\partial x}\left(-\frac{\mu_i}{T}\right)}_{X_{N_i}} \geq 0 \quad (12.11)$$

The resulting quantity, s'''_{gen}, is the volumetric (local) entropy-generation rate expressed in $W/K/m^3$. The same quantity is more frequently labeled σ in the irreversible-thermodynamics literature. Here we continue to use the symbol s'''_{gen}, which was used extensively in the thermodynamic design chapter [e.g., eq. (11.48)].

The intensive property gradients X_U and X_{N_i} defined on the right side of eq. (12.11) are the *conjugate forces* (called also *driving forces*, or *affinities*) associated with J_U and J_{N_i}, respectively. It is not the temperature gradient $\partial T/\partial x$ that serves as conjugate force for the energy flux J_U, but the gradient of the inverse thermodynamic temperature, $X_U = \partial(T^{-1})/\partial x$. Likewise, the conjugate force for the molal flux J_{N_i} is not the gradient of $-\mu_i$ but that of $-\mu_i/T$. If the flow through the partition is three-dimensional, then in place of eq. (12.11), an analogous vectorial analysis yields

$$s'''_{gen} = \mathbf{J}_U \cdot \nabla\left(\frac{1}{T}\right) + \sum_{i=1}^{n} \mathbf{J}_{N_i} \cdot \nabla\left(-\frac{\mu_i}{T}\right) \tag{12.12}$$

The local entropy-generation rate (12.11) is expressed in terms of thermodynamic properties such as T^{-1} and μ_i/T. Implicit in this formulation is the assumption of local thermodynamic equilibrium (p. 71) by which the state of the vanishingly thin system of thickness Δx is represented by one temperature and one chemical potential for each of its constituents. The local equilibrium assumption is essential because T and μ_i have been defined in the context of systems in equilibrium (see chapters 2 and 4). Looking at the greater picture painted in Fig. 12.1, we see two macroscopic subsystems—the left and the right halves—that are not in equilibrium with one another. The partition that separates these two subsystems can be viewed as a stack of a very large number of vertical sheets of thickness Δx, in which each sheet has its own state of equilibrium. The equilibrium property values of one sheet differ infinitesimally from the values of the adjacent sheet. In this arrangement, then, the local entropy-generation rate (12.11) is an expression of the irreversibility associated with the energy and mass transfer *interactions* between two adjacent sheets.

The general character of eq. (12.11) vis-à-vis the special s'''_{gen} expressions of chapter 11 becomes more evident if we replace the energy flux J_U by its two possible components (avenues), namely, the nonflow component called *heat flux* ($q'' = \dot{Q}/A$) and the n enthalpy-flow components of type ($\bar{h}_i J_{N_i}$):

$$J_U = q'' + \sum_{i=1}^{n} \bar{h}_i J_{N_i} \tag{12.13}$$

In place of eq. (12.11), we obtain:

$$s'''_{gen} = q'' \frac{\partial}{\partial x}\left(\frac{1}{T}\right) + \sum_{i=1}^{n} J_{N_i} \frac{\partial \bar{s}_i}{\partial x} \tag{12.14}$$

which, in the first term on the right side, reveals the entropy-generation contribution made by pure heat transfer in eq. (11.48). The connection with eq. (11.48) is even more evident if we recognize the three-dimensional analog of eq. (12.14):

$$s'''_{gen} = \mathbf{q}'' \cdot \nabla \left(\frac{1}{T} \right) + \sum_{i=1}^{n} \mathbf{J}_{N_i} \cdot \nabla(\bar{s}_i) \tag{12.15}$$

A more compact summary of the conclusion reached in eq. (12.11) is

$$s'''_{gen} = \sum_{i=0}^{n} J_i X_i \geq 0 \tag{12.16}$$

in which the new notation (J_i, X_i) stands for

$$J_i = J_U \quad X_i = X_U = \frac{\partial}{\partial x} \left(\frac{1}{T} \right) \quad (i = 0) \tag{12.17}$$

$$J_i = J_{N_i} \quad X_i = X_{N_i} = \frac{\partial}{\partial x} \left(-\frac{\mu_i}{T} \right) \quad (i = 1, \ldots, n) \tag{12.18}$$

The local entropy-generation rate emerges as the sum of $(n + 1)$ products of conjugate forces and fluxes. It takes at least one such product, that is, one pair (J_i, X_i) in which *both* partners are nonzero, for entropy to be generated at the point of interest. This observation is equivalent to comments made earlier in connection with individual irreversibility mechanisms (see p. 136 in chapter 3). If, on the other hand, all the driving forces X_i are zero, then the local entropy-generation rate is also zero. This last description applies to the two single-phase systems positioned to the left and right of the partition in Fig. 12.1, in which all the temperature and chemical potential gradients (hence, the X_is) are zero.

The conjugate fluxes J_i and forces X_i require special care [4]. The mere writing of eq. (12.16), that is, the identification of a number of products of type $J_i X_i$ that added together equal the local entropy-generation rate, does not mean that the flux J_i is the conjugate of the driving force X_i. In addition to eq. (12.16), the fluxes and forces must satisfy the definitions presented under eqs. (12.7) and (12.11), namely,

$$J_i = \frac{1}{A} \frac{da_i}{dt} \tag{12.19}$$

$$X_i = \frac{\partial}{\partial x} \left(\frac{\partial S}{\partial a_i} \right) \tag{12.20}$$

where the a_is are the $(n + 1)$ arguments of the homogeneous function S, eq. (12.1),

$$S = S(a_0, a_1, \ldots, a_n) \tag{12.21}$$

The conjugate fluxes are proportional to the time derivatives of the extensive properties that appear as arguments in the entropic fundamental relation ($a_0 = U$, $a_1 = N_1$, etc.). The driving forces are the gradients of the intensive-property coefficients that appear naturally in the writing of the differential form (12.1).

LINEARIZED RELATIONS

The practical experience gathered in the field of heat and mass transfer suggests that the rates of heat and mass transfer are finite only if the intensive-property gradients along which they occur are finite. Turned around, this first observation seems to suggest that all the J_i fluxes would be zero when all the X_i driving forces are zero. A second body of experimental evidence suggests that a particular flux (J_i) is driven in general not only by its conjugate force (X_i), but also by one or more of the remaining driving forces (X_j; $j \neq i$), that is, by the conjugate forces of other fluxes (J_j). This "cross-coupling" of the fluxes is illustrated by the phenomena discussed in the second half of this chapter.

The above observations are the basis for postulating that any flux of type J_i depends in general on all the driving forces evaluated at the point of interest (X_i; $i = 0, 1, \ldots, n$) as well as on all the intensive properties measured at the same point [5]:

$$J_i = J_i(X_0, X_1, \ldots, X_n; T, \mu_1, \ldots, \mu_n) \qquad (i = 0, 1, \ldots, n) \quad (12.22)$$

Furthermore, in view of the first observation of the preceding paragraph, we postulate that the J_i function (12.22) is such that J_i is zero when all the driving forces are zero:

$$(J_i)_{X_0 = X_1 = \cdots = X_n = 0} = 0 \qquad (i = 0, 1, \ldots, n) \qquad (12.23)$$

We expect, of course, a finite J_i in a case when not all the driving forces X_i are zero. The analytical relationship between J_i and the $(n + 1)$ driving forces, eq. (12.22), can be linearized in the limit of vanishing X_is by writing the Taylor expansion of J_i around the point $X_0 = X_1 = \cdots = X_n = 0$:

$$J_i = (J_i)_{X_0 = X_1 = \cdots = X_n = 0} + \frac{\partial J_i}{\partial X_0} X_0 + \frac{\partial J_i}{\partial X_1} X_1 + \cdots + \frac{\partial J_i}{\partial X_n} X_n$$

$$+ \text{ higher-order terms} \qquad (i = 0, 1, \ldots, n) \qquad (12.24)$$

Considering now the zero-flux situation (12.23), we conclude that in the limit of vanishingly small driving forces, each flux is related linearly to all the driving forces:

$$J_i = \sum_{k=0}^{n} L_{ik}X_k \qquad (i = 0, 1, \ldots, n) \qquad (12.25)$$

in which L_{ik} is shorthand for the partial-differential coefficients:

$$L_{ik} = \frac{\partial J_i}{\partial X_k} \qquad (12.26)$$

In the same limit, the local entropy-generation rate (12.16) assumes the form:

$$s'''_{gen} = \sum_{i=0}^{n} \sum_{k=0}^{n} L_{ik}X_iX_k \geq 0 \qquad (12.27)$$

ONSAGER'S RECIPROCITY RELATIONS

Equation (12.27) completes the formal introduction of the concepts of local entropy-generation rate, conjugate pairs of fluxes and driving forces, the flux postulate (12.22)–(12.23), and, finally, the linearized relations between fluxes and driving forces, eq. (12.25). It pays to review this analytical backbone by examining its form in the case when only two fluxes are present. Let the conjugate pairs of fluxes and forces be (J_0, X_0) and (J_1, X_1). The linear relations (12.25) reduce in this case to

$$J_0 = L_{00}X_0 + L_{01}X_1 \qquad (12.28)$$

$$J_1 = L_{10}X_0 + L_{11}X_1 \qquad (12.29)$$

The local entropy-generation rate expression (12.27) has now only three terms:

$$s'''_{gen} = L_{00}X_0^2 + (L_{01} + L_{10})X_0X_1 + L_{11}X_1^2 \geq 0 \qquad (12.30)$$

the second of which is due to the "coupling" of the two fluxes through the coefficients L_{01} and L_{10} of eqs. (12.28)–(12.29).

The Second Law of Thermodynamics is represented by the "$\geq$" sign appearing in eq. (12.30). This sign has something to say about the signs and relative sizes of the linearization coefficients L_{00}, L_{01}, L_{10}, and L_{11}. It is not difficult to show that the s'''_{gen} expression (12.30) can be written in two additional ways [5]:

$$s'''_{gen} = \left[L_{00} - \frac{(L_{01} + L_{10})^2}{4L_{11}} \right]X_0^2 + L_{11}\left(X_1 + \frac{L_{01} + L_{10}}{2L_{11}} X_0 \right)^2 \geq 0$$
$$(12.31a)$$

$$s'''_{gen} = L_{00}\left(X_0 + \frac{L_{01} + L_{10}}{2L_{00}} X_1\right)^2 + \left[L_{11} - \frac{(L_{01} + L_{10})^2}{4L_{00}}\right]X_1^2 \geq 0$$

$$(12.31b)$$

If s'''_{gen} is to remain nonnegative for all the values of X_0 and X_1, then all the coefficients of the squared quantities on the right side of eqs. (12.31a, b) must be nonnegative. This observation translates into only three inequalities:

$$L_{00} \geq 0 \tag{12.32a}$$

$$L_{11} \geq 0 \tag{12.32b}$$

$$L_{00}L_{11} \geq [\tfrac{1}{2}(L_{01} + L_{10})]^2 \tag{12.32c}$$

Note that, together, these three inequalities rule out the chance of seeing negative values in the two bracketed coefficients appearing in eqs. (12.31a, b).

The first two inequalites have a direct physical meaning in the realm of single-flux irreversible phenomena. For example, in the case of pure heat transfer through an isotropic medium, there is only one driving force, $X_0 = \partial(T^{-1})/\partial x$, and only one flux, J_0. Since in this case all the mass flows are zero, the lone flux J_0 (or J_U) is the same as the heat flux q'', eq. (12.13). If we write $J_0 = q''$ and invoke the Fourier law of heat conduction as the definition of the thermal conductivity coefficient k [6]:

$$J_0 = q'' = -k \frac{\partial T}{\partial x} \tag{12.33}$$

then the single-flux version of the linearized relation (12.25):

$$J_0 = L_{00} \frac{\partial}{\partial x} (T^{-1}) \tag{12.34}$$

allows us to conclude that

$$L_{00} = T^2 k \tag{12.35}$$

In conclusion, the L_{00} coefficient is not only proportional to k, but also has the same sign as k. As a manifestation of the second law, then, the inequality (12.32a) guarantees that the thermal-conductivity coefficient k cannot be negative.

Similar conclusions are reached by considering other examples of single-flux phenomena. In the case of pure electrical conduction through an isotropic medium, we can rely on Ohm's law (12.48) and eq. (12.32a) to show that the electrical conductivity k_e cannot be negative. Similarly, consideration of Fick's law [Ref. 6, p. 314] for the diffusion of species i

solely under the influence of the concentration gradient of i leads to the conclusion that the mass-diffusivity coefficient D_{ii} cannot be negative. We return to these particular phenomena in subsequent sections, in which the subject is the cross-coupling of two or more of such conjugate pairs of fluxes and forces.

Continuing with the two-flux phenomenon addressed by eqs. (12.28)–(12.32), we notice that the cross-coupling coefficients L_{01} and L_{10} are expressible in the same units. Specifically, the dimensions of both coefficients are the dimensions of s'''_{gen} divided by the dimensions of the product $X_0 X_1$. The empirical evidence that we already have on several two-flux irreversible phenomena suggests that the coefficients of type L_{01} and L_{10} have not only the same units, but also the same magnitude, in other words, that

$$L_{01} = L_{10} \tag{12.36}$$

The earliest observations of this kind were made by William Thomson [7] with respect to coupled thermal and electrical conduction (thermoelectric phenomena, or thermoelectricity), and by Stokes [8] in the analysis of heat conduction through an anisotropic medium. On these important engineering phenomena, we focus in detail beginning with the next section. Note at this stage that since $L_{01} = L_{10}$, the second-law requirement (12.32c) states that the absolute value of the cross-coupling coefficient cannot exceed the geometric average of the always nonnegative L_{00} and L_{11}:

$$(L_{00}L_{11})^{1/2} \ge |L_{01}| \tag{12.37}$$

Almost 100 years after William Thomson and Stokes, Onsager [1, 2] showed that the several examples of two-flux phenomena in which eq. (12.36) appears to hold are special manifestations of the more general statement that, for all is and js and in the absence of magnetic effects, reads:

$$L_{ij} = L_{ji} \tag{12.38}$$

According to Onsager's "reciprocity relations" (12.38), the matrix $\|L_{ij}\|$ associated with the linear system (12.25) is symmetric. Onsager derived these reciprocity relations based on statistical mechanical arguments. In the classical macroscopic framework of the present treatment, we have no choice but to adopt Onsager's relations as a new postulate—a law—whose validity remains to be supported or refuted by physical observations. The growing body of empirical evidence that supports Onsager's reciprocity relations was reviewed in 1960 by Miller [9]. A segment of this material forms the subject of the next three sections.

THERMOELECTRIC PHENOMENA

Formulations

The best-known example of coupled irreversible-flow phenomena is the phenomenon of thermoelectricity associated with the simultaneous conduction of thermal and electrical currents through a conducting material. An exhaustive review of the ideas associated with this subject has been published by Domenicali [10]. The following treatment is patterned after that of Callen [3]. Let $T(x)$ and $\phi(x)$ represent the distributions of temperature and electrostatic potential, respectively, along the one-dimensional conductor in Fig. 12.2. There are two fluxes that cross the unit cross-sectional area of the conductor, the internal energy flux $J_0 = J_U$, and the electron flux $J_1 = J_N$. Both fluxes are conserved from one cross-section to the next along the one-dimensional conductor. The role of the electrochemical potential μ in this problem is played by the potential energy per particle, $e\phi$, where e is the electric charge of one electron and ϕ is the electrostatic potential:

$$\mu = e\phi \tag{12.39}$$

In conclusion, in place of the two-flux relations (12.28)–(12.29), we can write immediately

$$J_U = L_{00}\frac{d}{dx}\left(\frac{1}{T}\right) + L_{01}\frac{d}{dx}\left(-\frac{e\phi}{T}\right) \tag{12.40}$$

$$J_N = L_{10}\frac{d}{dx}\left(\frac{1}{T}\right) + L_{11}\frac{d}{dx}\left(-\frac{e\phi}{T}\right) \tag{12.41}$$

This starting point is essential because it establishes the connection between the preceding formalism and the much older language of thermoelectricity, in which the preferred "energy" flux is the heat flux q'', not the internal energy flux J_U, and where the relevant gradients are those of

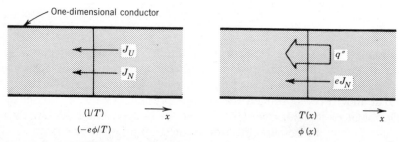

Figure 12.2   One-dimensional conductor housing the simultaneous flow of heat and electric current. Left side: conjugate fluxes and forces. Right side: formulation in terms of physically meaningful quantities.

temperature and electrostatic potential, not the driving forces listed in eqs. (12.40)–(12.41). The first step in the direction of translating eqs. (12.40)–(12.41) into old thermoelectricity consists of invoking eq. (12.13):

$$J_U = q'' + e\phi J_N \tag{12.42}$$

this, in order to replace J_U by q''. The new two-flux system that replaces eqs. (12.40)–(12.41) is

$$q'' = l_{00} \frac{d}{dx}\left(\frac{1}{T}\right) + \frac{l_{01}}{T}\frac{d}{dx}(-e\phi) \tag{12.43}$$

$$J_N = l_{10} \frac{d}{dx}\left(\frac{1}{T}\right) + \frac{l_{11}}{T}\frac{d}{dx}(-e\phi) \tag{12.44}$$

in which the new coefficients $(l_{00}, l_{01}, l_{10}, l_{11})$ are to be distinguished from the original "Onsager" set $(L_{00}, L_{01}, L_{10}, L_{11})$. The relationships between the two sets of coefficients are

$$l_{00} = L_{00} - e\phi(L_{01} + L_{10}) + (e\phi)^2 L_{11} \tag{12.45a}$$

$$l_{01} = L_{01} - e\phi L_{11} \tag{12.45b}$$

$$l_{10} = L_{10} - e\phi L_{11} \tag{12.45c}$$

$$l_{11} = L_{11} \tag{12.45d}$$

The new coefficients depend only on the local temperature $T(x)$ [5]. Worth noting is that if Onsager's reciprocity relation $L_{01} = L_{10}$ applies, then we also have $l_{01} = l_{10}$, that is, equal cross-coupling coefficients in the thermo-electric formulation of the two-flux problem. It is more instructive to postpone the use of the reciprocity relation until we take a closer look at the experimental basis for believing in a statement like $l_{01} = l_{10}$.

The second step in the thermoelectric translation of this two-flux problem consists of substituting directly measurable quantities in place of the four coefficients l_{00}, l_{01}, l_{10}, and l_{11}. For this second step, we need four relations (experiments). The first is the measurement of thermal conductivity, eq. (12.33), which in the present problem is carried out at *zero electric current*:

$$q'' = -k \frac{dT}{dx} \qquad (eJ_N = 0) \tag{12.46}$$

Eliminating q'' and $d(-e\phi)/dx$ at zero electric current between eqs. (12.43), (12.44), and (12.46), we obtain the first (thermal conductivity) relation:

$$k = \frac{1}{T^2}\left(l_{00} - \frac{l_{01}l_{10}}{l_{11}}\right) \tag{12.47}$$

Of course, this result matches eq. (12.35) in the case of $l_{01} = 0$, where the heat flux is the only flux or where q'' is uncoupled from J_N.

The second needed relation is the measurement of electrical conductivity, k_e, which is defined as the ratio between the electric current density (eJ_N) and the potential gradient, all in an *isothermal system*:

$$eJ_N = -k_e \frac{d\phi}{dx} \qquad \left(\frac{dT}{dx} = 0 \right) \tag{12.48}$$

Comparing this definition with the isothermal version of eq. (12.44), we find that k_e and l_{11} are directly proportional:

$$k_e = \frac{e^2}{T} l_{11} \tag{12.49}$$

The third needed relation comes from the *Seebeck effect* (p. 685), which represents a class of experimental observations that preceded William Thomson's paper [7]. These observations suggest that under conditions of *zero electric current* $(eJ_N = 0)$, a finite electromotive force is produced when the temperature gradient is finite. Rewriting eq. (12.44) for $J_N = 0$:

$$0 = l_{10} \frac{d}{dx}\left(\frac{1}{T} \right) + \frac{l_{11}}{T} \frac{d}{dx}(-e\phi) \tag{12.50}$$

we see that the electrostatic-potential difference registered for each unit temperature difference along the conductor is proportional to the second cross-coupling coefficient l_{10}:

$$\left(\frac{d\phi}{dT} \right)_{J_N = 0} = -\frac{l_{10}}{l_{11}eT} = -\epsilon(T) \tag{12.51}$$

The new temperature function $\epsilon(T)$ defined above is known as the *absolute thermoelectric power* of the conductor, or the *absolute Seebeck coefficient*.

Finally, the fourth relation needed for determining the "l" coefficients comes from the experimental finding known as the *Peltier effect* (p. 684). Equation (12.43) suggests the establishment of a finite "Peltier" heat current in an *isothermal* conductor with finite potential gradient along it:

$$q'' = \frac{l_{01}}{T} \frac{d}{dx}(-e\phi) \qquad \left(\frac{dT}{dx} = 0 \right) \tag{12.52}$$

Under the same isothermal conditions, the electron flux is, per eq. (12.44),

$$eJ_N = e\frac{l_{11}}{T} \frac{d}{dx}(-e\phi) \qquad \left(\frac{dT}{dx} = 0 \right) \tag{12.53}$$

The ratio of the two currents represents the *Peltier coefficient* of the particular conductor, π,

$$\frac{q''}{eJ_N} = \frac{l_{01}}{el_{11}} = \pi(T) \qquad \left(\frac{dT}{dx} = 0\right) \tag{12.54}$$

Combining now the four relations, (12.47), (12.49), (12.51), and (12.54), we can express the four coefficients $(l_{00}, l_{01}, l_{10}, l_{11})$ in terms of the physically more meaningful set (k, k_e, ϵ, π):

$$l_{00} = T^2(k + k_e \pi\epsilon) \qquad l_{01} = \frac{T}{e} k_e \pi$$

$$l_{10} = \frac{T^2}{e} k_e \epsilon \qquad l_{11} = \frac{T}{e^2} k_e \tag{12.55}$$

These results are arranged in the same array as in eqs. (12.43)–(12.44). It is now clear that the cross-coupling coefficients l_{01} and l_{10} are equal only if

$$\pi = T\epsilon \tag{12.56}$$

This last relation is supported by a long list of experimental reports [9]; therefore, $l_{01} = l_{10}$ and, according to the observation made under eq. (12.45d), $L_{01} = L_{10}$. The adoption of the reciprocity relation (12.56) means that the four "l" coefficients can now be expressed in terms of only k, k_e, and ϵ: substituting these new expressions into the two-flux linear system (12.43)–(12.44) yields, finally,

$$q'' = -(k + T\epsilon^2 k_e) \frac{dT}{dx} - T\epsilon k_e \frac{d\phi}{dx} \tag{12.57}$$

$$eJ_N = -\epsilon k_e \frac{dT}{dx} - k_e \frac{d\phi}{dx} \tag{12.58}$$

These two linear relations represent the end of the effort of converting the original statement (12.40)–(12.41) into one in terms of physically meaningful fluxes (q'', eJ_N), gradients $(dT/dx, d\phi/dx)$, and coefficients (k, k_e, ϵ). It is worth comparing these relations with the pure heat-conduction limit (12.33) and with the pure electric-conduction limit (12.48) in order to see better the coupling of the two phenomena. Another useful view of this coupling is provided by eliminating $d\phi/dx$ between the last two equations:

$$q'' = -k \frac{dT}{dx} + T\epsilon e J_N \tag{12.59}$$

This relation shows that in the presence of an electric current, the actual heat flux q'' exceeds the pure conduction (Fourier) flux by the amount $(T\epsilon eJ_N)$. This excess amount is what accounts for the Peltier heat flux in the case of an isothermal conductor with finite electrical current through it, eq. (12.52).

The Peltier Effect[†]

There are several important phenomena that can be described quantitatively by means of the phenomenological rewriting of this two-flux problem. The simplest is the release or absorption of a finite heat transfer rate at the junction between two constant-temperature electrical conductors made of different materials, α and β, Fig. 12.3. Since the entire system is isothermal, the heat fluxes on the two sides of the function are, cf. eq. (12.59),

$$q_\alpha'' = T\epsilon_\alpha eJ_N \qquad q_\beta'' = T\epsilon_\beta eJ_N \tag{12.60}$$

Note here the use of T for the temperature of the system and eJ_N for electric-current density. The latter is conserved as it passes through the junction. The Peltier heat fluxes, q_α'' and q_β'', are not necessarily equal because, generally speaking, two different materials have different absolute thermoelectric powers, $\epsilon_\alpha(T)$ and $\epsilon_\beta(T)$. We record these ideas by subtracting eqs. (12.60) and calculating the excess Peltier heat flux:

$$q_\beta'' - q_\alpha'' = T(\epsilon_\beta - \epsilon_\alpha)eJ_N \tag{12.61}$$

The question, next, is whether this excess heat flux leaves or enters the conducting path *laterally* at the junction, Fig. 12.3. There will be lateral heat transfer only if there is a change in the net energy current J_U that flows from right to left along the two-conductor system. Invoking eq. (12.42) for each conductor right at the junction, we write

$$J_{U,\alpha} = q_\alpha'' + e\phi J_N \qquad J_{U,\beta} = q_\beta'' + e\phi J_N \tag{12.62}$$

assuming, of course, the potential ϕ varies continuously across the junction. Equations (12.62) show that the excess Peltier flux is indeed equal to the step change in J_U across the junction:

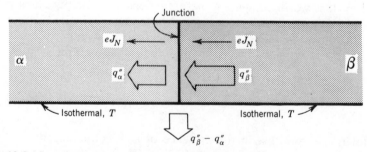

Figure 12.3   Junction between two one-dimensional isothermal conductors, showing the origin of the Peltier heat release.

[†]Discovered by French physicist Jean C. A. Peltier in 1834.

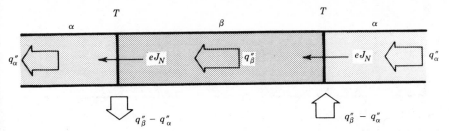

Figure 12.4 Two isothermal junctions at the same temperature, showing that the Peltier heat released by one is the same as the Peltier heat absorbed by the other.

$$q''_\beta - q''_\alpha = J_{U,\beta} - J_{U,\alpha} \qquad (12.63)$$

which means that $(q''_\beta - q''_\alpha)$ is transferred laterally, as shown in Fig. 12.3. In terms of the Peltier coefficient notation (12.54), eq. (12.61) is recognized as the *second Kelvin relation*:

$$\pi_\beta - \pi_\alpha = T(\epsilon_\beta - \epsilon_\alpha) \qquad (12.64)$$

where the difference $(\pi_\beta - \pi_\alpha)$ is heat transfer released by the junction per unit of electric current flowing from β to α:

$$\pi_\beta - \pi_\alpha = \frac{q''_\beta - q''_\alpha}{eJ_N} \qquad (12.65)$$

The actual sign of $(q''_\beta - q''_\alpha)$ depends on the sign of $(\epsilon_\beta - \epsilon_\alpha)$ and the direction of the electric current. One certain fact is that changing the direction of J_N changes also the sign of the lateral heating effect at the junction. Said another way, if we maintain the two junctions of Fig. 12.4 at the same temperature T, then the heat transfer released by the first junction equals the heat transfer received by the second junction, or vice versa. In Peltier's original discovery, α and β were antimony and bismuth, respectively. Peltier's junction released heat when the electric current passed from antimony to bismuth. The same junction produced a refrigeration effect when the electric current passed from bismuth to antimony [7].

The Seebeck Effect[†]

An even older manifestation of the two-flux phenomenon described by eqs. (12.57)–(12.58) is the principle that stands behind the *thermocouple* as a temperature-measuring device. Consider the two-conductor arrangement of Fig. 12.5, in which the two junctions are maintained at different tempera-

---

[†]Discovered by German physicist Thomas J. Seebeck in the 1820s.

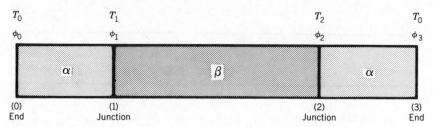

T_0 T_1 T_2 T_0
ϕ_0 ϕ_1 ϕ_2 ϕ_3

α β α

(0)   (1)   (2)   (3)
End   Junction   Junction   End

Figure 12.5 Two α–β junctions maintained at different temperatures (T_1, T_2) create also the end-to-end potential difference $(\phi_3 - \phi_0)$.

tures (T_1, T_2). The other distinguishing feature of the arrangement is that the electric current is zero $(J_N = 0)$. Based on our first encounter with the Seebeck effect, eq. (12.51), even in the absence of electric current, we know to expect a variation of the potential ϕ along the conducting path, because the temperature is not uniform. Let ϕ_0, ϕ_1, ϕ_2, and ϕ_3 denote the main values taken by the potential along the conducting path, Fig. 12.5. The extreme ends of the arrangement are at the same temperature T_0. The question we address next is whether the maintenance of the finite tempera-ture difference across the two junctions results in a finite potential difference across the entire arrangement.

The analysis consists of setting $J_N = 0$ in eq. (12.58) and integrating the remaining expression from station (0) to station (3). Or, we can start directly with eq. (12.51). The overall potential difference is

$$\phi_3 - \phi_0 = -\int_0^3 \epsilon(T)\, dT = -\left(\int_0^1 \epsilon_\alpha\, dT + \int_1^2 \epsilon_\beta\, dT + \int_2^3 \epsilon_\alpha\, dT \right) \quad (12.66)$$

The three integrals on the right side are all temperature integrals. Next, since $T_3 = T_0$, the first and third integrals can be combined:

$$\phi_3 - \phi_0 = -\left(-\int_{T_1}^{T_2} \epsilon_\alpha\, dT + \int_{T_1}^{T_2} \epsilon_\beta\, dT \right) = \int_{T_1}^{T_2} [\epsilon_\alpha(T) - \epsilon_\beta(T)]\, dT \quad (12.67)$$

In conclusion, an electromotive force is created between the ends (0) and (3) whenever a temperature difference is maintained across the junctions (1) and (2). The induced potential difference is clearly a function of only T_1, T_2, and the two materials used in the construction of the device. If in an existing device (fixed α, β), we also fix one of the junction temperatures (T_1), then the voltage measured between the extreme ends (0) and (3) becomes a function of only the temperature of the remaining junction, T_2. Note that the resulting $(\phi_3 - \phi_0)$ measurement does not depend on the temperature at the ends, T_0, or on the manner in which T varies along each conducting segment. When the device is used as a thermocouple, T_1 is usually fixed at 0°C by immersing junction (1) in a bath containing crushed ice and water.

The Thomson Effect

Finally, we consider the question of how much heat is generated and released by a conductor of both heat and electricity whose temperature distribution $T(x)$ is controlled by means of an entire sequence of temperature reservoirs $T_0(x)$, Fig. 12.6. A similar question formed the subject of the subsection on the Peltier effect, where the conductor temperature was maintained uniform presumably by contact with a single-temperature reservoir (e.g., the ambient). In the present problem, the imposed temperature distribution $T_0(x)$ is not arbitrary. The $T_0(x)$ distribution is the one that is established naturally when: (i) the electric current is zero, and (ii) the conductor does not exchange laterally any heat with the ambient. The temperature $T_0(x)$ is obviously the temperature of the conductor when it functions as a purely heat-conducting strut whose outer surface is perfectly insulated. Using the subscript "0" for this limiting conducting-strut condition, we note the special forms that are taken in this limit by eqs. (12.42) and (12.59):

$$J_{U,0} = q_0'' = -k\frac{dT_0}{dx} \tag{12.68}$$

Note further that in this limit, the energy flux is conserved, $dJ_{U,0}/dx = 0$.

Having defined the meaning of the controlled distribution $T_0(x)$, we focus our attention on the general case when J_N is finite and the conductor can exchange heat—at any x—with the neighboring reservoir $T_0(x)$. The lateral heat transfer interaction between the conductor and the reservoir is measured by the change in the total energy current J_U expressed by eq. (12.42). Keeping in mind that the electron flux is conserved along the conductor $(dJ_N/dx = 0)$, we begin with taking the derivative of eq. (12.42):

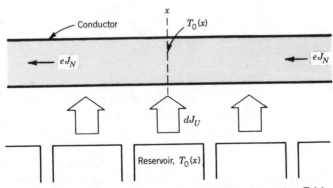

Sequence of temperature reservoirs matching the conductor temperature $T_0(x)$

Figure 12.6 One-dimensional conductor whose "zero-electric-current" temperature distribution $T_0(x)$ is maintained by contact with a sufficiently refined sequence of temperature reservoirs.

$$\frac{dJ_U}{dx} = \frac{dq''}{dx} + eJ_N \frac{d\phi}{dx} \tag{12.69}$$

and then use eqs. (12.59) and (12.58) in order to eliminate q'' and $d\phi/dx$, respectively:

$$\frac{dJ_U}{dx} = \frac{d}{dx}\left(-k\frac{dT_0}{dx} + T_0 \epsilon e J_N\right) + eJ_N\left(-\frac{eJ_N}{k_e} - \epsilon \frac{dT_0}{dx}\right)$$

$$= \frac{d}{dx}\left(-k\frac{dT_0}{dx}\right) + eJ_N T_0 \frac{d\epsilon}{dx} - \frac{(eJ_N)^2}{k_e} \tag{12.70}$$

The group in the first set of parentheses on the right side is equal to the energy flux in the limit of zero electric current, $J_{U,0}$. Because of the way in which we have selected $T_0(x)$, this limiting energy flux is x-independent and, consequently, the first of the three terms in eq. (12.70) is zero. The lateral heat transfer dJ_U/dx then has only two components:

$$\frac{dJ_U}{dx} = eJ_N T_0 \frac{d\epsilon}{dx} - \frac{(eJ_N)^2}{k_e} \tag{12.71}$$

the second of which the usual *Joule heat* generated by the electrical conductor. The first term on the right side represents the so-called *Thomson heat* absorbed by the conductor when the electric current proceeds along a finite temperature gradient (note that $d\epsilon/dx = \epsilon' \, dT/dx$, where $\epsilon' = d\epsilon/dT$). The Thomson term can be expressed alternatively as

$$eJ_N T_0 \frac{d\epsilon}{dx} = eJ_N \frac{dT_0}{dx} \tau(T_0) \tag{12.72}$$

where the new temperature function τ is the Thomson coefficient of the material:

$$\tau(T) = T \frac{d\epsilon}{dT} \tag{12.73}$$

The Thomson heat changes sign if the electric current (eJ_N) changes its direction. This means that if *two* parallel conductors are placed in contact with the $T_0(x)$ reservoirs and if the same (eJ_N) current flows in opposite directions through the two conductors, then the Thomson heat generated by one conductor matches exactly the Thomson heat absorbed by the other. The net heat transfer interaction with the $T_0(x)$ reservoirs consists of rejecting the two Joule heating rates produced by the two conductors. This is why in the first-law analysis of a cooled power cable (a pair of leads), eq. (10.62), the only heat-generation term is the one that accounts for the Joule heat, even though the absolute temperature varies dramatically along the cable.

Power Generation

The Peltier and Seebeck effects are at work simultaneously in the functioning of two important classes of energy-conversion devices known as thermoelectric power generators and refrigerators. The opportunity for constructing such devices out of as little as two $\alpha|\beta$ junctions held at different temperatures becomes apparent if we take one more look at Figs. 12.3–12.5. The Seebeck effect of Fig. 12.5 amounts to the observation that the ends of an open circuit consisting of two junctions at different temperatures (T_1, T_2) develop a finite potential difference, $(\phi_3 - \phi_0)$. In this open circuit, the Peltier heat interactions between the two junctions and the respective temperature reservoirs are zero because the electric current is zero, eq. (12.61). If, on the other hand, the circuit is closed by connecting the extreme ends through an external device (e.g., electrical resistance, R), the end-to-end potential difference will drive the finite current I through the entire circuit. When passing through the two junctions, this current triggers two Peltier heat interactions with the (T_1) and (T_2) reservoirs. These heat transfer interactions must be of opposite signs because the entire circuit represents a closed system that operates steadily in communication with only two heat reservoirs.[†]

Placing the circuit described above inside a black box, we see a closed system that absorbs heat steadily from one temperature reservoir, rejects heat steadily to a second reservoir, and delivers useful electrical power (size: I^2R) steadily to an external system. The black box has all the features required by a heat engine, which in this case generates electric power directly. Likewise, if the external system maintains the end-to-end potential difference and at the same time drives the current I against this difference, then all the energy transfer interactions around our system change sign, turning the two-junction system into a refrigerator or heat pump.

The conceptual design and optimization of thermoelectric energy-conversion devices is now the subject of a growing segment of the applied-thermodynamics literature [e.g., Refs. 10–17]. Thermoelectric power generators and refrigerators have two general characteristics, one negative, the other extremely advantageous. On the negative side, these devices are *inherently irreversible* because heat and electric currents must flow through the circuit during their operation. These inherent irreversibilities are the reason so much effort is being spent on optimizing, fine-tuning, and "matching" these devices to the larger systems (electric networks) in which they are to perform. The advantageous feature of thermoelectric power generators and refrigerators is that they do not require any moving parts. This makes them extremely reliable.

[†]Recall the second-law argument that the two heat transfer interactions of a closed system operating in an integral number of cycles in communication with two heat reservoirs must have opposing signs (chapter 2, p. 56).

In order to see how the internal irreversibility of heat and electric currents decreases the "heat-engine" efficiency of a thermoelectric power generator, consider the two-leg construction shown in Fig. 12.7. The two legs, α and β, are usually made out of different semiconductor materials. The hot junction is held at the high temperature T_H: the net heat transfer rate from the (T_H) reservoir into the junction is $\dot{Q}_H$. The figure shows the more common junction design in which the hot ends of α and β are shunted (tied together) thermally and electrically by a horizontal bar of a different and highly conductive material (e.g., copper). The "junction" is represented by the entire subsystem enclosed in the dashed-line box. It is easy to verify that the thermodynamics of this junction is equivalent to that of the simplest design, in which α and β are joined end to end (as in Fig. 12.5) and held at the temperature level T_H while heated at the rate $\dot{Q}_H$.

The cold end of the two-leg apparatus is designed similarly. The only differenced this time is that the cold ends of α and β are insulated electrically from each other. The potential difference generated between them drives now the total current I through the external resistance R. The net heat transfer rate to the low-temperature reservoir (T_L) is $\dot{Q}_L$.

The two legs (α, β) are one-dimensional conductors along which x is measured from T_H to T_L. The lateral surfaces of both legs are insulated thermally and electrically. The physical properties and geometry of the α leg

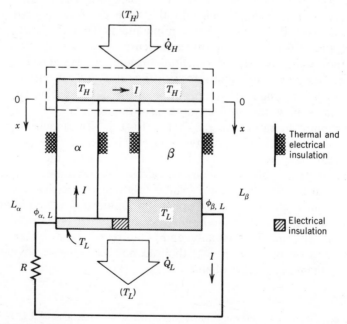

Figure 12.7 Thermoelectric power generator consisting of two differentially heated thermoelectric elements.

differ in general from the properties and geometry of the β leg. The geometry is represented by two lengths (L_α, L_β) and two cross-sectional areas (A_α, A_β).

An important parameter of the power generator in Fig. 12.7 is the energy-conversion or heat-engine efficiency:

$$\eta = \frac{I^2 R}{\dot{Q}_H} \qquad (12.74)$$

in which the numerator $I^2 R$ represents the useful power delivered by the device. From the first-law analysis of the (T_H) junction, we deduce the heat transfer rate intput:

$$
\begin{aligned}
\dot{Q}_H &= J_{U,\alpha,H} A_\alpha + J_{U,\beta,H} A_\beta \\
&= (q'' + \phi e J_N)_{\alpha,H} A_\alpha + (q'' + \phi e J_N)_{\beta,H} A_\beta
\end{aligned} \qquad (12.75)
$$

in which we have made use of eq. (12.42). Noting that the potential ϕ has the same value throughout the junction, $\phi_{\alpha,H} = \phi_{\beta,H}$, and that the current I proceeds in the positive x direction only along the β leg:

$$I = -(e J_N)_\alpha A_\alpha = +(e J_N)_\beta A_\beta \qquad (12.76)$$

the $\dot{Q}_H$ formula reduces to

$$
\begin{aligned}
\dot{Q}_H &= q''_{\alpha,H} A_\alpha + q''_{\beta,H} A_\beta \\
&= \left(-k \frac{dT}{dx} + T \epsilon e J_N\right)_{\alpha,H} A_\alpha + \left(-k \frac{dT}{dx} + T \epsilon e J_N\right)_{\beta,H} A_\beta \\
&= -k_\alpha A_\alpha T'_\alpha(0) - k_\beta A_\beta T'_\beta(0) + I \pi_{\alpha\beta}(T_H)
\end{aligned} \qquad (12.77)
$$

The last expression is the result of invoking eq. (12.59), writing $T'(0)$ as shorthand for dT/dx at $x = 0$ and, finally, using the $\pi_{\alpha\beta}(T_H)$ notation for the *Peltier coefficient of the junction* of temperature T_H:

$$\pi_{\alpha\beta}(T_H) = (\pi_\beta - \pi_\alpha)_{T_H} = T_H(\epsilon_\beta - \epsilon_\alpha)_{T_H} = T_H \epsilon_{\alpha\beta}(T_H) \qquad (12.78)$$

The next step in the analysis consists of evaluating the hot-end temperature gradients $T'_\alpha(0)$ and $T'_\beta(0)$. For this, we must solve the one-dimensional heat-conduction problem associated with each leg. The statement that J_U is conserved along each leg is equivalent to the following ordinary differential equation for the temperature distribution $T(x)$, Problem 12.4:

$$\frac{d}{dx}\left(k \frac{dT}{dx}\right) - e J_N \tau \frac{dT}{dx} + \frac{(e J_N)^2}{k_e} = 0 \qquad (12.79)$$

where $T(x)$ stands for either $T_\alpha(x)$ or $T_\beta(x)$. The results became manageable and expressible in closed form only if we assume that k, k_e, and ϵ are constant along each conductor [11], which also implies that $\tau = 0$. Solving eq. (12.79) leads to parabolic temperature distributions along both α and β, and, after invoking the temperature-boundary conditions listed on the figure, to

$$T'(0) = \frac{I^2 L}{2kk_e A^2} - \frac{T_H - T_L}{L} \tag{12.80}$$

The (L, A, k, k_e) quantities on the right side acquire the appropriate subscript "α" or "β" as $T'(0)$ refers to $T'_\alpha(0)$ or $T'_\beta(0)$, respectively.

The result of combining eqs. (12.80) and (12.77) into eq. (12.74) is the final efficiency formula [11]:

$$\eta = \frac{I^2 R}{\pi_{\alpha\beta}(T_H)I + \Delta T \dfrac{A_\alpha}{L_\alpha}(k_\alpha + Xk_\beta) - \dfrac{I^2}{2}\dfrac{L_\alpha}{A_\alpha}(k_{e,\alpha}^{-1} + X^{-1}k_{e,\beta}^{-1})} \tag{12.81}$$

where X is the ratio of the two geometric slenderness ratios:

$$X = \frac{(A/L)_\beta}{(A/L)_\alpha} \tag{12.82}$$

The current I depends on the potential difference generated across R and the electrical resistance posed by the entire circuit. We seek this relationship by integrating eq. (12.58) in the x direction along each leg:

$$\phi_{\alpha,L} - \phi_{\alpha,H} = -\left(\frac{eJ_N}{k_e}\right)_\alpha L_\alpha - \int_{T_H}^{T_L} \epsilon_\alpha \, dT \tag{12.83\alpha}$$

$$\phi_{\beta,L} - \phi_{\beta,H} = -\left(\frac{eJ_N}{k_e}\right)_\beta L_\beta - \int_{T_H}^{T_L} \epsilon_\beta \, dT \tag{12.83\beta}$$

And, since $\phi_{\alpha,H} = \phi_{\beta,H}$, subtracting these equations yields

$$\phi_{\beta,L} - \phi_{\alpha,L} = -I\left[\left(\frac{L}{k_e A}\right)_\alpha + \left(\frac{L}{k_e A}\right)_\beta\right] + \int_{T_L}^{T_H} (\epsilon_\beta - \epsilon_\alpha) \, dT \tag{12.84}$$

The potential difference appearing on the left side is simply IR; therefore, eq. (12.84) is the relation we have been seeking for electric current:

$$I = \frac{\int_{T_L}^{T_H} (\epsilon_\beta - \epsilon_\alpha) \, dT}{R + \dfrac{L_\alpha}{A_\alpha}(k_{e,\alpha}^{-1} + X^{-1}k_{e,\beta}^{-1})} \tag{12.85}$$

Between eqs. (12.81) and (12.85), the efficiency emerges as a function of two parameters, the geometric ratio X and the ratio between the external resistance R and the "internal" resistance represented, for example, by the group $(L_\alpha / k_{e,\alpha} A_\alpha)$. Sherman et al. [11] showed that η can be maximized with respect to both parameters and that the optimum values are

$$X_{\text{opt}} = \left(\frac{k_{e,\alpha} k_\alpha}{k_{e,\beta} k_\beta} \right)^{1/2} \tag{12.86}$$

$$\left(\frac{R}{L_\alpha / k_{e,\alpha} A_\alpha} \right)_{\text{opt}} = Z \left[1 + \left(\frac{k_{e,\alpha} k_\beta}{k_{e,\beta} k_\alpha} \right)^{1/2} \right] \tag{12.87}$$

where the new dimensionless group Z is defined as

$$Z^2 = 1 + \frac{1}{2} (T_H + T_L) \left[\frac{\epsilon_{\alpha\beta}}{(k_\alpha / k_{e,\alpha})^{1/2} + (k_\beta / k_{e,\beta})^{1/2}} \right]^2 \tag{12.88}$$

The optimum geometric ratio (12.86) shows that if α is a "better conductor" than β, then $X_{\text{opt}} > 1$ and, via eq. (12.82), $(A/L)_\beta > (A/L)_\alpha$. In other words, at the optimum, the leg that is a poorer conductor must have a more robust (less slender) outlook than the companion leg. The maximum efficiency that corresponds to the optimum conditions (12.86)–(12.87) is

$$\eta_{\text{max}} = \left(1 - \frac{T_L}{T_H} \right) \frac{Z - 1}{Z + T_L / T_H} \tag{12.89}$$

In conclusion, the efficiency cannot exceed a ceiling value that is *lower* than the Carnot efficiency associated with T_L and T_H; in other words,

$$\eta_{\text{max}} < 1 - T_L / T_H \tag{12.90}$$

Equation (12.89) shows that η_{max} approaches the Carnot limit as the dimensionless group Z becomes considerably greater than 1. Equation (12.88), on the other hand, shows that in order to approach the Carnot limit, the designer faces the very difficult task of finding materials with low thermal conductivity and—at the same time—high electrical conductivity. The efficiency ceiling η_{max} drops as the irreversible phenomena represented by (k_α, k_β) and $(k_{e,\alpha}^{-1}, k_{e,\beta}^{-1})$ intensify. For these reasons, the property group that appears in the brackets of eq. (12.88) is recognized as the *figure of merit* z of the thermoelectric device[†]:

[†]The figure of merit z, whose units are K^{-1}, should not be confused with the dimensionless quantity Z defined in eq. (12.88). The relationship between the two is

$$Z = (1 + T_{\text{avg}} z)^{1/2}$$

where $T_{\text{avg}} = (T_H + T_L)/2$.

$$z = \left[\frac{\epsilon_{\alpha\beta}}{(k_\alpha/k_{e,\alpha})^{1/2} + (k_\beta/k_{e,\beta})^{1/2}} \right]^2 \tag{12.91}$$

The efficiency of a thermoelectric power generator can be increased further by extracting electrical power not only from the cold end (as in Fig. 12.7), but also from every temperature level T between T_H and T_L. This can be done by bridging the temperature gap with a large number of small two-junction elements of the type considered in Fig. 12.7, such that each element covers the infinitesimal temperature range $(T) - (T + dT)$. Problem 12.5 shows that in this arrangement, the heat rejected by the cold junction of one element becomes the heat absorbed by the next (immediately colder) element. This method of continuous extraction of $d\dot{Q}$ between T_H and T_L is conceptually analogous to the continuous cooling of low-temperature installations, Figs. 10.9 and 10.14.

Refrigeration

The two-leg arrangement discussed until now can function also as a refrigerator, if modified as shown in Fig. 12.8. The device pulls the refrigeration load $\dot{Q}_L$ from the (T_L) reservoir: this is done at the expense of the external power supply, which delivers the net power $I(\phi_{\beta,H} - \phi_{\alpha,H})$. The coefficient of performance of the refrigerator:

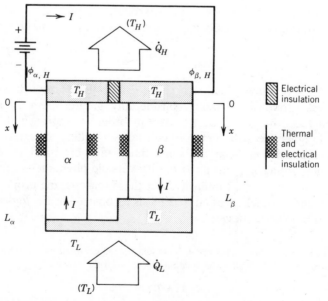

Figure 12.8   Thermoelectric refrigerator with two straight legs.

$$\text{COP} = \frac{\dot{Q}_L}{I(\phi_{\beta,H} - \phi_{\alpha,H})} \qquad (12.92)$$

can be expressed and maximized analytically by adopting the same simplifying assumptions as in the preceding subsection. Highlights of this analytical work are the relations:

$$\dot{Q}_L = k_\alpha A_\alpha T_\alpha'(L_\alpha) + k_\beta A_\beta T_\beta'(L_\beta) + I\pi_{\beta\alpha}(T_L) \qquad (12.93)$$

$$\phi_{\beta,H} - \phi_{\alpha,H} = I\left[\left(\frac{L}{k_e A} \right)_\alpha + \left(\frac{L}{k_e A} \right)_\beta \right] + \int_{T_L}^{T_H} (\epsilon_\alpha - \epsilon_\beta)\, dT \qquad (12.94)$$

$$T'(L) = -\frac{I^2 L}{2kk_e A^2} - \frac{T_H - T_L}{L} \qquad (12.95)$$

where we note again that in eq. (12.95), the quantities (L, A, k, k_e) receive the subscript "α" or "β" depending on whether the left side of the equation reads $T_\alpha'(L_\alpha)$ or $T_\beta'(L_\beta)$. It can be shown that the COP function derived based on eqs. (12.92)–(12.95) depends on two design parameters: the ratio of slenderness ratios, X, eq. (12.82), and the potential difference maintained by the external battery, $(\phi_{\beta,H} - \phi_{\alpha,H})$. The maximum coefficient of performance resulting from the two-parameter optimization process is [11]

$$\text{COP}_{\max} = \left(\frac{T_H}{T_L} - 1 \right)^{-1} \frac{Z - T_H/T_L}{Z + 1} \qquad (12.96)$$

It turns out that the optimum geometric ratio X necessary for achieving this $\text{COP}_{\max}$ is the same as the X_{opt} of eq. (12.86). The other optimized parameter is the potential difference [11]:

$$(\phi_{\beta,H} - \phi_{\alpha,H})_{\text{opt}} = \frac{Z}{Z-1} \int_{T_L}^{T_H} (\epsilon_\alpha - \epsilon_\beta)\, dT \qquad (12.97)$$

The $\text{COP}_{\max}$ formula above reveals once again the inherent irreversibility of a thermoelectric refrigerator. The maximum coefficient of performance is less than its carnot counterpart [eq. (3.25) with $\eta_{\text{II}} = 1$], and only in the large-Z limit, the difference between the two COPs disappears. According to eq. (12.88), the designer is faced again with the problem of procuring thermoelectric materials that are good electrical conductors and poor thermal conductors.

Another way of looking at the thermoelectric refrigeration problem is to ask the question: What is the largest temperature difference that the refrigerator can sustain in the vertical direction? Working with eqs. (12.93) and (12.95), one can show that one condition for maximum $(T_H - T_L)$ is that of zero refrigeration load, $\dot{Q}_L = 0$. Another is the optimum ratio of slenderness ratios, X_{opt}, listed in eq. (12.86). The maximum temperature difference is [11]

$$(T_H - T_L)_{max} = T_L \frac{Z^2 - 1}{1 + T_H/T_L} \qquad (12.98)$$

In summary, the preceding analyses of thermoelectric power generation and refrigeration were based on the following simplifying assumptions:

(i) one-dimensional thermoelectric flows,
(ii) elements (legs) with insulated lateral sides,
(iii) zero thermal and electrical contact resistances in all the junctions,
(iv) constant (x-independent) cross-sectional areas, A_α and A_β
(v) constant properties (k, k_e, ϵ), meaning also $\tau = 0$.

Considerably more sophisticated analyses and design optimizations in which these assumptions have been relaxed can be found in Refs. 11–17. The first treatment of the straight-leg geometry of Figs. 12.7–12.8 was reported in 1911 by Altenkirch [18].

HEAT CONDUCTION IN ANISOTROPIC MEDIA

Another interesting application of Onsager's reciprocity relation is found in the study of diffusion phenomena in anisotropic media. We focus here on the particular case of heat conduction as a diffusion phenomenon ("thermal" diffusion), keeping in mind that rewritten in a different notation, the same analysis applies to electric conduction through an anisotropic medium.

In the field of heat conduction, "anisotropy" means that the transport property of the medium—the thermal conductivity k—is not insensitive to the direction in which the transport-property measurement is being performed. Consider the measurement of the net heat flux driven by a temperature difference across a one-dimensional slab (the "sample"). In a wide variety of solid materials such as crystals, fiber-reinforced composites, and windings for electric machinery, the thermal-conductivity measurement depends on the way in which the "structure" of the material is oriented relative to the sample. Figure 12.9 shows an example of two-dimensional anisotropy, in which the ability of the material to transfer heat by conduction depends on the angle between the "fibers" of the material and the direction of interest (e.g., the axis x_1). The two-dimensional structure in Fig. 12.9 is what we might see in a perpendicular section through a stack of parallel sheets (laminae) of individually isotropic materials with highly different thermal conductivities, for example, copper and epoxy (k_a, k_b, respectively, Fig. 12.9).

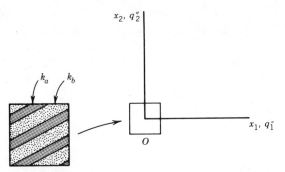

Figure 12.9 Two-dimensional anisotropic conducting medium and the general orientation of the system of coordinates (x_1, x_2).

Formulation in Two Dimensions

Consider now the heat flux in the x_1 direction, q_1'', and how this might depend on the x_1 and x_2 temperature gradients at point (O). The inclined orientation of the material structure suggests that if the temperature gradient $\partial T / \partial x_1$ is imposed in the x_1 direction, then heat will be conducted not only in the x_1 direction, but also in the perpendicular direction x_2. We record this observation by writing

$$-\frac{\partial T}{\partial x_1} = r_{11} q_1'' + r_{12} q_2'' \tag{12.99}$$

where q_2'' is the heat flux in the x_2 direction, and r_{11} and r_{12} are two thermal-resistivity coefficients. Based on the same reasoning, the temperature gradient imposed in the x_2 direction is capable of contributing to both q_2'' and q_1'':

$$-\frac{\partial T}{\partial x_2} = r_{21} q_1'' + r_{22} q_2'' \tag{12.100}$$

Equations (12.99) and (12.100) are equivalent to writing that, in general, each heat flux contains contributions from both temperature gradients:

$$q_1'' = -k_{11} \frac{\partial T}{\partial x_1} - k_{12} \frac{\partial T}{\partial x_2} \tag{12.101}$$

$$q_2'' = -k_{21} \frac{\partial T}{\partial x_1} - k_{22} \frac{\partial T}{\partial x_2} \tag{12.102}$$

The new coefficients k_{11}, k_{12}, k_{21}, and k_{22} are the four thermal conductivities associated with the cartesian frame (x_1, x_2). The relationships between these coefficients and the thermal resistivities of eqs. (12.99)–(12.100) can be established easily, however, they are not central to the objective of this section. The important feature is that eqs. (12.101)–(12.102), which are the classical starting point in the analysis of two-dimensional conduction in an anisotropic medium [8, 19], describe a coupled two-flux phenomenon in the same sense as eqs. (12.28)–(12.29). For the q_1'' and q_2'' relations listed above can easily be rewritten as

$$q_1'' = T^2 k_{11} \frac{\partial}{\partial x_1} \left(\frac{1}{T} \right) + T^2 k_{12} \frac{\partial}{\partial x_2} \left(\frac{1}{T} \right) \tag{12.103}$$

$$q_2'' = T^2 k_{21} \frac{\partial}{\partial x_1} \left(\frac{1}{T} \right) + T^2 k_{22} \frac{\partial}{\partial x_2} \left(\frac{1}{T} \right) \tag{12.104}$$

in order to show that the respective conjugate driving forces are $\partial(T^{-1})/\partial x_1$ and $\partial(T^{-1})/\partial x_2$. Note further that the Onsager coefficients $(L_{11}, L_{12}, L_{21}, L_{22})$ are related to the thermal conductivities via

$$L_{ij} = T^2 k_{ij} \tag{12.105}$$

This last statement is a generalization of what we wrote already in eq. (12.35).

To invoke Onsager's reciprocity relation $L_{12} = L_{21}$ at this stage means to postulate that the cross-coupling conductivity coefficients are equal:

$$k_{12} = k_{21} \tag{12.106}$$

In what follows, we postpone the adoption of the reciprocity relation (12.106) until we had a chance to examine the experimental evidence that supports its validity. We followed a similar strategy in the preceding section on thermoelectricity. In this way, we treat more justly the historical development of conduction heat transfer, for the relation $k_{12} = k_{21}$ was debated and tested in the laboratory long before Onsager's theoretical argument for writing $L_{12} = L_{21}$.

Principal Directions and Conductivities

The physical meaning of the coupling between q_1'' and q_2'' becomes evident if we consider a concrete heat transfer problem, that is, the analytical derivation of the isotherms and heat-flux lines associated with a certain heating configuration of the medium. The First Law of Thermodynamics for steady conduction without volumetric heat generation $(-w''')$, eq. (1.26), reduces to

$$\nabla \cdot \mathbf{q} = 0 \tag{12.107}$$

or, in the (x_1, x_2) system of coordinates:

$$\frac{\partial q_1''}{\partial x_1} + \frac{\partial q_2''}{\partial x_2} = 0 \qquad (12.107')$$

Substituting eqs. (12.101) and (12.102) into the so-called "energy-conservation" eq. (12.107′) yields the quadric:

$$k_{11} \frac{\partial^2 T}{\partial x_1^2} + (k_{12} + k_{21}) \frac{\partial^2 T}{\partial x_1 \partial x_2} + k_{22} \frac{\partial^2 T}{\partial x_2^2} = 0 \qquad (12.108)$$

According to a well-known theorem of second-order linear equations [20], there exists a suitable transformation of variables $(x_1, x_2) \rightarrow (\xi_1, \xi_2)$ such that eq. (12.108) is reduced to the *canonical form*:

$$k_1 \frac{\partial^2 T}{\partial \xi_1^2} + k_2 \frac{\partial^2 T}{\partial \xi_2^2} = 0 \qquad (12.109)$$

The special coordinates (ξ_1, ξ_2) that satisfy eq. (12.109) are the *principal axes* or the *principal directions* of the conducting material. The new coefficients (k_1, k_2) are recognized as *principal conductivities* and are to be distinguished from the coefficients associated with the (x_1, x_2) directions, namely, $(k_{11}, k_{12}, k_{21}, k_{22})$.

We could take eq. (12.109) and proceed directly toward a temperature-field solution, as done in a regular heat transfer course, however, we are especially interested in what that solution has to say about the original cross-coupling coefficients k_{12} and k_{21}. For this reason, it is important to establish first the relationship between the principal conductivities of eq. (12.109) and the original set used in eqs. (12.101)–(12.102). Let us assume that the new system of coordinates (ξ_1, ξ_2) is obained by rotating the old system (x_1, x_2) counterclockwise by the angle α (unknown) shown in Fig. 12.10. The transformation $(x_1, x_2) \rightarrow (\xi_1, \xi_2)$ is effected by the relations:

$$\xi_1 = x_1 \cos \alpha + x_2 \sin \alpha \qquad (12.110)$$

$$\xi_2 = -x_1 \sin \alpha + x_2 \cos \alpha \qquad (12.111)$$

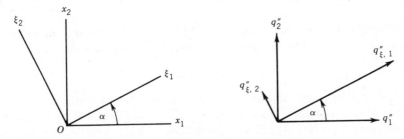

Figure 12.10 Principal directions and fluxes.

Omitting the algebra associated with going from eq. (12.108) to eq. (12.109) via eqs. (12.110)–(12.111), it is worth recording the resulting full-length expressions for the principal conductivities:

$$k_1 = k_{11} \cos^2 \alpha + (k_{12} + k_{21}) \sin \alpha \cos \alpha + k_{22} \sin^2 \alpha \qquad (12.112)$$

$$k_2 = k_{11} \sin^2 \alpha - (k_{12} + k_{21}) \sin \alpha \cos \alpha + k_{22} \cos^2 \alpha \qquad (12.113)$$

The blind transformation of eq. (12.108) into eq. (12.109) yields also a term in $\partial^2 T / \partial \xi_1 \, \partial \xi_2$: the coefficient of this term is zero [as required by the canonical form (12.109)] if

$$\tan 2\alpha = \frac{k_{12} + k_{21}}{k_{11} - k_{22}} \qquad (12.114)$$

Equation (12.114) pinpoints the orientation of the principal axes (ξ_1, ξ_2) relative to the original system (x_1, x_2). Let $q''_{\xi,1}$ and $q''_{\xi,2}$ be the heat fluxes that point in the principal directions, Fig. 12.10. Since the principal frame (ξ_1, ξ_2) is a special orientation of the original frame (x_1, x_2), we have every reason to assume that $q''_{\xi,1}$ and $q''_{\xi,2}$ are related to the directional temperature gradients $\partial T / \partial \xi_1$ and $\partial T / \partial \xi_2$ in the same way that the heat fluxes and temperature gradients are related in eqs. (12.101)–(12.102). Writing,

$$q''_{\xi,1} = -a_{11} \frac{\partial T}{\partial \xi_1} - a_{12} \frac{\partial T}{\partial \xi_2} \qquad (12.115)$$

$$q''_{\xi,2} = -a_{21} \frac{\partial T}{\partial \xi_1} - a_{22} \frac{\partial T}{\partial \xi_2} \qquad (12.116)$$

it is only a matter of algebraic manipulation to show that the new coefficients $(a_{11}, a_{12}, a_{21}, a_{22})$ are related to the original set via (Problem 12.7):

$$a_{11} = k_{11} \cos^2 \alpha + (k_{12} + k_{21}) \sin \alpha \cos \alpha + k_{22} \sin^2 \alpha \qquad (12.117)$$

$$a_{12} = k_{12} \cos^2 \alpha - k_{21} \sin^2 \alpha + (k_{22} - k_{11}) \sin \alpha \cos \alpha \qquad (12.118)$$

$$a_{21} = k_{21} \cos^2 \alpha - k_{12} \sin^2 \alpha + (k_{22} - k_{11}) \sin \alpha \cos \alpha \qquad (12.119)$$

$$a_{22} = k_{11} \sin^2 \alpha - (k_{12} + k_{21}) \sin \alpha \cos \alpha + k_{22} \cos^2 \alpha \qquad (12.120)$$

Comparing these results with eqs. (12.112)–(12.113), we see immediately that a_{11} and a_{22} are the principal conductivities of the material:

$$a_{11} = k_1 \qquad a_{22} = k_2 \qquad (12.121)$$

The less-expected conclusion is that the remaining coefficients, a_{12} and a_{21}, are not necessarily zero. Furthermore, adding eqs. (12.118) and (12.119), and invoking eq. (12.114) leads to the discovery that a_{12} and a_{21} add to zero, or that

$$a_{12} = -a_{21} \tag{12.122}$$

As summary to the preceding paragraph, the principal fluxes are related to the principal temperature gradients by the linear expressions:

$$q''_{\xi,1} = -k_1 \frac{\partial T}{\partial \xi_1} - a \frac{\partial T}{\partial \xi_2} \tag{12.123}$$

$$q''_{\xi,2} = +a \frac{\partial T}{\partial \xi_1} - k_2 \frac{\partial T}{\partial \xi_2} \tag{12.124}$$

in which we have substituted $a = a_{12}$ and $-a = a_{21}$. Note that for reasons that will be made clear by the right half of Fig. 12.11, the cross-coupling coefficient a is recognized as the *rotatory* conductivity coefficient. The pair of principal fluxes (12.123)–(12.124) satisfies the energy equation (12.107) regardless of the value of a:

$$\frac{\partial q''_{\xi,1}}{\partial \xi_1} + \frac{\partial q''_{\xi,2}}{\partial \xi_2} = 0 \tag{12.125}$$

Recognizing further the geometric relationship between the original fluxes (q''_1, q''_2) and the principal fluxes $(q''_{\xi,1}, q''_{\xi,2})$, Fig. 2.10,

$$q''_1 = q''_{\xi,1} \cos \alpha - q''_{\xi,2} \sin \alpha \tag{12.126}$$

$$q''_2 = q''_{\xi,1} \sin \alpha + q''_{\xi,2} \cos \alpha \tag{12.127}$$

it is possible to use eqs. (12.123)–(12.127) and the reverse of the transformation (12.110)–(12.111) in order to rederive the original flux relations (12.101)–(12.102). This operation forms the subject of Problem 12.7. Its purpose is to identify the relations that exist between the original conduc-

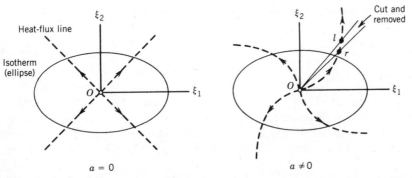

Figure 12.11 Isotherms and heat-flux lines in a two-dimensional anisotropic medium with concentrated heat source.

tivities $(k_{11}, k_{12}, k_{21}, k_{22})$ and the principal coefficients (k_1, k_2, a). These relations are

$$k_{11} = k_1 \cos^2 \alpha + k_2 \sin^2 \alpha \qquad (12.128)$$

$$k_{12} = (k_1 - k_2) \sin \alpha \cos \alpha + a \qquad (12.129)$$

$$k_{21} = (k_1 - k_2) \sin \alpha \cos \alpha - a \qquad (12.130)$$

$$k_{22} = k_1 \sin^2 \alpha + k_2 \cos^2 \alpha \qquad (12.131)$$

The important conclusion that becomes visible at this stage is that the cross-coupling coefficients k_{12} and k_{21} are equal only if

$$a = 0 \qquad (12.132)$$

The validity of Onsager's reciprocity relation, then, depends on whether eq. (12.132) is supported by experiment. To this aspect, we turn in the next subsection.

The Concentrated-Heat-Source Experiment

Consider the temperature field created if a heat source of strength q' (W/m) is buried right at the origin of the coordinate system (ξ_1, ξ_2). In the language employed in the field of conduction heat transfer, q' is a continuous (steady) line source oriented perpendicularly to the plane of Fig. 12.10. The temperature field $T(\xi_1, \xi_2)$ must satisfy the energy equation (12.109); therefore, a first conclusion is that T depends on the principal conductivities k_1 and k_2, but not on the disputed rotatory coefficient a. Indeed, the temperature-field solution is of the form [8, 19]:

$$T = -m \ln \left(\frac{\xi_1^2}{k_1} + \frac{\xi_2^2}{k_2} \right) \qquad (12.133)$$

where the leading constant is proportional to the strength of the heat source:

$$m = \frac{q'}{4\pi(k_1 k_2)^{1/2}} \qquad (12.134)$$

Holding T constant in eq. (12.133), we see that the isotherms are ellipsoidal (Fig. 12.11) and that their shape does not contain any information regarding the rotatory coefficient a. On the other hand, the principal heat fluxes and the shape of the heat-flux lines are affected by the rotatory coefficient. Combining eq. (12.133) with eqs. (12.123)–(12.124), we obtain

$$q''_{\xi,1} = 2m \left(\frac{\xi_1^2}{k_1} + \frac{\xi_2^2}{k_2} \right)^{-1} \left(\xi_1 + \frac{a}{k_2} \xi_2 \right) \qquad (12.135)$$

$$q''_{\xi,2} = 2m \left(\frac{\xi_1^2}{k_1} + \frac{\xi_2^2}{k_2} \right)^{-1} \left(\xi_2 - \frac{a}{k_1} \xi_1 \right) \tag{12.136}$$

and, from these:

$$\frac{q''_{\xi,1}}{q''_{\xi,2}} = \frac{k_1(k_2\xi_1 + a\xi_2)}{k_2(k_1\xi_2 - a\xi_1)} \tag{12.137}$$

Finally, invoking the principal flux relations (12.123)–(12.124) in order to eliminate $q''_{\xi,1}$ and $q''_{\xi,2}$ from eq. (12.137) leads to the differential equation for the heat-flux lines:

$$\frac{d\xi_2}{d\xi_1} = \frac{k_2(k_1\xi_2 - a\xi_1)}{k_1(k_2\xi_1 + a\xi_2)} \tag{12.138}$$

with the following solution [19]:

$$(k_1 k_2)^{1/2} \tan^{-1} \left(\frac{\xi_2 k_1^{1/2}}{\xi_1 k_2^{1/2}} \right) + \frac{a}{2} \ln (k_1 \xi_2^2 + k_2 \xi_1^2) = \text{constant} \tag{12.139}$$

Figure 12.11 shows the main characteristics of the heat-flux lines. If the coefficient a is zero, the heat-flux path solution (12.139) reduces to $\xi_2/\xi_1 =$ constant, that is, to a family of straight lines pointing radially away from the origin. In general, the heat-flux lines are not normal to the isotherms. This family of straight heat-flux lines is sketched on the left side of Fig. 12.11. The same family appears also as starting point (assumption) in Rohsenow and Choi's treatment of conduction in anisotropic media [21], where it is worth noting that the analysis leads directly to the conclusion that $k_{12} = k_{21}$, that is, to our eqs. (12.129)–(12.130) with $a = 0$ in them. Postulating straight heat-flux lines is synonymous with postulating $k_{12} = k_{21}$. The question we zero in on is whether k_{12} equals k_{21}, that is, whether the heat-flux lines are straight.

When the rotatory coefficient a is finite, the solution (12.139) represents the family of spirals shown on the right side of Fig. 12.11. The curvature of the spiral increases as the absolute value of a increases. If the sign of a changes, then the spiral changes only its sense of rotation.

That the heat-flux lines are straight and not spiral was demonstrated experimentally by Soret [22], among others. The experiment consisted of heating the two-dimensional anisotropic material (a thin-plate crystal) at a point and then cutting a slit radially away from the origin. This cut is shown also on the right side of Fig. 12.11. If the heat-flux lines are spiral, then those that are terminated by the cut will tend to raise the temperature of the right edge labeled $(O–r)$. Likewise, the spiral heat-flux lines that start from the cut will tend to depress the temperature of the left edge $(O–l)$. The net effect of the spiral would be the creation of a finite temperature difference between two points situated vis-à-vis across the cut. Soret did not detect any temperature change across the cut; therefore, the conclusion is that the heat-flux lines are straight, which means also that $a = 0$ and, finally, $k_{12} = k_{21}$. This conclusion and another experiment due to Voigt [23] are reviewed further by Miller [9].

Three-Dimensional Conduction

It is now easier to see the basis for the analytical formulation of the heat-conduction problem in a three-dimensional anisotropic solid. Let (x_1, x_2, x_3) be a cartesian system of coordinates not necessarily aligned with any of the "fibers" of the anisotropic material. Each heat flux is driven in general by a combination of all three temperature gradients:

$$q_1'' = -k_{11} \frac{\partial T}{\partial x_1} - k_{12} \frac{\partial T}{\partial x_2} - k_{13} \frac{\partial T}{\partial x_3} \tag{12.140}$$

$$q_2'' = -k_{21} \frac{\partial T}{\partial x_1} - k_{22} \frac{\partial T}{\partial x_2} - k_{23} \frac{\partial T}{\partial x_3} \tag{12.141}$$

$$q_3'' = -k_{31} \frac{\partial T}{\partial x_1} - k_{32} \frac{\partial T}{\partial x_2} - k_{33} \frac{\partial T}{\partial x_3} \tag{12.142}$$

where the nine conductivity coefficients k_{ij} are the components of the conductivity tensor (a second-order tensor). Now we know to rely from the start on the reciprocity relations, which in this case mean:

$$k_{12} = k_{21} \qquad k_{13} = k_{31} \qquad k_{23} = k_{32} \tag{12.143}$$

Let (ξ_1, ξ_2, ξ_3) be another cartesian system of coordinates with the same origin as the (x_1, x_2, x_3) system. The transformation from the old system to the new system is represented by the relations:

$$\xi_i = \sum_{j=1}^{3} c_{ij} x_j \qquad (i = 1, 2, 3) \tag{12.144}$$

where the c_{ij} coefficients represent the directional cosines (c_{ij} is the cosine of the angle formed by the axes ξ_i and x_j). The three heat fluxes that point along the axes of the new system can be written using the notation of eqs. (12.115)–(12.116):

$$q_{\xi,1}'' = -a_{11} \frac{\partial T}{\partial \xi_1} - a_{12} \frac{\partial T}{\partial \xi_2} - a_{13} \frac{\partial T}{\partial \xi_3} \tag{12.145}$$

$$q_{\xi,2}'' = -a_{21} \frac{\partial T}{\partial \xi_1} - a_{22} \frac{\partial T}{\partial \xi_2} - a_{23} \frac{\partial T}{\partial \xi_3} \tag{12.146}$$

$$q_{\xi,3}'' = -a_{31} \frac{\partial T}{\partial \xi_1} - a_{32} \frac{\partial T}{\partial \xi_2} - a_{33} \frac{\partial T}{\partial \xi_3} \tag{12.147}$$

The nine conductivity coefficients a_{ij} change as the cartesian system (ξ_1, ξ_2, ξ_3) rotates relative to the old system. Certain orientations of (ξ_1, ξ_2, ξ_3) are marked by the appearance of zeros in the matrix $\|a_{ij}\|$. For example, in a material whose structure consists of "fibers" oriented in three mutually perpendicular directions, it is possible to find that special orientation of the (ξ_1, ξ_2, ξ_3) frame for which all six cross-coupling coefficients vanish. In this position, the $\|a_{ij}\|$ matrix shows only three finite elements placed along the diagonal:

$$\begin{vmatrix} k_1 & 0 & 0 \\ 0 & k_2 & 0 \\ 0 & 0 & k_3 \end{vmatrix} \tag{12.148}$$

The surviving coefficients (k_1, k_2, k_3) are the principal conductivities of the anisotropic material. The special directions (ξ_1, ξ_2, ξ_3) that ensure eq. (12.148) are the principal axes or principal directions. This very common and important class of materials, whose conductivity matrices can be reduced to the form (12.148), are recognized as *orthorhombic solids*.

In the coordinate system formed by the principal directions, the three heat fluxes attain the much simpler forms:

$$q''_{\xi,1} = -k_1 \frac{\partial T}{\partial \xi_1} \tag{12.149}$$

$$q''_{\xi,2} = -k_2 \frac{\partial T}{\partial \xi_2} \tag{12.150}$$

$$q''_{\xi,3} = -k_3 \frac{\partial T}{\partial \xi_3} \tag{12.151}$$

The energy equation for steady conduction in three dimensions, eq. (12.107), reads now

$$k_1 \frac{\partial^2 T}{\partial \xi_1^2} + k_2 \frac{\partial^2 T}{\partial \xi_2^2} + k_3 \frac{\partial^2 T}{\partial \xi_3^2} = 0 \tag{12.152}$$

And, since this course is about to end and, perhaps, a new one in heat transfer about to begin, the reader may note that a useful step beyond eq. (12.152) is the definition of a set of compressed (or stretched) scales for the principal directions:

$$x = \xi_1 \left(\frac{k}{k_1}\right)^{1/2} \qquad y = \xi_2 \left(\frac{k}{k_2}\right)^{1/2} \qquad z = \xi_3 \left(\frac{k}{k_3}\right)^{1/2} \tag{12.153}$$

such that the energy equation reads

$$\frac{\partial^2 T}{\partial x^2} + \frac{\partial^2 T}{\partial y^2} + \frac{\partial^2 T}{\partial z^2} = 0 \tag{12.154}$$

In this way, the three-dimensional conduction problem of the anisotropic medium is transformed into one that is tractable by means of routine analytical methods now used in the realm of isotropic conducting media, eq. (12.154). The transformation (12.153) must be applied also to the boundary conditions of the problem statement, not just to the energy equation. The quantity k employed in the scale compression (12.153) is a constant whose units are W/m K: its role could be played by any of the principal conductivities, for example, $k = k_1$.

As irreversible thermodynamics attempts to bridge the gap between the classical fields of thermodynamics and transport processes, it is worth keeping in mind that the present analytical treatment of conduction in anisotropic media was used for 100 years before Onsager's formalism. Heat transfer analysts managed quite well using temperature gradients instead of T^{-1} gradients as "driving forces." When questions arose about the relationship between cross-coupling conductivity coefficients in crystals, heat transfer turned to clever experimental demonstrations such as those of Soret and Voigt. The present analytical treatment, which in essence is a generalization of Fourier's treatment of conduction in isotropic media, is due to Duhamel [24] and Stokes [8].

MASS DIFFUSION

Next to the thermoelectric and heat-conduction phenomena addressed in the preceding two sections, a third example that subscribes to the Onsager formalism is the phenomenon of "mass diffusion," or the migration of one or more species through a multicomponent system. It is common to expect that the diffusion of one species is intimately tied to the gradient shown by the concentration of that particular species in the mixture. This is, of course, true in the special circumstances in which the simplest form of *Fick's law*, eq. (12.179) applies, i.e., in isothermal binary mixtures. The mass-diffusion picture is considerably more complicated in the presence of temperature gradients, or in the presence of more than two diffusing species. In such cases, the mass flux of one species is driven not only by the concentration gradient of that species, but also by the temperature gradients and the concentration gradients of the other components.

Nonisothermal Diffusion of a Single Component

In order to examine the more general picture alluded to above, consider the simultaneous flow of energy and mass through the vertical partition in Fig. 12.1. All the flows in that early figure were oriented in the x direction: in the following analysis, the flow direction is labeled "z," whereas the "x" symbol is reserved for the mole fraction. Consider the case when there is just one component on both sides of the partition ($n = 1$). The fact that the intensive state of this single-phase pure substance changes as the observer's eyes move across the partition means that the partition maintains finite temperature and chemical-potential differences between the two parts of the system. This is the same as saying that the partition maintains finite-T and -P differences, because in the case of a single-component phase, the three intensities (μ, T, P) are related through the Gibbs–Duhem relation (4.33):

$$d\mu = -\bar{s}\,dT + \bar{v}\,dP \qquad (12.155)$$

In the present case, only two fluxes can penetrate the partition: the energy flux J_U, written also as J_0; and the molal flux—the migration of the lone component—J_N, written also as J_1. The linear relations (12.25) reduce to only two:

$$J_U = L_{00}\frac{d}{dz}\left(\frac{1}{T}\right) + L_{01}\frac{d}{dz}\left(-\frac{\mu}{T}\right) \qquad (12.156)$$

$$J_N = L_{10}\frac{d}{dz}\left(\frac{1}{T}\right) + L_{11}\frac{d}{dz}\left(-\frac{\mu}{T}\right) \qquad (12.157)$$

showing that the temperature gradient can contribute to the flow of mass and that the gradient of $-\mu/T$ can drive the flow of energy. These two cross-coupling effects, which later are adorned with the names of Soret and Dufour, respectively, are ruled by the Onsager coefficients L_{10} and L_{01}. We develop a better feel for what these coefficients represent if we focus on the situation in which the "differences" across the partition are such that the net flowrate is zero, namely, $J_N = 0$ or

$$0 = L_{10}\frac{d}{dz}\left(\frac{1}{T}\right) + L_{11}\frac{d}{dz}\left(-\frac{\mu}{T}\right) \qquad (J_N = 0) \qquad (12.158)$$

Eliminating $d\mu$ between eqs. (12.158) and (12.155) and using the single-phase and single-component identity $\mu = \bar{h} - T\bar{s}$ (i.e., $\mu = \bar{g}$) yields in a few steps [25]:

$$\bar{v}T\,dP = \left(\bar{h} - \frac{L_{10}}{L_{11}}\right)dT \qquad (J_N = 0) \qquad (12.159)$$

In conclusion, when the mass flow rate is zero, the finite temperature difference sustains a finite pressure difference across the partition, assuming of course that $L_{10}/L_{11} \neq \bar{h}$. The ratio L_{10}/L_{11} has a special physical meaning that is revealed by the isothermal limit of the flux relations (12.156)–(12.157):

$$J_U = L_{01}\frac{d}{dz}\left(-\frac{\mu}{T}\right) \qquad \left(\frac{dT}{dz} = 0\right) \qquad (12.160)$$

$$J_N = L_{11} \frac{d}{dz}\left(-\frac{\mu}{T}\right) \qquad \left(\frac{dT}{dz} = 0\right) \tag{12.161}$$

Dividing side by side and using the reciprocity relation $L_{01} = L_{10}$ yields

$$\left(\frac{J_U}{J_N}\right)_{dT/dz=0} = \frac{L_{01}}{L_{11}} = \frac{L_{10}}{L_{11}} \tag{12.162}$$

Therefore, L_{10}/L_{11} is equal to the ratio between the energy and molal fluxes in the case in which both sides of the partition are maintained at the same temperature.

Nonisothermal Binary Mixtures

Consider now the case of two differing species that cross the partition of Fig. 12.1 in the presence of a finite temperature gradient. The formalism that started this chapter suggests that there are three fluxes and three conjugate forces, all related through

$$J_U = L_{00} \frac{d}{dz}\left(\frac{1}{T}\right) + L_{01} \frac{d}{dz}\left(-\frac{\mu_1}{T}\right) + L_{02} \frac{d}{dz}\left(-\frac{\mu_2}{T}\right) \tag{12.163}$$

$$J_{N_1} = L_{10} \frac{d}{dz}\left(\frac{1}{T}\right) + L_{11} \frac{d}{dz}\left(-\frac{\mu_1}{T}\right) + L_{12} \frac{d}{dz}\left(-\frac{\mu_2}{T}\right) \tag{12.164}$$

$$J_{N_2} = L_{20} \frac{d}{dz}\left(\frac{1}{T}\right) + L_{21} \frac{d}{dz}\left(-\frac{\mu_1}{T}\right) + L_{22} \frac{d}{dz}\left(-\frac{\mu_2}{T}\right) \tag{12.165}$$

If the binary mixture, as a whole, does not move in the z direction, we also must account for the fact that the net volumetric flowrate is zero:

$$\bar{v}_1 J_{N_1} + \bar{v}_2 J_{N_2} = 0 \tag{12.166}$$

In this last statement, $\bar{v}_1$ and $\bar{v}_2$ are the partial molal volumes of constituents 1 and 2 at the location of interest. Note further that if the proportions of 1 and 2 are such that constituent 1 exists in very small amounts in the mixture (i.e., if constituent 2 plays the role of solvent), then the zero-net-flow condition reduces to $J_{N_2} = 0$. In any case, the two molal flux relations (12.164)–(12.165) can be combined with the zero-flow condition (12.166) in order to express $d(-\mu_2/T)/dz$ as a linear combination of $d(T^{-1})/dz$ and $d(-\mu_1/T)/dz$. Eliminating in this way $d(-\mu_2/T)/dz$ from the J_U and J_{N_1} linear laws above yields

$$J_U = L'_{00} \frac{d}{dz}\left(\frac{1}{T}\right) + L'_{01} \frac{d}{dz}\left(-\frac{\mu_1}{T}\right) \tag{12.167}$$

$$J_{N_1} = L'_{10} \frac{d}{dz}\left(\frac{1}{T}\right) + L'_{11} \frac{d}{dz}\left(-\frac{\mu_1}{T}\right) \tag{12.168}$$

The new Onsager coefficients L'_{00}, L'_{01}, L'_{10}, and L'_{11} are entirely different from the coefficients employed in eqs. (12.163)–(12.164). In conclusion, only two of the original fluxes and only two of the original driving forces are independent. The third flux can be deduced from the zero-flow condition (12.166). If needed, the third driving force $d(-\mu_2/T)/dz$ can be calculated from eq. (12.165).

It is possible to convert this new two-flux formulation into one in which the places of $d(T^{-1})/dz$ and $d(-\mu_1/T)/dz$ are taken by physically meaningful quantities such as dT/dz and dx_1/dz. If the binary mixture is an ideal gas, then μ_1 depends on T, P, and x_1 in the special way indicated by eq. (4.139):

$$\mu_1 = \bar{g}_1(T, P) + \bar{R}T \ln x_1 \qquad (12.169)$$

Assuming for the sake of analytical conciseness that the local pressure distribution is sufficiently uniform so that the function $\bar{g}_1/T$ varies primarily because of changes in T, the μ_1/T version of eq. (12.169) implies also

$$\frac{d}{dz}\left(-\frac{\mu_1}{T}\right) = \left(\frac{\bar{g}_1}{T^2} - \frac{1}{T}\frac{\partial \bar{g}_1}{\partial T}\right)\frac{dT}{dz} - \frac{\bar{R}}{x_1}\frac{dx_1}{dz} \qquad (12.170)$$

Substituting this into the J_U and J_{N_1} relations (12.167)–(12.168) yields two new linear relations in terms of temperature gradient and mole-fraction gradient:

$$J_U = \left[-\frac{L'_{00}}{T^2} - L'_{01}\left(-\frac{\bar{g}_1}{T^2} + \frac{1}{T}\frac{\partial \bar{g}_1}{\partial T}\right)\right]\frac{dT}{dz} - L'_{01}\frac{\bar{R}}{x_1}\frac{dx_1}{dz} \qquad (12.171)$$

$$J_{N_1} = \left[-\frac{L'_{10}}{T^2} - L'_{11}\left(-\frac{\bar{g}_1}{T^2} + \frac{1}{T}\frac{\partial \bar{g}_1}{\partial T}\right)\right]\frac{dT}{dz} - L'_{11}\frac{\bar{R}}{x_1}\frac{dx_1}{dz} \qquad (12.172)$$

The above system represents the starting point in the study of one-dimensional heat and mass diffusion through a bulk-stationary binary mixture. For this purpose, we can use the following shorthand notation:

$$J_U = -D'_{00}\frac{dT}{dz} - D'_{01}\frac{dx_1}{dz} \qquad (12.173)$$

$$J_{N_1} = -D'_{10}\frac{dT}{dz} - D'_{11}\frac{dx_1}{dz} \qquad (12.174)$$

in which D'_{00}, D'_{01}, D'_{10}, and D'_{11} are four "diffusivity" coefficients. More frequent than the use of molal flux J_{N_1} is the use of the mass flux:

$$J_{m_1} = M_1 J_{N_1} \qquad (12.175)$$

which is expressed in kilograms of constituent 1 per second and meter squared. Also, instead of the mole fraction x_1, it is customary to report the concentration C_1, in this case,

$$C_1 = \frac{M_1}{\bar{v}} x_1 \tag{12.176}$$

where $\bar{v}$ is the proper molal volume of the mixture at the z location of interest ($\bar{v}$ is the number of cubic meters per mole of mixture). Note further that the units of C_1 are kilograms of constituent 1 per cubic meter. In terms of J_{m_1}, C_1, and a new set of diffusivities, the two-flux relations (12.173)–(12.174) can be rewritten as

$$J_U = -D_{00} \frac{dT}{dz} - D_{01} \frac{dC_1}{dz} \tag{12.177}$$

$$J_{m_1} = -D_{10} \frac{dT}{dz} - D_{11} \frac{dC_1}{dz} \tag{12.178}$$

This version invites several comments. First, when the region of interest is isothermal, the mass flux of constituent 1 is due entirely to the concentration gradient of the same constituent:

$$J_{m_1} = -D_{11} \frac{dC_1}{dz} \quad \left(\frac{dT}{dz} = 0 \right) \tag{12.179}$$

This is Fick's law of diffusion [6] in which the *mass-diffusivity* coefficient D_{11} has the units m^2/s, i.e., the same units as the thermal and momentum diffusivities, α and ν, respectively, encountered in the field of heat and mass transfer. The second law guarantees that D_{11} is nonnegative.

The second observation is that the temperature gradient partially drives the mass flux J_{m_1}. When the mass flow ceases, $J_{m_1} = 0$, the temperature gradient sustains a concentration gradient:

$$\frac{dC_1}{dT} = -\frac{D_{10}}{D_{11}} \quad (J_{m_1} = 0) \tag{12.180}$$

in other words, the imposition of a temperature difference across the partition induces a concentration difference (a partial separation) of the constituents of the binary mixture. This diffusion phenomenon, whose magnitude is ruled by D_{10}, is recognized as the *Soret effect*, or Soret diffusion [26–28]. The D_{10} coefficient can have either sign.

Third, the J_U relation (12.177) shows that the concentration gradient contributes to the energy flux in proportion to the coefficient D_{01}. This aspect represents the *Dufour effect* first described in 1872 [29, 30]. The D_{01} coefficient can also be positive or negative.

Finally, although the D_{00} coefficient of the J_U relation (12.177) has units of thermal conductivity, W/mK, in general, it is not the same as the thermal conductivity k. The two are equal only if either the Soret effect or the Dufour effect are insignificant (Problem 12.9).

Isothermal Diffusion

The preceding discussion makes it easy to see that when the temperature gradient is zero, the diffusion of constituent 1 in a bulk-stationary binary mixture is ruled by the classical Fick's law (12.179). The same holds for constituent 2, because the two-flux formulation developed from eqs. (12.166)–(12.178) can be repeated identically for (J_U, J_{N_2}) as a flux pair.

The simplest example of isothermal diffusion in which cross-coupling effects are present is that of a stationary ternary mixture. In this case, $n = 3$ and, in place of eqs. (12.163)–(12.165), we could write three linear relations linking the three fluxes $(J_{N_1}, J_{N_2}, J_{N_3})$ to the conjugate forces $d(-\mu_1/T)/dz$, $d(-\mu_2/T)/dz$, and $d(-\mu_3/T)/dz$. The condition of zero volumetric flow:

$$\bar{v}_1 J_{N_1} + \bar{v}_2 J_{N_2} + \bar{v}_3 J_{N_3} = 0 \tag{12.181}$$

makes only two of the fluxes independent, for example, J_{N_1} and J_{N_2}. For these, the linear relations can be written the same way as in eqs. (12.167)–(12.168):

$$J_{N_1} = L'_{11} \frac{d}{dz}\left(-\frac{\mu_1}{T}\right) + L'_{12} \frac{d}{dz}\left(-\frac{\mu_2}{T}\right) \tag{12.182}$$

$$J_{N_2} = L'_{21} \frac{d}{dz}\left(-\frac{\mu_1}{T}\right) + L'_{22} \frac{d}{dz}\left(-\frac{\mu_2}{T}\right) \tag{12.183}$$

If the ternary mixture is an ideal gas and, in addition, if the pressure is sufficiently uniform in the region penetrated by the diffusion phenomenon, by invoking eq. (4.139), the above relations can be transformed into

$$J_{N_1} = -L'_{11} \frac{\bar{R}}{x_1} \frac{dx_1}{dz} - L'_{12} \frac{\bar{R}}{x_2} \frac{dx_2}{dz} \tag{12.184}$$

$$J_{N_2} = -L'_{21} \frac{\bar{R}}{x_1} \frac{dx_1}{dz} - L'_{22} \frac{\bar{R}}{x_2} \frac{dx_2}{dz} \tag{12.185}$$

In terms of mass fluxes and concentration gradients defined in the manner of eqs. (12.175)–(12.176), the same conclusion reads

$$J_{m_1} = -D_{11} \frac{dC_1}{dz} - D_{12} \frac{dC_2}{dz} \tag{12.186}$$

$$J_{m_2} = -D_{21} \frac{dC_1}{dz} - D_{22} \frac{dC_2}{dz} \tag{12.187}$$

Important to note is that the cross-coupling mass diffusivities D_{12} and D_{21} are in general not equal. The reciprocity relation applies to the starting formulation, eqs. (12.182)–(12.183), and reads $L'_{12} = L'_{21}$. Using this relation and identifying the quantities of eqs. (12.184)–(12.185) for which D_{12} and D_{21} have been devised as shorthand notation, it is easy to show that

$$\frac{D_{12}}{D_{21}} = \frac{M_1 C_1}{M_2 C_2} \tag{12.188}$$

SYMBOLS

a	rotatory conductivity coefficient
a_{ij}	principal conductivity coefficients [eqs. (12.115)–(12.116)]
A	cross-sectional area
C_i	concentration [kg of constituent i/m^3], eq. (12.176)]
COP	coefficient of performance of refrigerator
D_{ij}	thermal diffusivities [eqs. (12.186)–(12.187)]
D'_{ij}	coefficients [eqs. (12.173)–(12.174)]
e	electric charge of one electron
$\bar{g}$	specific Gibbs free energy, per mole
h	specific enthalpy, per mole
I	electric current
J_i	flux [eqs. (12.17)–(12.18)]
J_{m_i}	mass flux [kg/m^{2}s, eq. (12.175)]
J_{N_i}	molal flux [mol/m^{2}s]
J_U	energy flux [W/m^2]
k	thermal conductivity
k_e	electrical conductivity
k_{ij}	thermal-conductivity coefficients
k_1, k_2, k_3	principal thermal conductivities
L	length
L_{ik}	Onsager coefficient associated with J_i and X_k (several other sets of such coefficients are listed as l_{ik} and L'_{ik})
m	parameter, [eq. (12.134)]
M_i	molecular mass of constituent i
n	total number of constituents
N	number of moles
$\dot{N}$	molal flowrate [mol/s]
P	pressure
q''	heat flux [W/m^2] as distinct from J_U
$\dot{Q}$	heat transfer rate [W]
r_{ij}	thermal-resistivity coefficients [eqs. (12.99)–(12.100)]
R	electric resistance [eq. (12.74)]
$\bar{R}$	universal ideal gas constant

$\bar{s}$	specific entropy, per mole
s'''_{gen}	local entropy-generation rate [W/m^3 K]
S	entropy [J/K]
$\dot{S}_{gen}$	entropy-generation rate [W/K]
t	time
T	thermodynamic temperature
ΔT	temperature difference, $T_H - T_L$
U	internal energy
$\bar{v}$	specific volume, per mole
x	longitudinal coordinate for one-dimensional diffusion (Fig. 12.1)
x_i	mole fraction of constituent i [eq. (12.169)]
(x, y, z)	cartesian system of stretched coordinates [eqs. (12.153)]
(x_1, x_2, x_3)	cartesian system of coordinates
X	ratio of slenderness ratios [eq. (12.82)]
X_i	driving force, conjugate with the flux J_i
X_{N_i}	driving force, the conjugate of J_{N_i}
X_U	driving force, the conjugate of J_U
z	longitudinal coordinate for one-dimensional mass diffusion
z	figure of merit [eq. (12.91)]
Z	dimensionless group [eq. (12.88)]
α	angle of rotation (Fig. 12.10)
ϵ	absolute thermoelectric power or Seebeck coefficient [eq. (12.51)]
η	heat-engine efficiency
μ	chemical or electrochemical potential
(ξ_1, ξ_2, ξ_3)	cartesian system of principal directions
π	Peltier coefficient [eq. (12.54)]
$\pi_{\alpha\beta}$	Peltier coefficient of the $\alpha\|\beta$ junction [eq. (12.78)]
τ	Thomson coefficient [eq. (12.73)]
ϕ	electrostatic potential
$(\ )_H$	associated with the high-temperature end
$(\ )_L$	associated with the low-temperature end
$(\ )_{max}$	maximum
$(\ )_{opt}$	optimum
$(\ )_\alpha$	property of material α
$(\ )_\beta$	property of material β

REFERENCES

1. L. Onsager, Reciprocal relations in irreversible processes. I, *Phys. Rev.*, Vol. 37, 1931, pp. 405–426.
2. L. Onsager, Reciprocal relations in irreversible processes. II, *Phys. Rev.*, Vol. 38, 1931, pp. 2265–2279.

3. H. B. Callen, *Thermodynamics*, Wiley, New York, 1960, Chapter 17.

4. B. D. Coleman and C. Truesdell, On the reciprocal relations of Onsager, *J. Chem. Phys.*, Vol. 33, No. 1, 1960, pp. 28–31.

5. W. C. Reynolds, *Thermodynamics*, 2nd ed., McGraw-Hill, New York, 1968, p. 404.

6. A. Bejan, *Convection Heat Transfer*, Wiley, New York, 1984, p. 11.

7. W. Thomson, Thermo-electric currents, *Trans. R. Soc. Edinburgh*, Vol. 21, Part I, 1854; also *Mathematical and Physical Papers*, Vol. I, Cambridge University Press, London and New York, 1882, pp. 232–291.

8. G. G. Stokes, On the conduction of heat in crystals, *Cambridge Dublin Math. J.*, Vol. 6, 1851, pp. 215–238; also *Mathematical and Physical Papers*, Vol. III, Cambridge University Press, London and New York, 1901, pp. 203–227.

9. D. G. Miller, Thermodynamics of irreversible processes. The experimental verification of the Onsager reciprocal relations, *Chem. Rev.*, Vol. 60, 1960, pp. 15–37.

10. C. A. Domenicali, Irreversible thermodynamics of thermoelectricity, *Rev. Mod. Phys.*, Vol. 26, No. 2, 1954, pp. 237–275.

11. B. Sherman, R. R. Heikes, and R. W. Ure, Jr., Calculation of efficiency of thermoelectric devices, *J. Appl. Phys.*, Vol. 31, No. 1, 1960, pp. 1–16.

12. P. O. Gehloff, E. Justi, and M. Kohler, Verfeinerte Theorie der elektrothermischen Kälteerzeugung, *Abhl. Braunschw. Wiss. Ges.*, Vol. 2, 1950, pp. 149–164.

13. A. H. Boerdijk, Contribution to a general theory of thermocouples, *J. Appl. Phys.*, Vol. 30, No. 7, 1959, pp. 1080–1083.

14. W. H. Clingman, Entropy production and optimum device design, *Adv. Energy Convers.*, Vol. 1, 1961, pp. 61–79.

15. J. A. Brandt, Solutions to the differential equations describing the temperature distribution, thermal efficiency, and power output of a thermoelectric element with variable properties and cross sectional area, *Adv. Energy Convers.*, Vol. 2, 1962, pp. 219–230.

16. C. N. Rollinger, Convectively cooled thermoelements with variable cross-sectional area, *J. Heat Transfer*, Vol. 87, 1965, pp. 259–265.

17. K. Landecker, Some recent developments in thermoelectric junctions for refrigeration and power production, in K. R. Rao, ed., *Thermoelectric Energy Conversion*, IEEE, New York, 1976, pp. 150–154.

18. E. Altenkirch, Elektrothermische Kälteerzeugung und reversible elektrische Heizung, *Phys. Z.*, Vol. 12, 1911, pp. 920–924.

19. H. S. Carslaw and J. C. Jaeger, *Conduction of Heat in Solids*, 2nd ed., Oxford University Press, London and New York, 1959, p. 38.

20. K. Rektorys, ed., *Survey of Applicable Mathematics*, MIT Press, Cambridge, MA, 1969, p. 882.

21. W. M. Rohsenow and H. Y. Choi, *Heat, Mass and Momentum Transfer*, Prentice-Hall, Englewood Cliffs, NJ, 1961, Chapter 19.

22. C. Soret, Sur l'étude expérimentale des coefficients rotationnels de conductibilité thermique, *Soc. Phys. Hist. Nat. Geneve, Arch.*, Vol. 29, 1893, pp. 355–357.

23. W. Voigt, Fragen der Kristallphysik. I, *Nachr. Ges. Wiss. Goettingen, Math.-Phys. Kl.*, Vol. 87, No. 3, 1903, pp. 87–89.

24. J. M. C. Duhamel, *Journal de l'Ecole Polytechnique, Paris*, Vol. 13, Cahier 21, 1832, p. 356 (first presented in 1828); also Vol. 19, Cahier 32, 1848, p. 155.

25. K. G. Denbigh, *The Thermodynamics of the Steady State*, Methuen, London, 1951, p. 73.

26. C. Soret, Sur l'état d'équilibre que prend, au point de vue de sa concentration, une dissolution saline primitivement homogène dont deux parties sont portées a des températures différentes, *Ann. Chim. Phys.*, Ser. 5, Vol. 22, 1881, pp. 293–297.

27. E. D. Eastman, Thermodynamics of non-isothermal systems, *J. Am. Chem. Soc.*, Vol. 48, 1926, pp. 1482–1493.

28. E. D. Eastman, Theory of the Soret effect, *J. Am. Chem. Soc.*, Vol. 50, 1928, pp. 283–291.

29. L. Dufour, Abbreviated communication, *Arch. Sci. Phys. Nat.*, Ser. 5, Vol. 45, 1872, pp. 9–12.

30. L. Dufour, Ueber die Diffusion der Gase durch poröse Wände und die sie begleitenden Temperaturveränderungen, *Ann. Phys. (Leipzig)*, Ser. 2, Vol. 28, 1873, pp. 490–492.

PROBLEMS

12.1 Consider the single-flux phenomenon of pure electric conduction through the one-dimensional conductor of Fig. 12.2. Invoke the second law in order to prove that the electrical conductivity k_e cannot be negative.

12.2 Estimate the local entropy-generation rate s'''_{gen} for the coupled one-dimensional flow of energy and electric current discussed in connection with Fig. 12.2. Show that your result can be expressed as

$$s'''_{gen} = q'' \frac{d}{dx} \left(\frac{1}{T} \right) + \frac{1}{T} \frac{dq''}{dx}$$

in other words, that the irreversibility is due to: (i) the flow of heat down a finite temperature gradient, and (ii) the mere formation or appearance of heat current at the longitudinal position of interest.

12.3 Consider the junction between two one-dimensional conductors of heat and electric current, α and β (Fig. 12.3). Show that the Peltier coefficients, absolute thermoelectric powers, and Thomson coefficients of α and β are related through the so-called "first Kelvin relation" [3]:

$$\frac{d}{dT} (\pi_\beta - \pi_\alpha) = \epsilon_\beta - \epsilon_\alpha + \tau_\beta - \tau_\alpha$$

12.4   Show that the ordinary differential equation for the temperature distribution along a one-dimensional thermoelectric conductor is

$$\frac{d}{dx}\left(k\,\frac{dT}{dx}\right) - eJ_N\tau\,\frac{dT}{dx} + \frac{(eJ_N)^2}{k_e} = 0$$

An application of this result forms the subject of Problem 12.11.

12.5   The efficiency of a thermoelectric power generator operating between T_H and T_L can be improved by producing electric power *continuously*, that is, at every temperature level T between T_H and T_L. Conceptually, this is accomplished by a stack of infinitesimally tall two-junction elements of the type treated in Fig. 12.7. As shown in the figure for this problem, in this stack, the heat rejected by the cold junction of one element becomes the heat absorbed by the warm junction of the adjacent element. Show that the heat-engine efficiency of the entire stack is

$$\eta_{\text{stack}} = 1 - \exp\left(-\int_{T_L}^{T_H} \zeta\,\frac{dT}{T}\right)$$

where $\zeta = (Z-1)/(Z+1)$. Show also that if Z is independent of temperature, the efficiency simplifies further:

$$\eta_{\text{stack}} = 1 - \frac{T_L^{\zeta}}{T_H^{\zeta}}$$

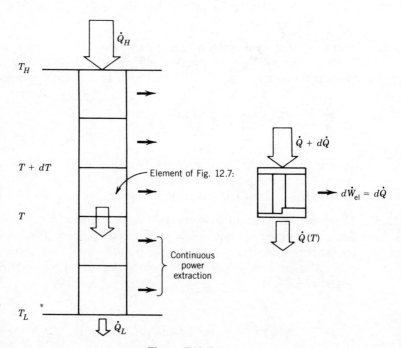

Figure P12.5

12.6   Demonstrate that the principal thermal conductivities (k_1, k_2) are related to the four coefficients $(k_{11}, k_{12}, k_{21}, k_{22})$ via eqs. (12.112)–(12.113). Show also that the angle between the x_1 axis and the principal direction ξ_1 in Fig. 12.10 must be

$$\alpha = \frac{1}{2} \tan^{-1} \frac{k_{12} + k_{21}}{k_{11} - k_{22}}$$

12.7   Show that the coefficients $(a_{11}, a_{12}, a_{21}, a_{22})$ that appear in the linear relations between principal fluxes and principal temperature gradients (12.115)–(12.116) are indeed equal to the expressions listed in eqs. (12.117)–(12.120). Prove also that a_{21} is the negative of a_{12}.

12.8   Here is an alternative to the method used in the text for showing that $k_{12} = k_{21}$ if $a = 0$, eqs. (12.129)–(12.130). Rely on the a_{12} expression derived above [or, simply start with eq. (12.118)] to show that

$$a_{12} = \tfrac{1}{2}(k_{12} - k_{21})$$

in other words, that $k_{12} = k_{21}$ if a_{12} (or a) is zero.

12.9   While examining the two flux relations of nonisothermal diffusion in a binary mixture, eqs. (12.177)–(12.178), note that the energy flux J_U is not necessarily equal to the heat flux q''. If the thermal conductivity k is defined under zero-mass-flux conditions:

$$q'' = -k \frac{dT}{dz} \qquad (J_{m_1} = 0)$$

prove that

$$k = D_{00} - \frac{D_{01} D_{10}}{D_{11}}$$

12.10   Consider the nonisothermal diffusion of a single-component fluid across the partition shown in the drawing. During a short enough time interval dN moles of fluid pass through the partition, that is, dN moles leave system I and enter system II. During this process, system I is maintained at constant temperature (T) and pressure (P) by contact with the temperature and pressure reservoirs shown in the sketch. The respective energy interactions with these reservoirs are δQ and δW, with the additional observation that $\delta W = P\, dV$, where dV is the volume swept by the piston (i.e. the change in the volume of system I).

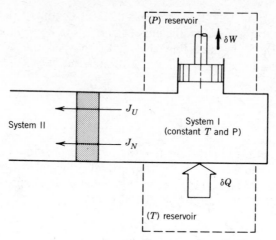

Figure P12.10

The above process provides a physical interpretation for the factor [25]

$$Q^* = \frac{L_{10}}{L_{11}} - \bar{h}$$

which appears in eq. (12.159),

$$\frac{dP}{dT} = -\frac{Q^*}{\bar{v}T} \qquad (J_N = 0)$$

Recall that the ratio L_{10}/L_{11} represents the "energy transport," i.e., the energy transfer from I to II, per mole and when the temperature is the same on both sides of the partition, eq. (12.162). This means that during the process described in the preceding paragraph the net energy transport across the partition is $(L_{10}/L_{11})\, dN$.

Invoke the first law for system I and show that

$$Q^* = \frac{\delta Q}{dN}$$

in other words, that the factor Q^* is the heat transfer absorbed by system I per mole of fluid transferred from I to II at constant T and P.

12.11 The two ends of the one-dimensional thermoelectric conductor of Problem 12.4 are maintained at the same temperature, T_0. The length of the conductor is L. The deviation of the conductor temperature $T(x)$ from the end temperature T_0 is sufficiently small so that k, k_e, and τ may be treated as three known constants. Show that the maximum conductor temperature occurs at

$$\frac{x}{L} = \frac{1}{B} \ln \left(\frac{e^B - 1}{B} \right)$$

where $B = eJ_N \tau L/k$. Show further that the temperature maximum passes through the midpoint of the conductor as the electrical current changes direction.

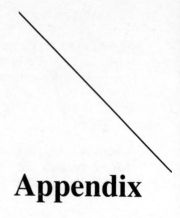

Appendix

CONSTANTS

Ideal gas (universal) constant

$$\bar{R} = 8314.3 \text{ J kmol}^{-1} \text{ K}^{-1}$$
$$= 1.9872 \text{ cal mol}^{-1} \text{ K}^{-1}$$
$$= 1.9872 \text{ Btu lbmol}^{-1} \text{ R}^{-1}$$
$$= 1545.33 \text{ ft lbf lbmol}^{-1} \text{ R}^{-1}$$

Boltzmann's constant $\quad k = (1.38054)10^{-23} \text{ J K}^{-1}$

Planck's constant $\quad h = (6.6256)10^{-34} \text{ J s}$

Speed of light $\quad c = (2.998)10^{8} \text{ m s}^{-1}$

Avogadro's number $\quad N = (6.02252)10^{23} \text{ mol}^{-1}$

Stefan–Boltzmann constant $\quad \sigma = (5.6693)10^{-12} \text{ W cm}^{-2} \text{ K}^{-4}$
$$= (0.1712)10^{-8} \text{ Btu hr}^{-1} \text{ ft}^{-2} \text{ R}^{-4}$$

Atmospheric pressure $\quad P_{\text{atm}} = 0.101325 \text{ MPa}$
$$= (1.01325)10^{5} \text{ N m}^{-2}$$

Gravitational acceleration $\quad g = 9.80665 \text{ m s}^{-2}$

Calorie $\quad 1 \text{ cal} = 4.184 \text{ J}$

Mole $\quad 1 \text{ mol} =$ sample containing $(6.02252)10^{23}$ elementary entities (e.g., molecules); abbreviated also as 1 gmol, or

$$1 \text{ mol} = 10^{-3} \text{ kmol}$$
$$= 10^{-3} \text{ kgmol}$$

$$= \frac{1}{453.6} \text{ lbmol}$$

MATHEMATICAL FORMULAS

Relations involving partial derivatives:

$$df = \left(\frac{\partial f}{\partial x}\right)_y dx + \left(\frac{\partial f}{\partial y}\right)_x dy \qquad \text{[total differential of function } f(x, y)\text{]}$$

720

$$\left(\frac{\partial f}{\partial x}\right)_y = \frac{1}{\left(\dfrac{\partial x}{\partial f}\right)_y} \qquad \text{(reciprocal relation)}$$

$$\left(\frac{\partial f}{\partial x}\right)_y = \left(\frac{\partial f}{\partial z}\right)_y \left(\frac{\partial z}{\partial x}\right)_y \qquad \text{(chain rule)}$$

$$\left(\frac{\partial f}{\partial x}\right)_y = \left(\frac{\partial f}{\partial z}\right)_y \left(\frac{\partial z}{\partial w}\right)_y \left(\frac{\partial w}{\partial x}\right)_y$$

$$\left(\frac{\partial x}{\partial y}\right)_z \left(\frac{\partial y}{\partial z}\right)_x \left(\frac{\partial z}{\partial x}\right)_y = -1 \qquad \text{(cyclical relation)}$$

$$\left(\frac{\partial x}{\partial y}\right)_z = \left(\frac{\partial x}{\partial y}\right)_w + \left(\frac{\partial x}{\partial w}\right)_y \left(\frac{\partial w}{\partial y}\right)_z$$

Jacobians, or functional determinants:

$$\frac{\partial(x, y, \ldots, z)}{\partial(\alpha, \beta, \ldots, \gamma)} = \begin{vmatrix} \dfrac{\partial x}{\partial \alpha} & \dfrac{\partial x}{\partial \beta} & \cdots & \dfrac{\partial x}{\partial \gamma} \\ \dfrac{\partial y}{\partial \alpha} & \dfrac{\partial y}{\partial \beta} & \cdots & \dfrac{\partial y}{\partial \gamma} \\ \dfrac{\partial z}{\partial \alpha} & \dfrac{\partial z}{\partial \beta} & \cdots & \dfrac{\partial z}{\partial \gamma} \end{vmatrix} \qquad \text{(definition)}$$

$$\frac{\partial(x, y, \ldots, z)}{\partial(\alpha, \beta, \ldots, \gamma)} = -\frac{\partial(y, x, \ldots, z)}{\partial(\alpha, \beta, \ldots, \gamma)} \qquad \begin{array}{l}\text{(interchanging the positions of } x \\ \text{and } y)\end{array}$$

$$\frac{\partial(x, y, \ldots, z)}{\partial(\alpha, \beta, \ldots, \gamma)} = \left[\frac{\partial(\alpha, \beta, \ldots, \gamma)}{\partial(x, y, \ldots, z)}\right]^{-1} \qquad \text{(reciprocal relation)}$$

$$\frac{\partial(x, y, \ldots, z)}{\partial(\alpha, \beta, \ldots, \gamma)} = \frac{\partial(x, y, \ldots, z)}{\partial(A, B, \ldots, C)} \frac{\partial(A, B, \ldots, C)}{\partial(\alpha, \beta, \ldots, \gamma)} \qquad \text{(chain rule)}$$

$$\left(\frac{\partial x}{\partial \alpha}\right)_{\beta, \ldots, \gamma} = \frac{\partial(x, \beta, \ldots, \gamma)}{\partial(\alpha, \beta, \ldots, \gamma)}$$

$$= \frac{\partial(x, \beta, \ldots, \gamma)}{\partial(A, B, \ldots, C)} \bigg/ \frac{\partial(\alpha, \beta, \ldots, \gamma)}{\partial(A, B, \ldots, C)}$$

Linear differential forms (Pfaffians), and conditions of integrability [1]:

$$\delta F = P(x, y, z)\, dx + Q(x, y, z)\, dy + R(x, y, z)\, dz$$

$$\left(\frac{\partial P}{\partial y}\right)_{x,z} = \left(\frac{\partial Q}{\partial x}\right)_{y,z}$$

$$\left(\frac{\partial P}{\partial z}\right)_{x,y} = \left(\frac{\partial R}{\partial x}\right)_{y,z}$$

$$\left(\frac{\partial Q}{\partial z}\right)_{x,y} = \left(\frac{\partial R}{\partial y}\right)_{x,z}$$

Definite integral entering the analytical derivation of the Stefan–Boltzmann law:

$$\int_0^\infty \frac{m^3 \, dm}{e^m - 1} = \frac{\pi^4}{15}$$

VARIATIONAL CALCULUS

The basic problem in variational calculus consists of determining, from among functions possessing certain properties, that function for which a given integral (functional) assumes its maximum or minimum value. The integrand of the integral in question depends on the function and its derivatives [2, 3]. Consider the many values of the integral

$$I = \int_a^b F(x, y, y') \, dx$$

where $y(x)$ is unknown, and $y' = dy/dx$. The special function y for which I reaches an extremum satisfies the *Euler equation*:

$$\frac{\partial F}{\partial y} - \frac{d}{dx}\left(\frac{\partial F}{\partial y'}\right) = 0$$

If, in addition to minimizing or maximizing I, the wanted function $y(x)$ satisfies an *integral constraint* of the type

$$C = \int_a^b G(x, y, y') \, dx$$

then $y(x, \lambda)$ satisfies the new Euler equation:

$$\frac{\partial H}{\partial y} - \frac{d}{dx}\left(\frac{\partial H}{\partial y'}\right) = 0$$

where

$$H = F + \lambda G$$

Constant λ is a Lagrange multiplier; its value is determined by substituting the $y(x, \lambda)$ solution into the integral constraint C.

In general, the integral may depend on the first n derivatives of the unknown function $y(x)$:

$$I = \int_a^b F(x, y, y', y'', \ldots, y^{(n)}) \, dx$$

In this case, the condition for I to reach an extremum is

$$\frac{\partial F}{\partial y} - \frac{d}{dx}\left(\frac{\partial F}{\partial y'}\right) + \frac{d^2}{dx^2}\left(\frac{\partial F}{\partial y''}\right) - \cdots + (-1)^n \frac{d^n}{dx^n}\left[\frac{\partial F}{\partial y^{(n)}}\right] = 0$$

The above differential equation for $y(x)$ is called the *Euler–Poisson equation*.

If the integral depends not on one but k unknown functions y_i ($i = 1, \ldots, k$) and their first derivatives, y_i',

$$I = \int_a^b F(x, y_1, \ldots, y_k, y_1', \ldots, y_k') \, dx$$

then the k functions $y_i(x)$ that minimize or maximize I satisfy the system of Euler equations:

$$\frac{\partial F}{\partial y_i} - \frac{d}{dx}\left(\frac{\partial F}{\partial y_i'}\right) = 0 \qquad (i = 1, \ldots, k)$$

PROPERTIES OF MODERATELY COMPRESSED-LIQUID STATES

The thermodynamic properties of pure substances in compressed (or sub-cooled) liquid form are usually tabulated in pressure–temperature matrices attached to the more refined tabulations of saturated-states and super-heated-vapor-states properties. One difficulty that is often encountered in the analysis of power and refrigeration cycles is that the P–T tabulation of compressed-liquid states starts with pressures well in excess of the highest pressure that occurs in the cycle analysis. In other words, the highest compressed-liquid pressure in the cycle analysis is too "moderate" and, consequently, the properties associated with this pressure are not tabulated.

The moderate-pressures domain is situated very close to the two-phase region. The analytical technique to be presented consists of expressing the wanted compressed-liquid properties in terms of the neighboring saturated-liquid properties associated with the same temperature. For example, for the enthalpy of the compressed-liquid state $(T\,P)$, we write [4]

$$h(T, P) = h_f(T) + (\text{correction term})$$

The correction term is easily identified by treating the compressed liquid as essentially *incompressible* ($v = v_f = \text{constant}$); hence, the approximate formula for enthalpy:

$$h(T, P) \cong h_f(T) + [P - P_{\text{sat}}(T)]v_f(T)$$

where $P > P_{\text{sat}}(T)$. The corresponding approximate formulas for internal energy and entropy are [4]

$$u(T, P) \cong u_f(T)$$
$$s(T, P) \cong s_f(T)$$

PROPERTIES OF SLIGHTLY SUPERHEATED-VAPOR STATES

Based on a geometric argument similar to the one used in the preceding section, the properties of a pure substance that is in a superheated-vapor state situated sufficiently close to the two-phase dome can be approximated in terms of the properties of the saturated-vapor state that has the same temperature. This time, the connection between the slighly superheated-vapor state (T, P) and the "g" state of temperature T is made through the ideal gas model:

$$h(T, P) \cong h_g(T)$$

$$u(T, P) \cong u_g(T)$$

$$s(T, P) \cong s_g(T) + R_g \ln \frac{P_{sat}(T)}{P}$$

where it is worth noting that $P_{sat}(T) > P$. The constant R_g is the ideal gas constant of saturated and slightly superheated vapor when viewed "locally" as an ideal gas, i.e., for sufficiently small departures from the saturated-vapor state of temperature T:

$$R_g = \frac{P_{sat}(T)v_g(T)}{T}$$

In the case of saturated and slightly superheated steam, for example, the local ideal gas model is particularly effective when the pressure is subatmospheric. For this reason, it is used routinely in air-conditioning calculations involving (air + water vapor) mixtures.

PROPERTIES OF COLD WATER NEAR THE DENSITY MAXIMUM

An approximate analytical model for the two equations of state of cold liquid water in the 1–100 atm pressure range was constructed by Thomsen and Hartka [5]:

$$v = v_0[1 + \lambda(T - T_0 + aP)^2 - \kappa_0 P]$$

where

$$v_0 = (1.00008)10^{-3} \, \text{m}^3/\text{kg}$$

$$\lambda = (8)10^{-6} \, \text{K}^{-2}$$

$$T_0 = 277 \, \text{K}$$

$$a = (2)10^{-7} \, \text{ms}^2 \, \text{K}/\text{kg}$$

$$\kappa_0 = (5)10^{-10} \, \text{ms}^2/\text{kg}$$

and

$$(c_P)_{P=1 \text{ atm}} = c_0 - b(T - T_0)$$

where

$$c_0 = (4.2057)10^3 \text{ J/kg/K}$$
$$b = 2.6 \text{ J/kg/K}^2$$

Other useful relations are

$$c_P(T, P) = c_0 - b(T - T_0) - 2\lambda v_0 TP \qquad \text{(Problem 4.9)}$$

$$\beta = \frac{1}{v}\left(\frac{\partial v}{\partial T}\right)_P = \frac{2\lambda(T - T_0 + aP)}{1 + \lambda(T - T_0 + aP)^2 - \kappa_0 P} \cong 2\lambda(T - T_0 + aP)$$

$$\kappa = -\frac{1}{v}\left(\frac{\partial v}{\partial P}\right)_T = \frac{\kappa_0 - 2\lambda a(T - T_0 + aP)}{1 + \lambda(T - T_0 + aP)^2 - \kappa_0 P} \cong \kappa_0 - 2\lambda a(T - T_0 + aP)$$

The approximate expressions listed for β and κ (and denoted by "$\cong$") are valid in the limit $|T - T_0| \ll T_0$. In the same limit, we have

$$c_v \cong c_0 - b(T - T_0) - 2\lambda v_0 TP - (4\lambda^2 v_0 T/\kappa_0)(T - T_0 + aP)^2$$

An extremely accurate model for density (or v), which is well suited for computer-aided simulations of the buoyancy-driven flow of cold pure water and saline water was developed by Gebhart and Mollendorf [6]. Their model is highly recommended for convection heat transfer applications.

ANALYSIS OF ENGINEERING-SYSTEM COMPONENTS

This section contains a summary [7] of the equations and inequalities obtained by applying the first law and the second law to the components encountered in engineering systems such as power cycles and refrigeration cycles. It is assumed that each component operates in the *steady state*.

Valve (throttle) or adiabatic duct with friction (Fig. A.1a):

First law $h_1 = h_2$
Second law $\dot{S}_{\text{gen}} = \dot{m}(s_2 - s_1) > 0$

Expander or *turbine* with negligible heat transfer to the ambient (Fig. A.1b):

First law $\dot{W}_t = \dot{m}(h_1 - h_2)$
Second law $\dot{S}_{\text{gen}} = \dot{m}(s_2 - s_1) \geq 0$

Efficiency $\eta_t = \dfrac{h_1 - h_2}{h_1 - h_{2,\text{rev}}} \leq 1$

Compressor or *pump* with negligible heat transfer to the ambient (Fig. A.1c):

First law $\dot{W}_c = \dot{m}(h_2 - h_1)$
Second law $\dot{S}_{gen} = \dot{m}(s_2 - s_1) \geq 0$

Efficiency $\eta_c = \dfrac{h_{2,rev} - h_1}{h_2 - h_1} \leq 1$

Nozzle with negligible heat transfer to the ambient (Fig. A.1d, where V = bulk speed):

First law $\frac{1}{2}(V_2^2 - V_1^2) = h_1 - h_2$
Second law $\dot{S}_{gen} = \dot{m}(s_2 - s_1) \geq 0$

Efficiency $\eta_n = \dfrac{V_2^2 - V_1^2}{V_{2,rev}^2 - V_1^2} \leq 1$

Diffuser with negligible heat transfer to the ambient (Fig. A.1e, where V = bulk speed):

First law $h_2 - h_1 = \frac{1}{2}(V_1^2 - V_2^2)$
Second law $\dot{S}_{gen} = \dot{m}(s_2 - s_1) \geq 0$

Efficiency $\eta_d = \dfrac{h_{2,rev} - h_1}{h_2 - h_1} \leq 1$

Heat exchangers with negligible heat transfer to the ambient (Figs. A.1f and A.1g):

First law $\dot{m}_{hot}(h_1 - h_2) = \dot{m}_{cold}(h_4 - h_3)$
Second law $\dot{S}_{gen} = \dot{m}_{hot}(s_2 - s_1) + \dot{m}_{cold}(s_4 - s_3) \geq 0$

Figures A.1f and A.1g show that a pressure drop always occurs in the direction of flow, in any heat-exchanger flow passage. Other heat-exchanger configurations are illustrated in chapter 11.

THE FLOW EXERGY OF GASES AT LOW PRESSURES

Figures 7.7 and 7.11 show the enthalpy and absolute entropy values of low-pressure gases. The enthalpy $\Delta \bar{h}$ measured on the ordinate of Fig. 7.7 is shorthand for

$$\Delta \bar{h}(T) = \bar{h}(T) - \bar{h}_f^{\circ} \qquad (A.1)$$

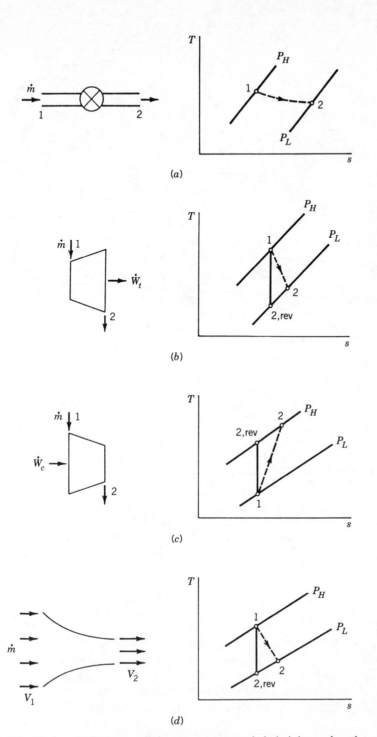

Figure A.1 Seven engineering-system components and their inlet and outlet states on the T–s plane (P_H = high pressure; P_L = low pressure).

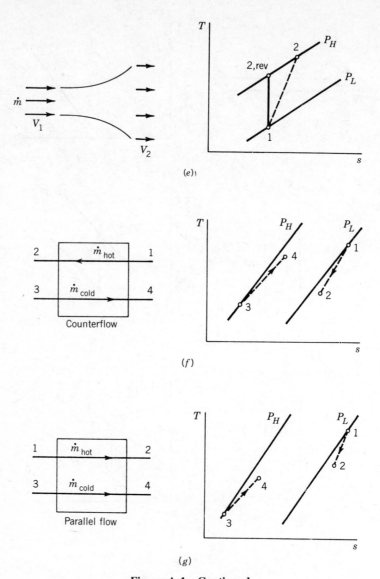

Figure A.1   Continued.

where $\bar{h}_f^{\circ}$ is the enthalpy of formation at $T_0 = 298.15\ \mathrm{K}$ (25°C) and $P_0 = 0.101325\ \mathrm{MPa}$ (1 atm). In writing eq. (A.1) it is being assumed that the pressure is sufficiently low so that $\bar{h}$ and $\bar{\Delta}h$ are functions of T only (in other words, the gas approaches the ideal gas limit).

Figure 7.11 shows the absolute entropy at atmospheric pressure,

$$\bar{s}^{\circ}(T) = s(T, P_0) \tag{A.2}$$

The observations made in the preceding paragraph apply here as well. The entropy at a state (T, P), where P is not necessarily the atmospheric pressure P_0, can be evaluated by invoking the ideal gas model,

$$\bar{s}(T, P) = \bar{s}^\circ(T) - \bar{R} \ln \frac{P}{P_0} \tag{A.3}$$

Figure A.2 shows the flow exergy of the 12 gases at atmospheric pressure. This particular chart is even more novel than Figs. 7.7 and 7.11 because tabulations of flow exergies for low pressure gases do not exist. Figure A.2 was constructed based on Figs. 7.7 and 7.11 by noting the general definition of flow exergy

$$\bar{e}_x(T, P) = \bar{h}(T, P) - T_0 \bar{s}(T, P) - [\bar{h}(T_0, P_0) - T_0 \bar{s}(T_0, P_0)] \tag{A.4}$$

Writing $\bar{e}_x^\circ$ for the $P = P_0$ version of eq. (A.4), we have

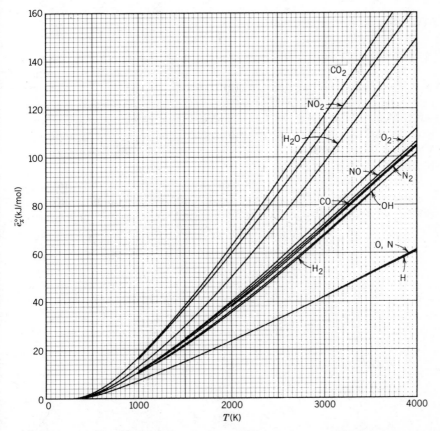

Figure A.2 Chart for the flow exergy of 12 gases at atmospheric pressure.

$$\bar{e}_x^{\,\circ}(T) = \Delta h(T) - T_0[\bar{s}^{\,\circ}(T) - \bar{s}^{\,\circ}(T_0)] \tag{A.5}$$

where $T_0 = 298.15\ \text{K}$ (25°C). The right-hand side of eq. (A.5) can be evaluated directly from Figs. 7.7 and 7.11 or from the tables on which those figures are based. The resulting quantity, $\bar{e}_x^{\,\circ}$, is reported in Fig. A.2. The observations listed in the paragraph above eq. (A.2) apply also to Fig. A.2. In addition, it is worth noting that each of the $\bar{e}_x^{\,\circ}$ curves has a minimum at $T = T_0$.

The flow exergy at a low pressure that is not necessarily equal to P_0 can be deduced from eqs. (A.3)–(A.5), that is, based on the ideal gas model,

$$\bar{e}_x(T, P) = \bar{e}_x^{\,\circ}(T) + \bar{R} T_0 \ln \frac{P}{P_0} \tag{A.6}$$

In conclusion, the flow exergy $\bar{e}_x(T, P)$ can be calculated by combining the reading of Fig. A.2 with the pressure correction formula (A.6).

TABLES

The tables assembled in this section contain the information necessary for determining the $P(v, T)$ relation and other properties of a large number of fluids encountered in engineering applications. For the $P(v, T)$ surface the reader can use the corresponding-states formulation presented in chapter 6 and Tables A.1 and A.2, or two empirical correlations whose coefficients are listed in Tables A.3 and A.4. The enthalpy, entropy, and fugacity can be calculated using Tables A.5–A.10, in the manner that was outlined in chapter 6.

Table A.11 and eq. (6.141) permit the calculation of the temperature dependence of the latent heat of vaporization of 20 common substances. Table A.11 reports also the critical-point and triple-point temperatures of these substances.

T_r \ P_r	0.010	0.050	0.100	0.200	0.400	0.600	0.800	1.000	1.200	1.500	2.000	3.000	5.000	7.000	10.000
0.30	0.0029	0.0145	0.0290	0.0579	0.1158	0.1737	0.2315	0.2892	0.3470	0.4335	0.5775	0.8648	1.4366	2.0048	2.8507
0.35	0.0026	0.0130	0.0261	0.0522	0.1043	0.1564	0.2084	0.2604	0.3123	0.3901	0.5195	0.7775	1.2902	1.7987	2.5539
0.40	0.0024	0.0119	0.0239	0.0477	0.0953	0.1429	0.1904	0.2379	0.2853	0.3563	0.4744	0.7095	1.1758	1.6373	2.3211
0.45	0.0022	0.0110	0.0221	0.0442	0.0882	0.1322	0.1762	0.2200	0.2638	0.3294	0.4384	0.6551	1.0841	1.5077	2.1338
0.50	0.0021	0.0103	0.0207	0.0413	0.0825	0.1236	0.1647	0.2056	0.2465	0.3077	0.4092	0.6110	1.0094	1.4017	1.9801
0.55	0.9804	0.0098	0.0195	0.0390	0.0778	0.1166	0.1553	0.1939	0.2323	0.2899	0.3853	0.5747	0.9475	1.3137	1.8520
0.60	0.9849	0.0093	0.0186	0.0371	0.0741	0.1109	0.1476	0.1842	0.2207	0.2753	0.3657	0.5446	0.8959	1.2398	1.7440
0.65	0.9881	0.9377	0.0178	0.0356	0.0710	0.1063	0.1415	0.1765	0.2113	0.2634	0.3495	0.5197	0.8526	1.1773	1.6519
0.70	0.9904	0.9504	0.8958	0.0344	0.0687	0.1027	0.1366	0.1703	0.2038	0.2538	0.3364	0.4991	0.8161	1.1241	1.5729
0.75	0.9922	0.9598	0.9165	0.0336	0.0670	0.1001	0.1330	0.1656	0.1981	0.2464	0.3260	0.4823	0.7854	1.0787	1.5047
0.80	0.9935	0.9669	0.9319	0.8539	0.0661	0.0985	0.1307	0.1626	0.1942	0.2411	0.3182	0.4690	0.7598	1.0400	1.4456
0.85	0.9946	0.9725	0.9436	0.8810	0.0661	0.0983	0.1301	0.1614	0.1924	0.2382	0.3132	0.4591	0.7388	1.0071	1.3943
0.90	0.9954	0.9768	0.9528	0.9015	0.7800	0.1006	0.1321	0.1630	0.1935	0.2383	0.3114	0.4527	0.7220	0.9793	1.3496
0.93	0.9959	0.9790	0.9573	0.9115	0.8059	0.6635	0.1359	0.1664	0.1963	0.2405	0.3122	0.4507	0.7138	0.9648	1.3257
0.95	0.9961	0.9803	0.9600	0.9174	0.8206	0.6967	0.1410	0.1705	0.1998	0.2432	0.3138	0.4501	0.7092	0.9561	1.3108
0.97	0.9963	0.9815	0.9625	0.9227	0.8338	0.7240	0.5580	0.1779	0.2055	0.2474	0.3164	0.4504	0.7052	0.9480	1.2968
0.98	0.9965	0.9821	0.9637	0.9253	0.8398	0.7360	0.5887	0.1844	0.2097	0.2503	0.3182	0.4508	0.7035	0.9442	1.2901
0.99	0.9966	0.9826	0.9648	0.9277	0.8455	0.7471	0.6138	0.1959	0.2154	0.2538	0.3204	0.4514	0.7018	0.9406	1.2835
1.00	0.9967	0.9832	0.9659	0.9300	0.8509	0.7574	0.6353	0.2901	0.2237	0.2583	0.3229	0.4522	0.7004	0.9372	1.2772
1.01	0.9968	0.9837	0.9669	0.9322	0.8561	0.7671	0.6542	0.4648	0.2370	0.2640	0.3260	0.4533	0.6991	0.9339	1.2710
1.02	0.9969	0.9842	0.9679	0.9343	0.8610	0.7761	0.6710	0.5146	0.2629	0.2715	0.3297	0.4547	0.6980	0.9307	1.2650
1.05	0.9971	0.9855	0.9707	0.9401	0.8743	0.8002	0.7130	0.6026	0.4437	0.3131	0.3452	0.4604	0.6956	0.9222	1.2481
1.10	0.9975	0.9874	0.9747	0.9485	0.8930	0.8323	0.7649	0.6880	0.5984	0.4580	0.3953	0.4770	0.6950	0.9110	1.2232
1.15	0.9978	0.9891	0.9780	0.9554	0.9081	0.8576	0.8032	0.7443	0.6803	0.5798	0.4760	0.5042	0.6987	0.9033	1.2021
1.20	0.9981	0.9904	0.9808	0.9611	0.9205	0.8779	0.8330	0.7858	0.7363	0.6605	0.5605	0.5425	0.7069	0.8990	1.1844
1.30	0.9985	0.9926	0.9852	0.9702	0.9396	0.9083	0.8764	0.8438	0.8111	0.7624	0.6908	0.6344	0.7358	0.8998	1.1580
1.40	0.9988	0.9942	0.9884	0.9768	0.9534	0.9298	0.9062	0.8827	0.8595	0.8256	0.7753	0.7202	0.7761	0.9112	1.1419
1.50	0.9991	0.9954	0.9909	0.9818	0.9636	0.9456	0.9278	0.9103	0.8933	0.8689	0.8328	0.7887	0.8200	0.9297	1.1339
1.60	0.9993	0.9964	0.9928	0.9856	0.9714	0.9575	0.9439	0.9308	0.9180	0.9000	0.8738	0.8410	0.8617	0.9518	1.1320
1.70	0.9994	0.9971	0.9943	0.9886	0.9775	0.9667	0.9563	0.9463	0.9367	0.9234	0.9043	0.8809	0.8984	0.9745	1.1343
1.80	0.9995	0.9977	0.9955	0.9910	0.9823	0.9739	0.9659	0.9583	0.9511	0.9413	0.9275	0.9118	0.9297	0.9961	1.1391
1.90	0.9996	0.9982	0.9964	0.9929	0.9861	0.9796	0.9735	0.9678	0.9624	0.9552	0.9456	0.9359	0.9557	1.0157	1.1452
2.00	0.9997	0.9986	0.9972	0.9944	0.9892	0.9842	0.9796	0.9754	0.9715	0.9664	0.9599	0.9550	0.9772	1.0328	1.1516
2.20	0.9998	0.9992	0.9983	0.9967	0.9937	0.9910	0.9886	0.9865	0.9847	0.9826	0.9806	0.9827	1.0094	1.0600	1.1635
2.40	0.9999	0.9996	0.9991	0.9983	0.9969	0.9957	0.9948	0.9941	0.9936	0.9935	0.9945	1.0011	1.0313	1.0793	1.1728
2.60	1.0000	0.9998	0.9997	0.9994	0.9991	0.9990	0.9990	0.9993	0.9998	1.0010	1.0040	1.0137	1.0463	1.0926	1.1792
2.80	1.0000	1.0000	1.0001	1.0002	1.0007	1.0013	1.0021	1.0031	1.0042	1.0063	1.0106	1.0223	1.0565	1.1016	1.1830
3.00	1.0000	1.0002	1.0004	1.0008	1.0018	1.0030	1.0043	1.0057	1.0074	1.0101	1.0153	1.0284	1.0635	1.1075	1.1848
3.50	1.0001	1.0004	1.0008	1.0017	1.0035	1.0055	1.0075	1.0097	1.0120	1.0156	1.0221	1.0368	1.0723	1.1138	1.1834
4.00	1.0001	1.0005	1.0010	1.0021	1.0043	1.0066	1.0090	1.0115	1.0140	1.0179	1.0249	1.0401	1.0747	1.1136	1.1773

TABLE A.2 Lee and Kesler's [8] Tables of $Z^{(1)}(T_r, P_r)$ Values

T_r \ P_r	0.010	0.050	0.100	0.200	0.400	0.600	0.800	1.000	1.200	1.500	2.000	3.000	5.000	7.000	10.000
0.30	-0.0008	-0.0040	-0.0081	-0.0161	-0.0323	-0.0484	-0.0645	-0.0806	-0.0966	-0.1207	-0.1608	-0.2407	-0.3996	-0.5572	-0.7915
0.35	-0.0009	-0.0046	-0.0093	-0.0185	-0.0370	-0.0554	-0.0738	-0.0921	-0.1105	-0.1379	-0.1834	-0.2738	-0.4523	-0.6279	-0.8863
0.40	-0.0010	-0.0048	-0.0095	-0.0190	-0.0380	-0.0570	-0.0758	-0.0946	-0.1134	-0.1414	-0.1879	-0.2799	-0.4603	-0.6365	-0.8936
0.45	-0.0009	-0.0047	-0.0094	-0.0187	-0.0374	-0.0560	-0.0745	-0.0929	-0.1113	-0.1387	-0.1840	-0.2734	-0.4475	-0.6162	-0.8606
0.50	-0.0009	-0.0045	-0.0090	-0.0181	-0.0360	-0.0539	-0.0716	-0.0893	-0.1069	-0.1330	-0.1762	-0.2611	-0.4253	-0.5831	-0.8099
0.55	-0.0314	-0.0043	-0.0086	-0.0172	-0.0343	-0.0513	-0.0682	-0.0849	-0.1015	-0.1263	-0.1669	-0.2465	-0.3991	-0.5446	-0.7521
0.60	-0.0205	-0.0041	-0.0082	-0.0164	-0.0326	-0.0487	-0.0646	-0.0803	-0.0960	-0.1192	-0.1572	-0.2312	-0.3718	-0.5047	-0.6928
0.65	-0.0137	-0.0772	-0.0078	-0.0156	-0.0309	-0.0461	-0.0611	-0.0759	-0.0906	-0.1122	-0.1476	-0.2160	-0.3447	-0.4653	-0.6346
0.70	-0.0093	-0.0507	-0.0744	-0.0148	-0.0294	-0.0438	-0.0579	-0.0718	-0.0855	-0.1057	-0.1385	-0.2013	-0.3184	-0.4270	-0.5785
0.75	-0.0064	-0.0339	-0.1161	-0.0143	-0.0282	-0.0417	-0.0550	-0.0681	-0.0808	-0.0996	-0.1298	-0.1872	-0.2929	-0.3901	-0.5250
0.80	-0.0044	-0.0228	-0.0487	-0.1160	-0.0272	-0.0401	-0.0526	-0.0648	-0.0767	-0.0940	-0.1217	-0.1736	-0.2682	-0.3545	-0.4740
0.85	-0.0029	-0.0152	-0.0319	-0.0715	-0.0268	-0.0391	-0.0509	-0.0622	-0.0731	-0.0888	-0.1138	-0.1602	-0.2439	-0.3201	-0.4254
0.90	-0.0019	-0.0099	-0.0205	-0.0442	-0.1118	-0.0396	-0.0503	-0.0604	-0.0701	-0.0840	-0.1059	-0.1463	-0.2195	-0.2862	-0.3788
0.93	-0.0015	-0.0075	-0.0154	-0.0326	-0.0763	-0.1662	-0.0514	-0.0602	-0.0687	-0.0810	-0.1007	-0.1374	-0.2045	-0.2661	-0.3516
0.95	-0.0012	-0.0062	-0.0126	-0.0262	-0.0589	-0.1110	-0.0540	-0.0607	-0.0678	-0.0788	-0.0967	-0.1310	-0.1943	-0.2526	-0.3339
0.97	-0.0010	-0.0050	-0.0101	-0.0208	-0.0450	-0.0770	-0.1647	-0.0623	-0.0669	-0.0759	-0.0921	-0.1240	-0.1837	-0.2391	-0.3163
0.98	-0.0009	-0.0044	-0.0090	-0.0184	-0.0390	-0.0641	-0.1100	-0.0641	-0.0661	-0.0740	-0.0893	-0.1202	-0.1783	-0.2322	-0.3075
0.99	-0.0008	-0.0039	-0.0079	-0.0161	-0.0335	-0.0531	-0.0796	-0.0680	-0.0646	-0.0715	-0.0861	-0.1162	-0.1728	-0.2254	-0.2989
1.00	-0.0007	-0.0034	-0.0069	-0.0140	-0.0285	-0.0435	-0.0588	-0.0879	-0.0609	-0.0678	-0.0824	-0.1118	-0.1672	-0.2185	-0.2901
1.01	-0.0006	-0.0030	-0.0060	-0.0120	-0.0240	-0.0351	-0.0429	-0.0223	-0.0473	-0.0621	-0.0778	-0.1072	-0.1615	-0.2116	-0.2816
1.02	-0.0005	-0.0026	-0.0051	-0.0102	-0.0198	-0.0277	-0.0303	-0.0062	0.0227	-0.0524	-0.0722	-0.1021	-0.1556	-0.2047	-0.2731
1.05	-0.0003	-0.0015	-0.0029	-0.0054	-0.0092	-0.0097	-0.0032	0.0220	0.1059	0.0451	-0.0432	-0.0838	-0.1370	-0.1835	-0.2476
1.10	-0.0000	0.0000	0.0001	0.0007	0.0038	0.0106	0.0236	0.0476	0.0897	0.1630	0.0698	-0.0373	-0.1021	-0.1469	-0.2056
1.15	0.0002	0.0011	0.0023	0.0052	0.0127	0.0237	0.0396	0.0625	0.0943	0.1548	0.1667	0.0332	-0.0611	-0.1084	-0.1642
1.20	0.0004	0.0019	0.0039	0.0084	0.0190	0.0326	0.0499	0.0719	0.0991	0.1477	0.1990	0.1095	-0.0141	-0.0678	-0.1231
1.30	0.0006	0.0030	0.0061	0.0125	0.0267	0.0429	0.0612	0.0819	0.1048	0.1420	0.1991	0.2079	0.0875	0.0176	-0.0423
1.40	0.0007	0.0036	0.0072	0.0147	0.0306	0.0477	0.0661	0.0857	0.1063	0.1383	0.1894	0.2397	0.1737	0.1008	0.0350
1.50	0.0008	0.0039	0.0078	0.0158	0.0323	0.0497	0.0677	0.0864	0.1055	0.1345	0.1806	0.2433	0.2309	0.1717	0.1058
1.60	0.0008	0.0040	0.0080	0.0162	0.0330	0.0501	0.0677	0.0855	0.1035	0.1303	0.1729	0.2381	0.2631	0.2255	0.1673
1.70	0.0008	0.0040	0.0081	0.0163	0.0329	0.0497	0.0667	0.0838	0.1008	0.1259	0.1658	0.2305	0.2788	0.2628	0.2179
1.80	0.0008	0.0040	0.0081	0.0162	0.0325	0.0488	0.0652	0.0816	0.0978	0.1216	0.1593	0.2224	0.2846	0.2871	0.2576
1.90	0.0008	0.0040	0.0079	0.0159	0.0318	0.0477	0.0635	0.0792	0.0947	0.1173	0.1532	0.2144	0.2848	0.3017	0.2876
2.00	0.0008	0.0039	0.0078	0.0155	0.0310	0.0464	0.0617	0.0767	0.0916	0.1133	0.1476	0.2069	0.2819	0.3097	0.3096
2.20	0.0007	0.0037	0.0074	0.0147	0.0293	0.0437	0.0579	0.0719	0.0857	0.1057	0.1374	0.1932	0.2720	0.3135	0.3355
2.40	0.0007	0.0035	0.0070	0.0139	0.0276	0.0411	0.0544	0.0675	0.0803	0.0989	0.1285	0.1812	0.2602	0.3089	0.3459
2.60	0.0007	0.0033	0.0066	0.0131	0.0260	0.0387	0.0512	0.0634	0.0754	0.0929	0.1207	0.1706	0.2484	0.3009	0.3475
2.80	0.0006	0.0031	0.0062	0.0124	0.0245	0.0365	0.0483	0.0598	0.0711	0.0876	0.1138	0.1613	0.2372	0.2915	0.3443
3.00	0.0006	0.0029	0.0059	0.0117	0.0232	0.0345	0.0456	0.0565	0.0672	0.0828	0.1076	0.1529	0.2268	0.2817	0.3385
3.50	0.0005	0.0026	0.0052	0.0103	0.0204	0.0303	0.0401	0.0497	0.0591	0.0728	0.0949	0.1356	0.2042	0.2584	0.3194
4.00	0.0005	0.0023	0.0046	0.0091	0.0182	0.0270	0.0357	0.0443	0.0527	0.0651	0.0849	0.1219	0.1857	0.2378	0.2994

TABLE A.3 Empirical Constants for Beattie and Bridgeman's Equation of State [9][a]

Gas	Formula	R (J/kg K)	A_0 (N m^4/kg^2)	$a \times 10^3$ (m^3/kg)	$B_0 \times 10^3$ (m^3/kg)	$b \times 10^3$ (m^3/kg)	$c \times 10^{-3}$ (m^3 K^3/kg)
Ammonia	NH_3	488.15	836.16	10.0012	2.0052	11.223	280.03
Air		286.95	157.12	0.66674	1.5919	-0.03801	1.498
Argon	Ar	208.14	81.99	0.58290	0.98451	0.0	1.499
n-Butane	C_4H_{10}	143.15	534.57	2.09388	4.2391	1.6225	60.267
Carbon dioxide	CO_2	188.93	262.07	1.62129	2.3811	1.6444	14.997
Carbon monoxide	CO	296.90	173.78	0.93457	1.8023	0.24660	1.498
Ethane	C_2H_6	276.62	659.89	1.95030	3.1283	0.63740	29.95
Ethylene	C_2H_4	296.58	793.50	1.77112	4.3370	1.2835	8.092
Ethyl ether	$C_4H_{10}O$	112.22	577.59	1.67748	6.1350	1.6137	4.499
Helium	He	2078.18	136.79	1.49581	3.5004	0.0	9.955
n-Heptane	C_7H_{16}	83.01	551.18	2.00398	7.0733	1.9159	39.92
Hydrogen	H_2	4115.47	4904.92	-2.50530	10.376	-21.582	0.24942
Methane	CH_4	518.60	897.96	1.15744	3.4854	-0.99013	8.003
Neon	Ne	411.98	52.89	1.08815	1.0207	0.0	50.10
Nitrogen	N_2	296.69	173.54	0.93394	1.8011	-0.24660	1.498
Nitrous oxide	N_2O	188.83	261.83	1.62005	2.3799	1.6431	14.98
Oxygen	O_2	259.79	147.56	0.80097	1.4452	-0.13148	1.498
Propane	C_3H_8	188.67	622.24	1.66125	4.1082	0.97452	27.19

[a]After Cravalho and Smith [10], with permission from the authors.

TABLE A.4 Empirical Constants for the Benedict–Webb–Rubin Equation of State [11][a]

Gas		$A_0 \times 10^{-2}$ (N m^4/kg^2)	$B_0 \times 10^3$ (m^3/kg)	$C_0 \times 10^{-7}$ (N m^4 K^2/kg^2)	a (N m^7/kg^2)	$b \times 10^5$ (m^6/kg^2)	$c \times 10^{-5}$ (N m^7 K^2/kg^3)	$\alpha \times 10^9$ (m^9/kg^3)	$\gamma \times 10^5$ (m^6/kg^2)
Methane	CH_4	7.31195	2.65735	0.889635	1.21466	1.31523	0.62577	30.1853	2.33469
Ethylene	C_2H_4	4.30550	1.98649	1.69071	1.19119	1.09451	0.97139	8.08173	1.17469
Ethane	C_2H_6	4.66269	2.08914	2.01509	1.28892	1.23191	1.22361	8.97220	1.30701
Propylene	C_3H_6	3.50217	2.02308	2.51642	1.05482	1.05806	1.39829	6.13014	1.03453
Propane	C_3H_8	3.58575	2.20855	2.65194	1.12224	1.15892	1.52759	7.09776	1.13317
i-Butane	C_4H_{10}	3.07308	2.36826	2.55256	1.00195	1.25806	1.47891	5.48279	1.00799
i-Butylene	C_4H_8	2.88571	2.06958	2.98871	0.97316	1.10774	1.58056	5.16963	0.941616
n-Butane	C_4H_{10}	3.02865	2.14127	2.98168	0.97334	1.18582	1.63610	5.62184	1.00799
i-Pentane	C_5H_{12}	2.49391	2.22006	3.40357	1.01546	1.28545	1.87887	4.53682	0.890805
n-Pentane	C_5H_{12}	2.37376	2.17426	4.13424	1.10159	1.28545	2.22807	4.83038	0.913893
n-Hexane	C_6H_{14}	1.97242	2.06498	4.53487	1.12913	1.47181	2.40013	4.40244	0.899353
n-Heptane	C_7H_{16}	1.77041	1.98756	4.79543	1.04602	1.51575	2.49275	4.33982	0.897754

[a]After Cravalho and Smith [10], with permission from the authors.

T_r \ P_r	0.010	0.050	0.100	0.200	0.400	0.600	0.800	1.000	1.200	1.500	2.000	3.000	5.000	7.000	10.000
0.30	6.045	6.043	6.040	6.034	6.022	6.011	5.999	5.987	5.975	5.957	5.927	5.868	5.748	5.628	5.446
0.35	5.906	5.904	5.901	5.895	5.882	5.870	5.858	5.845	5.833	5.814	5.783	5.721	5.595	5.469	5.278
0.40	5.763	5.761	5.757	5.751	5.738	5.726	5.713	5.700	5.687	5.668	5.636	5.572	5.442	5.311	5.113
0.45	5.615	5.612	5.609	5.603	5.590	5.577	5.564	5.551	5.538	5.519	5.486	5.421	5.288	5.154	4.950
0.50	5.465	5.463	5.459	5.453	5.440	5.427	5.414	5.401	5.388	5.369	5.336	5.270	5.135	4.999	4.791
0.55	0.032	5.312	5.309	5.303	5.290	5.278	5.265	5.252	5.239	5.220	5.187	5.121	4.986	4.849	4.638
0.60	0.027	5.162	5.159	5.153	5.141	5.129	5.116	5.104	5.091	5.073	5.041	4.976	4.842	4.704	4.492
0.65	0.023	0.118	5.008	5.002	4.991	4.980	4.968	4.956	4.945	4.927	4.896	4.833	4.702	4.565	4.353
0.70	0.020	0.101	0.213	4.848	4.838	4.828	4.818	4.808	4.797	4.781	4.752	4.693	4.566	4.432	4.221
0.75	0.017	0.088	0.183	4.687	4.679	4.672	4.664	4.655	4.646	4.632	4.607	4.554	4.434	4.303	4.095
0.80	0.015	0.078	0.160	0.345	4.507	4.504	4.499	4.494	4.488	4.478	4.459	4.413	4.303	4.178	3.974
0.85	0.014	0.069	0.141	0.300	4.309	4.313	4.316	4.316	4.316	4.312	4.302	4.269	4.173	4.056	3.857
0.90	0.012	0.062	0.126	0.264	0.596	4.074	4.094	4.108	4.118	4.127	4.132	4.119	4.043	3.935	3.744
0.93	0.011	0.058	0.118	0.246	0.545	0.960	3.920	3.953	3.976	4.000	4.020	4.024	3.963	3.863	3.678
0.95	0.011	0.056	0.113	0.235	0.516	0.885	3.763	3.825	3.865	3.904	3.940	3.958	3.910	3.815	3.634
0.97	0.011	0.054	0.109	0.225	0.490	0.824	1.356	3.658	3.732	3.796	3.853	3.890	3.856	3.767	3.591
0.98	0.010	0.053	0.107	0.221	0.478	0.797	1.273	3.544	3.652	3.736	3.806	3.854	3.829	3.743	3.569
0.99	0.010	0.052	0.105	0.216	0.466	0.773	1.206	3.376	3.558	3.670	3.758	3.818	3.801	3.719	3.548
1.00	0.010	0.051	0.103	0.212	0.455	0.750	1.151	2.584	3.441	3.598	3.706	3.782	3.774	3.695	3.526
1.01	0.010	0.050	0.101	0.208	0.445	0.728	1.102	1.796	3.283	3.516	3.652	3.744	3.746	3.671	3.505
1.02	0.010	0.049	0.099	0.203	0.434	0.708	1.060	1.627	3.039	3.422	3.595	3.705	3.718	3.647	3.484
1.05	0.009	0.046	0.094	0.192	0.407	0.654	0.955	1.359	2.034	3.030	3.398	3.583	3.632	3.575	3.420
1.10	0.008	0.042	0.086	0.175	0.367	0.581	0.827	1.120	1.487	2.203	2.965	3.353	3.484	3.453	3.315
1.15	0.008	0.039	0.079	0.160	0.334	0.523	0.732	0.968	1.239	1.719	2.479	3.091	3.329	3.329	3.211
1.20	0.007	0.036	0.073	0.148	0.305	0.474	0.657	0.857	1.076	1.443	2.079	2.807	3.166	3.202	3.107
1.30	0.006	0.031	0.063	0.127	0.259	0.399	0.545	0.698	0.860	1.116	1.560	2.274	2.825	2.942	2.899
1.40	0.005	0.027	0.055	0.110	0.224	0.341	0.463	0.588	0.716	0.915	1.253	1.857	2.486	2.679	2.692
1.50	0.005	0.024	0.048	0.097	0.196	0.297	0.400	0.505	0.611	0.774	1.046	1.549	2.175	2.421	2.486
1.60	0.004	0.021	0.043	0.086	0.173	0.261	0.350	0.440	0.531	0.667	0.894	1.318	1.904	2.177	2.285
1.70	0.004	0.019	0.038	0.076	0.153	0.231	0.309	0.387	0.466	0.583	0.777	1.139	1.672	1.953	2.091
1.80	0.003	0.017	0.034	0.068	0.137	0.206	0.275	0.344	0.413	0.515	0.683	0.996	1.476	1.751	1.908
1.90	0.003	0.015	0.031	0.062	0.123	0.185	0.246	0.307	0.368	0.458	0.606	0.880	1.309	1.571	1.736
2.00	0.003	0.014	0.028	0.056	0.111	0.167	0.222	0.276	0.330	0.411	0.541	0.782	1.167	1.411	1.577
2.20	0.002	0.012	0.023	0.046	0.092	0.137	0.182	0.226	0.269	0.334	0.437	0.629	0.937	1.143	1.295
2.40	0.002	0.010	0.019	0.038	0.076	0.114	0.150	0.187	0.222	0.275	0.359	0.513	0.761	0.929	1.058
2.60	0.002	0.008	0.016	0.032	0.064	0.095	0.125	0.155	0.185	0.228	0.297	0.422	0.621	0.756	0.858
2.80	0.001	0.007	0.014	0.027	0.054	0.080	0.105	0.130	0.154	0.190	0.246	0.348	0.508	0.614	0.689
3.00	0.001	0.006	0.011	0.023	0.045	0.067	0.088	0.109	0.129	0.159	0.205	0.288	0.415	0.495	0.545
3.50	0.001	0.004	0.007	0.015	0.029	0.043	0.056	0.069	0.081	0.099	0.127	0.174	0.239	0.270	0.264
4.00	0.000	0.002	0.005	0.009	0.017	0.026	0.033	0.041	0.048	0.058	0.072	0.095	0.116	0.110	0.061

TABLE A.6 Lee and Kesler's [8] Values of $(h^* - h)^{(1)}/RT_c$

T_r \ P_r	0.010	0.050	0.100	0.200	0.400	0.600	0.800	1.000	1.200	1.500	2.000	3.000	5.000	7.000	10.000
0.30	11.098	11.096	11.095	11.091	11.083	11.076	11.069	11.062	11.055	11.044	11.027	10.992	10.935	10.872	10.781
0.35	10.656	10.655	10.654	10.653	10.650	10.646	10.643	10.640	10.637	10.632	10.624	10.609	10.581	10.554	10.529
0.40	10.121	10.121	10.121	10.120	10.121	10.121	10.121	10.122	10.122	10.121	10.122	10.123	10.128	10.135	10.150
0.45	9.515	9.515	9.516	9.517	9.519	9.521	9.523	9.525	9.527	9.531	9.537	9.549	9.576	9.611	9.663
0.50	8.868	8.869	8.870	8.872	8.876	8.880	8.884	8.888	8.892	8.899	8.909	8.932	8.978	9.030	9.111
0.55	0.080	8.211	8.212	8.215	8.221	8.226	8.232	8.238	8.243	8.252	8.267	8.298	8.360	8.425	8.531
0.60	0.059	7.568	7.570	7.573	7.579	7.585	7.591	7.596	7.603	7.614	7.632	7.669	7.745	7.824	7.950
0.65	0.045	0.247	6.949	6.952	6.959	6.966	6.973	6.980	6.987	6.997	7.017	7.059	7.147	7.239	7.381
0.70	0.034	0.185	0.415	6.360	6.367	6.373	6.381	6.388	6.395	6.407	6.429	6.475	6.574	6.677	6.837
0.75	0.027	0.142	0.306	5.796	5.802	5.809	5.816	5.824	5.832	5.845	5.868	5.918	6.027	6.142	6.318
0.80	0.021	0.110	0.234	0.542	5.266	5.271	5.278	5.285	5.293	5.306	5.330	5.385	5.506	5.632	5.824
0.85	0.017	0.087	0.182	0.401	4.753	4.754	4.758	4.763	4.771	4.784	4.810	4.872	5.008	5.149	5.358
0.90	0.014	0.070	0.144	0.308	0.751	4.254	4.248	4.249	4.255	4.268	4.298	4.371	4.530	4.688	4.916
0.93	0.012	0.061	0.126	0.265	0.612	1.236	3.942	3.934	3.937	3.951	3.987	4.073	4.251	4.422	4.662
0.95	0.011	0.056	0.115	0.241	0.542	0.994	3.737	3.712	3.713	3.730	3.773	3.873	4.068	4.248	4.497
0.97	0.010	0.052	0.105	0.219	0.483	0.837	1.616	3.470	3.467	3.492	3.551	3.670	3.885	4.077	4.336
0.98	0.010	0.050	0.101	0.209	0.457	0.776	1.324	3.332	3.327	3.363	3.434	3.568	3.795	3.992	4.257
0.99	0.009	0.048	0.097	0.200	0.433	0.722	1.154	3.164	3.164	3.223	3.313	3.464	3.705	3.909	4.178
1.00	0.009	0.046	0.093	0.191	0.410	0.675	1.034	2.471	2.952	3.065	3.186	3.358	3.615	3.825	4.100
1.01	0.009	0.044	0.089	0.183	0.389	0.632	0.940	1.375	2.595	2.880	3.051	3.251	3.525	3.742	4.023
1.02	0.008	0.042	0.085	0.175	0.370	0.594	0.863	1.180	1.723	2.650	2.906	3.142	3.435	3.661	3.947
1.05	0.007	0.037	0.075	0.153	0.318	0.498	0.691	0.877	0.878	1.496	2.381	2.800	3.167	3.418	3.722
1.10	0.006	0.030	0.061	0.123	0.251	0.381	0.507	0.617	0.673	0.617	1.261	2.167	2.720	3.023	3.362
1.15	0.005	0.025	0.050	0.099	0.199	0.296	0.385	0.459	0.503	0.487	0.604	1.497	2.275	2.641	3.019
1.20	0.004	0.020	0.040	0.080	0.158	0.232	0.297	0.349	0.381	0.381	0.361	0.934	1.840	2.273	2.692
1.30	0.003	0.013	0.026	0.052	0.100	0.142	0.177	0.203	0.218	0.218	0.178	0.300	1.066	1.592	2.086
1.40	0.002	0.008	0.016	0.032	0.060	0.083	0.100	0.111	0.115	0.108	0.070	0.044	0.504	1.012	1.547
1.50	0.001	0.005	0.009	0.018	0.032	0.042	0.048	0.049	0.046	0.032	-0.008	-0.078	0.142	0.556	1.080
1.60	0.000	0.002	0.004	0.007	0.012	0.013	0.011	0.005	-0.004	-0.023	-0.065	-0.151	-0.082	0.217	0.689
1.70	0.000	0.000	0.000	-0.000	-0.003	-0.009	-0.017	-0.027	-0.040	-0.063	-0.109	-0.202	-0.223	-0.028	0.369
1.80	-0.000	-0.001	-0.003	-0.006	-0.015	-0.025	-0.037	-0.051	-0.067	-0.094	-0.143	-0.241	-0.317	-0.203	0.112
1.90	-0.001	-0.003	-0.005	-0.011	-0.023	-0.037	-0.053	-0.070	-0.088	-0.117	-0.169	-0.271	-0.381	-0.330	-0.092
2.00	-0.001	-0.003	-0.007	-0.015	-0.030	-0.047	-0.065	-0.085	-0.105	-0.136	-0.190	-0.295	-0.428	-0.424	-0.255
2.20	-0.001	-0.005	-0.010	-0.020	-0.040	-0.062	-0.083	-0.106	-0.128	-0.163	-0.221	-0.331	-0.493	-0.551	-0.489
2.40	-0.001	-0.006	-0.012	-0.023	-0.047	-0.071	-0.095	-0.120	-0.144	-0.181	-0.242	-0.356	-0.535	-0.631	-0.645
2.60	-0.001	-0.006	-0.013	-0.026	-0.052	-0.078	-0.104	-0.130	-0.156	-0.194	-0.257	-0.376	-0.567	-0.687	-0.754
2.80	-0.001	-0.007	-0.014	-0.028	-0.055	-0.082	-0.110	-0.137	-0.164	-0.204	-0.269	-0.391	-0.591	-0.729	-0.836
3.00	-0.001	-0.007	-0.014	-0.029	-0.058	-0.086	-0.114	-0.142	-0.170	-0.211	-0.278	-0.403	-0.611	-0.763	-0.899

T_r \ P_r	0.010	0.050	0.100	0.200	0.400	0.600	0.800	1.000	1.200	1.500	2.000	3.000	5.000	7.000	10.000
0.30	11.614	10.008	9.319	8.635	7.961	7.574	7.304	7.099	6.935	6.740	6.497	6.182	5.847	5.683	5.578
0.35	11.185	9.579	8.890	8.205	7.529	7.140	6.869	6.663	6.497	6.299	6.052	5.728	5.376	5.194	5.060
0.40	10.802	9.196	8.506	7.821	7.144	6.755	6.483	6.275	6.109	5.909	5.660	5.330	4.967	4.772	4.619
0.45	10.453	8.847	8.157	7.472	6.794	6.404	6.132	5.924	5.757	5.557	5.306	4.974	4.603	4.401	4.234
0.50	10.137	8.531	7.841	7.156	6.479	6.089	5.816	5.608	5.441	5.240	4.989	4.656	4.282	4.074	3.899
0.55	0.038	8.245	7.555	6.870	6.193	5.803	5.531	5.324	5.157	4.956	4.706	4.373	3.998	3.788	3.607
0.60	0.029	7.983	7.294	6.610	5.933	5.544	5.273	5.066	4.900	4.700	4.451	4.120	3.747	3.537	3.353
0.65	0.023	0.122	7.052	6.368	5.694	5.306	5.036	4.830	4.665	4.467	4.220	3.892	3.523	3.315	3.131
0.70	0.018	0.096	0.206	6.140	5.467	5.082	4.814	4.610	4.446	4.250	4.007	3.684	3.322	3.117	2.935
0.75	0.015	0.078	0.164	5.917	5.248	4.866	4.600	4.399	4.238	4.045	3.807	3.491	3.138	2.939	2.761
0.80	0.013	0.064	0.134	0.294	5.026	4.649	4.388	4.191	4.034	3.846	3.615	3.310	2.970	2.777	2.605
0.85	0.011	0.054	0.111	0.239	4.785	4.418	4.166	3.976	3.825	3.646	3.425	3.135	2.812	2.629	2.463
0.90	0.009	0.046	0.094	0.199	0.463	4.145	3.912	3.738	3.599	3.434	3.231	2.964	2.663	2.491	2.334
0.93	0.008	0.042	0.085	0.179	0.408	0.750	3.723	3.569	3.444	3.295	3.108	2.860	2.577	2.412	2.262
0.95	0.008	0.039	0.080	0.168	0.377	0.671	3.556	3.433	3.326	3.193	3.023	2.790	2.520	2.362	2.215
0.97	0.007	0.037	0.075	0.157	0.350	0.607	1.056	3.259	3.188	3.081	2.932	2.719	2.463	2.312	2.170
0.98	0.007	0.036	0.073	0.153	0.337	0.580	0.971	3.142	3.106	3.019	2.884	2.682	2.436	2.287	2.148
0.99	0.007	0.035	0.071	0.148	0.326	0.555	0.903	2.972	3.010	2.953	2.835	2.646	2.408	2.263	2.126
1.00	0.007	0.034	0.069	0.144	0.315	0.532	0.847	2.178	2.893	2.879	2.784	2.609	2.380	2.239	2.105
1.01	0.007	0.033	0.067	0.139	0.304	0.510	0.799	1.391	2.736	2.798	2.730	2.571	2.352	2.215	2.083
1.02	0.006	0.032	0.065	0.135	0.294	0.491	0.757	1.225	2.495	2.706	2.673	2.533	2.325	2.191	2.062
1.05	0.006	0.030	0.060	0.124	0.267	0.439	0.656	0.965	1.523	2.328	2.483	2.415	2.242	2.121	2.001
1.10	0.005	0.026	0.053	0.108	0.230	0.371	0.537	0.742	1.012	1.557	2.081	2.202	2.104	2.007	1.903
1.15	0.005	0.023	0.047	0.096	0.201	0.319	0.452	0.607	0.790	1.126	1.649	1.968	1.966	1.897	1.810
1.20	0.004	0.021	0.042	0.085	0.177	0.277	0.389	0.512	0.651	0.890	1.308	1.727	1.827	1.789	1.722
1.30	0.003	0.017	0.033	0.068	0.140	0.217	0.298	0.385	0.478	0.628	0.891	1.299	1.554	1.581	1.556
1.40	0.003	0.014	0.027	0.056	0.114	0.174	0.237	0.303	0.372	0.478	0.663	0.990	1.303	1.386	1.402
1.50	0.002	0.011	0.023	0.046	0.094	0.143	0.194	0.246	0.299	0.381	0.520	0.777	1.088	1.208	1.260
1.60	0.002	0.010	0.019	0.039	0.079	0.120	0.162	0.204	0.247	0.312	0.421	0.628	0.913	1.050	1.130
1.70	0.002	0.008	0.017	0.033	0.067	0.102	0.137	0.172	0.208	0.261	0.350	0.519	0.773	0.915	1.013
1.80	0.001	0.007	0.014	0.029	0.058	0.088	0.117	0.147	0.177	0.222	0.296	0.438	0.661	0.799	0.908
1.90	0.001	0.006	0.013	0.025	0.051	0.076	0.102	0.127	0.153	0.191	0.255	0.375	0.570	0.702	0.815
2.00	0.001	0.006	0.011	0.022	0.044	0.067	0.089	0.111	0.134	0.167	0.221	0.325	0.497	0.620	0.733
2.20	0.001	0.004	0.009	0.018	0.035	0.053	0.070	0.087	0.105	0.130	0.172	0.251	0.388	0.492	0.599
2.40	0.001	0.004	0.007	0.014	0.028	0.042	0.056	0.070	0.084	0.104	0.138	0.201	0.311	0.399	0.496
2.60	0.001	0.003	0.006	0.012	0.023	0.035	0.046	0.058	0.069	0.086	0.113	0.164	0.255	0.329	0.416
2.80	0.000	0.002	0.005	0.010	0.020	0.029	0.039	0.048	0.058	0.072	0.094	0.137	0.213	0.277	0.353
3.00	0.000	0.002	0.004	0.008	0.017	0.025	0.033	0.041	0.049	0.061	0.080	0.116	0.181	0.236	0.303
3.50	0.000	0.001	0.003	0.006	0.012	0.017	0.023	0.029	0.034	0.042	0.056	0.081	0.126	0.166	0.216
4.00	0.000	0.001	0.002	0.004	0.009	0.013	0.017	0.021	0.025	0.031	0.041	0.059	0.093	0.123	0.162

TABLE A.8 Lee and Kesler's [8] Values of $(s^* - s)^{(1)}/R$

T_r \ P_r	0.010	0.050	0.100	0.200	0.400	0.600	0.800	1.000	1.200	1.500	2.000	3.000	5.000	7.000	10.000
0.30	16.782	16.774	16.764	16.744	16.705	16.665	16.626	16.586	16.547	16.488	16.390	16.195	15.837	15.468	14.925
0.35	15.413	15.408	15.401	15.387	15.359	15.333	15.305	15.278	15.251	15.211	15.144	15.011	14.751	14.496	14.153
0.40	13.990	13.986	13.981	13.972	13.953	13.934	13.915	13.896	13.877	13.849	13.803	13.714	13.541	13.376	13.144
0.45	12.564	12.561	12.558	12.551	12.537	12.523	12.509	12.496	12.482	12.462	12.430	12.367	12.248	12.145	11.999
0.50	11.202	11.200	11.197	11.192	11.182	11.172	11.162	11.153	11.143	11.129	11.107	11.063	10.985	10.920	10.836
0.55	0.115	9.948	9.946	9.942	9.935	9.928	9.921	9.914	9.907	9.897	9.882	9.853	9.806	9.769	9.732
0.60	0.078	8.828	8.826	8.823	8.817	8.811	8.806	8.799	8.794	8.787	8.777	8.760	8.736	8.723	8.720
0.65	0.055	0.309	7.832	7.829	7.824	7.819	7.815	7.810	7.807	7.801	7.794	7.784	7.779	7.785	7.811
0.70	0.040	0.216	0.491	6.951	6.945	6.941	6.937	6.933	6.930	6.926	6.922	6.919	6.929	6.952	7.002
0.75	0.029	0.156	0.340	6.173	6.167	6.162	6.158	6.155	6.152	6.149	6.147	6.149	6.174	6.213	6.285
0.80	0.022	0.116	0.246	0.578	5.475	5.468	5.462	5.458	5.455	5.453	5.452	5.461	5.501	5.555	5.648
0.85	0.017	0.088	0.183	0.408	4.853	4.841	4.832	4.826	4.822	4.820	4.822	4.839	4.898	4.969	5.082
0.90	0.013	0.068	0.140	0.301	0.744	4.269	4.249	4.238	4.232	4.230	4.236	4.267	4.351	4.442	4.578
0.93	0.011	0.058	0.120	0.254	0.593	1.219	3.914	3.894	3.885	3.884	3.896	3.941	4.046	4.151	4.300
0.95	0.010	0.053	0.109	0.228	0.517	0.961	3.697	3.658	3.647	3.648	3.669	3.728	3.851	3.966	4.125
0.97	0.010	0.048	0.099	0.206	0.456	0.797	1.570	3.406	3.391	3.401	3.437	3.517	3.661	3.788	3.957
0.98	0.009	0.046	0.094	0.196	0.429	0.734	1.270	3.264	3.247	3.268	3.318	3.412	3.569	3.701	3.875
0.99	0.009	0.044	0.090	0.186	0.405	0.680	1.098	3.093	3.082	3.126	3.195	3.306	3.477	3.616	3.796
1.00	0.008	0.042	0.086	0.177	0.382	0.632	0.977	2.399	2.868	2.967	3.067	3.200	3.387	3.532	3.717
1.01	0.008	0.040	0.082	0.169	0.361	0.590	0.883	1.306	2.513	2.784	2.933	3.094	3.297	3.450	3.640
1.02	0.008	0.039	0.078	0.161	0.342	0.552	0.807	1.113	1.655	2.557	2.790	2.986	3.209	3.369	3.565
1.05	0.007	0.034	0.069	0.140	0.292	0.460	0.642	0.820	0.831	1.443	2.283	2.655	2.949	3.134	3.348
1.10	0.005	0.028	0.055	0.112	0.229	0.350	0.470	0.577	0.640	0.618	1.241	2.067	2.534	2.767	3.013
1.15	0.005	0.023	0.045	0.091	0.183	0.275	0.361	0.437	0.489	0.502	0.654	1.471	2.138	2.428	2.708
1.20	0.004	0.019	0.037	0.075	0.149	0.220	0.286	0.343	0.385	0.412	0.447	0.991	1.767	2.115	2.430
1.30	0.003	0.013	0.026	0.052	0.102	0.148	0.190	0.226	0.254	0.282	0.300	0.481	1.147	1.569	1.944
1.40	0.002	0.010	0.019	0.037	0.072	0.104	0.133	0.158	0.178	0.200	0.220	0.290	0.730	1.138	1.544
1.50	0.001	0.007	0.014	0.027	0.053	0.076	0.097	0.115	0.130	0.147	0.166	0.206	0.479	0.823	1.222
1.60	0.001	0.005	0.011	0.021	0.040	0.057	0.073	0.086	0.098	0.112	0.129	0.159	0.334	0.604	0.969
1.70	0.001	0.004	0.008	0.016	0.031	0.044	0.056	0.067	0.076	0.087	0.102	0.127	0.248	0.456	0.775
1.80	0.001	0.003	0.006	0.013	0.024	0.035	0.044	0.053	0.060	0.070	0.083	0.105	0.195	0.355	0.628
1.90	0.001	0.003	0.005	0.010	0.019	0.028	0.036	0.043	0.049	0.057	0.069	0.089	0.160	0.286	0.518
2.00	0.000	0.002	0.004	0.008	0.016	0.023	0.029	0.035	0.040	0.048	0.058	0.077	0.136	0.238	0.434
2.20	0.000	0.002	0.003	0.006	0.011	0.016	0.021	0.025	0.029	0.035	0.043	0.060	0.105	0.178	0.322
2.40	0.000	0.001	0.002	0.004	0.008	0.012	0.015	0.019	0.022	0.027	0.034	0.048	0.086	0.143	0.254
2.60	0.000	0.001	0.002	0.003	0.006	0.009	0.012	0.015	0.018	0.021	0.028	0.041	0.074	0.120	0.210
2.80	0.000	0.001	0.001	0.003	0.005	0.008	0.010	0.012	0.014	0.018	0.023	0.035	0.065	0.104	0.188
3.00	0.000	0.001	0.001	0.002	0.004	0.006	0.008	0.010	0.012	0.015	0.020	0.031	0.058	0.093	0.158

TABLE A.9 Lee and Kesler's [8] Values of $[\log_{10}(f/P)]^{(0)}$

$T_r \backslash P_r$	0.010	0.050	0.100	0.200	0.400	0.600	0.800	1.000	1.200	1.500	2.000	3.000	5.000	7.000	10.000
0.30	-3.708	-4.402	-4.696	-4.985	-5.261	-5.412	-5.512	-5.584	-5.638	-5.697	-5.759	-5.810	-5.782	-5.679	-5.461
0.35	-2.471	-3.166	-3.461	-3.751	-4.029	-4.183	-4.285	-4.359	-4.416	-4.479	-4.547	-4.611	-4.608	-4.530	-4.352
0.40	-1.566	-2.261	-2.557	-2.848	-3.128	-3.283	-3.387	-3.463	-3.522	-3.588	-3.661	-3.735	-3.752	-3.694	-3.545
0.45	-0.879	-1.575	-1.871	-2.162	-2.444	-2.601	-2.707	-2.785	-2.845	-2.913	-2.990	-3.071	-3.104	-3.063	-2.938
0.50	-0.344	-1.040	-1.336	-1.628	-1.912	-2.070	-2.177	-2.256	-2.317	-2.387	-2.468	-2.555	-2.601	-2.572	-2.468
0.55	-0.008	-0.614	-0.911	-1.204	-1.488	-1.647	-1.755	-1.835	-1.897	-1.969	-2.052	-2.145	-2.201	-2.183	-2.096
0.60	-0.007	-0.269	-0.566	-0.859	-1.144	-1.304	-1.413	-1.494	-1.557	-1.630	-1.715	-1.812	-1.878	-1.869	-1.795
0.65	-0.005	-0.026	-0.283	-0.576	-0.862	-1.023	-1.132	-1.214	-1.278	-1.352	-1.439	-1.539	-1.612	-1.611	-1.549
0.70	-0.004	-0.021	-0.043	-0.341	-0.627	-0.789	-0.899	-0.981	-1.045	-1.120	-1.208	-1.312	-1.391	-1.396	-1.344
0.75	-0.003	-0.017	-0.035	-0.144	-0.430	-0.592	-0.703	-0.785	-0.850	-0.925	-1.015	-1.121	-1.204	-1.215	-1.172
0.80	-0.003	-0.014	-0.029	-0.059	-0.264	-0.426	-0.537	-0.619	-0.685	-0.760	-0.851	-0.958	-1.046	-1.062	-1.026
0.85	-0.002	-0.012	-0.024	-0.049	-0.123	-0.285	-0.396	-0.479	-0.544	-0.620	-0.711	-0.819	-0.911	-0.930	-0.901
0.90	-0.002	-0.010	-0.020	-0.041	-0.086	-0.166	-0.276	-0.359	-0.424	-0.500	-0.591	-0.700	-0.794	-0.817	-0.793
0.93	-0.002	-0.009	-0.018	-0.037	-0.077	-0.122	-0.214	-0.296	-0.361	-0.437	-0.527	-0.637	-0.732	-0.756	-0.735
0.95	-0.002	-0.008	-0.017	-0.035	-0.072	-0.113	-0.176	-0.258	-0.322	-0.398	-0.488	-0.598	-0.693	-0.719	-0.699
0.97	-0.002	-0.008	-0.016	-0.033	-0.067	-0.105	-0.148	-0.223	-0.287	-0.362	-0.452	-0.561	-0.657	-0.683	-0.665
0.98	-0.002	-0.008	-0.016	-0.032	-0.065	-0.101	-0.142	-0.206	-0.270	-0.344	-0.434	-0.543	-0.639	-0.666	-0.649
0.99	-0.001	-0.007	-0.015	-0.031	-0.063	-0.098	-0.137	-0.191	-0.254	-0.328	-0.417	-0.526	-0.622	-0.649	-0.633
1.00	-0.001	-0.007	-0.015	-0.030	-0.061	-0.095	-0.132	-0.176	-0.238	-0.312	-0.401	-0.509	-0.605	-0.633	-0.617
1.01	-0.001	-0.007	-0.014	-0.029	-0.059	-0.091	-0.127	-0.168	-0.224	-0.297	-0.385	-0.493	-0.589	-0.617	-0.602
1.02	-0.001	-0.007	-0.014	-0.028	-0.057	-0.088	-0.122	-0.161	-0.210	-0.282	-0.370	-0.477	-0.573	-0.601	-0.588
1.05	-0.001	-0.006	-0.013	-0.025	-0.052	-0.080	-0.110	-0.143	-0.180	-0.242	-0.327	-0.433	-0.529	-0.557	-0.546
1.10	-0.001	-0.005	-0.011	-0.022	-0.045	-0.069	-0.093	-0.120	-0.148	-0.193	-0.267	-0.368	-0.462	-0.491	-0.482
1.15	-0.001	-0.005	-0.009	-0.019	-0.039	-0.059	-0.080	-0.102	-0.125	-0.160	-0.220	-0.312	-0.403	-0.433	-0.426
1.20	-0.001	-0.004	-0.008	-0.017	-0.034	-0.051	-0.069	-0.088	-0.106	-0.135	-0.184	-0.266	-0.352	-0.382	-0.377
1.30	-0.001	-0.003	-0.006	-0.013	-0.026	-0.039	-0.052	-0.066	-0.080	-0.100	-0.134	-0.195	-0.269	-0.296	-0.293
1.40	-0.001	-0.003	-0.005	-0.010	-0.020	-0.030	-0.040	-0.051	-0.061	-0.076	-0.101	-0.146	-0.205	-0.229	-0.226
1.50	-0.001	-0.002	-0.004	-0.008	-0.016	-0.024	-0.032	-0.039	-0.047	-0.059	-0.077	-0.111	-0.157	-0.176	-0.173
1.60	-0.000	-0.002	-0.003	-0.006	-0.012	-0.019	-0.025	-0.031	-0.037	-0.046	-0.060	-0.085	-0.120	-0.135	-0.129
1.70	-0.000	-0.001	-0.002	-0.005	-0.010	-0.015	-0.020	-0.024	-0.029	-0.036	-0.046	-0.065	-0.092	-0.102	-0.094
1.80	-0.000	-0.001	-0.002	-0.004	-0.008	-0.012	-0.015	-0.019	-0.023	-0.028	-0.036	-0.050	-0.069	-0.075	-0.066
1.90	-0.000	-0.001	-0.002	-0.003	-0.006	-0.009	-0.012	-0.015	-0.018	-0.022	-0.028	-0.038	-0.052	-0.054	-0.043
2.00	-0.000	-0.001	-0.001	-0.002	-0.005	-0.007	-0.009	-0.012	-0.014	-0.017	-0.021	-0.029	-0.037	-0.037	-0.024
2.20	-0.000	-0.001	-0.001	-0.001	-0.003	-0.004	-0.005	-0.007	-0.008	-0.009	-0.012	-0.015	-0.017	-0.012	0.004
2.40	-0.000	-0.000	-0.001	-0.001	-0.001	-0.002	-0.003	-0.003	-0.004	-0.004	-0.005	-0.006	-0.003	0.005	0.024
2.60	-0.000	-0.000	-0.000	-0.001	-0.000	-0.001	-0.001	-0.001	-0.001	-0.001	-0.001	0.001	0.007	0.017	0.037
2.80	0.000	-0.000	0.000	0.000	0.000	0.000	0.001	0.001	0.001	0.002	0.003	0.005	0.014	0.025	0.046
3.00	0.000	0.000	0.000	0.001	0.001	0.001	0.002	0.002	0.003	0.003	0.005	0.009	0.018	0.031	0.053
3.50	0.000	0.000	0.000	0.001	0.001	0.002	0.003	0.004	0.005	0.006	0.008	0.013	0.025	0.038	0.061
4.00	0.000	0.000	0.000	0.001	0.002	0.003	0.004	0.005	0.006	0.007	0.010	0.016	0.028	0.041	0.064

TABLE A.10 Lee and Kesler's [8] Values of $[\log_{10}(f/P)]^{(1)}$

T_r \ P_r	0.010	0.050	0.100	0.200	0.400	0.600	0.800	1.000	1.200	1.500	2.000	3.000	5.000	7.000	10.000
0.30	-8.778	-8.779	-8.781	-8.785	-8.790	-8.797	-8.804	-8.811	-8.818	-8.828	-8.845	-8.880	-8.953	-9.022	-9.126
0.35	-6.528	-6.530	-6.532	-6.536	-6.544	-6.551	-6.559	-6.567	-6.575	-6.587	-6.606	-6.645	-6.723	-6.800	-6.919
0.40	-4.912	-4.914	-4.916	-4.919	-4.929	-4.937	-4.945	-4.954	-4.962	-4.974	-4.995	-5.035	-5.115	-5.195	-5.312
0.45	-3.726	-3.728	-3.730	-3.734	-3.742	-3.750	-3.758	-3.766	-3.774	-3.786	-3.806	-3.845	-3.923	-4.001	-4.114
0.50	-2.838	-2.839	-2.841	-2.845	-2.853	-2.861	-2.869	-2.877	-2.884	-2.896	-2.915	-2.953	-3.027	-3.101	-3.208
0.55	-0.013	-2.163	-2.165	-2.169	-2.177	-2.184	-2.192	-2.199	-2.207	-2.218	-2.236	-2.273	-2.342	-2.410	-2.510
0.60	-0.009	-1.644	-1.646	-1.650	-1.657	-1.664	-1.671	-1.677	-1.684	-1.695	-1.712	-1.747	-1.812	-1.875	-1.967
0.65	-0.006	-0.031	-1.242	-1.245	-1.252	-1.258	-1.265	-1.271	-1.278	-1.287	-1.304	-1.336	-1.397	-1.456	-1.539
0.70	-0.004	-0.021	-0.044	-0.927	-0.934	-0.940	-0.946	-0.952	-0.958	-0.967	-0.983	-1.013	-1.070	-1.124	-1.201
0.75	-0.003	-0.014	-0.030	-0.675	-0.682	-0.688	-0.694	-0.700	-0.705	-0.714	-0.728	-0.756	-0.809	-0.858	-0.929
0.80	-0.002	-0.010	-0.020	-0.043	-0.481	-0.487	-0.493	-0.499	-0.504	-0.512	-0.526	-0.551	-0.600	-0.645	-0.709
0.85	-0.001	-0.006	-0.013	-0.028	-0.321	-0.327	-0.332	-0.338	-0.343	-0.351	-0.364	-0.388	-0.432	-0.473	-0.530
0.90	-0.001	-0.004	-0.009	-0.018	-0.039	-0.199	-0.204	-0.210	-0.215	-0.222	-0.234	-0.256	-0.296	-0.333	-0.384
0.93	-0.001	-0.003	-0.007	-0.013	-0.029	-0.048	-0.141	-0.146	-0.151	-0.158	-0.170	-0.190	-0.228	-0.262	-0.310
0.95	-0.001	-0.003	-0.005	-0.011	-0.023	-0.037	-0.103	-0.108	-0.114	-0.121	-0.132	-0.151	-0.187	-0.220	-0.265
0.97	-0.001	-0.002	-0.004	-0.009	-0.018	-0.029	-0.042	-0.075	-0.080	-0.087	-0.097	-0.116	-0.149	-0.180	-0.223
0.98	-0.000	-0.002	-0.004	-0.008	-0.016	-0.025	-0.035	-0.059	-0.064	-0.071	-0.081	-0.099	-0.132	-0.162	-0.203
0.99	-0.000	-0.001	-0.003	-0.007	-0.014	-0.021	-0.030	-0.044	-0.050	-0.056	-0.066	-0.084	-0.115	-0.144	-0.184
1.00	-0.000	-0.001	-0.003	-0.006	-0.012	-0.018	-0.025	-0.031	-0.036	-0.042	-0.052	-0.069	-0.099	-0.127	-0.166
1.01	-0.000	-0.001	-0.003	-0.005	-0.010	-0.016	-0.021	-0.024	-0.024	-0.030	-0.038	-0.054	-0.084	-0.111	-0.149
1.02	-0.000	-0.001	-0.002	-0.004	-0.009	-0.013	-0.017	-0.019	-0.015	-0.018	-0.026	-0.041	-0.069	-0.095	-0.132
1.05	-0.000	-0.001	-0.002	-0.002	-0.005	-0.006	-0.007	-0.007	-0.002	0.008	0.007	-0.005	-0.029	-0.052	-0.085
1.10	-0.000	-0.000	-0.001	-0.000	0.001	0.002	0.004	0.007	0.012	0.025	0.041	0.042	0.026	0.008	-0.019
1.15	0.000	0.000	0.001	0.002	0.005	0.008	0.011	0.016	0.022	0.034	0.056	0.074	0.069	0.057	0.036
1.20	0.000	0.001	0.002	0.003	0.007	0.012	0.017	0.023	0.029	0.041	0.064	0.093	0.102	0.096	0.081
1.30	0.000	0.001	0.003	0.005	0.011	0.017	0.023	0.030	0.038	0.049	0.071	0.109	0.142	0.150	0.148
1.40	0.000	0.002	0.003	0.006	0.013	0.020	0.027	0.034	0.041	0.053	0.074	0.112	0.161	0.181	0.191
1.50	0.000	0.002	0.003	0.007	0.014	0.021	0.028	0.036	0.043	0.055	0.074	0.112	0.167	0.197	0.218
1.60	0.000	0.002	0.003	0.007	0.014	0.021	0.029	0.036	0.043	0.055	0.074	0.110	0.167	0.204	0.234
1.70	0.000	0.002	0.004	0.007	0.014	0.021	0.029	0.036	0.043	0.053	0.072	0.107	0.165	0.205	0.242
1.80	0.000	0.002	0.003	0.007	0.014	0.021	0.028	0.035	0.042	0.052	0.070	0.104	0.161	0.203	0.246
1.90	0.000	0.002	0.003	0.007	0.013	0.020	0.027	0.034	0.041	0.050	0.068	0.101	0.157	0.200	0.246
2.00	0.000	0.002	0.003	0.007	0.013	0.019	0.025	0.034	0.040	0.047	0.066	0.097	0.152	0.196	0.244
2.20	0.000	0.002	0.003	0.006	0.012	0.018	0.024	0.032	0.038	0.044	0.062	0.091	0.143	0.186	0.236
2.40	0.000	0.002	0.003	0.006	0.011	0.017	0.023	0.030	0.036	0.042	0.058	0.086	0.134	0.176	0.227
2.60	0.000	0.001	0.003	0.005	0.011	0.016	0.021	0.028	0.034	0.039	0.055	0.080	0.127	0.167	0.217
2.80	0.000	0.001	0.003	0.005	0.010	0.015	0.020	0.027	0.032	0.037	0.052	0.076	0.120	0.158	0.208
3.00	0.000	0.001	0.003	0.004	0.009	0.013	0.018	0.025	0.030	0.033	0.049	0.072	0.114	0.151	0.199
3.50	0.000	0.001	0.002	0.004	0.008	0.012	0.016	0.022	0.026	0.029	0.043	0.063	0.101	0.134	0.179
4.00	0.000	0.001	0.002	0.004	0.008	0.012	0.016	0.020	0.023	0.025	0.038	0.057	0.090	0.121	0.163

TABLE A.11 Torquato and Smith's [12] Constants for the Latent Heat of Vaporization Formula (6.141)

	Ammonia	Argon	Carbon dioxide	Carbon monoxide	Ethane
T_c [K]	405.55	150.86	304.19	132.92	305.50
T_t [K]	195.42	83.78	216.55	68.14	89.88
$h_{fg}(T_t)$ [kJ/kg]	1473.90	161.80	347.30	235.52	601.19
a_1	0.47057	0.36102	1.35069	1.45169	−0.64121
a_2	3.04437	23.96446	−5.95311	−25.33722	36.15830
a_3	−8.04468	55.53528	−20.83607	−82.04570	66.16418
a_4	2.69778	−67.19804	21.89009	87.67934	−90.29894
a_5	5.37598	−17.88592	11.94955	32.14627	−14.05575
a_6	−2.82782	7.40818	−13.93713	−14.91833	4.03838
δ_m	0.33581	0.44605	0.15900	1.54475	0.49968
σ	0.00081	0.00085	0.00118	0.00325	0.00086

	Ethanol	Freon-12	Freon-22	Isooctane	Krypton
T_c [K]	521.35	384.95	369.28	544.25	209.39
T_t [K]	158.65	118.15	113.15	165.77	115.76
$h_{fg}(T_t)$ [kJ/kg]	1187.32	207.83	294.18	351.78	109.60
a_1	1.34686	0.94297	0.72254	0.67128	1.30503
a_2	−15.85366	−4.70274	2.95824	4.96085	−10.66454
a_3	−37.48493	−11.92221	3.42752	.7.92755	−32.53309
a_4	47.15897	15.46468	−5.36903	−10.87562	36.43934
a_5	7.37973	0.59654	−1.17015	−2.43041	9.54154
a_6	−1.70348	1.20085	0.81290	−0.91875	−2.80368
δ_m	0.26812	0.60933	0.24399	0.03399	0.56980
σ	0.00077	0.00063	0.00020	0.00031	0.00083

	Methane	Methanol	Neon	Nitrogen	n-Nonane
T_c [K]	190.55	513.15	44.40	126.25	595.15
T_t [K]	91.00	175.15	24.50	63.15	219.65
$h_{fg}(T_t)$ [kJ/kg]	543.40	1354.16	89.45	212.60	387.81
a_1	0.78206	0.29045	0.88154	1.41557	0.31647
a_2	0.84814	12.12910	9.79304	−10.94088	13.56551
a_3	−9.79152	14.11962	24.10286	−29.52156	22.12327
a_4	6.41067	−24.69802	−28.20568	34.51737	−31.61818
a_5	4.66431	1.28818	−8.03241	8.43211	−4.26098
a_6	−2.04333	−3.46706	2.70559	−3.10147	1.05594
δ_m	1.12302	0.74542	0.03090	0.59132	1.71695
σ	0.00073	0.00105	0.00007	0.00083	0.00207

	Oxygen	Propane	1-Propanol	Water	Xenon
T_c [K]	154.77	370.00	536.85	647.27	289.74
T_t [K]	54.35	85.46	147.15	273.16	161.36
$h_{fg}(T_t)$ [kJ/kg]	238.70	540.31	946.80	2501.00	96.98
a_1	1.24529	0.32838	1.15023	0.72241	1.13256
a_2	−4.05385	13.25739	−14.07059	5.33402	−3.08280
a_3	−6.31829	26.78045	−31.76606	8.97347	−8.18035
a_4	10.20602	−34.00190	40.45017	−11.93143	10.80348
a_5	−0.81329	−8.50819	−8.64866	−3.31206	−0.91775
a_6	0.98570	3.84203	4.26884	1.63257	2.42356
δ_m	0.84017	0.25917	1.09955	1.53773	0.17028
σ	0.00074	0.00064	0.00208	0.00028	0.00099

REFERENCES

1. J. Kestin, *A Course in Thermodynamics*, revised printing, Vol I, Hemisphere, Washington, DC, 1979, p. 91.

2. K. Rektorys, ed., *Survey of Applicable Mathematics*, MIT Press, Cambridge, MA, 1969, p. 1020.

3. F. B. Hildebrand, *Advanced Calculus for Applications*, Prentice-Hall, Englewood Cliffs, NJ, 1962, p. 355.

4. A. Bejan and H. M. Paynter, *Solved Problems in Thermodynamics*, Mechanical Engineering Department, Massachusetts Institute of Technology, Cambridge, MA, 1976, p. 9-2.

5. J. S. Thomsen and T. J. Hartka, Strange Carnot cycles; thermodynamics of a system with a density extremum, *Am. J. Phys.*, Vol. 30, 1962, pp. 26–33, 388–389.

6. B. Gebhart and J. C. Mollendorf, A new density relation for pure and saline water, *Deep-Sea Res.*, Vol. 24, 1977, pp. 831–848.

7. A. Bejan, Engineering thermodynamics, in M. Kutz, ed., *Mechanical Engineers' Handbook*, Wiley, New York, 1986, Chapter 54.

8. B. I. Lee and M. G. Kesler, A generalized thermodynamic correlation based on three-parameter corresponding states, *AIChE J.*, Vol. 21, No. 3, 1975, pp. 510–527.

9. J. A. Beattie and O. C. Bridgeman, A new equation of state for fluids, *Proc. Am. Acad. Arts Sci.*, Vol. 63, 1928, pp. 229–308.

10. E. G. Cravalho and J. L. Smith, Jr., *Engineering Thermodynamics*, Pitman, Boston, MA, 1981.

11. M. Benedict, G. B. Webb, and L. C. Rubin, An empirical equation for thermodynamic properties of light hydrocarbons and their mixtures. Constants for twelve hydrocarbons, *Chem. Eng. Prog.*, Vol. 47, No. 8, 1951, pp. 419–422.

12. S. Torquato and P. Smith, The latent heat of vaporization of a widely diverse class of fluids, *J. Heat Transfer*, Vol. 106, 1984, pp. 252–254.

Author Index

Duhem, P. M. M., 158–162
Dunning, J. W., Jr., 522, 527

Eastman, E. D., 710, 715
Edison, T. A., 405
Egolfopoulos, E., 454, 458
Ehnholm, G. J., 574, 581, 588
Ehrenfest, P., 353, 400
Ehrenfest-Afanassjewa, T., 353, 401
Ekström, J. P., 574, 581, 588
El-Masri, M. A., 449, 453, 457
El-Sayed, Y. M., 453, 458
Emmett, J. B., 42, 45
Epstein, P. S., 36, 45
Ericsson, J., 405
Esmaili, H., 453, 457
Euler, L., 32
Evans, R. B., 218, 234, 634, 663
Evans, W. H., 393, 401
Eyssa, Y. M., 555, 557, 586, 587

Fahrenheit, G. D., 16, 31
Fanger, P. O., 664
Ferdinand II, Grand Duke of Tuscany, 31
Ferrell, R. A., 328, 338
Feyerabend, P., 53, 100
Fisher, M. E., 306, 312, 313, 335
Fitch, J., 51
Fourier, J. B. J., 33, 36, 59
Fowler, R. H., 15, 44, 582, 588
Fox, R. F., 192, 210
Freeston, D. H., 454, 458
Freeth, F. A., 28, 45
Fujii, F., 320, 336
Fujiwara, M., 518, 526
Fulton, R., 51

Gaggioli, R. A., xi, 227, 229, 232, 234, 390,
    401, 638, 664, 665
Galilei, G., 22, 31
Gammon, R. W., 318, 336
Garceau, R. M., 454, 458
Garrabos, Y., 324, 337
Gay-Lussac, J. L., 32, 61, 105
Gebhart, B., 607, 662, 725, 742
Gehloff, P. O., 689, 696, 714
Georgescu-Roegen, N., 75, 101
Geskin, E. S., 651, 665
Giauque, W. F., 59, 100, 573, 574, 579, 581,
    587, 588
Gibbs, J. W., 12, 28, 44, 54, 75, 100, 101,
    147, 159, 209, 238, 263, 332, 343, 400
Gifford, P., 574
Gilbert, D., 51

Gillispie, C. C., 12, 44, 98, 102
Goldfrank, J. C., 287, 334
Golem, P. J., 639, 664
Goodenough, G. A., 28, 45, 118, 144
Gorrie, J., 531, 574
Gouy, G., 114, 144
Grabner, W., 328, 338
Grassmann, P., 547, 586
Green, M. S., 301, 322, 334
Greer, S. C., 320, 322, 336
Greer, W. L., 318, 336
Gribik, J. A., 467, 491, 492, 496, 499, 501,
    525
Griffiths, G. M., 408, 409, 457
Griffiths, R B., 306, 335
Grosjean, G., 506, 526
Guggenheim, E. A., 15, 44, 195, 196, 199,
    211, 244, 332, 349, 400, 582, 583, 588
Gunton, J. D., 306, 335
Gyftopoulos, E. P., xii

Hachette, J. N. P., 33
Hammond, R. P., 454, 458
Hanley, H. J. M., 181, 209
Harman, C. M., 454, 458
Harris, J., 518, 526
Hartka, T. J., 96, 102, 724, 742
Hartley, J. G., 65, 101
Hartnett, J. P., 526, 662
Haselden, G. G., 654, 665
Hastings, J. R., 318, 336
Hatsopoulos, G. N., 6, 12, 14, 28, 43, 96,
    102
Haught, A. F., 503, 506–509, 526
Hawthorne, W. R., 449, 457
Hayashi, A. K., 191, 210
Haywood, R. W., 65, 101, 181, 209, 404,
    405, 427, 431, 436, 438, 440, 456, 457
Hedley, 51
Heikes, R. R., 689, 692, 693, 695, 696, 714
Heisenberg, W., 313, 335
Helmholtz, H., 36, 37, 45
Hendricks, R. C., 522, 527
Henry, D. L., 328, 338
Herman, C. V., 622, 623, 663
Hesselmann, K., 638, 664
Heylandt, C. W. P., 538, 586
Hilal, M. A., 555, 586
Hilbert, D., 163, 209
Hildebrand, F. B., 722, 742
Hill, P. G., 177, 209, 419, 457, 460
Hirn, G. A., 33, 40
Hirschfelder, J. O., 654, 665
Hirst, T. A., 44

Subject Index

Absolute entropy, 365, 378
Absolute Seebeck coefficient, 682
Absolute temperature, 16
Absolute thermoelectric power, 682
Additivity of chemical equations, 358
Adiabatic boundary, 18, 468
Adiabatic demagnetization, 573, 578–579
Adiabatic flame temperature, 375
Adiabatic saturation process, 235–236
Advancement of the reaction, 346
Affinity, 347
Aftercooler, 529
Air-conditioning applications, 224–232
Amagat's law, 201
Andrews' diagram, 250, 266
Appendix, 720–742
Area constraint, 619–620, 668
Area and volume constraint, 621
Atmospheric air, 391
Augmentation of heat transfer, 137, 602–605
Availability:
   flow, 129, 381
   via heat transfer, 118
   nonflow, 125, 491
Available work, 115
Average high and low temperatures of cycle, 416
Axioms, 77–96
Azeotropic composition, 271

Beamed radiation, 472
Beattie–Bridgeman equation of state, 286, 733
Benedict–Webb–Rubin equation of state, 286, 287, 734
Binary mixtures, 196–197, 268–275, 708–711
Blackbody, 469
Blade and vane cooling, 448–450

Boiler, as separate component, 51, 427
Boundary, 2
Brayton cycle, 411–414, 614, 624–630
   refrigeration, 538–544
Bridgman's diagram, 266, 268
Bridgman's table, 182–185
Bubble point, 269
Bulk flow model, 71
Bulk-stationary mixture, 709
Burn-up domain, 560
Bushel, British imperial, 52

Cables, current leads, 557, 716
Caloric theory, 42
Calorimetry, 31
Canonical form, 699
Carathéodory's formulation, 39–41, 77–89, 95
Carnot cycle, 53–54, 478–479, 494
Celsius scale, 16–17
Charts, properties, 128, 161, 178, 179, 182, 205–207, 227, 254, 256, 257, 266, 267, 276–278, 281, 282, 288, 293, 294, 299, 319, 321, 323–327, 355, 374, 379, 580, 729
Chemical equation, 344
Chemical exergy, 218, 221, 223
   fuels, 390–394
Chemically reactive systems, 343–403
Chemical potential, 152
Chemical reactions, 343–346
Clapeyron relation (Clausius-Clapeyron relation), 262
Clapeyron's equation, 37, 61
Claude cycle, 538
Clausius equation of state, 285
Clausius–Clapeyron relation, 262
Closed system, 3

About the Author

Adrian Bejan received his B.S. (1972, Honors Course), M.S. (1972, Honors Course), and Ph.D. (1975) degrees in mechanical engineering, all from the Massachusetts Institute of Technology. He taught at M.I.T. until 1976 as a Lecturer and Research Associate. From 1976 until 1978 he was a Fellow of the Miller Institute for Basic Research in Science, at the University of California, Berkeley.

Adrian Bejan joined the faculty of the University of Colorado as an assistant professor in 1978 and was promoted to associate professor in 1981. Three years later he was appointed full professor with tenure at Duke University.

Professor Bejan is the author of 130 technical articles on a diversity of topics in natural convection, combined heat and mass transfer, convection through porous media, transition to turbulence, second law analysis and design, solar energy conversion, cryogenics, applied superconductivity, and energy policy. He is the author of two earlier graduate-level textbooks, *Entropy Generation through Heat and Fluid Flow* (Wiley, 1982) and *Convection Heat Transfer* (Wiley, 1984).